Time-Saver
Standards for
Site Planning

Time-Saver Standards for Site Planning

Joseph De Chiara

Lee E. Koppelman

McGraw-Hill Book Company

New York / St. Louis / San Francisco / Auckland / Bogotá / Hamburg
Johannesburg / London / Madrid / Mexico / Montreal / New Delhi
Panama / Paris / São Paulo / Singapore / Sydney / Tokyo / Toronto

Library of Congress Cataloging in Publication Data

De Chiara, Joseph, date.
  Time-saver standards for site planning.

  1. Building sites. I. Koppelman, Lee. E. II. De Chiara,
Joseph, date. Site planning standards.
NA2540.5.D4 1984   720   84-4333
ISBN 0-07-016266-2

  34567890   SEM/BKP   89876

ISBN 0-07-016266-2

The editors for this book were Joan Zseleczky and Esther Gelatt, the
designer was Mark E. Safran, and the production supervisor was Thomas
G. Kowalczyk. It was set in Optima by University Graphics, Inc.

Printed by Semline, Inc., and bound by The Book Press.

Parts of this book were adapted from *Site Planning Standards* by Joseph
De Chiara and Lee E. Koppelman, McGraw-Hill, 1978.

Contents

CONTENTS

Preface

Time-Saver Standards for Site Planning was conceived and organized to meet the needs of those who are involved in the planning, designing, or programming of individual buildings, complexes, and outdoor facilities. It provides basic design criteria, analytic methods, and construction details for all phases of site development. In format, the book follows the general path of the design process. It starts with the predesign stage of data inventory, environmental considerations, and ancillary surveys; continues through the analysis of site conditions; and finally deals with the design stages of plan alternatives through specific site details.

The material, as much as possible, is presented in graphic form for easy reference and utilization. It is intended for use primarily by the following groups:

1. Architects, landscape architects, engineers, and other related design professionals

2. Public agency planners and reviewers

3. Private organizations with environmental and developmental concerns

4. Developers, builders, and general contractors

5. Students and teachers of the design disciplines

Useful graphics and accompanying text are provided to assist these groups in establishing general concepts and review criteria. They also furnish a basis for further analysis and development in the complex planning process. Readers are reminded, however, that the material is not intended to provide definite design solutions, strict formulae, or design patterns that can be followed automatically; nor should it be used to supplement detailed engineering studies. The goal is to provide the broadest possible range of information and data relating to site development—to make possible site developments that are functional and well designed.

The material and examples in this book are generally applicable to typical conditions, reflecting planning and design principles and practices currently in vogue in the United States, and should be utilized judiciously.

The authors wish to take this opportunity to express their gratitude and sincere thanks to the many public agencies, private organizatiions, publications, and individual professionals who have allowed the use of their material.

JOSEPH De CHIARA
LEE E. KOPPELMAN

S E C T I O N

ONE

Preliminary Site Investigation and Analysis

The process of site planning begins with the gathering of basic data relating specifically to the site under consideration and the surrounding areas. The data should include such items as master plans and studies, zoning ordinances, base and aerial maps, surveys, topographic data, geological information, the hydrology of the area, types of soils, vegetation, and existing easements.

After all the available information has been obtained, it must be examined and analyzed. One of the first objectives is to establish the site's advantages and limitations. On the basis of these conclusions, it is then possible to determine whether the land is suitable for the proposed use. If the land is found suitable, the data must be analyzed further to establish other specific parameters of the site. These include such items as the best areas for building locations because of soil conditions, areas to avoid because of steep slopes, areas with soil erosion problems because of drainage patterns, or areas to be left in their natural state because of vegetation.

The use of computer applications in site planning has arrived in the form of computer graphics. Such techniques enable site designers to depict quickly a land area for pictorial and analytical purposes. This tool is suitable for perspective and isometric depictions of topography, slope analysis, cut-and-fill calculations, watershed analysis, and simulation studies. Illustrations of these techniques are included in this section. Since this approach is usually prepared by specialized consultants, it may be economically feasible only for large projects.

1.1 Site Inventory and Resource Analysis

Developing an environment requires a thorough knowledge of natural-resource systems, cultural features, and other relevant data. Only when this information has been gathered and analyzed can one proceed to determine possible end use allocations. Although many systems have been developed to inventory and analyze natural resources, almost all share the following three basic purposes:

1. Development of an understanding of separate ecosystem components (soils, vegetation, hydrology, and others)
2. Development of an understanding of the interrelationship of the various ecosystem components (soil and water, climate, vegetation, soils, and others)
3. Determination of the suitability of resource elements and resource aggregates for specified land uses and functions

Illustrated below is a process for realizing these purposes in a logical, sequential fashion.

Resource Category Determination The first step in resource planning is to determine which resources should be investigated. There are several factors that influence this choice.

The resources to be investigated should be relevant to the functions or land uses (outputs) being considered. The factors (soil, water, vegetation canopy, and so on) that determine the suitability of a site for a particular function may be assumed to be the critical factors for which data must be collected, mapped, and evaluated. For instance, if a planner is interested in locating the best sites for wood duck production, the first step is to determine the resource factors that best express the suitability of a site for that particular function.

Listing relevant factors shows which resources need to be investigated and the degree or detail of the data that must be collected. The relation of assumed factors to the example of wood duck production suggests that the resource information presented in the accompanying list should be collected and mapped.

RESOURCE CATEGORY FACTORS

Vegetation	1. Mature deciduous forest
	2. Medium-age deciduous forest
	3. Young deciduous forest
Shoreline features	1. Dense, overhanging woody vegetation
	2. Intermittent brush
	3. Emergent vegetation
Hydrology	1. Slow, meandering stream
	2. Slow- to medium-speed streams
	3. Ponds, 1 to 2 acres
	4. Ponds, 3 to 5 acres
	5. Large, open waters

The list shows that output factors greatly influence the sophistication of an inventory. If a planner is interested not only in locating optimum sites for wood duck production but also in locating optimum sites for interpretive centers, hiking trails, and automobile routes, the inventory will have to consist of all factors relating to the number of functions or outputs.

It is also imperative that resource planning activities be adjusted to respect local physiographic characteristics. Differences in terrain, climate, and vegetation greatly affect development constraints and opportunities. A function such as camping may be represented by one set of factors in the West and by another set in the Southeast. Thus no one set of factors can ever be used to represent accurately any function in all physiographic situations in the United States. All factors, however, may generally be accounted for within the framework of the following natural and cultural resource structure:

1. Soils
2. Vegetation
3. Hydrology
4. Climate
5. Topography
6. Aesthetics
7. Historical significance
8. Existing land use
9. Physiographic obstructions

Each of these structural components represents areas of major environmental influence for most functions. The influence of each component naturally depends upon the type and intensity of the proposed function. It should also be emphasized that the type of information needed about each of the components depends upon the particular function or land uses being considered. An inventory of soils relating to a function such as a visitors' center may focus on engineering capabilities, whereas information required for locating specific habitat types may focus on a soil's ability to support a certain type of plant growth. The following statements regarding the individual components suggest their importance as locational factors.

The result of the resource category determination phase should be a rather extensive list of natural and cultural resource features for which data must be collected, mapped, and evaluated. Although most of the information presented in the resource inventory will relate to on-site resources, care should be taken to record any off-site or regional factors that influence on-site phenomena. Important migration routes, watersheds, and vegetation patterns far from the actual site boundaries may greatly affect on-site conditions.

Soils An understanding of soil development, which depends upon (1) parent material, (2) topography, (3) climate, (4) biotic forces, and (5) time, provides insight into many phenomena associated with natural resources. A knowledge of soils is important not only in terms of engi-

neering capability but also in terms of its relevance to other natural-resource systems. An extensive knowledge of on-site soil conditions is helpful both in determining the suitability of a site to support buildings and roadways and in gaining an insight into existing plant communities and associated wildlife habitats. The following list presents soil information that could be considered part of the soil survey (example only).

1. Glacial till; shallow water table
2. Glacial till; deep, well drained
3. Glacial till; deep but pan
4. Glacial till; shallow bedrock, shallow water table
5. Stratified drift; deep, well drained
6. Stratified drift; shallow water table
7. Floodplain alluvium; shallow water table
8. Muck and peat
9. Floodplain alluvium; deep, well drained
10. Rock outcroppings

The following list represents soil inventory considerations.

1. Depth of horizon
2. Depth to seasonal high water
3. Depth to bedrock
4. Drainage characteristics
5. Suitability
   a. Septic tanks
   b. Excavation and grading
   c. Value as foundation material
6. Susceptibility to compaction
7. Susceptibility to erosion
8. pH rating
9. Soil fertility

Vegetation   Vegetative types and patterns represent a major visual, recreational, and ecological resource. Native vegetation types are closely related to soils as well as to microclimate, hydrology, and topography. This component is influential in determining the location of most natural-based outputs. The location of hiking trails, camping or picnicking sites, and, perhaps most important, wildlife habitats is greatly influenced by vegetation types and patterns. The following list presents vegetation features that might be studied as part of the vegetation survey (example only).

1. No vegetation
2. Emergent vegetation
3. Streamside and riverside vegetation
4. Vegetation beside ponds, lakes, and reservoirs
5. Wetlands; bog, and tree growth
6. Wetlands; brush growth

7. White pine and mixed northern hardwoods
8. White pine, northern hardwoods, and hemlock
9. Fir and spruce

The following list represents vegetation inventory considerations.

1. Density of canopy
2. Understory heights
3. Overstory heights

Hydrology   The type and quality of on-site water is a critical visual and recreational resource. Even more important is the overall consideration of the entire interlocking hydrological system. Surface water and drainage patterns greatly affect vegetation, wildlife, and even climatic systems. The capability of the hydrological system must be considered if it is to be utilized as a significant resource. The following list presents possible hydrological features that may be analyzed as part of the hydrology survey (example only).

1. Surface water
   a. First-order stream
   b. Second-order stream
   c. Third-order stream
   d. River
   e. Pond
   f. Lake
   g. Reservoir
2. Drainage basins (watersheds)
   a. First-order
   b. Second-order
   c. Third-order
3. Other
   a. Springs
   b. Pumping wells
   c. Artesian wells

The following list represents other hydrological considerations:

1. Runoff rates
2. Siltation
3. Oxygen content
4. Subsurface water characteristics

Climate   Overall precipitation and temperature variations affect the entire site, as do winds, cloud cover, and seasonal changes. It is important to consider both small- and large-scale climatic phenomena. Many on-site climatic changes are closely related to such on-site factors as changes in topography, slope orientation, vegetation, and the presence of

water. Climatic conditions are interconnected with overall regional climatological patterns as well as with smaller site characteristics. The following list presents possible climatological features that may be analyzed as part of the climate survey (example only).

1. Temperature
   a. Annual average temperature range
   b. Temperature extremes
   c. Monthly temperature averages
2. Precipitation
   a. Annual precipitation
   b. Monthly precipitation
3. Wind
   a. Intensity and duration
   b. Seasonal direction
   c. Frequency of damaging storms
4. Snowfall
   a. Annual snowfall
   b. Monthly snowfall
5. Other
   a. Killing-frost dates
   b. Length of growing season
   c. Sun angles (azimuth)

Topography The basic land form or topographic structure of a site is a visual and aesthetic resource that strongly influences the location of various land uses and recreational and interpretive functions. A complete understanding of the topographic structure not only gives an insight into the location of roadways and hiking trails but also helps reveal the spatial configuration of the site. This spatial structure is especially important when the visual aspects of the environment are being considered. The following list presents possible topographic features that may be analyzed as part of the topographic survey (example only).

1. Elevation above sea level
   a. 1520 to 1580 feet
   b. 1580 to 1620 feet
   c. 1620 to 1660 feet
   d. 1660 to 7700 feet
2. Topographic orientation
   a. Flat; less than 3 percent topographic slope
   b. East to southeast
   c. South to southwest
   d. North to northeast
   e. West to northwest
3. Topographic slope

a. 0 to 3 percent slope
b. 3 to 8 percent slope
c. 8 to 15 percent slope
d. 15 to 25 percent slope
e. 25 percent slope or higher

Aesthetics Aesthetic resources are largely responsible for locating sites for recreation and activities interpreting wildlife and wild land. These resources depend upon landform diversity, vegetation pattern, and surface waters as well as the spatial definition, views, vistas, and image of the site that stem from these features. The following list presents possible aesthetic features that may be analyzed as part of the aesthetic survey (example only).

1. Major spatial determinants. Landforms that serve as three-dimensional masses or barriers to define spaces from eye level
2. Promontory. Mountain peaks that serve as landmarks or points of reference; highly visible and identifiable landforms
3. Scenic vista. A visual panorama with particular scenic value illustrating a contrast between "closed views" and "open views" as found from an overlook or clearing
4. Orientation vista. A visual panorama with particular value as a locational reference for visitors; a site with a strong image
5. Tree cover. An area characterized by a dense vegetation canopy and limited views through woodlands
6. Flat grassland. An open expanse with long views and a high degree of exposure
7. Hilly grassland. An open expanse with short views and relatively hidden or concealed areas
8. Water image. A large area of surface water with a shallow shoreline profile and a strong sense of water

Historical Significance Any given area generally has significant historical landmarks. A knowledge of the location and importance of these landmarks is valuable in interpreting the overall management area and in locating interpretive displays or exhibits that may focus on them. The following list presents possible historical features that may be analyzed as part of the historical significance survey (example only).

1. Historic trails or passageways
2. Historic buildings or structures
3. Sites of particular significance

Existing Land Use A thorough knowledge of existing land uses on or adjacent to the site provides the planner with an understanding of constraints and opportunities. Existing land uses often represent significant expenditures and must be weighed accordingly. It is also important to document

functions that are not considered land uses per se but are associated with certain land uses, that is, roads, fences, and utilities. The following list presents possible existing land uses that may be analyzed as part of the land use survey (example only).

1. Conservation, forest, preservation
2. Recreation
3. Farm dwellings
4. Residences
5. Seasonal dwellings
6. Commercial use
7. Industrial use
8. Institutions
9. Air and railway facilities
10. Transmission lines
11. Water and sewage lines
12. Range fences
13. Transportation types
    a. Unimproved road
    b. Graded and drained road
    c. Gravel-surfaced road
    e. Paved road
    f. Divided highway with partial control of access
    g. Divided highway with full control of access

Physiographic Obstructions Physiographic obstructions are natural elements that obstruct or are hazardous to certain types of development. The elements of an obstruction are related to the output or function being considered. Such conditions as earthquake faults and flash-flood zones are physiographic obstructions that seriously restrict almost all public-use activities that require buildings. Most other obstructions are less restrictive. Floodplain zones, which may be viewed as an obstruction for facilities requiring intensive development, are nevertheless usable for picnicking, hiking, or other functions that cannot damage or be damaged by phenomena associated with such zones. The following list presents possible physiographic obstructions that may be analyzed as part of the physiographic obstruction survey (example only).

1. Fault zones
    a. Major fault zone
    b. Minor fault zone
2. Floodplains
    a. 10-year floodplains
    b. 50-year floodplains
    c. 100-year floodplains
3. Critical wildlife habitat areas
    a. Habitat of endangered or threatened species
    b. Critical migration routes
4. Aquifer recharge
5. Zones susceptible to storm damage
    a. Tornado
    b. Lightning
    c. Hurricane
6. Topography
    a. Low elevations susceptible to tidal inundation
    b. Area with a high water table
    c. Peat bogs
    d. Quicksand
7. Obstructions associated with wildlife or wild lands
    a. Poisonous snakes or reptiles
    b. Mosquitoes or other annoying insects
    c. Poison ivy, poison oak, and other poisonous plants

Data Collection Pertinent data may be collected in a variety of ways. A thorough inquiry into the recommended sources cited above, in combination with a comprehensive review of the literature, is essential. Information thus gathered may be complemented by field study. Again, the data to be collected should be the total resource data indicated by the output factors.

Data Mapping Most information on the location of natural and cultural resources will be represented on a variety of scales such as 1 inch = 400 feet or 1:24,000. An initial decision must be made on an appropriate scale for subsequent mapping. The determination of this scale generally depends upon the degree of detail needed to evaluate the resource in terms of previously developed output factors. For instance, a cursory review of data and needs may suggest that a U.S. Geological Survey map of a scale of 1:24,000 is the best mapping scale. If this is so, resource data collected on a different scale must be transposed to the 1:24,000 scale. The transposition may be accomplished by a grid system or by photographic techniques. A standard technique for evaluating resource data requires that individual resources (soil, hydrology) be mapped on single sheets of paper. Once all locational data have been delineated on appropriate drawings on a consistent scale, the drawings can be converted through photographic techniques to transparent overlays. These techniques provide an opportunity to superimpose several individual resource overlays so that combinations of resources or resource aggregates can be evaluated.

Resource Evaluation Once all the factors relating to the location of the proposed functions have been mapped and converted to transparent overlays, one can begin evaluating pertinent resources in terms of proposed functions or outputs. This process is accomplished by superimposing combinations of appropriate resource overlays to establish the locations of certain phenomena as well as to note pertinent

Output: Waterfowl Production — Wood Ducks

| Factors | Characteristics | | |
	Optimum	Acceptable	Minimum
Tree cover (nesting)	Mature deciduous forest	Medium-age deciduous forest	Young deciduous forest
Brood ponds	Slow, meandering stream and ponds of 1 to 2 acres, or both	Faster-moving stream and ponds of 2 to 5 acres, or both	Large, open water
Shoreline features	Dense, overhanging woody vegetation and emergent vegetation	Intermittent brush and emergent vegetation	Void of brush or limited emergent vegetation, or both
Proximity of brood to nesting site	Adjacent	Within a half mile	Over a half mile

relationships. For example, if a planner wants to locate a potential wood duck production habitat, the first step is to extract the transparent resource overlays pertaining to the location of wood duck production. The planner's chosen criteria for wood duck production are shown in the accompanying table.

Each of the output factors can be mapped as part of a composite drawing (see Figure 1.1a and b). Once the mapping is complete, the resources can be evaluated in terms of the ability of specific resource aggregates to meet specific output criteria; that is, optimum, first class; acceptable, second class; minimum, third class.

The result of an individual output composite drawing should be a drawing of the entire site, with output locations indicated in terms of suitability (optimum, acceptable, minimum). Individual output composite drawings can, in turn, be used to determine other land use decisions. For instance, the need to determine the optimum location for a multipurpose public-use area may require the development of a composite formed of individual composites relating to picnicking, camping, hiking, and other outdoor activities.

It should be emphasized that determination of the optimum or acceptable suitability of a particular site does not necessarily mean that the site should be developed for a given function. Suitability studies merely suggest the degree to which a site is suitable for a given function. For example, a site that may appear to be optimum for hiking may be located far from optimum camping facilities. If it is desirable to have camping and hiking facilities near each other, one must seek a lower-quality (acceptable or minimum) camping site near the hiking area, or vice versa. Decisions relating to the actual selection of final output locations can be made only after all relevant issues have been considered.

See also Figures 1.2, 1.3, and 1.4.

Figure 1.1a Map showing gross habitat types of wildlife areas. It is produced by superimposing the vegetation, soil, slope, and hydrology resource data.

Figure 1.1b Map showing suitable locations for four recreational activities. It is produced by superimposing the suitable resource overlays.

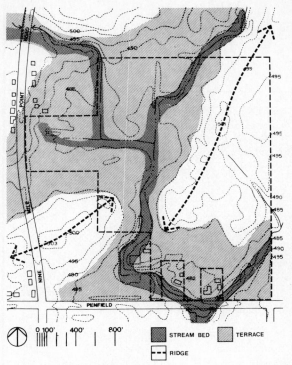

Figure 1.2 Landforms. (Penfield road study; Schnadelbach Associates.)

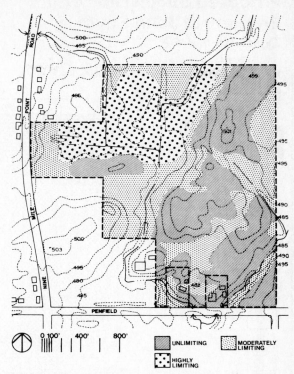

Figure 1.4 Relative soil limitations. (Penfield road study; Schnadelbach Associates.)

1.2 Site Considerations

Lay of the Land (Topography) Map

What the Developer Should Have Looked For (see Figure 1.5)

Level areas

High elevations:

    Crests

    Ridges

    Hilltops

Low elevations:

    Valleys

    Swales

    Depressions

    Wetlands

    Floodplains

Rocky areas:

    Exposed ledge

    Stony surface

Surface water:

    Lakes, ponds, rivers, streams, springs, seeps, bogs, wetlands, etc.

    Steep slopes

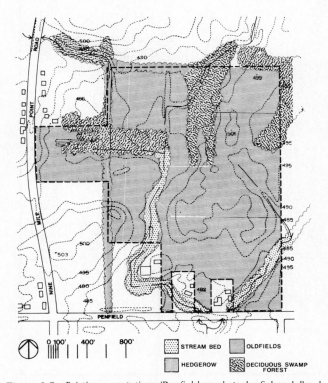

Figure 1.3 Existing vegetation. (Penfield road study; Schnadelbach Associates.)

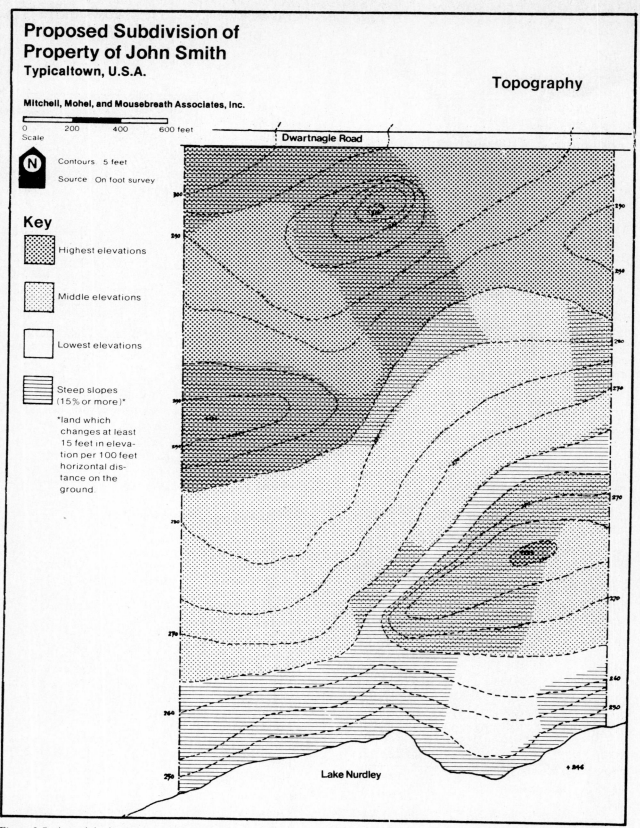

Figure 1.5 Lay of the land topography map.

Good views:

From high land

From low land

From site toward water

From water toward site

From site toward road or distance

From road or distance toward site

How the Developer Might Have Found It   On foot: the cheapest, most thorough and reliable source of information. Even though time-consuming on larger properties, it is worthwhile in the long run.

Aerial photographs: available from the nearest office of the Soil Conservation Service or directly from the federal government. You can tell a lot from these photographs; but thorough, accurate interpretation of them is a process which may require the services of a professional.

USGS maps: available from realtors, sporting goods stores, etc., or you can get them yourself from Washington. USGS maps are generally reliable as far as actual topography is concerned, but the scale and contour interval are usually too large for smaller properties and lots. Information about human activities is usually out of date.

Why It Is Important   Topographic information gives a fast overview of the character of the site — whether it is hilly or flat; sloping gradually or steeply; where one is likely to find wet spots or good views. A quick look at the lay of the land can suggest the best site for buildings, access location, plumbing, and activity areas. Alternatively, imposing conventional development patterns on variable topography can result in monotonous sites, sometimes dangerous conditions, loss of good views, and loss of good recreational opportunities.

Soils Map

What the Developer Should Have Looked For (see Figure 1.6)

Steep slopes

Obviously wet areas

Vegetation that indicates wetness, such as cattails, alders, and cedars

Rock outcroppings

Depth to bedrock

How the Developer May Have Found It

An on-foot visual check

SCS maps; SCS offices are located in many county seats

Test borings and pits, preferably conducted by a soils scientist

Why It Is Important   The suitability of soils for building sites depends, in part, on their capacity to drain water. Soil types are categorized by both wetness and degrees of slope,

as well as the texture and size of particles involved — all of which affect the ability of the soil to permit water to pass through it or to support the weight of buildings.

For example: Wetlands are found on soil types that have a low percolation rate — that is, the density of the soil particles dramatically slows the ability of water to seep through it (or it stops water from seeping through it altogether, as with clay). Water then remains close to, at, or above the surface. Wet soils thus require subdivision and building restrictions.

Dry, sandy soils with moderate to severe slopes also require building restrictions because such soils are highly susceptible to erosion. Erosion carries away topsoil and deposits excessive amounts of this material in streams and river floodplains (siltation).

A good soil to build on usually will have a moderate slope and will absorb water without serious erosion.

A good soils map, interpreted by an expert, can yield vast amounts of information to help plan a development consistent with the natural features of a site.

Water Systems (Hydrology) Map

What the Developer Should Have Looked For (see Figure 1.7)

Seeps, springs, wells, and other sources of water supply

Drainage patterns

Wet areas, bogs, etc.

Surface water, such as lakes, ponds, and streams

Water-table level

Vegetation typical of wet areas, such as cattails, alders, and cedar

How the Developer May Have Found It

On foot

Testing for location by boring test holes or digging pits

Testing for quality of water once location is established

Hydrologic maps (if available) from the U.S. Geological Survey, Water Resources Division

Dowsing (but be careful; have a contract)

Why It Is Important   Precipitation, in the form of rain and snow, seeps into the ground, moving through various permeable layers in a roughly vertical direction until it reaches bedrock or other impermeable strata. Water then begins to move in a roughly horizontal direction. This is called groundwater.

Sources of groundwater are important to dwellings with wells as the main source of drinking water.

The depth to groundwater, along with soil and other geologic data, can determine if a site will allow septic facilities.

Precipitation which does not drain into the soil, but remains or appears above the ground, is called surface water. Surface water includes lakes, ponds, rivers, seasonal and year-round streams, seeps, marshes, swamps, wetlands, and floodplains.

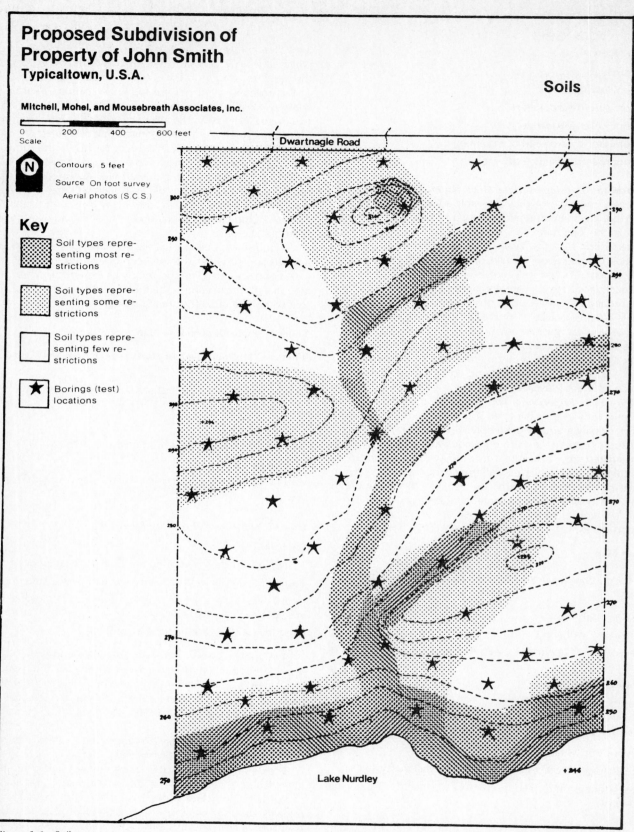

Figure 1.6 Soils map.

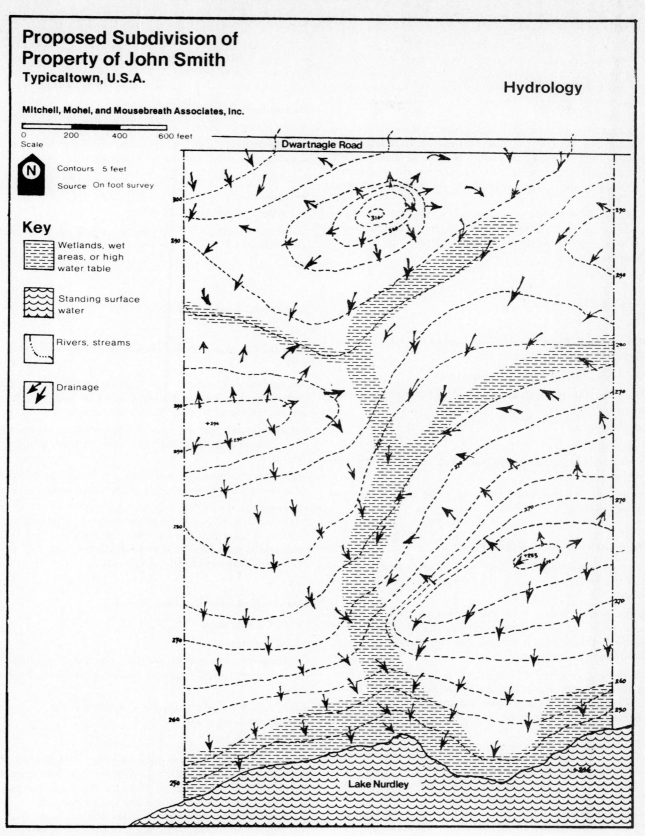

Figure 1.7 Water systems (hydrology) map.

The disruption of drainage patterns can result in erosion, siltation, and damage to the building. If large numbers of people are careless about water systems, the soil and water resources of an entire region may become polluted or destroyed.

Rainwater and snow melt which runs off the hard surfaces of city roads, parking lots, and roofs may strain drainage systems and lead to more frequent and more serious flooding. The pollution and trash which may accompany it also may affect the water's use for drinking, recreation, and other activities.

All these elements of groundwater and surface water quality and quantity are linked to one another, and modifying any one element will result in modification of all the other elements.

Vegetation Map

What the Developer Should Have Looked For (see Figure 1.8)

Areas made up primarily of mature trees

Areas made up primarily of younger trees (thin trunks, dense growth)

Cut-over areas

Open areas, such as fields

Unusual or unique specimens (such as a very old white pine of immense size)

Areas with no vegetation, such as ledge or outcrop, and dunes

How the Developer May Have Found It

On foot — the cheapest reliable way of getting information. It is the best way to spot details that aerial photography misses.

Aerial photography — a reliable means of gaining general information about vegetation on the site. A trained interpreter can determine species, age of the stand(s), and other information from the photographs.

USGS maps are generally too large in scale for accurate information and are usually outdated. The vegetated areas are indicated by a pale green color and are meant only to show vegetation which is large enough to hide a man (military use).

Why It Is Important   Aside from forests being harvested for timber and pulp, green areas are critical to our survival, since green plants are a source of oxygen, as well as the basic source of food for all living creatures.

Field, forest, and wetland alike provide both food and shelter for many different kinds of birds and animals. They are also useful to man for their educational, scientific, and recreational value.

There is no way to place a dollar value on such a resource — it is priceless.

Forest values other than wood products include:

Maintenance and enhancement of visual character

Ability to buffer noise

Wildlife habitat, such as wetlands, deer-wintering areas, eagle-nesting areas, and heron rookeries

Recreation

Soil stabilization

Watershed protection

Ability to moderate climate by screening sun and wind (microclimate)

Shade

Noise/Visual Impact Map

What the Developer Should Have Looked For (see Figure 1.9)

Natural features that will have a positive effect on the subdivision, such as near and distant views of bodies of water and interesting or dramatic landform features

Natural features that will have a negative effect on the subdivision, such as being in the shadow of a neighboring hill or mountain

Man-made features that will have a positive effect on the subdivision, such as a view of a nicely clustered old village

Man-made features that will have a negative effect on the subdivision, such as a view of an abandoned gravel pit; a view of a clear-cut area; a view of a commercial area/brightly lit signs; a view of a carelessly done subdivision; noise from a highway or industrial operation

How the Developer May Have Found It

On-foot exploration

Asking area residents

Checking on development plans for the area (i.e., roadway improvement usually leads to further development of an area)

Again, USGS maps are usually outdated as far as human changes are concerned

Why It Is Important   Being aware of the neighboring features of a site, both natural and man-made, can help the developer take advantage of those features felt to be pleasant and to avoid those features felt to be unpleasant.

Carefully siting and designing a development can minimize the impact of noise from both outside and within the development.

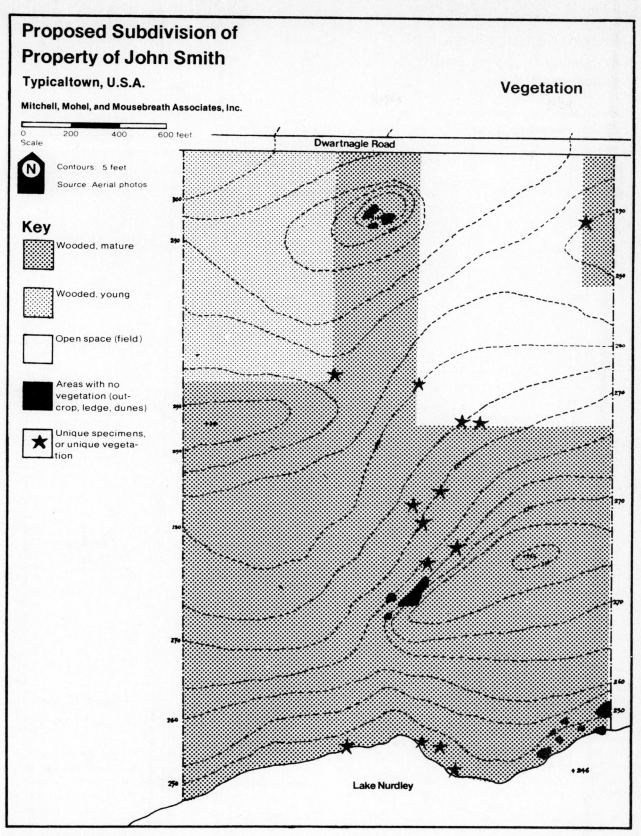

Figure 1.8 Vegetation map.

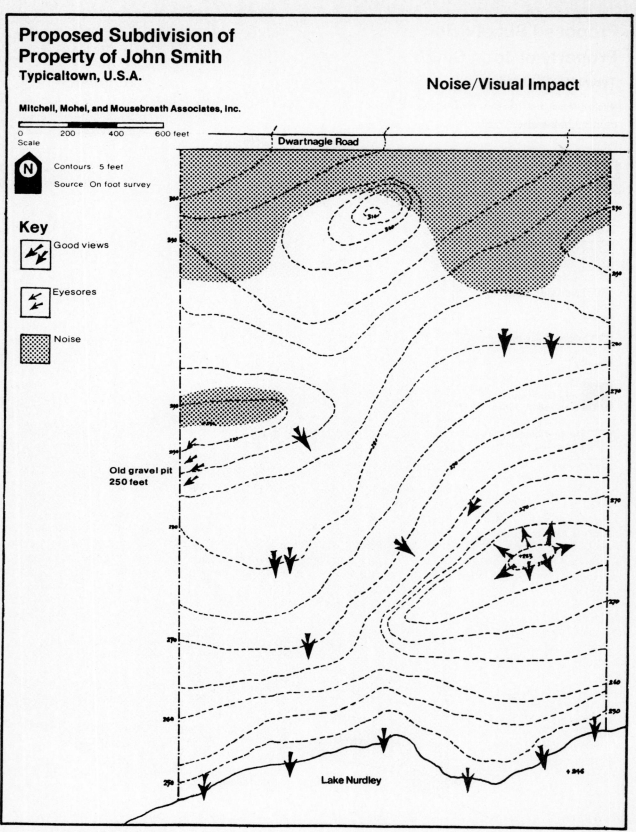

Figure 1.9 Noise/visual impact map.

1.3 Survey Systems

Every legal description is based upon a survey of the land. To prepare a legal description of a parcel or tract of land, someone at some time must have walked on the parcel or tract and made some sort of survey.

Metes-and-Bounds Descriptions A metes-and-bounds description is the oldest-known manner of describing land. Literally it means the measurements and boundaries of a tract of land. This method consists of beginning at some point in the boundary of the tract to be described and then reciting on the courses (the directions) and the distances from point to point entirely around the tract.

Two things are of prime importance in a metes-and-bounds description. First, the description must begin at some known point that can be readily identified; that is, the point must be substantial and so well established and witnessed that it can be relocated with certainty if the marker (or, as we say, the monument) that identifies the point is destroyed or removed. Second, the description must close; that is, if one follows the courses and distances of the description step by step from corner to corner, one must return to the place of beginning.

Monuments Monuments may be natural or artificial. Natural monuments are those created by nature, such as trees, rivers, and lakes. Artificial monuments are those created by man, such as highways, section corners, quarter corners, and boundaries or a stone or other permanent marker, properly located and witnessed.

Historical Background The metes-and-bounds method of description was followed by the settlers of the thirteen original colonies. An example of a legal description of property in Vermont contained in a deed of October 3, 1784, reads in part as follows: "Begin at the middle of a large white pine stump standing in the west side line of Simon Vender Cook's land and on the south side of the main road that leads to the new city, — and there is also a fence that stands a little to the west of Simon Vender Cook's barn, which said fence if it was to run cross the said road southerly, would run in the middle of said stump; and running thence north 2 degrees east 19 chains and 50 links to a small white oak tree," and so on. This particular description continues with courses that run to trees and to stakes and piles of stones, and concludes with a course that reads "thence north 9 chains 16 links to the middle of the stump where it first begun."

Such a description, although probably sufficient, lacked permanence both as to the monument that identified the place of beginning and as to the monuments that marked the ends of the various courses. The destruction or removal of such monuments would make a resurvey of the property difficult if not impossible.

This description is typical of the early descriptions used in the colonial states. References to monuments that lacked permanence, surveying of large and irregular tracts without regard to system or uniformity, and the failure of the surveyors to make their survey notes a matter of public record gave rise to frequent disputes and litigation over boundary lines.

After the Revolutionary War and the adoption of the federal Constitution the greater part of the land in the United States outside the original thirteen colonies became the property of the federal government, as a result of cession either by the original thirteen states or by trades, purchase, or treaty.

The federal government found itself with vast tracts of undeveloped and uninhabited land that had few natural characteristics suitable for use as monuments in metes-and-bounds descriptions. In any case the metes-and-bounds system was not satisfactory, and it was necessary to devise a new standard system of describing land that would make parcels readily and permanently locatable and easily available for land office sales. Even before the adoption of the Constitution, a committee headed by Thomas Jefferson evolved a plan for dividing the land into a series of rectangles, which the Continental Congress adopted on April 26, 1785. The system adopted was truly American. Designated the "rectangular system" or the "government system" of survey, it is in use today in thirty of the fifty states of the Union. The other twenty states, which include the original colonial states and other states in the New England area carved from them, the Atlantic Coast states except Florida, and the states of Hawaii, West Virginia, Kentucky, Tennessee, and Texas, retained their direction of the surveys of lands within their boundaries upon their admission to the Union and did not adopt the rectangular system. Florida is the only Atlantic Coast state in which the system is used.

Those states and parts of states shown as shaded areas on the map in Figure 1.10, and also the state of Hawaii, are those that are not controlled by a rectangular survey system.

Meridians and Base Lines To understand the rectangular system of survey one must have an understanding of meridians and base lines, which form the framework upon which the system is built. There are thirty-five principal meridians and thirty-two base lines in the United States, as shown in Figure 1.10. A meridian is a line that runs straight north and south. The dictionary defines it as "an imaginary line on the surface of the earth extending from the north pole to the south pole." A base line is a line that runs straight east and west. The government system of survey is based on a series of meridians and base lines run astronomically by surveyors, that is, by the same methods used by navigators to locate ships at sea or planes in the air.

Principal Meridians Under the government system of survey, as under any other surveying system, surveyors first had to find a substantial landmark from which a start could be made. Usually they selected a place that could readily be referred to, such as the mouth of a river. From this point they ran a line due north through the area to be surveyed. This north-and-south line was designated the principal meridian for that particular state or area. In addition to being marked and monumented, its location was fixed by a longitudinal reading, that is, as being so many degrees, minutes, and seconds west of the Greenwich meridian. As additional territories were opened and surveyed by the government, addi-

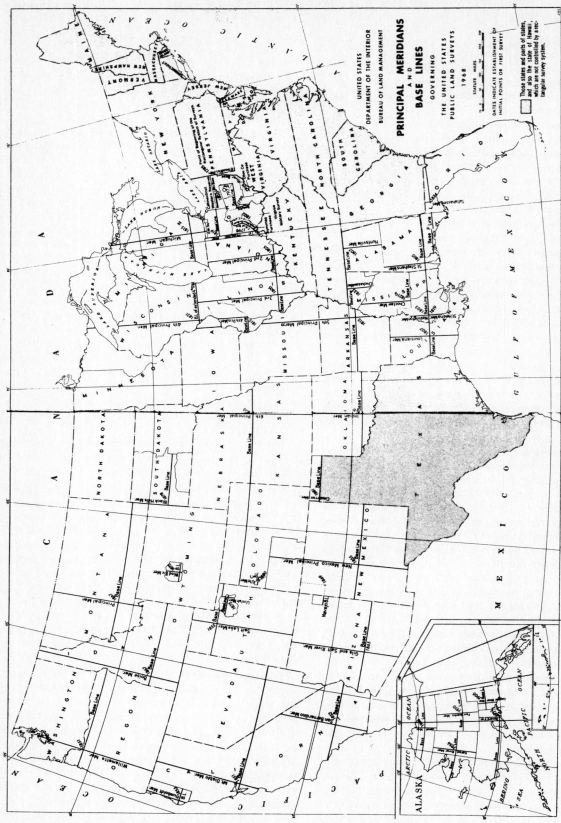

Figure 1.10 Principal meridians and base lines.

Figure 1.10 Principal meridians and base lines. Source: John S. Hoag, *Fundamentals of Land Measurements*, Chicago Title Insurance Company, 1971.

tional principal meridians were established, one being designated for each area so opened. Some of the principal meridians were numbered (first principal meridian, second principal meridian, and so on); while others were given names such as the Michigan meridian, which covers the survey of that state.

Base Lines Having fixed the line of the principal meridian (the north-and-south line) for a particular territory, the surveyors ascertained a point on the meridian from which they ran a line at right angles to the meridian, that is, a due east-and-west line. This line was designated the base line for the particular area, and its location (its latitude) was fixed astronomically as being so many degrees north of the equatorial line. In certain instances, when a new territory was opened up, the base line of an adjoining territory, previously surveyed, was extended to form the base line for the new territory. In Illinois, for example, the base line established for both the second principal meridian and the third principal meridian intersects the third principal meridian at a point about 10 miles south of Centralia and follows a parallel of latitude of 38 degrees, 28 minutes, and 20 seconds north of the equatorial line. The base line established for the fourth principal meridian follows a parallel of latitude of 40 degrees and 30 seconds north.

Correction Lines and Guide Meridians Having established a principal meridian and a base line, the surveyors proceeded with their survey, using the point of intersection of the principal meridian and the base line as their place of beginning. Their first step was to establish and locate east-and-west lines parallel to the base line at intervals of 24 miles measured along the meridian north and south of the base line. These lines were designated "correction lines."

Their next step consisted of establishing lines running due north and south at 24-mile intervals on each side of the principal meridian, commencing at the base line and extending to the first correction line. These lines were called "guide meridians." With the correction lines they divided the territory into squares measuring approximately 24 miles on each side. See Figure 1.11.

Township and Range Lines The 24-mile squares were then divided into smaller tracts of land by running east-and-west lines, called "township lines," at 6-mile intervals parallel to the base line. At the same time, north-and-south lines, called "range lines," were run parallel to the principal meridian at regular 6-mile intervals. The result of this crosshatching of lines was a grid of squares measuring approximately 6 miles on each side.

To locate a particular square in that grid, the government assigned to each square two numbers, a township number and a range number. To the first row of squares immediately adjacent to and parallel with the base line it assigned "township 1," to the second row it assigned "township 2," and so on. Thus each square in the first row north of the base line was called "township 1 north" of the base line, each in the second row, "township 2 north" of the base line, and so on. Likewise each square in the first row south of the base line was called "township 1 south," each in the second row, "township 2 south," and so on.

In the same manner, the government assigned a range number to each row of squares in the rows running parallel to the principal meridian, starting the numbering with the principal meridian. Thus the squares in the first row east of the principal meridian were numbered "range 1 east" of the particular principal meridian; those in the second row east, as "range 2 east," and so on. On the west side of the principal meridian those in the first row were numbered "range 1 west"; those in the second row, "range 2 west," and so on.

A glance at Figure 1.11, which shows a principal meridian, base line, correction lines, guide meridians, and township and range lines, makes it easy to understand the formation and numbering of townships and ranges, and the importance of these numbers in locating a particular tract of land.

Sections The act of 1785 creating the rectangular system of survey provided only for townships of 6 miles square; and in the early surveys only the outside boundaries of the townships were surveyed, although monuments were placed at every mile on the township lines. It soon became apparent that a 6-mile square was too large an area in which to describe and locate a given tract of land. Congress on May 18, 1796, passed an act directing that the townships theretofore surveyed be subdivided into thirty-six sections, each to be 1 mile square and containing "as nearly as may be" 640 acres; that each section corner be monumented; and that the sections be numbered consecutively from 1 to 36, beginning with 1 in the northeast corner of the township, proceeding west and east alternately through the township, and ending in the southeast corner with 36. This manner of numbering sections in a township has continued to the present time. See Figure 1.12.

By furnishing the section number, the township number north or south of the base line, and the range number east or west of the controlling principal meridian, a given tract of land can be located within the square mile of which it forms a part.

Purpose of Correction Lines You will recall that correction lines are east and west lines parallel to the base line and are established at intervals of 24 miles measured on the principal meridian. Each such correction line serves as a new base line for the townships that lie between it and the next correction line. It is necessary that such correction lines be established and used as new base lines if the townships and sections are to have substantially the size intended.

Because of the curvature of the earth it is necessary to compensate for the convergence of the meridians. If all the meridians were extended northward, they would meet at the north pole. The fact that they are constantly approaching each other is not observable from a point on the earth's surface without the aid of surveying instruments, but it is very real. An accurate survey of a township would show its north line to be about 50 feet shorter than its south line. Thus, in the case of the fourth township north of the base line, the difference is 4 times as great. The north line of the fourth township is 200.64 feet shorter than the south line of township 1.

To compensate for this convergence of the meridians, the south line of township 5 is measured the full distance of 6

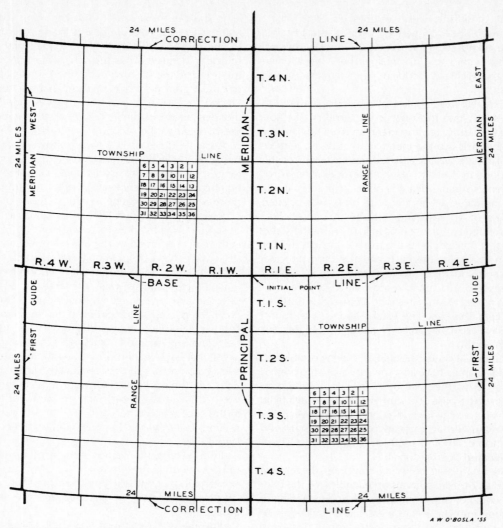

Figure 1.11   Correction lines and guide meridians.

miles on the correction line. The same practice is followed for every fourth township.

Subdivision of Sections   As time went on, Congress provided for the subdivision of sections into smaller units. An act passed in 1800 directed the subdivision of sections into east and west halves of 320 acres each "as nearly as may be" by running north-and-south lines through the center of the sections. Five years later, on February 11, 1805, Congress provided for the further division of sections into quarter sections by running east and west lines through the center of the sections and monumenting all quarter-section corners. This act further provided that all corners marked in the public surveys be established as the proper corners of the sections or quarter sections they were intended to designate and that "the boundary lines actually run and marked . . . shall be established as the proper boundary lines of the sections or subdi-

visions for which they were intended, and the length of such lines as returned by . . . the surveyor-general . . . shall be held and considered as the true length thereof." In other words, the original corners established by the government surveyors must stand as the true corners whether or not they were in the place shown by the field notes.

In 1820, Congress directed the further division of sections into half-quarter sections by the running of north-and-south lines through all quarter sections. Finally, on April 5, 1832, it directed the subdivision of all public lands into quarter-quarter sections by running east-and-west lines through the quarter sections. The quarter-quarter section of 40 acres is the smallest statutory division of regular sections.

Figure 1.12 shows a section divided into the fractional parts most frequently found in legal descriptions. The measurements of each typical fraction are noted. A little time

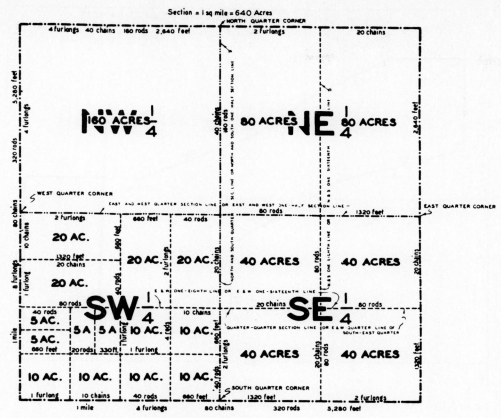

Figure 1.12 Section of land showing acreage and distances.
1 chain = 4 rods, or 66 feet
1 mile = 320 rods, or 5280 feet
1 section = 1 square mile, or 640 acres
1 acre = 160 square rods, or 43,560 square feet

spent in study of the typical subdivisions of a section and their dimensions will help one to understand more quickly plats of survey and description of parts of sections.

Legal Descriptions Legal descriptions of land that follow the regular subdivisions of a regular section are easily understood and present few difficult problems. Care and accuracy are required so that the terms used actually refer to the land intended. The section, township, and range must be given to complete the description of each fraction; for example, the southwest quarter of the northeast quarter of Section 6, Township 39, North, Range 13 east of the third principal meridian in Cook County, Illinois.

When one works with a description such as the north half of the northwest quarter of the southwest quarter of the southwest quarter of Section 6, it is somewhat easier to locate a tract of land if the description is read in reverse; that is, Section 6, southwest quarter, southwest quarter, northwest quarter, north half. Such a method must be used to avoid hopeless confusion.

A mental picture of a section or a chart such as Figure 1.12, together with familiarity with the usual measurements

and terms employed, plus care and thorough checking and rechecking, will produce descriptions that avoid trouble and loss of time.

Summary The foregoing discussion has outlined the government survey. As a result of the survey, a large unmarked area has been reduced to a series of small squares; basic lines of survey have been established by astronomical measurements; markers or permanent monuments have been located on the ground by the surveyors to establish quarter-quarter section corners; and careful survey notes of the description and location of all markers have been made and filed with the government. It is readily seen that any tract of land in the government survey can be described with certainty and identified to the exclusion of all other tracts by the proper numbering or naming of the principal meridian, the fixing of the base line, and the proper numbering of the section, township, and range.

TYPICAL METES-AND-BOUNDS DESCRIPTION

A typical description (see Figure 1.13) will read as follows:

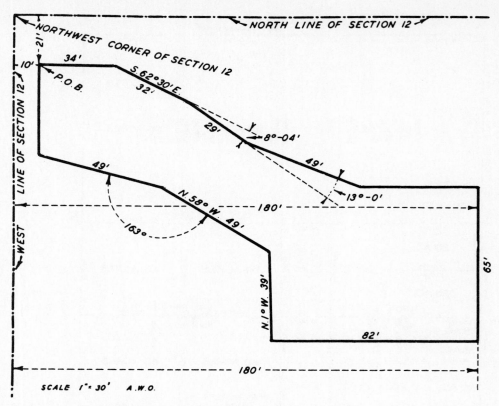

Figure 1.13 Sketch plat of a metes-and-bounds description.

Commencing at the northwest corner of Section 12, thence south along the section line 21 feet; thence east 10 feet for a place of beginning; thence continuing east 34 feet; thence south 62 degrees, 30 minutes, east 32 feet; thence southeasterly along a line forming an angle of 8 degrees, 4 minutes, to the right with a prolongation of the last-described course 29 feet; thence south 13 degrees, 0 minutes, to the left with a prolongation of the last-described line a distance of 49 feet; thence east to a line parallel with the west line of said section and 180 feet distant therefrom; thence south on the last-described line a distance of 65 feet; thence due west a distance of 82 feet to a point; thence north 1 degree west 39 feet; thence north 58 degrees west a distance of 49 feet; thence northwesterly along a line forming an angle of 163 degrees as measured from right to left with the last-described line a distance of 49 feet; thence north to the place of beginning.

Although today distances in the United States are generally expressed in feet and inches, land measure in the government survey is stated in chains, rods, and links, and these words should be understood.

A chain is 4 rods (66 feet) in length, divided into 100 links each 7.92 inches long. For years chain measurements were made with an actual metallic chain of 100 rings, or links, constructed in the early days from iron and later from heavy steel wire. Because the links tend to wear and stretch and the chain tends to twist and knot, accurate surveying often was difficult. Since about 1900 the surveyors have used steel ribbon tapes varying in length from 2 to 8 chains (from 166 to 528 feet).

A rod is 16½ feet. As the word implies, it was a pole of that length.

A link is the distance spanned by one loop of the chain formerly used by surveyors; it is 7.92 inches long.

1.4 Soil Classification and Surveys

Soils are classified and named, just as plants and animals are. Plants are identified by such characteristics as the structure of the flower and the form of the leaf. Soils are identified by such characteristics as the kinds and numbers of horizons, or layers, that have developed in them. The texture (the relative amounts of stones, gravel, sand, silt, and clay), the kinds of minerals present and their amounts, and the presence of salts and alkali help distinguish the horizons.

Most of the characteristics that identify soils can be determined in the field. A few can be determined only in the laboratory, but even without laboratory tests you often can get an accurate knowledge of soil characteristics from standard works on soils and geology. For example, you can estimate the amount of sand in a soil from its feel when you rub it

between your fingers, but for an accurate knowledge you must depend on laboratory analyses.

The type is the smallest unit in the natural classification of soils. One or a few types constitute a soil series. These are the common classification units seen on soil maps and survey reports.

A soil series is a group of soils having horizons that are essentially the same in the properties used to identify soils, with the exception of the texture of the surface soil and the kinds of layers that lie below what is considered the true soil. The names of soil series are taken from the towns or localities near the place where the soils were first defined.

The soil type, a subdivision of the soil series, is based on the texture of the surface soil. Stones, gravel, sand, silt, and clay have been defined as having the following diameters: gravel, from 3 inches down to 0.08 inch; sand, between 0.08 and 0.0002 inch; silt, between 0.0002 and 0.00008 inch; and clay, less than 0.00008 inch.

The full name of the soil type includes the name of the soil series and the textural class of the surface soil equivalent to the plow layer, that is, the upper 6 or 7 inches. Thus, if the surface of an area of the Fayette series is a silt loam, the name of the soil type is "Fayette silt loam."

The soil phase is not a part of the natural classification. It can be a subdivision of the soil type, soil series, or one of the higher units in the classification. Phases shown on soil maps commonly are subdivisions of soil types and are based on characteristics of the soil significant to its use for agriculture. Phases shown on large-scale soil maps generally have reflected differences in slope, degree of erosion, and stoniness, but other bases for defining phases include drainage and flood protection, climate, and the presence of contrasting layers below the soil.

The legends that accompany soil maps generally include such names for the units on the map as "Sharpsburg silty clay loam, eroded rolling phase," or "Fayette silt loam, 8–14 percent slopes, eroded." These names identify the soil series, the soil type, and the phase. They represent names of the most specific kinds of soil, comparable with the name of a practical subdivision of a variety of a plant, such as "old Jonathan apple trees."

The word "Fayette" in the second soil name mentioned above designates the soil series. This word, plus the words "silt loam," identifies the soil type, and the phase is identified by the words "8–14 percent slopes, eroded." In this name, the word "phase" is not used but is understood.

Higher units in the classification system include families, great soil groups, suborders, and orders. They are seldom used on any but small-scale soil maps.

Soil series, types, and phases do not occur at random in the landscape. They have an orderly pattern related to the landform; the parent material from which the soil was formed; and the influence of the plants that grew on the soils, the animals that lived on them, and the way humans have used them. On a given farm, the different kinds of soil commonly have a repeating pattern, associated with the slope.

The relationships between soils and landscapes vary in details in different parts of the country, but such relationships generally exist. Persons who are familiar with soils can visualize a landscape from a soil map; or if they see the landscape, they can predict where the boundaries are.

A soil survey includes determining which properties of soils are important, organizing the knowledge about the relations of soil properties and soil use, classifying soils into defined and described units, locating and plotting the boundaries of the units on maps, and preparing and publishing the maps and reports.

A soil survey report consists of a map that shows the distribution of soils in the area, descriptions of the soils, some suggestions as to their use and management, and general information about the area. Reports usually are prepared on the soils of one county, although a single report may cover several small counties or only parts of counties.

Soil surveys are made cooperatively by the Soil Conservation Service of the Department of Agriculture, agricultural experiment stations, and other state and federal agencies. Plans for the work in any area are developed jointly, and the reports are reviewed jointly before publication.

Soil maps have many uses, but generally they are made for one main purpose: to identify the soil as a basis for applying the results of research and experience to individual fields or parts of fields. Results from an experiment on a given soil can be applied directly to other areas of the same kind of soil with confidence. Two areas of the same kind of soil are no more identical than two oak trees, but they are so similar that, with comparable past management, they should respond to the same practices in a similar manner.

But many thousands of kinds of soil exist in the United States. Research can be conducted on only a few of them. The application of the research results must usually be based on the relationships of the properties of the soil on which the experiment was conducted to the properties of the soils shown on the maps. This can best be done by the soil classification system.

Significant properties that can be known from soil maps include physical properties, such as the amount of moisture that the soil will hold for plants, the rate at which air and water move through the soil, and the kinds and amounts of clays, all of which are important in drainage, irrigation, erosion control, maintenance of good tilth, and the choice of crops.

The soil map shows the distribution of specific kinds of soil and identifies them through the map legend. The legend is a list of the symbols used to identify the kinds of soil on the map. The commonest soil units shown on maps are the phases of soil types, but other kinds of units may also be shown.

Soil bodies, areas occupied by individual soil units, generally range from a few acres to a few hundred acres. Often within one soil body are small areas of other soils—series, types, or phases. If the included soils are similar, they are generally not identified unless they represent more than 10 to 15 percent of the soil body in which they are included. If the properties of the included soils differ markedly from those of

the rest of the soil body, they usually are indicated by special symbols.

Occasionally the individual parts of a unit are so small and so mixed with other units that they cannot be shown. Then the legend indicates the area occupied by the intricate mixture as a soil complex if all the included units are present in nearly every area.

A complex may consist of two or more phases of a soil type, but commonly it consists of two or more series. The names of complexes may carry a hyphen between the names of two soil types or phases, such as "Barnes-Buse loams." If several series or types are included in the complex, the names of one or two of the most important series or types will be followed by the word "complex," for example, the "Clarinda-Lagonda complex."

Two other kinds of units are common on soil maps: the undifferentiated group and the miscellaneous land type.

Two or more recognized kinds of soil that are not regularly associated in the landscape may be combined if their separation is costly and the differences between them are not significant for the objective of the soil survey. This kind of undifferentiated group is shown in the legend with the names of the individual units connected by a conjunction, for example, "Downs or Fayette silt loams."

Miscellaneous land types are used for land that has little or no natural soil. The map units then are given descriptive names, such as "steep, stony land," "gullied land," and "mixed alluvial land."

The relationships between the units that appear on the maps and legends and the use and management alternatives are explained in the text that accompanies the soil survey report.

The Department of Agriculture began making soil surveys about 1900. The purposes of the work were the same then as now, but there was no body of knowledge of how soils are formed or how nutrients become available to plants. The early definitions of the soil series therefore failed to take into account some important properties and overemphasized some of the more obvious but less important ones, such as color.

As scientists learned more about the relationships between soils and plants, ideas about the importance of the properties originally used to distinguish between soil series and types changed. The first soil series was split into two or more series. These in turn were often subdivided. Consequently many of the names shown on the older maps have been changed. Changes will continue to be made as long as we continue to learn new things about soil-plant relationships.

Soil maps are made by experienced soil scientists who are graduates of state agricultural colleges or other colleges or universities that offer courses in soil science.

Ordinarily soil scientists use aerial photographs as a base for plotting soil boundaries. The scientists go over the land and dig with a spade or auger as often as necessary to determine and evaluate the important characteristics of the entire profile. They identify the kind of soil, locate its boundaries in the field, plot the boundaries, and place the identification symbol of each soil mapping unit on the map.

In making detailed maps, they follow or see the boundaries between the kinds of soil through their entire length. In making reconnaissance surveys, they may not see the boundaries over their entire length; they merely identify one when they cross it and draw the boundary through to their next traverse, or crossing, on the basis of information obtained from aerial photographs.

Soil scientists make simple chemical tests in the field to determine the degree of acidity and the presence of lime, salts, and a few toxic compounds. They measure slopes with a hand level. Usually they take samples of a few representative soils during the survey and send them to the laboratory for detailed study. All stages of the work, from mapping to the contents of the report, are reviewed by the supervisors and representatives of the cooperating agencies.

Soil maps often are used before they are published. All cooperators of the soil conservation districts are furnished copies of the soil maps of their particular holdings.

County assessors and other users sometimes buy copies of such maps to use in their work before the publication of the completed survey. The local soils handbooks, available for reference at the Soil Conservation Service offices, give information needed to use and interpret the maps. Photographic copies of unpublished maps may be purchased through the Soil Conservation Service offices. These offices are usually located in county seats.

Soil maps are published by the Soil Conservation Service for all states except Illinois. In Illinois the University of Illinois Agricultural Experiment Station publishes them.

Copies of available published maps and reports may be obtained through a state extension service or the Soil Conservation Service offices. Files of unpublished maps are maintained in the Soil Conservation Service offices and may be examined there.

Interpretations of soil maps are physical and economic analyses of the alternative opportunities available to the users of the land. They indicate the capabilities of the soils for agricultural use, adapted crops, estimated yields of crops under defined systems of management, the presence of specific soil management problems, opportunities and limitations for various management practices, and problems in nonagricultural use.

The main bases for interpretations are yield estimates, related to specific combinations of practices for soils in their climatic setting. Yield estimates for a soil are predictions of the average production of specific crops that a group of farmers can expect during the next 10 or 15 years if they follow the defined system of soil management. These estimates apply less closely to individual farmers, whose skills are variable, than to averages of groups. Sources of information are the results of research, the experiences of farmers, ranchers, and others who grow plants, and observations of plants growing on different kinds of soils.

The definitions and descriptions of the kinds of soil shown on maps provide information on their characteristics. These are used to infer the qualities of soils such as productivity and erosion hazard. Predictions can be made about a soil whose behavior is unknown by comparing its characteristics with those of soils about which such basic information is known. Basic principles of soil management are another tool to help extend predictions of soil behavior and responses to all kinds of soils.

The Use of Soil Maps Soils may be grouped into land capability classes, subclasses, and units to help us use them properly. Of the eight classes, which normally do not all exist on any single farm or ranch, Classes I through IV are suited to cultivated crops, pasture or range, woodland, and wildlife. Classes V through VIII are suited to pasture or woodland and wildlife and are not generally recommended for cultivation. However, some kinds of soil in Classes V, VI, and VII may be cultivated safely with special management. Because several kinds of soil often occur in the same capability class on the same farm or ranch, the classes are divided into subclasses.

Four kinds of problems are recognized in the subclasses and are indicated by symbols: *e* — erosion and runoff; *w* — wetness and drainage; *s* — root zone and tillage limitations, such as shallowness, stoniness, droughtiness, and salinity; and

c — climatic limitations. The subclass therefore provides more specific information about the kind and the degree of limitation for the use of soil than does the capability class. See Figure 1.14.

The land capability unit is the most detailed and specific soil grouping of the capability classification. Soils that can be used in the same way and will give about the same crop yield are grouped into a capability unit. This unit is used most commonly for planning in specific areas, for it groups soils which are nearly alike in the features that affect plant growth and in their response to management.

Other interpretative soil groupings also are used in conservation planning. In extensive range areas, the mapping units are grouped into range sites, which give information about the kind and amount of vegetation that the area will produce when it is in its best condition. This grouping, together with range conditions, provides the basis needed for sound range planning.

On farms or ranches that are to be used for woodland, range or pasture, and cropland, the soil map is interpreted to show the suitability of the land for these uses. For areas that are to be planned as woodland, the mapping units are grouped into woodland sites and interpreted in terms of the kinds and amounts of wood crops that can be produced. See Figure 1.15.

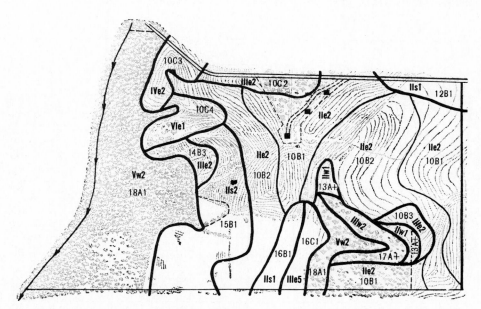

Figure 1.14 A map of the soil and its capability. The symbols pertain to various aspects of soil and topography. For example, 10B1 refers to the kind of soil, the number 10 to the soil type, the letter B to steepness of slope, and the number 1 to degree of erosion. The symbol IIe2 refers to the land capability unit. II designates the land capability class, e indicates the subclass, and 2 indicates the unit. Heavy lines indicate the boundaries of a capability unit.

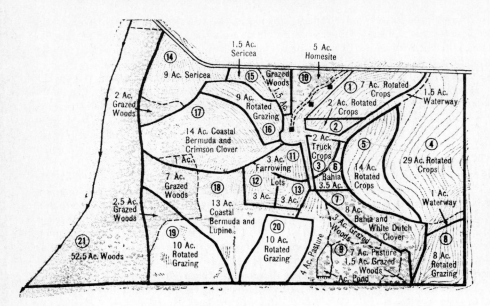

Figure 1.15 A conservation plan map. The decisions made by the farmer concerning the use and management of the land, together with the field unit arrangement, are recorded on this map. These decisions were based on the soil and capability map shown in Figure 1.14. The needs of the farmer relative to the farm enterprise also were considered.

1.5 Unified Classification and Properties of Soils

The basic outline of the unified classification system is presented in Figure 1.16. The system has several outstanding features:

1. It is simple. There are twelve materials with which technicians are normally concerned: four coarse-grained materials, four fine-grained materials, and four combined materials. In addition, three organics require special attention, making a total of fifteen items.

2. It provides important physical characteristics, such as size, gradation, plasticity, strength, brittleness, and consolidation potential.

3. It is reliable. The engineering properties implied by the classification are realistic.

TYPICAL NAMES	IMPORTANT PROPERTIES						UNIFIED SOIL CLASSES
	SHEAR STRENGTH	COMPRESS-IBILITY	WORKABILITY AS CONSTRUCTION MATERIAL	PERMEABILITY			
				WHEN COMPACTED	X CH. PER SEC.	X FT. PER DAY	
Well graded gravels, gravel-sand mixtures, little or no fines.	Excellent	Negligible	Excellent	Pervious	$X > 10^{-2}$	$X > 30$	GW
Poorly graded gravels, gravel-sand mixtures, little or no fines.	Good	Negligible	Good	Very Pervious	$X > 10^{-2}$	$X > 30$	GP
Silty gravels, gravel-sand-silt mixtures.	Good to Fair	Negligible	Good	Semi-Pervious to Impervious	$X - 3$ to 10^{-6}	$X - 3$ to 3×10^{-3}	GM
Clayey gravels, gravel-sand-clay mixtures.	Good	Very Low	Good	Impervious	$X - 10^{-6}$ to 10^{-8}	$X - 3 \times 10^{-3}$ to 3×10^{-5}	GC
Well graded sands, gravelly sands, little or no fines.	Excellent	Negligible	Excellent	Pervious	$X > 10^{-3}$	$X > 3$	SW
Poorly graded sands, gravelly sands, little or no fines.	Good	Very Low	Fair	Pervious	$X > 10^{-3}$	$X > 3$	SP
Silty sands, sand-silt mixtures	Good to Fair	Low Fair	to Impervious	Semi-Pervious to 10^{-6}	$X - 10^{-3}$ to 3×10^{-3}	$X - 3$ to 3×10^{-3}	SM
Clayey sands, sand-clay mixtures.	Good to Fair	Low	Good	Impervious	$X - 10^{-6}$ to 10^{-8}	$X - 3 \times 10^{-3}$ to 3×10^{-5}	SC
Inorganic silts and very fine sands, rock flour, silty or clayey fine sands or clayey silts with slight plasticity.	Fair	Medium to High	Fair	Semi-Pervious to Impervious	$X - 10^{-3}$ to 10^{-6}	$X - 3$ to 3×10^{-3}	ML
Inorganic clays of low to medium plasticity, gravelly clays, sandy clays, silty silty clays, lean clays.	Fair	Medium	Good to Fair	Impervious	$X - 10^{-6}$ to 10^{-8}	$X - 3 \times 10^{-3}$ to 3×10^{-5}	CL
Organic silts and organic silty clays of low plasticity.	Poor	Medium	Fair	Semi-Pervious to Impervious	$X - 10^{-4}$ to 10^{-6}	$X - 3 \times 10^{-3}$ to 3×10^{-3}	OL
Inorganic silts, micaceous or diatomaceous fine sandy or silty soils, elastic silts.	Fair to Poor	High	Poor	Semi-Pervious to Impervious	$X - 10^{-4}$ to 10^{-6}	$X - 3 \times 10^{-1}$ to 3×10^{-3}	MH
Inorganic clays of high plasticity, fat clays.	Poor	High to Very High	Poor	Impervious	$X - 10^{-6}$ to 10^{-8}	$X - 3 \times 10^{-3}$ to 3×10^{-5}	CH
Organic clays of medium to high plasticity, organic silts.	Poor	High	Poor	Impervious	$X - 10^{-6}$ to 10^{-8}	$X - 3 \times 10^{-3}$ to 3×10^{-5}	OH
Peat and other highly organic soils.	NOT SUITABLE FOR CONSTRUCTION						Pt

Figure 1.16 Unified classification system.

EMBANKMENTS

COMPACTION CHARACTERISTICS	STANDARD PROCTOR UNIT DENSITY LBS. PER CU. FT.	TYPE OF ROLLER DESIRABLE	RELATIVE CHARACTERISTICS		RESISTANCE TO PIPING	ABILITY TO TAKE PLASTIC DEFORMATION UNDER LOAD WITHOUT SHEARING	GENERAL DESCRIPTION & USE	UNIFIED SOIL CLASSES
			PERMEABILITY	COMPRESSIBILITY				
Good	125-135	crawler tractor or steel wheeled & vibratory	High	Very Slight	Good	None	Very stable, pervious shells of dikes and dams.	GW
Good	115-125	crawler tractor or steel wheeled &	High	Very Slight	Good	None	Reasonably stable, pervious shells of dikes and dams.	GP
Good with close control	120-125	rubber-tired or sheepsfoot	Medium	Slight	Poor	Poor	Reasonably stable, not well suited to shells but may be used for impervious cores or bankets.	GM
Good	115-120	sheepsfoot or rubber-tired	Low	Slight	Good	Fair	Fairly stable, may be used for for impervious core.	GC
Good	110-130	crawler tractor & vibratory or steel wheeled	High	Very Slight	Fair	None	Very stable, pervious sections, slope protection required.	SW
Good	100-120	crawler tractor & vibratory or steel wheeled	High	Very Slight	Fair to Poor	None	Reasonably stable, may be used to dike with flat slopes.	SP
Good with close control	110-125	rubber-tired or sheepsfoot	Medium	Slight	Poor to Very Poor	Poor	Fairly stable, not well suited to shells, but may be used for impervious cores or dikes.	SM
Good	105-125	sheepsfoot or rubber-tired	Low	Slight	Good	Fair	Fairly stable, use for impervious core for flood control structures.	SC
Good to Poor Close control essential	95-120	sheepsfoot	Medium	Medium	Poor to Very Poor	*Very Poor	Poor stability, may be used for embankments with proper control.	ML
Fair to Good	95-120	sheepsfoot	Low	Medium	to Poor	to Poor	*Varies with water content Stable, impervious cores and blankets.	CL
Fair to Poor	80-100	sheepsfoot	Medium to Low	Medium to High	Good to Poor	Fair	Not suitable for embankments.	OL
Poor to Very Poor	70-95	sheepsfoot	Medium to Low	Very High	Good to Poor	Good	Poor stability , core of hydraulic fill dam, not desirable in rolled fill construction.	MH
Fair to Poor	75-105	sheepsfoot	Low	High	Excellent		Fair stability with flat slopes, thin thin cores, blanket & dike sections.	CH
Poor to Very Poor	65-100	sheepsfoot	Medium to Low	Very High	Good to Poor	Good	Not suitable for embankments.	OH
DO NOT USE FOR EMBANKMENT CONSTRUCTION								Pt

CHANNELS LONG DURATION TO CONSTANT FLOWS. RELATIVE DESIRABILITY		FOUNDATION FOUNDATION SOILS. BRING UNDISTURBED, ARE INFLUENCED TO A GREAT DEGREE BY THEIR GEOLOGIC ORIGIN. JUDGEMENT AND TESTING MUST BE USED IN ADDITION TO THESE GENERALIZATIONS.					UNIFIED SOIL CLASSES
EROSION ASSISTANCE	COMPACTED EARTH LINING	0500 BEARING VALUE	RELATIVE DESIRABILITY		REQUIREMENTS FOR SEEPAGE CONTROL		
			SEEPAGE IMPORTANT	SEEPAGE NOT IMPORTANT	PERMANENT RESERVOIR	FLOOD WATER RETARDING	
1	-	Good	-	1	Position cutoff or blanket.	Control only within volume acceptable plus pressure relief if if required.	GW
2	-	Good	-	3	Positive cutoff or blanket.	Control only within volume acceptable plus pressure relief if required.	GP
4	4	Good	2	4	Core trench to none	None	GM
3	1	Good	1	6	None	None	GC
6	-	Good	-	2	Positive cutoff or upstream blanket & tow drains or wells.	Control only within volume acceptable plus pressure relief if required.	SW
7 if gravelly	-	Good to Poor depending upon density	-	5	Positive cutoff or upstream blanket & toe drains or wells.	Control only within volume acceptable plus pressure relief if required.	SP
8 if gravelly	5 erosion critical	Good to Poor depending upon density	4	7	Upstream blanket & toe drains or wells	Sufficient control to prevent dangerous seepage iping.	SM
5	2	Good to Poor	3	8	None	None	SC
-	6 erosion critical	Very Poor, susceptible to liquefication	6, if saturated or pre-wetted	9	Positive cutoff or upstream blanket & toe drains or wells.	Sufficient control to prevent dangerous seepage piping.	ML
9	3	Good to Poor	5	10	None	None	CL
-	7 erosion critical	Fair to Poor, may have excessive settlement	7	11	None	None	OL
-	-	Poor	8	12	None	None	MH
10	8 volume change critical	Fair to Poor	9	13	None	None	CH
-	-	Very Poor	10	14	None	None	OH
-	-	REMOVE FROM FOUNDATION					Pt

Figure 1.16 Unified classification system (*Continued*).

1.6 Soil Investigations

Geophysical Methods of Exploration

Name of method	Procedure or principle utilized	Applicability
Seismic methods:	Based on time required for seismic waves to travel from source of blast to points on ground surface, as measured by geophones spaced at intervals on a line at the surface	
Refraction	Refraction of seismic waves at the interface between different strata gives a pattern of arrival time vs. distance at a line of geophones	Utilized to determine depth to rock or other lower stratum substantially different in wave velocity from the overlying material. Generally limited to depths up to 100 feet of a single stratum. Used only where wave velocity in successive layers becomes greater with depth
Reflection	Geophones record travel time for the arrival of seismic waves reflected from the interface of adjoining strata	Generally applied only for depths exceeding 2000 feet
Electrical methods:		
Resistivity	Based on the difference in electrical conductivity or resistivity of strata. Resistivity of subsoils at various depths is determined by measuring the potential drop and current flowing between two current and two potential electrodes from a battery source. Resistivity is correlated to material type	Used to determine horizontal extent and depths up to 100 feet of subsurface strata. Principal applications for investigating foundations of dams and other large structures, particularly in exploring granular river channel deposits or bedrock surfaces
Drop in potential	Based on the determination of the ratio of potential drops between three potential electrodes as a function of the current imposed on two current electrodes	Similar to resistivity methods but gives sharper indication of vertical or steeply inclined boundaries and more accurate depth determinations. More susceptible than resistivity method to surface interference and minor irregularities in surface soils
Continuous vibration method	The travel time of transverse or shear waves generated by a mechanical vibrator consisting of a pair of eccentrically weighted disks is recorded by seismic detectors placed at specific distances from the vibrator	Velocity of wave travel and natural period of vibration gives some indication of soil type. Travel time plotted as a function of distance indicates depths or thicknesses of surface strata. Useful in determining dynamic modulus of subgrade reaction and obtaining information on the natural period of vibration for design of foundations of vibrating structures
Magnetic measurements	Magnetometer is used to measure the vertical component of the earth's magnetic field at closely spaced stations in an area	*Note: Not applicable to building foundation investigations*
Gravity measurements	Based on difference in density of subsurface materials as indicated by the vertical intensity or the curvature and gravitational field at various points being investigated	*Note: Rarely applicable to building foundation investigations*
Sonic method	The time of travel of sound waves reflected from the mud line beneath a body of water and a lower rock surface is computed by predetermining the velocity of sound in the various media	Currently used in shallow underwater exploration to determine position of mud line and depth to hard stratum underlying mud

On-Site Investigations

Type of investigation	Information sought	Recommended procedures
Reconnaissance and preliminary	General site conditions	Visual inspection Interpretation of agricultural maps Interpretation of geological maps Interpretation of air photos Study of other local developments Geophysical methods Field penetrometer and probings
Field exploration and sampling	Exploration hole logs	Exploration recording, ASTM Spec. Proc. 1970, (Sub. 2) Description of soils, ASTM D-2488 Classification of soils, ASTM D-2487
	Representative intact but disturbed samples	Split barrel sampling, ASTM D-1586
	1. Small samples	Diamond core drilling, ASTM D-2113 Any of (2) below
	2. Large samples	Auger borings, ASTM D-1452 Accessible excavations, USBR, E-1a
	Undisturbed samples	
	1. Soft to stiff cohesive soils	Thin-wall tube sampling, ASTM D-1587
	2. Stiff to hard cohesive or cemented soils and shales	Denison, Pitcher, Hollow Auger Samplers, USBR, E-2B
	3. Sand soils of low moisture	Piston or thin-wall tube samplers, USBR, E-2B
	4. Sand soils, wet or saturated	Piston samplers, USBR, E-2B
	5. All soils	Hand-cut samples, USBR, E-2A
	6. Rock	Diamond-core samples, ASTM D-2113
Field testing	In situ conditions	
	1. Moisture	Nuclear method, ASTM Spec. Proc., 1970 (Shepard)
	2. Density	Sand cone method, ASTM D-1556 Rubber balloon method, ASTM D-2167 Nuclear method, ASTM Spec. Proc. 1970 (Sub. 8) Penetration test, ASTM D-1586 (estimating relative density of sands)
	3. Permeability	FHA Minimum Property Standards for Single and Multi Family Dwellings Well permeameter test, USBR, E-19 Infiltration test, ASTM Spec. Proc. 1970 (Sub. 4) Borehole test, USBR, E-18
	4. Bearing capacity	Test for static load on spread footings, ASTM D-1194 Penetration test, ASTM D-1586 (estimating relative density of sands)
	5. Shear strength	Vane shear test, ASTM D-2573 (for soft cohesive soils) Penetration test, ASTM D-1586 (estimating relative density of soils)
	6. Pile load	Load-settlement relationship for vertically loaded single piles, ASTM D-1143

1.6 SOIL INVESTIGATIONS

Laboratory Investigations

Problem	Data required	Recommended test procedure
General information: Classification	Identification	Manual tests (unified system) ASTM D-2488 Consistency Liquid limit, ASTM D-423 Plastic limit, ASTM D-424 Shrinkage limit, ASTM D-427 Gradation, ASTM D-422 Related tests -200 fraction, ASTM D-1140 Dry preparation, ASTM D-421 Wet preparation, ASTM D-2217
Other properties	Specific gravity, void ratio, porosity, degree of saturation	Specific gravity: Fine soils, ASTM D-854 Gravelly soils, ASTM C-127
Building foundation: Bearing capacity	Strength	Unconfined compression, cohesive soils, ASTM D-2166 Direct shear, ASTM Spec. Proc. (by USBPR), USED, Appendix IX (S test) Undrained triaxial compression (Q test), ASTM D-2850 (without pore pressure measurement)
Settlement	Compressibility	Consolidation: Fine-grained soils: ASTM D-2435 Coarse-grained fills: USBR, E-14
Expansion—shrinkage	Swell and shrinkage Water content Consistency, gradation	Qualitative: FHA PVC meter, FHA Bull. 701 Quantitative: ASTM Spec. Proc. (by Holtz) ASTM D-2216; see identification test
Slope stability: (natural and fill)	Strength	Unconfined compression, cohesive soils, ASTM D-2166 Direct shear, sands ASTM Spec. Proc. (by USBPR) USED, Append. IX Triaxial shear, cohesive and cohesionless Unconsolidated-Undrained (Q test) ASTM D-2850 (without pore pressure measurement) Consolidated-Undrained (R test) USED, Appendix X; USBR, E-17 Consolidated-drained (S test) USED, Appendix X
Lateral earth pressure	Strength deformation and principal stresses	See Slope stability—strength
Earth fills	Unit weight	Compaction, ASTM D-698 or ASTM D-1557 Relative density, ASTM D-2049
	Water content	ASTM D-2216
	Compressibility	See Settlement—compressibility
	Strength	See Stability of slopes—strength
	Permeability	Granular soils, constant head, ASTM 2434 Cohesive and granular soils, constant head, USBR, E-13 Gravelly soils, constant head USBR, E-14

Sources of Geological Information

Series	Description of material
U.S. Geological Survey (USGS)	Consult "USGS Index to Publications" from Superintendent of Documents, Washington 25, D.C. Order publications from Superintendent of Documents. Order maps from USGS, Washington 25, D.C.
Geological Index map	Individual maps of each state showing coverage and sources of all published geological maps
Folios of the Geological Atlas of U.S.	Contains maps of bedrock and surface materials for many important urban and seacoast areas. When out of print, obtain folios through suppliers of used technical literature
Geological Quadrangle Maps of U.S.	This series supplants the older geological folios and includes areal or bedrock geology maps with brief descriptive text. Series is being extended to cover areas not previously investigated
Bulletins, professional papers, circulars, annual reports, monographs	General physical geology emphasizing mineral and petroleum resources. Areal and bedrock geology maps for specific locations included in many publications
Water supply papers	Series includes papers on groundwater resources in specific localities which are generally accompanied by description of subsurface conditions affecting groundwater plus observations of groundwater levels
Topographic maps	Topographic contour maps in all states, widespread coverage being continually expanded
U.S. Coast and Geodetic Survey (USC&GS) Nautical charts	Consult Index from Director, U.S. Coast and Geodetic Survey, Washington 25, D.C. Charts of coastal areas showing available soundings of sea bottom plus topographic and cultural features adjacent to the coast or waterways
U.S. Department of Agriculture (USDA), Soil Conservation Service Soil maps and reports	Consult Highway Research Board (HRB), BUL No. 22-R, "Agricultural Soil Maps, Status July 1957," for coverage by counties of USDA Soil Maps and Reports Surveys of surface soils described in agricultural terms. Physical geology summarized. Excellent for highway, airfield, *or subdivision* investigations. Coverage mainly in midwest, east, and southern United States
State geologists' bulletins, reports, and maps	Consult HRB Bull. No. 180, "Geologic Survey Mapping in the United States," for addresses of all state geological organizations. Most states provide excellent detailed local geological maps and reports covering specific areas or features in the publications of the state geologists
Geological Society of America (GSA)	Write for index to GSA, 419 West 117 Street, New York, N.Y.
Monthly bulletins, special papers, and memoirs	Texts cover specialized geological subjects and intensive investigations of local geology. Detailed geological maps are frequently included in the individual articles
Geologic maps	Publications include general geological maps of North and South America, maps of glacial deposits, and Pleistocene acollan deposits

Types of Test Borings

Boring method	Procedure utilized	Applicability
Displacement type	Repeatedly driving or pushing tube or spoon sampler into soil and withdrawing recovered materials. Changes indicated by examination of materials and resistance to driving or static force for penetration. No casing required	Used in loose to medium compact sands above water table and soft to stiff cohesive soils. Economical where excessive caving does not occur. Limited to holes < 3 inches in diameter
Auger boring	Hand- or power-operated augering with periodic removal of material. In some cases continuous auger may be used requiring only one withdrawal. Changes indicated by examination of material removed. Casing generally not used	Ordinarily used for shallow explorations above water table in partly saturated sands and silts, and soft to stiff cohesive soils. May be used to clean out hole between drive samples. Very fast when power-driven. Large-diameter bucket auger permits examination of hole
Wash-type boring for undisturbed or dry samples	Chopping, twisting, and jetting action of a light bit as circulating drilling fluid removes cuttings from hole. Changes indicated by rate of progress, action of rods, and examination of cuttings in drilling fluid. Casing used as required to prevent caving	Used in sands, sand and gravel without boulders, and soft to hard cohesive soils. Most common method of subsoil exploration. Usually can be adapted for inaccessible locations, such as over water, in swamps, on slopes, or within buildings

Types of Test Borings (*Continued*)

Boring method	Procedure utilized	Applicability
Rotary drilling	Power rotation of drilling bit as circulating fluid removes cuttings from hole. Changes indicated by rate of progress, action of drilling tools, and examination of cuttings in drilling fluid. Casing usually not required except near surface	Applicable to all soils except those containing much large gravel, cobbles, and boulders. Difficult to determine changes accurately in some soils. Not practical in inaccessible locations because of heavy truck-mounted equipment, but applications are increasing since it is usually most rapid method of advancing borehole
Percussion drilling (churn drilling)	Power chopping with limited amount of water at bottom of hole. Water becomes a slurry which is periodically removed with bailer or sand pump. Changes indicated by rate of progress, action of drilling tools, and composition of slurry removed. Casing required except in stable rock	Not preferred for ordinary exploration or where undisturbed samples are required because of difficulty in determining strata changes, disturbance caused below chopping bit, difficulty of access, and usual higher cost. Sometimes used in combination with auger or wash borings for penetration of coarse gravel, boulders, and rock formations
Rock core drilling	Power rotation of a core barrel as circulating water removes ground-up material from hole. Water also acts as coolant for core barrel bit. Generally hole is cased to rock.	Used alone and in combination with boring types to drill weathered rocks, bedrock, and boulder formations

1.7 Geological Investigations

Geological investigations are made to determine the geologic conditions that affect the design, safety, effectiveness, and cost of a proposed project. Insufficient geological investigations and faulty interpretations of results have been responsible for costly construction changes and could be the cause of the failure of a structure. The investigations are performed to determine the general geologic setting of the project, the geologic conditions that influence the selection of a site, the characteristics of the foundation soils and rocks, all other geologic conditions that influence design and construction, and sources of construction materials. The method employed for the investigations depends on the type of structures contemplated and the character and degree of accuracy of the information required. The extent of the investigations depends on the magnitude of the project and the simplicity or complexity of local geology.

Research This phase of investigation includes a careful search of published and unpublished papers, reports, maps, and records and consultation with federal, state, and local geological authorities for information pertinent to the project or problem. Thorough utilization of this source of information during preliminary investigations of projects cannot be overemphasized. Publications of federal and state agencies, such as those of the various geological surveys, in many instances contain information pertinent to the geology of the area and of the project site as well as data on mineral resources. Annual bulletins of the U.S. Department of Commerce titled *United States Earthquakes* are good reference material on earthquakes. Many state geological surveys and university libraries have well-drilling data, groundwater data, logs, and information on rock outcrops that are available for examination and study. Some state highway departments have geological profiles for cuts along existing or proposed highways and possess records of borings and construction data that can be studied. Copies of such unpublished data are obtainable in some instances. Exploration and construction data from work performed by private companies should be included in the research.

Maps Strictly speaking, the use of published maps in geologic investigations should fall under the category of research. It is discussed here separately for emphasis.

There are commonly available various types of published maps from which geologic information pertinent to a project can be obtained prior to exploration work. These include topographic maps, geologic maps, mineral resources maps, and soil maps. Such maps can be quite helpful for obtaining preliminary information and for planning subsequent reconnaissance and exploration.

Topographic Maps Most topographic maps published in the United States are prepared by the U.S. Geological Survey. These maps are available in 7½-minute quadrangle size plotted to scales of 1:24,000 and 1:31,680 and in 15-minute quadrangle size plotted to a scale of 1:62,500.

Topographic maps based on the United States military grid are published by the various agencies of the defense establishment, particularly the Army Map Service. Maps prepared by the Army Map Service are in 7½-minute quadrangle size plotted to a scale of 1:25,000, in 15-minute quadrangle size plotted to a scale of 1:50,000, and in 30-minute quadrangle size plotted to a scale of 1:125,000. Maps covering larger areas are prepared on scales of 1:250,000 and 1:500,000.

Certain information on engineering geology can be

inferred from topographic maps by a proper interpretation of the landforms and drainage patterns shown on them. Topography tends to reflect the geologic structure and composition of the underlying rocks. Geologic features, however, are not equally apparent on all topographic maps, and considerable skill is required to arrive at geologic interpretations from maps of some areas. Information of engineering significance that may be obtained or inferred from topographic maps may be classified as follows:

Physiography
    Significant topographic features
    Physiographic history pertinent to engineering
General rock types
    Crystalline or noncrystalline
    Massive or thin-bedded
    Alternating hard and soft rocks
    Glacial terrain
Rock structure
    Dip and strike
    Folding
    Faults
    Joints
    Slide areas
Soil types
    Glacial
    Alluvial
    Residual

Geologic Maps   Geologic maps are published by the U.S. Geological Survey, by various state geological surveys or equivalent state agencies, and by some of the geological societies. A partial list of readily available sources follows:

1. U.S. Geological Survey
    Geologic map of the United States; scale, 1:2,500,000
    Folios of the *Geologic Atlas of the United States* (no longer published; many available in libraries)
    Geologic quadrangle maps, in publication since 1945; scale, 1:31,680, 1:62,500, and 1:125,000 (maps titled ``Geology,'' ``Bedrock Geology,'' ``Surficial Geology,'' ``Economic Geology,'' and ``Engineering Geology'' published in this series)
    Bulletins and professional papers
2. State geological surveys or equivalent state agencies
    Geologic maps of most states; scale, 1:500,000
    Bulletins, county reports, and monographs
3. Geological Society of America
    Glacial map of North America

Glacial map of the United States east of the Rocky Mountains

Map of Pleistocene eolian deposits of the United States, Alaska, and parts of Canada

The published geologic maps show the general geology of an area and provide information useful in preliminary planning of investigations.

Mineral Resources Maps   Mineral resources maps are published by the U.S. Geological Survey and by state geological surveys or equivalent agencies. These may be helpful and should be consulted in determining the effect of a proposed project on the mineral resources of the area.

Aerial Photographs   Aerial photographs may be either vertical or oblique views. Both are used effectively in geologic studies. Geologic interpretations for engineering are based on conventional principles of photointerpretation of landforms, drainage patterns, and other surface features. The procedure is similar to making geologic interpretations from topographic maps, with the exception that many more details can be seen on photographs than are shown on corresponding maps. Definite criteria for complete interpretation from aerial photographs, however, are difficult to set forth. Much depends upon the experience and ability of persons making the interpretation and upon their general geologic knowledge of the area. Analytical factors used in making systematic studies of aerial photographs include (1) topography, (2) drainage and erosion, (3) color tones, and (4) vegetative cover.

Aerial photographs are valuable tools for geologists, particularly in connection with field reconnaissance and preliminary studies. They present an overall view of the land surface and of the features of bedrock geology that otherwise may be seen only in piecemeal fashion or escape notice entirely. They also cover areas that have difficult or limited access by ground. Study of the pictures in many instances reveals features to which special attention should be given during reconnaissance or in later studies.

Field Reconnaissance   A geological field reconnaissance is a preliminary investigation involving one or more trips to the project site or area for the purpose of gathering such information as is obtainable without subsurface exploration or detailed study. Before plans for detailed surface and subsurface investigations are prepared, the geologist responsible for planning such investigations should have some advance knowledge of the geology, physiography, and cultural features of the area involved. Some of this knowledge will be obtained from a study of published reports and maps, but field reconnaissance is required to orient the published information with actual field conditions. During field reconnaissance, the geologist will study and record all topographic and geologic features that may have a significant bearing on the suitability of the site for an engineering project. It is essential to obtain at this time information on landforms, soil types and thickness, bedrock types and structure, groundwater conditions, construction materials, and mineral resources that can

influence the type of structure or affect the cost of the project.

Geophysical Methods Geophysical exploration consists of making certain physical measurements by the use of special instruments to obtain generalized subsurface geologic information. Geophysical observations in themselves are not geologic facts but are statistical and orderly measurements. The required geologic information is obtained indirectly through analysis or interpretation of these measurements and is not subject to direct visual verification. Geophysical explorations lack the degree of accuracy provided by core drilling; therefore, boreholes or other direct geological explorations are needed for reference and control of measurements when geophysical methods are used. Geophysical explorations are not a substitute for core drilling, test pits, or other direct methods of subsurface exploration. They are appropriate, however, for a rapid though roughly approximate determination of certain geologic conditions, such as the depth to bedrock, when supplemented by core boring or other reliable exploration methods. The cost of geophysical explorations is generally low compared with the cost of core borings or test pits, and considerable savings may often be effected by a judicious use of this exploration method in conjunction with other more reliable methods.

The six major geophysical exploration methods are the seismic, electrical resistivity, sonic, electrical logging, magnetic, and gravity methods.

Subsurface Borings Borings of various kinds are the commonest methods of subsurface exploration. The boring methods most frequently used are discussed below.

Probings Probings ordinarily consist of driving a steel rod into the ground and noting variations in penetration resistance. Probe rods may be of any size. They are generally round or hexagonal in cross section and about ⅝ to ⅞ inch in diameter. Most are pointed or have a driving point attached. Penetration may be accomplished by driving the rod by manual or mechanical hammers.

Probings generally are intended to determine the presence or absence of bedrock within prescribed depths. The resistance to penetration can be interpreted roughly in terms of the character of the material penetrated. Probings, however, are not reliable. Penetration refusal, which is often erroneously interpreted as bedrock, may actually prove to be a cobble, a boulder of a cemented soil condition. Probings should never be used for other than preliminary exploration purposes, and interpretations from them should always be checked by other boring methods.

Wash Borings Wash boring consists essentially of forcing a wash pipe or a hollow drill rod through the overburden by chopping and jetting while water is pumped through the boring device. Usually, a chopping bit is attached to the bottom end of the wash pipe. The displaced soil is washed to the surface, where it may be caught in a bucket or other container for sampling. A casing is required in soft or loose soils, but it may not be needed in such materials as stiff clay or glacial till.

Wash borings are used principally to advance holes through overburden materials between zones of drive sampling or to obtain a bore through overburden to rock preparatory to rock drilling. They are not reliable for determining subsurface conditions, and the information obtained from them is nearly always misleading. Definite identification of soils cannot be made. Fine materials in the cuttings are lost in the wash water, and coarser materials, such as gravel or cobbles, either may not be forced up out of the hole or may be broken into fine chips that would not show in the cuttings. Boulders and cobbles may easily be taken for bedrock when arbitrary depths are selected for these borings.

Core Drilling Core drilling involves the cutting and recovery of cylindrical cores of subsurface materials. The term includes certain methods of soil drilling and sampling but is most commonly applied to the drilling and recovery of continuous cores from bedrock.

Core drilling in bedrock is accomplished with rotary drills. These are obtainable in various sizes and models to meet different requirements related to drilling location, hole size and depth, and purpose of the borings. Drills may be post-, carriage-, skid-, or truck-mounted and may be powered by air or electric motors or by diesel or gasoline engines. For drills employed in cutting small-diameter cores, maintenance of pressure on and advancement of the bit are accomplished by a screw-feed or hydraulic-feed device in the swivel head of the drill.

Drilling is performed by rotating a tubular bit in a hole by means of a series of drill rods driven by a machine at the top of the hole. The cutting edge of the bit may be diamond- or alloy-set, or the bit may be of the calyx type or have alloy steel teeth. All coring bits are fastened to the bottom of a core barrel. The bit cuts an annular hole around a central core, and the barrel passes down over the core as the bit advances. The operation continues until the core barrel is filled, at which time the barrel with the core is recovered from the hole and the core samples are arranged in boxes for logging. The drilling tools then are returned to the hole, and the drilling operation is resumed. Water or other drilling fluid is circulated through the drill rods and core barrel to the bit and thence back up the hole for the purpose of removing rock cuttings and keeping the bit cool.

Exploratory core borings are indispensable for determining the character of the overburden materials, the depth to bedrock, and the character and condition of bedrock materials. They should be employed on all foundation studies subsequent to preliminary examination, since the information obtained from them is more accurate than that obtained from most other exploration methods except excavation of test pits, shafts, trenches, and tunnels. Core drilling is the only practical method of sampling rock materials short of these more costly methods. Good core recovery, however, may be difficult in soft or weathered rock, in rock containing closely spaced fractures and cleavage planes, and in rock of widely varying hardness. As a general rule, the larger the diameter of the core, the higher the percentage of core

recovered and the more accurate the geological observations on the character and condition of the rock.

Calyx Core Drilling Calyx drilling differs from other coring methods chiefly in the type of bit used and in the means of obtaining the cutting action from the bit. The bit consists of a core barrel of soft steel with one or more slots cut in the bottom rim or of a short slotted tube of mild steel attached to the bottom edge of a single-tube core barrel. Steel shot, which is fed to the bottom edge of the barrel by drill water, is the cutting medium. Under the weight of the rotating barrel, the shot cuts an annular groove around the core without disturbing the wall of the borehole. Rock cuttings are removed by circulating water and are caught in a sludge barrel on top of the core barrel. Cores as small as 2 inches in diameter may be obtained by the calyx drilling method, but the greatest use of this method is for drilling large-diameter holes into which a person can be lowered to examine the rock in place. Such calyx holes usually are 36 inches or larger in diameter and afford opportunities to observe the actual condition of the rock through which the holes were drilled. Calyx drilling can be performed only in a vertically downward direction, because an even distribution of the shot at the base of the slotted bit cannot be achieved in inclined borings. Drilling progress is slow, and the drilling of large-diameter holes is expensive. Where extremely open or fractured water-bearing rocks are to be explored, rock voids must be consolidated by colored grout if inspection of the calyx hole is to be successful.

Borehole Photography Borehole photography consists of photographing the interior surfaces of boreholes and studying the photographs to obtain information on the materials through which the borings have penetrated. The photographs are taken with special borehole cameras.

Borehole photography provides more extensive information from small-diameter borings than is obtainable from the cores alone and is much less expensive than the excavation of shafts and tunnels or the drilling of large-diameter calyx holes. The borehole photographs show existing conditions when no core is recovered or when the core is too badly broken to give reliable information. They show the conditions of weathering and the details and orientation of rock structures such as fractures, joints, and bedding planes. They can also show the effectiveness of pressure grouting of foundation openings.

Churn Drilling This exploration method consists of drilling holes with cable tool drills or well drills. The hole is drilled by the impact and cutting action of a heavy chisel-edged drilling tool that is alternately raised and dropped by means of a cable. The up-and-down motion of the cable and the attached drilling bits may be accomplished by a tripod-winch-cathead arrangement or by a crank or spudding arm that is operated by motorized units and used in conjunction with a hoisting mast. Such a drilling rig as the latter may be skid- or truck-mounted. The hole is kept partially filled with water during chopping, and the cuttings are cleaned from the hole periodically with a bailer at intervals of 5 or 10 feet. Casing is required when drilling through unstable materials.

Churn drilling is used principally in advancing holes through overburden to rock. The cuttings from churn drilling represent only the average disturbed composition of the materials penetrated, and an accurate determination of the undisturbed character of the subsurface materials is impossible. If the character of the subsurface materials must be ascertained, undisturbed samples are taken with a soil-sampling head. Churn drilling has been used effectively for groundwater studies in regions where wells are needed to observe water table elevations and fluctuations in limestone formations.

Wagon and Jackhammer Drilling Wagon drills and jackhammers are percussion-type drills operated by compressed air or by a self-contained gasoline engine. As the name implies, the wagon drill is mounted on a wheeled carriage. The jackhammer drill is hand-held. Explorations with these drills consist of drilling holes into rock and observing the rate of and the resistance to penetration of the drill as well as the color of the cuttings from the hole. These observations are then interpreted in terms of bedrock condition.

Explorations with wagon drills or jackhammers may be useful as supplementary investigations for searching out shallow cavities, solution channels, and soft zones after the overburden has been stripped from foundation rock. Because such borings are extremely unreliable for determining either the composition or the engineering properties of subsurface materials, they cannot be substituted for core borings.

Test Pits, Trenches, and Tunnels

Test Pits Test pits are openings excavated vertically from the ground surface to expose the subsurface materials for examination in place. They may also be excavated to take undisturbed samples of soil materials. Their greatest use is in connection with soil exploration and testing. They may also be employed to study the character of the overburden bedrock contact and the position, character, and condition of the bedrock surface.

Trenches Trenches are functionally somewhat similar to test pits except that they are usually limited to relatively shallow depths below the ground surface. They are particularly useful for the continuous exploration, examination, and sampling of soil foundations of earth embankments and for examining and correlating bedrock surface conditions that elude accurate identification by conventional drilling and sampling methods. Combined with test pits, exploratory trenches are the only reliable method of determining the occurrence, composition, distribution, structure, and stability of unsatisfactory materials in deep alluvial and residual soil foundations for high dams.

Tunnels Tunnels and drifts are nearly horizontal underground passages or openings. They are excavated by common mining methods, which vary with the type of material being tunneled. Their principal function as an exploratory device is to permit detailed examination of the composition and geometry of such rock structures as joints, fractures,

faults, shear zones, and solution channels when these conditions affect foundation excavation and treatment. Excavation of exploration tunnels generally is slow and expensive and should be used only when other methods are inadequate for supplying information. Tunnels are especially useful in the proper exploration of foundation conditions in the abutments of high concrete gravity and arch dams. Logs of exploration tunnels should be made concurrently with their excavation whenever possible. Sampling, if required, also should be done during the excavation.

Groundwater Observations Groundwater observations include observations and measurements of flows from springs and of water levels in existing wells, boreholes, selected observation wells, and piezometers. Such observations are made in connection with studies to determine water table elevations and profiles, fluctuations in water table elevations, the possible existence and location of perched water tables, depths of probable water-bearing horizons, artesian flow, and locations of possible leakage.

Springs The method employed for measuring the flow from natural springs generally depends on the size and location of the springs and on the shape of the outlet. Spring flow may be estimated when the quantity involved is too small to make measurements practicable, but in most instances actual measurements are practicable and should be made. Flows from small springs often are measured by directing the water into a container of known capacity and noting the time required to fill the container. Measurements also can be made by installing a weir across the channel along which the water flows from the spring. Flows are computed from measurements of the depths of water flowing over the notch in the weir by means of formulas that vary with the cross-sectional shape and size of the weir. Flows from large springs are computed from current velocity measurements in the outlet channel and from measurements of the size together with a consideration of the shape of the channel.

Existing Wells Water levels in existing wells usually are obtained from the well owners. If the owners do not have such information, measurements may be obtained by the methods described below.

Boreholes The depth to water level in a borehole may be determined by means of a chalked tape, a ''plunker,'' or an electrical water-level indicator. Because of the influence of the drilling fluid on the water levels in the holes, time should elapse after the fluid circulation has stopped and the tools have been removed before measurements are made; the water in the hole can thus adjust to its static level. If measurements are desired before a hole is completed, a common practice is to fill the hole with water at the end of the workday and to measure the depth to water level at the start of the next workday. Measurements after completion of a hole generally are made 24 hours or more after drilling has stopped. Numerous subsequent check readings often are required to obtain accurate water table elevations. Because of seasonal and other fluctuations in groundwater levels, all records of water-level measurements should contain the date on which the measurements were made. Records of measurements on uncompleted boreholes also should contain the time of day and the depth of hole at which the measurements were made in order that the measurements may be correlated with the progress of the borings.

Observation Wells These wells are employed when observations of groundwater levels are to be continued over a more or less prolonged period of time during which detailed observations are desired. Either existing wells and boreholes or holes drilled especially for groundwater observation purposes may be used. Although casing often is left in boreholes used as observation wells, the conversion of these holes to observation wells by the installation of piezometers or the use of specially constructed observation wells is preferable. Piezometers usually are composed of a well point or section of porous pipe attached to the lower end of a metal or plastic pipe. This is placed in a hole and surrounded by a filter of well-graded sand or gravel. A seal of compacted clay or bentonite is placed above the filter material in the annular space around the piezometer pipe to prevent the inflow of surface water and to seal off any portion of the hole above the zone for which waterhead observations are desired. Water-level measurements may be obtained by the same methods as described above under ''Boreholes.''

1.8 Soil Bearing Values

The requirements for the determination of soil bearing values should be in accordance with the following data or with the American Standards Association's *Building Code Requirements for Excavations and Foundations.* Where the bearing value of soil is determined by field loading tests and where other bearing values are established by local practice and experience or because of special conditions, soil bearing values should not exceed those given in the tables on undisturbed soil on pages 35, 36, and 37.

Modification of Bearing Value

Variation in Underlying Soils Where the bearing materials directly under a foundation overlie strata with a smaller allowable bearing value, such smaller value should not be exceeded at the top level of such strata. The computation of the vertical pressure in the bearing materials at any depth below a foundation should be made on the assumption that the load is spread uniformly at an angle of 1 horizontal to 2 vertical.

Loosened Bearing Materials Wherever bearing material is loosened or disturbed by a flow of water, the bearing value is to be reduced to the allowable bearing value of the loosened material, unless the loosened material is removed. Where the flow of water is controlled by well points or by another method so that the bearing material is not disturbed or loosened, the full bearing value of the unloosened material may be assumed.

Foundations on Laterally Supported Soil   The presumptive unit bearing values given in the following tables may be increased for a load on soil where, because of the depth below ground level and permanent lateral support of the bearing soil, greater bearing values are justified.

Soil Bearing Load Test   Tests should be made and interpreted so as to take into account all significant factors, such as the presence of soft underlying strata, variations in the size of footings, and the compressibility of the soils encountered. When the size of proposed footings varies substantially, loading tests should be made on several areas of different sizes as a guide in the determination of allowable bearing values for the various footing sizes.

Tests should be made where surface-water conditions and groundwater conditions are representative of the bearing soil and when the soil tested is free from frost. Tests should be made on leveled but otherwise undisturbed portions of foundation bearing material. Where tests are made materially below the ground level, any material immediately adjoining the test location should be removed to eliminate the effect of surcharge or reinforcing.

The test assembly should consist of a vertical timber or post, with or without braced timber footing, resting upon the soil to be tested and supporting a platform on which the test loads are to be placed. The exact area resting upon the soil should be ascertained; it should be not less than 1 square foot for bearing materials of Classes 1 to 4 inclusive as indicated in the table "Presumptive Unit Soil Bearing Values" and not less than 4 square feet for other bearing materials. The platform should be symmetrical in respect to the post and as close to the bearing soil as is practicable. The post should be maintained in a vertical position by guys or wedges. The load may be any convenient material that can be applied in the increments required, such as cement or sand in bags, or pig iron or steel in bars. In applying the load, precautions should be taken to prevent jarring or moving the post.

Settlement readings should be taken at least once every 24 hours at a point that should remain undisturbed during the test, and the settlement should be plotted against time. The proposed allowable load per square foot should be applied and allowed to remain undisturbed until there has been no settlement for 24 hours.

Frost Penetration   The effect of freezing and thawing is much greater upon soil than upon other materials, such as brick or concrete. Footings should be carried below the frost line, the depth that frost penetrates below the grade. Many building codes require the footing to be carried 1 foot below the frost line. If the footings are above the frost line, they are likely to heave (move) as a result of soil pressures caused by extreme temperature change. It is evident from the maps shown in Figures 1.18 and 1.19 that maximum frost penetration differs in various sections of the United States. Local building codes usually specify the footing depth. Depths below grade are determined by the general drainage conditions and extreme temperatures in a locality.

Underpinning, Shoring, and Sheetpiling   Figures 1.17 and 1.21 demonstrate methods of underpinning abutting foundations and of preventing cave-ins in excavations by means of shoring and sheetpiling.

Nominal Values of Allowable Bearing Pressures for Spread Foundations

Type of bearing material	Consistency in place	Allowable bearing pressure (tons per square foot)	
		Ordinary range	Recommended value for use
Massive crystalline igneous and metamorphic rock: granite, diorite, basalt, gneiss, thoroughly cemented conglomerate (sound condition allows minor cracks)	Hard, sound rock	60–100	80
Foliated metamorphic rock: slate, schist (sound condition allows minor cracks)	Medium-hard sound rock	30–40	35
Sedimentary rock: hard cemented shales, siltstone, sandstone, limestone without cavities	Medium-hard sound rock	15–25	20
Weathered or broken bedrock of any kind except highly argillaceous rock (shale)	Soft rock	8–12	10
Compaction shale or other highly argillaceous rock in sound condition	Soft rock	8–12	10
Well-graded mixture of fine- and coarse-grained soil: glacial till, hardpan, boulder clay (GW–GC, GC, SC)	Very compact	8–12	10
Gravel, gravel-sand mixtures, boulder-gravel mixtures (GW, GP, SW, SP)	Very compact	7–10	8
	Medium to compact	5–7	6
	Loose	3–6	4

Nominal Values of Allowable Bearing Pressures for Spread Foundations (*Continued*)

Type of bearing material	Consistency in place	Allowable bearing pressure (tons per square foot)	
		Ordinary range	Recommended value for use
Coarse to medium sand, sand with little gravel (SW, SP)	Very compact	4–6	4
	Medium to compact	3–4	3
	Loose	2–3	2
Fine to medium sand, silty or clayey medium to coarse sand (SW, SM, SC)	Very compact	3–5	3
	Medium to compact	2–4	2.5
	Loose	1–2	1.5
Fine sand, silty or clayey medium to fine sand (SP, SM, SC)	Very compact	3–4	3
	Medium to compact	2–3	2
	Loose	1–2	1.5
Homogeneous inorganic clay, sandy or silty clay (CL, CH)	Very stiff to hard	3–6	4
	Medium to stiff	1–3	2
	Soft	0.5–1	0.5
Inorganic silt, sandy or clayey silt, varved silt-clay–fine sand (ML, MH)	Very stiff to hard	2–4	3
	Medium to stiff	1–3	1.5
	Soft	0.5–1	0.5

Under the Casagrande system of classification, C = clay, G = gravel, H = high compressibility, L = low to medium compressibility, M = silt, P = poorly graded, S = sand, and W = well graded.

Bearing Powers of Materials

Material	English system		Metric system	
	lb/in^2	tons/ft^2	kg/cm^2	tons/m^2
Rock, solid	350	24	24.6	240
Rock, semishattered	70	5	4.9	50
Clay:				
Dry	55	4	3.9	40
Damp	27	2	1.9	20
Wet	14	1	1.0	10
Gravel, cemented	110	8	7.7	80
Sand:				
Dry, compacted	55	4	3.9	40
Dry, clean	27	2	1.9	20
Quicksand and alluvial soil	7	0.5	0.5	5

Presumptive Unit Soil Bearing Values

Class	Material	Allowable bearing value, tons per square foot*
1	Massive crystalline bedrock, such as granite, gneiss, or traprock, in sound condition	100
2	Foliated rocks, such as schist and slate, in sound condition	40
3	Sedimentary rocks, such as hard shales, silt-stones, or sand-stones, in sound condition	15
4	Exceptionally com-pacted gravels or sands	10
5	Compact gravel sand-gravel mixtures	6
6	Loose gravel; com-pact coarse sand	4
7	Loose coarse sand; loose sand-grav-el mixtures; com-pact fine sand; wet coarse sand (confined)	3
8	Loose fine sand; wet fine sand (confined)	2
9	Stiff clay	4
10	Medium-stiff clay	2
11	Soft clay	1
12	Fill, organic materi-al, or silt	†

*Presumptive bearing values apply to loading at the surface or in cases where permanent lateral support for the bearing soil is not provided.
†Where, in the opinion of the enforcement officer, the bearing value is ade-quate for light frame structures, fill material, organic material, and silt are deemed to be without presumptive bearing value. The bearing value of such material may be fixed on the basis of tests or other satisfactory evidence.

Approximate Angles of Repose of Materials*

Material	Slope ratio, horizontal:vertical	Angle of repose*
Ashes, coal	1.0:1	45
Cinders, coal	1.0:1	45
Clay:		
Dry	1.3:1	38
Damp	2.0:1	27
Coal, broken	1.4:1	36
Earth:		
Dry	1.3:1	38
Damp	2.0:1	27
Gravel:		
Round	1.7:1	30
Angular	1.3:1	38
Rock, broken:		
Soft	1.5:1	34
Hard	1.3:1	38
Rock, weathered:		
Residuals and weathered rock	1.5:1	34

*Angle of repose is the angle between the horizontal and the slope of a heaped pile of material.

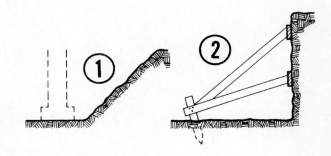

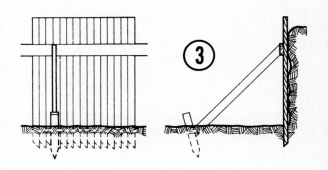

Figure 1.17 Shoring and sheetpiling: (1) *Sloped bank:* In this type of excavation the earth takes its natural angle of repose where the soil lacks the stability to stand vertically when cut. Such an excava-tion is undesirable and is frequently forbidden in specifications because the undisturbed earth remaining creates a bowl for the col-lection of water (both before and after backfilling) and an undesira-bly large amount of soil removal and backfilling is required.
(2) *Braced bank:* If the soil has some stability but will not stand unaided, very simple bracing may be sufficient.
(3) *Sheetpiling:* In very fluid soils sheetpiling is driven and braced and may be used as the outside form for poured foundations. Wood, steel, or concrete sheetpiling is available.

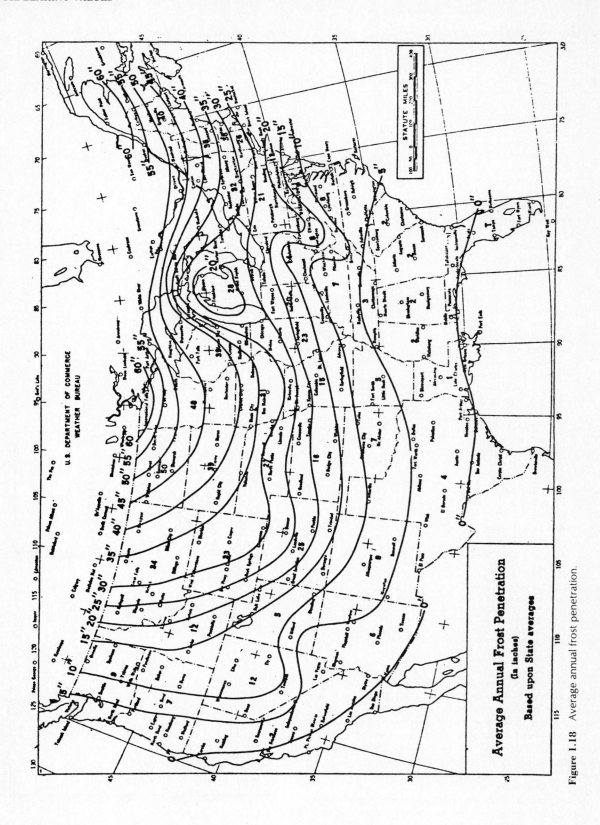

Figure 1.18 Average annual frost penetration.

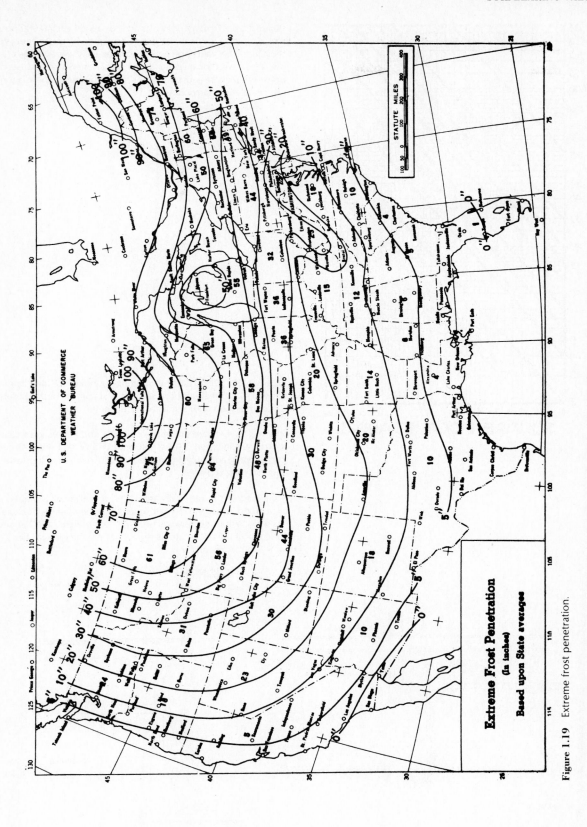

Extreme Frost Penetration
(In inches)
Based upon State averages

Figure 1.19 Extreme frost penetration.

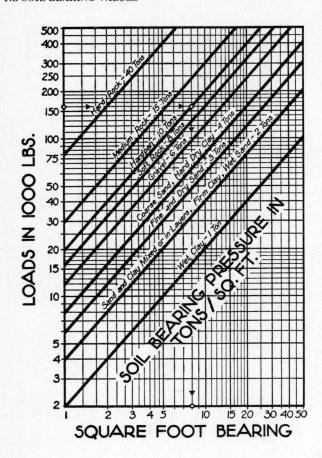

LOADS IN 1000 LBS.

500
400
300
250
200
150
100
75
50
40
30
20
15
10

5
4
3
2

Hard Rock - 40 Tons
Medium Rock - 15 Tons
Hardpan - 10 Tons
Soft Rock - 8 Tons
Gravel - 6 Tons
Coarse Sand, Hard Dry Clay - 4 Tons
Fine and Dry Sand - 3 Tons
Sand and Clay Mixed or in Layers, Firm Clay, Wet Sand - 2 Tons
Wet Clay - 1 Ton

SOIL BEARING PRESSURE IN TONS/SQ.FT.

1   2   3 4 5   10   15 20   30 40 50

SQUARE FOOT BEARING

Figure 1.20 Footing design chart. For a load of 160,000 pounds on hardpan, the chart shows that a footing of 8 square feet would be required. The values given for various soils are averages and may not agree with local codes. Local requirements should be checked before this chart is used. Values falling between the diagonal lines can readily be interpolated.

Figure 1.21 Underpinning abutting foundations. Needling: Where old walls are weak and the soil is not stable, or both, underpinning is accomplished with the aid of needling.

Shoring: Sockets are cut in the old wall, and shores (also called spur braces) are inserted. These rest on a crib of timbering. Shores prevent slipping and bulging and reduce the load to be supported while underpinning is placed.

Sectioning: If the old walls are sound and the soil is stable, a short excavation is made and a 6-foot length of new wall is built under the old wall. When this new section will bear weight, another section is added, and the procedure is repeated until the old wall has a continuous foundation under it.

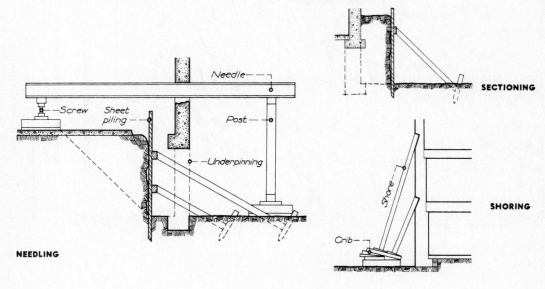

Needle

Screw   Sheet
piling

Post

Underpinning

SECTIONING

Shore

Crib

SHORING

NEEDLING

1.9 Piles

Piles are used to transmit foundation loads to strata of adequate bearing capacity and to eliminate settlement from the consolidation of overlying materials. Figure 1.22 lists nine principal pile categories of the three structural materials: wood, steel, and concrete. No exact criteria for the applicability of the various pile types can be given. The selection of types should be based on factors listed in the figure and on comparative costs.

Pile type	Timber	Steel
Consider for length of....	30-60 ft	40-100 ft
Applicable material specifications.	TS-2P3	TS-P67
Maximum stresses.	Measured at most critical point, 1200 psi for Southern Pine and Douglas Fir. See U.S.D.A. Wood Handbook No. 72 for stress values of other species.	12,000 psi.
Consider for design loads of.	10-50 tons	40-120 tons.
Disadvantages.	Difficult to splice. Vulnerable to damage in hard driving. Vulnerable to decay unless treated, when piles are intermittently submerged.	Vulnerable to corrosion where exposed. BP section may be damaged or deflected by major obstructions.
Advantages..	Comparatively low initial cost. Permanently submerged piles are resistant to decay. Easy to handle.	Easy to splice. High capacity. Small displacement. Able to penetrate through light obstructions.
Remarks ...	Best suited for friction pile in granular material.	Best suited for endbearing on rock. Reduce allowable capacity for corrosive locations.
Typical illustrations.	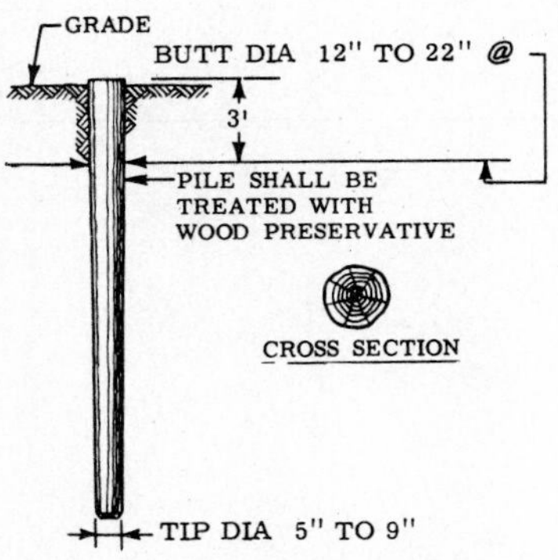 GRADE / BUTT DIA 12" TO 22" @ 3' / PILE SHALL BE TREATED WITH WOOD PRESERVATIVE / CROSS SECTION / TIP DIA 5" TO 9"	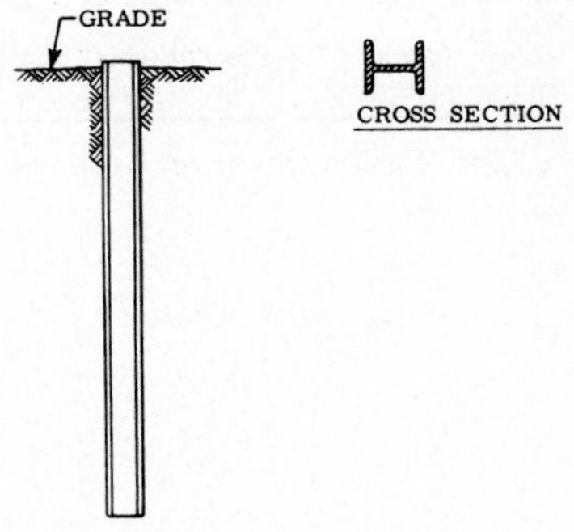 GRADE / CROSS SECTION

See General Notes on last page of table.

Figure 1.22 Design criteria for bearing piles.

Pile type	Precast concrete (including prestressed)	Cast-in-place concrete (thin shell driven with mandrel)
Consider for length of...	40-50 ft for precast. 60-100 ft for prestressed.	100 ft.
Applicable material spec- ifications.	TS-P57	ACI Code 318—For Concrete.
Maximum stresses.	For precast—15% of 28-day strength of con- crete, but no more than 700 psi. For prestressed—20% of 28-day strength of concrete, but no more than 1,000 psi in excess of prestress.	25% of 28-day strength of concrete with 1,000 psi maximum, measured at midpoint of length in bearing stratum.
Specifically designed for a wide range of loads.	....	
Disadvan- tages.	Unless prestressed, vulnerable to handling. High initial cost. Considerable displacement. Prestressed difficult to splice.	Difficult to splice after concreting. Redriving not recommended. Thin shell vulnerable during driving. Considerable displacement.
Advantages..	High load capacities. Corrosion resistance can be attained. Hard driving possible.	Initial economy. Tapered sections provide higher bearing resistance in granular stratum.
Remarks ...	Cylinder piles in particular are suited for bending resistance.	Best suited for medium load friction piles in granular materials.
Typical illustrations.		

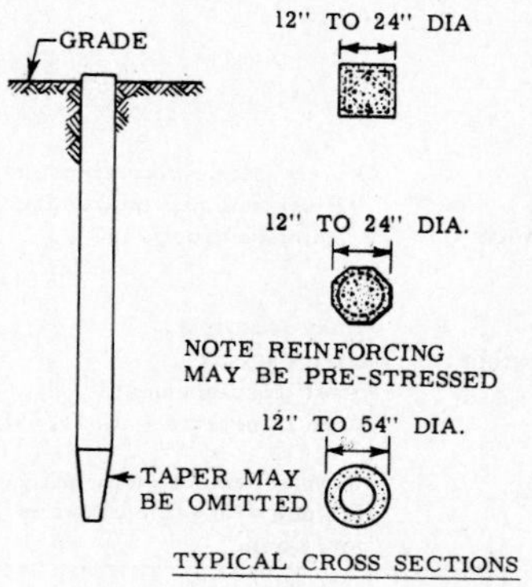

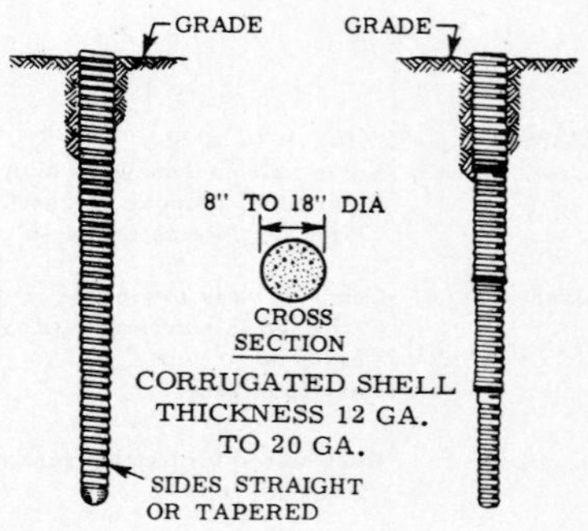

See General Notes on last page of table.

Figure 1.22 Design criteria for bearing piles (*Continued*).

Pile type	Cast-in-place concrete piles (shells driven without mandrel)	Pressure injected footings
Consider for length of...	30-80 ft	10 to 60 ft.
Applicable material specifications.	ACI Code 318	TS-F16.
Maximum stresses.	25% of 28-day strength of concrete with maximum of 1,000 psi measured at midpoint of length in bearing stratum. 9,000 psi in shell.	25% of 28-day strength of concrete, with a maximum of 1,000 psi. 9,000 psi for pipe shell if thickness greater than 1/8".
Consider for design loads of. . .	50-70 tons	60-120 tons.
Disadvantages.	Hard to splice after concreting. Considerable displacement.	*Base* of footing cannot be made in clay. When clay layers must be penetrated to reach suitable material, special precautions are required for *shafts* if in groups.
Advantages..	Can be redriven. Shell not easily damaged.	Provides means of placing high capacity footings on bearing stratum without necessity for excavation or dewatering. Required depths can be predicted accurately. High blow energy available for overcoming obstructions. Great uplift resistance if suitably reinforced.
Remarks ...	Best suited for friction piles of medium length.	Best suited for granular soils where bearing is achieved through compaction around base. Minimum spacing 4'-6" on center. For further design requirements see Philadelphia Building Code 4-1710.
Typical illustrations.	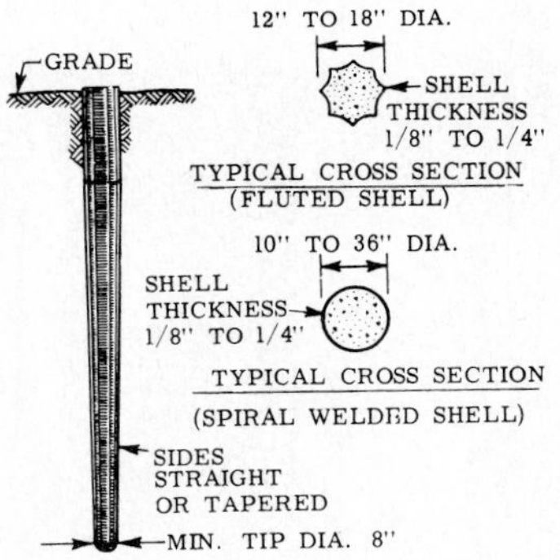	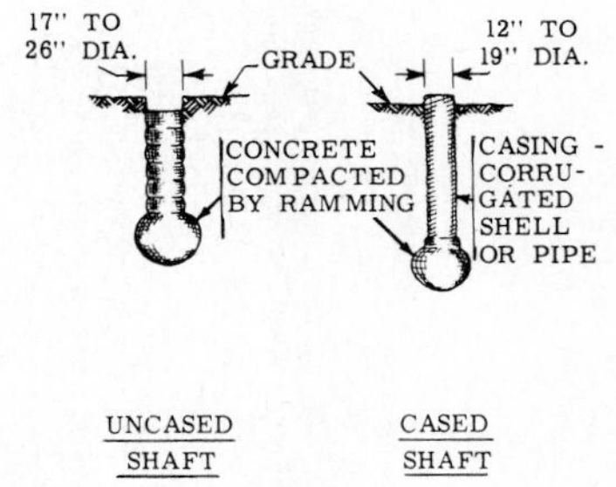

See General Notes on last page of table.

Figure 1.22 Design criteria for bearing piles (*Continued*).

Pile type	Concrete filled steel pipe piles	Composite piles
Consider for length of...	40-120 ft	60-120 ft.
Applicable material specifications.	ASTM A7—for Core. ASTM A252—for Pipe. ACI Code 318—for Concrete.	ACI Code 318—for Concrete. ASTM-36—for Structural Section. ASTM A252—for Steel Pipe. TS-P2—for Timber.
Maximum stresses.	9,000 psi for pipe shell. 25% of 28-day strength of concrete with a maximum of 1,000 psi. 12,000 psi on Steel Cores.	25% of 28-day strength of concrete with 1,000 psi maximum. 9,000 psi for structural and pipe sections. Same as timber piles for wood composite.
Consider for design load of...	80-120 tons without cores. 500-1,500 tons with cores.	30-80 tons.
Disadvantages.	High initial cost. Displacement for closed end pipe.	Difficult to attain good joint between two materials.
Advantages..	Best control during installation. No displacement for open end installation. Open end pipe best against obstructions. High load capacities. Easy to splice.	Considerable length can be provided at comparatively low cost.
Remarks ...	Provides high bending resistance where unsupported length is loaded laterally.	The weakest of any material used shall govern allowable stresses and capacity.
Typical illustrations.	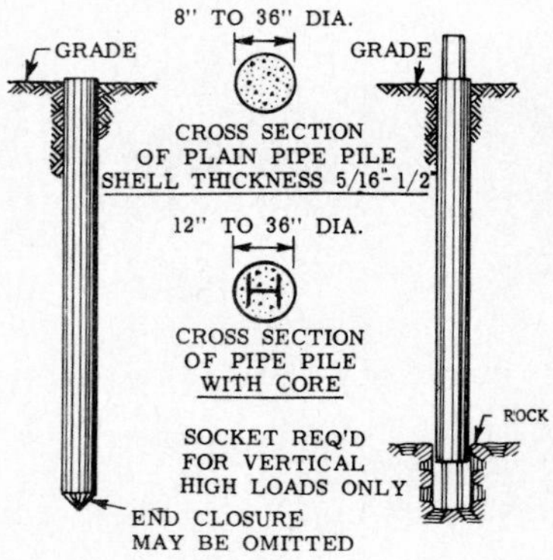	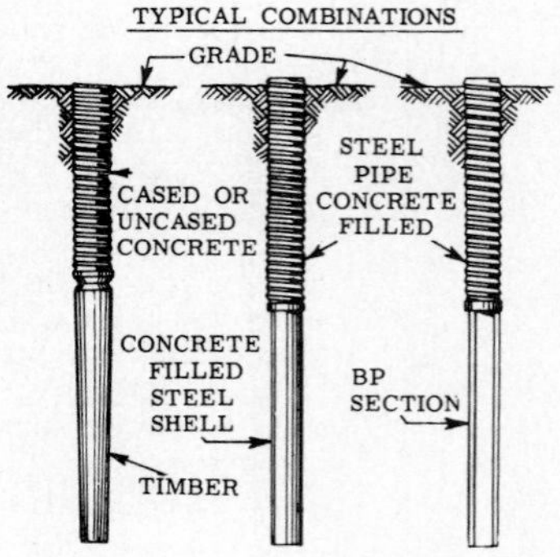

See General Notes on last page of table.

Figure 1.22 Design criteria for bearing piles (*Continued*).

Pile type	Auger-placed, pressure-injected concrete piles	— General Notes —
Consider for length of ...	30-60 ft	1. Stresses given for steel piles are for noncorrosive locations. For corrosive locations, estimate possible reduction in steel cross section or provide protection from corrosion. 2. Lengths and loads indicated are for feasibility guidance only. They generally represent current practice. 3. Design load capacity should be determined by soil mechanics principles limiting stresses in piles and type and function of structure. See Section 3.
Applicable material specifications.	TS-2P69............................. ..	
Maximum stresses.	25% of 28-day strength of concrete with maximum of 1,000 psi.	
Consider for design load of. . .	35-70 tons	
Disadvantages.	More than average dependence on quality workmanship. Not suitable thru peat or similar highly compressible material.	
Advantages..	Economy. Completely nondisplacement. No driving vibration to endanger adjacent structures. High skin friction. Good contact on rock for end bearing. Convenient for low-headroom underpinning work. Visual inspection of augered material. No splicing required.	
Remarks ...	Process patented.........................	
Typical illustrations.	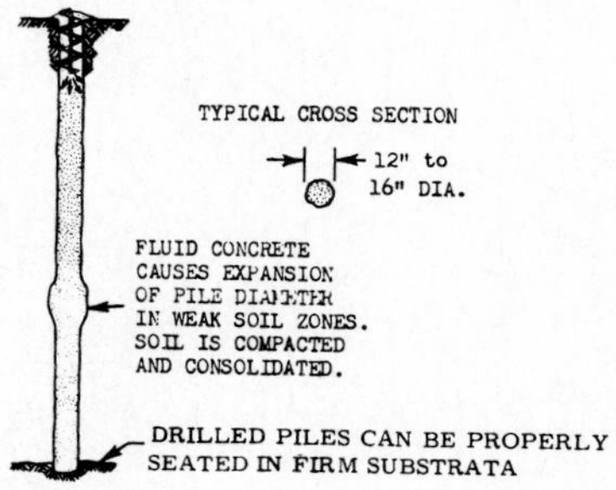 TYPICAL CROSS SECTION 12" to 16" DIA. FLUID CONCRETE CAUSES EXPANSION OF PILE DIAMETER IN WEAK SOIL ZONES. SOIL IS COMPACTED AND CONSOLIDATED. DRILLED PILES CAN BE PROPERLY SEATED IN FIRM SUBSTRATA	

Figure 1.22   Design criteria for bearing piles (*Continued.*)

1.10 Grading (FHA Requirements)

Grading Design Grading design should be considered in the early planning stages with the following principal objectives:

1. Development of attractive, suitable, and economical building sites.

2. Provision of safe, convenient, and functional access to all areas for use and maintenance.

3. Disposal of surface runoff from the site area without erosion or sedimentation, or its collection as needed for water features, debris basins, or irrigation storage.

4. Diversion of surface and subsurface flow away from buildings and pavements to prevent undue saturation of the subgrade that could damage structures and weaken pavements.

5. Preservation of the natural character of the site by minimum disturbance of existing ground forms and meeting of satisfactory ground levels at existing trees to be saved.

6. Optimum on-site balance of cut and fill; stockpiling for reuse of existing topsoil suitable for the establishment of ground cover or planting.

7. Avoidance of filled areas that will add to the depth or instability of building foundations and pavement subgrades.

8. Avoidance of wavy profiles in streets and walks and of steps in walks.

9. Avoidance of earth banks requiring costly erosion control measures, except where these are needed in place of costly retaining walls.

10. Keeping finished grades as high as practicable where rock will be encountered close to the surface, thus reducing the cost of utility trenching and other excavation and improving growing conditions for vegetation.

11. Avoidance of runoff water over roadways. Ice forms during freezing weather and a hazardous driving situation results.

Areas Adjacent to Buildings Unpaved areas adjacent to buildings should be sloped to direct surface water and roof drainage, including snow melt, away from buildings at a minimum slope of 6 inches in the first 10 feet of horizontal distance and not across sidewalks. Surfaces paved with portland cement concrete should have a slope of not less than 0.5 percent, and bituminous pavements a slope of not less than 1.5 percent, to assure adequate drainage without ponding or "birdbaths."

Unoccupied Site Areas Portions of the site not occupied by buildings or pavement should have adequate continuous slopes to drain toward watercourses, drainage swales, roadways, and the minimum necessary storm drainage inlets. Drainage swales or channels should be sized and sloped to accommodate design runoff. The runoff should be carried under walkways in pipes with diameters of not less than 8 inches or of larger sizes if clogging by debris or grass cuttings is a problem. Swales should be used to intercept water at the top and bottom of banks where large areas are drained. To provide positive drainage, a slope of not less than 2 percent for turfed areas is usually desirable, but more permeable soils may have adequate drainage with a lesser slope. Turf banks, where required, should be graded to permit the use of gang mowers, providing a maximum slope of 1 vertical to 3 horizontal and, if feasible, a slope of 1 vertical to 4 horizontal. The tops and bottoms of all slopes should be gently rounded in a transition curve for optimum appearance and ease of maintenance.

Grading Procedure The site areas shall be roughly graded to comply with the foregoing criteria on grading design and criteria on fine grading. The subgrade should be established parallel to the proposed finish grade and at elevations to allow for the thickness of topsoil or other surface. In fill areas, all topsoil, debris, and other noncompatible materials should be removed, and all tree stumps removed or cut out 18 inches below grade. On sloping areas to be filled, the original ground should be scarified to provide bond for fill material where the original ground is clay. Fill material should be free from debris and have a moisture content near the optimum when it is placed. Fill should be compacted to a density that would avoid damaging settlement to drainage structures, walks, or other planned improvements.

See also Figures 1.23, 1.24, and 1.25.

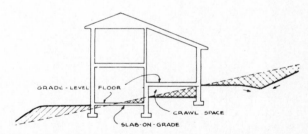

Figure 1.23 A three-level house that creates two levels at grade and involves a minimum of excavation.

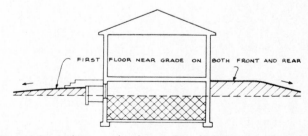

Figure 1.24 A house on level ground in which the excavated earth is reused to raise the grade immediately around the house. This procedure reduces required excavation and provides drainage away from the house.

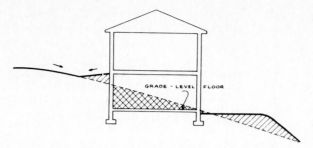

Figure 1.25 A house on a slope in which the slope is used to create a ground floor, the earth being reused to establish a level terrace and to care for the drainage.

1.11 Cut and Fill

Cut and fill to create the desired surface in land grading may be determined by the profile adjustment method or the plane method.

PROFILE ADJUSTMENT METHOD

The procedure for computing cuts and fills should be easier to follow than the plane method. Although it is not as accurate as the plane method, it should be adequate for surface drainage.

The field should be surveyed by the grid method. The size of the grids is not critical, but 100 feet is customary. Calculations are simpler when the first line of grid points in each direction is started at half the grid size from the boundaries (see Figure 1.26). Now each grid point becomes the center of a square, and the field is made up of such squares.

The elevations of the grid points should be plotted along each grid line on the direction of greatest slope or the direction in which row drainage is desired. A profile should be drawn for the existing land surface along the grid line. Limits within which the slope of the surface may be allowed to vary should be adopted. By the use of these limits, a new profile can be drawn in for each grid line so as to have more cutoff area than area to be filled in and thus allow for the cut-fill ratio. Existing topography should be followed as closely as possible to keep soil movement to a minimum.

When this has been done for each grid line, the cuts and fills should be added for each line. The total cuts and the total fills can then be calculated. The ratio

$$\frac{\text{Total cuts}}{\text{Total fills}} \times 100$$

can be determined. If this ratio is greater or smaller than that required for the soil, the cuts and fills must be adjusted by raising or lowering the new surface profiles by an amount that will give the desired cut-fill ratio.

The grid survey will probably have been done to the nearest 0.05 foot. This means that calculations to any greater

accuracy than to the nearest 0.05 foot are time-wasting. Grid points should all be raised or lowered by 0.05 foot until the cut-fill ratio closest to the desired ratio is obtained. This is accurate enough for drainage work. If necessary, one grid line can be adjusted at a time until an overall cut-fill ratio close to that desired is achieved.

At this point the profiles across the field perpendicular to those already drawn should be plotted. This is a check to see whether the cross slopes exceed the limits that have been decided upon. These limits need not be the same as those chosen for the row grade; frequently greater variation is allowed for the cross slope than for the row grade. If the cross slopes do exceed the permissible variations, the row grade profiles must be further adjusted until the cross slopes are acceptable.

Example   Figure 1.27 is an example of a field which is to be graded and which has been surveyed on a 100-foot grid system. Profiles along the lines *A, B, C, D,* and *E* have been plotted and shown in Figure 1.27.

Assume that the row grade may vary between 0.05 and 0.3 percent and that the cross slope may vary by 0.5 percent. The cut-fill ratio is to be 150 percent. Now new profile lines are sketched in as shown.

The cuts and fills are determined for each station and put on the original survey sheet. These are the figures to the right and below each grid point in Figure 1.27. Now the cuts and fills are totaled.

The cut-fill ratio is

$$\frac{2.20}{1.65} \times 100 = 133 \text{ percent}$$

This is too low and therefore must be increased. Line *A* is the only line in which cut does not exceed fill. Line *A* is now lowered by 0.05 foot, keeping to the same grade, and the new cuts and fills are totaled. Now the cut-fill ratio is

$$\frac{2.40}{1.55} = 154 \text{ percent}$$

No further adjustment is necessary.

The cross slopes have been plotted in Figure 1.28. They are within the previously determined limits.

PLANE METHOD

As in the profile adjustment method, the field should be surveyed on a grid system. The size of the grids is not critical, but 100 feet is customary. Calculations are simpler when the first line of grid points in each direction is started at half the grid size from the boundaries, thus placing the "origin" outside the field by 50 feet in each direction.

The equation

$$E = a + S_x X + S_y Y$$

where E = elevation at any point (X, Y)
a = elevation of origin
S_x, S_y = slope in X and Y directions, respectively

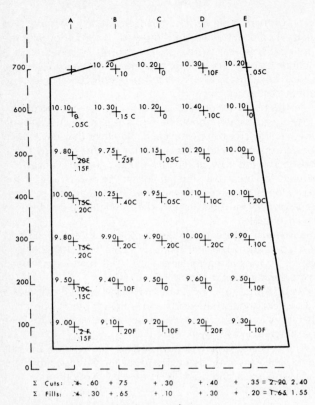

Σ Cuts: .60 + 75 + .30 + .40 + .35 = 2.20 2.40
Σ Fills: .30 + .65 + .10 + .30 + .20 = 1.65 1.55

Figure 1.26 Topographical survey sheet.

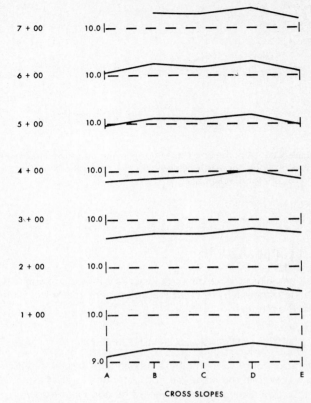

CROSS SLOPES

Figure 1.28 Cross-slope profiles. The distance between horizontal lines is equal to 1 foot.

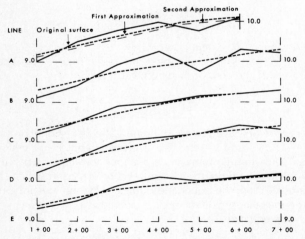

Figure 1.27 Profiles in row direction. The distance between horizontal lines is equal to 1 foot.

will give the elevation of any point in a plane once S_x and S_y have been determined.

However, this equation will not give a plane that will balance the cut area with the fill or best fit the surface. To give equal cut and fill, the plane must pass through the centroid. The position of the centroid of the field can be determined as follows. Multiply the number of grid positions in each grid line by the distance of that line from the origin in each direction. Total these products in each direction and divide each figure thus obtained by the number of grid positions in that direction. This will give two figures, X_c and Y_c, representing the distance along each boundary from the origin, which will locate the centroid in the field.

The elevation of the centroid will be

$$E_c = \frac{\text{sum of all elevations}}{\text{total number of grid positions}}$$

The elevation of the origin will be

$$a = E_c - S_x X_c - S_y Y_c$$

With known X and Y slopes, the grid point elevation for any plane can be calculated so that total cut equals total fill. From the new elevations the cut and fill can be determined.

To avoid unnecessary earth moving, it is of value to know the best-fitting plane to any particular field. For rectangular fields the best-fitting slope in the X direction, S_x, can be found from

$$S_x = \frac{\Sigma(XE) - nX_cE_c}{\Sigma X^2 - nX_c^2}$$

where n = number of grid positions (stations)
E = elevation at any station
X = station number in X direction
X_c = X distance of centroid

To find S_y, substitute Y and Y_c for X and X_c, respectively.

For nonrectangular fields, the problem can be solved more accurately by using the following simultaneous equations:

$$S_x(\Sigma X^2 - nX_c^2) + S_y[\Sigma(XY) - nX_cY_c] = \Sigma XE - nX_cE_c$$

$$S_y(\Sigma Y^2 - nY_c^2) + S_x[\Sigma(XY) - nX_cY_c] = \Sigma YE - nY_cE_c$$

Example See Figure 1.29.
In the field in Figure 1.29, which is nonrectangular,

$$n = 34$$

$$X_c = \frac{104}{34} = 3.1$$

$$X_c^2 = 9.6$$

$$Y_c = \frac{133.0}{34} = 3.9$$

$$Y_c^2 = 15.2$$

$$E_c = \frac{335.2}{34} = 9.9$$

$\Sigma XY = 1(1 + 2 + 3 + 4 + 5 + 6) + 2(1 + 2 + 3 + 4 + 5 + 6 + 7) + \cdots + 7(1 + 2 + 3 + 4 + 5 + 6 + 7) = 413$

$\Sigma X^2 = 6(1) + 7(4) + 7(9) + 7(16) + 7(25) = 384$
$\Sigma Y^2 = 5(1) + 5(4) + 5(9) + 5(16) + 5(25) + 5(36) + 4(49) = 651$

$\Sigma XE = 58.2 + 2(68.9) + 3(69.2) + 4(69.8) + 5(69.1) = 1028.3$

$\Sigma YE = 45.8 + 2(47.5) + 3(49.5) + 4(50.4) + 5(49.5) + 6(51.1) + 4(41.0) = 1332$

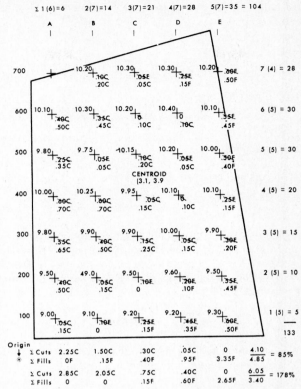

Figure 1.29 Cut-and-fill sheet for the plane method of grading.

Substitute these figures in the simultaneous equations

$S_x(384 - 326) + S_y(413 - 411)$
$\qquad = 1028.3 - 1044.0 = -13.7$
$S_y(651 - 517) + S_x(413 - 411)$
$\qquad = 1332.0 - 1312.0 = 20.0$

$$58S_x + 2S_y = -13.7$$
$$2S_x + 134S_y = 20.0$$
$$58S_x + 2S_y = -13.7$$
$$58S_x + 3890S_y = 580$$
$$3888S_y = 593.7$$
$$S_y = 0.153 \text{ foot}/100 \text{ feet}$$
$$S_x = 0.236 \text{ foot}/100 \text{ feet}$$

If the slopes of the best-fitting plane are outside the limits imposed by erosion hazard, they will indicate at which end of the limits the slopes should be chosen. The best-fitting plane, once determined, can be varied or lowered until the difference between cut and fill volumes is appropriate to give the desired cut-fill ratio. Thus the best-fitting plane would have a slope in the Y direction of 0.15 foot/100 feet and a slope in the X direction of 0.24 foot/100 feet. Since the plane

must pass through the centroid, the elevation of the origin can be computed:

$$a = E_c - S_x X_c' - S_y Y_c$$

$$= 9.9 - (0.24)(3.1) - (0.15)(3.9) = 8.6$$

Now the best-fitting plane is

$$E = 8.6 + 0.24\,X + 0.15\,Y$$

Each new elevation is computed and the cut or fill is noted (see Figure 1.29).

The cuts and fills are totaled, and the ratio is determined:

$$\frac{\text{Total cuts}}{\text{Total fills}} = \frac{4.10}{4.85} = 85 \text{ percent}$$

The figure is not 100 percent because the numbers have been rounded off to the nearest 0.05 foot, which is the limit of accuracy for this work. However, we need a cut-fill ratio of 150 percent. If the whole plane is lowered by 0.05 foot, the new cuts and fills will give a ratio as follows:

$$\frac{5.00}{4.05} = 123 \text{ percent}$$

This is not adequate if we assume that a 150 percent cut-fill ratio is needed. Lowering the plane by a further 0.05 foot will give

$$\frac{6.05}{3.40} = 178 \text{ percent}$$

This is too much. For drainage work it is not necessary to maintain an exact plane; so simply lower half of the points by 0.10 foot and the remainder by 0.05 foot, which would result in the desired 150 percent. This should be done across the slope rather than along the length of the slope so as not to interfere with the desired row grade.

Earthwork Volume Calculations When grading land for drainage, it is almost impossible to determine precisely the volume of soil being moved. This is due to the degree of accuracy used in surveying.

When the grid points are positioned so that they are in the center of a square, the following formula is used:

Volume of cut (V_c)

$$= \frac{(\Sigma \text{ cuts}) \times \text{area of grid square}}{27} \text{ cubic yards}$$

By the profile adjustment method the volume of soil moved would be

$$V_c = \frac{2.40 \times 100 \times 100}{27} = 888 \text{ cubic yards}$$

By the plane method the volume would be

$$V_c = \frac{5.50 \times 100 \times 100}{27} = 2037 \text{ cubic yards}$$

1.12 Soil Erosion Control

There are two kinds of erosion and sediment control measures, mechanical and vegetative. The most widely used of these measures are discussed here from the standpoint of their general use and purpose. Detailed information and standards and specifications developed for local conditions can be obtained from local offices of the Soil Conservation Service (SCS) or from conservation districts. SCS can also give technical advice. Erosion control measures must be properly designed, installed, and maintained if they are to accomplish their intended purpose.

Mechanical Measures Mechanical measures are used to reshape the land to intercept, divert, convey, retard, or otherwise control runoff.

Land Grading Grading only areas going into immediate construction, as opposed to grading the entire site, helps immensely in controlling erosion. On large tracts, to avoid having a large area bare and unprotected, units of workable size can be graded one at a time. As construction is completed on one area, grading proceeds to another.

On some sites only the street rights-of-way are graded, until storm sewers have been installed. This leaves only limited areas exposed to erosion. Usually the adjacent undisturbed areas can be used as temporary outlets for diversions, or berms, built to protect the graded street rights-of-way. After the storm sewers have been installed, other areas can be graded and the runoff can be directed to the streets and storm sewers.

As a general rule, grading should be held to the minimum that makes the site suitable for its intended purpose without appreciably increasing runoff. Wherever possible, only undesirable trees should be removed. In some areas heavy cutting, filling, or reshaping of the natural topography is needed to increase the percentage of usable land. Heavy grading almost always increases erosion hazards and should be accompanied by the maximum use of appropriate erosion control measures.

The grading plan should show the location, slope, and elevation of the areas to be graded and the measures to be used for disposing of runoff and for erosion control. Constructed slopes should be limited to a degree of steepness that will provide stability and allow easy maintenance. Retaining walls may be required.

Stumps and other decayable material should not be used in fills. Soft, mushy soil material is not suitable for fills that are to be used to support buildings or other structures.

Bench Terraces Bench terraces constructed across the slope of the land and fitted to the natural terrain are used to break long slopes and slow the flow of runoff. In some areas the terraces are constructed wide enough to be used as residential sites. Since the cut-and-fill slopes of the bench terraces are always steeper than the natural slope of the land, landslides may be a threat. Engineering studies should always be made to guide the design of the slopes and ensure a reasonable degree of slope stability and safety.

Small bench terraces are sometimes used on the face of cut-and-fill slopes to help control runoff and erosion and establish vegetation.

Subsurface Drains   Subsurface drains are sometimes required at the base of fill slopes to remove excess groundwater. In heavy grading it may be necessary to fill natural drainage channels; subsurface drains may have to be installed below the newly filled areas to prevent the accumulation of groundwater.

Subsurface drains may be needed in vegetated channels to lower a high water table that prevents establishing an effective plant cover.

Diversions   Diversions intercept and divert runoff so that it will not cause damage; they consist of a channel and a ridge constructed across the slope. Diversions need a stable outlet to dispose of water safely.

In many places diversions are placed above critical slopes to divert runoff. Runoff over such slopes would cause serious erosion. Diversions can be used in the same way to protect construction areas. They can also be used on long slopes, in a series if necessary.

Permanent diversions should be seeded to the same grasses that cover the surrounding areas. If they are built to protect open spaces, they should blend into the landscape for both better appearance and ease of maintenance.

Berms   Berms are a type of diversion. They are compacted earth ridges on a slight grade and have no channels, and they may be permanent or temporary.

Berms can be used to protect newly constructed slopes until the slopes are stabilized with permanent vegetation. They can be constructed across graded rights-of-way in a series and at intervals needed to intercept runoff. The side slopes of the berms are made flat enough to allow work vehicles to cross them.

Berms too must have stable outlets. Well-stabilized upgraded areas adjacent to the street rights-of-way are often used as temporary outlets. In many places half-channel flumes, sod, or other material can be used to make temporary outlets.

Storm Sewers   Storm sewers dispose of runoff from the streets and adjacent lots. Temporary diversions may be needed to control runoff on the lots and convey it safely to the streets and storm sewers.

The use of storm sewers for runoff disposal does not prevent sediment from being deposited downstream. To reduce the sediment load carried by runoff through storm sewers during construction, some developers have improvised small sediment basins adjacent to sewer inlets. The sediment collected in the basins is removed following each runoff-producing rain.

Storm sewers should discharge at a place where the grade is stable. Generally an energy dissipater is needed to slow the force of the flow at the point of discharge.

Outlets   Most outlets are grassed waterways, either natural or artificial, and serve to dispose safely of water from diversions and from parking lots, highways, and other areas.

Natural waterways, or swales, can be improved by grading, reshaping, and revegetating. Artificial outlets should have flat side slopes and a wide bottom so that they can be easily maintained. They should have adequate capacity.

Grass protects a channel against erosion by reducing the velocity of flow. The most suitable grass species are those that produce a dense uniform cover near the soil surface, are long-lived, can withstand small amounts of sedimentation, and provide protection during all seasons of the year. The species selected should be adapted to the locality and the site.

Jute netting or fiber glass can be used as a channel liner to protect the channel from erosion until vegetation becomes established. Liners may not be needed if the runoff can be diverted from the channel during the establishment period.

Waterway Stabilization Structures   A waterway needs a stabilizing structure if its slope is so steep that the velocity of runoff exceeds the limit of protection that the vegetation alone gives. Grade stabilization structures, special culverts, and various kinds of pipe can be used in combination with vegetation. Energy dissipaters may be required. The structures should be designed and built to provide permanent stabilization.

Lined Channels   The alternative to using vegetated waterways with grade stabilization structures is using lined channels. Such channels (paved ditches and valley gutters, for example) have many uses in urban areas where slopes are too steep or soils too unstable for control by vegetation alone. Fiber glass mats can be used as temporary lining for ditches and channels.

Sediment Basins   The function of a sediment basin is to detain runoff and trap sediment, thus preventing damage to areas downstream. By detaining runoff, sediment basins also reduce peak flows. Basins can be excavated or formed by a combination of dam and excavation. Earth dams can be constructed across waterways to form basins. Under some conditions a highway embankment can serve as a dam. Sediment basins are almost always temporary structures. They are graded into the surrounding landscape after construction has been completed and the area has been stabilized. But they can be designed as permanent structures if there is a permanent need for them. Some industrial firms have preferred permanent basins that can be used later to protect downstream areas from accidentally released materials that would cause pollution.

The location, design, and construction of a sediment basin should be such that serious damage to areas downstream would be avoided if the basin failed. If the minimum storage requirement cannot be met, excavation to enlarge the basin and periodic cleanout may be necessary.

Sediment basins are constructed to discharge on stable ground below the dam. Emergency spillways should be added to increase safety. Exposed areas of the embankment and the emergency spillway should be protected by mulching and seeding.

Stream Channel and Bank Stabilization   The increased

runoff from construction sites may make it necessary to stabilize the stream channel below.

Stream channels can be stabilized by installing grade control structures or by paving. The undercutting of banks can be controlled by measures that withstand the flow, such as concrete structures or rock riprap built along the toe and lower facing of the bank, or by measures that dissipate the energy of the flow, such as jetties, piling, and fencing built into or along the channel. Realigning the channel may be desirable or necessary in many places, but it creates the risk of starting a new erosion cycle.

Stabilizing stream channels and stream banks is usually complex and costly. Control measures should be undertaken only on the basis of thorough engineering studies and plans. If a stream runs along or through a floodplain that is to be developed for parks and recreation, for example, aesthetic values may determine the methods of improvement.

Vegetative Measures Vegetative measures provide temporary cover to help control erosion during construction and permanent cover to stabilize the site after construction has been completed. The measures include the use of mulches and temporary and permanent cover crops.

Erosion can be controlled with less difficulty on some sites than on others during construction, and permanent cover is easy to establish on some sites and difficult on others. Establishing and maintaining good plant cover is easy in areas of fertile soil and moderate slopes. Usually such areas can be stabilized by using plants and cultural methods common in the community.

Sites that are difficult to stabilize because of exposed subsoil, steep slopes, a droughty exposure, and other conditions, require special treatment. Such sites are called critical areas because they erode severely and are the source of much sediment if they are not well stabilized.

Mulch Straw mulch can be used to protect constructed slopes and other areas brought to final grade at an unfavorable time for seeding. The areas can be seeded when the time is favorable without removing the mulch.

Mulch is essential in establishing good stands of grasses and legumes on steep cut-and-fill slopes and other areas where it is difficult to establish plants. By reducing runoff, mulch allows more water to infiltrate the soil. It also reduces the loss of soil moisture by evaporation, holds seed, lime, and fertilizer in place, and reduces seedling damage from the heaving of the soil caused by freezing and thawing.

The materials most widely used in mulching are small-grain straw, hay, and certain processed materials. Grain straw is easily applied and generally is more readily available than hay, and it costs less. In some places, certain hays are preferred because they are a source of seed of plants that can be used for stabilization. Straw and hay mulches are usually applied at the rate of 1½ tons per acre.

A number of processed mulches are available, and some show promise of greater use under specific conditions. Hydromulching, in which seed, fertilizer, and mulch are applied as a slurry, is a fast, all-in-one operation that requires little labor. Hydromulching may not be successful if it is carried out during a period of high-intensity storms.

Straw and hay mulches must be anchored to keep them from blowing or washing away. Anchoring methods include spraying the mulch with asphalt, tucking the mulch into the soil with a straight-blade disk, stapling netting over the mulch, and driving pegs into the mulched area at intervals of about 4 feet and interlacing them with twine.

Temporary Cover Temporary cover crops can be used when cover is needed for a few months or a year or two. If construction is delayed on a site that has been cleared and graded, temporary cover crops can be employed to protect the site from erosion. And they can be planted at a time of year that is unfavorable for seeding and establishing permanent cover.

Rapidly growing plants, such as annual rye grass, small grain, sudan grass, and millet, are most often used for temporary cover. Plants that are adapted to the locality and the season of the year during which protection is needed should always be used.

Permanent Cover Special care should be taken in selecting plants for permanent cover. There are many grasses and legumes, trees, shrubs, vines, and ground covers from which to choose in most humid areas of the United States, but only a few in most dry regions. Final choice should be based on the adaptation of the plants to the soils and climate, ease of establishment, suitability for a specific use, longevity or ability to self-reseed, maintenance requirements, aesthetic values, and other special qualities.

The best plants are those that are well adapted to the site and to the purpose for which they are to be used. For example, grasses used for waterway stabilization must be able to withstand submergence and provide a dense cover to prevent scouring of the channel. In playgrounds, grasses must be able to withstand trampling. In some places, such as south-facing cut-and-fill slopes, the plants needed are those that are adapted to droughty areas. In other places, plants must be able to tolerate shade. Some plants can beautify as well as stabilize an area.

Maintenance may be the most important factor in selecting plants for permanent stabilization. Most tame grasses and legumes require much maintenance, and they gradually give way to native grasses, shrubs, and weedy plants if they are not mowed and fertilized regularly. In some areas native plants are preferred. On steep slopes and other inaccessible areas, it is preferable to select plants that require little or no maintenance. Sericea lespedeza, crown vetch, and honeysuckle, for example, are long-lived and provide good erosion control with a minimum of maintenance. Most native grasses, trees, and shrubs grow well with little or no maintenance.

Fibrous Materials A number of fibrous materials have special uses in erosion control. Jute netting, a coarse, open-mesh, weblike material, can be applied directly on the soil to protect newly seeded channels until vegetation becomes established. It can also be used in repairing outlets and diversions where gullies have cut the channel. In some places it can be used to hold down straw mulch. Cotton netting and paper netting are both lightweight; they can hold straw mulch in place and prevent it from blowing or washing away.

Solid, heavy-duty fiber glass matting can be used as a tem-

porary channel liner where the water velocity is too high for the use of vegetation or where vegetation is not wanted. Impregnating the mat with asphalt prolongs its life. Perforated fiber glass matting can be used in the same way as jute netting to protect newly seeded channels. It can be used as a transition apron, lining the head of a channel to protect it from runoff, especially runoff from road culverts, which tends to cut a gully down the channel. And used as erosion stops, perforated matting can check rilling.

Fiber glass erosion stops have certain advantages over rigid stops of masonry or wood. Soil often settles around rigid structures, causing a turbulent and erosive flow. The fiber glass stops are flexible and conform to the channel. Also, they are porous (water can seep through them), and subsurface drainage is improved.

Stabilizing Cut-and-Fill Slopes The first requirement in stabilizing cut-and-fill slopes is to prevent runoff from flowing over the face of the slopes. Temporary diversions, berms, shoulder dikes, or other measures should be used to intercept and divert the runoff. Permanent structures such as brow ditches and valley gutters are required where the areas contributing to runoff are large, for example, areas of highway construction. Small benches, interceptor ditches, or other measures can be used to protect long slopes from runoff originating on the slope itself.

Methods of establishing vegetation vary for different parts of the United States and depend on the plants used and on the soil and climate. In selecting plants and in getting them established, it is best to be guided by the methods commonly used and recommended in the particular area.

As a general rule, when seeding grasses and legumes it is advisable to prepare as good a seedbed as site conditions permit. Applications of lime and fertilizer should be based on local standards or on soil tests.

Usually a good stand can be obtained by broadcasting, drilling, or hydroseeding if other conditions are met. Mulching after seeding and then anchoring the mulch are essential in most areas. Irrigation is needed in many places.

Sodding is more costly than seeding, but it provides immediate protection. It should be used where the concentration of runoff is such that other methods of stabilization will not be effective. Sod can be laid whenever the soil is not frozen. Sod responds to a good seedbed and to lime and fertilizer. The strips of sod should always be laid across the slope, anchored to the soil, and watered. Some grasses can be established by sprigging and chunk sodding.

If trees are used to stabilize steep slopes, they are usually planted in pure stands. Mulching is important. Vines are generally established by transplanting individual plants or crowns.

Site Analysis The term "analysis" in this case means drawing conclusions about the information gathered during the site investigation. The objective of the analysis is to pinpoint site areas that are especially vulnerable to erosion or sedimentation because of existing topography, soils, vegetation, or drainage patterns. These characteristics must be interrelated in assessing the hazards of erosion and sedimentation for different site areas. The inventory and analysis

information is of little value, however, if it is not used as a basis for later site planning decisions. After an assessment of erosion vulnerability has been made, an earth change plan can be developed so that the disturbance of vulnerable or critical areas will be minimized.

The process of making specific conclusions about critically erodible areas involves interpretation of the inventory information already gathered for topography and soils (specifically, a slope interpretation and a soils survey). The following list illustrates the site conditions or combinations of conditions that determine critically erodible areas:

Condition 1: highly erodible soil = critical area

Condition 2: high-erosion-hazard slope = critical area

Condition 3: moderately erodible soil, moderate erosion hazard slope = critical area

Conclusions The considerations that should be addressed in an analysis of the site investigation information on topography, vegetation, soils, and existing drainage for site development are outlined below. Specific conclusions about these issues should be noted in the analysis.

1. Indicate where soils, topography, and vegetation combine to create critically erodible areas.

2. Indicate how the site or corridor relates to surrounding streams, drainageways, or other bodies of water. Assess the vulnerability to erosion and sediment damage of these drainageways and surface waters and all off-site areas.

3. Indicate where storm-water runoff crosses the site boundaries. Indicate the potential options for disposing of storm-water runoff by including potential locations of sediment control structures.

4. Indicate how areas disturbed by construction might be protected from increasing surface storm-water runoff.

The map in Figure 1.30 provides a graphic example of what a site analysis might look like. By using the conclusions noted in the analysis above to guide the process of preparing an earth change plan, a plan that is strongly related to site conditions and minimizes soil erosion can be developed. An earth change plan developed in this way will be a direct response to soil erosion and sedimentation as well as other considerations of function, economics, engineering feasibility, and aesthetics.

1. Erosion must be controlled to maintain an effective and clear drainage system with a minimum of maintenance, and to reduce hazardous dust conditions. Erosion may occur at any point where the force of moving water exceeds the cohesive strength of the material with which the water is in contact. Initial design of side slopes in cut-and-fill sections, based on soil type, will reduce the need for extensive erosion control measures. However, additional control is usually required. Most methods of control are based on either dissipating the energy of water or providing an erosion-resistant surface.

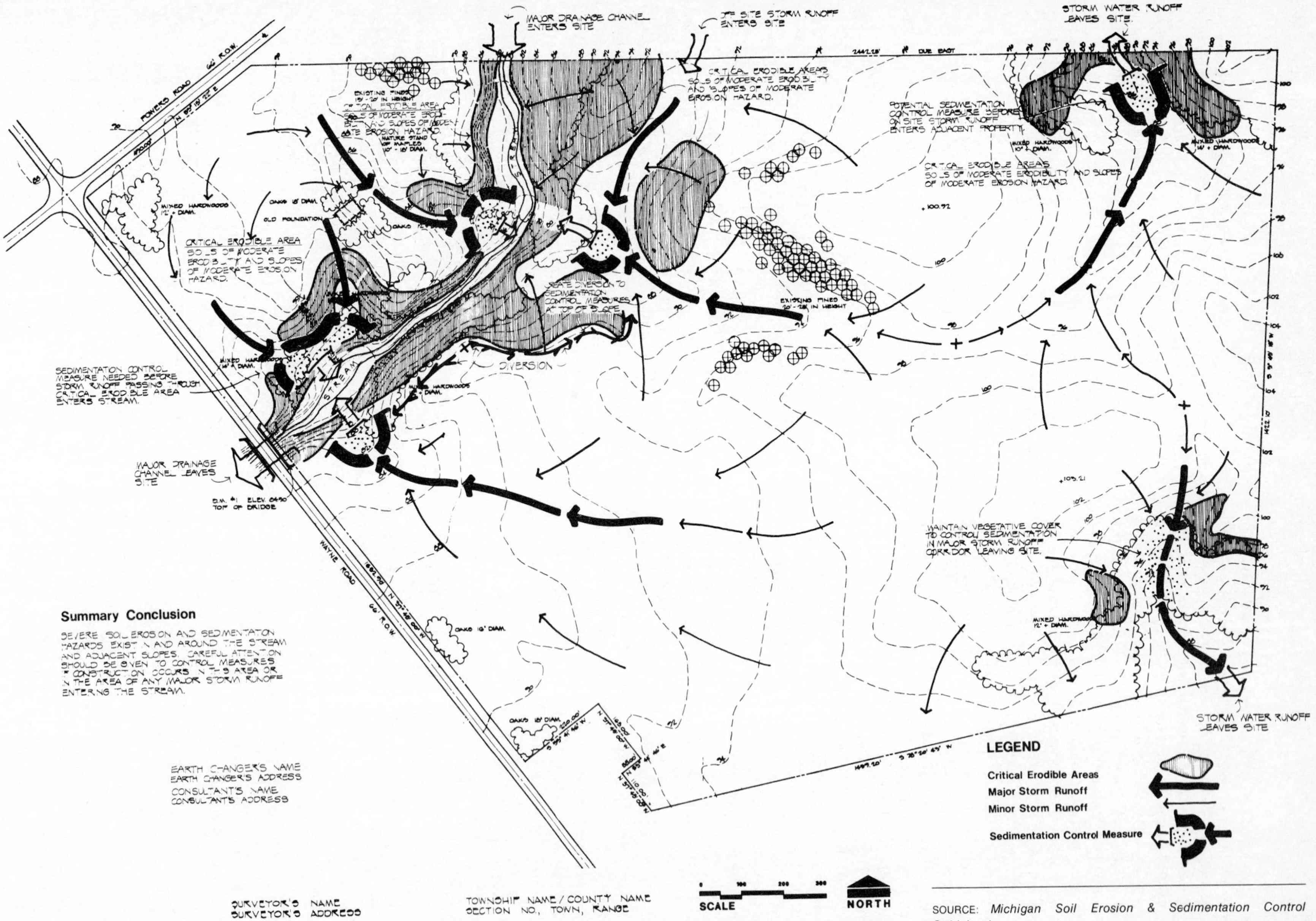

Figure 1.30 Site analysis.

2. Terracing is a control measure designed to dissipate the energy of overland flow in nonuse areas. A terrace consists of a low, broad-based earth levee constructed approximately parallel to the contours. It is designed to intercept and hold the water until it infiltrates into the soil or to conduct it as overland flow to a suitable discharge point. A hardy, vigorous turf should be established to hold the disturbed soil in place. Vertical spacing and longitudinal gradients of terraces are given in Figure 1.31.

3. Turfing, placing riprap, and spreading rubble are control methods designed to cause turbulence and increase the retardance in order to dissipate the energy of flow in channels, ditches, and pipe outfalls. Ditches are often protected by placing strips of sod held in place by wooden boards or stakes perpendicular to the path of flow at intervals along the ditch as shown in Figure 1.32.

4. Paving with either asphaltic or portland cement concrete provides superior erosion-resistant linings in gutters, ditches, and outfall structures. Ditches having grades in excess of 5 percent usually require a paved or gunite lining. Paving where $S = 5$ percent must be extended downslope at least to the point in the ditch section at which the erosive energy of the water is controlled or absorbed without erosion damage. The expense in terms of manpower, time, materials, and equipment usually limits the use of paved ditch linings to only the most demanding conditions in theater of operations airfield construction.

5. Gunite lining of ditches controls erosion effectively. Gunite is a mixture of portland cement and sand to which water is added just before being sprayed from a high-pressure nozzle onto the surface which is to be pro-

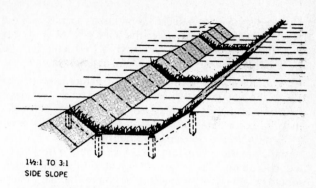

1½:1 TO 3:1
SIDE SLOPE

A. PERSPECTIVE OF DRAINAGE DITCH WITH EROSION CHECKS.

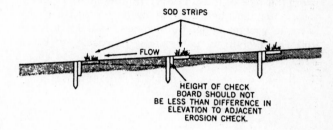

SOD STRIPS

FLOW

HEIGHT OF CHECK BOARD SHOULD NOT BE LESS THAN DIFFERENCE IN ELEVATION TO ADJACENT EROSION CHECK.

B. PROFILE OF EROSION CHECKS.

Figure 1.32 Erosion control checks.

pressure nozzle onto the surface which is to be protected. The gunite lining is formed over steel mesh which has been placed over the bottom and sides of the ditch.

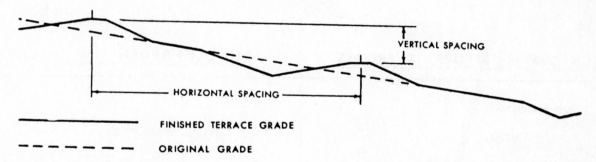

VERTICAL SPACING

HORIZONTAL SPACING

――――――― FINISHED TERRACE GRADE

― ― ― ― ― ORIGINAL GRADE

VERTICAL SPACING OF TERRACES		
AVERAGE LAND SLOPE (PERCENT)	HORIZONTAL SPACING (FEET)	VERTICAL SPACING (FEET)
2	125	2.50
4	75	3.00
6	58	3.50
8	50	4.00
10	45	4.50
12	42	5.00
14	39	5.50

LONGITUDINAL GRADIENTS FOR TERRACES	
LENGTH OF TERRACE (FEET)	TERRACE CHANNEL GRADE (PERCENT)
0 - 300	0.10
300 - 600	0.15
600 - 900	0.20
900 - 1200	0.30
1200 - 1500	0.40

Figure 1.31 Terrace spacing and gradients.

Gunite is sprayed to a thickness of 1 to 1½ inches with the steel mesh located midway in the thickness. As with paving, this erosion control method is restricted in use because of expense and special equipment required.

6. Checkdams are the most common structure used to dissipate the energy of water flowing in ditches that have longitudinal grades not exceeding 5 percent. Normally they are constructed of timber and should extend at least 24 inches into the bottom and sides of the ditch. The top of the ditch should be at least 12 inches above the top of the checkdam. The effective height of the checkdam should be at least 12 inches, but not more than 36 inches. An apron of rubble or riprap should extend at least 4 feet from the face of the checkdam on the discharge side. A weir notch must be cut in the top of the checkdam with a capacity large enough to discharge the anticipated run-off. The method for determining the spacing of check-

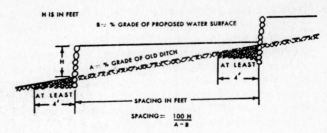

Figure 1.33 Spacing of checkdams.

dams is given in Figure 1.33. It is considered uneconomical to space checkdams closer than 50-feet intervals. An alternate solution should be considered for shorter intervals, such as use of concrete, timber, or boulder linings.

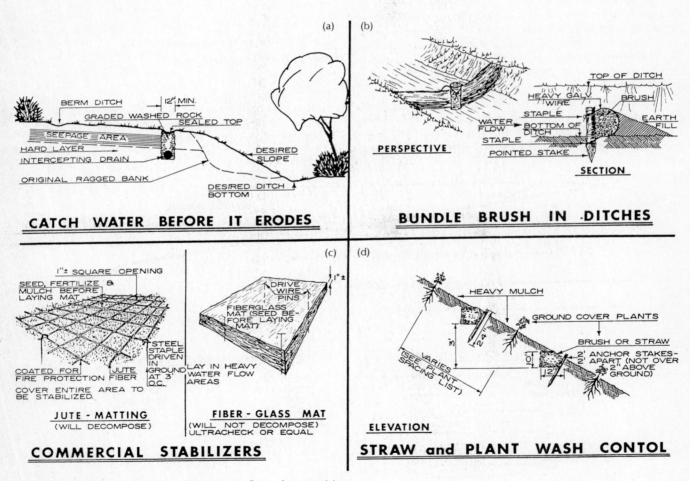

Figure 1.34 (*a*) Intercept water; (*b*) interrupt flow of water; (*c*) earth stabilizers; (*d*) straw and plant cover.

1.13 Sources of Surveying and Geological Data

The governmental agencies in the accompanying table are sources of surveying and geological data.

Guide to Map Scales   The table on page 58 is a guide to the conversion of map scales to linear and area measurements.

Agency	Nature of survey data
U.S. Geological Survey (USGS, Department of the Interior)	Topographic maps and indexes; bench mark locations, level data, and tables of elevations; streamflow data; water resources; geologic maps; horizontal control data; monument location
National Ocean Survey (Department of Commerce)	Topographic maps; coastline charts; topographic and hydrographic studies of inland lakes and reservoirs; bench mark locations, level data, and tables of elevations; horizontal control data; state tables for lambert and transverse Mercator projections; tide and current tables; coast pilots' information; seismological studies; magnetic studies; aeronautical charts
Bureau of Land Management	Township plots, showing land divisions; state maps, showing public land and reservations; survey progress map of the United States, showing the progress of public-land surveys
Army Map Service (Department of the Army)	Topographic maps and charts
Corps of Engineers (Department of the Army)	Topographic maps; charts of Great Lakes and connecting waters
Board of Engineers for Rivers and Harbors (Department of the Army)	Maps and charts of ports and permits for construction of bridges, piers, and so on in navigable rivers and harbors
Coastal Engineering Research Center (Department of the Army)	Beach erosion data
Mississippi River Commission (Department of the Army)	Hydraulic studies and flood control information
Soil Conservation Service (SCS, Department of Agriculture)	Soil charts and maps; index
U.S. Forest Service (Department of Agriculture)	Forest reserve maps including topography and culture and vegetation classification
U.S. Postal Service	Rural free delivery maps by counties, showing roads, streams, and so on
Naval Oceanographic Office (Department of the Navy)	Nautical charts; navigational manuals; aeronautical charts
International Boundary Commission, United States and Canada	Topographic maps for ½ to 2½ miles on either side of the United States–Canadian boundaries
Local municipalities: county, town, village, city	Street maps; zoning maps, drainage maps; horizontal and vertical control data; utility maps

Scale	Feet per inch	Inches per 1000 feet	Inches per mile	Miles per inch	Meters per inch	Acres per square inch	Square inches per acre	Square miles per square inch
1:500	41.67	24.00	126.72	0.008	12.70	0.040	25.091	0.00006
1:600	50.00	20.00	105.60	0.009	15.24	0.057	17.424	0.00009
1:1,000	83.33	12.00	63.36	0.016	25.40	0.159	6.273	0.00025
1:1,200	100.00	10.00	52.80	0.019	30.48	0.230	4.356	0.00036
1:1,500	125.00	8.00	42.24	0.024	38.10	0.359	2.788	0.00056
1:2,000	166.67	6.00	31.68	0.032	50.80	0.638	1.568	0.00100
1:2,400	200.00	5.00	26.40	0.038	60.96	0.918	1.089	0.0014
1:2,500	208.33	4.80	25.34	0.039	63.50	0.996	1.004	0.0016
1:3,000	250.00	4.00	21.12	0.047	76.20	1.435	0.697	0.0022
1:4,000	333.33	3.00	15.84	0.063	101.60	2.551	0.392	0.0040
1:5,000	416.67	2.40	12.67	0.079	127.00	3.986	0.251	0.0062
1:6,000	500.00	2.00	10.56	0.095	152.40	5.739	0.174	0.0090
1:7,920	660.00	1.515	8.00	0.125	201.17	10.000	0.100	0.0156
1:8,000	666.67	1.500	7.92	0.126	203.20	10.203	0.098	0.0159
1:9,600	800.00	1.250	6.60	0.152	243.84	14.692	0.068	0.0230
1:10,000	833.33	1.200	6.336	0.158	254.00	15.942	0.063	0.0249
1:12,000	1,000.00	1.000	5.280	0.189	304.80	22.957	0.044	0.0359
1:15,000	1,250.00	0.800	4.224	0.237	381.00	35.870	0.028	0.0560
1:15,840	1,320.00	0.758	4.000	0.250	402.34	40.000	0.025	0.0625
1:19,200	1,600.00	0.625	3.300	0.303	487.68	58.770	0.017	0.0918
1:20,000	1,666.67	0.600	3.168	0.316	508.00	63.769	0.016	0.0996
1:21,120	1,760.00	0.568	3.000	0.333	536.45	71.111	0.014	0.1111
1:24,000	2,000.00	0.500	2.640	0.379	609.60	91.827	0.011	0.1435
1:25,000	2,083.33	0.480	2.534	0.395	635.00	99.639	0.010	0.1557
1:31,680	2,640.00	0.379	2.000	0.050	804.67	160.000	0.006	0.2500
1:48,000	4,000.00	0.250	1.320	0.758	1,219.20	367.309	0.003	0.5739
1:62,500	5,208.33	0.192	1.014	0.986	1,587.50	622.744	0.0016	0.9730
1:63,360	5,280.00	0.189	1.000	1.000	1,609.35	640.000	0.0016	1.00
1:100,000	8,333.33	0.120	0.634	1.578	2,540.00	1,594.225	0.0006	2.49
1:125,000	10,416.67	0.096	0.507	1.973	3175.01	2,490.980	0.0004	3.89
1:126,720	10,560.00	0.095	0.500	2.000	3,218.69	2,560.000	0.0004	4.00
1:250,000	20,833.33	0.048	0.253	3.946	6,350.01	9,963.907	0.0001	15.57
1:253,440	21,120.00	0.047	0.250	4.000	6,437.39	10,244.202	0.0001	16.00
1:500,000	41,666.67	0.024	0.127	7.891	12,700.02	39,855.627	0.$\overset{4}{-}$25	62.27
1:750,000	62,500.00	0.016	0.084	11.837	19,050.04	89,675.161	0.$\overset{4}{-}$11	140.12
1:1,000,000	83,333.33	0.012	0.063	15.783	25,400.05	159,422.507	0.$\overset{5}{-}$62	249.10
FORMULAS	$\dfrac{\text{Scale}}{12}$	$\dfrac{12{,}000}{\text{Scale}}$	$\dfrac{63{,}360}{\text{Scale}}$	$\dfrac{\text{Scale}}{63{,}360}$	Feet per inch $\times$ 0.3048006	$\dfrac{\text{Scale}^2}{43{,}560 \times 144}$	$\dfrac{43{,}560 \times 144}{\text{Scale}^2}$	$\dfrac{\text{Feet per inch}^2}{5{,}280^2}$

1.14 Mapping of Physical Environment

MASTER PLAN

Many basic data concerning the physical characteristics of a site can be obtained from local planning and assessors' offices. These include topographic, soil, climatic, and property maps (see Figure 1.35). Although the scale in some cases may not be readily suitable for small parcels, the maps nevertheless can help to establish boundary conditions around the site.

The following pages give examples of the most generally used sources and map forms.

TOPOGRAPHIC MAPS

The U.S. Geological Survey continues to publish the familiar and very widely used topographic maps on a variety of scales (see Figure 1.36). Quadrangles covering 7½ minutes of latitude and longitude are published on a scale of 1:24,000 (1 inch represents 2000 feet). Quadrangles covering 15 minutes of latitude and longitude are published on a scale of 1:62,500

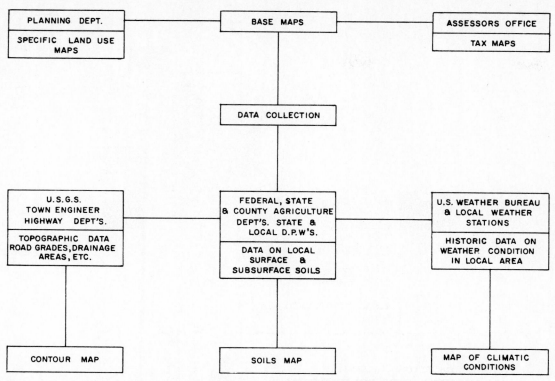

Figure 1.35 Source and type of site data available from maps of a local master plan.

(1 inch represents approximately 1 mile), and quadrangles covering 2 degrees of longitude by 1 degree of latitude are published on a scale of 1:250,000 (1 inch represents approximately 4 miles). Each quadrangle is designated by the name of a city, town, or prominent natural feature within it.

The maps are printed in five colors. The cultural features such as roads, railroads, and civil boundaries and lettering are in black. Water bodies are shown in blue. Green is used to identify areas of woodland. Red is used to identify developed urban areas and to classify roads. Features of terrain relief are shown by brown contour lines. The contour intervals differ according to the scale of the map and the amount of local relief.

Most topographic maps are available either with or without the green woodland overprint. Also, a special printing on which both green woodland tint and contours are omitted is available for certain quadrangles. These quadrangles are identified on the Geological Survey index maps.

As a means of reducing the time required to revise its maps, the Geological Survey has begun a new type of map updating, called interim revision. Using this technique, revisions to published maps are compiled in the office from recent aerial photography and then are printed on the published map in a distinctive heliotrope color. No field checking is performed under this system, and production time is approximately one-sixth of that required for standard revision.

An index map identifying published maps and giving their publication date is published by the Geological Survey. This map contains details on other maps published by the Geological Survey as well as details on how to purchase maps. Free copies of the index map may be obtained from the Map Information Office, U.S. Geological Survey, Sunrise Valley Drive, Reston, Virginia 22092.

HYDROGRAPHIC CHARTS

The navigable waters in and around the United States are shown on a variety of maps and charts. The more common of these, such as topographic and county maps, generally do not include much detail on hydrographic features. Therefore, persons needing information on such features as navigational aids, underwater hazards, lights, and harbor facilities must consult specialized hydrographic charts. The principal agencies producing such charts are the National Ocean Survey and the Lake Survey Center.

National Ocean Survey The National Ocean Survey, previously known as the U.S. Coast and Geodetic Survey, publishes a wide variety of charts of United States coastal waters (see Figure 1.37.)

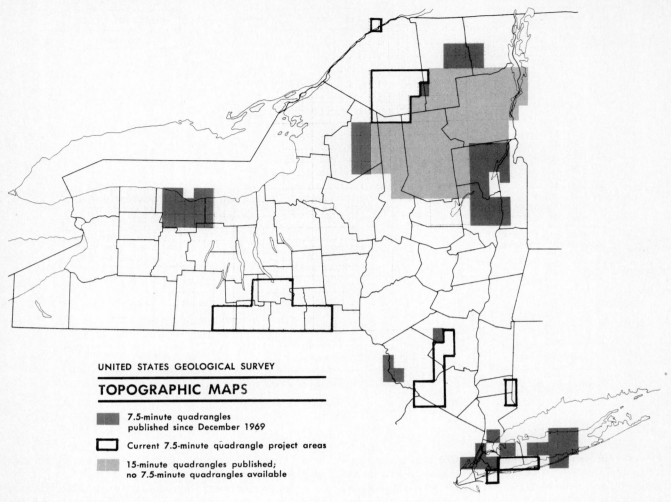

UNITED STATES GEOLOGICAL SURVEY

TOPOGRAPHIC MAPS

■ 7.5-minute quadrangles published since December 1969

▢ Current 7.5-minute quadrangle project areas

▨ 15-minute quadrangles published; no 7.5-minute quadrangles available

Figure 1.36 Topographic maps.

Small Craft Charts These compact, accordion-folded charts are specifically designed for cockpit use and include special features such as large-scale harbor area insets, locations of facilities, lists of services and supplies, tide tables, current tabulations, weather data, course bearings, and illustrated whistle and distress signals. They are constructed from the same source material and to the same high standards as conventional charts and are issued with protective covers or jackets.

Harbor Charts These charts are intended for navigation and anchorage in harbors and smaller waterways and are published at scales of 1:50,000 and larger, depending on the size and importance of the harbor and the number and types of existing dangers.

Coast Charts These charts are intended for coastwise navigation inside offshore reefs and shoals, for entering bays and harbors of considerable size, and for navigating certain inland waterways. Their scales range from 1:50,000 to 1:100,000.

General Charts These charts are intended for the navigation of vessels whose positions can be fixed by landmarks, lights, buoys, and characteristic soundings but whose courses are well offshore. Their scales range from 1:100,000 to 1:600,000.

Sailing Charts These charts are plotting charts for offshore sailing between coastal ports and for approaching coasts from the open ocean. They are published at scales smaller than 1:600,000.

CENSUS MAPS

Population data published by the federal Bureau of the Census are summarized for many different types and sizes of areas. Many of the areas coincide with political boundaries, but these areas are frequently unsatisfactory for meaningful interpretation of the data. The Bureau of the Census therefore delineates areas that correlate more closely with settlement patterns and centers of activity.

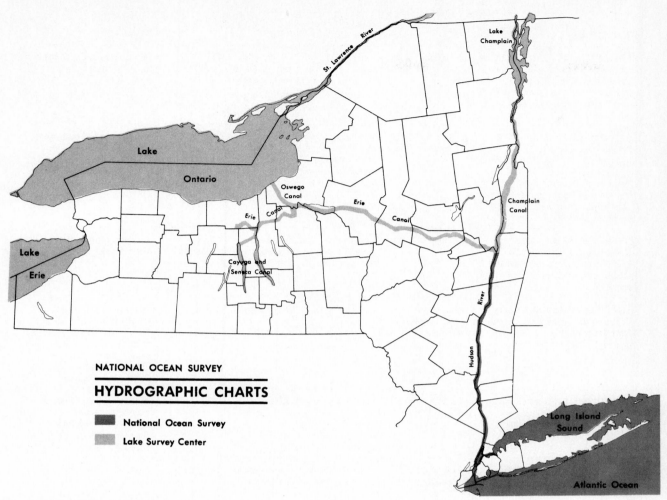

Figure 1.37 Hydrographic charts.

To aid the census user in interpreting and using the various summaries of data, the Bureau of the Census publishes a variety of maps that delineate the limits of the statistical areas it uses. Figure 1.38 summarizes these maps and includes information on how to obtain copies. Some of the maps listed are published as part of census reports, and these are identified in the figure. Individual copies of the maps are usually available, but they generally cost more when they are purchased separately. Reproducible copies (either diazo mylars or photographic negatives or positives) are available for all census maps. The cost of these copies is higher than the cost of paper prints.

SANBORN MAPS

Utilization The Sanborn fire insurance map serves numerous diversified industries and purposes. This accurately scaled visual depiction of street layout, building location and construction, exposures, occupancies, fire hazards, and fire protection is a valuable guide to the fire and casualty underwriter. Fire insurance companies maintain complete nationwide files of Sanborn maps in their principal offices. These form a graphic record for review and underwriting of individual risks as well as block liability as daily reports are received from the field. See Figure 1.39 for a diagram of a residential block.

Mapping, the noting of liability on the face of the map, is an important function in underwriting. Insurance companies utilize the map as a source of original entry on individual risks, generally recording such information as policy numbers, expiration dates, amounts of net retention, and distribution of liability between buildings and contents, thus completing a visual record of current commitments.

Local fire insurance agents are equipped with Sanborn maps of their respective cities and towns. The maps' careful identification of risks by map volume, page, block, and house number on daily reports renders a helpful, time-saving service. Maps are kept up to date by correction slips made from

MAP SERIES	SCALE	NUMBER OF MAP SHEETS	PRICE	SHEET SIZE	AVAILABILITY
US COUNTY OUTLINE MAP: Contains boundaries and names of all counties in the US. State boundaries are in black, county boundaries and names in black or blue.	1:5,000,000 (1" = approx. 79 mi.)	1 map sheet	$0.20 for black edition, $0.25 for black/blue edition	26" x 41"	Available from Government Printing Office
COUNTY SUBDIVISION MAPS: State maps showing 1970 boundaries of counties and county subdivisions, the location of all incorporated places and those unincorporated places for which population figures are published (Replaces Minor Civil Division Map)	Generally 1:750,000 (1" = approx. 12 mi.)	1 per state, except for each of the combinations noted below: Mass.-Conn.-R.I.-Vt.-N.H. D.C.-Del.-Md.	$0.20 per map ($9.40 for entire set)	36" x 48"	Available from Government Printing Office
COUNTY MAPS (Unpublished): Contain boundaries of MCD-CCD's, incorporated places, tracts (where established), and enumeration districts.	Generally 1" = 2 mi.	1 per county except for very large counties	$1.00 and up depending on map size	18" x 24" and larger	Available from Central Users' Service
TRACT OUTLINE MAPS: Show tract boundaries and incorporated limits for places over 25,000 population for each SMSA.	Varies depending on size and complexity of the area	1 to 4 per area	Varies depending on size of report	Varies	Available as part of tract reports (PHC(1) Series) from Government Printing Office. Preliminary maps may be obtained from the Central Users' Service
URBANIZED AREA MAPS: Show extent of and all areas included within 1970 Urbanized Areas.	1:250,000 (1" = approx. 4 miles)	1 to 6 per area	Varies	9" x 11⅓"	Available as part of final population reports (PC(1)A Series) from Government Printing Office
METROPOLITAN MAPS (MMS): Published for Urbanized Areas for which block statistics are available, contains all Census boundaries down to block level.	1" = 2000' with a few enlargements at 1" = 1000'	2 to 144 depending on size of area	Varies depending on size of area	18" x 24"	Available as part of Block Statistics reports (HC(3) Series) from Government Printing Office (Also available in microfiche form (4 maps to a sheet) from the National Technical Information Service, US Dept. of Commerce, Springfield, Va. 22151
PLACE MAPS (Unpublished): Prepared for incorporated and unincorporated places not shown on MMS, contains boundaries of tracts (where established), enumeration districts, and wards (where reported to the Bureau).	Varies with the size of the area, range from 1" = 400' to 1" = 1500'	Generally 1 per place	$1.50 and up depending on size of the map	Varies	Available from Central Users' Service

Figure 1.38 United States census maps available for public use.

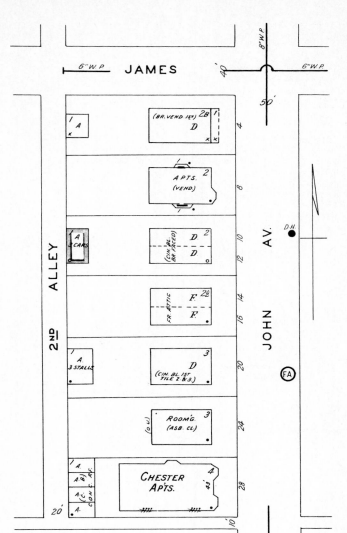

Figure 1.39 Diagram of a residential block. (*Source: The Sanborn Map,* Sanborn Map Company, Inc., Pelham, N.Y.)

periodic field surveys covering changes since the previous inspection of the territory.

Federal, state, county, and municipal governmental agencies rely on Sanborn maps to save costly field inspections and to keep a permanent record of valuable information. The maps are widely used by municipal and county departments such as building, education, engineering, health and sanitation, highway, planning and zoning, public libraries, public works, sewer, tax assessment, and water and by city managers. Utility companies find them invaluable as an original record of their outside plant data. Banks and mortgage and life insurance companies utilize them in underwriting mortgage loans, recording their commitments much in the manner of the fire insurance companies.

In recent years the Sanborn Map Company has developed special market analysis maps, converting for marketing studies the wealth of detail of individual communities appearing in the standard fire insurance map. Sanborn's nationwide corps of field engineers and publishers has accumulated an unparalleled record of the physical growth and current status of municipalities throughout the United States. The company facilities are available for special surveys on request.

Narrative Description of a Mercantile Block   The scale of the block shown in Figure 1.40 is 50 feet to the inch, and the direction is indicated by the meridian in the right margin.

The block is bounded on the north by Center Street, on the east by Atlantic Avenue, on the south by Main Street, and on the west by Pacific Avenue. All thoroughfares are 50 feet in width, as indicated by the figures at the extremity of each street. This block is elevated 42 feet above the city datum, as indicated by the figure in the circle at the intersection of Pacific Avenue and Center Street.

The number of the block is 28, as indicated by the prominent figure in the center of the 10-foot alley that divides the block into equal portions. This number represents the official designation for the block provided by the city or an arbitrary designation supplied by the Sanborn Map Company in the absence of a suitable official number.

5 Main Street   The size of the building is 100 by 100 feet. As indicated by the name, it is a hotel building.

On the first floor in the southwest corner of the building there is a drugstore *S,* the extent of which is indicated by a broken line, 50 by 15 feet. Adjoining this to the east is a florist shop *S,* 30 by 15 feet. At the southeast corner there is a haberdashery *S,* 40 by 30 feet. In the northeast corner of the building there is a restaurant *Rest.,* 50 by 50 feet, the kitchen of which is directly underneath in the basement shown.

With the exception of the 50- by 50-foot section in the northeast corner, the main part of the building contains ten stories, as indicated by the figure in the southwest corner, and is 106 feet in height from the street level to the roof level, as indicated by the figure in the center of the Main Street side of the building.

The 50- by 50-foot section in the northeast corner of the building contains only one story, as indicated by the figure in the northwest corner of this section, and is 14 feet in height from the street level to the roof level.

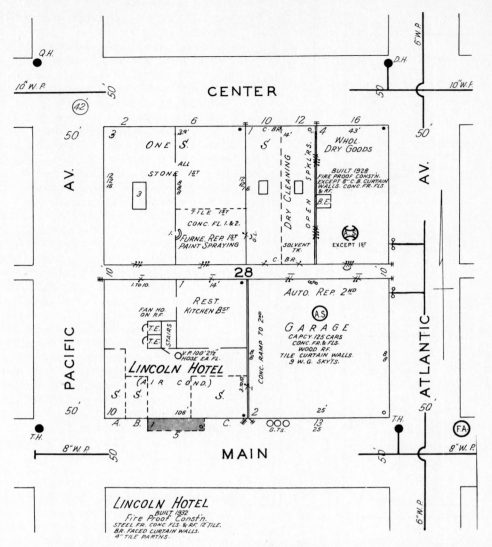

Figure 1.40 Diagram of a mercantile block. (*Source: The Sanborn Map*, Sanborn Map Company, Inc., Pelham, N.Y.)

1.15 Utility Easements

Underground installations in new subdivisions are usually located within easements along rear lot lines. The easement plan in Figure 1.41, prepared by the Detroit Edison Company and the Michigan Bell Telephone Company, is typical. A utility easement guide prepared jointly by the two companies suggests the following design criteria for underground service lines:

In general, the locations of easements for underground service lines are similar to those required for overhead service lines, in that easements should be located along rear or side lot lines, or should be provided across lots in some cases. In addition, easements along front lot lines will be required at some locations.

Subdivision layout, topography, natural obstructions, and the use of these easements for water, sewer, and drainage facilities will influence the design of an easement system adequate for an underground distribution system.

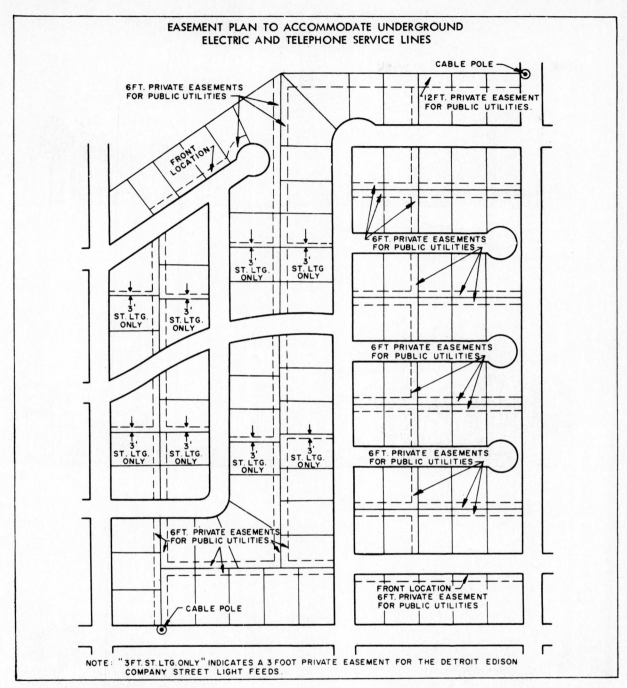

Figure 1.41 Utility easement plan.

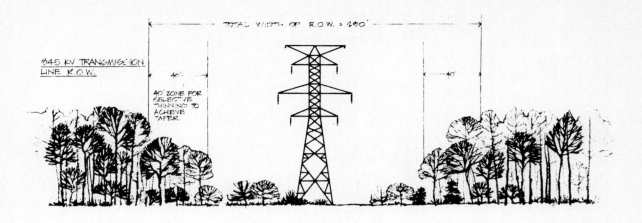

MINIMUM GROUND CLEARANCE = 30'
MINIMUM BRUSH CLEARANCE = 20'

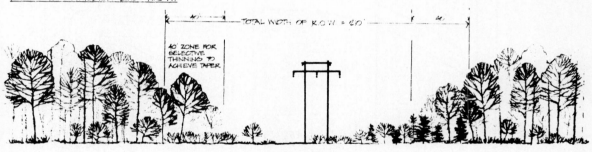

MINIMUM GROUND CLEARANCE = 25'
MINIMUM BRUSH CLEARANCE = 12'

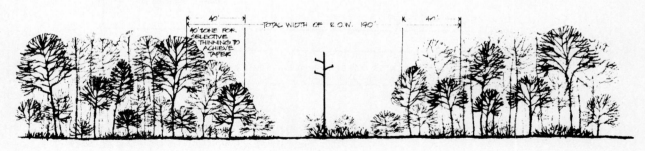

MINIMUM GROUND CLEARANCE = 22'
MINIMUM BRUSH CLEARANCE = 10'

Figure 1.42   Right-of-way requirements for single lines.

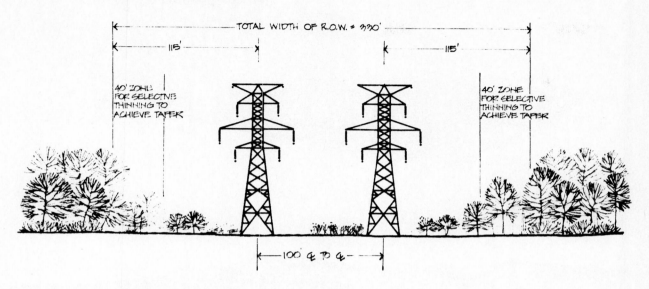

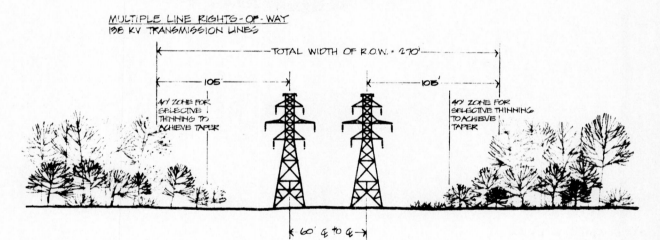

Figure 1.43   Right-of-way requirements for multiple lines.

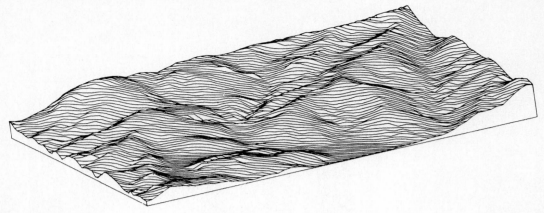

Figure 1.44

Figure 1.45

1.16 Graphic Analysis

VIEWS: VISUAL PERSPECTIVES AND ISOMETRICS

Computer-drawn illustrations can communicate the visual feelings of a land area. A given site can quickly be observed in a series of drawings shown from various directions of view. By using these drawings, an individual can extract and analyze a great deal of information in a short time.

Large land areas can easily be illustrated. Figure 1.44 is a ski resort in Park City, Utah. The visual perspective communicates a quick impression of the mountains and valleys and their distinct topographic values for skiing purposes.

The computer produces both two-dimensional and three-dimensional cross sections accurately and inexpensively. Sections, at any desired point or along any line, can easily be generated and plotted.

Figure 1.45 displays the proposed grading changes for a regional park, Prado Dam, in Southern California. The visual effects of the proposed park are quickly communicated.

GRADING

As shown in Figures 1.46 and 1.47 the appearance of a particular subdivision design before and after grading can be simulated by visual perspectives and isometrics.

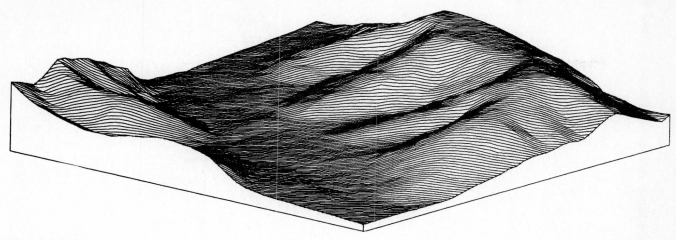

Figure 1.46 Before grading.

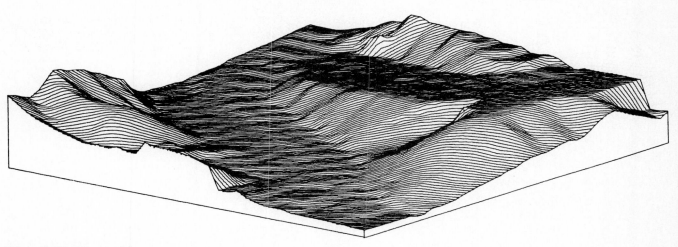

Figure 1.47 After grading.

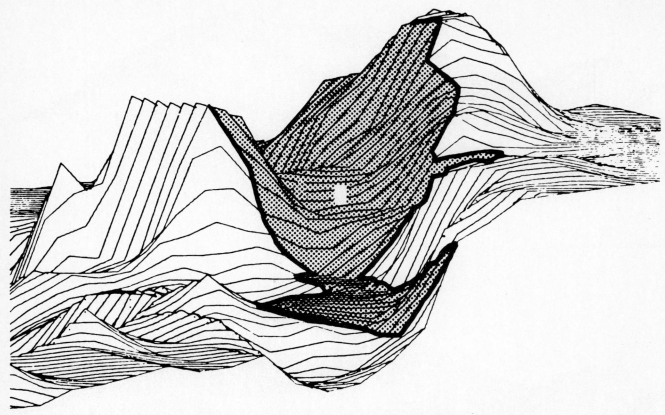

Figure 1.48 Dark areas exposed — and see opposite page: (a) point exposure; (b) relative exposure.

EXPOSURE

The visual exposure of a single point or the relative exposure of a land area can be analyzed by the computer. Figure 1.48a is a single-point analysis that displays with dark grid cells areas that can be seen in 360 degrees when one is standing at a point of observation. This analysis of visual exposure for a given viewing point can be accomplished by using a standard computer bank.

Figure 1.48b displays the relative visual exposure of a proposed park site near San Diego. In this example, the darker cells are more visible than the other cells, whereas the lighter shades of gray symbolize the less visible cells.

The single-point exposure has been used in locating vista points, restaurants, water towers, and refuse sites. The relative-exposure analysis is beneficial in the allocation of land uses. Open spaces might be located in high-exposure areas, while industry, power lines, and transportation might be located in low-exposure areas.

SLOPES

Topographic slope analysis is an important factor in land evaluation. This is an example of a computer service that analyzes graphically the relative percentage of slope for landscape terrain. This analysis has been extremely effective in evaluating the slope constraints on land developments and road designs. Normally, such an analysis is extremely time-consuming and suffers from inaccuracy. With the use of the computer technique, however, slopes can be rapidly located with a minimum of cost and effort.

Figure 1.49 compares a normal map of topography with a computer-produced slope map. Changes in tone represent changes in relative gradient. The key shows the gray tones that relate to the varying percentages of slope.

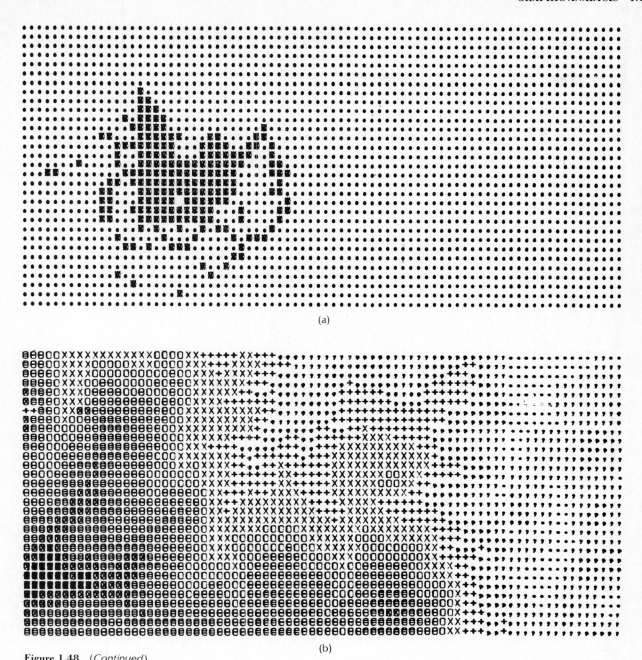

(a)

(b)

Figure 1.48 (*Continued*)

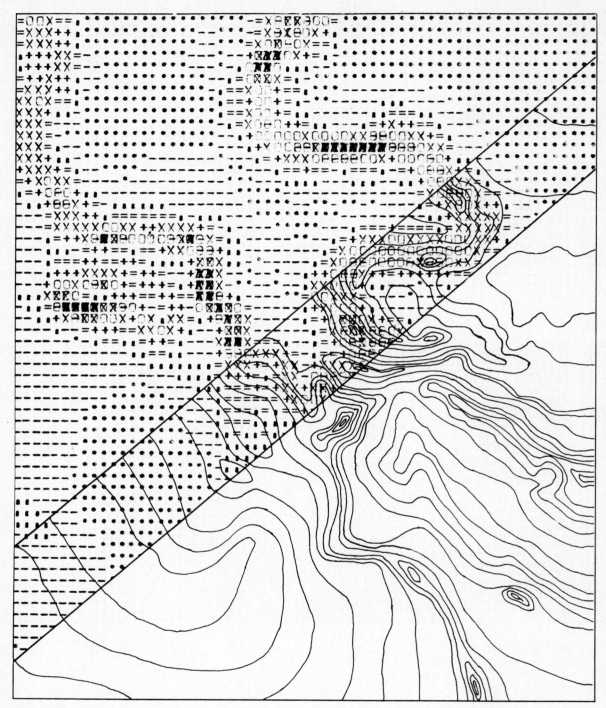

Figure 1.49

MAGNITUDE OF GRADIENT

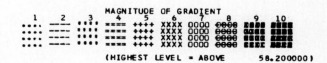

(HIGHEST LEVEL = ABOVE   58.200000)

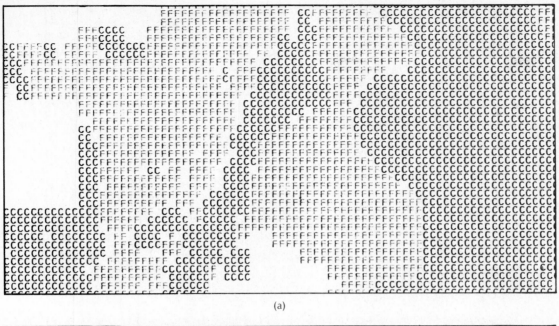

(a)

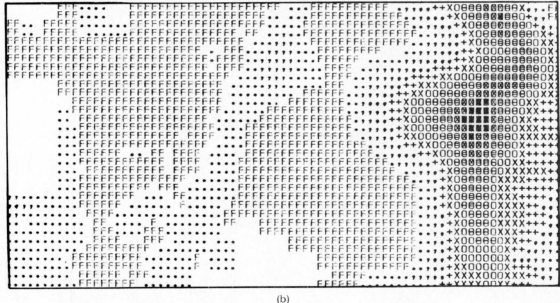

(b)

Figure 1.50   (*a*) Map of cut-and-fill areas; (*b*) map of relative cut (the darkest area indicates the greatest cut).

CUT AND FILL

Accurate and inexpensive cut-and-fill analysis can be done by the computer. This analysis uses data banks of existing and proposed topography to calculate the differences in each grid cell (normally a 10-foot square). Five separate computer printouts are prepared:

1. Map of cut-and-fill areas
2. Map of relative cuts
3. Map of relative fills
4. Cut-and-fill balance
5. Elevation and volume differences at each grid cell

Two of the printouts are shown in Figure 1.50*a* and *b*.

Figure 1.51 Topographic map.

TOPOGRAPHIC MAPS

Extremely accurate topographic maps can be prepared from field survey data or other sources of point information. The computer program uses a highly advanced interpolation technique for producing contour lines. As shown in Figure 1.51, topographic elevations are indicated along given contour lines.

WATERSHED AREAS

Land planning tools include a complete watershed analysis. This analysis provides statistical, schematic, and graphic information on natural boundaries and subwatersheds. Watershed analysis produces two separate outputs. The first is a printout of the general watershed areas illustrating graphically the ridgelines that separate them (see Figure 1.52a). The

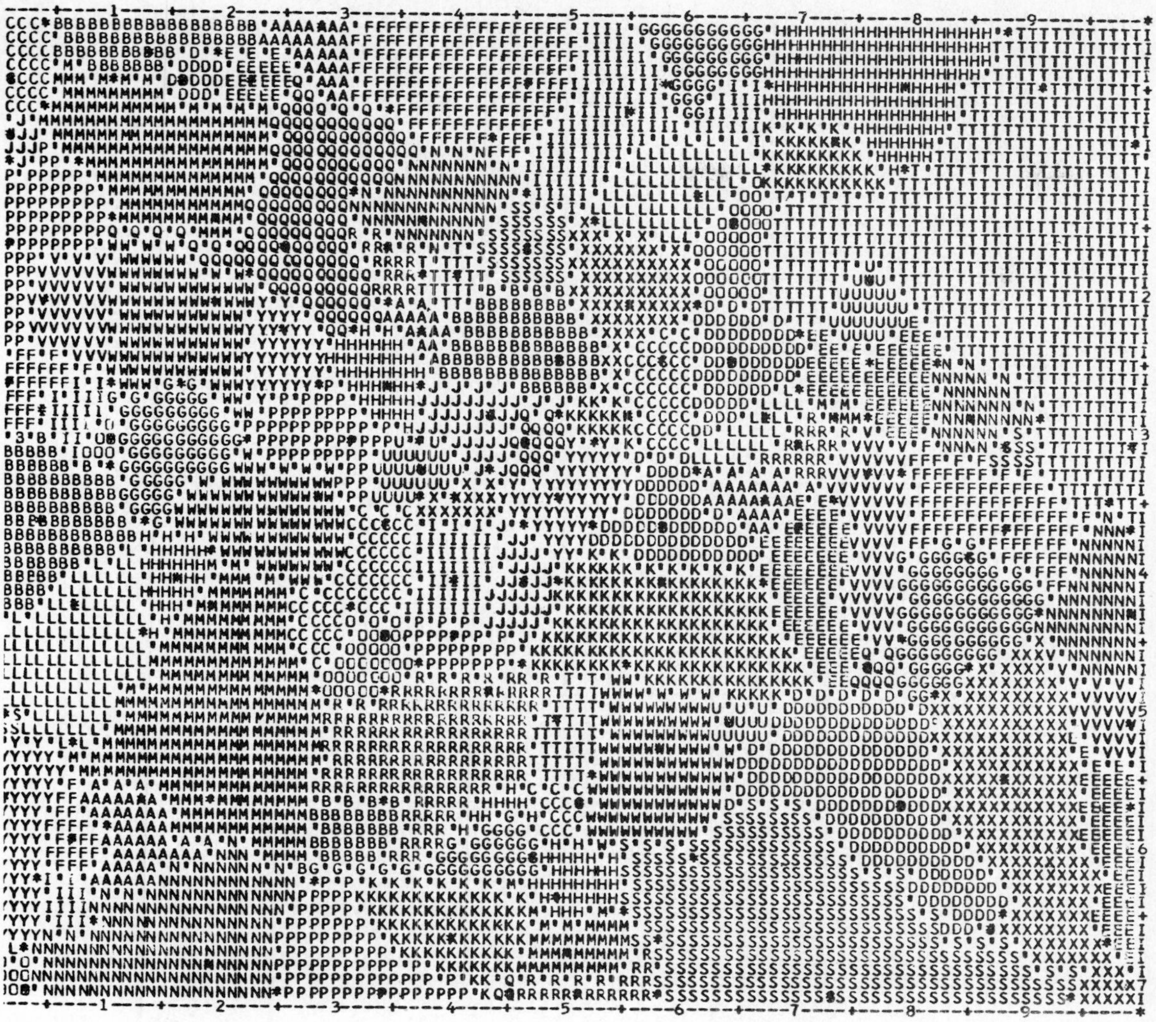

LIST OF PITS

NUMBER	ROW	COLUMN	SYMBOL	AREA
(1)	1	28	A	33
(2)	3	13	B	54
(3)	5	1	C	26
(4)	5	16	D	16
(5)	5	22	E	22
(6)	5	46	F	155
(7)	5	58	G	54
(8)	5	79	H	149
(9)	7	55	I	127
(10)	9	1	J	10
(11)	9	73	K	49
(12)	13	61	L	80
(13)	15	19	M	174
(14)	15	37	N	74
(15)	15	64	O	47
(16)	17	1	P	71
(17)	17	25	Q	185
(18)	17	34	R	21
(19)	17	46	S	44
(20)	19	40	T	20
(21)	19	76	U	33
(22)	21	4	V	51
(23)	21	19	W	127
(24)	21	55	X	84
(25)	23	25	Y	45
(26)	23	37	Z	20
(27)	25	49	A	94
(28)	25	58	B	50
(29)	25	64	C	71
(30)	25	76	D	69
(31)	27	1	E	29
(32)	27	16	F	82
(33)	27	34	G	43
(34)	29	4	H	25
(35)	29	43	I	51
(36)	29	55	J	18
(37)	29	67	K	29
(38)	29	73	L	8
(39)	29	85	M	46
(40)	31	10	N	8

LIST OF PITS

NUMBER	ROW	COLUMN	SYMBOL	AREA
(41)	31	31	P	64
(42)	31	46	Q	21
(43)	31	70	R	24
(44)	31	88	S	11
(45)	31	95	T	623
(46)	33	37	U	29
(47)	33	76	V	80
(48)	35	28	W	104
(49)	35	40	X	17
(50)	35	49	Y	63
(51)	35	67	Z	36
(52)	37	4	A	101
(53)	37	34	B	71
(54)	37	58	C	70
(55)	37	70	D	81
(56)	37	83	E	104
(57)	39	85	F	106
(58)	41	16	G	37
(59)	41	40	H	49
(60)	41	46	I	41
(61)	41	58	J	189
(62)	43	7	K	132
(63)	43	19	L	285
(64)	43	95	M	74
(65)	45	34	N	36
(66)	45	40	O	32
(67)	47	76	P	11
(68)	49	43	Q	178
(69)	51	1	R	4
(70)	51	49	S	41
(71)	51	64	T	17
(72)	51	95	U	39
(73)	53	58	V	130
(74)	55	88	W	237
(75)	57	1	X	66
(76)	57	13	Y	69
(77)	57	34	Z	38
(78)	57	51	A	18
(79)	57	76	B	216
(80)	57	96	C	62

(a)

(b)

Figure 1.52 (a) Subwatershed analysis; (b) acreage calculations.

Figure 1.53 Watercourse map.

acreages for each watershed are calculated and included as part of the output. An example of this output appears in Figure 1.52b.

The computer also produces a pen-line drawing of water "courses" or "flows." An example of this drawing is presented in Figure 1.53. The asterisks represent the topographic peaks, whereas the boxes are topographic pits, and the arrows delineate the watercourse lines.

Some of the uses of this analysis are potential runoff calculations, flooding simulations, site analysis, and storm drain layout. Such analyses, although difficult to perform manually, are easily produced from the basic data bank.

Communities offer local services in response to different needs through varying authorities and regulations. In this respect they require study and comparison on an individual case basis. Figure 1.54 will help as a reminder to bring together some of the more important site service, site design factors that should be regarded. Whenever possible, services should be coordinated and unified.

Local Services which require consideration in the design of a site include	Major Site Design considerations relating to Local Services include																
	SIZE	SHAPE	COLOR	SCALE	LOCATION	UTILITY CONNECTION	GROUND AREA	PAVING	APPEARANCE	VISIBILITY	ACCESSIBILITY	GRAPHICS	STRUCTURE	SCREEN	COST	LOCAL REGULATIONS	MATERIAL SELECTION
Street Construction and Maintenance	x			x	x		x	x	x	x	x				x	x	x
Sidewalk Construction and Maintenance	x			x	x		x	x	x	x	x				x	x	x
Street Planting and Maintenance	x	x	x	x	x			x		x	x	x		x	x	x	x
Street Lighting Maintenance					x					x	x				x	x	
Trash, Garbage, and Ash Removal					x			x		x				x	x	x	
Snow Removal					x		x			x					x	x	
Fire Protection					x	x				x	x				x	x	
Police Protection					x					x	x				x	x	
Ambulance Service					x					x					x	x	
Water Service	x			x	x	x			x	x					x	x	x
Gas Service	x			x	x	x			x	x					x	x	x
Electric Service	x			x	x	x			x	x	x			x	x	x	x
Sewer Service	x			x	x	x			x	x			x		x	x	x
Mail Delivery					x					x						x	
Household Deliveries					x			x		x	x				x		
Local Park, Playground, and Community Space and Maintenance	x	x	x	x	x	x	x	x	x	x	x	x	x	x	x	x	x
Public Transportation Stops and Shelters	x	x	x	x	x	x	x	x	x	x	x	x	x	x	x	x	x
Community Social and Family Services	x	x	x	x	x	x	x		x	x	x		x		x	x	x
Street Furniture, Installation and Maintenance																	
Traffic Signals	x	x	x	x	x	x			x	x	x	x			x	x	x
Police Call Boxes	x	x	x	x	x				x	x	x	x			x	x	x
Fire Alarm Boxes	x	x	x	x	x	x			x	x	x	x			x	x	x
Seating	x	x	x	x	x				x	x	x				x	x	x
Planting Tubs and Boxes	x	x	x	x	x				x						x	x	x
Hydrants	x	x	x	x	x	x			x	x	x				x	x	x
Trash Receptacles	x	x	x	x	x				x	x	x			x	x	x	x
Lighting Posts and Fixtures	x	x	x	x	x	x			x	x					x	x	x
Signs	x	x	x	x	x	x			x	x			x		x	x	x
Railings, Parapets and Safety Barriers	x	x	x	x	x				x	x			x		x	x	x
Pedestrian Direction Facilities	x	x	x	x	x	x			x	x	x	x			x	x	x
Mail Collection Boxes	x	x	x	x	x				x	x	x	x			x	x	x

Figure 1.54 Design considerations for services.

1.18 Conversion Factors

Linear Linear measure is used to express distances and to indicate the differences in elevation. The standard units of linear measure are the foot and the meter. In surveying operations, both units are frequently divided into tenths, hundredths, and thousandths for measurements. When long distances are involved, the foot is expanded into the statute mile, and the meter into the kilometer. The accompanying table shows the conversion factors for the most commonly used linear measurements.

1	Inches	Feet	Yards	Statute miles	Nautical miles	Millimeters
Inch	1	0.0833	0.0277			25.40
Foot	12	1	0.333			304.8
Yard	36	3	1	0.00056		914.4
Statute mile	63,360	5,280	1,760	1	0.8684	
Nautical mile	72,963	6,080	2,026	1.1516	1	
Millimeter	0.0394	0.0033	0.0011			1
Centimeter	0.3937	0.0328	0.0109			10
Decimeter	3.937	0.328	0.1093			100
Meter	39.37	3.2808	1.0936	0.0006	0.0005	1,000
Decameter	393.7	32.81	10.94	0.0062	0.0054	10,000
Hectometer	3,937	328.1	109.4	0.0621	0.0539	100,000
Kilometer	39,370	3,281	1,094	0.6214	0.5396	1,000,000
Myriameter	393,700	32,808	10,936	6.2137	5.3959	10,000,000

1	Centimeters	Decimeters	Meters	Decameters	Hectometers	Kilometers	Myriameters
Inch	2.540	0.2540	0.0254	0.0025	0.0003		
Foot	30.48	3.048	0.3048	0.0305	0.0030	0.0003	
Yard	91.44	9.144	0.9144	0.0914	0.0091	0.0009	
Statute mile	160,930	16,093	1,609	160.9	16.09	1.6093	0.1609
Nautical mile	185,325	18,532	1,853	185.3	18.53	1.8532	0.1853
Millimeter	0.1	0.01	0.001	0.0001			
Centimeter	1	0.1	0.01	0.001	0.0001		
Decimeter	10	1	0.1	0.01	0.001	0.0001	
Meter	100	10	1	0.1	0.01	0.001	0.0001
Decameter	1,000	100	10	1	0.1	0.01	0.001
Hectometer	10,000	1,000	100	10	1	0.1	0.01
Kilometer	100,000	10,000	1,000	100	10	1	0.1
Myriameter	1,000,000	100,000	10,000	1,000	100	10	1

Example 1:
Problem: Reduce 76 centimers to ? inches

$$76 \text{ cm} \times 0.3937 = 29 \text{ inches}$$

Answer: There are 29 inches in 76 centimeters.

Example 2:
Problem: How many feet are there in 2.74 meters?

$$\frac{2.74}{0.3048} = 9 \text{ feet}$$

Answer: There are approximately 9 feet in 2.74 meters.

Linear Measure

Measures of length		Nautical units	
12 inches	= 1 foot	6080.20 feet	= 1 nautical mile
3 feet	= 1 yard	6 feet	= 1 fathom
5½ yards = 16½ feet	= 1 rod, pole, or perch	120 fathoms	= 1 cable length
		1 nautical mile per hour	= 1 knot
40 poles = 220 yards	= 1 furlong	**Surveyor's or Gunter's Measure**	
8 furlongs = 1760 yards } = 5280 feet	= 1 mile	7.92 inches	= 1 link
3 miles	= 1 league	100 links = 66 ft = 4 rods	= 1 chain
4 inches	= 1 hand	80 chains	= 1 mile
9 inches	= 1 span	33⅓	= 1 vara (Texas)

Length equivalents

Centimeters	Inches	Feet	Yards	Meters	Chains	Kilometers	Miles
1	0.3937	0.03281	0.01094	0.01	0.0_34971	10^{-5}	0.0_56214
2.540	1	0.08333	0.02778	0.0254	0.001263	0.0_4254	0.0_41578
30.48	12	1	0.3333	0.3048	0.01515	0.0_33048	0.0_31894
91.44	36	3	1	0.9144	0.04545	0.0_39144	0.0_35682
100	39.37	3.281	1.0936	1	0.04971	0.001	0.0_36214
2012	792	66	22	20.12	1	0.02012	0.0125
100000	39370	3281	1093.6	1000	49.71	1	0.6214
160935	63360	5280	1760	1609	80	1.609	1

Subscripts after any figure, 0_3, 9_4, etc., mean that the figure is to be repeated the indicated number of times.

Volumetric Measure

Measures of volume

1728 cubic inches	= 1 cubic foot
27 cubic feet	= 1 cubic yard
1 cord of wood	= 128 cu ft
1 perch of masonry	= 16½ to 25 cu ft

Liquid or fluid measure

4 gills	= 1 pint
2 pints	= 1 quart
4 quarts	= 1 gallon
7.4805 gallons	= 1 cubic foot

(There is no standard liquid barrel; by trade custom, 1 bbl of petroleum oil, unrefined = 42 gal)

Dry measure

2 pints	= 1 quart
8 quarts	= 1 peck
4 pecks	= 1 bushel

1 std. bbl. for fruits and vegetables = 7056 cu in or 105 dry quarts, struck measure

Board measure

1 board foot = $\begin{cases} 144 \text{ cu in} = \text{volume} \\ \text{of board 1 ft sq and} \\ 1 \text{ in thick.} \end{cases}$

No. of board feet in a log = $[1\frac{1}{4}(d - 4)]^2 L$, where d = diam. of log (usually taken inside the bark at small end), in., and L = length of log, ft. The 4 in deducted is an allowance for slab. This rule is variously known as the Doyle, Conn. River, St. Croix, Thurber, Moore and Beeman, and the Scribner rule.

Volume and capacity equivalents

Cubic inches	Cubic feet	Cubic yards	U.S. apothecary liquid ounces	U.S. quarts Liquid	U.S. quarts Dry	U.S. gallons Liquid	U.S. gallons Dry	Bushels U.S.	Liters (l)
1	0.0_35787	0.0_42143	0.5541	0.01732	0.01488	0.0_24329	0.0_23720	0.0_34650	0.01639
1728	1	0.03704	957.5	29.92	25.71	7.481	6.429	0.8036	28.32
46656	27	1	25853	807.9	694.3	202.0	173.6	21.70	764.6
1.805	0.001044	0.0_43868	1	0.03125	0.02686	0.007813	0.006714	0.0_38392	0.02957
57.75	0.03342	0.001238	32	1	0.8594	0.25	0.2148	0.02686	0.9464
67.20	0.03889	0.001440	37.24	1.164	1	0.2909	0.25	0.03125	1.101
231	0.1337	0.004951	128	4	3.437	1	0.8594	0.1074	3.785
268.8	0.1556	0.005761	148.9	4.655	4	1.164	1	0.125	4.403
2150	1.244	0.04609	1192	37.24	32	9.309	8	1	35.24
61.02	0.03531	0.001308	33.81	1.057	0.9081	0.2642	0.2270	0.02838	1

Subscripts after any figure, 0_3, 9_4, etc., mean that that figure is to be repeated the indicated number of times.

Measures of Area

144 square inches	= 1 square foot
9 square feet	= 1 square yard
30¼ square yards	= 1 square rod, pole or perch

160 square rods
= 10 square chains
= 43,560 sq ft $\Big\}$ = 1 acre
= 5645 sq varas (Texas)

640 acres = 1 square mile = 1 "section" of U.S. government surveyed land

Area equivalents

Square meters	Square inches	Square feet	Square yards	Square rods	Square chains	Roods	Acres	Square miles or sections
1	1550	10.76	1.196	0.0395	0.002471	0.0_39884	0.0_32471	0.0_63861
0.0_96452	1	0.006944	0.0_37716	0.0_42551	0.0_51594	0.0_66377	0.0_61594	0.0_92491
0.09290	144	1	0.1111	0.003673	0.0_32296	0.0_49184	0.0_42296	0.0_72491
0.8361	1296	9	1	0.03306	0.002066	0.0_38264	0.0002066	0.0_63228
25.29	39204	272.25	30.25	1	0.0625	0.02500	0.00625	0.0_59766
404.7	627264	4356	484	16	1	0.4	0.1	0.0001562
1012	1568160	10890	1210	40	2.5	1	0.25	0.0_33906
4047	6272640	43560	4840	160	10	4	1	0.001562
2589998		27878400	3097600	102400	6400	2560	640	1

1 hectare = 100 acres = 10,000 centiares or square meters.
Subscripts after any figure 0_3, 9_4, etc. mean that that figure is to be repeated the indicated number of times.

Measures of Weight

Weights
(The grain is the same in all systems)
Avoirdupois weight

16 drams = 437.5 grains	= 1 ounce
16 ounces = 7000 grains	= 1 pound
100 pounds	= 1 cental
2000 pounds	= 1 short ton
2240 pounds	= 1 long ton
1 std lime bbl, small	= 180 lb net
1 std lime bbl, large	= 280 lb net

Also (in Great Britain):

14 pounds	= 1 stone
2 stone = 28 lb	= 1 quarter
4 quarters = 112 lb	= 1 hundred-weight (cwt.)
20 hundredweight	= 1 long ton

Troy weight

24 grains		= 1 penny-weight (dwt.)
20 pennyweights	= 480 grains	= 1 ounce
12 ounces	= 5760 grains	= 1 pound

1 Assay Ton = 29,167 milligrams, or as many milligrams as there are troy ounces in a ton of 2000 lb. avoirdupois. Consequently, the number of milligrams of precious metal yielded by an assay ton of ore gives directly the number of troy ounces that would be obtained from a ton of 2000 lb. avoirdupois

Apothecaries' weight

20 grains		= 1 scruple
3 scruples = 60 grains		= 1 dram ℨ
8 drams		= 1 ounce ℨ
12 ounces	= 5760 grains	= 1 pound

Mass equivalents

		Ounces		Pounds		Tons		
Kilograms	Grains	Troy and apoth.	Avoir-dupois	Troy and apoth.	Avoir-dupois	Short	Long	Metric
1	15432	32.15	35.27	2.6792	2.205	0.0_21102	0.0_39842	0.001
0.0_46480	1	0.0_22083	0.0_22286	0.0_31736	0.0_31429	0.0_7143	0.0_76378	0.0_76480
0.03110	480	1	1.09714	0.08333	0.06857	0.0_43429	0.0_43061	0.0_43110
0.02835	437.5	0.9115	1	0.07595	0.0625	0.0_43125	0.0_42790	0.0_42835
0.3732	5760	12	13.17	1	0.8229	0.0_34114	0.0_33673	0.0_33732
0.4536	7000	14.58	16	1.215	1	0.0005	0.0_34464	0.0_34536
907.2	140_6	29167	320_3	2431	2000	1	0.8929	0.9072
1016	15680_4	32667	35840	2722	2240	1.12	1	1.016
1000	15432356	32151	35274	2679	2205	1.102	0.9842	1

Subscripts after any figure, 0_3, 9_4, etc., mean that that figure is to be repeated the indicated number of times.

English Units to English Units

This unit	Multiplied by	Equals this unit
Feet (ft)	12	Inches (in)
Yards (yd)	3	Feet (ft)
Fathoms	6	Feet (ft)
Rods, poles, or perches	16.5	Feet (ft)
Furlongs (fur)	660	Feet (ft)
Miles, statute	5280	Feet (ft)
Miles, statute	1760	Yards (yd)
Miles, nautical	6080	Feet (ft)
Miles, nautical	2027	Yards (yd)
Square feet (ft^2)	144	Square inches (in^2)
Square yards (yd^2)	9	Square feet (ft^2)
Square miles (mi^2)	27,878,400	Square feet (ft^2)
Square miles (mi^2)	3,097,600	Square yards (yd^2)
Acres	43,560	Square feet (ft^2)
Acres	4840	Square yards (yd^2)
Gallons (gal)	231	Cubic inches (in^3)
Gallons, U.S. (gal)	0.833	Imperial gallon
Cubic feet (ft^3)	1728	Cubic inches (in^3)
Cubic feet (ft^3)	7.48	Gallons (gal)
Cubic yards (yd^3)	27	Cubic feet (ft^3)
Cubic yards (yd^3)	202	Gallons (gal)
Acre-feet (acre·ft)	43,560	Cubic feet (ft^3)
Acre-feet (acre·ft)	1613	Cubic yards (yd^3)
Acre-feet (acre·ft)	325,829	Gallons (gal)
Pounds, lb	16	Ounces (oz)
Tons, short	2000	Pounds (lb)
Tons, long	2240	Pounds (lb)
Miles per hour (mi/h)	88	Feet per minute (ft/min)
Miles per hour (mi/h)	1.47	Feet per second (ft/s)
Horsepower (hp)	33,000	Foot-pounds per minute (ft·lb/min)
Horsepower (hp)	550	Foot-pounds per second, standard (ft·lb/s)
Horsepower (hp)	0.746	Kilowatts (kW)
British thermal units (Btu)	778	Foot-pounds (ft·lb)
British thermal units (Btu)	0.0236	Horsepower (hp)
Equals this unit	Divided by	This unit

Metric Units to Metric Units

This unit	Multiplied by	Equals this unit
Meters (m)	100	Centimeters (cm)
Kilometers (km)	1000	Meters (m)
Square meters (m²)	10,000	Square centimeters (cm²)
Hectares	10,000	Square meters (m²)
Square kilometers (km²)	1,000,000	Square meters (m²)
Cubic meters (m³)	1,000,000	Cubic centimeters (cm³)
Cubic meters (m³)	1000	Liters (L)
Kilograms (kg)	1000	Grams (g)
Quintals	100	Kilograms (kg)
Tons, metric	1000	Kilograms (kg)
Kilometers per hour (km/h)	16.7	Meters per minute (m/min)
Kilometers per hour (km/h)	0.278	Meters per second (m/s)
Equals this unit	Divided by	This unit

English Units to Metric Units and Metric Units to English Units

This unit	Multiplied by	Equals this unit
Inches (in)	2.54	Centimeters (cm)
Feet (ft)	0.305	Meters (m)
Yards (yd)	0.914	Meters (m)
Miles, statute	1.609	Kilometers (km)
Square inches (in²)	6.45	Square centimeters (cm²)
Square feet (ft²)	0.0929	Square meters (m²)
Square yards (yd²)	0.836	Square meters (m²)
Acres	0.405	Hectares
Square miles (mi²)	2.590	Square kilometers (km²)
Cubic inches (in³)	16.4	Cubic centimeters (cm³)
Cubic feet (ft³)	0.0283	Cubic meters (m³)
Cubic yards (yd³)	0.765	Cubic meters (m³)
Gallons, U.S. (gal)	3.79	Liters (L)
Ounces (oz)	28.4	Grams (g)
Pounds (lb)	0.454	Kilograms (kg)
Tons, short	0.907	Tons, metric
Tons, long	1.016	Tons, metric
Feet per minute (ft/min)	0.305	Meters per minute (m/min)
Feet per second (ft/s)	0.305	Meters per second (m/s)
Miles per hour (mi/h)	1.609	Kilometers per hour (km/h)
Pounds per square inch (lb/in²)	0.0703	Kilograms per square centimeter (kg/cm²)
Pounds per square foot (lb/ft²)	4.887	Kilograms per square meter (kg/m²)
Pounds per cubic yard (lb/yd³)	0.593	Kilograms per cubic meter (kg/m³)
Foot-pounds or pound-feet (ft·lb)	0.138	Kilogram-meter (kg·m)
Horsepower (hp)	0.746	Kilowatts (kW)
British thermal units (Btu)	0.252	Calories (cal)
Equals this unit	Divided by	This unit

Units of Measurement

Area   Surveys are often made to obtain measured data from which area can be computed. The names of area units are frequently derived from the linear units, since area is the product of two linear measurements. The accompanying table shows the exact comparative values of some of the units. See also Figure 1.55.

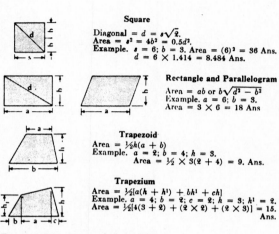

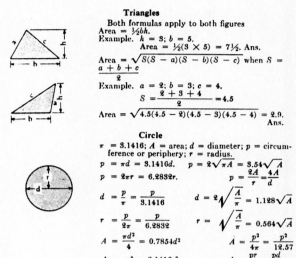

Square

Diagonal = $d = s\sqrt{2}$.
Area = $s^2 = 4b^2 = 0.5d^2$.
Example. $s = 6$; $b = 3$. Area = $(6)^2 = 36$ Ans.
$d = 6 \times 1.414 = 8.484$ Ans.

Rectangle and Parallelogram

Area = ab or $b\sqrt{d^2 - b^2}$
Example. $a = 6$; $b = 3$.
Area = $3 \times 6 = 18$ Ans

Trapezoid

Area = $\frac{1}{2}h(a + b)$
Example. $a = 2$; $b = 4$; $h = 3$.
Area = $\frac{1}{2} \times 3(2 + 4) = 9$. Ans.

Trapezium

Area = $\frac{1}{2}[a(h + h^1) + bh^1 + ch]$
Example. $a = 4$; $b = 2$; $c = 2$; $h = 3$; $h^1 = 2$.
Area = $\frac{1}{2}[4(3 + 2) + (2 \times 2) + (2 \times 3)] = 15$. Ans.

Triangles

Both formulas apply to both figures
Area = $\frac{1}{2}bh$.
Example. $h = 3$; $b = 5$.
Area = $\frac{1}{2}(3 \times 5) = 7\frac{1}{2}$. Ans.
Area = $\sqrt{S(S - a)(S - b)(S - c)}$ when $S = \frac{a + b + c}{2}$
Example. $a = 2$; $b = 3$; $c = 4$.
$S = \frac{2 + 3 + 4}{2} = 4.5$
Area = $\sqrt{4.5(4.5 - 2)(4.5 - 3)(4.5 - 4)} = 2.9$. Ans.

Circle

$\pi = 3.1416$; A = area; d = diameter; p = circumference or periphery; r = radius.
$p = \pi d = 3.1416d$.
$p = 2\pi r = 6.2832r$.
$p = 2\sqrt{\pi A} = 3.54\sqrt{A}$
$p = \frac{2A}{r} = \frac{4A}{d}$
$d = \frac{p}{\pi} = \frac{p}{3.1416}$
$d = 2\sqrt{\frac{A}{\pi}} = 1.128\sqrt{A}$
$r = \frac{p}{2\pi} = \frac{p}{6.2832}$
$r = \sqrt{\frac{A}{\pi}} = 0.564\sqrt{A}$
$A = \frac{\pi d^2}{4} = 0.7854d^2$
$A = \frac{p^2}{4\pi} = \frac{p^2}{12.57}$
$A = \pi r^2 = 3.1416r^2$
$A = \frac{pr}{2} = \frac{pd}{4}$

Figure 1.55   Areas of plane figures.

	Square inch	Square link	Square foot	Square vara (California)	Square vara (Texas)	Square yard	Square meter	Square rod, pole, or perch	Square chain	Rood	Acre	Square kilometer	Square mile (statute)
Square inch	1	0.01594	0.00694										
Square link	62.7264	1	0.4356	0.0576	0.05645	0.0484	0.04047	0.0016					
Square foot	144	2.29568	1	0.13223	0.1296	0.11111	0.0929	0.00367	0.00174				
Square vara (California)	1089	17.3611	7.5625	1	0.9801	0.84028	0.70258	0.02778					
Square vara (Texas)	1,111.11	17.7136	7.71605	1.0203	1	0.85734	0.71685	0.02834	0.00177				
Square yard	1,296	20.6612	9	1.19008	1.1664	1	0.83613	0.03306	0.00207				
Square meter	1,549.80	24.7104	10.7639	1.42332	1.395	1.19599	1	0.03954	0.00247				
Square rod, pole, or perch		625	272.25	36	35.2836	30.25	25.2930	1	0.0625	0.025	0.00625		
Square chain		10,000	4,356	576	564.538	484	404.687	16	1	0.4	0.1		
Rood		25,000	10,890	1,440	1,411.34	1,210	1,011.72	40	2.5	1	0.25	0.00101	
Acre		100,000	43,560	5,760	5,645.38	4,840	4,046.87	160	10	4	1	0.00405	0.00156
Square kilometer							1,000,000	39,536.7	2,471.044	988.418	247.104	1	0.3861
Square mile (statute)							102,400	6,400	2,560	640	2.59	1	

TWO

Environmental Considerations of the Site

Environmental considerations have always been an important aspect of the site design process. They may include the analysis of microclimates and macroclimates, ecosystems and their interrelationships, surface and subsurface hydrology, vegetation, and subsurface soil conditions. All these considerations require extensive and detailed study to produce meaningful conclusions. For sites that are strongly influenced by any of these factors, such studies are essential. Such sites most likely would include those on the shoreline, on mountainous terrain, or near floodplain areas. This section treats considerations dealing with climate, floodplains, flood control, drainage, water supply, and the location and design of sanitary landfills.

Over the years, general criteria for the selection of sites for different kinds of uses have evolved from a variety of sources. These criteria encompass the total environment, both regional and local. With such a guide or checklist, it is possible to evaluate most proposed sites and determine their suitability for proposed uses. However, care should be used in applying these criteria. Few sites ever meet all the ideal conditions suggested by the criteria. Special local considerations must be balanced with general requirements. It is also of increasing importance to provide adequate conservation measures for developing areas. Good site design may mitigate the negative impacts of construction to some extent. It is also possible to provide for wildlife in the site design process.

Included also in this section are a general discussion and some of the requirements for environmental impact statements. In recent years such statements have become extremely important, if not critical, for new large-scale developments of any kind.

2.1 United States Climatic Conditions

The map in Figure 2.11 delineates regional climatic zones for the continental United States; the site assessment criteria suggest factors to look for in evaluating sites in each zone. A word of caution: the map boundaries are only approximations, and sites that fall on or near them should be considered on a case-by-case basis.

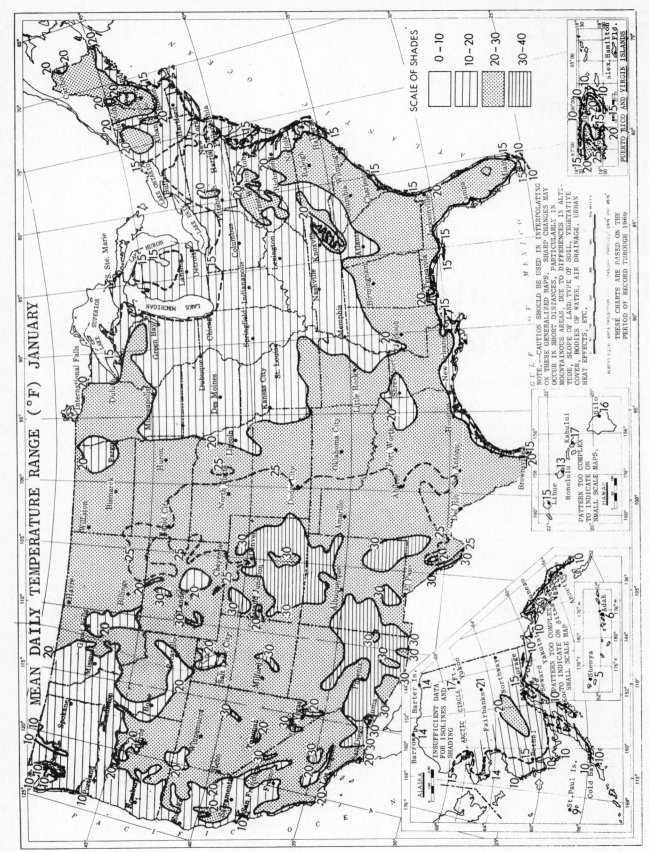

MEAN DAILY TEMPERATURE RANGE (°F) JANUARY

SCALE OF SHADES

0–10 10–20 20–30 30–40

NOTE.—CAUTION SHOULD BE USED IN INTERPOLATING ON THESE GENERALIZED MAPS. SHARP CHANGES MAY OCCUR IN SHORT DISTANCES, PARTICULARLY IN MOUNTAINOUS AREAS, DUE TO DIFFERENCES IN ALTITUDE, SLOPE OF LAND, TYPE OF SOIL, VEGETATIVE COVER, BODIES OF WATER, AIR DRAINAGE, URBAN HEAT EFFECTS, ETC.

ALBERS EQUAL AREA PROJECTION. STANDARD PARALLELS 29½° AND 45½°

THESE CHARTS ARE BASED ON THE PERIOD OF RECORD THROUGH 1969

PUERTO RICO AND VIRGIN ISLANDS

HAWAII

ALASKA

Figure 2.1

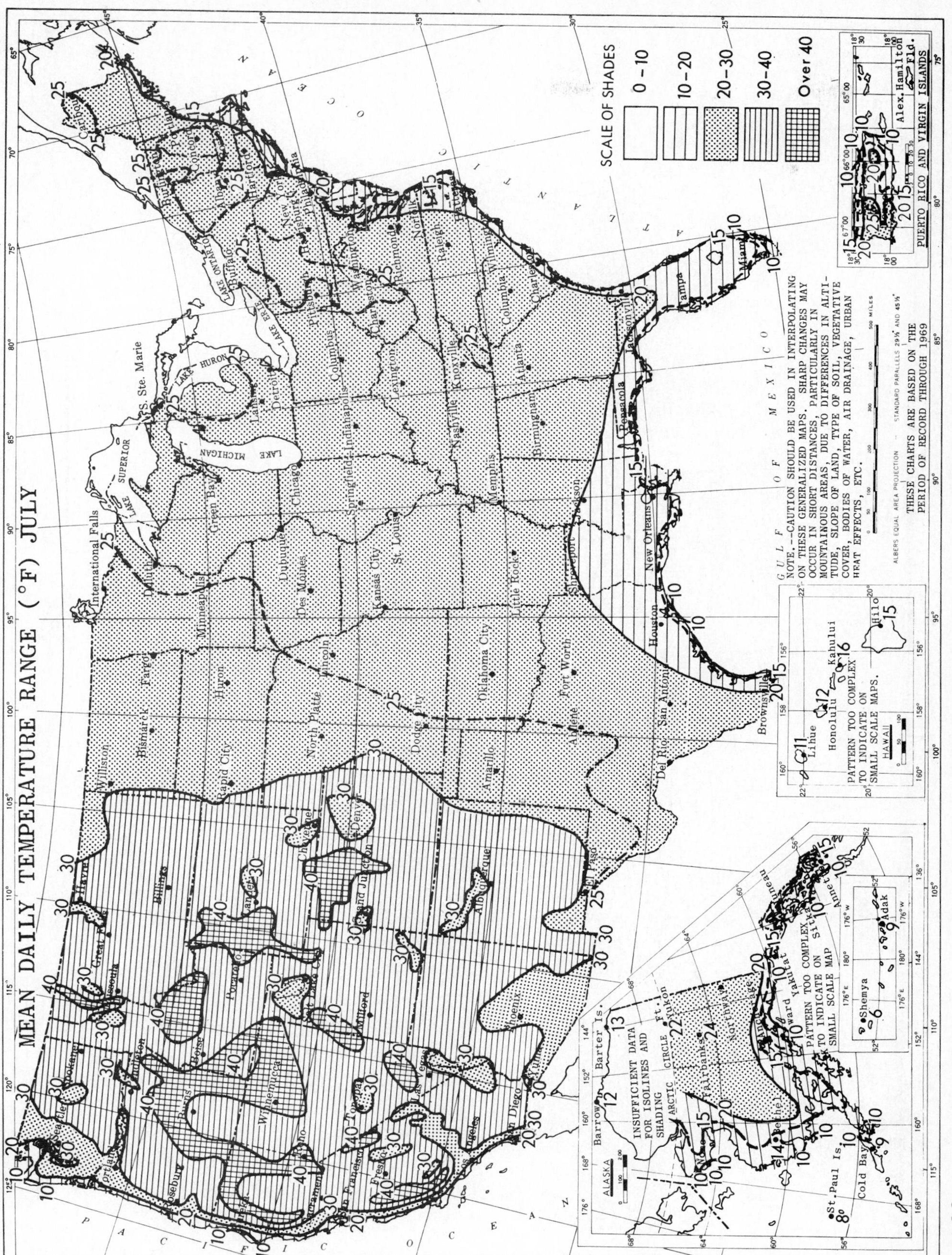
MEAN DAILY TEMPERATURE RANGE (°F) JULY
SCALE OF SHADES
0–10
10–20
20–30
30–40
Over 40
NOTE.--CAUTION SHOULD BE USED IN INTERPOLATING ON THESE GENERALIZED MAPS. SHARP CHANGES MAY OCCUR IN SHORT DISTANCES, PARTICULARLY IN MOUNTAINOUS AREAS, DUE TO DIFFERENCES IN ALTITUDE, SLOPE OF LAND, TYPE OF SOIL, VEGETATIVE COVER, BODIES OF WATER, AIR DRAINAGE, URBAN HEAT EFFECTS, ETC.
ALBERS EQUAL AREA PROJECTION — STANDARD PARALLELS 29½ AND 45½
THESE CHARTS ARE BASED ON THE PERIOD OF RECORD THROUGH 1969
PUERTO RICO AND VIRGIN ISLANDS
HAWAII
PATTERN TOO COMPLEX TO INDICATE ON SMALL SCALE MAPS.
ALASKA
INSUFFICIENT DATA FOR ISOLINES AND SHADING
PATTERN TOO COMPLEX TO INDICATE ON SMALL SCALE MAP
ARCTIC CIRCLE
PACIFIC OCEAN
ATLANTIC OCEAN
GULF OF MEXICO
LAKE SUPERIOR
LAKE MICHIGAN
LAKE HURON
LAKE ERIE
LAKE ONTARIO

Figure 2.2

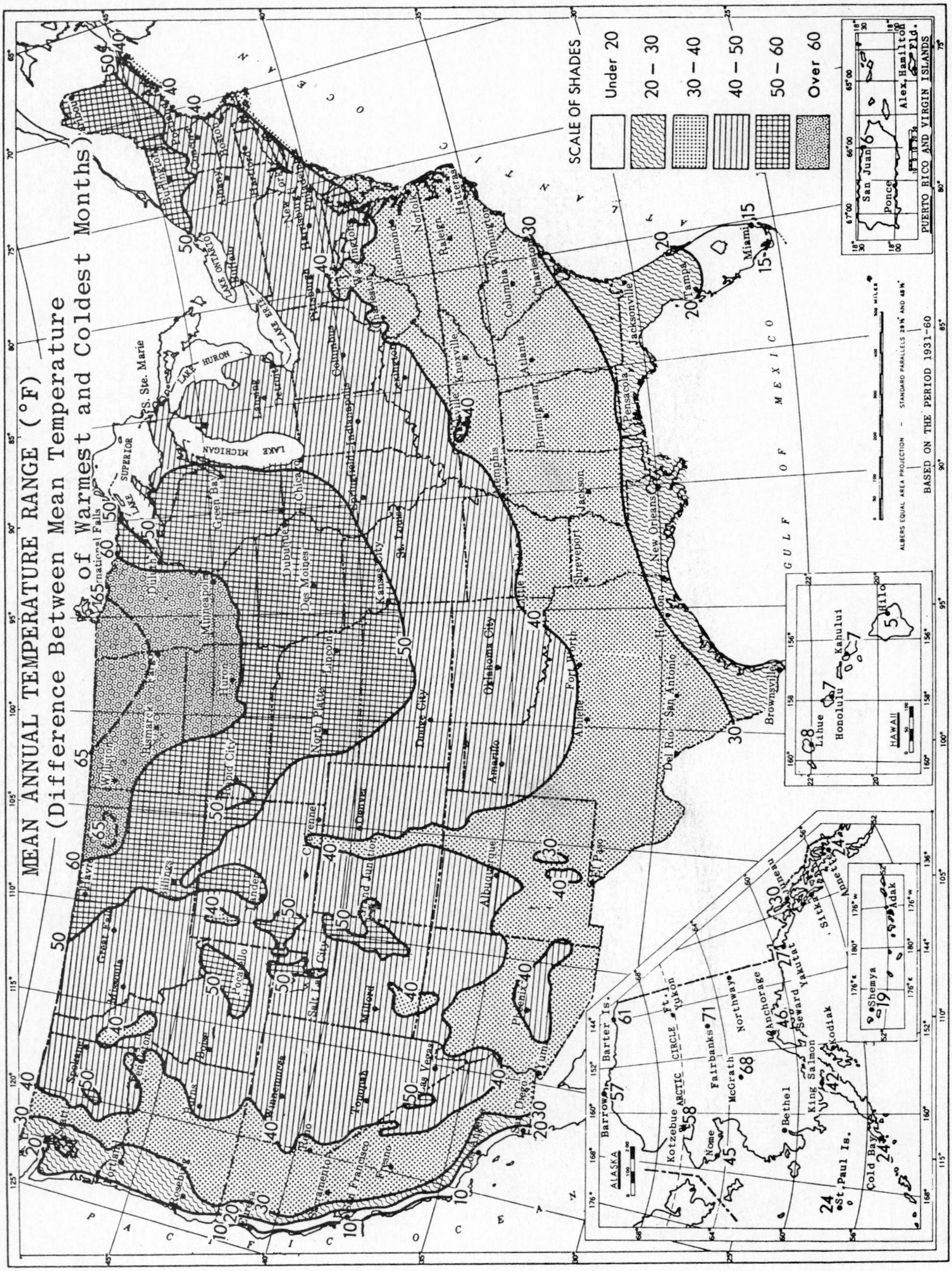

MEAN ANNUAL TEMPERATURE RANGE (°F)
(Difference Between Mean Temperature
of Warmest and Coldest Months)

SCALE OF SHADES

Under 20   20 – 30   30 – 40   40 – 50   50 – 60   Over 60

PUERTO RICO AND VIRGIN ISLANDS

BASED ON THE PERIOD 1931-60

ALBERS EQUAL AREA PROJECTION – STANDARD PARALLELS 29½°N AND 45½°N

HAWAII

ALASKA

Figure 2.3

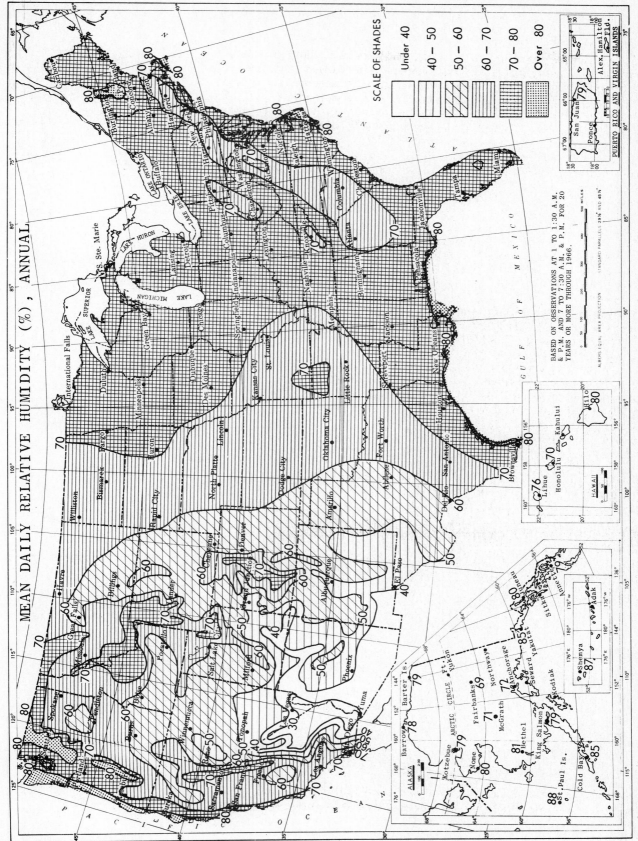

MEAN DAILY RELATIVE HUMIDITY (%), ANNUAL

SCALE OF SHADES

| Under 40 | 40 – 50 | 50 – 60 | 60 – 70 | 70 – 80 | Over 80 |

PUERTO RICO AND VIRGIN ISLANDS

BASED ON OBSERVATIONS AT 1 TO 1:30 A.M.
& P.M. AND 7 TO 7:30 A.M. & P.M. FOR 20
YEARS OR MORE THROUGH 1966.

ALBERS EQUAL AREA PROJECTION

STANDARD PARALLELS 29½° AND 45½°

HAWAII

ALASKA

Figure 2.4

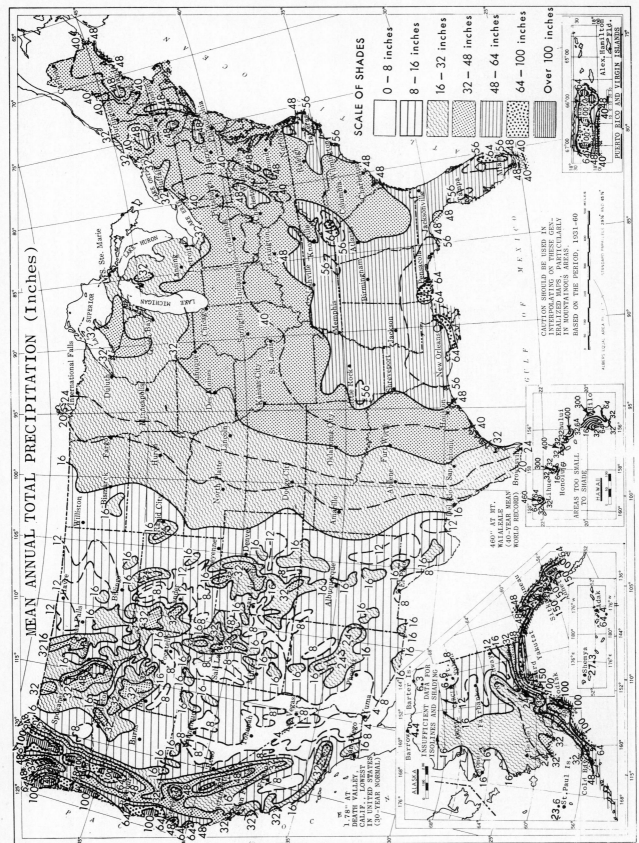

MEAN ANNUAL TOTAL PRECIPITATION (Inches)

Figure 2.5

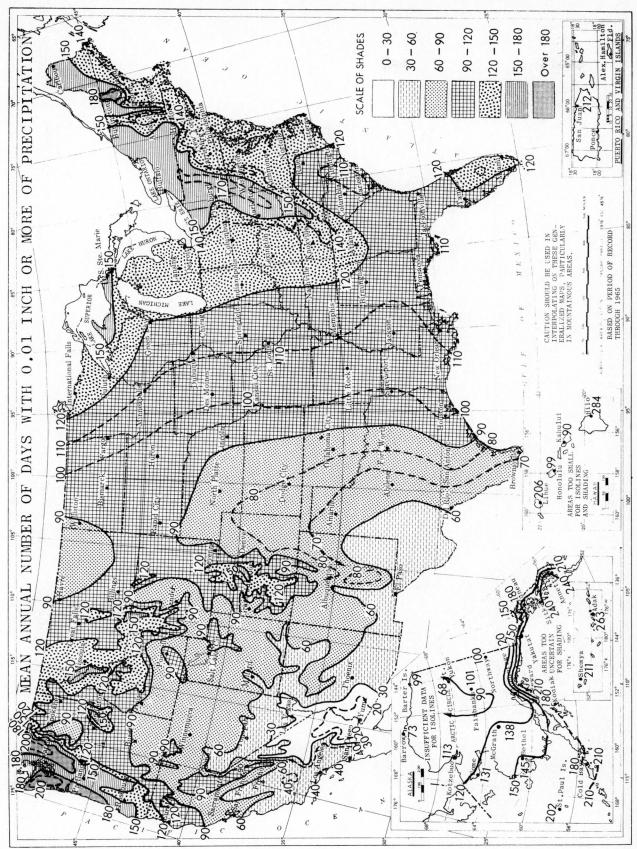

MEAN ANNUAL NUMBER OF DAYS WITH 0.01 INCH OR MORE OF PRECIPITATION

Figure 2.6

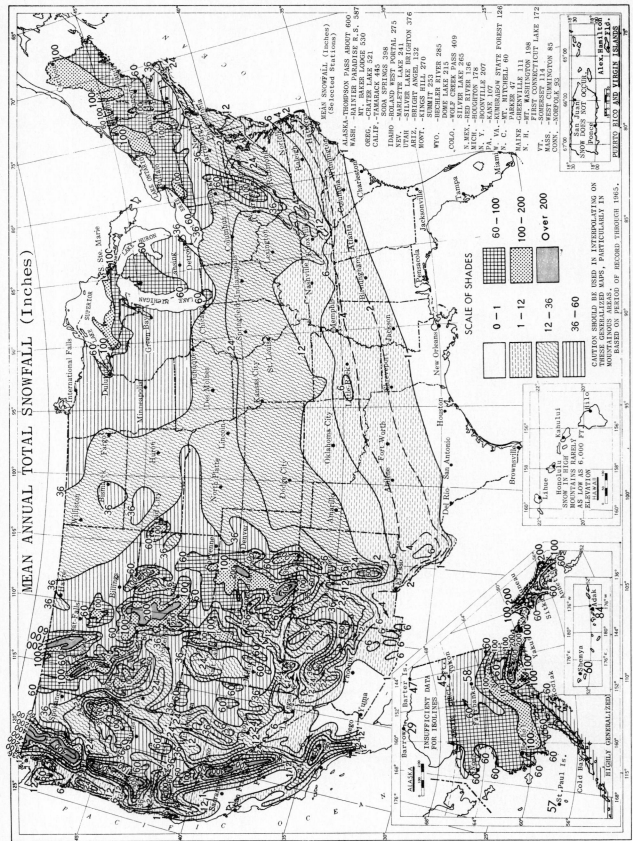

MEAN ANNUAL TOTAL SNOWFALL (Inches)

MEAN SNOWFALL, (Inches)
(Selected Stations)

ALASKA–THOMPSON PASS ABOUT 600
WASH. –RAINIER PARADISE R.S. 587
       –MT. BAKER LODGE 530
OREG. –CRATER LAKE 521
CALIF. –TAMARACK 445
       –SODA SPRINGS 398
IDAHO –ROLAND WEST PORTAL 275
NEV. –MARLETTE LAKE 241
UTAH –SILVER LAKE BRIGHTON 376
ARIZ. –BRIGHT ANGEL 132
MONT. –KINGS HILL 270
       –SUMMIT 253
WYO. –BECHLER RIVER 285
       –DOME LAKE 215
COLO. –WOLF CREEK PASS 409
       –SILVER LAKE 265
N.MEX. –RED RIVER 136
MICH. –HOUGHTON 178
N. Y. –BOONVILLE 207
PA. –KANE 107
W. VA. –KUMBRABOW STATE FOREST 126
N. C. –MT. MITCHELL 60
       –PARKER 47
MAINE –GREENVILLE 111
N. H. –MT. WASHINGTON 198
VT. –FIRST CONNECTICUT LAKE 172
MASS. –SOMERSET 114
CONN. –WEST CUMMINGTON 85
       –NORFOLK 93

SCALE OF SHADES

0 – 1	1 – 12	12 – 36	36 – 60	60 – 100
				100 – 200
				Over 200

CAUTION SHOULD BE USED IN INTERPOLATING ON
THESE GENERALIZED MAPS, PARTICULARLY IN
MOUNTAINOUS AREAS.
BASED ON PERIOD OF RECORD THROUGH 1965.

HAWAII

Honolulu
SNOW IN HIGH
MOUNTAINS RARELY
AS LOW AS 6,000 FT.
ELEVATION

PUERTO RICO AND VIRGIN ISLANDS
SNOW DOES NOT OCCUR.

ALASKA

HIGHLY GENERALIZED

INSUFFICIENT DATA
FOR ISOLINES

Figure 2.7

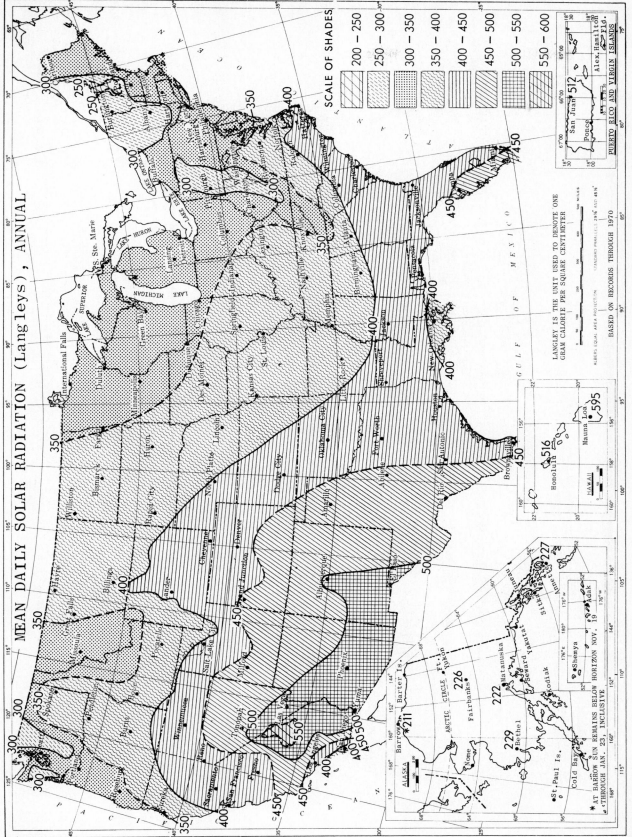

MEAN DAILY SOLAR RADIATION (Langleys), ANNUAL

SCALE OF SHADES

200 – 250	400 – 450
250 – 300	450 – 500
300 – 350	500 – 550
350 – 400	550 – 600

PUERTO RICO AND VIRGIN ISLANDS

LANGLEY IS THE UNIT USED TO DENOTE ONE
GRAM CALORIE PER SQUARE CENTIMETER

ALBERS EQUAL AREA PROJECTION   (STANDARD PARALLELS 29½° AND 45½°)

BASED ON RECORDS THROUGH 1970

HAWAII

ALASKA

* AT BARROW SUN REMAINS BELOW HORIZON NOV. 19
THROUGH JAN. 23, INCLUSIVE

Figure 2.8

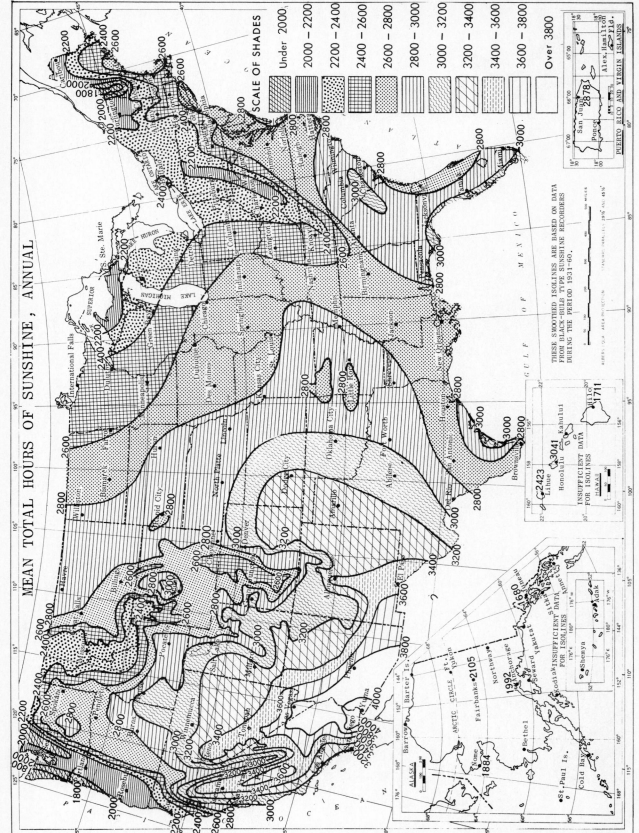

Figure 2.9

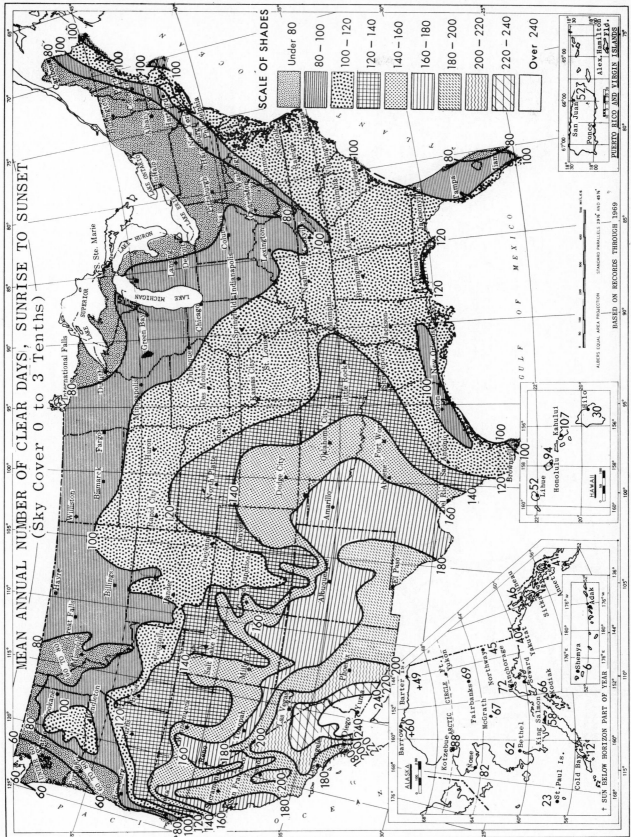

MEAN ANNUAL NUMBER OF CLEAR DAYS, SUNRISE TO SUNSET
(Sky Cover 0 to 3 Tenths)

SCALE OF SHADES

Under 80
80 — 100
100 — 120
120 — 140
140 — 160
160 — 180
180 — 200
200 — 220
220 — 240
Over 240

BASED ON RECORDS THROUGH 1969

ALBERS EQUAL AREA PROJECTION
STANDARD PARALLELS 29½° AND 45½°

PUERTO RICO AND VIRGIN ISLANDS

HAWAII

ALASKA

+ SUN BELOW HORIZON PART OF YEAR

Figure 2.10

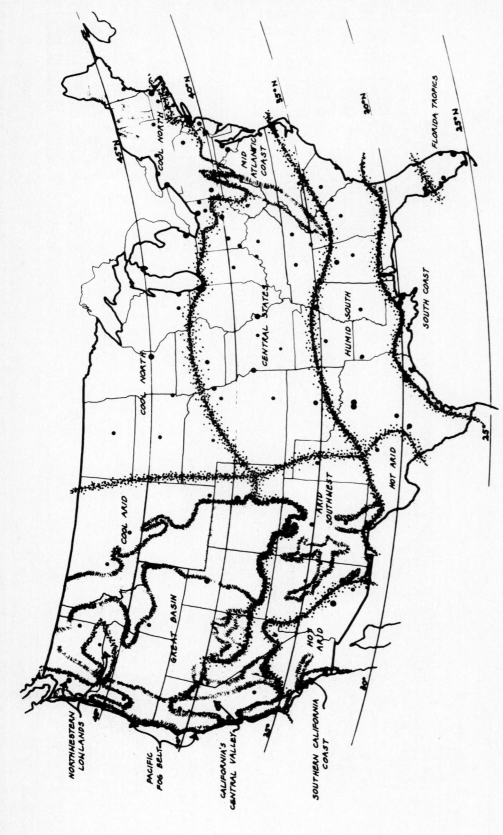

Figure 2.11 Regional climate zone map. (See also Site Assessment Criteria table, page 97.)

Site Assessment Criteria

Region	Siting guidelines			Notes
	Best	Good	Poor	
Pacific fog belt	Sheltered sites. Medium slopes facing southwest to southeast. Look for sites sheltered from fog and winds, both of which have regular patterns	Flat sites with wind shelter. Shallow slopes in any direction (except northwest, northeast, and north in Oregon and Washington). On slopes over 15 percent southeast to southwest slopes will allow east-west roads	Windy ridges and hilltops, especially in Oregon and Washington	Sunlight characteristics change in short distances. Make sure this is your climate. Vegetation can indicate fog and wind incidence. For example, redwoods indicate fog; sheared cypress and bay indicate wind
Northwestern lowlands	Midway up southwest to southeast slopes, sheltered by trees and topography	Southerly slopes, flat sites with good solar access and gentle east and west slopes	Northwest to northeast slopes, frost hollows, exposed ridges, steep west slopes	Beware of sites with evergreen tree cover: they limit solar access
Great Basin, cold arid	Sites on lower, sheltered south to southeast slopes 5 to 15 percent	Sites with south orientation and wind shelter	North slopes, cold air drainage, and exposed ridges	
Arid: Southwest, California's Central Valley	Lower south and southeast slopes for early cool-season warmth	Upper south and southeast slopes; flat land	North slopes (difficult solar access) and west slopes (afternoon overheating problems)	
Southern California coast	Any south-facing site on moderate slopes	Any south-facing site on moderate slopes	Avoid only steep slopes to north, northeast, and northwest; on more inland sites avoid steep east or west slopes	This is the least exacting of any U.S. climate. Determine whether site is in a summer fog belt or in the summer inversion layer
Hot arid	South to southeast slopes, flat land, shallow north slopes	Flat land, cool air drainages, cool ridges	Southwest and northwest slopes, steep north slopes, hot valley bottoms	Sites near water tend to be cooler
Cool north	Sheltered sites on gently south-facing slopes	Sheltered sites on flat ground or any slope southeast to southwest	Exposed ridges, hillcrests, north slopes, steep west slopes, frost hollows, windy sites	South orientation and wind shelter are the keys
Central United States, Mid-Atlantic Coast	Gentle southeast to southwest slopes with scattered, mature trees	Flat sites, wooded sites, sheltered slopes, steep slopes south to southeast	Windy ridges, steep north, northwest, or west slopes, unventilated depression	Look for winter sun and wind shelter, summer shade and breezes
Humid south	Mature deciduous woodland on gentle south or north slopes	Mature deciduous woodland on gentle slope in any direction; flat sites; scattered trees on steeper north or south slope; breezy ridges	Sites without breezes; steep slopes; treeless sites	Look for a balance of sun and shade, good summer breezes, and mature deciduous trees
South coast	Mature deciduous woodland on flat land or gentle slopes. Ridgetops with good breezes	Mature deciduous woodland on steeper north, east, or west slopes	Steep or sheltered slopes any direction, unventilated pockets, treeless sites	Mature trees and air movement are the keys to cooling here
Florida tropics	Mature broadleaf woodland essential, ridgetops, north slopes, gentle south slopes	Mature broadleaf trees on flat sites	Treeless sites, steep east and west slopes, airless hollows without breezes	Through breezes and mature shade trees are more important than solar access

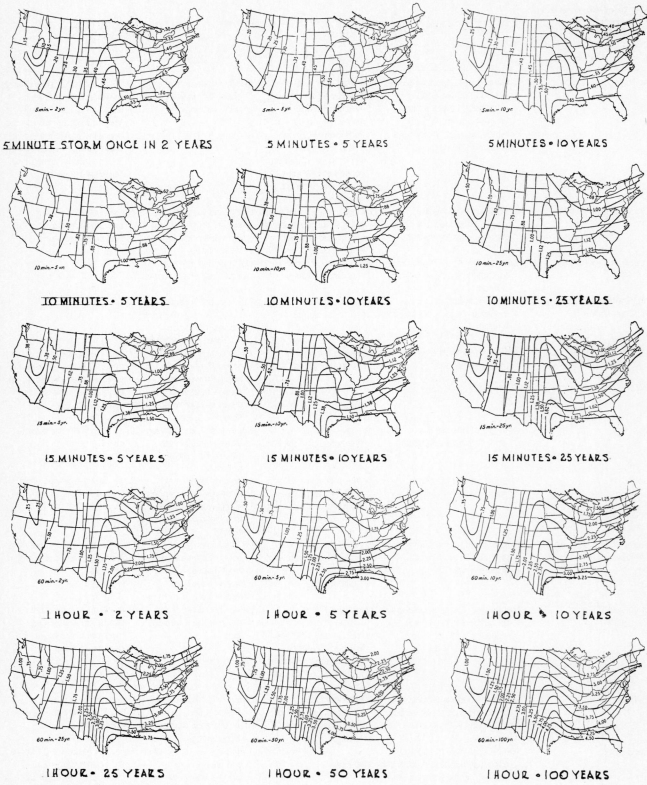

Figure 2.12   Rainfall intensity — frequency data.

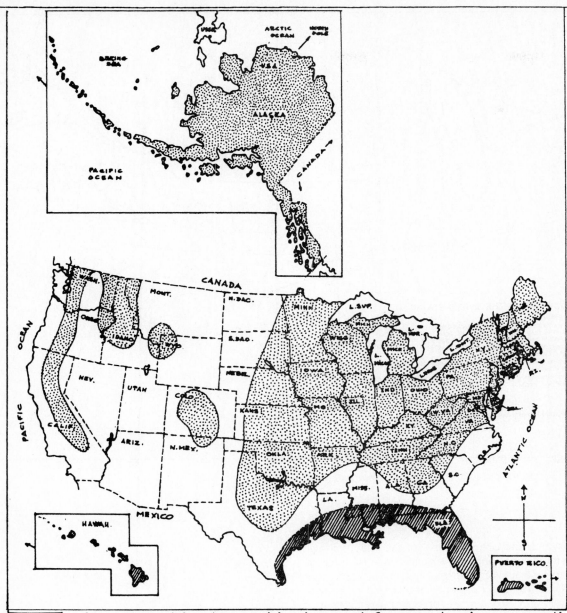

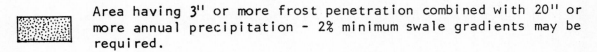

Area not subject to combined ground frost and moisture conditions - 1% minimum swale gradients acceptable.

Area having 3" or more frost penetration combined with 20" or more annual precipitation - 2% minimum swale gradients may be required.

Area having intense rainfall - 2% minimum swale gradients required - special drainage design may be required.

Figure 2.13   Areas of combined ground frost and moisture conditions.

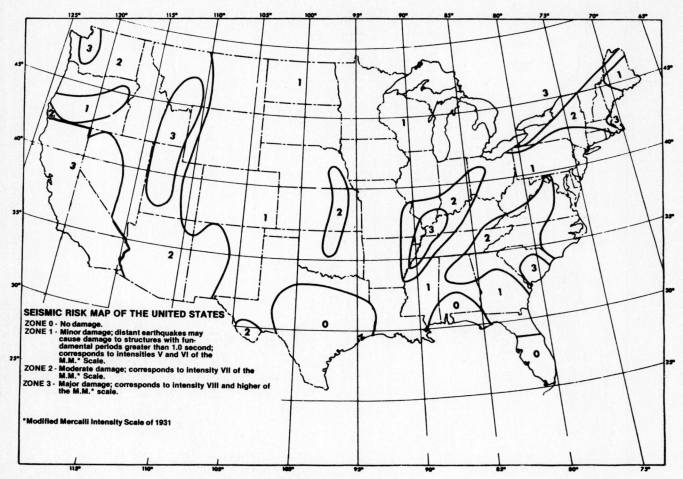

Figure 2.14 Seismic risk map of the contiguous United States.

The division of the United States into the regions shown on the map is only one approach to defining relevant regional climates. A recent publication from HUD, for example, uses slightly different climatic categories and design criteria for buildings. Research into the use of regional climates as a design consideration has just begun and is likely to lead to further refinements.

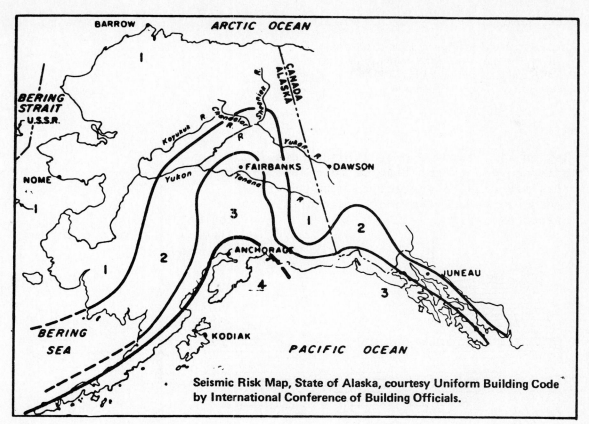

Seismic Risk Map, State of Alaska, courtesy Uniform Building Code by International Conference of Building Officials.

Figure 2.15

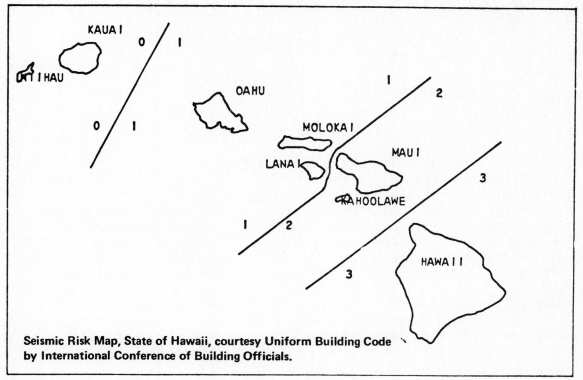

Seismic Risk Map, State of Hawaii, courtesy Uniform Building Code by International Conference of Building Officials.

Figure 2.16

2.2 Termite Zones

Figure 2.17 indicates the general geographic distribution of termite infestation. The lines defining the various regions are approximate only, and local conditions may be more or less severe than indicated by regional classification.

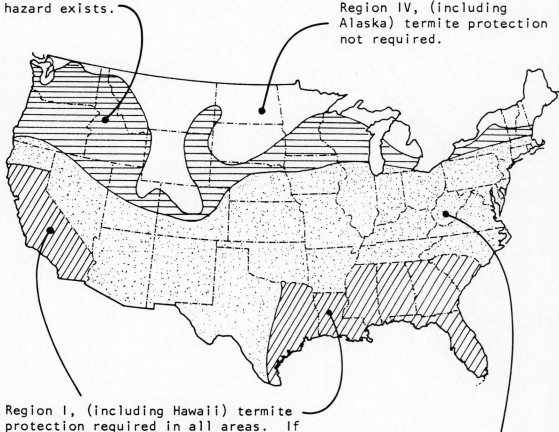

Region III, termite protection generally not required except that certain localities may require protection when determined by the HUD field office that a hazard exists.

Region IV, (including Alaska) termite protection not required.

Region I, (including Hawaii) termite protection required in all areas. If construction is of slab-on-ground type a chemical barrier should be used, except that monolithic slab design may be considered as termite protection when acceptable to the HUD field office.

Puerto Rico is an area of severe infestation, all lumber should be pressure treated per AWPI Standards.

Region II, termite protection generally required. Specific areas may be exempted as determined by the HUD field office. If construction is slab-on-ground type, use a chemical barrier or a monolithic concrete slab design (where permitted).

Figure 2.17 Geographic distribution of termite infestation.

2.3 Floodplain Classifications

Data on floods and their applicability to various areas and situations are presented in Figure 2.18.

SUMMARY OF FLOOD-DATA CLASSIFICATIONS AND APPLICABILITY

Classification	Determination of Three Hydraulic Factors Essential to Regulations			Desirable Areas of Application	Disadvantages	Advantages	Suggested Zoning Districts
	Profile	Floodway	Flood Plain				
Class A, Exact Flood Data Based on Hydraulic Calculations	Determined initially from engineering study	Determined initially from engineering study	Determined initially from engineering study	1) Areas of high development pressures for urban growth 2) Areas of intense existing development 3) Areas of high land values 4) Areas where zoning is the only management tool	Very costly and sometimes takes many years before studies are complete	1) Provide sound legal base for zoning 2) Expedites evaluation of proposed developments or amendments to adopted plans 3) Minimize hardships to applicant of proposed development 4) Contribute to flood emergency preparedness plan	1) Floodway 2) Flood fringe 3) Basement
Class B, Interpreted Flood Data Based on Known High-water Marks	Extrapolation from past floods records	Normal depth analysis (see footnote p. 35) on a case-by-case basis	Location by elevations from extrapolated profile on 1) topographic maps, 2) street-sewer maps that show ground elevations or 3) field surveys	1) Small amount of existing development affected 2) Little development pressure 3) Strong sanitary subdivision controls and public policy 4) Much land under public ownership 5) Ongoing program to acquire Class A data	1) Legal base questionable 2) Requires technical assistance to evaluate floodway 3) Applicant has greater burden to provide survey information	1) Low cost 2) Discourage land speculation 3) Valuable river basin 4) Guide to public facilities and transportation 5) Can be made available in short time to serve immediate need	1) General flood-plain district
Class C, Interpreted Flood Data Based on Nonhydraulic Calculations	Determined case-by-case (see footnote on hydraulic analysis p. 35)	Determined case-by-case (see footnote on hydraulic analysis p. 35)	Experienced flood maps, aerial photo examination, or detailed soil maps that have been correlated with engineering studies on similar streams	1) Rural areas little potential for development 2) Large portion of area under public ownership 3) Land-easement or acquisition program 4) Storng sanitary and subdivision regulations 5) Ongoing program to acquire Class A data	1) Frequency of mapped flood unknown 2) All 3 hydraulic factors unknown 3) Weak legal base 4) Source of technical assistance is needed 5) Burden to applicant to furnish surveys	1) Low cost 2) Readily expected by local people 3) Identifies pressure areas 4) Discourages land speculation	1) General flood-plain district

Figure 2.18 Floodplain classification.

2.4 Floodplain Districts

The three illustrations in Figure 2.19 show how a floodway can be developed to accommodate varying flood heights.

Plans for a single flood district and for two flood districts are shown in Figures 2.20 and 2.21.

Addition of Flood Basement Distribution to Flood Fringe and Floodway Districts Figure 2.22 presents a three-district approach involving floodway, flood fringe, and flood basement districts for urban flood hazard areas with detailed engineering data that indicate serious basement flooding problems (Class A data). This approach is applicable to communities where soil conditions allow flood waters to seep into basements located beyond the floodplain, a problem usually caused by protracted floods.

Common Elements in Floodplain Regulations Floodplain regulations, whether they are included in comprehensive zoning ordinances or in a separate ordinance, should incorporate the following elements, which correspond approximately to those in the examples shown.

1. Finding of facts
2. Objectives
3. Establishment of floodplain zoning map
   a. Rules for the interpretation of district boundaries
   b. Warning and disclaimer of liability
4. Floodplain district regulations
   a. Floodway use standards
   b. Flood fringe use standards
5. Administrative provisions
   a. Variances
   b. Mapping disputes
   c. Special-use permits
      (1) Procedure to be followed in passing on special-use permits
      (2) Factors upon which decisions are to be based
      (3) Conditions attached
6. Nonconforming uses
7. Definitions

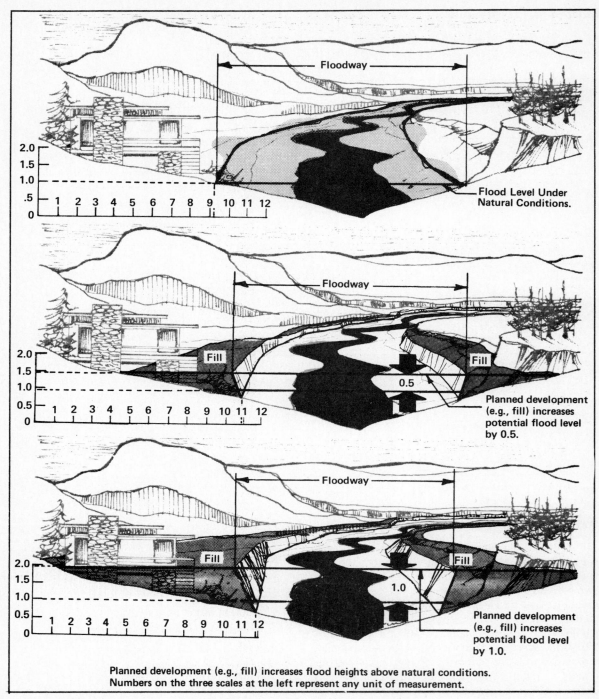

Figure 2.19 Floodway planned for the same discharge with varying increase in flood heights.

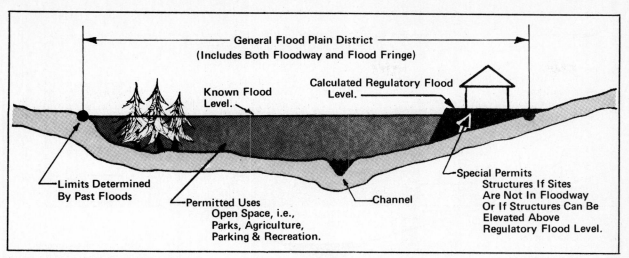

Figure 2.20 One floodplain district.

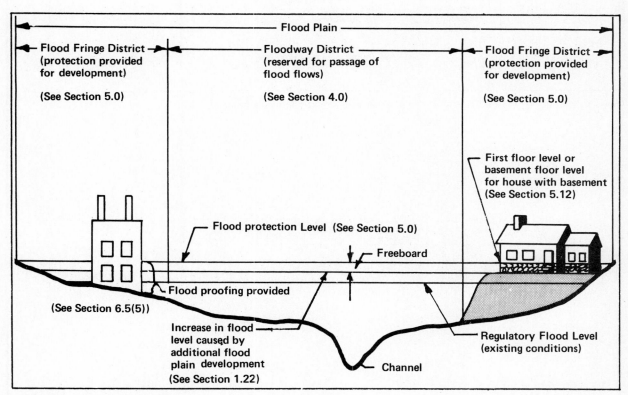

Figure 2.21 Two floodplain districts.

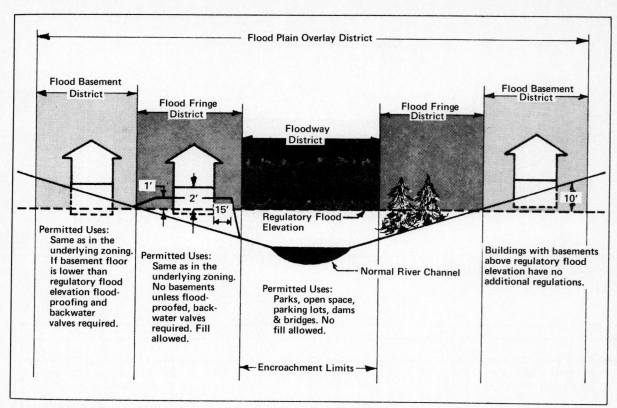

Figure 2.22 Addition of flood basement districts to flood fringe and floodway districts.

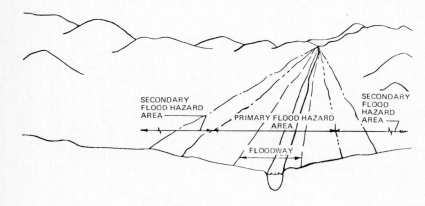

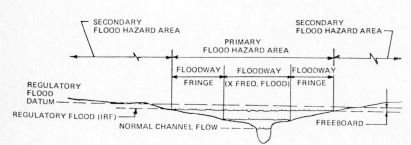

Figure 2.23 Flood hazard areas and regulatory flood datum.

2.5 Drainage Design and Flood Hazard Exposure

Planning for storm drainage should be coordinated with local area drainage plans which can be expected to have an impact on the site drainage system.

Permanence of location and maintenance for off-site outfall drainage ways should be assured by public rights-of-way, easements, or other satisfactory means, and acceptance of maintenance responsibility for local public or private jurisdictions.

Careful analysis should be made to determine that no probable drainage backup exists or is being created which will adversely affect site facilities, site surroundings, or the drainage system.

Most locations have two separate and distinct storm drainage systems. The first is the designed system and the second is the route followed by flood or runoff waters when the designed system is inoperable or inadequate due to blockage or lack of capacity. The second system should be arranged and designed with close attention to prevent serious erosion and damage to major site facilities when it is in operation.

Storm drainage mains should be designed and located so that permanent dedication to an appropriate local jurisdiction for maintenance and repair is possible.

Where possible, storm and sanitary sewers should be in separate trenches.

Where foreseeable land development will cause probable downstream flooding or require additional drainage structures, design techniques to moderate peak runoff rates should be used. Detention storage arrangements for this purpose can range from controlled discharge from flat building roofs to ponds or ponding areas with extra storage capacity and controlled outlet flows. Where detention ponds are used, there should be provision for permanent maintenance.

Existing storm drainage patterns and systems should be used to the maximum extent possible.

Drainage design should be arranged to avoid the downstream deposit of sediment.

Drainage from large surfaced areas should discharge onto paved or natural drainage ways. Where it becomes necessary to discharge paved area runoff onto planted areas, quantities should be kept small, runoff speed slow, and drainage should be evenly diffused and not concentrated at any point.

It is desirable when efficient and economical to direct surface drainage to the street drainage system.

Runoff should not be directed to pockets or low areas where drainage system failure would cause damage to buildings or create erosion. Precautions such as overflow channels or emergency outlets should be designed to prevent serious damage in case of failure in the drainage system.

Open Channels Open channels should be designed upon the basis of the flow friction factor, Manning's *n*, and in a manner which will result in self-scouring with no maintenance.

Splash Blocks Splash blocks should have a minimum width of 12 inches and a minimum length of 30 inches.

Splash blocks should be arranged to discharge runoff without causing erosion. Where this cannot be assured, consideration should be given to other methods of handling downspout discharge such as connection of downspouts to underground storm drains.

Catch Basins Catch basins in grassed areas should be set in small depressions about 3 inches below the surrounding ground level to help ensure positive flow into the inlet opening.

The use of grate inlets and trapped catch basins should be avoided to reduce incidents of inlet blockage and high maintenance costs.

Manholes Removable inlet parts should be designed to minimize opportunities for vandalism.

Manholes or junction boxes should be provided at maximum intervals of approximately 400 feet.

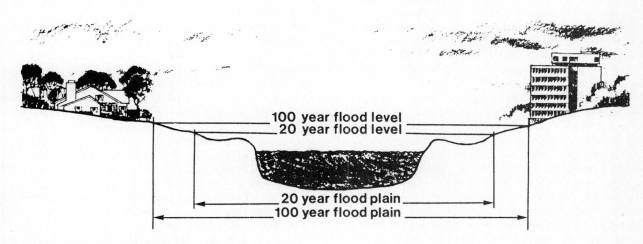

Figure 2.24 Floods and floodplains.

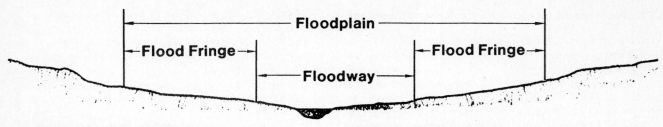

Floodway Land Uses

Prohibited or Discouraged:
Most structures and fills.

Permitted:
Farming, pasture, forestry
open space, recreation,
wildlife preserves.

Flood Fringe Uses

Prohibited or Discouraged:
Storage of toxic materials.
Hospitals and other vital facilities.

Permitted:
All other uses with developments
elevated on fill or otherwise
protected against damage.

Figure 2.25 Floodplain regulation can be employed to limit uses in an area to those which are consistent with the flood hazard.

2.6 Flood Control

Floods and Floodplains The size of a flood is usually described in terms of its statistical frequency of occurrence. For example, a "20-year flood" is one likely to be equaled or exceeded on the average of only once every 20 years. However, such averages are only meaningful over decades or centuries. Several floods of that size or larger could occur in a particular year or short period of years.

Floodplains are typically low areas adjacent to a river, ocean, or other body of water which are subject to flooding. They are described in a fashion similar to floods. For instance, the land flooded by the 20-year flood is called the "20-year floodplain."

The 100-year flood and 100-year floodplain are particularly significant. The boundary of the 100-year flood has been widely used in federal and state programs for identifying areas where the risk of flooding is significant. For example, floodplain zoning laws and other regulations controlling building in flood hazard areas are often applied to the 100-year floodplain.

Dams and Reservoirs Reservoirs capture and temporarily hold floodwaters upstream of flood-prone areas. Excess water is gradually released after the flood threat has passed. Use of dams requires sites capable of storing sufficient amounts of water which are upstream but fairly close to the area to be protected. In addition, the location must normally be "available" in the sense that it does not contain significant urban development or sites warranting preservation for historical or other reasons.

Reservoirs have the potential for reducing flooding in several downstream communities rather than protecting single areas as with some flood-modifying measures. Another advantage is their potential for serving other uses, such as recreation, hydroelectric power generation, and water supply.

Dams and reservoirs are most frequently used on small and moderate-sized streams. The large areas of land required to store the flood flows of major rivers are usually no longer available. Because of their cost, large dams and reservoirs are usually feasible only where they protect urban areas or high-value agricultural lands. For some heavily developed urban areas, dams and reservoirs may provide the only practical means for significantly reducing flood damages.

The protection afforded by a dam is greatest in the area immediately downstream. Protection farther downstream is reduced by flows from tributaries and by runoff from lands next to the stream. Protection can also decrease over time if the reservoir gradually fills with sediment.

A major disadvantage of dams and reservoirs is that downstream residents may not recognize that they only control floods up to some particular size for which they were designed. Such a false sense of security encourages further development and encroachment on downstream floodplains. The value of protection provided by a dam and reservoir may be reduced unless land use controls are enforced to prevent such development. Dams and reservoirs can also have large social and environmental effects and require significant amounts of maintenance. A remote possibility also exists that a dam will fail, causing catastrophic damage and loss of life.

Dams can also play a role in the management of coastal floodplains. If constructed across the lower end of streams, they can prevent storm surges from traveling up the stream valley and flooding inland areas.

Levees and Walls The principal purpose of levees and floodwalls is to confine floodwaters to the stream and a

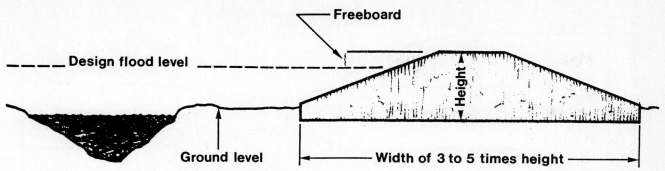

Figure 2.26 Earth levees require significant space because of their width.

selected portion of the floodplain. Seawalls serve much the same purpose in stopping ocean or lake waves. Each of these measures protects only the area immediately behind it and only against the selected flood height for which it was designed.

Levees are normally constructed of earth. The width of a levee's base must be several times greater than the levee's height for stability. As a result, levees more than a few feet in height require that significant space be available between the stream and the property to be protected. Significant space may also be required for setting levees back far enough from the channel to provide adequate flow carrying capacity. Levees are best suited for use along large rivers where their construction is less likely to be a severe encroachment on the floodplain. Floodwalls and seawalls are usually constructed of concrete or steel and take up far less room. They are more suitable for use in congested areas.

A principal advantage of levees, floodwalls, and seawalls is the flexibility they offer to protect either a specific site or a larger area. Unlike dams and reservoirs, they can be used to protect a single community or a portion of a community. However, as with dams and reservoirs, they sometimes create a false sense of security about the protection provided. Floods exceeding the level for which levees and walls are designed can cause disastrous losses of life and property.

Levees and floodwalls may increase flooding in other areas unless designed as part of a comprehensive program. Blocking off a flood's natural course on one side of a stream normally causes more widespread flooding on the other side. Containing a flood between levees or walls on both sides of a stream can create a bottleneck that raises flood levels in upstream areas and increases the speed of water flow immediately downstream. Seawalls in coastal areas also have disadvantages. They are highly susceptible to erosion and, if not properly placed, may also increase erosion in adjacent areas.

Levees and walls often have other undesirable environmental aspects such as blocking access to and views of the stream or ocean, or distruction of wildlife habitat. For communities straddling a river, high levees or walls may make expensive construction necessary to permit travel between sections of the community.

Levees, floodwalls, and seawalls can also block natural drainage from the protected area to the stream or ocean. Preventing flooding of the low-lying areas behind the protective walls by blocked drainage usually requires equipping the structures with a pumping system or other provisions for getting rid of unwanted water. Areas behind a wall or levee may need to be set aside for storing drainage until it can be pumped or discharged through or over the wall.

Levees and walls cannot be used in as many ways as reservoirs. However, they can sometimes be used as roads or paths for access to riverfront or beach areas, fishing sites, and boat launching lanes. They may also be used as trails and even road rights-of-way for recreational purposes.

A different type of wall, called a "jetty" or "groin," is sometimes used in coastal areas. Their purpose is usually to break the force of waves or control erosion. However, they disrupt natural processes and sometimes have unanticipated adverse effects.

Channel Alterations Channel alterations reduce flooding by increasing the flow carrying capacity of a stream's channel. The various types of alterations include straightening, deepening, or widening the channel; removing debris; paving the channel; raising or enlarging bridges and culverts that restrict flow; and removing dams that interfere with flow. Underground conduits can also be installed to carry part or all of a small stream's flow.

Shortening a stream improves its flow carrying capacity and reduces flooding. Cutoffs must be carefully designed so that increased water velocities do not cause erosion and greater flooding in downstream areas. Channel alterations often have undesirable effects on fish and wildlife.

All these channel alterations contribute toward reducing the height of a flood. It is sometimes possible, by extensively reconstructing a stream channel, to contain major floods within its banks. Unfortunately, such alterations sometimes increase flooding downstream by accelerating the flow of flood waters.

Channel deepening is not very well suited to major streams because sediments can quickly fill in the excavated area. Frequent redredging often is necessary to maintain the deeper channel even on smaller streams. This can become a

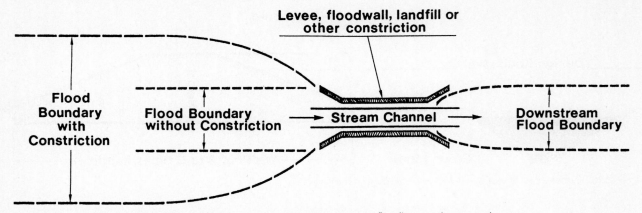

Figure 2.27 A bottleneck in a stream which causes or aggravates upstream flooding may be created by a narrow bridge opening, levees or floodwalls too close to the channel, or other constrictions.

significant expense to local governments. Care must also be taken to avoid causing erosion when changing a channel.

Channel alterations are like levees and floodwalls in that they can be used to protect a specific site or region. However, they differ in that they are not subject to sudden or disastrous failure. Channel alterations for flood control can sometimes be used for other purposes such as navigation and recreation. For example, features such as boat launching facilities can be included in projects to deepen a channel.

The environmental impact of altering a stream channel depends on the specific techniques used. Some, such as reconstructing bridges and culverts, usually have only a temporary effect during construction. However, widening, deepening, or paving of channels may destroy fish and wildlife habitat and other natural values for several years, decades, or perhaps even permanently.

Diversions Diversions intercept flood flows upstream of a damage-prone area and route them around the area through an artificial channel. Diversions may either completely reroute a stream or only collect and transport flows that exceed the normal capacity of the channel or that would cause damage.

Diversions sometimes offer the advantage of protecting several nearby communities with one major facility. A negative aspect is the false sense of security which may prevail in the ''unprotected areas.''

Diversions are particularly well suited for protecting developed areas because they do not require land acquisition or construction within the protected area. However, opportunities for diversions are often limited by the nature of local land formations and soil conditions. There must also be a stream channel available into which the diversion can empty. The receiving channel must have enough capacity to carry the flow disposed of through the diversion without causing flooding.

Also, the use of diversions may be limited in some states by laws prohibiting transfer of water between basins or watersheds.

Land Treatment Land treatment measures are used to reduce runoff of water to streams or other areas. Techniques of land treatment include maintenance of trees, shrubbery and vegetative cover, terracing, slope stabilization, grass waterways, contour plowing, and strip farming. These measures reduce water flow by improving infiltration of rainfall into the soil, slowing and reducing runoff, and reducing the sedimentation that can clog stream channels or storage reservoirs. While the effect of any individual measure is small, extensive land treatment programs can effectively reduce flooding in small headwater areas. Land treatment measures are less effective in downstream areas subject to larger floods.

Land treatment measures are most commonly used in agricultural areas. In areas with steep slopes and unstable soils, maintaining a good growth of grass and other vegetation is the most practical way of reducing runoff and erosion. Several land treatment measures involve little or no additional cost to the farmer, and some, such as no till or minimum tillage practices, actually reduce costs. Land treatment measures may be undertaken as either a public or private effort. Practices requiring significant expenditures by the landowner are frequently encouraged by providing technical and financial assistance from public sources.

Dune stabilization is the most frequently used type of land treatment in coastal areas. It includes protecting or establishing plant cover on existing dunes and constructing replacement dunes. The objective of these measures is to preserve the dunes as barriers to large waves and storm surges.

On-Site Detention Flooding can be increased significantly by the runoff from lands which have been stripped of vegetation or covered with buildings, pavements, and other impervious materials. The main objective of on-site detention is to prevent excessive runoff from such areas. A secondary benefit is that on-site detention traps pollutants and so sometimes improves water quality.

The principal on-site detention measures are those restricting land clearing and providing for temporarily storing

SUNKEN PARKING LOT

Figure 2.28   On-site detention can be provided for in the design of parking lots, rooftops, and other such areas. Slow release of the trapped water avoids aggravating flood problems downstream from large new developments.

some or all runoff from a property. Use of the measures may be voluntary or required by regulatory or permit programs. Regulations requiring on-site detention are often part of zoning or other broad programs controlling land use and development in upland areas.

Small ponds can provide on-site detention on land in open space uses. These ponds sometimes take the form of shallow grass-covered basins that can be used during dry periods as athletic fields, parking lots, or for other purposes. Detention basins can sometimes also be created by excavation during sand and gravel mining operations. Controls on clearing land are most applicable to sites under construction. On-site detention measures in urban areas are usually design provisions which slow runoff. These may include equipping roofs or parking lots for temporarily storing at least a part of the water which falls on them, designing streets in hilly areas to prevent rapid runoff, incorporating small retention basins into landscaping, using rock-filled pits to catch gutter runoff, and using pavements that let water seep through into the ground below.

The cost of individual on-site detention measures is usually not high. However, the cost often falls on the owner of the land where flood waters arise, while most other flood control measures are paid for by the people in the protected area or by the general public.

On-site detention ponds or reservoirs can lose their effectiveness over time if they are not regularly cleaned and maintained. This can involve a significant cost. Another potential problem with on-site detention measures is the lack of unified control over patterns of flow drainage. However, this problem can be handled through broad-scale planning of the overall system.

2.7   Floodplain Management

The table on pages 112 to 113 provides an overview of possible management tools for floodplain control. It indicates relative cost as well as advantages and limitations.

An Overview of Floodplain Management Tools

Tool	Purpose	Approach to flooding threat	Incidence of costs	Advantages	Limitations
Land use regulations	1. Foster health and safety 2. Prevent nuisances 3. Prevent fraud 4. Promote wisest use of lands throughout a community	1. Require individual adjustment of uses to the flooding threat	1. Landowner must bear cost of adjustment. Community bears cost of adoption and administration of regulations	1. Low costs 2. Promote economic and social well-being 3. Promote most suitable use of lands 4. Can be put into effect immediately 5. May remain effective for long periods if adequately enforced	1. Must not violate state and federal constitutional provision 2. Cannot prevent all losses 3. Generally do not apply to governmental uses 4. Limited application to existing uses
Dams, reservoirs, levees	1. Reduce flood losses, protect safety, promote economic well-being 2. Protect existing uses 3. Promote navigation, water recreation 4. Make new sites available for development, increase tax base	1. Adjust flooding threat to land use needs	1. Generally public at large pays for benefits which accrue to landowners, local communities	1. Reduce wide range of flood losses 2. Protect existing uses 3. Promote navigation and recreation 4. Permit regional approach to problems	1. Federal subsidy leads to private gains 2. High costs 3. Construction may take many years 4. May not be consistent with community plans, environmental quality 5. Maintenance required 6. Sedimentation may reduce effectiveness 7. Catastrophic losses may result from failure of dam or levee 8. No site may be available for dam, or levee; geology wrong
Land treatment (to retain precipitation)	1. Prevent future increases in flood heights. Reduce existing levels 2. Promote water and soil conservation	1. Reduce existing flood conditions, prevent future increases in flood heights in frequent floods	1. Expense largely public; however, landowners may bear portion of costs	1. Limited cost 2. Attack flood problem where it begins 3. May be consistent with broad community needs	1. Not applicable in many instances 2. Effectiveness limited to relatively frequent, small floods

–112–

Measure	Objectives	Adjustments to flood	Who pays	Advantages	Disadvantages
Public open space acquisition for parks, wildlife areas, floodways	1. Reduce flood losses 2. Achieve broader community recreation and conservation goals	1. Adjust use to threat	1. Public pays but receives multiple benefits	1. Multiple benefits 2. No problem of constitutionality 3. Permanent 4. Active public use of lands possible 5. Federal grants may be available for open space acquisition 6. Particularly attractive in urban areas	1. Acquisition costly 2. Flood losses to open space uses (e.g., campgrounds) remain 3. Sites not always suitable for recreation, wildlife 4. May create shortage of land needed for businesses, industry, etc. 5. Creates public land management requirements
Flood insurance (National Flood Insurance program)	1. Promote flood regulations 2. Promote long-term cost bearing by individual occupant	1. Require individual cost bearing 2. Adjust use to threat	1. Public pays, in part, for subsidized insurance 2. Private landowner pays for unsubsidized insurance	1. Spread cost of flood losses 2. Promote regulation 3. Encourage consideration of flood costs in private decision making	1. Subsidized insurance may promote continued use at primarily public rather than private expense 2. May undercut floodway regulations to abate existing uses
Warning systems	1. Warn property owners of impending threats 2. Permit advance evacuation, installation of temporary flood abatement measures	1. Adjust use to threat	1. Public bears costs (usually)	1. Can permit adjustment to threat 2. Useful in combination with regulations	1. Of no use unless floodplain occupants are willing and able to take necessary protection measures 2. Systems must be adequately operated and maintained

These and other floodplain management tools such as permanent evacuation and relocation, floodproofing, and flood emergency and recovery measures are usually used in combinations.

2.8 Urban, Urbanizing, and Rural Floodplain Contexts

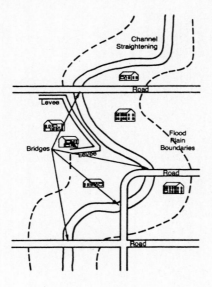

Figure 2.29 Urban areas.

Conditions

1. Intensive existing development of flood plain.
2. Agriculture, forestry and other open space uses often uneconomic.
3. High land values and taxes.
4. Levees and channel straightening.
5. Sewer and water.
6. Bridge openings act as restraints upon flood flows.
7. Blighted flood plain areas.
8. Recreation and open space demands.
9. Land use demands for commercial, industrial, and residential uses.
10. Often adopted zoning, subdivision controls, and building codes without flood provisions.

Common Land Use Management Goals

1. Combine flood plain management tools to reduce flood losses to existing uses and prevent losses to new uses.
2. Preserve floodway areas; require flood protection for new uses in flood fringe areas.
3. Provide park and other active recreation areas.
4. Redevelop blighted areas.
5. Provide levees, fill, flood-proofing and other protection for existing uses.
6. Prevent subdivision of unsuitable lands.
7. Permit (in some instances) residential, commercial and industrial uses.

Tool	Purpose
1. Regulations (overlay more comprehensive zoning; must be combined with other approaches.)	
a. Zoning: Two-District	Preserve floodway, define protection elevations for new uses.
b. Subdivision	Insure suitability of lands, installation of flood-protected facilities.
c. Building Code	Establish minimum protection elevations flood-proofing requirement.
d. Housing Code	Require flood-proofing for certain types of existing development (apartments, motels, etc.).
2. Public Acquisition	Fee acquisition for recreation areas; easements for floodway areas.
3. Urban Renewal	Clear certain flood blighted areas and replace them by open space or restricted new uses. Relocation of uses.
4. Levees, Channel Straightening	Protect existing intensive use areas, improve suitability of lands for new uses.
5. Public Facilities Planning	Withhold public facilities from flood areas or insure adequate protection for facilities.
6. Flood Warning Systems	Permit emergency evacuation, flood-proofing measures.

Figure 2.30 Urban areas.

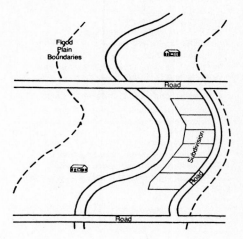

Figure 2.31 Urbanizing area.

Conditions

1. Partial development of flood plain.
2. Transition from agriculture and forestry to residential and commercial uses.
3. Rapidly increasing land values and taxes.
4. Subdivision and land speculation common.
5. Some sewer and water facilities.
6. Often limited existing zoning and subdivision controls.
7. Increasing recreation demands.

Common Land Use Management Goals

1. Preserve floodway areas; prevent development in selected flood fringe areas with special values.
2. Preserve agricultural uses (in some areas).
3. Provide recreation, wildlife and scenic areas.
4. Prevent subdivision of unsuitable lands.
5. Prevent water pollution and nuisances.
6. Prevent extension of public facilities into flood prone areas (in some circumstances).

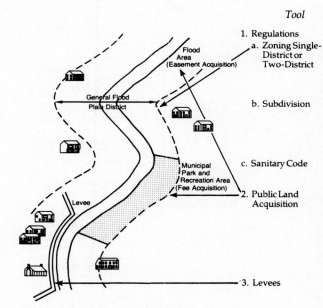

Figure 2.32 Urbanizing area.

Tool	Purpose
1. Regulations	
a. Zoning Single-District or Two-District	Preserve floodway; define protection elevations for new uses. (Case-by-case approach for data gathering if a single-district ordinance is used.)
b. Subdivision	Insure suitability of lands, installation of flood-protected services consistent with municipal systems. May be exercised extraterritorially.
c. Sanitary Code	Prohibit onsite waste disposal where public sewers are not provided.
2. Public Land Acquisition	Fee acquisition for park areas; easements for floodway and flood stor-areas. (Note: Consistent policies should be applied to the compensated acquisition of some lands versus uncompensated regulations of others.)
3. Levees	Protect existing intensive use areas.

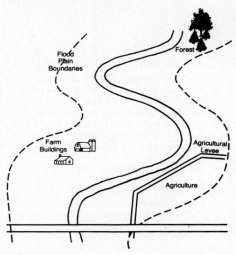

Figure 2.33 Rural area.

Conditions

1. Low intensity flood plain use and little development pressure (except for recreation areas). No established development pattern.
2. Forestry, agriculture, recreation and other low intensity uses, economic.
3. Onsite waste disposal systems.
4. Wildlife and scenic values.
5. Little flood data.
6. Often no zoning, subdivision control; little planning and engineering expertise.

Common Land Use Management Goals

1. Guide development to unflooded sites.
2. Preserve floodway areas and required flood storage areas.
3. Maintain viable agricultural and open space uses.
4. Prevent water pollution and nuisances.
5. Protect ecological and scenic values.
6. Rely primarily upon regulations to guide flood plain uses.

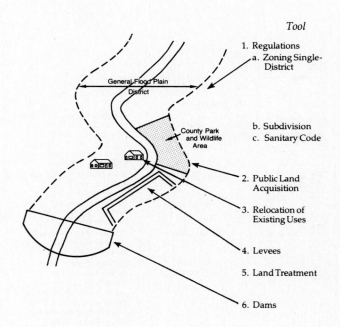

Figure 2.34 Rural area.

Tool	Purpose
1. Regulations a. Zoning Single-District	Preserve broad floodway; define protection elevations for new uses. (Case-by-case approach for data gathering and elevation of individual uses very common.)
b. Subdivision	Insure suitability of lands.
c. Sanitary Code	Prohibit onsite waste disposal in flood plain. (May apply to most single structural uses.)
2. Public Land Acquisition	Fee or easement acquisition for parks, ponding areas.
3. Relocation of Existing Uses	Relocation program may be attractive where little development exists in flood plain.
4. Levees	Provide protection for agricultural uses.
5. Land Treatment	Prevent increases in flood threats; combine with agricultural practices.
6. Dams	Provide protection for downstream properties, improve navigation, provide water recreation.

2.9 Floodproofing

Many thousands of structures and potential building sites are located in floodplains and thus are susceptible to flooding. Although flood control projects have partially protected some of these structures and building sites through reduction of the flood threat, the residual threat to these sites and the total threat to unprotected sites remain as major problems. Evidence of this situation is given every year by the millions of words and hundreds of headlines that dramatically describe floods and their resulting damage and loss of life. When floods strike developed areas, whole cities may be disrupted and their productive capacities impaired. Strategic transportation lines are cut. Public-service facilities are sapped, homes and crops are destroyed, and soils are eroded. Yet flood-vulnerable lands are the setting for continued urban growth in the United States.

Studies of floodplain use show that some encroachment is undertaken in ignorance of the hazard, some occurs in anticipation of increased federal protection, and some takes place because shifting the cost of the hazard to society makes it profitable for private owners to undertake such development. Even if full information on the flood hazard were available to all owners or users of floodplain property, there would still be conscious decisions for some reason or other to build in areas subject to flooding. To escape this dismal cycle of losses, partial protection, further induced development and new unnecessary losses, old attitudes must be transformed into positive actions.

Primary among these actions is the revision of development policies and the enaction of a regulatory program to encourage the direction of growth or change necessary to achieve floodplain management objectives. Information programs are essential to this revision. They foster the development of more appropriate policies and involve the gathering and dissemination of data on past floods, estimates of future floods, and alternative ways of dealing with flood losses in areas where intensive development has taken place or is anticipated. Such programs have led to an expanded approach to flood damage reduction and prevention that recognizes the need to control or regulate the use of lands adjacent to watercourses and the need to provide guidance in the design of floodplain structures through the planned management and development of the flood hazard areas.

Regulation of the use of floodplain lands is a responsibility of state and local governments and can be accomplished by a variety of means, such as the establishment of designated floodways and encroachment lines, zoning ordinances, subdivision regulations, and building codes. These land use controls, most often known as floodplain regulations, do not attempt to reduce or eliminate flooding but are intended to guide and regulate floodplain development to lessen the adverse effects of floods. Floodplain regulations are being adopted by communities and used as legal tools to control the extent and type of development permitted on floodplains.

Floodproofing standards applied through building codes and regulations to floodplain structures can permit economic development in the lower-risk areas by holding flood damages and other adverse effects within acceptable limits. Floodproofing requires adjustments both to structures and to building contents and involves keeping water out as well as reducing the effects of water entry. Such adjustments can be applied individually or as part of collective action either when buildings are under construction or during the remodeling or expansion of existing structures. They may be permanent or temporary.

Floodproofing, like other methods of adjusting to floods, has its limitations. For example, in addition to reducing loss potentials, a main purpose of floodproofing habitable structures is to provide for early return to normalcy after floods have receded rather than for continuous occupancy. See also Figure 2.23.

The concept of floodproofing (or more accurately, flood-resistant construction practices) is to modify buildings, their sites, or their contents to keep water out or reduce the damage caused by water entry. Floodproofing also can be used to reduce disruption to activities, to maintain vital services in operation during a flood, and to permit faster recovery from flooding.

Unlike dams and other measures protecting large areas or long sections of streams, floodproofing is used to protect individual buildings or small groups of buildings. Floodproofing is most easily incorporated in new construction since it may often be included at little or no additional cost. However, its addition to existing structures may be economical in some cases. A large number of floodproofing techniques are available. The most common ones are described below.

The most frequently used method of floodproofing is the elevation of buildings above expected flood levels. Existing structures can sometimes be raised and the original foundation extended upward with walls, piers, or posts. These measures are best suited for smaller structures with basements or crawl spaces. A common way to elevate a new structure above flood levels is to raise the elevation of the site with earth fill. New structures can also be raised above the ground on columns or walls.

Small floodwalls or levees can be used to protect single buildings or small groups of properties. They have the advantage of protecting the whole enclosed area rather than just the building. These measures are also useful for protecting buildings for which other floodproofing measures cannot be used because of the building's size or lack of structural strength. However, small walls and levees sometimes may be unattractive, may require substantial maintenance, are subject to failure and/or overtopping, or even may intensify flood problems on adjacent property by redirecting flood flows.

Structures with walls and foundations which are generally impermeable to water can sometimes be made almost completely watertight by permanently blocking unused door-

Figure 2.35 Small walls can be an attractive addition to property as well as providing protection against floods. However, they must be properly designed and built to avoid failure. Openings in the walls for access during nonflood periods can be provided as needed.

ways and windows and providing temporary covers which can be installed over the remaining openings during flooding. Seepage can be reduced by applying a sealant to walls and floors. If temporary closures are to be useful, early flood warnings must be available.

Any contents of a building which are particularly vulnerable to water damage should be moved above the expected flood level. Depending on the potential depth of flooding, items such as appliances may need to be temporarily or permanently raised only a few inches above floors or moved to an upper story. This technique can also be applied to industrial equipment, commercial stocks, electrical wiring, and other damageable items.

The amount of flood damage and the cost of cleanup after a flood may also be reduced by using water-resistant materials for constructing and finishing buildings. The water resistance of paints, paneling, insulation, floor coverings, cabinet materials, and other items varies widely.

The major expense of floodproofing is usually the original cost. This cost varies depending on the type of property being protected and the floodproofing technique selected.

Decisions on what techniques to use should take into consideration that some approaches, such as elevation of a building, require no further attention after completion. Other techniques, such as putting temporary covers over doors and windows or closing gates in small walls, require action each time a flood occurs. Dependence on these actions requires availability of advance warning of floods as well as people trained to take the necessary steps.

Some floodproofing measures may cause the structural failure of buildings and increased damages if not properly designed and used. For example, blocking the entry of flood-

waters into a building prevents equalization of inside and outside pressures and results in forces of a type most structures are not designed to withstand. If floodwater on the outside of the structure is deep enough, its pressure can buckle or even break walls and floors. Similarly, raising buildings may cause their foundations to become unstable. These and most other floodproofing measures require professional design and supervision of construction for successful use. Their improper use can increase flood damage.

Laying Out the Site The practice of "clustering" buildings is prevalent in planned-unit developments. This clustering permits buildings to be attractively grouped on parts of a site which are above flood levels and reserves the low-lying sections as landscaped green areas and parking facilities (Figure 2.36).

The use of the higher ground for development allows streams and other natural lowland features to be kept intact as scenic elements and fish and wildlife habitats. Many of the natural features of the valley can be made more useful for recreational puposes. For example, low-lying swampy areas can be transformed into permanent lakes that provide opportunities for water-oriented recreation and modest amounts of floodwater storage as well.

Alternate Methods of Floodproofing

Site Selection The one method of assuring complete flood hazard protection of a building or structure is to select a site or structure location which places all spaces in the structure above the "floodplain flood." This could apply to sites both inside and outside the floodplain limits. Locating a structure outside the floodplain would eliminate the need to consider floodwater loads in the building design. The building could be located in the floodplain and be protected to design flood level by dikes, levees, or floodwalls; also eliminating the need for flood load consideration in the building design for flooding to a design flood level.

Floodproofing by Elevating the Building

Natural Terrain Structures constructed above the regulatory flood datum (RFD) and outside the regulatory floodplain will not be subject to loads from regulatory floodwaters if basements are not used. The effect of soil saturation on basement walls and foundations may still have to be considered. Natural slopes should be investigated for stability and scour potential if the structure is to be built at the regulatory-flood runout line on the ground surface. A building located outside the regulatory floodplain is shown at the left side in Figure 2.37.

Building on Fill Buildings can be located within the floodplain or primary flood hazard area on a fill constructed to an elevation above the RFD. This method of protection can be accomplished by constructing an earth fill either partially or entirely within the floodplain, as also shown in Figure 2.37. Such a design should provide assurances that the fill does not restrict or obstruct the flow of floodwaters or

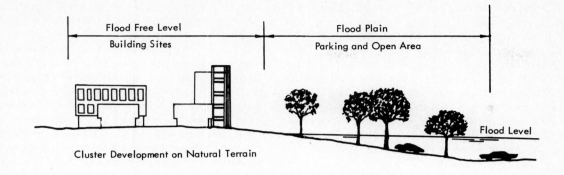

Flood Free Level
Building Sites

Flood Plain
Parking and Open Area

Flood Level

Cluster Development on Natural Terrain

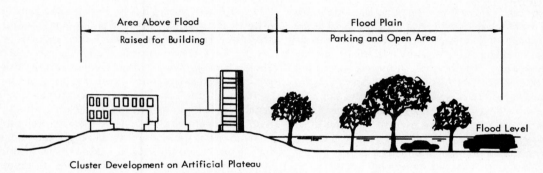

Area Above Flood
Raised for Building

Flood Plain
Parking and Open Area

Flood Level

Cluster Development on Artificial Plateau

Figure 2.36   A planned-unit development on a floodplain site. The example at the top illustrates a valley location with only a part of the site subject to flooding. The lower example shows a site entirely within a pondage area and subject to low-stage flooding. In this case, a portion of the site was artificially raised to be above flood levels.

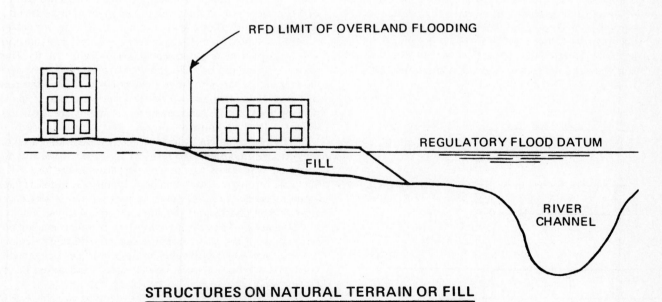

RFD LIMIT OF OVERLAND FLOODING

REGULATORY FLOOD DATUM

FILL

RIVER
CHANNEL

STRUCTURES ON NATURAL TERRAIN OR FILL

Figure 2.37   Structures on natural terrain or fill.

reduce the hydraulic efficiency of the channel, which in turn could cause floodwater backup and resultant higher floodwater elevations upstream of the filled building site.

The fill material should be suitable for the intended purpose as determined by an investigation of the soil properties. The earth fill should be compacted to provide the necessary permeability and resistance to erosion or scour. Where velocities of floodwaters are such as to cause scour, adequate slope protection should be provided with vegetation or stone protection as required. Slope stability should be analyzed by an experienced soils engineer to assure its adequacy.

Where the fill is partially within the floodplain, access and utilities should be provided from the "dry" side. If the fill is entirely in the floodplain, access and utilities could be provided by constructing an access road or bridge to an elevation above the RFD.

Building on Stilts   Often it is geographically undesirable or economically not feasible to locate a structure outside the floodplain. Available land areas are being developed rapidly and communities are finding it necessary to permit construction in the fringe areas of floodways. In these areas, structures can be built which place all functional aspects above the RFD by building on "stilts" as shown in Figure 2.38.

In elevating a building on stilts, piles, columns, piers, and walls, or other similar members are used to raise the functional floors or spaces of the building above the RFD elevation. The design should consider the loads that result from possible debris blockage between supporting members and impact of floating debris.

The open space created at ground level below the functional floors could be used as a plaza, parking area, materials handling, or recreational area, or for storage of special non-damageable materials, equipment, etc. This open space would be essentially free from the damaging effects of floodwater, except that lobbies and entrance would have to be protected by some approved floodproofing method.

The equipment necessary to maintain building functions should be located safely above the RFD. If access to the building were provided from a location above the RFD, the

normal building activities would not be disrupted and the building could continue to function during the flood emergency.

Protection by Dikes, Levees, and Floodwalls   As an alternate to providing flood protection through building or structure modifications, the necessary protection may be achieved by detached dikes, levees, or floodwalls. The primary purpose of these constructions is to prevent the flood from reaching the structure and associated functional land areas. The choice of using a dike or floodwall is made on the basis of economic considerations when compared with structural floodproofing modifications, the ability of a structure to be structurally modified, and the degree of protection to be provided. The type of protection barrier depends on location, availability of material, foundation conditions, and right-of-way restrictions. Floodwalls would be used in tight, restricted areas where foundation conditions are favorable. Dikes or levees would be used where adequate space and material are available. The dike or floodwall may not have to completely surround a structure. Protection may be required only on the low sides as illustrated in Figures 2.39 and 2.40. The ends of the works would be tied into the existing high ground or to the structure, depending on local conditions.

Dikes   If used, dikes should be constructed to a section capable of supporting the imposed loads and providing the required impermeability. Suitable material preferably should be available at the site and should be tested and approved for use prior to constructing the dike. An investigation should also be made of the foundation material to determine the presence of, location, and extent of unsuitable materials and necessity for drainage of cutoff provisions.

At locations where the foundation material has a high degree of permeability, an impervious cutoff may be necessary to reduce seepage through the in situ foundation materials. The cutoff may be a sheet pile wall, compacted barrier of impervious soil, fabric-reinforced membrane, concrete wall, or grouted cutoff. As no cutoff is totally impermeable, provisions should be made to collect the excess seepage and any seepage from less permeable soils without cutoffs. The excess seepage can be collected with drainage blankets, pervious trenches, or perforated pipe drains placed at the toe of the embankment and on the dry landward side. Typical dike sections, cutoffs, and drainage provisions are shown in Figures 2.41, 2.42, and 2.43.

If any drainpipes or related structures are within a dike, they should be designed to resist all applicable loads and be provided with gates to prevent backflow to the dry side. Backflow through conduits can be prevented by installing flap gates, manually operated valves, or slide gates that would be closed when floodwaters would reach critical elevations.

Floodwalls   A floodwall is subject to hydraulic loading on one side with little or no earth loading as a resisting force on the opposite side. Floodwalls can be constructed as cantilever I-type sheetpiling walls, cellular walls, buttress walls, or gravity walls.

The walls should be founded on and keyed into rock where suitable rock is encountered reasonably close to the

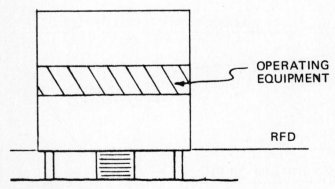

Figure 2.38   Building on stilts.

OPERATING EQUIPMENT

RFD

founding elevations. Where the soil provides inadequate bearing capacity and removal of unsuitable material and replacement is costly, an adequately designed system of piling should be considered. Cutoffs and drains should be used to intercept seepage. Drainpipes should not be placed directly under the wall base, and any drainage provided should not be considered as a factor for reduction of uplift pressures. The problem of scour should be further investigated and corrective measures provided where necessary.

Drainage features through floodwalls should be equipped with the necessary devices to prevent backflow. Typical sections of various floodwall types are shown on page 123.

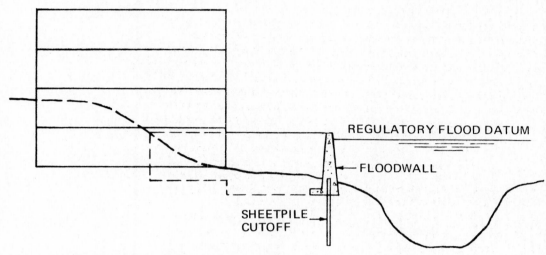

Figure 2.39 Flood protection with floodwalls.

Dike or Levee Protection

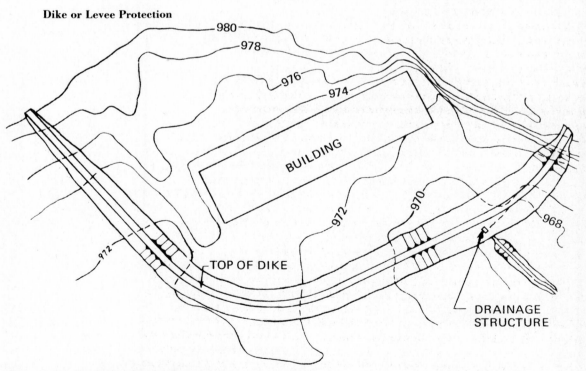

Figure 2.40 Flood protection by dikes. Structure with restricted use.

DIKE OR LEVEE PROTECTION

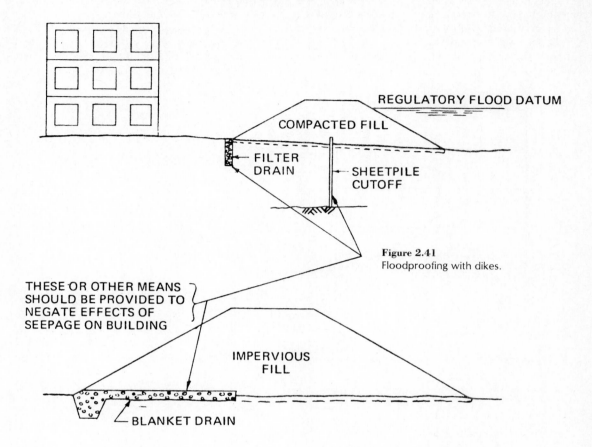

Figure 2.41
Floodproofing with dikes.

Figure 2.42
Dike with blanket drain.

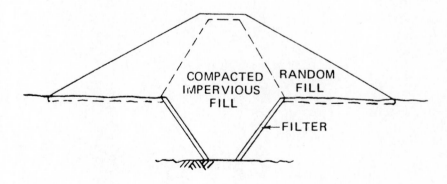

Figure 2.43
Dike with impervious core.

Various Floodwall Types

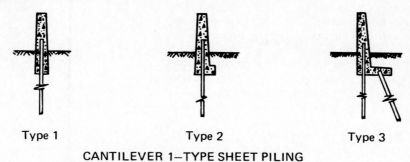

Type 1   Type 2   Type 3

CANTILEVER 1—TYPE SHEET PILING

Figure 2.44 Floodwalls: cantilever 1−type sheetpiling.

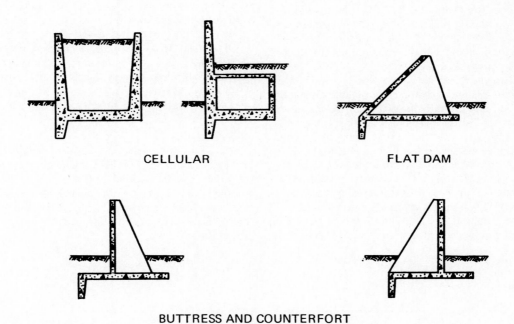

CELLULAR   FLAT DAM

BUTTRESS AND COUNTERFORT

Figure 2.45 Floodwalls: cellular, flat dam, buttress, and counterfort.

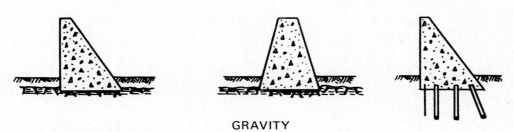

GRAVITY

Figure 2.46 Floodwalls: gravity.

Controlled or Intentional Flooding   In many situations, the basement walls and floor slab(s) of existing buildings and structures lack the structural strength required to withstand flood loadings. The expense of reinforcing an existing structure or replacement with a new structure at the same location to withstand such flood loadings is, in most cases, not justified. As an alternate means of floodproofing these structures, provisions may be made for flooding of the structure interior to balance the external flood pressures on the building components. This intentional flooding would have to be accomplished in such a manner as to keep the unbalanced hydrostatic pressures safely within the load-carrying capacity of the slab and walls. Provisions must be made for interconnections through and around all floors and partitions in order to prevent unbalanced filling of chambers or spaces within the structures.

Flooding   Flooding should be with potable water from a piping or storage system of adequate capacity to fill the basement at a rate consistent with the anticipated floodwater rise. The provisions should be such as to keep the internal water surface as nearly even with the outside as possible. All spaces should be provided with air vents to prevent the trapping of air by the rising water surface.

Draining   Outlets to drain the water as floodwaters recede should be located to completely drain the structure and all spaces at a uniform rate corresponding to that of the receding waters. The water level in all interior spaces should be kept even and all spaces should be completely drained. Upper spaces and levels should be drained before the lower spaces. All watertight walls should be designed for an internal hydrostatic pressure resulting when waters trapped in the building are higher than those of the receding floodwaters outside, a possibility with malfunction of required drains.

Use   Where provisions are made for internal flooding, all floors and spaces below the RFD should be restricted as to

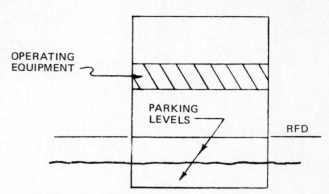

STRUCTURE WITH RESTRICTED USE

Figure 2.47   Structure on natural terrain or fill.

types of use permitted. Examples of controlled flooding of structures with restricted use are shown in Figures 2.47 and 2.48.

Backflow   Where intentional flooding with potable water is proposed (or where floodwater backflow through the sewer system may occur), backflow preventers should be installed in the sewer lines.

Total Approach

General   The design and implementation of floodproofing systems and procedures requires a total approach. No element or item, regardless of how minute it might appear, should be overlooked or left to chance. The most elaborate, extensive, and expensive floodproofing system may be rendered useless by a minor omission or by the failure of a weak link in the system.

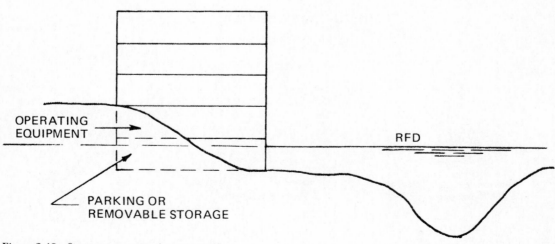

Figure 2.48   Structure on natural terrain or fill.

2.10 Drainage (FHA Requirements)

Surface and subsurface drainage systems should be provided, as appropriate, for the collection and disposal of storm drainage and subsurface water. These systems should provide for the safety and convenience of occupants and the protection of dwellings, other improvements, and usable lot areas from water damage, flooding, and erosion. Where storm drainage flow is concentrated, permanently maintainable facilities should be provided to prevent significant erosion and other damage or flooding on the site or on adjacent properties

Drainage Design and Exposure to Flood Hazards Drainage should be designed to accommodate storm runoff, calculated on the basis of the ultimate foreseeable developed conditions of contributory site and off-site drainage areas.

The minimum grades at buildings and at openings into basements should be at elevations that will prevent adverse effects by water or water entering basements from flood levels equivalent to a 50-year return frequency. The floor elevations of all habitable space should be above flood levels equivalent to a 100-year return frequency.

Provision should be made for the best available routing of runoff water to assure that buildings or other important facilities will not be endangered by a major emergency flood runoff that would become active if the capacity of the site's storm drainage system were exceeded.

Streets should be usable during runoff equivalent to a 10-year return frequency. Where drainage outfall is inadequate to prevent runoff equivalent to a 10-year return frequency from ponding more than 6 inches deep, streets should be made passable for local commonly used emergency vehicles during runoff equivalent to a 25-year return frequency except where an alternate access street not subject to such ponding is available.

Site drainage should be routed to permanent surface or subsurface outfall adequate to dispose of present and future anticipated runoff from the site and from contributing off-site watershed areas except where such water is necessary for controlled irrigation.

Drainage swales should not carry runoff across walks in quantities that will make them undesirable to use. Walks should not be designed as drainageways.

Developed portions of a site that can be adversely affected by a potentially high groundwater table should be drained where possible by subsurface drainage facilities adequate for the disposal of excess groundwater.

Storm-water drainage should be connected only to outfalls approved by the local jurisdiction.

Primary Storm Sewer The pipe size for the primary storm sewer system should have an inside diameter based on design analysis but not less than 15 inches. The minimum gradient should be selected to provide for self-scouring of the conduit under low-flow conditions and for the removal of foreseeable sediments from the drainage area.

Secondary Drains Pipe drains of adequate size from minor runoff concentration points should be provided and connected to appropriate disposal lines when analysis indicates that they are necessary.

Drainage Swales and Gutters Paved gutters should have a minimum grade of 0.5 percent. Paved gutters and unpaved drainage swales should have adequate depth and width to accommodate the maximum foreseeable runoff without overflow. Swales and gutters should be seeded, sodded, sprigged, or paved as appropriate to minimize potential erosion.

Open Channels Channels should be protected from erosion by appropriate vegetative cover, lining, or other treatment indicated as necessary by analysis. Earthen channel side slopes should be no steeper than 2 to 1 and should be flatter to prevent erosion where analysis indicates the need.

Open channels with lining should have a maximum gradient on side slopes of 67 percent (1½: 1), with adequate provisions for weep hole drainage. Channel side slopes steeper than 67 percent should be designed as structural retaining walls with provision for live and dead surcharge load.

2.11 Drainage Systems

The following are four commonly used methods for providing site drainage:

1. Surface drainage system
2. Enclosed underground drainage system
3. Enclosed underground drainage system with on-site storage
4. Combination system with enclosed drainage for paved areas and surface drainage for unpaved areas

The selection of a particular drainage system has a direct effect on the control of erosion and sedimentation.

Surface Drainage System Under this system the runoff from paved and unpaved areas is collected and conveyed off the site in surface drainage channels (see Figure 2.49). Channels must be designed so that channel erosion does not occur.

The surface roughness of the vegetative channel lining slows the velocity of the runoff. Such a reduction of velocity is desirable, but under certain conditions surface channels must be paved to prevent erosion within the channel. The outlets of paved surface channels must control the runoff and sediment load at discharge sites. Where structural lining is necessary in only a small percentage of the total surface of the drainage system, cost advantages will make the use of a surface drainage system a preferred alternative.

Surface drainage systems can release runoff to off-site surface drainageways or streams, to a street or municipal storm drainage system, or to an on-site sediment basin. In some cases, an on-site retention basin will be necessary to control the velocity of the runoff released from the site.

The principal disadvantage of the surface drainage system

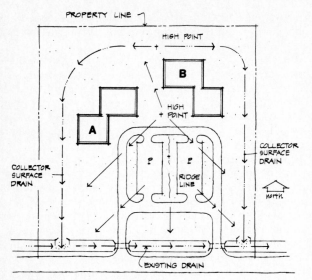

Figure 2.49 Surface drainage system.

is its potential for on-site erosion. Such erosion will occur if channels are not adequately designed, stabilized, and maintained. Where the runoff flows from paved areas to grassy areas, the potential for erosion is high. Careful stabilization of the areas that will receive runoff from paved areas is necessary.

Enclosed Underground Drainage System An enclosed drainage system intercepts runoff from both paved and unpaved areas and conveys it to an outlet at the edge of the site (surface drainageway or stream), a municipal storm drainage system, or an on-site sedimentation and storage basin (see Figure 2.50).

The major advantage of the enclosed drainage system is

that the increased volume and velocity of runoff generated by development can be intercepted before runoff can cause on-site erosion damage. The principal disadvantage is that the velocity of runoff is increased and sediment often is not filtered from runoff. As a result, the points at which runoff is released from the system are subject to erosion and sedimentation. Thus, while the potential for erosion damage to the site itself is minimized, the erosion and sedimentation damage to off-site areas may be increased.

Where the enclosed drainage system empties into a municipal storm system, sediment must be removed before it reaches the municipal system outlet. Where the system outlet is at a site boundary or within the site, the potential for erosion and sedimentation damage that may be caused must be controlled by an earth changer. The velocity of runoff released at these outlets must be maintained at nonerosive levels. Sediment must be removed from runoff before it is released. An on-site storage and sediment basin may be included in the permanent erosion control plan if the enclosed drainage system within the site does not empty into an enclosed, underground municipal storm drainage system.

Enclosed Underground Drainage with On-Site Storage This kind of alternative drainage system (see Figure 2.51) has the advantages of the enclosed drainage system of on-site erosion control; yet it avoids off-site damage. Instead of merely delaying the erosion and sedimentation impact of the enclosed drainage system, the on-site, storage-controlled runoff release system largely eliminates this impact.

Combination Drainage System Under this system runoff from open spaces is collected in surface drainageways, while runoff from paved areas is collected in an enclosed drainage system (see Figure 2.52). Because the enclosed drainage system intercepts runoff from only a limited area, the erosion and sedimentation risk at its outfall is likely to be less than if an enclosed system is used to drain the entire site. The runoff

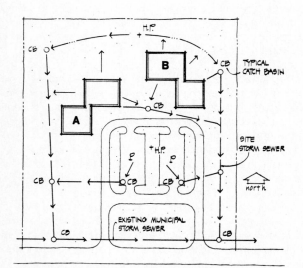

Figure 2.50 Enclosed underground drainage system.

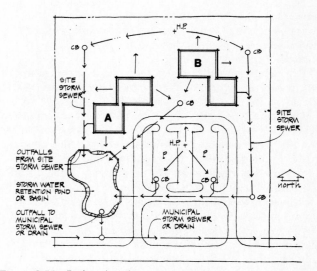

Figure 2.51 Enclosed underground drainage system with on-site storage.

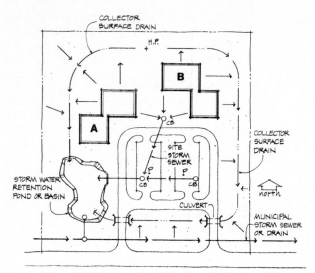

Figure 2.52 Combination drainage system.

from the enclosed drainage system can be let out into the surface drainage system, and in some cases a permanent sediment basin may not be necessary. See Figure 2.53.

This system of dual drainage has the advantage of ensuring that no erosion will occur in vegetated areas adjacent to paved surfaces. Surface channels, especially if they must accommodate runoff released from the enclosed drainage system, must be carefully designed and stabilized to prevent channel erosion. A permanent sediment and storage basin will be required if either the content of runoff sediment or the runoff velocity is high. Careful design and maintenance are essential when this dual system is used.

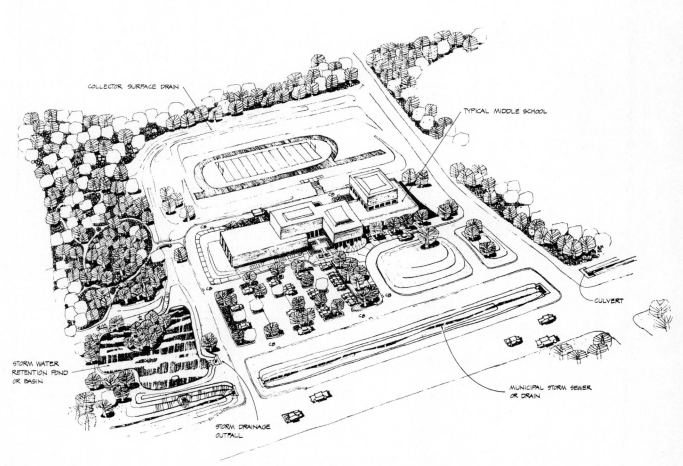

Figure 2.53 A dual drainage system can slow runoff velocity and trap sediment on the site.

2.12 Utility Systems

Water The water distribution system is that part of the total water system which transmits the water from distribution reservoirs to the consumer. The layout of the distribution system depends on the street plan, topography, and the location of the supply and storage facilities. There are two basic types of distribution systems: a branching pattern, and a gridiron or network pattern. These types are illustrated in Figure 2.54. The branching pattern is mainly used in suburban areas. One obvious drawback of this type of pattern is the dead ends where the possibility of stagnant water conditions exist. The gridiron or network pattern is more common in built-up areas where a more reliable supply for fire-fighting capability is needed; however, it is often required in subdivisions by zoning regulations. The network pattern allows water to be supplied to a given point from two directions and avoids dead ends.

In general, with the network system each street will contain at least one water main with the mains interconnected at each street intersection. Valves are installed at pipe junctions to provide the ability to shut off a leaking or broken line until repairs can be made. Depending on the design of the system, as many as three valves are generally installed at each intersection. Other valves in a water distribution system are located on each service connection and at each fire hydrant.

Valves are normally installed with a valve box attached to the top to allow the valve to be operated from the surface. Larger valves may be installed in pits or manholes for access. Repairs to valves directly buried in the ground, of course, require excavation for access to the valve itself. The valves for small service connections (corporation cocks) are almost always directly buried, and access to these valves requires excavation. A typical service connection is shown in Figure 2.55.

Since a major function of a city water system is to provide fire protection, hydrants are located at predetermined points. Fire protection requirements also determine the pipe diameter and water flow rate to be provided. The National Board of Fire Underwriters requires 8-inch pipe as the minimum diameter but will permit 6-inch pipes in gridiron systems if the length of pipe between interconnections is less than 600 feet. If fire protection is not required, 4-inch or less pipe sizes can be installed. This practice is normally followed only in low-density suburban areas. In areas where fire-protection capability is required, the National Board of Fire Underwriters standard is for hydrants to be spaced to serve areas within a radius of 200 feet. These hydrants are normally installed at street intersections, although in large city blocks they may be required at other locations.

Sewers There are two basic types of sewer systems used in the United States. The first of these is the combined sewer system in which waste waters from domestic and industrial sources are collected along with storm water runoff in a single collection system. The second type consists of a collection system for domestic and industrial wastes, and a separate system of storm sewers. Construction economy favors the installation of combined sewers; however, emphasis is currently being directed to the use of separate sewers to avoid hydraulic overloading of sewage treatment facilities caused by the peak flows from storms. In the past this problem was overcome by bypassing the waste which could not be handled in the treatment facility directly to the outfall. Recently, it has been found that the pollutants present in storm-water runoff are significant and that treatment of this waste is often necessary. Installation of separate sewers allows sanitary sewage and storm water to be collected separately and treated before release.

In the design of sewer systems every effort is normally directed toward a system which will allow gravity flow. Because of the gravity flow requirements, sewers must be installed with definite grades which, because of topographical features, results in their being at various depths. Pumping systems or force mains may be installed where topographic

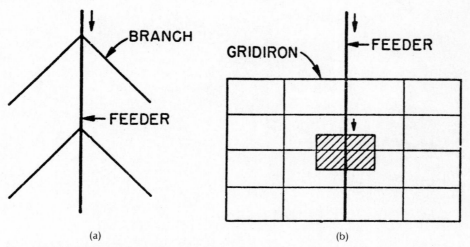

(a)   (b)

Figure 2.54 Typical water distribution system patterns: (*a*) branching pattern with dead ends; (*b*) gridiron pattern with central feeder.

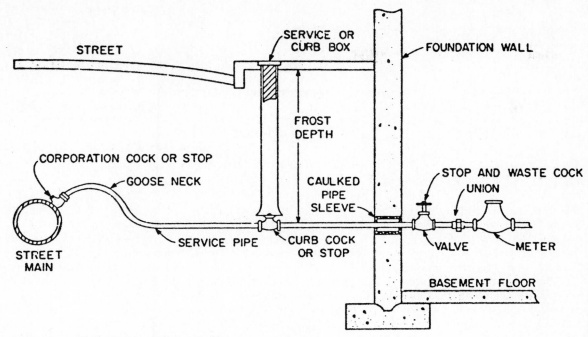

Figure 2.55   Typical water service connection.

features require deep excavation. Many of the construction problems encountered in the installation of sewers for sanitary waste could be eliminated by the use of pumped or pressurized systems which would eliminate grade considerations and the need for deep excavation in some areas. However, the energy costs for pumping and the possibility of power and pump failures tend to reduce the advantages of the pumped sewer concept. The application of pumped sewers for storm sewers or combined sewers is difficult because of the extreme variability of waste flows in these systems.

In the case of combined sewers, each sewer must be connected to the waste producer as well as to each catch basin located in the street for collecting storm water. Manholes must be installed at frequent intervals to allow the sewers to be cleaned and to make connections between other sewers. Current design practice calls for manholes at each sewer junction, at each change in grade or change in pipe size, or at intervals of about 300 feet. Valves and other special fittings are not normally found in sewer systems.

In most new installations, branch fittings for service connections are installed as the sewer is constructed, and the user later connects to these fittings. If a minimum number of street cuts is desired, lines may be installed from the main sewer to the property line at the time the sewer is built.

Natural Gas   Natural gas systems can have both transmission and distribution functions. However, consumers do not normally receive gas directly from transmission systems. Upon entering the distribution system from the transmission system, the gas is reduced in pressure. Both low-pressure (about 0.5 pounds per square inch) and intermediate-pres-

sure distribution systems (up to 50 pounds per square inch) are employed. Where intermediate-pressure systems are used, additional pressure reduction is generally provided for each customer at the meter. An odorizing agent is commonly added to assist in detecting leaks in the distribution system and on the customer's premises. Valving is provided to localize service outages and to stop the flow of gas in the event of a break or leak. Since corrosion of metal gas mains is a major concern, distribution mains constructed of steel pipe generally require protective coatings and/or cathodic protection. The feasibility of plastic pipe for gas distribution systems has been demonstrated, and its use is undergoing rapid growth. In the next year or two, it is expected that the use of plastic pipe will increase to about 25 percent of the total annual installation mileage.

Valves used in distribution mains are generally located in vaults or valve boxes to provide protection and permit proper maintenance and operation. In low-pressure service and on smaller lines (i.e., 2 inches and under), plug-type valves are most commonly used. As shown in Figure 2.56, there are generally two valves on a gas service line, one at the connection to the main and the other at the customer's meter. Where taps are made on the mains for service connections, full-opening plug or gate valves are preferred.

Electric Power   Underground electric power systems are different from aerial systems in that the conductors are larger and are located closer together. Because of the close proximity of the conductors, and because heat transfer from the cable is restricted, thermal considerations become the main design problem. These factors thus limit the number of

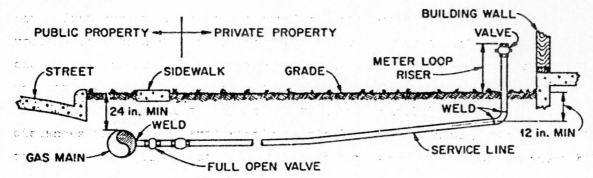

Figure 2.56 Typical gas service connection.

power cables that can be incorporated in a single installation without excessive reduction of the current-carrying capacity of the cables.

As in the case of water systems, a number of wiring patterns (e.g., radial, loop, and network) are used to distribute electricity to the consumer. The choice of wiring pattern used depends on the degree of reliability required.

The lowest voltages in the distribution system generally occur at the consumer's utilization level. Typically, this is 120/240-volt, single-phase, three-wire for residential and small consumers, and 120/208- or 265/460-volt, three-phase, four-wire for commercial areas and industrial users. Higher voltages exist in the distribution system depending on the size of the individual loads, the total load, the load density, size of the area, and the distribution pattern used. The trend in distribution voltages in high-density urban areas currently is toward the use of high voltages similar to those used in aerial transmission for long distances. This has been brought about by the development of improved high-voltage underground cables.

Underground cables can be buried directly in the soil or placed in underground conduits or ducts which are installed separately. The use of a duct system allows the cable to be installed at a later date when needed, or allows existing cables to be removed and replaced without excavation. When conduit banks are installed, extra ducts are normally included to provide space for future expansion. The conduit system, with its inherent flexibility for cable installation and replacement without excavation is commonly used in dense urban areas. Ducts are generally constructed of vitrified clay tile, or fiber, cement-asbestos, concrete, or plastic pipe.

Installation of cables in ducts is performed by pulling the cable between manholes which are usually located at each street intersection. Not only are the manholes utilized for installing and removing cables, they also are the major points in the distribution system where the cables can be spliced together to form the network used. Because the manholes are the only points at which the cable can be serviced without removal, they must be located at each point where two cables are connected together. Much smaller openings (hand holes) are normally installed at points where connections are made between the main system and the wires to the con-

sumer. In addition, it is necessary to have space and access available for transformers and switching devices. In urban areas these units are normally installed in underground vaults adjacent to the building served or in the building basement.

In less-dense areas, underground cable is installed by burying it directly in the ground. Manholes are not normally used in this kind of installation.

Telephone The basic components of a telephone system are telephones, teletypes, and other customer communications equipment; wire system connections to central offices; switching equipment in the central office; and trunk-line systems between central offices. Only the connecting line systems from the individual user equipment to the exchanges and the trunk lines between exchanges are of concern here. Both systems generally consist of cables containing as many as 2700 pairs of conductors in lead or plastic sheathing. The cables are buried directly in the earth or installed in a duct system similar to that described for electric power. The principal differences between telephone and power duct systems are: (1) thermal effects are not a problem in the low-voltage telephone system; so more ducts may be grouped together into one large duct bank, and (2) the cables must be protected from moisture to a greater degree than power cables. This is done by passing dry air through the cable sheath. In order to supply sufficient air at large distances from the central stations, pipelines are run with the cables, or portable supplies (bottled compressed air) are used. Cable connections are made in manholes or in specially designed connection cabinets located above ground.

In order to improve the transmission efficiency of telephone cables, the inductance of the telephone line is generally increased by placing low-resistance coils with magnetic alloy cores (called "loading coils") in each phone line every 2000 to 6000 feet depending on the line application. The large number of loading coils required for each cable is installed by sealing banks of these coils in containers to handle small systems, or placing large banks of coils in cabinets for large applications. Working space must be provided for access to cabinets when they are used.

Since buried telephone cables are permanently installed and are not amenable to rapid large-scale changes, system reliability can be achieved by providing extra conductors, by

using multiple cables, and by providing multiple cable routes. When this added reliability is provided at the consumer's level for critical applications, there are attendant extra charges.

Community Antenna Television (CATV) Community antenna television (CATV) systems are increasing in popularity in areas where standard television broadcast reception is poor; they are also capable of augmenting the amount of programming available to subscribers by originating programs for distribution on available channels for transmission and reception.

Basically, these systems consist of a central receiving antenna and facilities for amplifying and transmitting the signals over a distribution network. This network generally consists of coaxial cables, with amplifiers located at regular intervals. Amplifier spacing is determined by attenuation and aging characteristics of the cable but is normally 1500 to 2000 feet.

Coaxial cables can be installed overhead on existing power or telephone poles, placed in underground conduit, or buried directly in the ground.

Central Heating In some dense urban areas, buildings are heated by steam or hot water transmitted from a central heating plant. The basic differences between steam and hot water distribution systems are: (1) steam systems must be equipped with traps to drain liquid (condensate) from the steam lines, (2) steam lines are larger in diameter than hot water lines of the same energy-carrying capacity and operating temperature, (3) steam systems are frequently equipped with pressure-reducing stations, and (4) steam systems normally employ only one pipe. Condensate is normally discharged to sewers, but it can be returned to the central plant, depending on the economics of the operation and the likelihood of contamination of the condensate. Hot water systems almost always return the water after removal of the useful heat.

Heat exchangers are sometimes used by each customer to prevent contamination of the main circulating water and to allow higher distribution pressures and temperatures to be used than the customer requires. The distribution system is made up of insulated pipes which may be in concrete envelopes buried directly in the ground or installed in utility tunnels. If pipes are installed in conduit systems, they are sloped to drain so that the insulation can be kept dry. Expansion loops or joints are placed every few hundred feet and insulated pipe guides and supports are used to prevent concentrated bearing loads due to thermal expansion of the pipes. Expansion loops are preferred since they are relatively maintenance-free, but bellows, slip-joint, and ball-and-socket joints are used where space is limited. Expansion joints are generally not buried directly in the ground but are located in manholes or vaults. Unless the pipes are installed in tunnels, manholes are provided at frequent intervals to permit access to traps, expansion joints, and valves for maintenance. Drainage from conduit systems or traps is removed by sump pumps or steam ejectors.

2.13 Water Supply

Quantity of Water One of the first steps in the selection of a suitable water supply source is determining the demand that will be placed on it. The essential elements of water demand include the average daily water consumption and the peak rate of demand. The average daily water consumption must be estimated for these purposes:

1. To determine the ability of the water source to meet continuing demands over critical periods when surface flows are low and groundwater tables are at minimum elevations

2. To ascertain the quantities of stored water that will sustain demands during these critical periods

The peak demand rates must be estimated to determine plumbing and pipe sizing, pressure losses, and the storage facilities necessary to supply sufficient water during periods of peak demand.

Average Daily Water Use Many factors influence water use for a given system. For example, the mere fact that water under pressure is available stimulates its use for watering lawns and gardens, for washing automobiles, for operating air-conditioning equipment, and for performing many other utility activities at home and on the farm. Modern kitchen and laundry appliances, such as food waste disposers and automatic dishwashers, contribute to a higher total water use and tend to increase peak demands. Since water requirements will influence all features of an individual development or improvement, they must figure prominently in plan preparation. The table "Planning Guide for Water Use" presents a summary of average water use as a guide in preparing estimates, local adaptations being made where necessary.

Peak Demands The rate of water use for an individual water system will vary directly with activity in the home or with the operational farm program. Rates are generally highest in the home near mealtimes, during midmorning laundry periods, and shortly before bedtime. In the intervening daytime hours and at night, water use may be virtually nil. Thus, the total amount of water used by a household may be distributed over only a few hours of the day, during which the actual use is much greater than the average rate determined from the table "Planning Guide for Water Use."

Simultaneous operation of several plumbing fixtures will determine the maximum peak rate of water delivery for the home water system. For example, a shower, an automatic dishwasher, a lawn sprinkler system, and a flush-valve toilet all operated at the same time would probably produce a near-critical peak. It is true that not all these facilities are usually operated together; but if they exist on the same system, there is always a possibility that a critical combination may result, and for design purposes this method of calculation is sound. Rates of flow are summarized in the table "Rates of Flow for Certain Plumbing, Household, and Farm Fixtures."

Planning Guide for Water Use

Types of establishments	Gallons per day
Airports (per passenger)	3 to 5
Apartments, multiple-family (per resident)	60
Bathhouses (per bather)	10
Camps:	
Construction, semipermanent (per worker)	50
Day with no meals served (per camper)	15
Luxury (per camper)	100 to 150
Resorts, day and night, with limited plumbing (per camper)	50
Tourist with central bath and toilet facilities (per person)	35
Clubs:	
Country (per resident member)	100
Country (per nonresident member present)	25
Cottages with seasonal occupancy (per resident)	50
Courts, tourist, with individual bath units (per person)	50
Dwellings:	
Boardinghouses (per boarder)	50
Additional kitchen requirements for nonresident boarders	10
Luxury (per person)	100 to 150
Multiple-family apartments (per resident)	40
Rooming houses (per resident)	60
Single-family (per resident)	50 to 75
Estates (per resident)	100 to 150
Factories (gallons per person per shift)	15 to 35
Highway rest areas (per person)	5
Hotels with private baths (two persons per room)	60
Hotels without private baths (per person)	50
Institutions other than hospitals (per person)	75 to 125
Hospitals (per bed)	250 to 400
Laundries, self-service (gallons per washing, that is, per customer)	50
Livestock (per animal):	
Cattle (drinking)	12
Dairy (drinking and servicing)	35
Goat (drinking)	2
Hog (drinking)	4
Horse (drinking)	12
Mule (drinking)	12
Sheep (drinking)	2
Steer (drinking)	12
Motels with bath, toilet, and kitchen facilities (per bed space)	50
With bed and toilet (per bed space)	40
Parks:	
Overnight with flush toilets (per camper)	25
Trailers with individual bath units, no sewer connection (per trailer)	25
Trailers with individual baths, connected to sewer (per person)	50
Picnic facilities:	
With bathhouses, showers, and flush toilets (per picnicker)	20
With toilet facilities only (gallons per picnicker)	10
Poultry:	
Chickens (per 100)	5 to 10
Turkeys (per 100)	10 to 18

Types of establishments	Gallons per day
Restaurants with toilet facilities (per patron)	7 to 10
Without toilet facilities (per person)	2½ to 3
With bars and cocktail lounge (additional quantity per patron)	2
Schools:	
Boarding (per pupil)	75 to 100
Day with cafeteria, gymnasiums, and showers (per pupil)	25
Day with cafeteria but no gymnasiums or showers (per pupil)	20
Day without cafeteria, gymnasiums, or showers (per pupil)	15
Service stations (per vehicle)	10
Stores (per toilet room)	400
Swimming pools (per swimmer)	10
Theaters:	
Drive-in (per car space)	5
Movie (per auditorium seat)	5
Workers:	
Construction (per person per shift)	50
Day (school or offices per person per shift)	15

Rates of Flow for Certain Plumbing, Household, and Farm Fixtures

Location	Flow pressure* (pounds per square inch)	Flow rate (gallons per minute)
Ordinary basin faucet	8	2.0
Self-closing basin faucet	8	2.5
Sink faucet, ⅜-inch	8	4.5
Sink faucet, ½-inch	8	4.5
Bathtub faucet	8	6.0
Laundry tub faucet, ½-inch	8	5.0
Shower	8	5.0
Ball cock for closet	8	3.0
Flush valve for closet	15	15 to 40†
Flushometer valve for urinal	15	15.0
Garden hose (50 feet; ¾-inch sill cock)	30	5.0
Garden hose (50 feet; ⅝-inch outlet)	15	3.33
Drinking fountains	15	0.75
Fire hose (1½ inches; ½-inch nozzle)	30	40.0

*Flow pressure is the pressure in the supply near the faucet or water outlet while the faucet or water outlet is wide-open and flowing.
†The wide range is due to variations in the design and type of closet flush valves.

Special Water Considerations

Lawn Sprinkling The amount of water required for lawn sprinkling depends upon the size of the lawn, the type of sprinkling equipment, climate, soil, and water control. In dry or arid areas the amount of water required may equal or exceed the total used for domestic or farmstead needs. For estimating purposes, a rate of approximately ½ inch per hour of surface area is reasonable. This amount of water can be applied by sprinkling 30 gallons of water per hour over each 100 square feet. For example:

$$\tfrac{1000}{100} \times 30 = 300 \text{ gallons per hour, or 5 gallons per minute}$$

A lawn of 1000 square feet would require 300 gallons per hour.

When possible, the water system should have a minimum capacity of 500 to 600 gallons per hour. A water system of this size may be able to operate satisfactorily during a peak demand. Peak flows can be estimated by adding lawn sprinkling to peak domestic flows but not to fire flows.

Fire Protection In areas of individual water supply systems, effective fire fighting depends upon the facilities provided by the property owner. The National Fire Protection Association has prepared a report that outlines and describes ways to utilize available water supplies.

The most important factors in successful fire fighting are early discovery and immediate action. For immediate protection, portable fire extinguishers are desirable. Such first-aid protection is designed only for the control of fires in the early stage; therefore, a water supply is desirable as a second line of defense.

The use of gravity water supplies for fire fighting presents certain basic problems. These include (1) the construction of a dam, farm pond, or storage tank to hold the water until needed and (2) the determination of the size of pipeline installed from the supply. The size of the pipe is dependent upon two factors, the total fall or head from the point of supply to the point of use and the length of pipeline required.

A properly constructed well tapping a good aquifer can be a dependable source for both domestic use and fire protection. If the well is to be relied upon for fire protection without supplemental storage, it should demonstrate, by a pumping test, a minimum capacity of 8 to 10 gallons per minute continuously for a period of 2 hours during the driest time of the year.

An installation is more dependable when the motor, controls, and power lines are protected from fire. A high degree of protection is achieved when all the electrical elements are located outside at the well and a separate power line bypasses other buildings.

Numerous factors determine the amount of fire protection that should be built into a water system. Publications of the National Fire Protection Association provide further information on this subject.

The smallest individual pressure systems commercially available provide about 210 gallons per hour (3½ gallons per minute). While this capacity will furnish a stream, through an ordinary garden hose, of some value in combating incipient fires or in wetting down adjacent buildings, it cannot be expected to be effective on a fire that has gained any headway. When such systems are already installed, connections and hose should be provided. When a new system is being planned or a replacement of equipment made, it is urged that a capacity of at least 500 gallons an hour (8⅓ gallons per minute) be specified and the supply increased to meet this demand. If necessary, storage should be added. The additional cost of the larger unit necessary for fire protection is offset partially by the increased quantities of water available for other uses.

Sanitary Survey The importance of a sanitary survey of water sources cannot be overemphasized. With a new supply, the sanitary survey should be made in conjunction with the collection of initial engineering data covering the development of a given source and its capacity to meet existing and future needs. The survey should include the detection of all health hazards and an assessment of their present and future importance. Persons trained and competent in public-health engineering and the epidemiology of waterborne diseases should conduct the sanitary survey. With an existing supply, the survey should be made as often as the control of health hazards and the maintenance of a good sanitary quality require.

The information furnished by the sanitary survey is essential to a complete interpretation of bacteriological and, frequently, chemical data. This information should always accompany the laboratory findings. The following outlines cover the essential factors to be investigated or considered in a sanitary survey. Not all the items are pertinent to any one supply, and items not on the lists would be important additions in some situations.

Groundwater Supplies

1. Character of local geology; slope of ground surface

2. Nature of soil and underlying porous strata; whether clay, sand, gravel, or rock (especially porous limestone); coarseness of sand or gravel; thickness of water-bearing stratum, depth to water table; location, log, and construction details of local wells

3. Slope of water table, preferably as determined from observational wells or as indicated, presumptively but not certainly, by slope of ground surface

4. Extent of drainage area likely to contribute water to the supply

5. Nature, distance, and direction of local sources of pollution

6. Possibility of surface drainage water's entering the supply and of wells becoming flooded; methods of protection

7. Methods used for protecting the supply against pollution by means of sewage treatment, waste disposal, and the like

8. Well construction (see Figures 2.57, 2.58, and 2.59.
   a. Total depth of well
   b. Casing: diameter, wall thickness, material, and length from surface
   c. Screen or perforations: diameter, material, construction, locations, and lengths
   d. Formation seal: material (cement, sand, bentonite, and so on), depth intervals, annular thickness, and method of placement

9. Protection of well at top: presence of sanitary well seal, casing height above ground, floor, or flood level, protection of well vent, protection of well from erosion and animals

10. Pump house construction (floors, drains, and so on),

capacity of pumps, drawdown when pumps are in operation

11. Availability of an unsafe supply, usable in place of normal supply and hence involving danger to the public health

12. Disinfection: equipment, supervision, test kits, or other types of laboratory control

Surface-Water Supplies

1. Nature of surface geology: character of soils and rocks

2. Character of vegetation, forests, cultivated and irrigated land, including salinity, effect on irrigation water, and so on

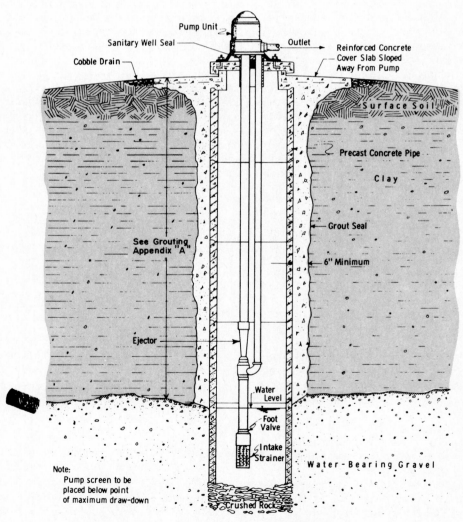

Figure 2.57 A dug well with a two-pipe jet pump installation.

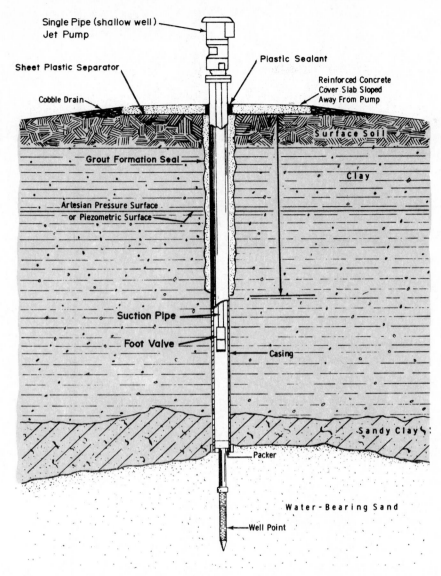

Figure 2.58   A hand-bored well with a driven well point and a "shallow well" jet pump.

3. Population and sewered population per square mile of catchment area

4. Methods of sewage disposal, whether by diversion from watershed or by treatment

5. Character and efficiency of sewage treatment works on watershed

6. Proximity of sources of fecal pollution to intake of water supply

7. Proximity, sources, and character of industrial wastes, oil field brines, acid mine waters, and so on

8. Adequacy of supply as to quantity

9. For lake or reservoir supplies: wind direction and velocity data, drift of pollution, sunshine data (algae)

10. Character and quality of raw water: coliform organisms, algae, turbidity, color, objectionable mineral constituents

11. Nominal period of detention in reservoir or storage basin

12. Probable minimum time required for water to flow from sources of pollution to reservoir and through reservoir intake

13. Shape of reservoir, with reference to possible currents of water, induced by wind or reservoir discharge, from inlet to water supply intake

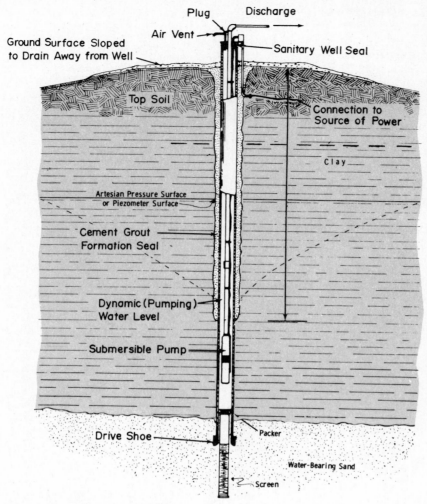

Plug
Discharge
Air Vent
Ground Surface Sloped
to Drain Away from Well
Sanitary Well Seal
Top Soil
Connection to
Source of Power
Clay
Artesian Pressure Surface
or Piezometer Surface
Cement Grout
Formation Seal
Dynamic (Pumping)
Water Level
Submersible Pump
Drive Shoe
Packer
Water-Bearing Sand
Screen

Figure 2.59   A drilled well with a submersible pump.

14. Protective measures in connection with the use of watershed to control fishing, boating, landing of airplanes, swimming, wading, ice cutting, permitting of animals on marginal shore areas and in or upon the water, and so on

15. Efficiency and constancy of policing

16. Treatment of water: kind and adequacy of equipment, duplication of parts, effectiveness of treatment, adequacy of supervision and testing, contact period after disinfection, free chlorine residuals carried

17. Pumping facilities: pump house, pump capacity and standby units, and storage facilities

The selection and use of surface-water sources for individual water supply systems require the consideration of additional factors not usually associated with groundwater sources. When small streams, open ponds, lakes, or open

reservoirs must be used as sources of water supply, the danger of contamination and of the consequent spread of enteric diseases such as typhoid fever and dysentery is increased. As a rule, surface water should be used only when groundwater sources are unavailable or inadequate. Clear water is not always safe, and the old saying that "running water purifies itself" to drinking-water quality within a stated distance is false. The physical and bacteriological contamination of surface water makes it necessary to regard such sources of supply as unsafe for domestic use unless reliable treatment, including filtration and disinfection, is provided.

The treatment of surface water to ensure a constant, safe supply requires diligent attention to operation and maintenance by the owner of the system. When groundwater sources are limited, consideration should be given to their development for domestic purposes only. Surface-water sources can then provide water needed for stock and poultry watering, gardening, fire fighting, and similar purposes. Treat-

ment of surface water used for livestock is not generally considered essential. There is, however, a trend to provide stock and poultry with drinking water free from bacterial contamination and certain chemical elements.

Sources of Surface Water The principal sources of surface water that may be developed include controlled catchments, ponds or lakes, surface streams, and irrigation canals. Except for irrigation canals, where discharges depend on irrigation activity, these sources derive water from direct precipitation over the drainage area. Because of the complexities of the hydrological, geological, and meteorological factors affecting surface-water sources, it is recommended that in planning the development of natural catchment areas of more than a few acres, engineering advice be obtained.

To estimate the yield of the source, it is necessary to consider the following information pertaining to the drainage area:

1. Total annual precipitation
2. Seasonal distribution of precipitation
3. Annual or monthly variations of rainfall from normal levels
4. Annual and monthly evaporation and transpiration rates
5. Soil moisture requirements and infiltration rates
6. Runoff gauge information
7. All available local experience records

Much of the required data, particularly that concerning precipitation, can be obtained from publications of the National Weather Service. Essential data such as soil moisture and evapotranspiration requirements may be obtained from local soil conservation and agricultural agencies or from field tests conducted by hydrologists.

Controlled Catchments In some areas groundwater is almost inaccessible or is so highly mineralized that it is not satisfactory for domestic use. In these cases the use of controlled catchments and cisterns may be necessary. A properly located and constructed controlled catchment and cistern, augmented by a satisfactory filtration unit and adequate disinfection facilities, will provide safe water.

A controlled catchment is a defined surface area from which rainfall runoff is collected. It may be a roof or a paved ground surface. The collected water is stored in a constructed covered tank called a cistern or in a reservoir. Ground surface catchments should be fenced to prevent unauthorized entrance by people or animals. There should be no possibility of the mixture of undesirable surface drainage and controlled runoff. An intercepting drainage ditch around the upper edge of the area and a raised curb around the surface will prevent the entry of any undesirable surface drainage.

For these controlled catchments, simple guidelines to determine water yield from rainfall totals can be established. When the controlled catchment area has a smooth surface or is paved and the runoff is collected in a cistern, water loss due to evaporation, replacement of soil moisture deficit, and

infiltration is small. As a general rule, losses from ground catchments covered with smooth concrete or asphalt average less than 10 percent; for shingled roofs or tar and gravel surfaces losses should not exceed 15 percent, and for sheet-metal roofs the loss is negligible.

A conservative design can be based on the assumption that the amount of water recoverable for use is three-fourths of the total annual rainfall. See Figure 2.60.

Location A controlled catchment may be suitably located on a hillside near the edge of a natural bench. The catchment area can be placed on a moderate slope above the receiving cistern.

The location of the cistern should be governed by both convenience and quality protection. A cistern should be as close to the point of ultimate use as is practicable. It should not be placed closer than 50 feet from any part of a sewage disposal installation and should be on higher ground.

A cistern collecting water from a roof surface should be located adjacent to the building but not in a basement subject to flooding. It may be placed below the surface of the ground to keep the water from freezing in cold climates and to keep water temperatures low in warm climates, but it should be situated on the highest ground practicable, with the surrounding area graded to provide good drainage.

Size The size of a cistern will depend on the size of the family and the length of time between periods of heavy rainfall. Daily water requirements can be estimated from the table "Planning Guide for Water Use." The size of the catchment or roof will depend on the amount of rainfall and the character of the surface. It is desirable to allow a safety factor for lower-than-normal rainfall levels. Designing for two-thirds of the mean annual rainfall usually will result in a catchment area of adequate capacity.

The following example illustrates the procedure for determining the size of the cistern and required catchment area. Let us assume that the minimum drinking and culinary requirements of a family of four persons are 100 gallons per day* (4 persons × 25 gallons per day = 100 gallons) and that the effective period† between rainy periods is 150 days. The minimum volume of the cistern required will be 15,000 gallons (100 × 150). This volume could be held by a cistern 10 feet deep and 15 feet square. If the mean annual rainfall is 50 inches, then the total design rainfall is 33 inches (50 × ⅔). In Figure 2.60 the catchment area required to produce 36,500 gallons (365 days × 100 gallons per day), the total annual requirement, is 2400 square feet.

Construction Cisterns should be of watertight construction with smooth interior surfaces. Manhole or other covers should be tight to prevent the entrance of light, dust, surface water, insects, and animals.

A manhole opening should have a watertight curb with

---

*Twenty-five gallons per person per day if it is assumed that other uses are supplied by water of poorer quality.

†The effective period is the number of days between periods of rainfall during which there is negligible precipitation.

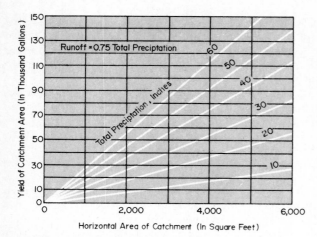

Figure 2.60 Yield of impervious catchment area.

Underground cisterns can be built of brick or stone, although reinforced concrete is preferable. If brick or stone is used, it must be low in permeability and laid with full portland cement mortar joints. Brick should be wet before laying. High-quality workmanship is required, and the use of unskilled labor for laying brick or stone is not advisable. Two ½-inch plaster coats of 1:3 portland cement mortar on the interior surface will aid in providing waterproofing. A hard, impervious surface can be made by troweling the final coat before it is fully hardened.

Figures 2.61 and 2.62 show a suggested design for a cistern of reinforced concrete. A dense concrete should be used to obtain watertightness and should be vibrated adequately during construction to eliminate honeycomb. All masonry cisterns should be allowed to wet-cure properly before being used.

The cistern should be disinfected with chlorine solutions. Initial and periodic water samples should be taken to determine the bacteriological quality of the water supply. Chlorination may be required on a continuing basis if the bacteriological results indicate that the quality is unsatisfactory.

Ponds or Lakes   A pond or a lake should be considered as a source of water supply only after groundwater sources and controlled catchment systems are found to be inadequate or unacceptable. The development of a pond as a supply source depends on several factors: (1) the selection of a watershed that permits only water of the highest quality to enter the pond, (2) usage of the best water collected in the pond, (3) filtration of the water to remove turbidity and reduce bacteria, (4) disinfection of filtered water, (5) proper storage of the treated water, and (6) proper maintenance of

edges projecting a minimum of 4 inches above the level of the surrounding surface. The edges of the manhole cover should overlap the curb and project downward a minimum of 2 inches. The cover should be provided with a lock to minimize the danger of contamination and accidents.

Provision can be made for diverting initial runoff from paved surfaces or rooftops before the runoff is allowed to enter the cistern (see Figure 2.61). Inlet, outlet, and waste pipes should be effectively screened. Cistern drains and waste or sewer lines should not be connected.

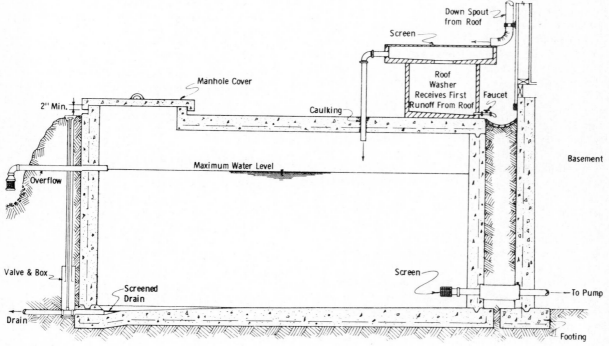

Figure 2.61   Cistern.

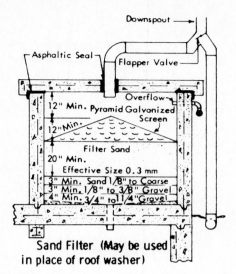

Figure 2.62 Sand filter for a cistern.

the entire water system. Local authorities may be able to furnish advice on pond development.

The value of a pond or lake development is its ability to store water during wet periods for use during periods of little or no rainfall. A pond should be capable of storing a minimum of a year's supply of water. It must be of sufficient capacity to meet water supply demands during periods of low rainfall with an additional allowance for seepage and evaporation losses. The drainage area (watershed) should be large enough to catch sufficient water to fill the pond or lake during wet seasons of the year.

Careful consideration of the location of the watershed and pond site reduces the possibility of chance contamination. The watershed should:

1. Be clean, preferably grassed
2. Be free from barns, septic tanks, privies, and soil absorption fields
3. Be effectively protected against erosion and drainage from livestock areas
4. Be fenced to exclude livestock

The pond should:

1. Be not less than 8 feet deep at the deepest point
2. Be designed to have the maximum possible water storage area more than 3 feet in depth
3. Be large enough to store at least a year's supply
4. Be fenced to keep out livestock
5. Be kept free of weeds, algae, and floating debris

In many instances pond development requires the construction of an embankment with an overflow or spillway. Assistance in designing a storage pond may be available from

federal, state, or local health agencies, from the U.S. Soil Conservation Service, and in publications of state or county agricultural, geological, or soil conservation departments. For specific conditions, engineering or geological advice may be needed.

Intake A pond intake must be properly located so that it may draw water of the highest possible quality. When the intake is placed too close to the pond bottom, it may draw turbid water or water containing decayed organic material. When it is placed too near the pond surface, the intake system may draw floating debris, algae, and aquatic plants. The depth at which it operates best varies with the season of the year and the layout of the pond. The most desirable water is usually obtained when the intake is located between 12 and 18 inches below the water surface. An intake located at the deepest point of the pond makes maximum use of stored water.

Pond intakes should be of the type illustrated in Figure 2.63. This is known as a floating intake. It consists of a flexible pipe attached to a rigid conduit that passes through the pond embankment.

In accordance with applicable specifications, gate valves should be installed on the main line below the dam and on any branch line to facilitate control of the rate of discharge.

Treatment The pond water treatment facility consists of four general parts. See Figure 2.64.

1. *Settling basin.* The first unit is a settling basin. The purpose of the basin is to allow the large particles of turbidity to settle. This may be accomplished adequately in the pond. When this settling is not completely effective, a properly designed settling basin with provision for coagulation may be needed. The turbid water is mixed with a suitable chemical such as alum. Alum and other chemical aids speed the settling rate of suspended materials present in the water. This initial process helps to reduce the turbidity of the water to be passed through the filter. The addition of alum will lower the pH, which may have to be readjusted with lime if the distribution piping becomes corroded.

2. *Filtration unit.* After settling, the water moves to a second compartment, where it passes through a filter bed of sand and gravel. The suspended particles that have not been removed by settlement or flocculation are now removed.

3. *Clear water storage.* After the water leaves the filter, it drains into a clear well, cistern, or storage tank.

4. *Disinfection.* After the water has settled and has been filtered, it must be disinfected. Proper disinfection is the most important part of pond water treatment. The continuous operation and high-quality performance of the equipment are very important. When the water is chlorinated, livestock unaccustomed to chlorinated water may refuse to drink the water for several days, but they usually become accustomed to it within a short period of time.

Bacteriological Examination After the treatment and disinfection equipment have been checked and are operating

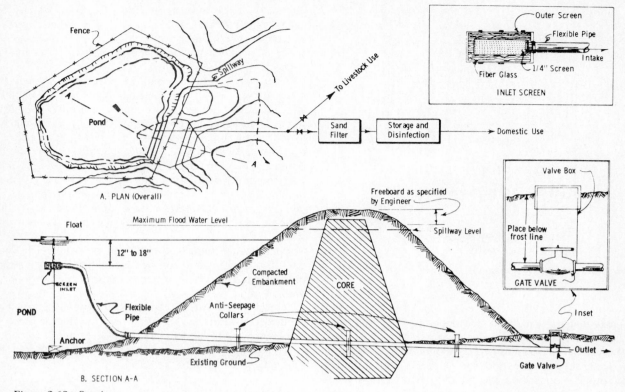

Figure 2.63 Pond.

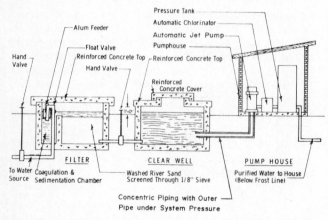

Figure 2.64 Schematic diagram of a pond water treatment system.

satisfactorily, a bacteriological examination of a water sample should be made. Before a sample is collected, the examining laboratory should be asked to furnish recommendations. These recommendations should include the type of container to be used and the method and precautions to take during collection, handling, and mailing. Water should not be used for drinking and culinary purposes until the results of the bacteriological examination show it to be safe.

The frequency of subsequent bacteriological examinations should be based on any breakdown or changes made in the sanitary construction or protective measures associated with the supply. A daily determination and record of the chlorine residual is recommended to ensure that proper disinfection is accomplished.

Plant Maintenance The treatment facility should be inspected daily. The disinfection equipment should be checked to make sure that it is operating satisfactorily. When chlorine disinfection is practiced, the chlorinator and the supply of chlorine solution should be checked. The water supply should be checked daily for its chlorine residual. The water may become turbid after heavy rains, and the quality may change. Increases in the amount of chlorine and coagulates used will then be required. The performance of the filter should be watched closely. When the water becomes turbid or the available quantity of water decreases, the filter should be cleaned or backwashed.

Protection from Freezing Protection from freezing must be provided unless the plant is not operated and is drained during freezing weather. In general, the filter and pump room should be located in a building that can be heated in winter. If the topography is suitable, the need for heat can be eliminated by placement of the pump room and filter under-

ground on a hillside. Gravity drainage from the pump room must be possible to prevent flooding. No matter what the arrangement, the filter and pump room must be easily accessible for maintenance and operation.

Tastes and Odors Surface water frequently develops musty or undesirable tastes and odors. These are generally caused by the presence of microscopic plants called algae. There are many kinds of algae. Some occur in long, threadlike filaments that are visible as large green masses of scum; others may be separately free-floating and entirely invisible to the unaided eye. Some varieties may grow in great quantities in the early spring, others in summer, and still others in the fall. Tastes and odors generally result from the decay of dead algae. This decay occurs naturally as plants pass through their life cycle.

Tastes and odors in water can usually be satisfactorily removed by passing the previously filtered and chlorinated surface water through activated carbon filters. These filters may be helpful in improving the taste of small quantities of previously treated water used for drinking or culinary purposes. They also absorb excess chlorine. Carbon filters are commercially available and require periodic servicing.

Carbon filters should not be expected to be a substitute for sand filtration and disinfection. Their area is insufficient to handle raw surface water, and they clog very rapidly when filtering turbid water.

Weed Control The growth of weeds around a pond should be controlled by cutting or pulling. Before weed killers are used, the advice of the local health department should be obtained, since herbicides often contain compounds that are highly toxic to humans and animals. Algae in the pond, particularly the blue-green types that produce scum and objectionable odors and that, in unusual instances, may harm livestock, should be controlled.

Streams Streams receiving runoff from large uncontrolled watersheds may be the only source of water supply. The physical and bacteriological quality of surface water varies and may impose unusually or abnormally high loads on the treatment facilities.

Stream intakes should be located upstream from sewer outlets or other sources of contamination. The water should be pumped when the silt load is low. A low-water stage usually means that the temperature of the water is higher than normal and that the water is of poor chemical quality. Maximum silt loads, however, occur during maximum runoff. High-water stages shortly after storms are usually the most favorable for diverting or pumping water to storage. These conditions vary, and the best time should be determined for the particular stream.

Irrigation Canals If properly treated, irrigation water may be used as a source of domestic water supply. Water obtained from irrigation canals should be treated like water from any other surface-water source.

When return irrigation (tail water) is practiced, the water may contain large concentrations of undesirable chemicals, including pesticides, herbicides, and fertilizers. Whenever water from return irrigation is used for domestic purposes, a

periodic chemical analysis should be made. Because of the poor quality of this water, it should be used only if no other water source is available.

2.14 Types of Wells

Wells may be classified with respect to construction methods as dug, bored, driven, drilled, and jetted. Drilled wells may be drilled by either the rotary or the percussion method. Each type of well has distinguishing physical characteristics and is best adapted to meet particular water development requirements as shown in the table on pages 142–143. The following factors should be considered when choosing the type of well to be constructed in a given situation.

1. Characteristics of the subsurface strata to be penetrated and their influence upon the method of construction
2. Hydrology of the specific situation and hydraulic properties of the aquifer; seasonal fluctuations of water levels
3. Degree of sanitary protection desired, particularly as this is affected by well depth
4. Cost of construction work and materials

Location from Sources of Pollution

Protective Horizontal Distances Minimum horizontal protective distances between the well and common sources of pollution should not be less than those specified in the accompanying table. For wells terminating in creviced formations, or where the overlaying soil formation is highly permeable, greater distances may be required by the FHA.

Protective Depths The well should be watertight to the depth necessary to seal off water-bearing formations that are or may be polluted or have undesirable characteristics.

Minimum Protective Distances for Wells*

Distance from	Type of well			
	Dug†	Bored	Driven	Drilled
Property line	100 ft	50 ft	50 ft	20 ft
Improperly abandoned well or sinkhole		Unacceptable		200 ft
Seepage pit	200 ft	200 ft	200 ft	100 ft
Disposal field or bed	200 ft	200 ft	200 ft	100 ft
Industrial lagoon		Unacceptable		‡
Watertight sewer lines	50 ft	50 ft	50 ft	10 ft
Other sewer lines	100 ft	100 ft	100 ft	50 ft

*Any distances specified above should be increased if necessary to meet the minimum requirements of the health authority.
†Radial water collectors, springs, and infiltration galleries should comply with the minimum distances specified for dug wells.
‡The minimum safe distances from a lagoon should be specified depending upon the nature of the water, geological surface and subsurface conditions, and the recommendations of the FHA and health authority.

Water Well Construction Methods and Applications

Method	Materials for which the well is best suited	Water table depth for which best suited (feet)	Usual maximum depth (feet)	Usual diameter range	Usual casing material	Customary use	Yield* (gallons per minute)	Remarks
Driven wells: hand, air hammer	Silt, sand, gravel less than 2 inches	5 to 15	50	1¼ to 4 inches	Standard-weight pipe	Domestic, drainage	3 to 40	Limited to shallow water table; no large gravel
Jetted wells: light portable rig	Silt, sand, gravel less than 1 inch	5 to 15	50	1½ to 3 inches	Standard-weight pipe	Domestic, drainage	3 to 30	Limited to shallow water table; no large gravel
Drilled wells: cable tools	Unconsolidated and consolidated medium-hard and hard rock	Any depth	1500†	3 to 24 inches	Steel or wrought-iron pipe	All uses	3 to 3000	Effective for water exploration; requires casing in loose materials; mud scow and hollow rod bits developed for drilling unconsolidated fine to medium sediments
Hydraulic rotary	Silt, sand, gravel less than 1 inch; soft to hard consolidated rock	Any depth	1500†	3 to 18 inches	Steel or wrought-iron pipe	All uses	3 to 3000	Fastest method for all except hardest rock; casing usually not required during drilling; effective for gravel envelope wells
Reverse hydraulic rotary	Silt, sand, gravel, cobble	5 to 100	200	16 to 48 inches	Steel or wrought-iron pipe	Irrigation, industrial, municipal	500 to 4000	Effective for large-diameter holes in unconsolidated and partially consolidated deposits; requires large volume of water for drilling; effective for gravel envelope wells

Method	Material	Usual depth	Maximum depth	Diameter	Casing material	Use	Yield*	Remarks
Air rotary	Silt, sand, gravel less than 2 inches, soft to hard consolidated rock	Any depth	2000†	12 to 20 inches	Steel or wrought-iron pipe	Irrigation, industrial, municipal	500 to 3000	Now used in oil exploration; very fast drilling; combines rotary and percussion methods (air drilling); cuttings removed by air; would be economical for deep-water wells
Augering: hand auger	Clay, silt, sand, gravel less than 1 inch	5 to 30	35	2 to 8 inches	Sheet metal	Domestic, drainage	3 to 50	Most effective for penetrating and removing clay; limited by gravel over 1 inch; casing required if material is loose
Power auger	Clay, silt, sand, gravel less than 2 inches	5 to 50	75	6 to 36 inches	Concrete, steel, or wrought-iron pipe	Domestic, irrigation, drainage	3 to 100	Limited by gravel over 2 inches; otherwise the same as for hand auger

*Yield influenced primarily by geology and availability of groundwater.
†Greater depths reached with heavier equipment.

2.15 Types of Springs

The various types of springs and their characteristics are shown in the accompanying table.

Types of Springs and Their Characteristic Features

Characteristic	Gravity				Artesian	
	Depression springs	Contact springs	Fracture and tubular springs	Aquifer outcrop springs		Fault springs
Location	Along outcrop of the water table at the edges or in bottom of valleys, basins, and depressions in moraines filled with alluvium (stream deposits of gravel, sand, silt, and clay) and in valleys cut in massive, permeable sandstone	Possibly present on hillsides or in valleys wherever the outcrop of an impermeable layer beneath a water-bearing permeable layer occurs	On hillsides or in valleys or wherever land surface is below the water table	Possibly present in any topographic position along outcrop of aquifer		Possibly present at any location along a fault or related fractures
Type of opening and water-bearing material	Irregular spaces between grains of the permeable material	Openings in sand or gravel irregular, intergranular spaces; openings in rocks joints or fractures; openings possibly tubular in limestone, gypsum, and basalt	Fractures in all kinds of rocks and sometimes tubular openings in limestone, gypsum, and lava Water-bearing material; fractured or jointed rocks	Depending upon nature of water-bearing material; if aquifer is sandstone, water possibly seeping from spaces between grains or from joints or tubular openings Water-bearing material; sandstone, limestone, or jointed basalt	Depending upon the nature of materials at the land surface; if surface is alluvium, water issuing from spaces between grains; if the surface is rock, water issuing from fractures Water-bearing material; possibly sandstone, limestone, or basalt; surface material may not indicate nature of the aquifer	
Yield	Depending upon permeability of water-bearing material and size of tributary area; flow possibly ranging from less than 1 to several gallons per minute	Volume of flow possibly ranging from less than 1 gallon per minute to several thousand gallons per minute, depending upon permeability of the water-bearing material, the volume of aquifer tributary to the spring, and conditions of water intake	Flow possibly ranging from less than 1 to hundreds of gallons per minute, depending upon the extent of fracturing or joint system tributary to the opening	Flow possibly ranging from a few to several thousand gallons per minute		Volume of flow possibly ranging from a few to several thousand gallons per minute

Type of flow	May be either perennial or intermittent, depending upon rise or fall of the water table; if the contributing area is small, the flow will depend on local precipitation	Usually perennial for contact springs supplied by the area water table; if the contact spring is supplied by a perched water table, the flow possibly intermittent	Usually perennial; possibly fluctuating with precipitation if the contributing area is small	Perennial, usually constant; quickly affected by wells drawing from same aquifer; possibly affected by long droughts	Perennial, constant, and only affected by long periods of drought; quickly affected by pumping from wells drawing upon the source aquifer
Quality of water	Usually fair to excellent but may be mineralized if the aquifer contains soluble substances	Usually fair to excellent but may be mineralized if water-bearing material is soluble	Usually good to excellent; possibly hard, because of calcium carbonate if spring issues from or percolates through limestone	Usually good to excellent; water possibly hard if aquifer is limestone	Usually good to excellent; water possibly hard if aquifer is limestone
Features produced	Usually none in valleys; in windswept arid and semiarid basins the wetted area and the vegetation growing around the spring may cause deposition of material resulting in a mound of loess and organic matter	Travertine (calcium carbonate) deposited as described under "Fracture and tubular springs"	Travertine deposited about the spring opening if warmer than the mean annual temperature and has percolated through limestone on its way to the point of discharge; water from other materials usually producing no surface features		

If water is from a limestone aquifer and is warmer than the mean annual temperature, travertine will be deposited; the reduction of pressure in an artesian aquifer that occurs as the water reaches the surface also causes deposition of dissolved solids about the spring opening

2.16 Sanitary Landfill

Soils and Geology A study of the soils and geologic conditions of any area in which a sanitary landfill may be located is essential to understanding how its construction might affect the environment. The study should outline the limitations that soils and geologic conditions impose on safe, efficient design and operation.

A comprehensive study identifies and describes the soils present, their variation, and their distribution. It describes the physical and chemical properties of bedrock, particularly as it may relate to the movement of water and gas (Figure 2.65). Permeability and workability are essential elements of the soil evaluation, as are stratigraphy and structure of the bedrock.

Rock materials are generally classified as sedimentary, igneous, or metamorphic. Sedimentary rocks are formed from the products of erosion of older rocks and from the deposits of organic matter and chemical precipitates. Igneous rocks derive from the molten mass in the depths of the earth. Metamorphic rocks are derived from both igneous and sedimentary rocks that have been altered chemically or physically by intense heat or pressure.

Sands, gravels, and clays are sedimentary in origin. The sedimentary rocks, sometimes called aqueous rocks, are often very permeable and therefore represent a great potential for the flow of groundwater. If leachate develops and enters the rock strata, contaminant travel will usually be greatest in sedimentary formations. Other rocks commonly classed as sedimentary are limestone, sandstone, and conglomerates. Fracturing and jointing of sedimentary formations are common, and they increase permeability. In fact, the most productive water-bearing strata for wells are formations of porous sandstone, highly fractured limestone, and sand and gravel deposits. Siltstones and shales, which are also of sedimentary origin, usually have a very low permeability unless they have been subjected to jointing and form a series of connected open fractures.

Igneous and metamorphic rocks, such as shist, gneiss, quartzite, obsidian, marble, and granite, generally have a very low permeability. If these rocks are fractured and jointed, however, they can serve as aquifers of limited productivity. Leachate movement through them should not, therefore, be categorically discounted.

Information concerning the geology of a proposed site may be obtained from the U.S. Geological Survey, the U.S. Army Corps of Engineers, state geological and soil agencies, university departments of soil sciences and geology, and consulting soil engineers and geologists.

Soil Cover The striking visual difference between a dump and a sanitary landfill is the use of soil cover at the latter. Its compacted solid waste is fully enclosed within a compacted earth layer at the end of each operating day, or more often if necessary.

The cover material is intended to perform many functions at a sanitary landfill (see the accompanying table); ideally, the soil available at the site should be capable of performing all of them.

The cover material controls the ingress and egress of flies, discourages the entrance of rodents seeking food, and prevents scavenging birds from feeding on the waste. Tests have demonstrated that 6 inches of compacted sandy loam will prevent fly emergence. Daily or more frequent application of soil cover greatly reduces the attraction of birds to the waste and also discourages rodents from burrowing to get food. The cover material is essential for maintaining a proper appearance of the sanitary landfill.

Many soils, when suitably compacted, have a low permeability, will not shrink, and can be used to control moisture that might otherwise enter the solid waste and produce leachate.

Control of gas movement is also an essential function of the cover material. Depending on anticipated use of the completed landfill and the surrounding land, landfill gases can be either blocked by or vented through the cover material. A permeable soil that does not retain much water can serve as a good gas vent. Clean sand, well-graded gravel, or crushed stone are excellent when kept dry. If gases are to be prevented from venting through the cover material, a gas-impermeable soil with high moisture-holding capacity compacted at optimum conditions should be used.

Enclosing the solid waste within a compacted earth shell offers some protection against the spread of fire. Almost all soils are noncombustible; thus the earth sidewalls and floor help to confine a fire within the cell. Top cover over a burning cell offers less protection because it becomes undermined and caves in, thus exposing the overhead cell to the fire. The use of a compactible soil of low permeability is an excellent fire-control measure, because it minimizes the flow of oxygen into the fill.

To maintain a clean and sightly operation, blowing litter must be controlled. Almost any workable soil satisfies this requirement, but fine sands and silts without sufficient binder and moisture content may create a dust problem.

The soil cover often serves as a road bed for collection vehicles moving to and from the operating area of the fill. When it is, it should be trafficable under all weather conditions. In wet weather, most clay soils are soft and slippery.

In general, soil used to cover the final lift should be capable of growing vegetation. It should therefore contain adequate nutrients and have a large moisture-storage capacity. A minimum compacted thickness of 2 feet is recommended.

Comparison of the soil characteristics needed to fulfill all of these functions indicates that some anomalies exist. To serve as a road base, the soil should be well drained so that loaded collection vehicles do not bog down. On the other hand, it should have a low permeability if water is to be kept out of the fill, fire is to be kept from spreading, and gas is not to be vented through the final cover. These differences can be solved by placing a suitable road base on top of the normally low-permeability-type cover material. A reverse situation occurs when landfill gases are to be vented uniformly through the cover material. The soil should then be gas-permeable, have a small moisture-storage capacity, and not be highly compacted. As before, the criteria for moisture and

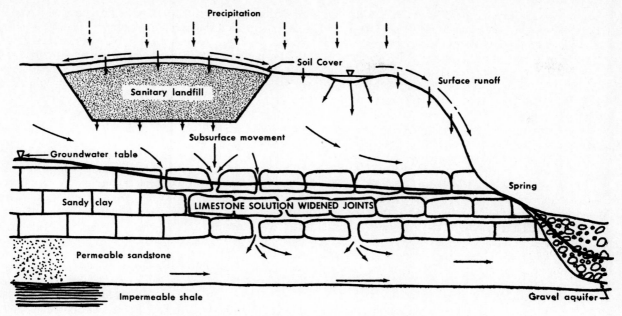

Figure 2.65 Leachate and infiltration movements are affected by the characteristics of the soil and bedrock.

Suitability of General Soil Types as Cover Material*

Function	Clean gravel	Clayey-silty gravel	Clean sand	Clayey-silty sand	Silt	Clay
Prevent rodents from burrowing or tunneling	G	F-G	G	P	P	P
Keep flies from emerging	P	F	P	G	G	E†
Minimize moisture entering fill	P	F-G	P	G-E	G-E	E†
Minimize landfill gas venting through cover	P	F-G	P	G-E	G-E	E†
Provide pleasing appearance and control blowing paper	E	E	E	E	E	E
Grow vegetation	P	G	P-F	E	G-E	F-G
Be permeable for venting decomposition gas‡	E	P	G	P	P	P

*E, excellent; G, good; F, fair; P, poor.
†Except when cracks extend through the entire cover.
‡Only if well drained.

fire control require the soil to have a low permeability. Leachate collection and treatment facilities may be required if a highly permeable soil is used to vent gas uniformly through the cover materials; if this is not done, an alternative means of venting gas through the cover material must be sought.

There are many soils capable of fulfilling the functions of cover material. Minor differences in soil grain size or clay mineralogy can make significant differences in the behavior of soils that fall within a given soil group or division. In addition, different methods of placing and compacting the same soil can result in a significantly different behavior. Moisture content during placement, for example, is a critical factor—it influences the soil's density, strength, and porosity.

The soils present at proposed sites should be sampled by augering, coring, or excavating, and then be classified. The volume of suitable soil available for use as cover material can then be estimated and the depth of excavation for waste disposal can be determined. Specific information on the top 5 feet of the soil mantle can often be obtained from the Soil Conservation Service, U.S. Department of Agriculture.

Sanitary landfilling is a carefully engineered process of solid waste disposal that involves appreciable excavating,

hauling, spreading, and compacting of earth. When manipulating soils in this manner, the Unified Soil Classification System (USCS) is useful. Although recommendations for soil to be used at a landfill are often expressed in the U.S. Department of Agriculture textural classification system as seen in Figure 2.66, the USCS is preferred because it relates in more detail the workability of soils from an engineering viewpoint. (See Figure 2.67.)

Clay soils are very fine in texture even though they commonly contain small to moderate amounts of silt and sand. They vary greatly in their physical properties, which depend not only on the small particle size but on the type of clay minerals and soil water content. When dry, a clay soil can be almost as hard and tough as rock and can support heavy loads. When wet, it often becomes very soft, is sticky or slippery, and is very difficult to handle. A clay soil swells when it becomes wet, and its permeability is very low.

Many clay soils can absorb large amounts of water but, after drying, usually shrink and crack. These characteristics make many clays less desirable than other soils for use as a cover material. The large cracks that usually develop allow water to enter the fill and permit decomposition gases to escape. Rats and insects can also enter or leave the fill through these apertures.

Clay soil can, however, be used for special purposes at a landfill. If it is desirable to construct an impermeable lining or cover to control leachate and gas movement, many clays can be densely compacted at optimum moisture. Once they are in place, it is almost always necessary to keep them moist so they do not crack.

The suitability of coarse-grained material (gravel and sand) for cover material depends mostly on grain size distribution (gradation), the shape of grains, and the amount of clay and silt fines present. If gravel, for example, is poorly graded and relatively free of fines, it is not suitable as cover material for moisture, gas, or fly control. It cannot be compacted enough, and the gravel layer will be porous and highly permeable; this would allow water to enter the fill easily. Flies would have little difficulty emerging through the loose particles. On the other hand, a gravel layer no more than 6 inches deep would probably discourage rats and other rodents from burrowing into the fill and would provide good litter control. If gravel is fairly well graded and contains 10 to 15 percent sand and 5 percent or more fines, it can make an excellent cover. When compacted, the coarse particles maintain grain-to-grain contact, because they are held in place by the binding action of the sand and fines and cohesion of the clays. The presence of fines greatly decreases a soil's permeability. A well-graded, sandy, clayey gravel does not develop shrinkage cracks. It can control flies and rodents, provide odor control, can be worked in any weather, and supply excellent traction for collection trucks and other vehicles.

Many soils classified as sand (grain size generally in the range of 4.0 to 0.05 millimeters) contain small amounts of silt and clay and often some gravel-size material as well. A well-graded sand that contains less than 3 percent fines usually has good compaction characteristics. A small increase in fines,

particularly silt, usually improves density and allows even better compaction. A poorly graded sand is difficult to compact unless it contains abundant fines. The permeability of clean sand soils is always high, even when compacted, and they are not, therefore, suitable for controlling the infiltration of water. They are also ineffective in constraining flies and gases.

A well-drained sandy soil can be easily worked even if temperatures fall below freezing, while a soil with a large moisture-storage capacity will freeze.

Practically the only soils that can be ruled out for use as cover material are peat and highly organic soils. Peat is an earthy soil (usually brown to black) and is composed largely of partially decomposed plant matter. It usually contains a high amount of voids, and its water content may range from 100 to 400 percent of the weight of dried solids. Peat is virtually impossible to compact, whether wet or dry. Peat deposits are scattered throughout the country but are most abundant in the states bordering the Great Lakes. Highly organic soils include sands, silts, and clays that contain at least 20 percent organic matter. They are usually very dark, have an earthy odor when freshly turned, and often contain fragments of decomposing vegetable matter. They are very difficult to compact, are normally very sticky, and can vary extremely in their moisture content.

Many soils contain stones and boulders of varying sizes, especially those in glaciated areas. The use of soils with boulders that hinder compaction should be avoided.

Soil surveys prepared by the Soil Conservation Service of the U.S. Department of Agriculture are available for a major portion of the country. Local assistance in using and interpreting them is available through soil conservation districts located in some 3000 county seats throughout the United States. The surveys cover such specific factors as natural drainage, hazards of flooding, permeability, slope, workability, depth to rock, and stoniness. They are commonly used to locate potential areas for sanitary landfills. They also can serve as the basis for designing effective water management systems and selecting suitable plant cover to control runoff and erosion during and after completion of fill operations. Sanitary landfill owners and their consultants can avoid costly investigations of unsuitable sites by using soil surveys to select areas for which detailed investigations appear warranted. Using soil surveys for the foregoing purposes does not, however, eliminate the need for making detailed site investigations.

Land Forms A sanitary landfill can be constructed on virtually any terrain, but some land forms require that extensive site improvements be made and expensive operational techniques followed. Flat or gently rolling land not subject to flooding is best, but this type is also highly desirable for farming and industrial parks, and this drives up the purchase price.

Depressions, such as canyons and ravines, are more efficient than flat areas from a land use standpoint since they can hold more solid waste per acre. Cover material may, however, have to be hauled in from surrounding areas. Depressions usually result when surface waters run off and erode

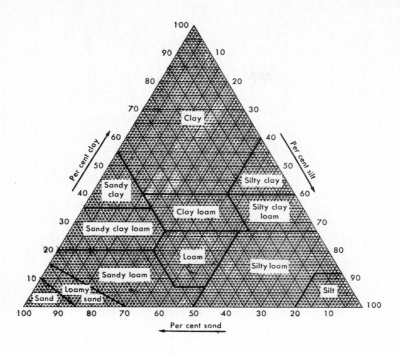

Sand—2.0 to 0.05 mm. diameter
Silt—0.05 to 0.002 mm. diameter
Clay—smaller than 0.002 mm. diameter

COMPARISON OF PARTICLE SIZE SCALES

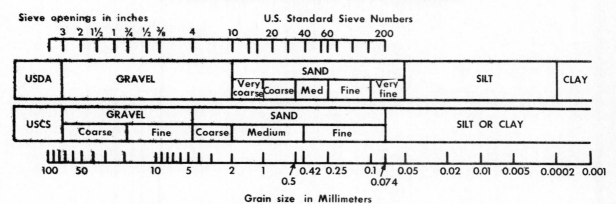

Figure 2.66 Textural classification chart (U.S. Department of Agriculture) and comparison of particle size scales.

UNIFIED SOIL CLASSIFICATION SYSTEM AND CHARACTERISTICS PERTINENT TO SANITARY LANDFILLS

Major Divisions	Symbol — Letter	Symbol — Color	Name	Potential Frost Action	Drainage Characteristics*	Value for Embankments	Permeability cm per sec	Compaction Characteristics	Std AASHO Max Unit Dry Weight lb per cu ft†	Requirements for Seepage Control‡
COARSE-GRAINED SOILS — GRAVEL AND GRAVELLY SOILS	GW	RED	Well-graded gravels or gravel-sand mixtures, little or no fines	None to very slight	Excellent	Very stable, pervious shells of dikes and dams	$k > 10^{-2}$	Good, tractor, rubber-tired steel-wheeled roller	125-135	Positive cutoff
	GP	RED	Poorly graded gravels or gravel-sand mixtures, little or no fines	None to very slight	Excellent	Reasonably stable, pervious shells of dikes and dams	$k > 10^{-2}$	Good,-tractor, rubber-tired steel-wheeled roller	115-125	Positive cutoff
	GM	YELLOW	Silty gravels, gravel-sand-silt mixtures	Slight to medium	Fair to poor	Reasonably stable, not particularly suited to shells, but may be used for impervious cores or blankets	$k = 10^{-3}$ to 10^{-6}	Good, with close control, rubber-tired, sheepsfoot roller	120 -135	Toe trench to none
	GC	YELLOW	Clayey gravels, gravel-sand-clay mixtures	Slight to medium	Poor to practically impervious	Fairly stable, may be used for impervious core	$k = 10^{-6}$ to 10^{-8}	Fair, rubber-tired, sheepsfoot roller	115-130	None
COARSE-GRAINED SOILS — SAND AND SANDY SOILS	SW	RED	Well-graded sands or gravelly sands little or no fines	None to very slight	Excellent	Very stable, pervious sections slope protection required	$k > 10^{-3}$	Good, tractor	110-130	Upstream blanket and toe drainage or wells
	SP	RED	Poorly graded sands or gravelly sands, little or no fines	None to very slight	Excellent	Reasonably stable. may be used in dike section with flat slopes	$k > 10^{-3}$	Good, tractor	100-120	Upstream blanket and toe drainage or wells
	SM	YELLOW	Silty sands, sand-silt mixtures	Slight to high	Fair to poor	Fairly stable, not particularly suited to shells, but may be used for impervious cores or dikes	$k = 10^{-3}$ to 10^{-6}	Good, with close control, rubber-tired, sheepsfoot roller	110-125	Upstream blanket and toe drainage or wells
	SC	YELLOW	Clayey sands, sand-clay mixtures	Slight to high	Poor to practically impervious	Fairly stable, use for impervious core for flood control structures	$k = 10^{-6}$ to 10^{-8}	Fair, sheepsfoot roller, rubber-tired	105-125	None
FINE-GRAINED SOILS — SILTS AND CLAYS LL IS LESS THAN 50	ML	GREEN	Inorganic silts and very fine sands rock flour, silty or clayey fine sands or clayey silts with slight plasticity	Medium to very high	Fair to poor	Poor stability, may be used for embankments with proper control	$k = 10^{-3}$ to 10^{-6}	Good to poor, close control essential, rubber-tired roller, sheepsfoot roller	95-120	Toe trench to none
	CL	GREEN	Inorganic clays of low to medium plasticity, gravelly clays, sandy clays, silty clays.lean clays	Medium to high	Practically impervious	Stable, impervious cores and blankets	$k = 10^{-6}$ to 10^{-8}	Fair to good, sheepsfoot roller, rubber-tired	95-120	None
	OL	GREEN	Organic silts and organic silt-clays of low plasticity	Medium to high	Poor	Not suitable for embankments	$k = 10^{-4}$ to 10^{-6}	Poor to fair, sheepsfoot roller	80-100	None
FINE-GRAINED SOILS — SILTS AND CLAYS LL IS GREATER THAN 50	MH	BLUE	Inorganic silts, micaceous or diatomaceous fine sandy or silty soils, elastic silts	Medium to very high	Fair to poor	Poor stability, core of hydraulic dam, not desirable in rolled fill construction	$k = 10^{-4}$ to 10^{-6}	Poor to very poor, sheepsfoot roller	70-95	None
	CH	BLUE	Inorganic clays of high plasticity, fat clays	Medium	Practically impervious	Fair stability with flat slopes, thin cores, blankets and dike sections	$k = 10^{-6}$ to 10^{-8}	Fair to poor, sheepsfoot roller	75-105	None
	OH	BLUE	Organic clays of medium to high plasticity, organic silts	Medium	Practically impervious	Not suitable for embankments	$k = 10^{-6}$ to 10^{-8}	Poor to very poor, sheepsfoot roller	65-100	None
HIGHLY ORGANIC SOILS	Pt	Orange	Peat and other highly organic soils	NOT RECOMMENDED FOR SANITARY LANDFILL CONSTRUCTION						

*Values are for guidance only; design should be based on test results.

†The equipment listed will usually produce the desired densities after a reasonable number of passes when moisture conditions and thickness of lift are properly controlled.

‡Compacted soil at optimum moisture content for Standard AASHO (Standard Proctor) compactive effort.

Figure 2.67

the soil and rock. By their nature, they require special measures to keep surface waters from inundating the fill. Permeable formations that intersect the sidewalls or floor of the fill may also have to be lined with an impervious layer of clay or other material to control the movement of fluids.

There are also numerous man-made topographic features scattered over the country—strip mines, worked-out stone and clay quarries, open-pit mines, and sand and gravel pits. In most cases, these abandoned depressions are useless, dangerous eyesores. Many of them could be safely and economically reclaimed by utilizing them as sanitary landfills. Clay pits, for example, are located in most impermeable formations, which are natural barriers to gas and water movement. Abandoned strip mines also are naturally suited for use as sanitary landfills. Most coal formations are underlaid by clays, shales, and siltstones, that have a very low permeability. When permeable formations, such as sandstones, are encountered near an excavation, impermeable soil layers can be constructed from the nearby abundant spoil. Abandoned limestone, sandstone, siltstone, granite, and traprock quarries and open-pit mines generally require more extensive improvements because they are in permeable or often open-fractured formations. The pollution potential of sand and gravel pits is great, and worked-out pits consequently require extensive investigation and probably expensive improvements to control gas movement and water pollution.

Marsh and tidal lands may also be filled, but they are less desirable from an ecological point of view. They have little value as real estate, but possess considerable ecological value as nesting and feeding grounds for wildlife. Filling of such areas requires, however, the permanent lowering of the groundwater or the raising of the ground surface to keep organic and soluble solid waste from being deposited in standing water. Roads for collection vehicles are also needed, and cover material generally has to be hauled in.

Sanitary Landfill Design The designing of a sanitary landfill calls for developing a detailed description and plans that outline the steps to be taken to provide for the safe, efficient disposal of the quantities and types of solid wastes that are expected to be received. The designer outlines volume requirements, site improvements (clearing of the land, construction of roadways and buildings, fencing, utilities), and all the equipment necessary for day-to-day operations of the specific landfilling method involved. The designer also provides for controlling water pollution and the movement of decomposition gas. The sanitary landfill designer should also recommend a specific use of the site after landfilling is completed. Finally, the designer should determine capital costs and projected operating expenditures for the estimated life of the project.

Volume Requirements If the rate at which solid wastes are collected and the capacity of the proposed site are known, its useful life can be estimated. The ratio of solid waste to cover material volume usually ranges between 4:1 and 3:1; it is, however, influenced by the thickness of the cover used and cell configuration. If cover material is not excavated from the fill site, this ratio may be compared with

the volume of compacted soil waste and the capacity of a site determined, as seen in Figure 2.68. For example, a town having a 10,000 population and a per capita collection rate of 5 pounds per day must dispose of, in a year, 11 acre-feet of solid waste if it is compacted to 1000 pounds per cubic foot. If it were compacted to only 600 pounds per cubic yard, the volume disposed of in a year would occupy 19 acre-feet. The volume of soil required for the 1000-pound density at a solid waste-to-cover ratio of 4:1 would be 2.75 acre-feet; the 600-pound density waste would need 4.75 acre-feet. A density of 800 pounds per cubic yard is easily achievable if the compacting of a representative municipal waste is involved. A density of 1000 pounds per cubic yard can usually be obtained if the waste is spread and compacted according to recommended procedures.

The number of tons to be disposed of at a proposed sanitary landfill can be estimated from data recorded when solid wastes are delivered to disposal sites. The daily volume of compacted solid waste can then be easily determined for a large community (Figure 2.69) or for a small community (Figure 2.70). The volume of soil required to cover each day's waste is then estimated by using the appropriate solid waste-to-cover ratio.

The terms used to report densities at landfills can be confusing. Solid waste density (field density) is the weight of a unit volume of solid waste in place. Landfill density is the weight of a unit volume of in-place solid waste divided by

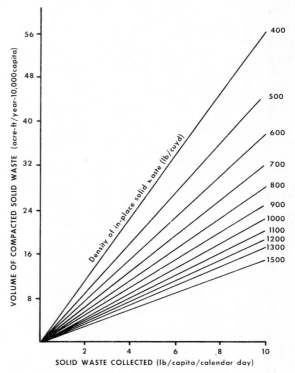

Figure 2.68 Determining the yearly volume of compacted solid waste generated by a community of 10,000 people.

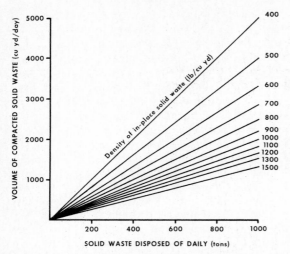

Figure 2.69 Determining the daily volume of compacted solid waste generated by large communities.

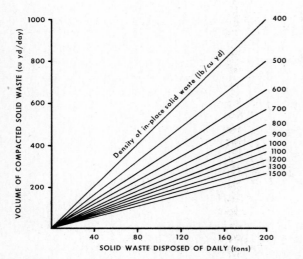

Figure 2.70 Determining the daily volume of compacted solid waste generated by small communities.

the volume of solid waste and its cover material. Both methods of reporting density are usually expressed as pounds per cubic yard, on an in-place weight basis, including moisture, at time of the test, unless otherwise stated.

Site Improvements The plan for a sanitary landfill should prescribe how the site will be improved to provide an orderly and sanitary operation. This may simply involve the clearing of shrubs, trees, and other obstacles that could hinder vehicle travel and landfilling operations, or it could involve the construction of buildings, roads, and utilities.

Clearing and Grubbing Trees and brush that hinder landfill equipment or collection vehicles must be removed. Trees that cannot be pushed over should be cut as close as possible to the ground so that the stumps do not hinder compaction or obstruct vehicles. Brush and tall grass in working areas can

be rolled over or grubbed. A large site should be cleared in increments to avoid erosion and scarring of the land. If possible, natural windbreaks and green belts of trees or brush should be left in strategic areas to improve appearance and operation.

Roads Permanent roads should be provided from the public road system to the site. A large site may have to have permanent roads that lead from its entrance to the vicinity of the working area. They should be designed to support the anticipated volume of truck traffic. In general, the roadway should consist of two lanes (total minimum width, 24 feet), for two-way traffic. Grades should not exceed equipment limitations. For loaded vehicles most uphill grades should be less than 7 percent and downhill grades less than 10.

Temporary roads are normally used to deliver wastes to the working face from the permanent road system, because the location of the working face is constantly changing. Temporary roads may be constructed by compacting the natural soil present and by controlling drainage or by topping them with a layer of a tractive material, such as gravel, crushed stone, cinders, broken concrete, mortar, or bricks. Lime, cement, or asphalt binders may make such roads more serviceable.

If fewer than 25 round trips per day to the landfill are expected, a graded and compacted soil will usually suffice. More than 50 round trips per day generally justifies the use of calcium chloride as a dust inhibitor or such binder materials as soil cement or asphalt. A base course plus a binder is desirable if more than 100 to 150 round trips per day are anticipated.

Scales Recording the weights of solid waste delivered to a site can help regulate and control the sanitary landfill operation as well as the solid waste collection system that serves it.

The scale type and size used will depend on the scope of the operation. Portable scales may suffice for a small site, while an elaborate system employing load cells, electronic relays, and printed output may be needed at a large sanitary landfill. Highly automated electronic scales and recorders cost more than a portable, simple, beam scale, but their use may often be justified, because they are faster and more accurate. The platform or scale deck may be constructed of wood, steel, or concrete. The first type is the least expensive, but also the least durable.

The scale should be able to weigh the largest vehicle that will use the landfill on a routine basis; 30 tons is usually adequate. Generally, the platform should be long enough to weigh all axles simultaneously. Separate axle-loading scales (portable versions) are the cheapest, but they are less accurate and slower-operating. The scale platform should be 10 by 34 feet to weigh most collection vehicles. A 50-foot platform will accommodate most trucks with trailers.

The accuracy and internal mechanism of the scale and the recording device should meet the commercial requirements imposed by the state and any other jurisdiction involved, particularly if user fees are based on weight. Recommended

scale requirements have been outlined by the National Bureau of Standards.

Buildings   A building is needed for office space and employee facilities at all but the smallest landfill; it can also serve as a scale house. Since a landfill operates in wet and cold weather, some protection from the elements should be provided. Operational records may also be kept at a large site. Sanitary facilities should be provided for both landfill and collection personnel. A building should also be provided for equipment storage and maintenance.

Buildings on sites that will be used for less than 10 years should be temporary types and, preferably, be movable. The design and location of all structures should consider gas movement and differential settlement caused by the decomposing solid waste.

Utilities   All sanitary landfill sites should have electrical, water, and sanitary services. Remote sites may have to extend existing services or use acceptable substitutes. Portable chemical toilets can be used to avoid the high cost of extending sewer lines, potable water may be trucked in, and an electric generator may be used instead of having power lines run into the site.

Water should be available for drinking, fire fighting, dust control, and employee sanitation. A sewer line may be called for, especially at large sites and at those where leachate is collected and treated with domestic wastewater. Telephone or radio communications are also desirable.

Fencing   Peripheral and litter fences are commonly needed at sanitary landfills. The first type is used to control or limit access, keep out children, dogs, and other large animals, screen the landfill, and delineate the property line. If vandalism and trespassing are to be discouraged, a 6-foot-high fence topped with three strands of barbed wire projecting at a 45-degree angle is desirable. A wooden fence or a hedge may be used to screen the operation from view.

Litter fences are used to control blowing paper in the immediate vicinity of the working face. As a general rule, trench operations require less litter fencing because the solid waste tends to be confined within the walls of the trench. At a very windy trench site, a 4-foot snow fence will usually suffice. Blowing paper is more of a problem in an area operation; 6- to 10-foot litter fences are often needed.

Control of Surface Water   Surface water courses should be diverted from the sanitary landfill. Pipes may be used in gullies, ravines, and canyons that are being filled to transmit upland drainage through the site and open channels employed to divert runoff from surrounding areas.

Groundwater Protection   It is a basic premise that groundwater and the deposited solid waste not be allowed to interact. It is unwise to assume that a leachate will be diluted in groundwater because very little mixing occurs in an aquifer since the groundwater flow there is usually laminar.

When issuing permits or certificates, many states require that groundwater and deposited solid wastes be 2 to 30 feet apart. Generally, a 5-foot separation will remove enough readily decomposed organics and coliform bacteria to make the liquid bacteriologically safe. On the other hand, mineral pollutants can travel long distances through soil or rock formations. In addition to other considerations, the sanitary landfill designer must evaluate the (1) current and projected use of the water resources of the area, (2) effect of leachate on groundwater quality, (3) direction of groundwater movement, (4) interrelationship of this aquifer with other aquifers and surface waters.

Sanitary Landfilling Methods   The designer of a sanitary landfill should prescribe the method of construction and the procedures to be followed in disposing of the solid waste, because there is no "best method" for all sites. The method selected depends on the physical conditions involved and the amount and types of solid waste to be handled.

The two basic landfilling methods are trench and area; other approaches are only modifications. In general, the trench method is used when the groundwater is low and the soil is more than 6 feet deep. It is best employed on flat or gently rolling land. The area method can be followed on most topographies and is often used if large quantities of solid waste must be disposed of. At many sites, a combination of the two methods is used.

Cell Construction and Cover Material   The building block common to both methods is the cell. All the solid waste received is spread and compacted in layers within a confined area. At the end of each working day, or more frequently, it is covered completely with a thin, continuous layer of soil, which is then also compacted. The compacted waste and soil cover constitute a cell. A series of adjoining cells, all of the same height, makes up a lift (Figure 2.71). The completed fill consists of one or more lifts.

The dimensions of the cell are determined by the volume of the compacted waste, and this, in turn, depends on the density of the in-place solid waste. The field density of most compacted solid waste within the cell should be at least 800 pounds per cubic yard. (It should be considerably higher if large amounts of demolition rubble, glass, and well-compacted inorganic materials are present.) The 800-pound figure may be difficult to achieve if brushes from bushes and trees, plastic turnings, synthetic fibers, or rubber powder and trimmings predominate. Because these materials normally tend to rebound when the compacting load is released, they should be spread in layers up to 2 feet thick, then covered with 6 inches of soil. Over this, mixed solid waste should be spread and compacted. The overlying weight keeps the fluffy or elastic materials reasonably compressed.

An orderly operation should be achieved by maintaining a narrow working face (that portion of the uncompleted cell on which additional waste is spread and compacted). It should be wide enough to prevent a backlog of trucks waiting to dump, but not be so wide that it becomes impractical to manage properly—never over 150 feet.

No hard-and-fast rule can be laid down regarding the proper height of a cell. Some designers think it should be 8 feet or less, presumably because this height will not cause severe settlement problems. On the other hand, if a multiple lift operation is involved and all the cells are built to the same height, whether 8 or 16 feet, total settlement should not dif-

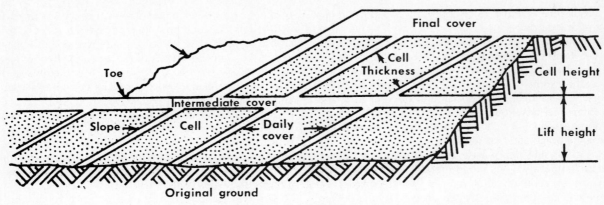

Figure 2.71 The cell is the common building block in sanitary landfilling. Solid waste is spread and compacted in layers within a confined area. At the end of each working day, or more frequently, it is covered completely with a thin, continuous layer of soil, which is then also compacted. The compacted waste and soil constitute a cell. A series of adjoining cells makes up a lift. The completed fill consists of one or more lifts.

fer significantly. If land and cover material are readily available, an 8-foot height restriction might be appropriate, but heights up to 30 feet are common in large operations. Rather than deciding on an arbitrary figure, the designer should attempt to keep cover material volume at a minimum while adequately disposing of as much waste as possible.

Cover material volume requirements are dependent on the surface area of waste to be covered and the thickness of soil needed to perform particular functions. As might be expected, cell configuration can greatly affect the volume of cover material needed. The surface area to be covered should therefore be kept minimal.

In general, the cell should be about square, and its sides should be sloped as steeply as practical operation will permit. Side slopes of 20 to 30 degrees will not only keep the surface area, and hence the cover material volume, at a minimum but will also aid in shredding and obtaining good compaction of solid waste, particularly if it is spread in layers not greater than 2 feet thick and worked from the bottom of the slope to the top.

Trench Method Waste is spread and compacted in an excavated trench. Cover material, which is taken from the spoil of the excavation, is spread and compacted over the waste to form the basic cell structure (Figure 2.72). In this method, cover material is readily available as a result of the excavation. Spoil material not needed for daily cover may be stockpiled and later used as a cover for an area fill operation on top of the completed trench fill.

Cohesive soils, such as glacial till or clayey silt, are desirable for use in a trench operation because the walls between the trenches can be thin and nearly vertical. The trenches can therefore be spaced very closely. Weather and the length of time the trench is to remain open also affect soil stability and must be considered when the slope of the trench walls is being designed. If the trenches are aligned perpendicularly to the prevailing wind, this can greatly reduce the amount of

blowing litter. The bottom of the trench should be slightly sloped for drainage, and provision should be made for surface water to run off at the low end of the trench. Excavated soil can be used to form a temporary berm on the sides of the trench to divert surface water.

The trench can be as deep as soil and groundwater conditions safely allow, and it should be at least twice as wide as any compacting equipment that will work in it. The equipment at the site may excavate the trench continuously at a rate geared to landfilling requirements. At small sites, excavation may be done on a contract basis.

Area Method In this method, the waste is spread and compacted on the natural surface of the ground, and cover material is spread and compacted over it (Figure 2.73). The area method is used on flat or gently sloping land and also in quarries, strip mines, ravines, valleys, or other land depressions.

Combination Methods A sanitary landfill does not need to be operated by using only the area or trench method. Combinations of the two are possible, and flexibility is, therefore, one of sanitary landfilling's greatest assets. The methods used can be varied according to the constraints of a particular site.

One common variation is the progressive slope or ramp method, in which the solid waste is spread and compacted on a slope. Cover material is obtained directly in front of the working face and compacted on the waste (Figure 2.74). In this way, a small excavation is made for a portion of the next day's waste. This technique allows for more efficient use of the disposal site when a single lift is constructed than the area method does, because cover does not have to be imported, and a portion of the waste is deposited below the original surface.

Both methods might have to be used at the same site if an extremely large amount of solid waste must be disposed of. For example, at a site with a thick soil zone over much of it

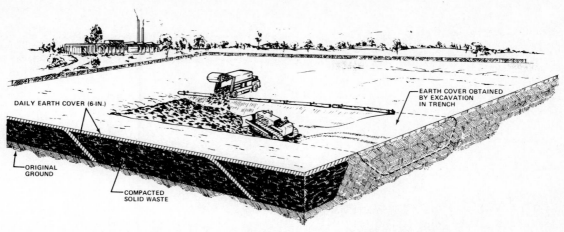

Figure 2.72 In the trench method of sanitary landfilling, the collection truck deposits its load into a trench where a bulldozer spreads and compacts it. At the end of the day, the trench is extended, and the excavated soil is used as daily cover material.

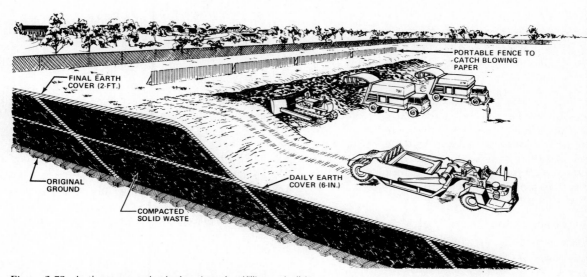

Figure 2.73 In the area method of sanitary landfilling, a bulldozer spreads and compacts the waste on the natural surface of the ground, and a scraper is used to haul the cover material at the end of the day's operations.

but with only a shallow soil over the remainder, the designer would use the trench method in the thick soil zone and use the extra spoil material obtained to carry out the area method over the rest of the site. When a site has been developed by either method, additional lifts can be constructed using the area method by having cover material hauled in.

The final surface of the completed landfill should be so designed that ponding of precipitation does not occur. Settlement must therefore be considered. Grading of the final surface should induce drainage but not be so extreme that the cover material is eroded. Side slopes of the completed surface should be 3:1 or flatter to minimize maintenance.

Finally, the designer should consider completing the sanitary landfill in phases so that portions of it can be used as parks and playgrounds, while other parts are still accepting solid wastes.

Summary of Design Considerations The final design of a sanitary landfill should describe in detail (1) all employee and operational facilities; (2) operational procedures and their sequence, equipment, and manpower requirements; (3) the pollution potential and methods of controlling it; (4) the final grade and planned use of the completed fill; (5) cost estimates for acquiring, developing, and operating the proposed site.

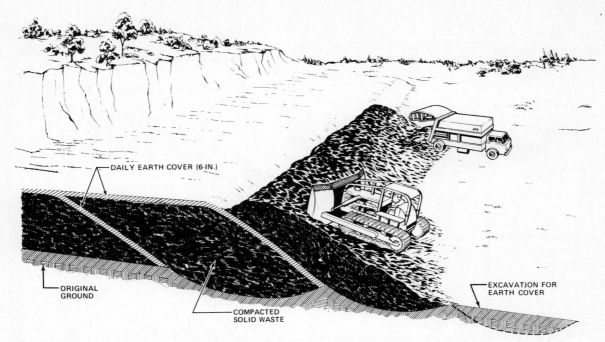

Figure 2.74 In the progressive slope or ramp method of sanitary landfilling, solid waste is spread and compacted on a slope. Cover material is obtained directly in front of the working face and compacted on the waste.

The designer should also provide a map that shows the location of the site and the area to be served and a topographic map covering the area out to 1000 feet from the site. Additional maps and cross sections should also be included that show the planned stages of filling (startup, intermediate lifts, and completion). They should present the details of:

1. Roads on and off the site
2. Buildings
3. Utilities above and below ground
4. Scales
5. Fire protection facilities
6. Surface drainage (natural and constructed) and groundwater
7. Profiles of soil and bedrock
8. Leachate collection and treatment facilities
9. Gas control devices
10. Buildings within 1000 feet of property (residential, commercial, agricultural)
11. Streams, lakes, springs, and wells within 1000 feet
12. Borrow areas and volume of material available
13. Direction of prevailing wind
14. Areas to be landfilled, including special waste areas, and limitations on types of waste that may be disposed of
15. Sequence of filling
16. Entrance to facility
17. Peripheral fencing
18. Landscaping
19. Completed use

Completed Sanitary Landfill Reclaiming land by filling and raising the ground surface is one of the greatest benefits of sanitary landfilling. The completed sanitary landfill can be used for many purposes, but all of them must be planned before operations begin.

Characteristics Designers should know the proposed use of the completed sanitary landfill before they begin to work. Unlike an earthfill, a sanitary landfill consists of cells containing a great variety of materials having different physical, chemical, and biological properties. The decomposing solid waste imparts characteristics to the fill that are peculiar to sanitary landfills. These characteristics require that the designer plan for gas and water controls, cell configuration, cover material specifications (as determined by the planned use), and the periodic maintenance needed at the completed sanitary landfill.

Decomposition Most of the materials in a sanitary landfill will decompose, but at varying rates. Food wastes decompose readily, are moderately compactible, and form organic acids that aid decomposition. Garden wastes are resilient and difficult to compact but generally decompose rapidly. Paper products and wood decay at a slower rate than food wastes. Paper is easily compacted and may be pushed into voids, whereas lumber, tree branches, and stumps are difficult to

compact and hinder the compaction of adjacent wastes. Car bodies, metal containers, and household appliances can be compacted and will slowly rust in the fill with the help of organic acids produced by decomposing food wastes. Glass and ceramics are usually easily compacted but do not degrade in a landfill. Plastics and rubber are resilient and difficult to compact; rubber decomposes very slowly, most plastics not at all. Leather and textiles are slightly resilient but can be compacted; they decompose, but at a much slower rate than garden and food wastes. Rocks, dirt, ashes, and construction rubble do not decompose and can be easily worked and compacted.

Density   The density of solid waste in a landfill is quite variable. One that is well constructed can have an in-place density as great as 1500 pounds per cubic yard, while that of poorly compacted solid waste may be only 500. Generally, 800 to 1000 pounds per cubic yard can be achieved with a moderate compactive effort. Soft and hard spots occur within the fill as a result of different decomposition rates and compaction densities. Density influences such other characteristics as settlement and bearing capacity.

Settlement   A sanitary landfill will settle as a result of waste decomposition, filtering of fines, superimposed loads, and its own weight. Bridging that occurs during construction produces voids. As the waste decomposes, fine particles from the cover material and overlying solid waste often sift into these voids. The weight of the overhead waste and cover material helps consolidate the fill, and this development is furthered when more cover material is added or a structure or roadway is constructed on the fill.

The most significant cause of settlement is waste decomposition, which is greatly influenced by the amount of water in the fill. A landfill will settle more slowly if only limited water is available to decompose the waste chemically and biologically. In Seattle, where rainfall exceeds 30 inches per year, a 20-foot fill settled 4 feet in the first year after it was completed. In Los Angeles, where less than 15 inches of rain falls per year, 3 years after a landfill had been completed a 75-foot high area had settled only 2.3 feet, and another section that had been 46 feet high had settled a mere 1.3 feet.

Settlement also depends on the types of wastes disposed of, the volume of cover material used with respect to the volume of wastes disposed of, and the compaction achieved during construction. A fill composed only of construction and demolition debris will not settle as much as one that is constructed of residential solid wastes. A landfill constructed of highly compacted waste will settle less than one that is poorly compacted. If two landfills contain the same types of wastes and are constructed to the same height, but one has a waste-to-cover volume ratio of 1:1 and the other a ratio of 4:1, the first will settle less. Because of the many factors involved, a fill may settle as much as 33 percent.

Settling can produce wide cracks in the cover material that expose the wastes to rats and flies, allow water to infiltrate, and permit gas to escape. Differential settling may form depressions that permit water to pond and infiltrate the fill. Settling may also cause structures on the landfill to sag and possibly collapse; the underground utility lines that serve these buildings or traverse the site may then shear. Because every landfill settles, its surface should be periodically inspected and soil should be added and graded when necessary.

Bearing Capacity   The bearing capacity of a completed sanitary landfill is the measure of its ability in pounds per square foot to support foundations and keep them intact. Very little information is available on the subject, but a few investigators place the bearing capacity of a completed landfill between 500 and 800 pounds per square foot; higher values have, however, been noted. Since there is no definite procedure for interpreting the results of solid waste bearing tests, any value obtained should be viewed with extreme caution. Almost without exception, the integrity and bearing capacity of soil cover depend on the underlying solid waste. Most bearing strength tests of soil are conducted over a short period—several minutes for granular materials to a maximum of 3 days for clay having a high moisture or air content. During the test, the soil adjusts to its limits under the load imposed and conditions of confinement. Solid waste, on the other hand, does not follow this pattern of deformation but continues to alter its structure and composition over a long period of time. Natural soils, which are not as heterogeneous as solid waste, produce test values that fall within a predictable range. Moreover, repeated tests of the soil will produce similar results—similar relationships have not been established for solid waste.

Landfill Gases   Landfill gases continue to be produced after the landfill is completed and can accumulate in structures or soil, cause explosions, and stunt or kill vegetation. Placement of a thick, moist, vegetative, final cover may act as a gastight lid that forces gases to migrate laterally from the landfill. If the site is converted into a paved parking lot, this may also prevent the gases from venting into the atmosphere. Design of gas controls should therefore conform with the planned use of the completed fill.

Corrosion   The decomposing material in a landfill is very corrosive. Organic acids are produced from food, garden, and paper wastes, and some weak acids are derived from ashes. Unprotected steel and galvanized pipe used for utility lines, leachate drains, and building foundations are subject to severe and rapid pitting. All structural materials susceptible to corrosion should be protected. Acids present in a sanitary landfill can deteriorate a concrete surface and thus expose the reinforcing steel; this could eventually cause the concrete to fail.

Uses   There are many ways in which a completed sanitary landfill can be used; it can, for example, be converted into a green area or be designed for recreational, agricultural, or light construction purposes. The landfill designer should evaluate each proposal from a technical and economic viewpoint. More suitable land is often available elsewhere that would not require the expensive construction techniques required at a sanitary landfill.

Green Area   The use of a completed sanitary landfill as a green area is very common. No expensive structures are

built, and a grassed area is established for the pleasure of the community. Some maintenance work is, however, required to keep the fill surface from being eroded by wind and water. The cover material should be graded to prevent water from ponding and infiltrating the fill. Gas and water monitoring stations, installed during construction, should be periodically sampled until the landfill stabilizes. Gas and water controls and drains also require periodic inspection and maintenance.

If the final cover material is thin, only shallow-rooted grass, flowers, and shrubs should be planted on the landfill surface. The decomposing solid waste may be toxic to plants whose roots penetrate through the bottom of the final cover. An accumulation of landfill gas in the root zone may interfere with the normal metabolism of plants. This can be avoided by selecting a cover material having a low water-holding capacity, but this type of soil provides poor support for vegetation. On the other hand, a moist soil does not allow decomposition gas to disperse, and consequently gas venting must be considered.

The most commonly used vegetation is grass. Most pasture and hay grasses are shallow-rooted and can be used on a landfill having only 2 feet of final cover, but alfalfa and clover need more than this. The soil used for final cover influences the choice of vegetation. Some grasses, such as tall meadow oat grass, thrive well on light sand or gravelly soils, while others, such as timothy grass, do better in such heavier soils as clays and loams.

Climate also influences the selection of grasses. Bermuda is a good soil binder and thrives in southern states. Perennial rye does best where the climate is cool and moist and winter is mild; it roots rapidly but dies off in 2 to 3 years if shaded. Redtop and bent grass thrive almost anywhere except in drier areas and the extreme south. The selection of the grass or mixture of grasses depends, therefore, on climate, depth of the root system, and soil used for cover material.* Mowing and irrigating requirements should also be considered. In general, it is not advisable to irrigate the landfill surface, because the water may infiltrate and leach the fill.

Agriculture A completed sanitary landfill can be made productive by turning it into pasture or crop land. Many of the grasses mentioned above are suitable for hay production. Corn and wheat usually have 4-foot roots, but the latter occasionally has longer ones. The depth of the final cover must therefore be increased accordingly.

If cultivated crops are used, the final cover should be thick enough that roots or cultivating do not disturb its bottom foot. If the landfill is to be cultivated, a 1- to 2-foot layer of relatively impermeable soil, such as clay, may be placed on top of the solid waste and an additional layer of agricultural soil placed above to prevent the clay from drying out. Excessive moisture will also be prevented from entering the fill. Such a scheme of final cover placement must also provide for gas venting via gravel trenches or pipes.

---

*Information on the grasses mainly used in a landfill area is available from county agricultural agents and the U.S. Soil Conservation Service.

Construction A foundations engineering expert should be consulted if plans call for structures to be built on or near a completed sanitary landfill. This is necessary because of the many unique factors involved—gas movement, corrosion, bearing capacity, and settlement. The cost of designing, constructing, and maintaining buildings is considerably higher than it is for those erected on a well-compacted earth fill or on undisturbed soil. The most problem-free technique is to preplan the use of islands to avoid settlement, corrosion, and bearing-capacity problems. Ideally, the islands should be undisturbed soils that are bypassed during excavating and landfilling operations. Settlement would then be governed by the normal properties of the undisturbed soil. Alternatively, truck loads of rocks, dirt, and rubble could be laid down and compacted during construction of the landfill at places where the proposed structure would be built.

The decomposing landfilled waste can be excavated and replaced with compacted rock or soil fill, but this method is very expensive and could prove hazardous to the construction workers. The decomposing waste emits a very putrid smell, and hydrogen sulfide, a toxic gas, may be present with methane, an explosive gas. These two gases should be monitored throughout the excavating operation. Gas masks may have to be provided for the workers and no open flames should be permitted.

Piles can also be used to support buildings when the piles are driven completely through the refuse to firm soil or rock. Some of the piles should be battered (angled) to resist lateral movement that may occur in the fill. Another factor to consider is the load imposed on the piles by solid wastes settling around them. The standard field penetration resistance test is used to determine the strength of the earth material in which the piles are to be founded. During this test, penetration will be resisted by the solid waste, but as the refuse decomposes and settling occurs, it may no longer resist and will more likely create a downward force on the pile. There are no data for established procedures for predicting this change in force.

Several peculiar problems arise when piles are used to support a structure over a landfill. The decomposing waste is very corrosive; so the piles must be protected with corrosion-resistant coatings. It may be very difficult to drive the piles through the waste if large bulky items, such as junked cars and broken concrete, are in the fill where the structure is to be located. The fill underlying a pile-supported structure may settle, and voids or air spaces may develop between the landfill surface and the bottom of the structure. Landfill gases could accumulate in these voids and create an explosion hazard.

Light, one-story buildings are sometimes constructed on the landfill surface. The bearing capacity of the landfill should be determined by field investigations in order to design continuous foundations. Foundations should be reinforced to bridge any gaps that may occur because of differential settling in the fill. Continuous floor slabs reinforced as mats can be used, and the structure should be designed to accommodate settlement. Doors, windows, and partitions should be able to adapt to slight differential movement between

them and the structural framing. Roads, parking lots, sidewalks, and other paved areas should be constructed of a flexible and easily repairable material, such as gravel or asphaltic concrete.

Consolidating the landfill to improve its bearing capacity and reduce settlement by surcharging it with a heavy layer of soil does not directly influence the decomposition rate. If the surcharge load is removed and the structure is built before the waste has stabilized, settlement will still be a problem, and the bearing capacity may not be as great as expected.

None of the methods for supporting a structure over a landfill are problem-free. A common difficulty is keeping landfill gases from accumulating in the structure. Even buildings erected on undisturbed islands of soil must be specially designed to prevent this from developing. A layer of sand can be laid over the proposed structural area and then be covered by two or more layers of polyvinyl chloride sheeting. An additional layer of sand can then be emplaced. If the bottom layer of sand is not saturated, it will act as a gas-permeable vent, and the sheeting will prevent the gas from entering or collecting under the structure. The top layer of sand protects the sheeting from being punctured. Another approach is to place an impermeable membrane of jute and asphalt under all below-grade portions of the structure. A gravel or sand layer must underlie the jute-asphalt membrane and be vented to the atmosphere. The most reliable method is to construct a ventilated false basement to keep gas from accumulating.

Utility connections must be made gasproof if they enter a structure below grade. If the building is surrounded by filled land, utility lines that traverse the fill must be flexible, and slack should be provided so the lines can adjust to settlement. Flexible plastic conduits are more expensive than other materials but would probably work best, because they are elastic and resist corrosion. Gravity waste-water pipelines may develop low points if the fill settles. Liquid wastes should be pumped to the nearest sewer unless the grade from the structure to the sewer prevents low points from forming. Shearing of improperly designed water and waste-water services caused by differential settlement can occur where they enter the structure or along the pipeline that traverses the fill.

Recreation Completed landfills are often used as ski slopes, toboggan runs, coasting hills, ball fields, golf courses, amphitheaters, playgrounds, and parks. Small, light buildings, such as concession stands, sanitary facilities, and equipment storage sheds, are usually required at recreational areas. These should also be constructed to keep settlement and gas problems at a minimum. Other problems encountered are ponding, cracking, and erosion of cover material. Periodic maintenance includes regrading, reseeding, and replenishing the cover material.

Registration The completed landfill should be inspected by the government agency responsible for ensuring its proper operation. Following final acceptance of the site, a detailed description, including a plat, should be recorded with the proper authority in the county where the site is located. This provides future owners or users with adequate information regarding the previous use of the site. The description should therefore include type and general location of wastes, number and type of lifts, and details about the original terrain.

2.17 Site Comparison: Checklist

Sites may be examined and compared with the following checklist.

1. Conformance with urban pattern
   a. Conformance with accepted urban development plans, tentative plans, or probable trends in land use
   b. Present zoning: possible changes
   c. Approval of city planning bodies
   d. Possibility of closing existing streets and dedicating new streets
   e. Effect of building codes and possibility of modification

2. Slum clearance considerations
   a. Number, character, and condition of existing buildings on site
   b. Number of families housed at present
   c. Relocation of present residents
   d. Equivalent elimination

3. Characteristics of site and environment
   a. Area of site compared with area needed for buildings and project facilities
   b. Shape of site, parcels necessarily excluded, deed restrictions, and easements
   c. Topography as it affects livability of the site plan; favorable features such as existing shade trees, pleasing outlook, and desirable slopes
   d. Quality of neighborhood: extent of nonresidential land use, suitability of neighborhood for dwelling type desired
   e. Effect of project on neighborhood
   f. Hazards: possibility of flooding, slides, or subsidence; proximity to railroads, high-speed trafficways, high embankments, unprotected bodies of water; presence of insect or rodent breeding places; high groundwater level that might cause dampness in building
   g. Nuisances: nearness to industrial plants, railroads, switchyards, heavy-traffic streets, airports, and so on, causing noise, smoke, dust, odor, or vibrations

4. Availability of special municipal services
   a. Garbage and rubbish collection
   b. Fire protection as affected by site location and street access
   c. Streets: lighting, cleaning, maintenance, snow removal, tree planting and maintenance, and so on

 d. Police protection and other municipal services

5. Civic and community facilities

 a. Public transportation facilities: means, routes, adequacy and expense of transportation to employment, schools, central business district, and so on

 b. Accessibility to paved thoroughfares

 c. Amount and character of employment within walking distance and within reasonable travel radius

 d. Stores and markets: kinds and locations; need for additional facilities as part of project development

 e. Schools (grade, junior high, and high): locations, capacities, adequacy; probability of enlargement if needed

 f. Parks and playgrounds: locations, facilities provided, adequacy, maintenance and supervision supplied; possible additions

 g. Churches, theaters, clinics

6. Appropriateness of project design to site, with reference to livability

 a. Type or types of dwellings

 b. Project density

 c. Utility selection

7. Elements of project development cost

 a. Land costs, including site acquisition, expense, and unpaid special assessments

 b. Effect of soil conditions, topographic features, project density appropriate to the neighborhood, availability of utilities, extent of existing street improvements, recreational facilities and additions to be provided by municipality or utility companies, and so on

 c. Building types, utility selection, site conditions, and requirements for nondwelling structures

8. Project maintenance and operating costs

 a. Differences in costs of utilities appropriate to the respective sites

 b. Differentials in grounds maintenance costs due to topography

 c. Differences in payments in lieu of taxes

2.18   Residential Site Selection

Controlling Importance of Site Selection   The purpose of selecting a site for residential development may be summarized as follows: to procure a site which is suitable for physical development, including the installation of utilities, and for the provision of dwellings, a circulation system, and neighborhood community facilities in a well-planned relation (all within the economic means of a definitely visualized group of families), and which is free from any grossly unfavorable environmental factors.

 The selection of a site for a neighborhood, housing project, or subdivision is an irrevocable step that often makes the difference between success and failure. It is for this reason that site selection assumes such critical importance.

 The perfect site seldom exists. Judgment must be made, on the one hand, as to the limitations that wholly preclude satisfactory development and, on the other hand, as to the site defects that can be brought within satisfactory limits or must, in a given case, be accepted as minor but necessary evils. The lines of demarcation are not always definite, and much depends on cost. If adverse conditions make the choice doubtful, the developer must determine three fundamental conditions:

1. Are the necessary improvements technically feasible?

2. If remedial action is the responsibility of an outside agency, is there a guarantee that defects will be corrected within a reasonable period of time?

3. If such improvements fall upon the developer, are they feasible within the economic range of the projected development?

Many problems to be judged in the course of site selection will of course be evident to lay people on their first inspection of a site; others lie below the surface of the land and will be handled by proper specialists. A third type of problem, however, should be stressed: the problem that varies seasonally (or in any other cycle). A piece of land that is bone-dry in August may be regularly flooded by a rise of groundwater in March. The odors of a nearby hog farm may not be evident on a still day, but the prevailing summer breezes might waft them to every open window on the site.

 Essential Physical Characteristics of the Site   The following conditions for healthful development and maintenance must be borne in mind in the selection of a site.

 Soil and Subsoil Conditions   Soil and subsoil conditions must be suitable for excavation and site preparation, for the location of utility connections, and for grading and planting. Subsoil conditions should afford suitable bearing capacity for the economical construction of buildings of the types contemplated. Bearing capacity will be affected if the site contains muck, peat, poorly compacted fill, shifting sand, or quicksand. Test borings will normally be needed as a check on these and other characteristics. For economical construction subsoil should contain no ledge, hardpan, or other obstruction to efficient excavation for the necessary utilities, foundations, or basements.

 Groundwater and Drainage   Essential factors in site selection include a water table low enough to protect the buildings against basement flooding and interference with sewerage, the absence of swamps or marshes, and sufficient slope to permit surface drainage of normal rainfall and a free flow of sanitary sewers. Periodic flooding due to the high groundwater table should disqualify a site unless preventive measures can be applied.

If dwelling basements are contemplated, the groundwater table should be below basement floors. Even where basements are not used, high groundwater may cause dampness in crawl spaces beneath the buildings. Such dampness has caused serious problems and expense in many housing developments.

Flooding due to groundwater may occur not only through conditions on the site itself but also from present or future drainage onto the site from adjacent areas. Groundwater observation may be needed in several seasons, or the testimony of those familiar with the site throughout the year should be sought. In addition, possible future developments in adjoining areas should be taken into account.

Freedom from Surface Floods The development area should be free from danger of surface flooding by streams, lakes, or tidal waters. Significant floods are those that inundate buildings, make them unusable by drowning utilities, or impede circulation within or to and from the development area.

Ideally, no land should be included in a development area that has been flooded at any time of record unless flood control measures have subsequently removed the danger. As a practical matter, locally varying compromises may be made below this standard. It would seem reasonable to insist, however, that land be excluded from development areas if it shows a history of flooding at intervals of less than 25 years unless the source of the flood has subsequently been controlled.

Suitability for Siting of Projected Buildings Land should not be too steep for satisfactory grading in relation to dwelling construction. Building sites should not have elevations above those at which normal water pressure for domestic use and fire fighting can be obtained.

The orientation of slopes may affect the possibility of good development. Southerly slopes will favor exposure of dwellings to the winter sun (steep northerly slopes may be undesirable on this count alone).

Suitability for Access and Circulation Topography should permit adequate vehicular and pedestrian access to and circulation within the development area. It should permit grading so that streets and walks conform to grade standards.

Suitability for Development of Open Areas Land to be reserved for private yards or gardens, play lots, playgrounds, and neighborhood parks should permit grading and development in conformance with specifications.

Freedom from Topographic Accident Hazards The development area should be free from, or the plan should assure correction of, topographic conditions that might be a serious cause of bodily accidents. Under this heading would come bluffs or precipices, open pits, and hazardous shorelines.

If there is a reasonable expectancy of a major earth movement that may cause loss of life or serious damage to structures or utilities, every attempt should be made to avoid sites within the area affected. If this is impossible, special consideration should be given to placing and constructing the buildings so as to reduce the hazard to a minimum. Hazards of this type include landslides, earth settlement above disused mine workings, and earthquake slippage along known geological fault lines.

Availability of Sanitary and Protective Services

Water Supply and Sanitary Sewage Disposal Without the assurance of water for their overnight needs, campers will not pitch their tents. A healthful neighborhood can be developed only on a site with a water supply that is adequate and certain as to amount, that will not be a means of conveying disease, and that is reasonably free from chemical and physical impurities. Equally important is the collection and ultimate disposal of human excreta without sanitary hazard.

Where an existing municipal water supply and sewerage system can be used, the needed safeguards will usually have been assured by local authorities. However, where a new supply must be developed or new disposal facilities provided, the usability of a site may be determined wholly by the problems of such an installation. Under no circumstances should a site lacking public water supply and sewerage systems be accepted without a binding assurance that these problems can be solved.

A point that can hardly be overstressed is that water and sewerage systems must be visualized as long-term sanitary services and not merely as physical installations. Advance approval by the proper health authorities is equally imperative for on-site sewage disposal facilities and for proposed extensions of existing water or sewer lines to serve a site.

A public water supply, with assured maintenance of official health standards, is generally preferable to individual or community private supplies on grounds both of safety and of economy. Preference should therefore be given to a site having access to a public system. Where a public supply is unavailable, it is necessary to be certain that a local supply can be developed at a reasonable cost.

In any case, under standard conditions water of safe quality must be available in each dwelling under pressure, and the general supply must be adequate in amount to provide for fire fighting and other special needs. Under certain semirural conditions, it may temporarily be necessary to accept supplies of safe quality from individual wells.

Removal of bodily wastes from a dwelling by a water carriage system is accepted by urban and semiurban America as standard practice. From a public health standpoint, a public sewerage system that is properly designed and operated affords the greatest safety. A site will usually be suitable for development if every residence can be connected to a public sewerage system of sufficient capacity. In the absence of a public sewerage system, an on-site community system is generally safer than septic tanks or other individual installations, provided adequate maintenance is assured.

The technical considerations in the development and operation of water and sewer systems should be studied before any neighborhood site is selected, for even if public water and sewer systems exist near the site, they may be inadequate or unavailable.

Removal of Refuse It is essential that a projected site have facilities for the effective removal from the neighborhood of domestic wastes (notably garbage but also inflammable and noncombustible rubbish) or that ultimate disposal on the site can be provided without sanitary hazard. Garbage should be regularly collected at time intervals varying from daily in hot weather to weekly in cold weather. If rubbish is collected separately, it may be removed at somewhat less frequent intervals.

If regular municipal collection services can be tapped, no site selection problem of this kind will generally exist. The necessity of facilities within or near the site for burial, incineration, or chemical reduction will call for study of precautionary measures. Major problems will be a segregated location for disposal, avoidance of wind-borne smoke or odors, and the use of disposal methods that eliminate rat harborage and insect breeding.

Power, Fuel, and Communications Electricity is essential in every home, but since electric service can usually be extended to any development of more than a few families and can even be generated on the site if necessary, it seldom offers a serious problem in site selection. Reasonable electricity rates, especially in outlying areas where, in the absence of gas, electricity may be used for cooking and water heating, may be an important factor in achieving a reasonable total housing cost.

Gas is not considered an essential utility. If a domestic gas supply is desired beyond the reach of distribution mains, portable high-pressure tanks may serve the need.

Telephone service, like electricity, can be extended to most sites offering a demand. Its availability is seldom important as a site selection factor.

Fire and Police Protection Since water requirements for fire fighting normally set the peak-load demand for a community supply, this factor in the adequacy of fire protection will automatically be checked as a part of other site selection judgments.

The availability of fire-fighting crews and equipment from the larger community will depend on both the location of the new neighborhood and the administrative relations between the two. When these facilities must be supplied on the site by the developer, they will be an important factor of operating cost.

The feasibility of police protection is little affected by location, but like fire protection it may involve special costs in isolated neighborhoods.

Freedom from Local Hazards and Nuisances The site should be entirely free from grave hazards to life or health and as free as possible from minor hazards and nuisances. Adequate techniques for measuring the seriousness of specific nuisances and standards for site selection in regard to them do not exist. Research on this whole problem, especially on such factors as minimum distances of dwellings from railroad lines, is badly needed. Some guides to specific standards for new development can be obtained from the pro-

cedures for the evaluation of existing neighborhoods given in a standard housing appraisal method.*

Certain nuisances and hazards may not be apparent in a single inspection of the site. Investigation must take into account the fact that many nuisances depend on season of the year, the time of day, the wind direction, or other weather factors. Nuisances and hazards common to all parts of a city, such as smoke in Knoxville or steep hills in San Francisco, must naturally be tolerated except where the degree of seriousness is markedly above the citywide norm.

Accident Hazards Major accident hazards are collision with moving vehicles, fire and explosions, falls, and drowning. The chief causes of collision are street traffic and railroads, with crash landings of aircraft to be considered near an airport. Sources of fire and explosion hazards include bulk storage of petroleum, gasoline, or gas; rifle ranges and other places where firearms are used under potentially dangerous conditions; dumps and rubbish piles; large expanses of brushland or cutover woodland from which the slash has not been cleared, especially in dry climates; and certain industries. Falls and drowning may occur with unprotected bodies of water, quarries, pits, junkyards, and so on.

Housing should not be located within the influence range of fire and explosion hazards from industrial sources. Safe distances from airports, as specified in airport zoning ordinances, should be maintained.

With respect to fire and explosion hazards, the "safe" distance must be determined for each specific hazard. Fire hazards not involving explosions may be partially controlled by the provision of adequate firebreaks.

It is desirable for both safety and noise protection to avoid housing construction on sites adjacent to heavy street traffic or railroads. When this is impossible, the following precautions against accidents must be enforced:

1. No grade crossings of railroads in or near the development area without a 24-hour guard. Rail lines at grade or depressed should be completely fenced or otherwise closed off so that children cannot wander onto them.
2. Adequate control of traffic hazards on all streets.

Accident hazards such as quarries, junkyards, docks, swamps, or hidden ponds are particularly dangerous to children. Sites near or including these hazards should not be used for housing without fencing off such features.

Noise and Vibration Excessive noise, sometimes with appreciable vibration, is commonly produced by railroads, airports, street traffic, heavy industry, boat whistles, foghorns, and the like. The site and the surrounding area should be investigated for such potential sources of noise. Where they exist, their distance from the source of the site and the presence of sound barriers should be determined. In the case of an undeveloped site in open country, distant noise of moderate intensity will tend to be masked by the general noise level of the new development itself.

Persons concerned with the operation of nearby airports, railroads, or industry and persons familiar with the site should be consulted as to the timing of noises and the carrying effect

---

*An Appraisal Method for Measuring the Quality of Housing, part 3, Appraisal of Neighborhood Environment, American Public Health Association, Committee on the Hygiene of Housing, New York, 1950.

of prevailing winds. In investigating railroads, the character of the railroad (frequency of passenger and freight service, existence of switchyards) and the design of rights-of-way (tracks depressed, at grade, or elevated) should be considered. For airports, the type of traffic (commercial, military, or private) and its frequency should be determined.

Street noise should be considered not only as to volume of traffic but as to traffic congestion and steep hills, or stop intersections that necessitate gear shifting, braking, and the use of horns.

Housing should not be located on sites where excessive and uncontrollable noise regularly occurs, especially at night. Steady noises of moderate intensity, such as those of a highway or a major traffic street, can sometimes be brought within tolerable limits by barrier planting or by deep setbacks of dwellings from the source.

Odors, Smoke, and Dust The commonest sources of objectionable odors are the following:

1. Industrial plants, especially slaughterhouses, tanneries, and other animal product factories; rubber, chemical, or fertilizer plants; textile dyeing, bleaching, and finishing plants; paper, soap, or paint factories; and gasworks
2. Refuse dumps, especially when the disposal process involves burning
3. Streams polluted by sewage, sewer outfalls, or poorly operated sewage disposal plants
4. Farm animals, especially pigs or goats when kept under crowded or insanitary conditions; also, under some circumstances, any farm animals, manure, and fertilizer
5. Fumes from heavy motor traffic and from coal-burning railroads

Common sources of smoke and dust include industry, railroads, dumps, and incinerators. Dust may also come from open untreated dirt such as vacant lots, unplanted farmland and recreation areas, or other large expanses of dirt.

The seriousness of these nuisances will depend on their intensity and frequency. Investigation should cover not only the distance of the site from potential sources but the direction of winds prevailing in all seasons. Smoke, dust, and odors may be a serious nuisance without being evident in a single investigation of the site, owing to the absence of wind or the season or time of day. Inquiries should be made of impartial persons familiar with the site over a long period. If there is any doubt as to the seriousness of smoke, odors, or dust, public-health officials should be consulted.

Excessive localized smoke, odor, or dust, unless it can be controlled, should disqualify a site for residential use. Because standards for measuring smoke are inadequate for site selection purposes and standards for odors are nonexistent, decisions as to the seriousness of the nuisance must be based on the judgment of qualified investigators. It should be noted that a continuous low-intensity odor is less likely to be objectionable than a periodic odor of equal intensity. However, an odor with unpleasant associations (from sewage disposal or

pigsty) causes greater annoyance than stronger odors without unpleasant associations.

Control of smoke and industrial odors can usually be obtained only by legal regulation, perhaps on a citywide basis. When ordinances exist, enforcement for the protection of a particular site may be possible.

2.19 Site Selection for Multifamily Housing

The following topics should be considered in analyzing sites for multifamily housing:

1. Marketability
   a. Demand for multifamily housing
      (1) At what rents?
      (2) Distribution (no bedroom or one, two, or three bedrooms)
      (3) Size of rooms
   b. Existing population and potential growth
   c. Type of existing tenants living in apartments
      (1) With children and how many?
      (2) Without children
      (3) Elderly
      (4) Single-occupancy
      (5) Income brackets
      (6) Age brackets
   d. Industries in the area and their future plans
2. Pertinent information of the surrounding area
   a. Existing street layout and how it may affect the parcel in question
   b. Proposed street changes
      (1) Widening
      (2) Elimination of streets
      (3) Map changes
   c. Location of main arteries (parkways, freeways, and highways)
   d. Mobility from site in all directions
   e. Zoning and proposed changes
   f. Kind of buildings
      (1) Single-family
      (2) Multifamily
      (3) Commercial and industrial
   g. Appearance and general character
      (1) Design of exteriors
      (2) Condition of buildings, grounds, and streets
   h. Off-street parking
      (1) Is the surrounding area provided with adequate off-street parking?

(2) Are the existing streets wide enough for street parking and easy access for cars and service vehicles?

   i. Proximity of parks, public playgrounds, other recreation areas, and waterways, if any, to site

   j. Hazards

      (1) Noise

      (2) Proximity of airports, railroads, and trucking highways

      (3) Smoke and fumes

      (4) High-tension wires

      (5) Ravines

   k. General trend

      (1) Stability of area

      (2) Building expansion

      (3) Deterioration

3. Transportation available

   a. Other than automobile

      (1) Rapid transit

      (2) Bus

      (3) Railroad

      (4) Taxis and other vehicles, such as helicopters, hydrofoils, ferryboats, and airplanes

   b. Time of travel to center of city and to job location

   c. Automobile travel

      (1) To center city

      (2) To job location

   d. Cost of daily traveling

      (1) Daily fares

      (2) Gas and oil and parking charges if by car

   e. Schedule of transport services

4. Zoning of the site

   a. Density coverage, height, yard requirements, and parking

   b. Proposed changes, if any

5. Planning boards

   a. Rules and regulations that control land development other than zoning

6. Deed restrictions

7. Community facilities (distance from the site and methods of getting there)

   a. Schools

      (1) Public, parochial, or preschool

      (2) Elementary, junior high, and high school

   b. Shopping

      (1) Necessities

      (2) All others

   c. Religious buildings

      (1) Denominations

   d. Recreation

      (1) Theaters

      (2) Playgrounds, beaches, swimming pools, bowling alleys, and others

   e. Hospitals, medical centers, and clinics

   f. Cultural

      (1) Libraries, art galleries, museums, and other cultural facilities

8. Community services

   a. Garbage and refuse collection

   b. Police and fire protection

   c. Snow removal

   d. Street cleaning

   e. Street maintenance

   f. Street lighting

9. Size and shape

   a. If irregular, can plot be utilized efficiently?

   b. If small, can an economical project be built?

10. Topography

   a. Rugged, gently sloping, or flat terrain

   b. Rock exposure or filled-in land

   c. Type of surface soil

   d. Surface drainage and groundwater

   e. Natural features

      (1) Trees, streams, lakes, adjoining parks, and rock outcroppings

11. Subsurface conditions (information usually received from borings)

   a. Composition of soil

   b. Evidence of rock or filled-in land

   c. Soil bearing capacity (necessity of piling)

   d. Underground streams

   e. Water level

   f. Percolation of soil

12. Utilities

   a. Storm and sanitary

      (1) Combined or separate

      (2) Nearby body of water or drainage ditch

      (3) Depth of sewers

      (4) Adequacy of sizes and pitch for additional loads anticipated

      (5) Public or private

   b. Water supply

(1) Pressure

(2) Reservoir, well, or other

(3) Rates

(4) Who pays for installation (from what point to what point)?

c. Gas

(1) High- or low-pressure

(2) Natural or manufactured

(3) Rates

(4) Who pays for installation (from what point to what point)?

d. Electricity

(1) Overhead or underground

(2) Current available

(3) Rates

(4) Who pays for installation (from what point to what point)?

e. Telephone service

13. Features

a. Views

b. Trees, streams, lakes, and parks

14. Cost of site

a. Potential yield (number of families)

(1) Land cost per family

(2) Land cost per room

b. Rent limitations for area

(1) Rent per room per month

(2) Relation of total rent to cost of site

c. Cost of abnormal site conditions

(1) Excessive fill and grading

(2) Piling

(3) Rock excavation

(4) Possible retaining walls

(5) Cost of bringing utilities to site

2.20 Site Selection for Schools

In choosing a school site from several parcels of land under consideration, school officials may wish to rate each site with the aid of a scorecard that carries the criteria found by experience to be most significant. Such a scorecard is shown in the accompanying table.

With the use of this scorecard, ratings for school sites under consideration may be given on a 1000-point scale. Although these ratings will reflect the subjective judgments of the scorers, they ought to fall within a fairly narrow range if the basic criteria for selection are accepted in equal degree by all the scorers. An average of three carefully worked-out scores may be expected to give a fair rating to a site. Sites with ratings below 750 to 800 points can be considered poor choices for a school. To rate a site, scores should be placed to the right of the numbers in column 1 representing the perfect or maximum scores for each item. The totals in each section should then be entered to the right of the ideal totals in column 2. The total of the column 2 scores represents the overall score for the site.

Scorecard for Selection of School Building Sites
Location of site under consideration_____

Item	Description	1	2
1. Present and future environment:			75
a. Nature of present surroundings		50	
(1) Character of nearby residential housing	General locality offering only the most favorable social influences		
(2) Freedom from business distractions	Not near commercial centers or shops that take on undesirable characteristics		
(3) Freedom from noise, odors, dust, and traffic of industry	Set distinctly apart from industry and its inconveniences; prevailing winds fully considered		
(4) Remoteness from railroads, landing fields, and docks	Without impact of disturbing conditions from these traffic centers		
(5) Remoteness from heavily traveled highways	Sufficiently protected from highway noises and hazards		
b. Protection from present and possible future air travel routes	Location approved after careful study of takeoff and landing practices	10	
c. Future prospect for surroundings	Conservation of an attractive community setting apparently assured	15	
2. Integration with community planning:			75
a. Acceptability in complete community plan	Satisfying requirements of the comprehensive community plan and contributing its share of values	50	

(Continued)

Scorecard for Selection of School Building Sites (*Continued*)
Location of site under consideration _____

Item	Description	1	2
b. Noninterference with other community projects	Sufficiently remote from hospital, church, and other community zones so that they will suffer no disturbance from large groups of children	15	
c. Value for extensive community use	Accessible and readily adjusted to adult use	10	
3. Role in comprehensive school-building plan:			100
a. Scientific determination of location with respect to present and future population	Objective techniques used to measure population in all aspects contributing to best choice	25	
b. Integration with existing schools	Serving a territory without overlapping or duplication with existing schools that have the promise of permanence	25	
c. Place in ultimate school program	Permanent dedication to education insofar as foreseeable	25	
d. Official approval of general location	Satisfactory to board of education and approved by current faculty	25	
4. Size of site:			300
a. Conformity to present and future educational programs	Making for satisfactory educational use and for educational expansion	50	
b. Compliance with following suggestions as the minimum in each case:	The minimum to be met; characteristics of locality and costs affecting final decision	150	
(1) 10 acres for a primary or elementary school and 1 acre for each 100 pupils	15 acres not necessarily excessive		
(2) 20 acres for a middle or junior high school and 1 acre for each 100 pupils	A defensible minimum because of present and future middle or junior high school programs		
(3) 30 acres for a senior high school and 1 acre for each 100 pupils	Acreage in excess of this minimum usually a good purchase		
c. Safeguarding of future educational extensions	Vision in selection encompassing all foreseeable extension needs	50	
d. Provision for present and future play areas for all groups	Character of land and orientation ensuring play and recreational facilities for all	50	
5. Accessibility:			100
a. Accessibility for general public	Free from approach and exit hazards; no dangerous gradients	25	
b. Optimum travel distances for children		25	
(1) 1½ to 2 miles for senior high school	Based upon national practices and not in conflict with local traditions; distances measured as the crow flies; travel routes protected by traffic lights and with police cooperation (these distances usually make possible schools of acceptable enrollments)		
(2) 1 mile for junior high or middle school			
(3) ½ mile for elementary school			
(4) ¼ to ½ mile for home-school units			
c. Feasibility of approaches	Pedestrian and vehicular approaches possible without congestion	25	
(1) Pedestrian	Attractive and readily traversable		
(2) Bicycle	Minimum of intersections with other traffic		
(3) Automobile	Minimum of intersections and no excessive grades		
(4) School bus	Easy access to loading center possible		
d. Safety of approaches	Safety the first consideration	25	
(1) Freedom from hazardous crossroads	Entrance and exit routes unhampered by conflicting traffic		
(2) Provision of sidewalks and good roads	Assurance of sidewalks and preferred road approaches		
(3) Elimination of conflicting travel currents	Freedom from heavy travel at school opening and closing hours		
(4) Provision of underpasses and pedestrian bridges	Artificial protection from crossing heavy or through traffic lines		

(Continued)

Scorecard for Selection of School Building Sites (*Continued*)
Location of site under consideration_____

Item	Description	1	2
6. Site characteristics:			200
a. Shape of site	Square or rectangular preferred over very irregular or "shoestring" sites	50	
b. Present utilization	Site to be free of structures involving high costs for removal	25	
c. Aesthetic value of site	Maximum capitalization of views at a distance and at close range	25	
d. Influence of site on building design	Stimulation of community-acceptable design through characteristics of site	10	
e. Possibility of preferred orientation for all rooms and all game areas	Dimensions of site offering no restriction to freedom of planning	25	
f. Prevalence of characteristics usable to educational advantage	Abundance of natural resources such as trees, water, and elevations	15	
g. Ease of surface adaptation for buildings, play areas, and parking	Surface and near-surface conditions offering no known handicap to planning	25	
h. Subsoil conditions	No excessive fill, rock, quicksand, or subsurface water conditions known	25	
7. Utility services:			50
a. Proximity of utility connections	Ready access to utilities possible	25	
(1) Water connections	Excessive trenching not required		
(2) Sewage connections	Reasonably near connections possible		
(3) Gas	Distance for gas connections reasonably short		
b. Feasibility of making serviceable utility connections	Freedom from undesirable subsurface conditions	25	
8. Costs:			100
a. Cost of land	Favorable comparison with other nearby land costs per acre	50	
b. Cost of site preparation	No unusual site features necessitating excessive costs	25	
(1) General adjustment of land contours for building and play areas	Site characteristics lending themselves to complete and distinctive planning		
(2) Sufficient elevation for safeguarding drainage at reasonable cost	Sufficiently commanding location for buildings and reasonable adjustment for play areas		
(3) Freedom from drainage from contiguous land	Proposed site, rather than adjoining land, controlling drainage problem		
(4) Ease of preparation of parking areas, entrances, and service roads	Parking areas feasible for teachers, visitors, and students; ready creation of roads possible		
(5) Additional charges for piling, rock excavation, tree removal, and the like	Site conditions causing no serious costs for these items		
(6) Removal or razing of existing buildings	Salvage value of existing structures establishing low cost		
c. Cost of utility connections	Reasonably low	15	
(1) Length of trench work necessary	Not excessive		
(2) Extent of pumping needs	Not beyond average expectation		
d. Cost of new improvements adjoining and approaching site	Much of this cost not chargeable to the school	10	
(1) New street paving required	Payment following local practice		
(2) New sidewalk installations	This requirement entailing costs chargeable to school building budget		
Maximum possible score		1000	1000

NOTE: Use the second half of column 1 for the scores on the lettered subdivisions when a specific site is being rated. The second half of column 2 permits summation of these scores.

2.21 Industrial Site Selection

Noncost Factor Evaluation   Under normal location procedures, particularly for a small or medium-sized installation, it is extremely difficult, if not impossible, to establish reliable cost values for all the factors applicable in the location study. For factors which it is desirable to consider relative to the plant location but for which, because of required expenditures, required time, or lack of facilities, it is impractical to establish reliable costs, a noncost evaluation system may be designed, and each location may be evaluated on the basis of the cumulative effect of the factor values for the particular location.

As with any situation requiring subjective evaluation, it is necessary to design an evaluation system whereby each factor is assigned a proportional value relative to all other factors under consideration, while at the same time providing a means whereby a value for each factor may be assigned to each location according to the degree or quality of that factor existing at the particular location under consideration. For plant location purposes, the steps by which to accomplish this are as follows:

1. Prepare a list of the factors considered important to the location of the individual plant that are not being evaluated on a cost basis. In the initial list all possible factors should be included. If after the start of the evaluation procedure it is found that certain factors tend to be insignificant, they can be eliminated.

2. Establish relative values for each factor. The relative value assigned to each factor should be based upon its value under the most desirable conditions. It is convenient to express such relative values in terms of assigned points. A percentage of total worth may be used under certain conditions, but when a large number of factors are being evaluated, the individual percentage values assigned are numerically so small that difficulties are created in later steps in the evaluation procedure. If desired, a predetermined total number of points to be distributed among the various factors can be established. This provides a convenient reference value during an evaluation but is not necessary.

3. Establish a number of degrees for each factor. Each factor which has been listed and for which relative values have been established may be present at any particular location in varying degrees. It is necessary to establish means of measuring variations in the presence of each factor at the various locations. Each degree represents a measured variation. The number of degrees assigned to each factor should vary from four to six. The use of four to six degrees permits relative evaluation of the factor present at individual locations without requiring unrealistic differentiation between slight variations in factor presence.

4. Define the degree. The definition of each degree is required so that all personnel involved in the evaluation use the same reference points in selecting the degree of presence of the factor at an individual location. The definition of the maximum degree for each factor would be identical to the definition of the desirable level of presence of the factor at an individual location, as established in step 2. The definition of the lowest degree of each factor will be "not present" or "present in insufficient amounts."

5. Assign point values to the degree. In step 4 we defined the maximum degree of each factor as that level for which the relative point values were assigned in step 2. Therefore, these relative point values can now be assigned as the maximum degree value for each factor. In step 4 we also defined the lowest degree of each factor as being either not present or present in insufficient amounts. Therefore, the value of this lowest degree for the location under consideration must be zero. This establishes the minimum and the maximum point values assignable to the individual factor. It is now possible by either linear or curvilinear methods to establish point values for the intermediate degrees. If curvilinear methods are used, the equation of the curve used for distribution should be the same for each factor. Linear assignment of degree values is the most common and can be accomplished in the following manner:

   a. Prepare a table, listing each factor with columns for each degree.

   b. Enter zero in the lowest degree column and the point values assigned in step 2 to the maximum degree for each factor.

   c. Calculate degree point increments. Increment value is equal to the maximum point value assigned for the individual factor divided by the number of degrees for the factor minus 1. The second degree value will be equal to the degree increment, the third degree value equal to 2 times the degree increment, and so on for each factor.

6. Designate mandatory factors. There will be certain factors within the list whose presence at some level is mandatory at the final location. These factors and the factor degree that is mandatory (which will normally be the second degree) must be designated on the evaluation form. Later, during the course of the actual evaluation, if a particular location fails to have the factor present at the mandatory level, that location is immediately eliminated from further consideration.

7. Evaluate all locations. Considering each location individually, proceed through the list of factors, selecting for each factor the degree that best defines the presence of the factor at the individual location under consideration.

8. Assign points to each location factor. Using the degree evaluations as established in step 7, return to the point value table established in step 5, and enter on the location evaluation sheet the assignable point values for each degree. Summarize the factor points earned by each location under consideration.

9. Select the location. From the standpoint of location selection on the basis of noncost factors, the location which has the highest total number of points and which does not fail to have any of the mandatory factors is selected as the best location for the particular installation.

The location selected on a noncost factor basis can now be compared with the location selected on a cost basis. If the location selected by the two methods agrees, there is a high degree of probability that the most advantageous location for the particular installation has been indicated. If the location selected by the noncost factors differs from that selected on a cost basis, it is normally wise to choose finally the location that offers cost advantages unless the differentiation of the cost advantage over the noncost evaluation is so slight as to be offset by noncost considerations. This, however, is the exception rather than the rule and must be treated as such.

Valuation of Noncost Factors   The following is a sample noncost evaluation system. Twenty-one factors with degree definition and assigned degree point values are included. An evaluation of three alternative locations is made below in the table "Converta Speed-King Co.: Summary of Noncost Rating."

1. *Nearness to market.* This factor takes into consideration the speed with which orders can be received and filled. The speed of delivery is of primary importance in building and developing good customer relations.

Degree	Description	Point assignment
0	Very inaccessible to the market	0
1	All markets relatively far from the plant	56
2	Many of the markets relatively far from the plant	112
3	Various distances to the markets fairly well distributed as to being near or distant from the plant	168
4	Majority of the market areas relatively close to the plant	224
Maximum	Location such that weighted distances to the markets are minimized	280

2. *Nearness to unworked goods.* This factor is instrumental in production planning and scheduling. Unworked goods should be available when needed with minimum delay and cost.

Degree	Description	Point assignment
0	Unworked goods practically inaccessible because of excessive travel distance	0

Degree	Description	Point assignment
1	All sources relatively far from the plant	44
2	Many sources relatively far from the plant	88
3	Various distances to the sources fairly well distributed as to being near or distant from the plant	132
4	Majority of sources relatively close to the plant	176
Maximum	Location such that weighted distances to the sources are minimized	220

3. *Availability of power.* Power should meet present and future needs. Interruptions to any extent should be nonexistent.

Degree	Description	Point assignment
0	Unavailable	0
1	Available but not of correct nature (DC, 220–110 AC, etc.)	6
2	Available and of correct nature but in insufficient quantity	12
3	Available and of correct nature and in sufficient quantity but not dependable and unable to meet future demands	18
4	Available and of correct nature and in sufficient quantity to meet all future proposed demands	24
Maximum	Available and of correct nature and in sufficient quantity to meet all future proposed demands; also excellent consulting service facilities	30

4. *Climate.* The climate provides a pleasant atmosphere for employees to live and work in.

Degree	Description	Point assignment
0	Unlivable or prohibitive to planned manufacture; corrective measures unable to change conditions	0
1	Extreme variations in climate conditions; susceptible to violent, destructive storms, floods, etc.	6

Degree	Description	Point assignment
2	Wide climate variation; infrequent destructive climatic forces	12
3	Wide climate variation; little likelihood of destructive climatic forces	18
4	Moderate climatic variations; very livable; corrective measures needed for limited periods of the year	24
Maximum	Ideal for both living and manufacturing; limited climatic variations	30

5. *Availability of water.* Water is in sufficient amount and pressure to meet drinking, heating, cleaning, and sprinkler system requirements.

Degree	Description	Point assignment
0	Unavailable	0
1	Available in small quantities at premium prices; of dubious purity for manufacturing process	2
2	Available in sufficient quantities for households but not for manufacturing processes	4
3	Available in sufficient quantities for manufacturing but highly treated	6
4	Available in sufficient quantities and pure enough for proposed manufacturing process	8
Maximum	Abundant for proposed usage; of a very pure nature	10

6. *Capital availability.* It is relatively easy to acquire capital for construction, expansion, mortgages, payroll, or other needs by loans or other means.

Degree	Description	Point assignment
0	Unavailable	0
1	Available but at exorbitant rates; very hard to obtain	12
2	Available at exorbitant rates	24
3	Equitable rates of return but hard to obtain	36
4	Equitable rates of return and relatively easy to obtain	48
Maximum	Available at low rates and in sufficient quantities to encourage location	60

7. *Momentum of early start.* There is prior availability of service facilities, markets, labor, materials, and capital, established by similar industries in the general area.

Degree	Description	Point assignment
0	No similar industry present and none coming	0
1	No similar industry present but some coming	2
2	Similar industry present	4
3	Similar industry present and more coming	6
4	Similar industry large portion of industry in area	8
Maximum	Center for given industry in consideration	10

8. *Fire protection.* Adequate facilities are available for protecting plant and employees against the hazard of fire, thereby allowing reduced insurance rates.

Degree	Description	Point assignment
0	No fire protection facilities and many fire hazards	0
1	Few fire protection facilities but no extreme hazards	2
2	Fire hazards present but excellent fire protection facilities existing	4
3	Excellent fire protection facilities available	6
4	Excellent fire protection facilities available; proposed plan to maintain protection at its present quality	8
Maximum	Excellent fire protection facilities available; proposed plan to maintain protection at its present quality; absence of fire hazards	10

9. *Police protection.* Protection is of such a nature as to prevent harm from theft or destruction of property.

Degree	Description	Point assignment
0	No police protection available; theft and property damage common	0
1	Little police protection available; theft and property damage common	4

Degree	Description	Point assignment
2	Theft and property damage occurring but excellent police protection in existence	8
3	Excellent police protection available	12
4	Excellent police protection available; proposed plans to maintain this status	16
Maximum	Excellent police protection available; proposed plans to maintain this status; low crime rates	20

10. *Schools and colleges*. There are adequate educational facilities for employees' children of all ages, for continuing adult education, and for the provision of skilled labor.

Degree	Description	Point assignment
0	No schools	0
1	Only low-quality public schools through the high school level	4
2	Only low-quality public schools through the high school level but good private schools	8
3	High-quality public schools through the high school level	12
4	High-quality public schools through the high school level; excellent private, vocational, and junior colleges; colleges or universities very near	16
Maximum	High-quality public schools through the high school level; excellent private, vocational, and junior colleges; colleges or universities very near; comprehensive plan for further adult education available	20

11. *Union activity*. This factor includes the existence of unions and their methods of attaining goals, influence in the industry, general attitude, and influence in the locality.

Degree	Description	Point assignment
0	Powerful, aggressive unions organized through the national and international level	0
1	Powerful, aggressive local unions	12
2	Weak, aggressive local unions	24
3	Unions nonexistent	36
4	Weak cooperative unions	48
Maximum	Cooperative unions	60

12. *Churches and religious facilities*. These are provided for all denominations and faiths.

Degree	Description	Point assignment
0	Nonexistent	0
1	Few denominations or faiths	3
2	Variety of denominations and representative faiths	6
Maximum	Excellent facilities for all faiths	10

13. *Recreation opportunities*. Employees should have access to numerous kinds of recreation.

Degree	Description	Point assignment
0	Nonexistent	0
1	A few poor-quality facilities	5
2	A few poor-quality facilities and a few good-quality facilities	10
3	Many good-quality facilities and a few poor-quality facilities; would meet a large range of interests	15
Maximum	Many excellent facilities to meet almost any interest	20

14. *Housing*. Housing is available in various types and in sufficient quantities at reasonable costs.

Degree	Description	Point assignment
0	Nonexistent	0
1	Largely unavailable and of poor quality	2
2	Largely available but of poor quality	4
3	Available, of acceptable quality, and at reasonable rates	6
4	Of excellent quality, in a limited range of types, and at reasonable rates	8
Maximum	Of excellent quality, in a wide range of types, and at reasonable rates	10

15. *Vulnerability to air attack.* The site is near areas that might be considered primary targets by an aggressor force.

Degree	Description	Point assignment
0	Almost assured of being a target area in case of attack	0
1	Probable target area in case of attack	2
2	Possible target area in case of attack	5
3	Improbable target area if there is an attack	7
Maximum	Nonexistent present potential target	10

16. *Community attitude.* The community is willing to accept industry as part of the community; city officials should be receptive and helpful; the general attitude should not be parasitic.

Degree	Description	Point assignment
0	Hostile, bitter, and non-cooperative	0
1	Parasitic in nature	15
2	Noncooperative	30
3	Cooperative	45
Maximum	Friendly and more than co-operative	60

17. *Local ordinances.* Are existing or proposed ordinances likely to prevent planned operation of the plant? Do they impose undue costs in meeting their provisions?

Degree	Description	Point assignment
0	Of such a nature as to prohibit location	0
1	Very restrictive and burdensome	10
2	Enforced biasedly	20
3	Regulated with discretion	30
4	Regulated with discretion and not generally burdensome in nature	40
Maximum	Nonexistent or not burdensome	50

18. *Labor laws.* Do existing or proposed laws have any derogatory effect upon the functioning of labor regulations or proposed employment policies?

Degree	Description	Point assignment
0	Strict and rigidly enforced	0
1	Strict but not rigidly enforced	8
2	Working no hardship on employment policy	15
3	Very few and not troublesome	23
Maximum	Nonexistent or of such a nature as to be conducive to good relations	30

19. *Future growth of community.* Is it likely to parallel the firm's growth and development and provide for the firm's increased demands?

Degree	Description	Point assignment
0	Community dying	0
1	Community growth stagnant	6
2	Community growing very slowly	12
3	Community growing quite rapidly	18
4	Community growing quite rapidly and not experiencing growing pains	24
Maximum	Community prospering and growing in such a manner as to enhance location	30

20. *Medical facilities.* Facilities are adequate to meet and maintain a high overall level of health.

Degree	Description	Point assignment
0	Nonexistent	0
1	Few and of poor quality	2
2	Adequate but of poor quality	4
3	Adequate and of good quality	6
4	Adequate and of good quality; proposed plans for maintenance of this high standard	8
Maximum	Full range of facilities at reasonable cost; proposed plans for maintenance of this standard	10

21. *Employee transportation facilities.* Facilities allow the convenient and rapid movement of employees to and from work as needed.

Degree	Description	Point assignment
0	Nonexistent	0
1	Of poor quality; little available	4
2	Of poor quality but variety of types available	8
3	A few good-quality transportation facilities available and at reasonable rates	12
4	Many good-quality facilities available at reasonable rates; cooperative	16
Maximum	Almost any kind of facilities available at reasonable rates; cooperative	20

Converta Speed-King Co.: Summary of Noncost Rating

Factor	Name	ATWN		CHO		TEO	
		Rating	Points	Rating	Points	Rating	Points
1.	Nearness to market	3	168	3	168	maximum	280
2.	Nearness to unworked goods	1	44	2	88	4	176
3.	Availability of power	4	24	4	24	4	24
4.	Climate	3	18	3	18	3	18
5.	Availability of water	3	6	3	6	3	6
6.	Capital availability	4	48	4	48	maximum	60
7.	Momentum of early start	1	2	1	2	1	2
8.	Fire protection	4	8	3	6	4	8
9.	Police protection	3	12	3	12	3	12
10.	Schools and colleges	3	12	3	12	4	16
11.	Union activity	0	0	2	24	0	0
12.	Churches	maximum	10	2	6	maximum	10
13.	Recreation	maximum	20	3	15	maximum	20
14.	Housing	4	8	4	8	4	8
15.	Vulnerability to air attack	2	5	3	7	1	2
16.	Community attitude	maximum	60	maximum	60	maximum	60
17.	Local ordinances	3	30	3	30	3	30
18.	Labor laws	3	23	3	23	3	23
19.	Future growth of community	3	18	2	12	2	12
20.	Medical facilities	4	8	3	6	3	6
21.	Transportation facilities	4	16	3	12	maximum	20
	Totals		528		577		793

2.22 Site Design for Wildlife

Planning for wildlife can be integrated readily into conventional planning approaches. For purposes of this discussion, let us consider planning as taking place at two levels: (1) site-level design and (2) public or large-area — regional — planning.

We are defining site design as being oriented to a specific residential development project, with or without a mix of industrial/commercial facilities. At the site design scale, a more detailed approach to wildlife planning is possible, and consideration is given to the relationship of infrastructure and other design components to wildlife.

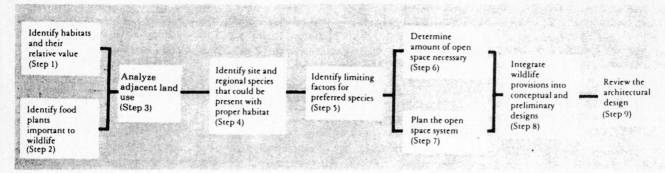

Figure 2.75 Flow diagram of procedures for integrating wildlife considerations into site design.

Larger-than-site-scale level includes all planning activities from the smallest municipal efforts to the largest regional plan. This includes city, county, state, interstate, river basin and watershed planning, and so forth. Planning for wildlife at this level is conducted on a much broader scale and focuses on the preservation and incorporation of regionally limited and/or unique habitat types in a continuous open space network.

The successful integration of wildlife is possible at both levels of planning.

This section will discuss site design considerations, since they form the basis for the discussion on integrating wildlife at the regional scale. Integration of wildlife considerations into the site design process has been emphasized purposely, since this is the planning level at which the authors feel the greatest results can be achieved.

Two basic approaches to residential site design are commonly practiced. One is the conventional subdivision with no provision for open or common space (sometimes called "lot-by-lot" development). The other is cluster or planned unit development, more commonly referred to as PUD. Projects of this type vary in size from a few acres to the multithousand-acre new-town developments.

The PUD planning approach significantly increases the potential for wildlife amenities within the community because of the amount of open space typically provided. It also allows greater opportunities for integrating wildlife because of the flexibility in the design process and general requirements for open space, preservation of natural features, and so forth. Little opportunity for wildlife planning is provided with the conventional subdivision type of development, other than in the landscaping of individual lots. For these reasons, the guidance presented for integrating wildlife into the site design process applies only to the PUD type of development.

Planning by individual residents is not considered, although much of the information presented here can be applied by homeowners.

Ecological considerations have been incorporated into the site planning process for many years and refined through the environmental planning efforts of Lewis, McHarg, and others.* Wildlife has been included as one of the ecological

parameters in the overall land development planning process. However, with a few exceptions, this has not been a totally integrated process resulting in communities that fully maximize benefits to be derived from wildlife.

Wildlife planning must go beyond the identification of important habitats and their incorporation into the open space system. Consideration also must be given to design of the open space system, the types and locations of all design components (including infrastructure), and proposed management policies and practices.

The following is an outline of a procedure for incorporating wildlife considerations into the site design process — one that does not require significant readjustment of conventional planning approaches. It begins with the identification and analysis of relative habitat types and continues through the siting of design components and architectural considerations. A deliberate attempt is made to prescribe an approach that can be accommodated by the normal complement of site designers, landscape architects, and environmental planners employed in most planning firms involved in land development.

The types of techniques and methodologies requiring a consulting wildlife biologist have been minimized or modified where possible. However, planners are urged to obtain first-hand assistance from wildlife biologists to ensure a well-integrated planning effort.

Steps in Wildlife Planning The proposed methodology for wildlife planning involves the same steps as the overall site planning process: an inventory of existing conditions followed by an analysis of the findings and their incorporation into the structure of the open space system and other components of the conceptual and preliminary designs. The steps in the procedure are detailed on the pages following and summarized in Figure 2.75.

*Philip H. Lewis, "Nature in Our Cities," pp. 23–27 in *Man and Nature in the City*, U.S. Department of the Interior, Washington, D.C., 1968. Ian L. McHarg, *Design with Nature*, Doubleday/Natural History Press, New York, 1969.

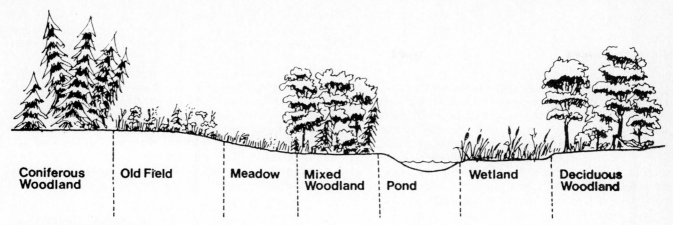

Figure 2.76 Habitat types.

Step 1. Identify Habitats on the Site and Determine Their Relative Value for Wildlife The first step in integrating wildlife into the site design process is to determine the types of habitats that exist on the site, their relative values, and their potential for habitat development (see Figure 2.76).

Each fish and wildlife species is adapted to living in a specific environment or habitat. Some species have broad tolerance to variations in soil, water, vegetation, and climate and occur in wide geographic ranges; others are restricted to rather specific ranges or habitat types. There are aquatic species and terrestrial species; species that have small home ranges; and those that require large areas to survive. Some are adapted to living in wooded areas; others to grasslands; and many prefer edges or ecotones between one type of vegetation and another. There are species or communities that are adapted for desert or semidesert areas, the high mountains, or floodplain areas.

Biotic communities of streams and rivers, lakes and reservoirs, estuaries and bays differ among themselves and from communities found on agricultural, forest, or range lands.

Habitat types for site planning can be identified, in part, from good vegetation maps. Wildlife habitats can be categorized in many ways. Individual trees can be considered as separate habitats, as can broad areas of forest or other vegetative types that may extend over several states.

The purpose of defining habitat types on a site is to distinguish between areas that differ with respect to the types of wildlife species present and/or their value to wildlife in general. Unique biological communities, particularly those containing threatened or endangered species, should be identified with the help of experts. They should be mapped and every effort made to preserve them.

Other sensitive areas, such as streams and creeks, aquifers, wetlands, woodlands, and hillsides, should be identified for protection. The American Society of Planning Officials has recommended ways that these can be protected by using the police powers invested in municipal and county governments.*

Representative habitat types are listed below to provide examples of the level of detail at which habitat differentiation should be defined for site planning purposes. All are typical of undeveloped sites, and each could feasibly become a component of the planning design.

Not all of the examples exist on any one site, and modifications would have to be made to this list for different geographical areas. Further information relative to forest cover types of North America is available from the Society of American Foresters.† In the West, for example, rangelands or prairie grassland might be added.

1. Coniferous woodlands
   a. natural
   b. plantation
2. Deciduous woodlands
   a. beech-maple woodlands
   b. oak-hickory woodlands
3. Mixed coniferous-deciduous woodlands
4. Old fields
5. Meadows
6. Watercourses
7. Impoundments

---

*Charles Thurow, William Toner, and Duncan Erley, *Performance Controls for Sensitive Lands: A Practical Guide for Local Administrators*, Planning Advisory Service Report 307/308, American Society of Planning Oficials, 1313 E. 60th St., Chicago, Ill., 60637, 1975, 156 pp.

†*Forest Cover Types of North America (Exclusive of Mexico)*, Society of American Foresters, 5400 Grosvenor Lane, Washington, D.C. 20014, 1967, 67 pp.

8. Marshes

9. Wooded swamps

10. Agricultural land

The relative value of the different habitat types on a site can be determined in many ways. One basic approach is to compare those present with habitat types in the general area (¼ to ½ mile in all directions from the edge of the site). Because of the importance of habitat diversity, those that are limited with respect to the general area should be considered of high value to wildlife. For larger sites—i.e., new-town developments—those habitat types that are limited compared with all types present on the site should be given special consideration.

Some indication of the relative value of different habitat types also can be obtained by listing the species in the region that could exist in the new community and the habitat types for which they show a strong affinity. Information on species and habitat can be obtained from secondary sources, such as national and regional field guides and other pertinent literature, plus consultation with local wildlife biologists.

Although the methodology based on the use of secondary sources may be helpful, whenever possible the planner should obtain the assistance of biologists in defining wildlife habitats present, determining their relative importance, and identifying which species may be able to be retained on the site. Often such assistance can be obtained from the regional or district wildlife biologists of the state fish and game departments, from the fisheries and wildlife department of the state university, or from the local Audubon or natural history societies.

Step 2. Identify Plant Species of Importance to Wildlife as Food Sources Plants of all types are important to wildlife, from the single-celled bacteria and algae to trees. However, in connection with urban wildlife planning, the seed-producing species are particularly important in that they provide food, cover, and nesting sites.

In step 1, guidance was given for determining the relative value of various habitats. Identification of important habitat types forms the basis for wildlife planning, but this should be supplemented with information on specific plant species. Their value to wildlife can then be integrated with habitat selection in determining which plant species to retain and which are preferable for supplemental planting and landscaping.

The choice of habitat types must include consideration of the importance of particular plant species to wildlife. This section focuses on the food value of individual plant species. All plant parts—roots, bark, leaves, fruit, seeds, and twigs—are utilized as food by wildlife. Some biologists believe that one of the major factors in attracting seedeaters, scavengers, and birds of mixed diet to urban and suburban areas is the larger amount of available food.

This is particularly true during periods when their normal food supply is limited—e.g., during winter. Waste, refuse, and bird feeders supply food directly, at the same time encouraging the presence of small animals that become the food source for larger species. Maintenance of lawns and gardens also provides an additional food supply as well as drinking water during dry seasons.

The availability of a food source near cover should be a consideration in the identification and selection of areas for producing and raising new generations of most wildlife species. After habitat types have been selected and preserved and plant species that provide good food value to wildlife have been chosen, a community richly endowed with wildlife amenities cannot be far behind.

Many studies have been conducted by wildlife biologists on food preferences of different wildlife species.

Step 3. Analyze Adjacent Land Uses Planning for wildlife in relation to site development must take into consideration what exists adjacent to the site as well as on the site itself. Many additional wildlife benefits can be derived from adjoining open space areas that can serve as potential habitats or refuge areas for species on the site. Maintaining movement corridors through the site will not only enhance the potential wildlife amenities within the proposed development but also will help realize the potential for maintaining the wildlife amenities of adjacent residential areas and the region as a whole.

An existing site in its undeveloped state can not only act as a refuge area for wildlife but provide access for wildlife to adjacent areas if left as part of a connective open space system. Such sites need to be identified and an open space system planned so as to maintain a continuous wildlife corridor. By tying the development to adjacent open space areas, species requiring larger home ranges can be accommodated. Such open space areas could include national or state forests, natural areas, wildlife management areas, golf courses, cemeteries, parks, estates, wooded stream corridors, and so forth.

Step 4. Identify Species on the Site and in the Region That Could Be Present If Proper Habitat Were Provided Although various methods have been devised for determining the types and populations of wildlife existing in an area, their application usually requires trained biologists or people working under the supervision of such biologists. When budgetary and scheduling constraints do not permit the use of biologists, secondary data sources can be used, and local naturalists or fish and wildlife biologists can be contacted regarding the availability of source materials and for help in interpreting the information.

Regional and national species lists and field guides exist for birds, reptiles, amphibians, mammals, and fish. These give broad geographical ranges for each species along with the description of their habitat types. The process then basically becomes one of listing those species whose geographical range includes the proposed site and for which preferred habitats exist either on or adjacent to the site or which can be provided (e.g., ponds and lakes).

For many sites, it may suffice to list only those species identified as common or frequent inhabitants. The planner should be alert, however, to the possibility that the devel-

opment may further endanger threatened species and should take every precaution possible to preserve their habitat.

The conversion of an undeveloped area into a residential area will further decrease the amount and types of critical habitat available. However, when site planning provides for development of ponds, lakes, or other habitats that presently are limited in the area, species adapted to such habitats can be benefited.

Step 5. Identify Limiting Factors for Preferred Species Although identifying and retaining habitats for preferred species can result in their increased production, it also can result in a decrease of other wildlife. Each individual species or group of species has its own particular requirements that must be satisfied to ensure its retention within the proposed development.

Performing the previous steps does not guarantee that specific species will be retained in desired densities or that they will be retained at all. In order to ensure retention of individual species or groups of species, the existing conditions within the site must be analyzed in relation to the specific requirements of the species. Any limiting factors must be identified and methods prescribed for their incorporation into the design or management system for the proposed development.

Step 6. Determine How Much Open Space Is Necessary The amount of open space required is difficult to determine. As indicated earlier, each species has a home range within which it satisfies all of its requirements. Although the home range of many species in the wild has been defined, little knowledge presently exists on how much area these same species would require in order to be retained within developed areas.

One of the main factors influencing home range requirements within residential areas is the effect of human disturbance. Species such as the raccoon, many songbirds, and the gray squirrel have adapted well to the presence of humans; many other species are more sensitive. To help ensure the retention of the latter species, acreage greater than that required under natural conditions should be provided to buffer against human disturbance.

Unfortunately, it is impractical for many reasons to list the home range for all wildlife species. For one, data either do not exist for all species or the samples taken have been too few to be conclusive. Determinations made for a species in one geographical area may not be applicable for other parts of the country, and the size of a home range for any one species in a given area may vary greatly according to the conditions present.

While no absolute answer can be given as to how much area is required, it can only be assumed that the larger and more diverse the open space is within a developed area, the more wildlife there is likely to be. As indicated previously, this acreage can be maximized by connecting the open space system with undeveloped areas adjacent to the site.

Also, size requirements can be compensated for to a certain degree by concentrating efforts toward providing high-quality habitat. This can be achieved by proper selection of habitat types to be retained in communal open space areas and by taking into consideration the limiting factors of preferred species. It should be apparent, also, that the way residents of an area manage the vegetation on their individual lots will have a bearing on the wildlife in the community.

Step 7. Plan the Open Space System Once the existing conditions that affect wildlife have been identified on site and within the adjacent area, one can proceed to integrate wildlife into the design of the open space system. In most cases, it will be impossible to incorporate all of the habitat types identified earlier. Frequently, these habitats would occupy greater acreage than it is economically feasible to retain. Also, they may coincide with the areas that are preferable for development or use as recreation areas.

Therefore, the integration of wildlife habitats into the open space design must begin with those areas that normally would be retained, irrespective of wildlife. These include floodplains, steep slopes, wetlands, utility line and road rights-of-way, and other areas intrinsically unsuitable or economically undesirable for development.

However, once the basic structure of the open space system is formulated, additional habitat types can be integrated, either as recreational/aesthetic components (e.g., ponds and lakes) or as part of the more detailed planning associated with the design components. Also, special landscaping techniques can be applied to large and small open space areas to increase their value to wildlife.

The areas most likely to remain undeveloped must form the spatial framework for wildlife. The task now becomes one of analyzing this acreage to see what modifications and additions can be made in light of the knowledge obtained in steps 1 through 6. It is recognized that certain limitations exist that bear on the amount and location of open space that can be retained. The areas defined in step 1 as valuable to wildlife must be integrated into this framework as well as possible within the limitations imposed. This task naturally becomes more difficult as the amount of land most likely to remain undeveloped approaches the percent of open space allowable.

Before we discuss potential modifications to the open space framework, let us consider the value of the land remaining undeveloped in residential areas. One should not assume that these areas have limited value to wildlife simply because wildlife is not a consideration in their retention. Quite the opposite is true, particularly in the case of wooded or semiwooded floodplains, which are typical in most areas.

Most species of wildlife show a stronger affinity for woodlands than for any other habitat. This would hold true for wooded or semiwooded floodplains as well. In many ways, wooded floodplains are of even greater value to wildlife than upland woodlands since the water provides an additional habitat type, thereby increasing the potential wildlife.

The ecotone formed between the aquatic and woodland habitats provides additional benefits to wildlife. The soil moisture available in floodplains frequently allows for more luxurious growth of understory species, thereby providing additional cover and screening from human disturbance.

In utilizing the undevelopable land, the first thing is to see whether it is a continuous open space/wildlife corridor system that is connected with adjacent open space parcels. If it is not, interconnecting parcels of open space should be incorporated. The most valuable habitats and groups of plant species should be used. As many corridors as possible should be provided through the site. Frequently, many are possible because of the existence of streams and related floodplains.

Where these are minimal or nonexistent, primary consideration should be given to maintaining continuity with any corridor presently existing within the immediate area and contiguous with the site.

Where possible, efforts should be directed toward developing or retaining at least two major corridor systems that are generally perpendicular to each other and meeting near the center of the site. The purpose of this type of system is to make an equitable distribution of wildlife amenities throughout the proposed development.

All corridors do not have to be equal in width, but at least one major corridor should be provided. It should be selected as the one that connects with the largest undeveloped tracts of land adjacent to the site. The primary corridor can then be supplemented with secondary and tertiary corridors of small size. These ancillary corridors extend from the major system into and through the site, encouraging the movement of wildlife into and through residential areas. The tertiary corridors may be only a row of trees or shrubs along a road right-of-way or a drainage swale allowed to undergo natural succession. They may be of value primarily to squirrels and songbirds.

The corridors should be predominantly wooded or have dense shrubbery to help buffer the effects of human disturbance. Portions that presently are not wooded, particularly those bordering watercourses, should be planted with tree species selected from the list in Section 3.8 or allowed to undergo natural succession to mature woodland. Often the latter is preferable.

Only the primary and possibly some of the secondary corridors can be defined at the early planning stages. The tertiary corridors and at least some of the secondary corridors will have to be defined more practically as the preliminary design plans are developed. Their integration probably will continue into the final design stages.

Planning the open space system of a site for which an open space framework already exists — stream corridors with attendant floodplains, wetlands, steep slopes, and so forth — calls for a different approach than a site where few preexisting determinants are present. In this first case, little opportunity may exist for incorporating additional preferred habitats.

As mentioned previously, the amount of latitude available is dependent upon what percentage the predetermined open space represents compared with the amount allowable economically. Any difference should be augmented with the most valuable habitats that remain. If the total acreage allocated for open space is already appropriated, a readjustment in unit densities can be justified if the additional acreage can

be demonstrated to improve wildlife amenities significantly for the development.

For all wooded components of the open space system, the importance of understory density, discussed in step 1, should be kept in mind. Where it is not practical to incorporate wooded areas with a relatively dense understory, the shrub layer can be increased either through supplemental plantings or by opening the overstory canopy if light restriction is the causal factor. Often, evergreen shrubs are preferable for supplemental planting because they provide cover and screening all year. Additionally, evergreens are effective at reducing noise levels in urban areas.

Once the basic open space system is established, many opportunities exist for integrating additional habitats into the design process. Some of these are quite obvious and are frequently included as part of the overall design; others are much less obvious and frequently are overlooked.

Ponds, lakes, and retention basins are examples of the more obvious design components. If planned and designed properly, they can increase potential wildlife benefits within a community significantly by increasing habitat diversity.

For maximum utilization by wildlife, impoundments should be incorporated into or adjacent to the open space system. Those surrounded completely by developed areas offer few wildlife amenities. To provide use by both humans and wildlife, medium- or high-density development can be accommodated along one side of the impoundment if the open space system or very low density housing borders the opposite side. The utilization of impoundments by wildlife is also dependent upon design.

It may be difficult to justify the retention of old fields in developing site plans. Typically, these represent prime areas for development. However, old fields are very valuable to many wildlife species. Because of the ecotone provided, their value is significantly increased when they are adjacent to wooded areas.

Many opportunities exist for integrating old fields within the site. Road rights-of-way adjacent to the open space system are an excellent example. Allowing these areas to undergo natural succession not only benefits wildlife but also reduces maintenance costs. A border of approximately 20 feet adjacent to all other portions of the wooded open space system also should be allowed to develop naturally into the vegetation typical of old fields.

The same practice can apply to utility rights-of-way and borders of drainage swales if mowing or selective herbiciding is done only once every 2 to 5 years. Even though these areas will be relatively narrow, their value to wildlife is significant.

The landscaping efforts normally performed by the planning firm also can significantly supplement benefits derived from the open space system. The presence of properly selected trees on the borders of roads, for example, can increase songbird diversity and numbers.

Step 8. Integrate Wildlife Considerations into the Conceptual and Preliminary Designs Thus far, guidance has been given on how to structure the open space system to derive

maximum benefits for wildlife. Although this may be considered as a major accomplishment, the process should not end here.

The relationship of the design components to the open space also is an important consideration. Many of the wildlife benefits anticipated in the planning stage may not be realized if wildlife considerations are not carried through into the design process.

In this section, basic wildlife planning concepts are discussed with respect to typical design components—roads, commercial and industrial facilities, outdoor recreation, and housing units. The level of detail at which each is discussed varies according to its relative importance with respect to wildlife and the amount of information available.

Two major areas of concern must be addressed with respect to the design components. These are (1) minimizing adverse physical impacts on components of the open space system that are important to wildlife and (2) minimizing the effects of human disturbance.

Minimizing Physical Impacts As a very basic rule of thumb, all efforts should be made to minimize adverse impacts on any component of the open space system. Particular care must be taken to minimize any breaks in the open space/wildlife corridor system, particularly those caused by roads and highways. It is recognized that most roads designed in association with site planning serve some type of development on both sides. In some cases, such as entrance roads to the development, freeway extensions, and segments of major arterials, efforts should be made to align road segments in such a way as to help buffer human disturbance to refuge areas and to restrict access.

Where crossing the open space system cannot be avoided, roads should be planned in areas where the open space system is narrow. Large wooded areas should be kept intact to maximize their utility as refuge or sanctuary areas. Overpasses or underpasses should be considered to facilitate wildlife movement.

Road width plays an important role with respect to wildlife. A study on the effects of roads on the movement of small and medium-sized mammals in forested areas suggested that divided highways of 90 or more meters (293 feet) in width may provide a barrier to movement comparable with a water body of 180 meters (585 feet). Therefore, all efforts should be made to minimize crossing the open space system with major roads.

Where roads cross watercourses, culverts also may act as barriers to movement by reptiles, amphibians, fish, and small mammals. Where the use of culverts cannot be avoided, they should be oversized or located so as to permit animal passage.

Minimizing Disturbances The reactions of wildlife to human disturbances such as noise and physical activities vary. Continuous noise and movement has little effect on most species, whereas discontinuous or intermittent types of disturbance may greatly affect wild animals. For example, most species can be approached quite closely by a moving vehicle.

They will remain nearby as long as the motor continues to run and the occupants do not make sudden moves. But, as soon as the motor is shut off and someone gets out of the vehicle or the car door is slammed, the animal will make a quick exit.

In planning the areas surrounding an open space system, those components generating disturbances of an intermittent and discontinuous nature are of greatest concern, whereas those whose disturbance factor is more or less continuous are of less concern. Housing units and outdoor recreation activities are examples of the former; roads and highways represent the latter, even though traffic causes considerable wildlife mortality.

Different housing types generate different levels of disturbance. It is safe to assume that the major cause of disturbance is associated with children and dogs and cats. Therefore, housing types producing the greatest concentration of children and pets represent the major area of concern. The high-bedroom-count units (single-family detached and large town houses) typically contain the largest number of children and pets per unit. Low-bedroom-count units and units for the elderly generate the least disturbance.

Garden and high-rise apartments, although typically containing much smaller bedroom-count units, may produce similar concentrations of children and pets simply because of the larger number of units per area. However, most of the children in this kind of housing are typically much younger and therefore more confined in their activities. It also has been observed that people of all ages in apartments, particularly high-rise, tend to make less use of adjacent natural open space than those in single-family homes.

In addition, although the concentration of pets per unit also may be higher, they generally are more confined. When allowed outdoors, their activities are more likely controlled. In single-family-home developments, there is more tendency to allow pets to roam freely.

Concern for human disturbance therefore focuses on high-bedroom-count single-family units. Town houses with comparable bedroom counts are of greater concern than single-family detached homes, particularly those on larger lots, because of the greater number of individuals per acre. Also, single-family detached homes typically have more property, which helps to buffer the effects of disturbances.

Given the fact that certain types of housing developments do not mix well with wildlife, it is desirable for planners and developers to consider what actions can be taken. Enactment and/or enforcement of laws to prohibit pets from roaming and educational campaigns to inform people about the impact they may have on wildlife are possibilities.

The planting of thorny vegetation along borders may be sufficient to discourage activities of children in areas of particular value to wildlife. At the same time, well-developed nature trails may enable children and their parents to derive benefits from the wildlife. Trails should be elevated over wet areas to avoid disturbance of the vegetation. The posting of signs explaining why tortoises and other animals should not be collected for pets would be helpful.

Most types of commercial, office, and industrial facilities are design components that pose less disturbance for wildlife in the adjacent open space system. Although much activity may be generated at commercial areas, the noise level is generally low; and little of the activity affects adjacent open space areas. Commercial development therefore may help buffer the larger refuge components of the open space system.

The effects on wildlife can be minimized for small commercial facilities by placing all parking in front of the buildings. For larger commercial areas — mini-malls and malls — it is preferable to separate the buildings from the open space system with parking. The fringes of the parking areas in these larger commercial facilities are used infrequently and therefore act as buffers for the open space.

Industrial facilities vary greatly in the amount of disturbance generated. Those typically integrated into planned-unit developments are of the "clean industry" variety and can be considered in the same light as the commercial facilities discussed above. Because of their buffering effect, employee parking areas can be placed between the open space system and the buildings.

Those types of recreation units — i.e., playgrounds or baseball lots — that have the greatest potential for generating human disturbance should not be located adjacent to the most sensitive components of the open space system. It is difficult to say what parts of an open space system would be more sensitive to human disturbance. The existing literature provides little guidance in this area. However, it seems reasonable to conclude that the larger wooded areas are the most sensitive since they have the greatest potential for serving as refuge areas for wildlife. The more they can be protected, the more likely they will be to function in this capacity.

The narrower portions will support some songbirds and small mammals and also will function as travel corridors for other species.

Step 9. Review the Architectural Design Although architectural design is not normally considered as part of the site planning process, for wildlife planning purposes certain principles should be considered. Architectural design is related to the bird population of any developed area. Even if full consideration is given to wildlife in planning the open space system, the design components, and site improvements, a bird population dominated by such pest species as pigeons, starlings, and house sparrows may still result. For this reason, architectural design also should be considered in the wildlife planning process.

Among the reasons suggested for the success of pigeons, starlings, and house sparrows are that they are all birds characteristic of urban areas in the Old World, where they had time to acclimate to such conditions; and they all are either hole nesters or they build their nests in crevices of buildings, where they are protected from predation and accidental disturbance of the nest. They generally are nonmigratory, which gives them an advantage over migratory species in the com-

petition for nest sites; and there is less seasonal fluctuation in the availability of food in cities than in other habitats.

The ability of these birds to use buildings can be attributed sometimes to the design of the buildings and sometimes to poor construction. Some commercial and residential buildings have proved to be excellent roosting and/or nesting sites. Certain sections of new communities, with buildings of one type of design or construction, may harbor one or more of these three species, whereas other sections of the same new community may have few of them.

Buildings with unboxed eaves having small openings beneath the roof afford starlings and house sparrows access to attractive nest sites. A recent study conducted in Columbia, Maryland, showed that these birds were able to remove some of the fine wire screening that covered air vents and use these protected sites for nesting. Heavy screening on vent holes kept the birds out, but frequently building flaws allowed continued access elsewhere on some buildings. Louvered air vents also resulted in increased house sparrow populations, and vents of fans were associated with large numbers of nesting starlings.

Pigeons also moved into some apartment and commercial areas in Columbia in great numbers. Overhanging portions of some of these buildings were supported by large exposed I-beams that provided roosting and nesting sites. The population of these nuisance birds could have been reduced significantly if wildlife had been considered in developing the architectural design.

Some researchers believe that the increased use of reflective plate glass on buildings may account for the death of many birds. When reflective glass is used on buildings in wooded areas, birds in flight have the illusion of additional trees or woods where the buildings stand because of the reflection of the trees.

Review of Development Plans Municipal planners also can help preserve wildlife amenities through review of site development plans. It is at the site plan review stage that the greatest concessions favoring wildlife can be obtained, since developers are likely to be willing to accommodate changes in order to obtain permits.

When reviewing development plans, municipal planners should use the steps outlined above for incorporating wildlife into the site design process, with one addition. If regional planning efforts have included consideration of wildlife, the site plan should be reviewed in light of these plans. This includes determining whether the site contains habitats of threatened or endangered species and regionally limited or unique habitat types and whether there is any linkage with regionally defined open space corridors.

The plans then should be reviewed to ensure that valuable habitats on the site have not been eliminated in favor of others of lesser value, that a continuous open space system has been provided wherever possible, and that the design is sensitive to the effects of human disturbances on wildlife.

Finally, the detailed design should be reviewed for such things as tree and shrub plantings of value to wildlife. Such

plantings would substitute, in part, for natural forest areas that have been eliminated or made less valuable by having the understory cleared. The review should involve consideration of impoundments of benefit to fish and wildlife. The developer should be alerted to the possible detrimental effects of developments on streams.

The proposed architectural design can be discussed also, to make the developer aware of the potential effects it can have on nuisance bird populations.

2.23 Waterfront Considerations of Site Development

Integrating aquatic resource considerations into the planning process offers opportunities to (1) maintain or protect high-quality or unique aquatic ecosystems, (2) enhance or rehabilitate disturbed systems, and (3) create new aquatic resources. This section suggests a procedure for incorporating helpful steps to ensure that water bodies in urban areas suitable for urban fishing or other desired uses are part of the plan.

Aquatic ecosystems are complex, involving not only the water and living organisms within or dependent on it, but also physicochemical components and their interrelationships with plants, animals, and land use on watersheds. Planning should focus on the entire aquatic ecosystem. Aquatic

resource optimization depends on site planning that is sensitive and responds to the natural functioning of the aquatic environment. Figure 2.77 shows suggested steps for aquatic resources planning.

Step 1. Identify Existing Aquatic Resources   An inventory of existing aquatic resources should be made early in the planning process and serves as the basis for planning and management efforts designed to protect or restore an aquatic resource. It should focus on the following:

Delineation of Aquatic Resources and Description of Their Structural and Functional Characteristics   Specificity and topics covered in an inventory can vary considerably, but the more detailed it is, the more useful it is. An inventory simply identifying the name and location of a lake or wetland is less useful than one giving size, physico-chemical-biological properties, and relationship to other water bodies and surrounding land uses. A detailed inventory supplies data useful in identifying functional characteristics, outstanding features, limitations, and special management-protection measures.

The planner must recognize different water body types. Because wetlands are transition areas, it is often difficult to delineate their margins. Streams and rivers are easier to define and delineate, but the inventory should be sensitive to changing physical, chemical, and biological characteristics throughout their course. Typical inventory outlines for lakes, ponds, wetlands, streams, and rivers are listed. A detailed inventory for estuarine and marine areas might include gen-

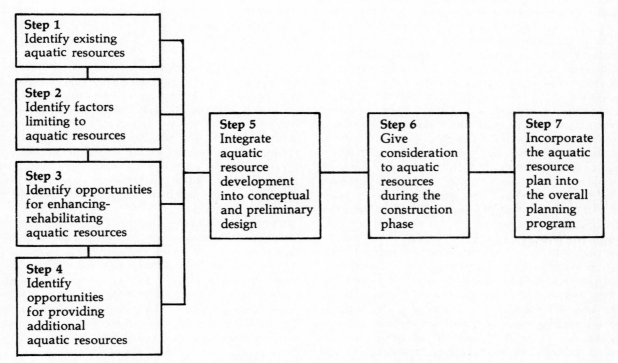

Figure 2.77   Flow diagram of procedures for integrating aquatic resources into site design.

eral information on longitudinal, vertical, and lateral features with data on the specific site.

Evaluating Functional Characteristics of the Aquatic Resource An inventory should also pay heed to fin and shellfish spawning and nursery areas, migratory fish passageways, fishery support areas, wildlife usage, unique biotic communities, scenic, cultural, scientific, and aesthetic values, recreation and educationally important resources, and other water uses. The inventory should reveal not only the present use, but what uses, like fishing, could be supported after a resource's restoration.

Determine If Critical or Outstanding Areas Exist Some factors to be considered in determining water bodies deserving special protection or attention are area use, relative scarcity of the aquatic resource, its proximity or accessibility for human use, and its vulnerability.

Identify Factors Influencing the Importance of Resource Functions These factors may vary from an area's size and location, habitat diversity, water quality, and substrate composition with respect to fish and wildlife presence and abundance, to access and availability for recreation.

The inventory should provide a basic framework for future protection and enhancement for the site and larger-than-site planning scales. By evaluating structural and functional characteristics, critical areas and specific water quality criteria needed for enhancement and protection measures can be identified. These measures and critical area designations can be incorporated directly into performance standards and land use control schemes. Criteria should serve, too, as a basis for sound planning and decision making during the conceptual and preliminary design stages of construction.

Methods of Inventory and Evaluation Inventory information can be obtained from federal, state, and private sources. Recent aerial photographs or U.S. Geological Survey topographic maps are useful in determining a water body's size, drainage patterns, surrounding land uses, proximity to other water bodies, and similar features. County planning commissions are a convenient source for information on land uses and some aquatic resource features. River basin commissions and regional planning commissions may be tapped for information. For species verification, occurrence, and distribution in a water body, or for other technical information, employ or seek the services of specialists. Maps having overlays depicting existing and proposed land uses, vegetation, erodible soils, aquifers, and steep slopes are useful to understand resource limitations and interrelationships between land and water resources.

Step 2. Identify Existing Limitations to Aquatic Resources Most water bodies in urban and suburban areas have been disturbed by man. In many instances, water quality is the major limiting factor, and, in others, hydrological or structural modifications limit the aquatic resource. Water bodies can be rehabilitated if corrective action is taken, thereby enhancing these resources for multiple use. It is important to identify the source of disturbance, and to develop strategies for eliminating or minimizing it. The resource inventory can be used to identify limiting influences on a body of water.

Water Quality Considerations Common water quality limitations include depressed oxygen levels, elevated temperatures, turbidity, nutrient levels (including phosphorus, nitrogen, and carbonaceous matter), and contamination of bottom sediments by metals and other toxic materials. Depressed oxygen levels often occur near the outfalls of sewage treatment and industrial plants. Since oxygen solubility is strongly affected by water temperature (the higher the temperature, the lower the solubility), oxygen depletion may occur locally, also, in the heated effluent waters of electric power stations or industrial plants.

Turbidity is caused by suspended organic and inorganic matter and, in some cases, by suspended microscopic plants. Sediment loading in runoff from the watershed and erosion within the stream channel are primary causes of excessive turbidity, but dredging, mining activities, navigation, and recreational uses also contribute to elevated turbidity. High turbidity and suspended fine-particle sediments drastically reduce the numbers and kinds of organisms present in an aquatic ecosystem.

High concentrations of dissolved nutrients encourage nuisance algae blooms in water bodies. Bloom death and decay exerts high oxygen demands, resulting in fish kills on a massive scale. High concentrations of ammonia can also poison aquatic organisms. Sediments contaminated by heavy metals, pesticides, and other toxic materials have a devastating effect on aquatic species' diversity and abundance. Hence, anything that can be done to promote adequate waste treatment and control urban runoff of materials like deicing salts helps ensure water bodies suitable for fish and water-related recreation. Methods for identifying and evaluating the nature and extent of pollution and contaminants entering urban waters from streets and other sources are obtainable from the EPA.

Structural and Hydrological Alterations Channelization, highway and residential construction, navigation, and other projects limit many aquatic ecosystems. Bridge crossings, culverts, and highway drainage can all affect river channel dynamics and the aquatic community. Channel modification sometimes eliminates valuable inshore river shallows by dredging, filling, bulkheading, or other operations affecting streams. Channel alteration can cause major changes in stream or river erosion and sediment patterns which ultimately affect the substrate on which many aquatic life forms depend.

Dams and impoundments placed across rivers should be identified and evaluated concerning fish movement effects, downstream receiving waters' chemical quality, flow regimes, and habitat suitability for aquatic organisms. In tidal areas, note any barriers like dikes and levees across wetlands and estuaries. Also note any undersized culverts impeding or preventing sufficient fresh and salt water exchange, because changes in the salinity regime can alter the wetland/estuarine ecosystem. Many species in these areas exhibit definite salinity sensitivity.

In western states stream flow diversion into impoundments, canals, and drainage ditches is common. Loss of flow,

or even seasonal modification of a river's flow can alter an aquatic ecosystem drastically.

Aquatic ecosystems are highly susceptible to disturbances involving water quality and physical habitat alterations. If aquatic resources are to be enhanced or protected, planners must evaluate existing and potential resource limitations owing to these disturbances and identify their sources.

Step 3. Identify Opportunities for Enhancing and Rehabilitating Aquatic Resources Limitations identified in step 2 should be carefully reviewed to determine opportunities for enhancing or rehabilitating degraded aquatic resources. Aquatic habitat surveys in most developed areas, and often in those yet to be developed, will indicate need for improvement. In most instances water bodies can be improved by better solid erosion and pollution control, or changes in land use on watersheds. Planners, assisted by biologists, may identify causes of aquatic habitat degradation and recommend solutions to the problems.

Site planners, in cooperation with biologists and engineers, may devise ways for improving water quality without federal or state assistance by enactment and enforcement of local ordinances or by other means. Key enhancement measures to improve the aquatic environment are discussed below.

Erosion Control at Construction Sites Erosion and sedimentation can be controlled if the following principles are used in the treatment of land-using soils suited for development: leave the soil bare for the shortest time possible; reduce the velocity and control runoff flow; detain runoff on the site to trap sediment; and release runoff safely to downstream areas.

Stream Bank Stabilization and Protection Erosion from bank instability of rivers and other water bodies and within river channels themselves can be reduced by stabilization through vegetation and by artificial measures. However, vertical-walled structures like sheetpilings and bulkheads are less well suited for aquatic life; sloped natural stone riprap construction is preferable.

One of the most effective means for enhancing aquatic resources is to maintain vegetative buffers along water bodies and by developing such buffers where they do not occur. When composed of grasses, shrubs, and trees, buffers not only help stabilize stream banks, but entrap sediment, aid in adsorbing pollutants in overland runoff, and provide food and cover for fish and wildlife. Soil Conservation Service offices can provide plant listings suitable for different localities. Generally, a border extending at least 15 meters from the high-water mark with a minimum of 36 meters total is recommended. (See Figures 2.78 and 2.79.)

Removal of Sediment Where massive amounts of sediment have accumulated and are known to be a major factor in limiting the health of an aquatic ecosystem or the use of the water for water-based recreation, excessive sediment can be removed. It should be recognized, however, that this is a temporary means for enhancing lakes and ponds because they will fill in again quickly if accelerated erosion on the watershed is not prevented. Unless care is used in the process, sediment removal operations can contribute to turbid-

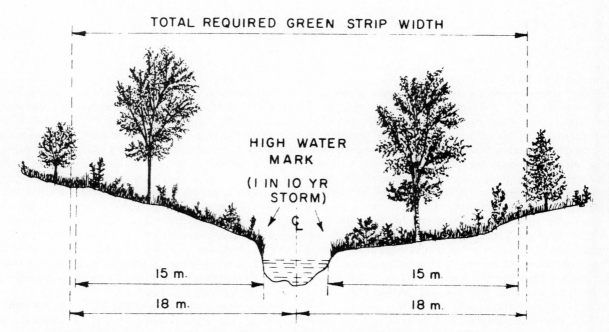

Figure 2.78 A watercourse with a well-defined high-water mark. In this example, 18 horizontal meters from the stream centerline is greater than 15 horizontal meters from the high-water mark.

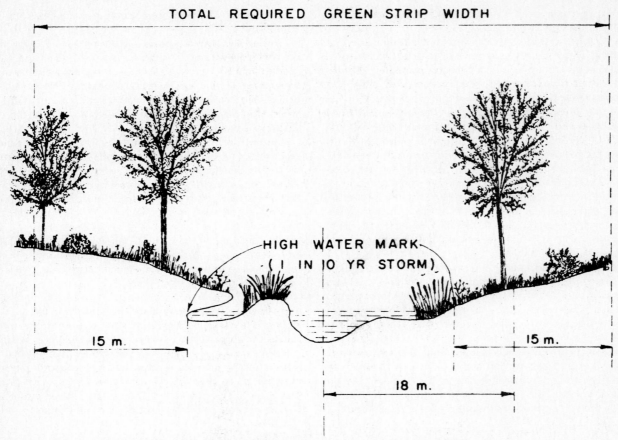

Figure 2.79 A watercourse with an ill-defined high-water mark. In this example, 15 horizontal meters from the high-water mark is greater than 18 horizontal meters from the stream centerline.

ity; dissolved oxygen levels can be reduced through oxidation of organic and other sediment-bound substances; nutrients, metals, and other toxic substances can be released from stirred-up sediments, altering water quality. Therefore, suction dredges or other methods for sediment removal reducing resuspension to a minimum should be used. Information about dredging methods, its undesirable effects, disposal of dredged materials on aquatic organisms, and alternative dredging techniques is available.

Rehabilitation of Lakes Excess sediment removal from lakes is but one of many methods available for rehabilitating lakes and reservoirs. A common objective of lake or reservoir improvement is the control of nuisance growths of aquatic plants interfering with recreation and which have other undesirable effects. Planners and municipal decision makers are reminded that excessive nuisance plant growth removal will only be temporary if they exert no control over nutrient quantities entering the lakes.

Rehabilitation of Rivers and Streams In addition to sediment removal, stream rehabilitation may involve removing channel obstructions or installing fish ladders and the like to permit fish passage. Debris dams and other obstructions, while often considered to be a major factor in fish migration, flooding, or eliminating spawning areas, sometimes should be left intact because they serve as important habitats for aquatic life. Channelized streams can often be improved by installing stone or log deflectors, gabions, and low dams. Many urban streams and rivers may be improved insofar as aquatic resources are concerned by restoring inshore shallows via removing hydrological obstacles like revetments. Where bulkheads are in disrepair, they can be removed and replaced by sloping riprap to offer a more hospitable habitat than vertical-walled structures.

Wetlands Restoration Although dikes may be necessary to manage water levels in some wetlands effectively, many of the dikes are in disrepair or obsolete. Often their removal

can rejuvenate a wetland system and surrounding water bodies dependent on wetland functions. By restoring tidal flushing in coastal wetlands, deteriorated water quality caused by sediment and toxic substance accumulations can be reduced. Simultaneously, marsh productivity and usefulness as shelter, fish and waterfowl feeding, and a fish spawning ground will be increased. By furnishing connections between wetlands, corridors valuable for fish movement are created. Water circulation can be improved, too, by replacing undersized culverts with ones that can better accept runoff and assure sufficient tidal flow.

Removal of Undesirable Species   In addition to controlling undesirable growths of vegetation, control of fish species such as carp may be warranted in some situations because they contribute to high turbidity levels with consequent effects on a water system's productivity.

Step 4. Identify Opportunities for Providing Additional Aquatic Resources   In some settings new water areas can be created to provide additional recreational opportunities and, in some cases, serve as a trade-off for other aquatic resources that have been disrupted or destroyed. Treatment effluents may be useful for creating new aquatic areas as well. Siting and design considerations are important, whether the water area to be developed is a wetland, pond, lake, or other water body, even when it is intended specifically for recreation or multiple use.

Ponds and Reservoirs   Ponds or reservoirs created by damming streams or drainages often require expensive diversion structures and large-scale dams to accommodate watershed runoff, and in large reservoirs, wind-generated waves can cause erosion. Therefore, many siting and engineering design aspects must be considered to ensure a valuable aquatic resource following construction. Among factors to be considered are the amount and quality of runoff and influent waters, size and configuration of the impoundment, need for water level control devices, and downstream impoundments effects created by river damming.

Wetlands   Most wetland establishment schemes involve dredge disposal sites, extant sea level habitats such as tidal flats, or excavating depressions in upland habitat into which water is introduced. Wetland rehabilitation usually involves water level and vegetation control along with occasional seeding or transplanting of aquatic plants. Though newly created and rehabilitated wetlands differ from those that have existed for thousands of years, the continual organic matter buildup in the substrate of newly vegetated wetland affords new opportunities for colonizing and establishing benthic organisms and habitats for fish and wildlife as wetlands mature.

Ditches and Canals   Drainage ditches have reduced wetlands in both urban and rural areas. While more drainage is not encouraged, present roadside ditches establish aquatic habitats in areas where they are lacking. Existing ditches might favor aquatic organisms by maintaining vegetated strips along them. Canals are important for recreation. Water area diversity and extent is achieved by widening canal portions to form a lake, varying the canal bank's slope, and

diversifying the vegetation, to encourage different fish and wildlife populations.

Step 5. Integrate Aquatic Resource Development into Conceptual and Preliminary Design   A comprehensive planning format needs an integrated approach to land development and aquatic resource control. Municipalities' and developers' initiative and interest in integrating aquatic resource development into conceptual and preliminary design phases of a land development project can help preserve and enhance the resource during site construction. Planning and design guidelines for accomplishing this follow.

Siting Considerations   Preliminary evaluation of a site's suitability for a particular land use or construction activity should be based on an inventory of key natural resource features. Features to be considered are:

1. Base soil limitations and suitabilities on erodibility, water filtration capacity, on geology, slope length, and gradient. Avoid development in those areas prone to non-point-source loading. Areas having steep slopes (greater than 15 percent) and short slope length or where soils are highly erodible should be excluded from development. Maintain them instead for permanent open space because their development is likely to affect aquatic ecosystems adversely, requiring expensive sediment and storm-water control.

2. Consider critical aquatic resources or systems highly vulnerable to disturbance, like wetlands harboring threatened or endangered species of fish and wildlife. Identify areas overlying underground aquifers. Rivers — especially those included in state or federal scenic river lists, floodplains, and areas representing important ecological, educational, and recreational opportunities should be examined.

3. Note hydrologic and hydraulic features (surface and subsurface). Sites where development would impact on water bodies having hydrologic significance (e.g., water supply, flood control) should be excluded from development.

4. Perceive any watershed problems upstream or downstream from proposed construction sites. Unless methods surely control disturbances from proposed construction, it is best to avoid areas where such impacts exacerbate drainage problems.

Vegetative buffer zone installation along water courses and exclusion of sewage disposal systems proximal to their margins need identification and control during preliminary design phases, thus preventing future adverse environmental impact. Where water bodies support cold water fisheries, it is important to identify and plan for preservation or development such that wooded areas and springs supplying cool water will continue to do so.

Power plants must be carefully sited to avoid disturbing habitats for threatened and endangered species. Research has resulted in screening devices at cooling water intakes

which now help prevent unnecessary aquatic animal mortality through screen impingement, and stop entrainment.

Planning for Water Pollution and Storm-Water Control Once a suitable site has been determined, an effective water pollution abatement plan should be devised to protect aquatic resources both at the site and within the watershed. The plan should reflect planning principles with regard to the landscape unit addressed above and, relative to water resource protection, should consider the following: groundwater disturbance; construction on or near potential landslide or mudslide area; stream crossing structures; land fill, culvert, dike, and building encroachment on stream flow; influences on storm-water runoff imposed by an increased area of impervious streets, parking lots, and buildings; changes in drainage caused by diversions and gradings, sediment spoil and other solid wastes disposal; floodplain excavation work; stream channel modification; petroleum waste; pesticide, and other chemical disposal; control of dust; access and haul road construction; sewage treatment; construction site proximity to streams, lakes, and other vulnerable water bodies; vegetation alteration; wetland modification; natural circulation pattern interference with respect to tide; and sediment erosion.

A plan for sediment and storm-water control, which also relates to pollution control, should be devised for all development sites. This plan should include means for:

1. Controlling water runoff speed and volume on the construction site: Options for accomplishing this objective include natural drainage systems, grassed swales instead of concrete ditches along roadsides in certain types of housing developments, diversionary structures to delay runoff delivery to watercourses, rooftop or parking lot ponding or use of recreation areas for temporary water storage, lawn and golf course aeration increasing infiltration, inline storage in sewers, instream, side-channel and off-channel watercourse storage, and use of dry detention basins or small permanent sediment ponds.

   Sediment basins and storm-water control ponds can be valuable for fish and wildlife if allowed to remain in place after construction.

2. Minimizing pollutant loadings: Urban street routing and design can govern pollutant quantities entering urban water bodies. Polluted water from urban streets contains toxic contaminants from many sources, including automobiles and lawns where insecticides have been used. Serious damage to aquatic organisms may occur as a result, particularly during initial pollution loadings resulting from storms. Some EPA suggestions for reducing nutrients, toxic, and oxygen-demanding substances introduced from street runoff during a storm are:

- Select roadway sites minimizing the area draining directly into the receiving water body.
- Use low curbs where the road joins flat unpaved areas or those sloping gently away from the street surface. This will

ease dust and dirt deposition into grass and gravel and reduce the deposition rate of runoff water.

- Consider using porous pavement where climate and soil types will permit it.
- Intensify and improve street-cleaning operations to reduce urban runoff effects.
- Design curbs and gutters to ease concentration and collection of particulate material.

Where possible, routing runoff water through a wetlands area before it enters a river or lake helps remove sediment and suspended solids.

3. Determining sediment control measures at the site: A variety of sediment control measures are considered in the next step (6). However, careful consideration should be given in the preliminary design stage to ensure cost-effective and environmentally sound measures for entrapping sediment at the development site.

4. Determining soil stabilization practices in relation to seasons: When possible, construction should be scheduled to avoid heavy rainfall months so prompt reseeding and sodding can reduce erosive soils exposure.

Planning Tools Effective site plans are built primarily by compiling and analyzing data derived from local, county, state, and federal records, supplemented as necessary by special studies or surveys. Maps constitute a valuable tool. Base maps having a scale of 2000 feet to the inch (contour intervals = 20 feet) are useful for generalized site planning purposes, but scales ranging from 50 feet to the inch (contour interval = 5 feet) are more suitable for average-sized sites, storm-water, and sediment control plans.

Map and other data availability does not eliminate the need for detailed on-site studies by specialists to provide guidance on identifying critical areas and designing storm-water, sediment, and erosion control methods. Depending on the development site, agronomists, soil scientists, geologists, hydrologists, engineers, landscape architects, economists, site planners, and biologists might participate. Composite maps depicting specific natural resource limitations as they relate to aquatic resource protection are particularly valuable during preliminary design stages to identify optimal development sites and areas having moderate to severe restrictions on development.

Step 6. Give Consideration to Aquatic Resources During the Construction Phase Although judicious development site selection and layout design avoids many erosion-sedimentation and pollution problems, controls must be instituted during the construction phase for any development project. Some of these have been alluded to in the previous section; others will be addressed here.

Staging of Construction Construction projects and associated grading and revegetation operations should be staged so that a minimum amount of soil surface is exposed at any time. For large-scale developments that take some time to complete, it is not necessary to denude and grade the entire tract at one time. Schedule earth-moving operations at sea-

sons when seeding or planting can be done to revegetate the disturbed areas quickly.

Erosion Control Surface roughening involving scarification and serration of exposed slopes reduces runoff velocity and thereby the extent of erosion. With slopes roughened, seed and fertilizers are less apt to wash out and vegetation can become better established. Various methods for intercepting and diverting runoff and establishing vegetation have already been described. Many local SCS offices can provide listings of plants suitable for specific sites, based on slopes, soil condition, and maintenance expectancy. They can provide information, too, on structural methods for stabilizing soils.

In building siting, and construction, natural vegetation, including shrubs and ground cover, should be left intact insofar as possible. Trees provide shade and protection from wind, add to residential property value, attract wildlife, and reduce erosion. The Agricultural Research Service has developed criteria for determining whether trees are valuable enough to justify removal from construction sites for transplanting elsewhere, and guides to methods that can be used to protect them during construction.

Control of Sedimentation Sediment abatement controls for a long-term project's life should be instituted during the construction phase, also. As indicated previously, maintaining vegetation buffer strips along waterways is a useful approach. Structural controls, including gravel inlet filters, sediment traps, permanent wet sediment basins, and other sediment control practices can be incorporated directly into the storm-water management plan for a construction site.

Other Pollutants Many pollutants are kept from escaping the construction site by effective storm-water and erosion-sediment controls and good "housekeeping" practices in disposing of excess paints, asphalt products, pesticide containers, etc.

Hydrological and In-Channel Modification Hydrological and in-channel modification during the construction phase should be kept minimal. Riparian vegetation should be protected during development, and heavy equipment should be kept out of streams. To stabilize shore lines, use native vegetation or gabions rather than impervious or vertical wall structures like bulkheads. If bulkheads are used, they should be located no farther waterward than mean high tide. Sloping riprap construction, using appropriately sized rock, is preferable to concrete structures.

During some construction projects, obstructions and debris like fallen trees or limbs are often removed because they are believed to contribute to flooding or present an "untidy" appearance. Unless such debris and obstructions are a major factor in flooding or interfere with recreation, they should be left within the channel because they serve as important habitats for aquatic organisms.

Consideration should be given, also, to the potential effects of development on the hydrology of adjacent or nearby water bodies. Projects should be designed to ensure natural circulation patterns, salinity regimes, and nutrient dis-

tributions so that aquatic life is not altered in wetlands, estuaries, and marine settings. Thus, docks and piers should be designed so they do not restrict circulation and should be located in areas where there are existing channels or where initial and maintenance dredging will be minimal. In tidal areas where wetlands can be affected by development, culverts should be large enough to accommodate tidal flow along streams.

Step 7. Incorporate the Aquatic Resource Plan into the Overall Planning Program Aquatic resources planning is one important component of a larger planning process in which many other socioeconomic, legal, and natural resource elements must be considered. Before implementing a water resources plan, it should be reviewed to see how well it fits into the overall plan. Sound land use planning principles involve comprehensive consideration of an area's natural resources and the restrictions or limitations they place on certain uses, including construction. The aquatic resource is an integral part of this larger interacting system. Land management practices recognizing suitable land uses and sound pollution control methods and standards based on natural constraints, protect water bodies found in that system. Even in areas where there are water shortages and many competing uses for water, there are opportunities for enhancing and protecting aquatic ecosystems. In reviewing the water resources plan and integrating it into the larger one, additional opportunities for remodeling urban fishing and waterfront recreation may be identified; if so, they can be incorporated in the plan.

Some Lake and Pond Inventory Considerations This is a partial listing of general considerations for inventory of water bodies. It is intended to alert the planner to some measurements that may be beneficial in planning for urban fishing. However, the planner should consult with state and local agencies for detailed information.

General

Location: Drainage, state, county, geographic landmarks, latitude, longitude

Elevation: Feet above sea level, uppermost point to lowermost point

Climate: Mean temperature, rainfall

Origin/age: Glacial, impoundment, etc., approximate age

Physical

Size: Area in acres, volume in acre-feet

Depth profile: Depth contours

Fluctuation: High pool to low pool and frequency of change

Sources and discharges: Streams entering or leaving and annual discharge of each

Substrate: Substrate sizes, silt, sand, peat, rubble, etc., percent and area of each

Littoral zone size: Percent of total surface area and in acres

Shoreline configuration: Length of shoreline compared with surface area

Stratification and turnover: Does seasonal stratification occur? Number of turnovers

Water Quality

Transparency/turbidity: Secchi disk or NTUs or JTUs seasonally, after rain and during dry period

Dissolved oxygen profile: Seasonally, during maximum temperature, in milligrams per liter

Temperature profile: Seasonally, during stratification if it occurs

Other

Land use practices: Land use classification, residential density, method of sewage disposal, proximity

Shoreline development: Docks, houses, industry, etc. Also bank stabilization or filling and dredging

Recreation: Public or private, fishing, boating, swimming, hunting, etc.

Point discharges: Location and type

Nuisance species: Algal blooms, rough fish, etc.

Special species: Rare and endangered plants or animals

pH: Seasonally, at several locations and depths

Total dissolved solids: Seasonally, after rain if tributaries are present, at several locations and depths

Nutrients: Include total phosphorus, ammonia, nitrate, nitrite, total Kjeldahl nitrogen, seasonally, at several locations and depths

Chemical oxygen demand (COD): Seasonally, at several locations and depths

Biological

Primary production: Carbon-14 fixation, chlorophyll *a* concentration, several locations

Phytoplankton: Dominant forms, phytoplankton, filamentous, densities (numbers per milliliter). Seasonally

Submergents: Species, categorize as rare, common, or abundant. Seasonally

Emergents: Same as previous

Floating-leafed: Same as previous

Zooplankton: Dominant forms, densities (numbers per cubic meter), seasonally

Fish: Species and numbers, general condition, tolerance, dominance, biomass

Waterfowl and other wildlife: Species and abundance, waterfowl, furbearers, small and big game, reptiles and amphibians, seasonally

Fecal coliform: Seasonally, several locations which reference potential contamination sources

Biochemical oxygen demand (BOD): Same as previous

Critical habitats: Spawning areas, migration routes, etc.

Intake points and volumes: Use, e.g., cooling water, municipal water supply, pump storage, etc. Entrainment and impingement potential

Subsurface geology: Surficial soil infiltration rates, texture, bedrock fractures, chemical composition, depth, extent and type

Subsurface hydrology: Groundwater flow rate, wells (location and pumping rate)

Other aspects of surrounding land uses: Slope, surrounding bank vegetation, amount and concentration of nonpoint pollutants resulting from land uses and distances from water body

Some Wetland Inventory Considerations

General

Location: Drainage, state, county, geographic landmarks, latitude, longitude

Elevation: Feet above sea level, uppermost point to lowermost point

Climate: Mean temperature, rainfall

Type: Shrubs swamp, bog, tidal marsh, etc.

Physical

Size: Surface area in acres

Configuration: Boundary length compared with total surface area

Hydrological location: Lakeside, streamside, estuary, deltaic, isolated, etc.

Open water areas: In acres and as a percent of total wetland area

Water source: If stream, discharge, drainage, etc.

Shoreline profile: Categorize as steep, medium, flat

Flow: Retention and rejuvenation of water, direction of movement, stagnation

Water Quality

Salinity: Report salinity from several locations in parts per thousand. If water movement is significant measure seasonally

Turbidity: Seasonally in NTUs or JTUs

pH: Seasonally

Special species: Rare and endangered plants and animals

Critical habitats: Spawning areas, migration routes, etc.

Subsurface geology: Surficial soil infiltration rates, texture, bedrock fractures, chemical composition, depth, extent and type

Subsurface hydrology: groundwater flow rate, wells (location and pumping rate)

Biological

Algae: Dominant forms, general abundance

Submergents: Species; categorize as rare, common, or abundant, seasonally

Emergents: Same as previous

Floating-leafed: Same as previous

Cover type: Brush, trees, grasses, etc., relative proportion in percent and acreage

Vegetative interspersion: "Edge" lengths between cover types, open water, etc., compared with area

Fish: Species and abundance, general condition, tolerance, biomass, breeding use

Waterfowl and other wildlife: Species and abundance, waterfowl, furbearers, small and big game, reptiles and amphibians, seasonally

Other

Land use practices: Land use classification, residential density, method of sewage disposal, proximity

Filling or draining: Extent

Physical alterations/hydrological barriers: Dikes and highways which may affect water movement

Recreation: Public or private, fishing, hunting, etc.

Point discharges: Location and type

Nuisance species: Algal blooms, fish, etc.

Other aspects of surrounding land uses: Slope, surrounding bank vegetation, amount and concentration of nonpoint pollutants resulting from land uses and distances from water body

Some Stream and River Inventory Considerations

General

Location: Drainage, state, county, geographic landmarks, latitude, longitude

Elevation: Feet above sea level, uppermost point to lowermost point

Climate: Mean temperature, rainfall

Stream order: Based on number of tributaries, from 1 to *n*

Physical

Basin size: Square miles

Discharge, fluctuation: Mean annual, by month, cubic feet per second

Floodplain size: Area inundated in square miles or acres by yearly high water and 25-, 50-, and 100-year floods

Channel width and cross-sectional area: Width in meters and cross section in square meters, average depth times width or by segments

Flooding history: 25-, 50-, and 100-year flood discharges in cubic feet per second

Pool and riffle: Using several transects, the percent of each

Substrate: Boulder, rubble, cobble, gravel, sand, or silt, percent of each on several transects

Bank stability: Eroding or stable, extent of each

Gradient: Change in elevation per unit of stream length

Bank vegetation and shading: Trees, shrubs, herbs, grasses, percent of each, overhang extent

Macroinvertebrates: Number of taxa, abundance of each, general tolerance of common taxa

Fish: Species and numbers, general condition, tolerance, dominance, biomass

Waterfowl and other wildlife: Species and abundance, waterfowl, furbearers, small and big game

Fecal coliform: Seasonally, several locations which reference point and nonpoint discharges

Biochemical oxygen demand (BOD): Same as previous

Other

Land use practices: Agriculture, urban, residential, natural, industrial; percent of each and square miles

Stream bank uses: Structures, docks, etc., or natural

Stream alterations: Channelization, damming, dredging, bank stabilization—reasons for each

Water Quality

Temperature: Average annual and seasonal or monthly in degrees Celsius. Large rivers measured at several depths and locations

Dissolved oxygen: Consider backwaters and channel areas. In milligrams per liter

Turbidity: During low-flow and high-flow (after rain) conditions, seasonally, in NTUs or JTUs

Specific conductance: During low-flow and high-flow conditions, seasonally, in micromhos per centimeter

Total dissolved solids: During low-flow and high-flow conditions, seasonally, in milligrams per liter

Alkalinity: Same as previous

Chemical oxygen demand (COD): During low-flow and high-flow conditions, seasonally

pH: Same as previous

Nutrients: Include total phosphorus, ammonia, nitrate, nitrite, total Kjeldahl nitrogen, during low flow and high flow, seasonally

Biological

Primary production: Carbon-14 fixation, chlorophyll *a* concentration

Algae: Dominant forms — filamentous, diatoms, periphyton, planktonic; indicate abundance of each form

Macrophytes: Types and locations, abundance

Recreation: Sport or commercial fishing, boating, swimming, hunting, etc.

Point discharges: Location and type (municipal, industrial, type of industry, etc.)

Nuisance species: Algal blooms, mosquitoes, rough fish, etc.

Special species: Rare and endangered plants or animals

Intake points and volumes: Use, e.g., cooling water, irrigation, municipal water supply. Potential entrainment and impingement problems

Critical habitats: Spawning areas, migration routes

Subsurface geology: Surficial soil infiltration rates, texture, bedrock fractures, chemical composition, depth, extent and type

Subsurface hydrology: Groundwater flow rate, wells (location and pumping rate)

Other aspects of surrounding land uses: Slope, surrounding bank vegetation, amount and concentration of nonpoint pollutants resulting from land uses and distances from water body.

2.24 Conservation Measures for a Developing Area

Increasingly, farms, forests, and other open spaces are being converted to urban uses. Land disturbances associated with residential, industrial, and commercial developments and supporting activities (building streets, sewer lines, power transmission lines, and airports) contributes to a serious nationwide problem of soil erosion and sediment damage. More than 25,000 tons of soil may be eroded from a square mile of developing area and find its way into marshes, streams, ponds, rivers, lakes, and marine estuaries. Silted ponds, lakes, and reservoirs have less room to store storm water, and thus water supplies are damaged and flood hazards increased. Government and industry spend millions of dollars to remove sediment from water to be used in homes, hospitals, and factories. Sediment destroys spawning grounds for fish and is a health hazard because disease germs, pesti-

cides, and other unwanted materials attached to it are transported from one community to another.

Many villages, towns, townships, cities, and counties are enacting ordinances or issuing rules and regulations designed to protect the public from unnecessary and destructive soil erosion and sedimentation. In many areas, conservation plans must be prepared by developers and approved by conservation districts before land-disturbing permits are granted. Such conservation plans show, for instance, how housing developers will use their land and control erosion and sedimentation during and after construction.

Soil surveys provide developers with information on the location and extent of the different kinds of soil in an area and show the soil limitations for selected uses. Once the soil limitations are known, streets, homes, utilities, and other kinds of construction often can be planned on selected soils that are stable, dry, and generally free of problems. In some places soils with moderate or severe limitations must be used. Soil interpretations show the kinds of soil problems so that engineers and developers can investigate them in detail and plan and design structures to reduce or overcome them. In this way decisions can be made early about selecting areas for specific uses, and maintenance costs can be held to a minimum. Constructing homes and other buildings on desirable soils increases opportunities for landscaping with a variety of plants, both for beauty and for erosion control.

A conservation plan shows the steep land that is to be protected with vegetation. Waterways are preserved and protected, and recreation areas are located on suitable sites. Land subject to overflow from streams also is preserved and protected to curtail flooding and to provide additional open space for wildlife and recreation.

Equally important, a conservation plan shows the location of conservation measures, such as dikes, water diversions, terraces, dams, reservoirs, water conduits, grassed waterways, and plantings of grass, trees, and shrubs. A plan accounts for the timing and sequence of installing conservation measures to provide maximum control of erosion and sedimentation.

Measures to Control Erosion and Sedimentation   To control erosion and sedimentation in this area during and after construction, the developers agree to:

1. Disturb only the areas needed for construction. At the present time, natural vegetation covers this area, and there is little erosion. The stream bed and stream banks are stable. The vegetation on the floodplain and on the adjacent slopes will contribute to the aesthetic and environmental quality of the development.

2. Remove only those trees, shrubs, and grasses that must be removed for construction; protect the rest to preserve their aesthetic and erosion control values.

3. Stockpile topsoil, and protect it with anchored straw mulch.

4. Install sediment basins and diversion dikes before disturbing the land that drains into them. Diversion dikes in

the central part of the development may be constructed after streets have been installed but before construction has been started on the lots that drain into them.

5. Install streets, curbs, water mains, electric and telephone cables, storm drains, and sewers in advance of home construction.

6. Install erosion and sediment control practices as indicated in the plan and according to soil conservation district standards and specifications. The practices are to be maintained in effective working condition during construction and until the drainage area has been permanently stabilized.

7. Temporarily stabilize each segment of graded or otherwise disturbed land, including the sediment control devices not otherwise stabilized, by seeding and mulching or by mulching alone. As construction is completed, permanently stabilize each segment with perennial vegetation and structural measures. Both temporary and permanent stabilization practices are to be installed according to soil conservation district standards and specifications.

8. "Loose-pile" material that is excavated for home construction purposes. Keep it loose-piled until it is used for foundation backfill or until the lot is ready for final grading and permanent vegetation.

9. Stabilize each lot within 4 months after work has been started on home construction.

10. Backfill, compact, seed, and mulch trenches within 15 days after they have been opened.

11. Level diversion dikes, sediment basins, and silt traps after areas that drain into them have been stabilized. Establish permanent vegetation on these areas. Sediment basins that are to be retained for storm-water detention may be seeded to permanent vegetation soon after they have been built.

12. Discharge water from outlet structures at nonerosive velocities.

13. Design and retain two debris basins as detention reservoirs so that peak runoff from the development area is no greater than the peak runoff before the development was established.

The following tables interpret the data in the soil map shown in Figure 2.80.

Legend and Soil Descriptions

Symbol	Soil	Brief description
CrA CrB2	Crosby silt loam, 0 to 2 percent slopes Crosby silt loam, 2 to 6 percent slopes, eroded	Light-colored to moderately dark-colored, deep, somewhat poorly drained, slowly permeable soils on nearly level to gently sloping areas in uplands; developed from firm glacial till
Es	Eel silt loam	Moderately dark-colored, deep, moderately well drained, moderately permeable soils on nearly level areas in bottom lands; developed from friable alluvium
HeF	Hennepin soils, 18 to 35 percent slopes	Light-colored, deep, well-drained, moderately permeable soils on steep slopes in uplands; developed from firm glacial till
MnB2 MnC MnC2 MnD MnD2	Miami silt loam, 2 to 6 percent slopes, eroded Miami silt loam, 6 to 12 percent slopes Miami silt loam, 6 to 12 percent slopes, eroded Miami silt loam, 12 to 18 percent slopes Miami silt loam, 12 to 18 percent slopes, eroded	Light-colored, deep, well-drained, moderately permeable soils on gentle slopes to moderately steep slopes in uplands; developed from firm glacial till

Soil Interpretations

| Symbol | Soil | Erosion hazard | Dwellings (three stories or less) | | Septic tank absorption fields | Location of roads and streets | Parks or nature trails |
			With basement	Without basement			
CrA	Crosby	Slight	Severe; wetness	Moderate; wetness	Severe; wetness	Moderate; wetness	Moderate; wetness
CrB2	Crosby	Moderate; sheet erosion	Severe; wetness	Moderate; wetness	Severe; wetness	Moderate; wetness	Moderate; wetness
Es	Eel	Moderate; stream bank erosion	Severe; floods	Severe; floods	Severe; floods	Severe; floods	Moderate; floods
HeF	Hennepin	Severe; sheet and gully erosion	Severe; slope	Severe; slope	Severe; slope	Severe; slope	Severe; slope
MnB2	Miami	Moderate; sheet erosion	Slight	Slight	Slight	Slight	Slight
MnC, MnC2	Miami	Severe; sheet and gully erosion	Moderate; slope	Moderate; slope	Moderate; slope	Moderate; slope	Moderate; slope
MnD, MnD2	Miami	Severe; sheet and gully erosion	Severe; slope	Severe; slope	Severe; slope	Severe; slope	Severe; slope

Soils rated as slight have few or no limitations for the use. Soils rated as moderate have limitations that reduce to some degree their desirability for the purpose being considered. They require some corrective measures. Soils rated as severe have unfavorable soil properties or features that severely restrict their use and desirability for the purpose. A severe rating does not mean that the soil cannot be used for a specific purpose, because many of the problems can be corrected.

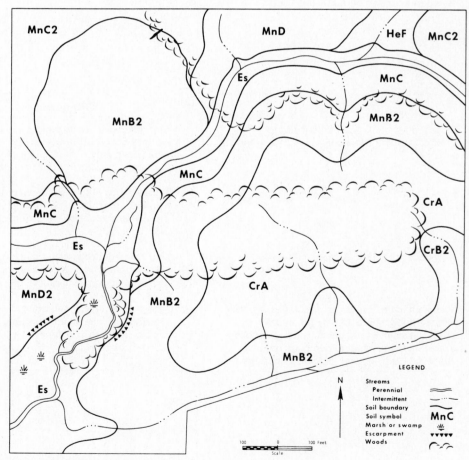

Figure 2.80 Soil map.

Land Use   The land use pattern proposed by the developers contributes to the control of erosion and sedimentation and to the maintenance of the environmental and aesthetic values of this area.

1. The floodplain and adjacent steep slopes are reserved for public use. Trails or other projects for recreation or environmental improvement may be developed by the local unit of government.

2. Public access lands connect each street with the open space dedicated to public use.

3. Storm sewers carry street and lot runoff to stable outlets or water detention reservoirs.

4. Street patterns conform to land contours and are designed for pedestrian safety and abatement of traffic noises.

5. Building lots are laid out to conform to the contours of the land in order to reduce land disturbance.

6. The area is platted so that a maximum number of lots share the wooded sectors.

7. Natural watercourses are preserved and protected.

The Plan   This example of developers' conservation plans for a housing development provides for an attractive environment based on careful use of soil, water, and plant resources. The plan is based on interpretations of the soils in the area. A pictorial map (Figure 2.81) is used here to show the developers' plan. More often, however, a plat map that shows topography, lot measurements, street widths, and other features is used.

Detailed designs of conservation measures, although not shown in this example, are a necessary part of conservation plans. Designs must comply with standards and specifications adopted by the governmental entity that has responsibility for reviews and approval.

Conservation plans for developing areas that have different climate, soils, and topography require different land use and conservation measures. For instance, in some areas a high water table is as great a problem as erosion and sedimentation.

Conservation plans can be prepared by engineers, developers, building contractors, or other technically qualified resource planners. If needed, Soil Conservation Service technical assistance can be made available through the local soil conservation district.

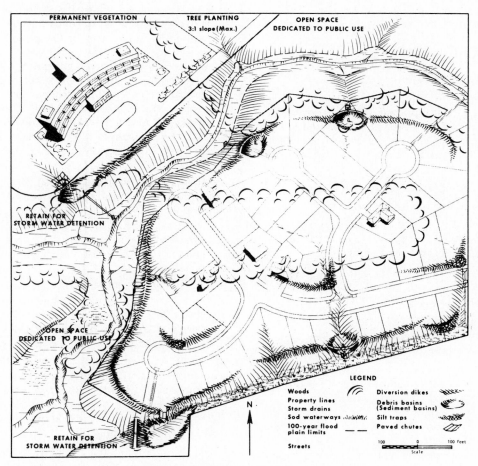

Figure 2.81   Planned land use and conservation treatment.

2.25 Environmental Impact Statements

General Content of an Environmental Impact Statement The following outline presents general topics usually required in an impact statement. Various topics may be expanded, modified, reduced, added, or eliminated to accommodate a specific project, to apply to a particular type of site, or to satisfy the guidelines of the particular agency to which the statement must be submitted.

Section 1. Existing condition of the region and the project site (resource inventory section): The resource inventory is the most important phase of the environmental assessment process. Without a detailed, scientifically accurate analysis of the conditions of the site, there is no basis, or only a faulty one, for the subsequent decisions on the rational use of the site. Some of the general items to be included are:

a. Landform, topography, physiography, location, and size

b. Vegetation and land use

c. Soils, geology, and groundwater

d. Streams and water quality

e. Wildlife

f. Climatology

g. Air quality

h. Noise

i. Socioeconomics, demography, transportation, utilities

j. Archaeology and historic sites

Section 2. Description of the proposed project: In this section, which is the most important part of an environmental assessment, the proposed project is described in a semitechnical manner that emphasizes the way in which the project will interface with the environment. Some aspects to be included are:

a. Need, public benefits, and economic feasibility

b. Facilities to be placed on the site

c. Gaseous, liquid, and solid wastes to be generated from project; project noise levels, ability to meet air and water quality standards and noise limitations

d. Energy and water requirements

e. Transportation requirements (raw material imports, product exports, employee traffic, and customer traffic)

f. Manpower requirements (expected local recruitment, expected transfer of employees from other areas, and number of new employees to be attracted to the region)

Section 3. The environmental impact assessment: This section, which is the most important part of the process, is unique to each project. It is an evaluation based on an overlay of the proposed project on the resource inventory of the site and the region. One must determine the goodness of fit, identify environmental and socioeconomic conflicts, and compare project requirements with resource supplies.

In regard to air and water quality, for example, the applicant must consider and explain the direct and secondary impact of his or her project during construction and during operation, including anticipated rundown periods, during times of facility upsets, and in the event of accident. These considerations must include not only average conditions but also worst-case conditions of weather and facility performance. Such evaluations properly are a phase of careful environmental planning and usually result in minor to very major revisions in the project. The most suitable solutions should be reflected in the final proposed plan. Discarded concepts and less-than-best solutions may be listed in Section 4 as alternatives.

Section 4. Alternatives to the proposed project

a. Alternative locations for project siting. What are the determinants in site selection (size of tract, transportation facilities, availability of large volumes of water or high-quality water, labor force, or others)? Why was the proposed site selected? What unique qualities does it provide for the project? What other sites were considered? Why were they rejected?

b. Alternative processes or methods. In what other way can the goals of this project be achieved? Why was an alternative process or method not proposed? Can the proposed project be modified further to avoid or minimize some adverse impacts? If the project were an electric generating station powered by coal, one should explain why a nuclear or oil-fired generator was not proposed or whether hydroelectric generation is possible. If once-through water is proposed for cooling, the applicant must explain why evaporative cooling towers or dry towers cannot be used. In the case of a road crossing of a marshland, the alternatives of construction on fill, on piles, or on open structures of short and long spans should be discussed.

c. Alternative configurations within the site. Sensitive areas on the site often can be preserved by a careful arrangement of facilities. A slight alteration of ingress and egress points may reduce potential traffic problems. Aesthetic impacts may be reduced by shifting the positions of some components. These aspects should be explained in this subsection.

d. The no-action alternative. What would be the consequences of not implementing the proposed project? What public benefits would be lost? What alternative use is likely to be made of the site? What adverse impacts would be avoided?

Section 5. Adverse impacts that cannot be avoided: In this summary section, the adverse environmental and socioeconomic effects of the proposed project are listed in the approximate order of magnitude and described briefly. No discussion of possible justifications, mitigations, or trade-offs should be included here.

This listing is intended to red-flag critical points for the consideration of regulatory agencies and other interested

parties. It serves to focus the attention of reviewers who may be able to suggest methods to reduce the adverse effects.

Review, Comment, and Revision   This is the heart of the environmental assessment process. It establishes an interface for the free exchange of information, criticisms, and suggestions between the following:

1. The applicant and his or her physical planners (engineers, architects, and so on)

2. Environmental and socioeconomic resource analysts

3. Regulatory agency personnel

4. Government organizations with jurisdiction or expertise, or both, citizen groups, and individuals

Most state and local ordinances, laws, or regulations establish an appointed official or a particular agency as the authority with the power to determine the acceptability of an environmental impact statement. This authority extends to the form, content, adequacy of analysis, and suitability of environmentally protective measures described in **the** statement.

Public informational meetings and public hearings are important components of the environmental assessment process. Especially in publicly financed projects, public informational meetings should be initiated early, while project planning still is uncommitted and flexible. Interim meetings should be convened at critical moments in the planning process. These will provide forums to receive criticism, additional information, and new ideas.

2.26 Inventory of Site Data

Inventory of On-Site Factors   After the program has been thrashed out, the designer turns to gathering facts about the site, securing information from maps and personal inspections of the area under study. Such data could include the location of or knowledge about existing

1. Artificial elements
   a. Legal and physical boundaries, private holdings, and public easements
   b. Buildings, bridges, and other structures including those of historical and archaeological significance
   c. Roads, walks, and other transportation ways
   d. Electric lines, gas mains, and other utilities
   e. Land uses: agriculture, industrial, recreation, and so on
   f. Applicable ordinances such as zoning regulations and health codes
2. Natural resources
   a. Topography, including high and low points (Figure 2.82), gradients (Figure 2.83), and drainage patterns (Figure 2.84).

b. Soil types, by name if available, for clues regarding ground surface permeability, stability, and fertility (Figure 2.85).

c. Water bodies, including permanence, fluctuations, and other habits

d. Subsurface matter: geology of the underlying rock including the existence of commercially or functionally valuable material such as sand and gravel, coal, and water

e. Vegetation types (mixed hardwoods, pine forest, prairie grassland, and so on) and individual specimens of consequence (Figure 2.86)

f. Wildlife including the existence of desirable habitat as low cover for pheasants, caves for bears, berries for birds, and so on

3. Natural forces (including both macroclimate as generally found over the entire site and microclimate characteristics

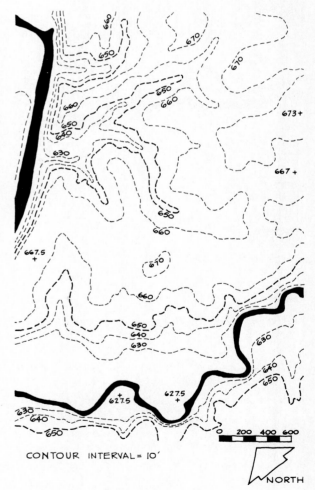

CONTOUR INTERVAL = 10'

Figure 2.82   Site data: topography.

Figure 2.83 Site data: slope gradient.

Legend for Figure 2.83:
- 0-2%
- 2-4%
- 4-10%
- 10-20%
- OVER 20%

0 200 400 600

NORTH

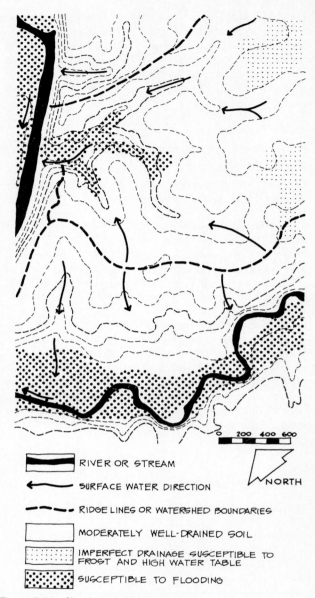

Figure 2.84 Site data: drainage patterns.

Legend for Figure 2.84:
- RIVER OR STREAM
- SURFACE WATER DIRECTION
- RIDGE LINES OR WATERSHED BOUNDARIES
- MODERATELY WELL-DRAINED SOIL
- IMPERFECT DRAINAGE SUSCEPTIBLE TO FROST AND HIGH WATER TABLE
- SUSCEPTIBLE TO FLOODING

0 200 400 600

NORTH

or changes from the norm as experienced in isolated patches)

a. Temperature (air and water), especially day, night, and seasonal norms, extremes, and their durations

b. Sun angles at various seasons and times of the day

c. Sun pockets such as may be found in forest clearings; frost pockets that may be in low places where the wind that sweeps away the morning dew is blocked

d. Wind directions and intensities as they occur daily and seasonally

e. Precipitation: rain, snow, and sleet seasons and accumulations; storm frequencies and intensities

4. Perceptual characteristics

a. Views into and from the site; significant features

b. Smells and sounds and their sources

c. Spatial patterns

d. Lines, forms, textures, and colors and scales that give the site its peculiar character

e. General impressions regarding the experience potential of the site and its parts

Inventory of Off-Site Factors The designer must also accumulate information about the artificial, natural, and perceptual elements on the properties that surround or otherwise affect the site. These might include both existing and anticipated

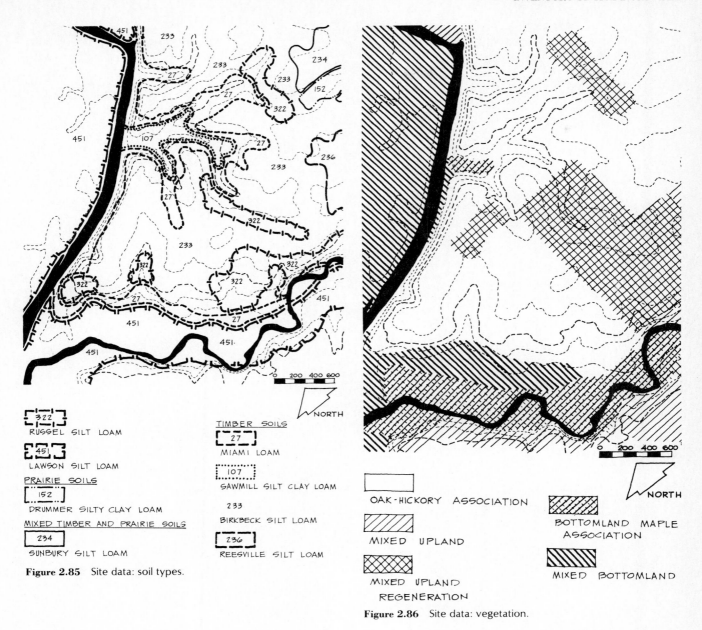

Figure 2.85 Site data: soil types.

Figure 2.86 Site data: vegetation.

1. Land use patterns
2. Stream and drainage sources
3. Visuals, smells, and sounds
4. Neighboring aesthetic character
5. Public utility locations and capacities
6. Transportation ways and systems

Each step in the survey phase begins in isolation, the first facts collected being merely those that are immediately handy. Soon all steps become intertwined, each giving direction to another to turn general notions of what might be needed into specific requirements and ensure that nothing that could affect the design's outcome has been left out of consideration. Designers find themselves working back and forth between program and inventory. Program items suggest to them not only what information must be collected but also what is inconsequential. The fact that a playground is being dealt with sends the landscape architect after data about sun angles and the peripheral traffic situation. At the same time, it suggests little urgency to seek out a map showing the location of bear dens and pheasant cover. In a complementary fashion, data garnered might point out modifications necessary in the program.

S E C T I O N

THREE

Factors Affecting Building Location and Orientation

The location of a building on a site or its relationship to other buildings is extremely important. If properly situated, the building achieves harmony with the topography, livability is enhanced, drainage problems are minimized, and the building's functional efficiency is increased. If the building is not properly situated, many problems that cannot easily be corrected can and will ensue.

Orientation of the building to sun, wind, and vistas is a basic consideration. Under most conditions it is desirable to be protected from the hot summer sun and exposed to the sun's rays during the cold winter months. Taking advantage of summer breezes can reduce or eliminate the need for air conditioning. In winter maximum protection against northerly winds can substantially reduce heat loss and heating costs.

Siting a building to conform with its topography will result in a minimum of necessary grading, reduce initial construction costs, and eliminate continuous drainage problems.

Exposure of a site to noise pollution has become a serious problem in both urban and suburban areas. This problem can be controlled or minimized by the judicious placement of the structure on the site.

Plant materials, both trees and shrubs, are an integral part of site design. They serve not only as aesthetic elements but as buffer strips, screens, and dividers.

Creating greater physical security and street safety through site planning has become in the past few years a significant new development. The concept, which is known as "defensible space," deals with the placement of buildings, tight control of access to and egress from the site, and visual surveillance of all public areas.

Sewage disposal requirements also impose constraints on site development and building location, particularly for buildings served by on-site water wells. Well-drained soils with adequate depth to groundwater may be

served by cesspools. Less well-drained soils or areas in which the groundwaters are near the surface may require septic tanks and leaching fields. In some instances subdivision tract housing may be better served by a communal sewerage system.

3.1 Sun Orientation

Solar Utilization

Climate Solar radiation, wind, temperature, humidity, and many other factors shape the climate of the United States. Basic to using solar energy for space heating and domestic hot water heating is understanding the relationship of solar radiation, climate, and dwelling design.

The amount and type of solar radiation varies between and within climate regions: from hot, dry climates where clear skies enable a large percentage of direct radiation to reach the ground, to temperate and humid climates where up to 40 percent of the total radiation received may be diffuse sky radiation, reflected from clouds and atmospheric dust, to cool climates where snow reflection from the low winter sun may result in a greater amount of incident radiation than in warmer but cloudier climates.

As a result of these differences in the amount and type of radiation reaching a building site, as well as in climate, season, and application—heating or domestic hot water—the need for and the design of solar system components will vary in each locale (see Figure 3.1).

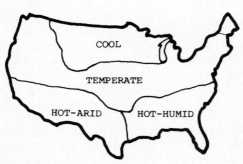

Figure 3.1

The Solar Window Imagine the sky as a transparent dome with its center at the solar collector of a house. The path of the sun can be painted (projected) onto the dome, as can be the outline of surrounding houses and trees. The morning and afternoon limits of useful solar collection (roughly 9 A.M. and 3 P.M.) and the sun's path between those hours throughout the year scribes a "solar window" on the dome. Almost all of the useful sun that reaches the collector must come through this window except for the added effect of diffuse radiation. If any of the surrounding houses, trees, etc., intrude into this "solar window," the intrusion will cast a shadow on the collector. The isometric drawing in Figure

3.2 illustrates the "solar window" for a latitude 40°N. The solar window will change for different latitudes. See Figures 3.3 to 3.5.

Orientation for sun and prevailing summer breezes always merits consideration. Latitude largely determines the former, local conditions the latter. Orientation for sunlight is most successful when sunshine is made available in kitchens on winter mornings and when some sun reaches living rooms in afternoons. When ideal conditions cannot be secured, a desirable minimum is considered to be achieved if some sun is available in each room at some time of day.

Although no absolute rules can be laid down, it is generally recognized that in Northern states buildings should lie with their long axes running from northeast to southwest, at an angle of from 30 to 60 degrees off north (see Figure 3.6a). Such an orientation will enable some sun to melt snow and to dry ground on the northerly side of the building. Where winters are shorter and there is less snow, this approximate angle diminishes until, in the extreme South, a true east-west alignment is usually preferred.

Under some conditions insolation (the sun's heat) becomes important, both negatively in summer and positively at other seasons. This factor, considered alone, points to southerly exposures in winter, southwesterly in summer; the building is aligned from northwest to southeast. The axis should not swing more than from 15 to 20 degrees from an east-west line (see Figure 3.6b).

Local atmospheric conditions also affect orientation. In the vicinity of New York, for example, prevalent morning mists make southwesterly exposures the sunniest.

Satisfactory orientation for sunlight thus becomes a compromise between conflicting factors. Local conditions furnish the basis for choice or for evaluating one factor against another. See also Figure 3.7.

Although the study of orientation is a comprehensive and detailed science, the knowledge and use of even a few basic rules will mean comfort and economy for the homeowner and a better house for the developer. The latitude belt between 35 and 45 degrees across the United States includes many of the most densely populated areas. This set of solar conditions thus has a wide application. In most locations below this latitude belt capturing summer breezes and protection from intense sun heat are the main consideration. Above the latitude belt shown in the map, protection from cold winter winds and utilization of the warming rays of the sun in winter are the prime objective. When the solar conditions of the belt between 35 and 45 degrees are used (except as noted), it is well to remember that the information presented in Figure 3.8 becomes more general and less exact as the distance increases from the midpoint of 40 degrees. See also Figure 3.9.

SOLAR HEAT ON BUILDINGS

In the north temperate zone, heavy radiation loads will act most decisively on the roof and on the east and west exposures during the summer. South exposures permit moder-

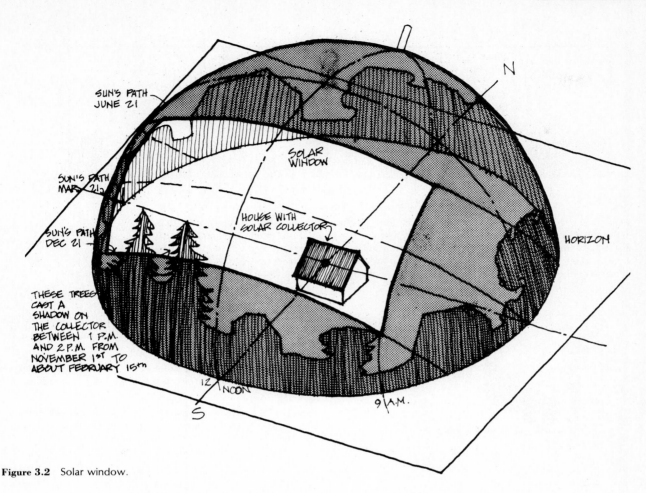

SUN'S PATH
JUNE 21

SUN'S PATH
MAR 21

SUN'S PATH
DEC 21

N

SOLAR
WINDOW

HOUSE WITH
SOLAR COLLECTOR

HORIZON

THESE TREES
CAST A
SHADOW ON
THE COLLECTOR
BETWEEN 1 P.M.
AND 2 P.M. FROM
NOVEMBER 1ST TO
ABOUT FEBRUARY 15TH

12 NOON

S

9 A.M.

Figure 3.2 Solar window.

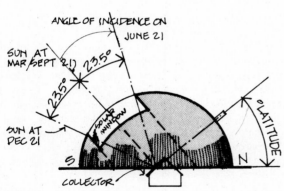

ANGLE OF INCIDENCE ON
JUNE 21

SUN AT
MAR/SEPT 21

23.5°

23.5°

SUN AT
DEC 21

SOLAR WINDOW

°LATITUDE

S

N

COLLECTOR

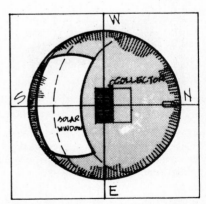

W

S

COLLECTOR

N

SOLAR
WINDOW

E

Figure 3.3 Side view of sky dome with "solar window": A side view of the sky dome from the east illustrates the relative position and angle of the sun throughout the year that defines the boundaries of the "solar window." Angle of incidence, a term often used in solar collector design, is the angle measured from the normal of the collector surface to the line indicating the sun's altitude at a particular time. The diagram specifically identifies the angle of incidence for June 21.

Figure 3.4 Plan view of sky dome with "solar window": Viewed from above the sky dome, the seasonal path of the sun can be plotted, thus defining the boundaries of the "solar window." This is easily accomplished by the use of a standard sun path diagram for the proper latitude. Sun path diagrams are widely reproduced and used for determining the azimuth and altitude of the sun at any time during the year, and give the points which can be plotted to determine the solar window.

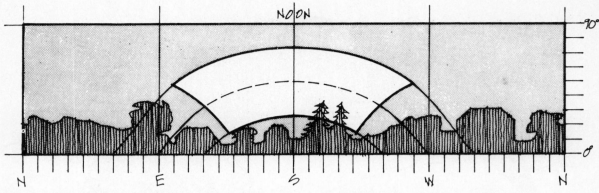

Figure 3.5 Panorama of the sky dome: As with the spherical earth, the spherical sky dome with its "solar window" can be mapped using a Mercator projection, in which all latitude and longitude lines are straight lines. Such a map is very useful for comparing the site surroundings with the "solar window" outline, since both can be easily plotted on the map. Any elements surrounding the site that intrude into the "solar window" will cast shadows on the collector.

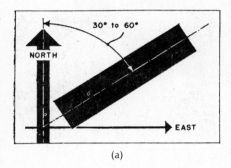

(a)

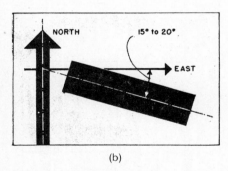

(b)

Figure 3.6 (a) Orientation for sunlight in Northern states. The building angle approaches an east-west line as one goes south. (b) Secondary favorable zone giving good early-afternoon sun in winter and protection from afternoon sun in summer.

ately significant heat gains during the summer, but they allow very significant heat gains during the winter. North exposures receive minimal radiation throughout the year. To be somewhat more specific:

1. If walls facing north of east or north of west receive direct radiation at all, they will tend to receive this radiation only in the late spring and early summer.

2. Walls facing south of east or south of west will tend to receive maximum direct radiation in the late fall and early winter.

3. Walls facing east of north or east of south will tend to receive maximum direct radiation at sunrise or during the morning hours.

4. Walls facing west of north or west of south will tend to receive maximum direct radiation during the afternoon hours or at sunset.

See Figure 3.10 for azimith and altitude angles and Figures 3.11 and 3.12 for the effects of solar heat through glassed openings in winter and summer.

LANDSCAPING FOR THERMAL CONTROL

Site vegetation and landforms can influence the immediate thermal environment of a building. These influences generally involve (1) the diversion of storm winds, (2) the channeling of cooling summer breezes, and (3) sun shading. The typical locations of basic landscape elements for thermal control on an open site are indicated in Figure 3.13. However, the optimum positioning of these elements may vary with local variations in prevailing wind patterns.

Under north temperate climatic conditions, it is desirable to follow these steps in utilizing sites:

1. To facilitate maximum exposure to the sun during prolonged winter periods, utilize warm slopes for building sites in colder regions.

2. Where it is desirable to provide natural summer cooling, utilize the lower portion of windward slopes. To induce penetration of prevailing summer breezes, openings should be placed to admit ventilation air on the windward

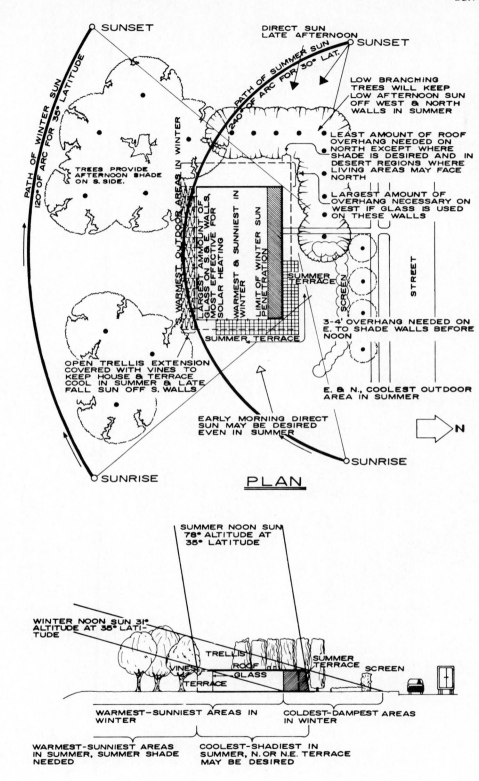

PLAN

ELEVATION

Figure 3.7

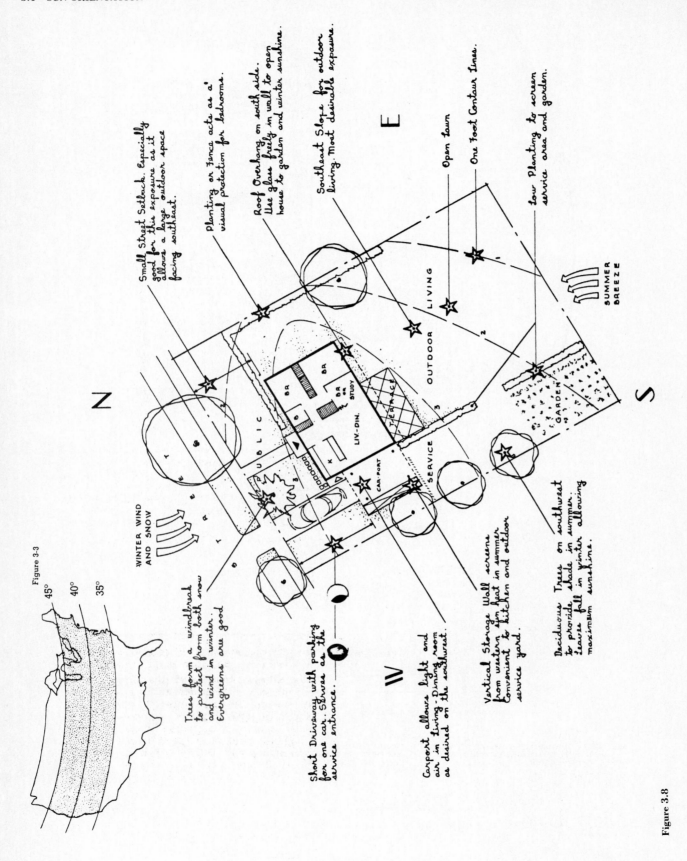

Figure 3-3

45°
40°
35°

WINTER WIND
AND SNOW

Trees form a windbreak to protect from both snow and wind in winter. Evergreens are good.

Short Driveway with parking for one car. Serves as the service entrance.

Carport allows light and air in living-Dining room as desired on the southwest.

Vertical Storage Wall screens from western sun heat in summer. Convenient to kitchen and outdoor service yard.

Deciduous Trees on southwest to provide shade in summer, leaves fall in winter allowing maximum sunshine.

Small Street Setback. Especially good for this exposure as it allows a large outdoor space facing southeast.

Planting on fence acts as a' visual protection for bedrooms.

Roof Overhang on south side. Use glass freely in wall to open house to garden and winter sunshine.

Southeast Slope for outdoor living. Most desirable exposure.

Open Lawn

One foot Contour lines.

Low Planting to screen service area and garden.

SUMMER BREEZE

N

E

S

PUBLIC
PRIVATE

BR
BR
BR
STUDY
LIV-DIN.
K
CAR-PORT
TERRACE

OUTDOOR LIVING
SERVICE
GARDEN

Figure 3.8

—204—

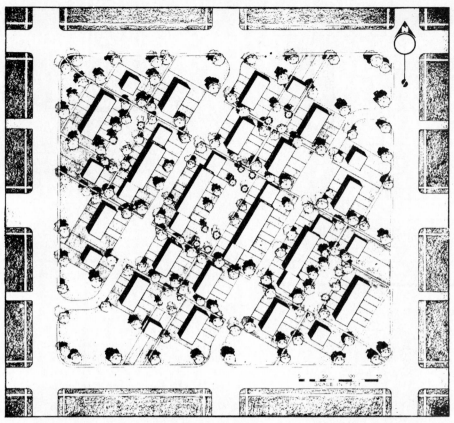

Figure 3.9   This site plan was developed from a project in a small Midwestern city. Since the ground is comparatively flat, orientation for sunlight was the principal consideration.

side of the building, exhaust outlets being placed on the leeward side.

As a related factor, to facilitate natural internal cooling action during warm periods, the blockage of prevailing summer breezes should be minimized. Usually this means that dense site screening should be curtailed on the south and southwest exposures.

3. Wind screening is desirable on the windward side of the building during cold periods. Usually this consideration applies to north and northwest exposures.

4. Utilize evergreens for wind-screening purposes. Utilize deciduous trees for sun-shading purposes.

5. If possible, locate the building so that available fully developed shade trees will provide shading on the east and west sides of low buildings. Similar considerations apply to the location of outdoor living areas.

6. Paving should be minimized immediately adjacent to the building. Where possible, vegetation should be used in this location to absorb rather than reflect solar energy. The critical west and southwest exposures are the most likely to produce significant energy during periods of peak solar heat gain.

7. Walks should be shielded from winter winds and summer sun.

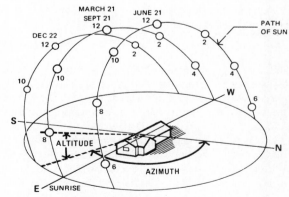

Figure 3.10 . Azimuth and altitude angles.

(a)

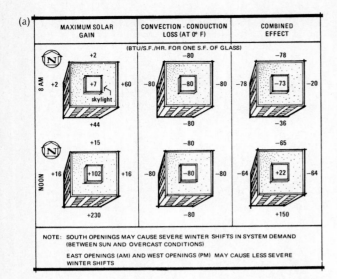

NOTE: SOUTH OPENINGS MAY CAUSE SEVERE WINTER SHIFTS IN SYSTEM DEMAND (BETWEEN SUN AND OVERCAST CONDITIONS)

EAST OPENINGS (AM) AND WEST OPENINGS (PM) MAY CAUSE LESS SEVERE WINTER SHIFTS

(b)

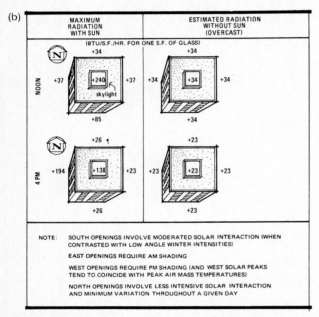

NOTE: SOUTH OPENINGS INVOLVE MODERATED SOLAR INTERACTION (WHEN CONTRASTED WITH LOW ANGLE WINTER INTENSITIES)

EAST OPENINGS REQUIRE AM SHADING

WEST OPENINGS REQUIRE PM SHADING (AND WEST SOLAR PEAKS TEND TO COINCIDE WITH PEAK AIR MASS TEMPERATURES)

NORTH OPENINGS INVOLVE LESS INTENSIVE SOLAR INTERACTION AND MINIMUM VARIATION THROUGHOUT A GIVEN DAY

Figure 3.11 (a) Effect of solar heat gain through glassed openings in peak winter periods; (b) effect of solar heat gain through glassed openings in peak summer periods.

Landscaping Considerations Site design for solar buildings requires special attention to landscaping. The location of plants and trees can be planned to enhance energy conservation and still permit unobstructed access to direct sunlight. Landscaping is often used to buffer wind and thus to conserve energy in the building, but trees and other vegetation must be located so they do not shade solar collectors.

Balancing energy-conservation and solar access needs can mean that the landscaping around a solar building will differ

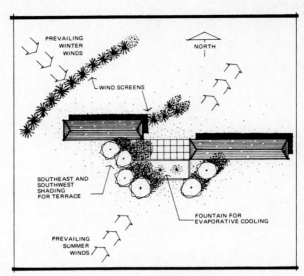

Figure 3.12 Landscape elements for thermal control.

from that of a conventional building. Often, vegetation that is close to the building will be concentrated near the east, west, and north walls, leaving the south wall and roof open to the sun. This practice means that the east and west walls, which receive the most solar radiation in the summer, are properly shaded.

Placing vegetation, especially evergreen species, near the north and west walls buffers the building from cold winter winds. To avoid shading collectors, tall vegetation usually is not located in the area of a lot from the southwest to the southeast of the house. Deciduous trees, which allow light in winter and shade in summer, can be located quite close to the south wall without impairing solar access too much. (See Figure 3.13.)

In warmer climates, vegetation can be used to channel cooling breezes. Often, buildings that use solar energy are designed to take advantage of natural ventilation. Figure 3.14 shows how hedges can serve both to divert breezes through buildings and to decorate the site.

Berming Like vegetation, berms—built-up masses of earth—can be used to insulate or buffer a solar building from the elements. The berms are piled up against the walls of the building and then usually sodded. Conventional houses sometimes have this type of treatment as a result of construction or landscaping practice. For a solar building, the most likely locations for berms are the north, east, and west walls. (See Figure 3.15.)

Figure 3.16 shows a berm that is used as a base for a ground-mounted collector. The berm is constructed so that it provides the collector with proper orientation—that is, facing south and tilted up at a certain angle. The berm may be landscaped on the north, east, and west sides to blend into the rest of the lot landscaping.

Figure 3.13   Tall trees on south exposures.

Figure 3.14   Landscaping used to channel winds.

Figure 3.15 Using berms for insulation.

Figure 3.16 Using a berm as a collector mounting.

3.2 Solar Access: Preliminary Site Planning

Solar Access, Density, and Environmental Protection
Local regulations establish development goals for both the community and the developer. It is likely that once barriers to the use or installation of solar energy equipment are identified and eliminated, conflicts between solar access needs and local development standards will diminish. In most cases, good solar access site planning (i.e., planning that takes into account environmental factors affecting the availability of sunlight as well as energy conservation) will surpass the minimum development standards set forth in local regulations. The result will be a unique, innovative, well-planned development that is an asset to the community.

Designing a development for solar access does not have to conflict with conventional development standards. Density, energy conservation, environmental protection, land-scaping, and other objectives can be accommodated easily while planning to promote solar access.

Density concerns need not rule out solar access planning. Depending on latitude and topographic conditions, it is likely that full buildout, as authorized in local regulations, can be maintained while solar access is still protected. For example, the recent study of a development indicates that densities of over eight dwelling units per acre are possible even in latitudes as far north as New York City, provided that streets are oriented east and west. If town house or multifamily development is proposed, densities can go even higher without sacrificing south-wall access.

Density becomes a crucial issue when the site has rolling topography and relatively steep slopes. Density objectives can change dramatically, depending on slope gradient and direction. For example, higher densities may be possible on south-facing slopes because shadow lengths are shorter. Conversely, north-facing slopes, especially steep ones, may

require reduced densities if solar access is to be protected. In PUDs, these differences in density can be balanced off — south slopes on a site can be designed with densities higher than those authorized, while north slopes can be more sparsely developed. In areas with poor solar access potential, solar access goals may have to be modified to achieve other development objectives — south-wall access, for example, may be infeasible, leaving roof access as the next desirable objective.

Environmental protection is also compatible with solar access protection. Although solar energy use is not so closely tied to environmental goals as energy conservation, the design options need not conflict with practices for controlling runoff, open-space preservation, and related conservation practices.

When it is necessary for streets to follow topography, it is still possible to site lots and buildings for proper solar orientation and access. With good site planning, areas restricted from development can be dedicated for open space, and this open space can facilitate both the environmental objective and solar access protection.

Another environmental concern involves protecting steep slopes from development. Although south slopes are particularly desirable from a solar access standpoint, soil conditions or the severity of the slope may entirely restrict or preclude development. As in conventional development, building on such slopes, when permitted at all, is likely to be difficult and expensive.

Meeting landscape standards may require compromise. Protecting solar access will sometimes mean selectively removing trees, especially in forested areas — not clear cutting, but selective removal to the south of buildings. Heavily forested sites probably should not be developed for solar energy use.

Creative planning and design can accommodate these and other environmental restrictions both imaginatively and economically.

Conventional Versus Planned Unit Development Local ordinances often give developers a choice. They can follow prescriptive regulations that establish minimum building or development standards, such as those in conventional zoning and subdivision ordinances. Or, they may choose the planned unit development (PUD) option, which enables them to negotiate with the community in designing a project and developing a parcel of land. The PUD approach offers greater flexibility at the cost of increased preplanning effort and negotiation time. Sometimes PUD ordinances contain incentives, such as density bonuses, to encourage innovative site design or to secure amenities for the community.

Preliminary Site Planning Procedures Based on the analysis of the site, the developer or site planner can proceed to allocate the major land uses of the project, including housing, open space, and transportation. In practice, site analysis and preliminary site planning are accomplished more or less simultaneously but are presented separately here for simplification.

Assessing a site for solar access and energy conservation is only a part of preliminary site planning. Developers must also decide which areas of a site are more desirable for building from an environmental standpoint. Some areas, for example, have more stable soils than others. The goal is to identify suitable areas for buildings, open space, streets, and other components of the project.

In analyzing the site, solar access and energy-conserving features of the site should be identified and mapped on a base map. The base map should be an aerial photograph of the site, a topographic map, or a land feature sketch. (See Figure 3.17.) The site can then be evaluated according to the site analysis checklist.

Site Analysis Checklist

1. Map topographic and major site features:
   Indicate slopes and flat areas.
   Indicate existing trees and buildings.
   Mark site elevations and contours.
   Mark all significant natural features, such as water courses or historic sites.

2. Map all potential solar access obstructions.
   Indicate individual trees, noting species, height, and whether evergreen or deciduous.
   Indicate all tall objects on the site or on adjoining property that can cast shadows on the site; estimate location and height.
   Indicate all north slopes or other areas with poor solar access, such as fog pockets. Sketch shadow patterns of major tall obstructions on the plan.

3. Map all energy-conserving factors of the site.
   Indicate seasonal wind directions and features that can influence wind flows.
   Mark possible frost or fog pockets.
   Note bodies of water.
   Note air quality.
   Indicate ground surfaces, such as bare soil, pavement, grass. Note reflective surfaces such as sand, water, concrete.

4. Discuss the terrain and site limitations with neighbors and other people familiar with the area.

A Solar Site Planning Example How would the principles and techniques described be used in actual practice? Let us take a hypothetical case to see how it is done.

First, the development objectives established in local zoning and subdivision regulations are compared with the base map developed as part of the site analysis. The easiest method for assessing these objectives and comparing them with both solar energy use and site constraints is to use overlay maps. Information is mapped on the overlays and placed over the base map and bubble diagram map to determine the suitability or unsuitability of site areas.

The first overlay should clearly show site constraints that prohibit the economic development or that threaten the environmental resources of the site. This overlay can be labeled "Exclusions," as in Figure 3.18. Excluded areas can be

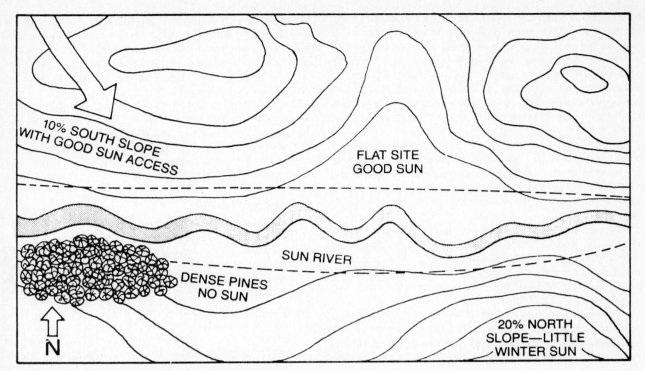

Figure 3-17 Base maps should be analyzed for solar access.

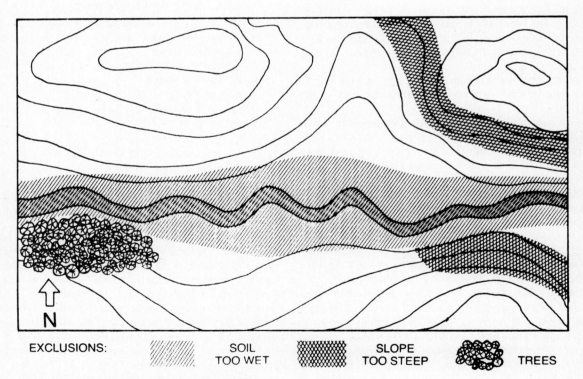

Figure 3-18 Site exclusions marked on the base map.

considered for other uses that are consistent with the development objectives.

Next, the solar access potential of the site should be mapped on an overlay and placed over the base map. The overlay should also suggest the solar energy objectives within the development, the constraints indicated by the site analysis checklist, and the regional building climate criteria. This map should be overlaid on the exclusion map and base map to indicate areas of optimal development potential and solar access, as in Figure 3.19.

The site should then be examined for energy-conserving site features, such as sheltered areas or trees that dam cold wind flows down slopes. Poor energy-conserving features should be identified on the base map, including exposed ridges or frost-prone areas at the foot of north-facing slopes. These frost pockets require greater care in building siting and may require larger collector areas to compensate for heat loss. (See Figure 3.20.)

The area that remains on the site map now has both good development potential and good solar access characteristics. This buildable area, shown in Figure 3.21, should be analyzed and housing allocations derived for the site.

At this point, preliminary site planning is finished. Housing and land uses are broadly allocated on the site according to environmental constraints and solar access requirements. The next step is to examine specific strategies for developing the site according to the development objectives and the site plan, so that solar access is maintained in those areas with good solar energy potential. This means designing land uses and housing that best use solar energy and minimize shading problems.

Collector Orientation and Tilt   Solar collectors must be oriented and tilted within prescribed limits to receive the optimum level of solar radiation for system operation and performance. See Figure 3.22.

Shading of Collector   Another issue related to both collector orientation and tilt is shading. Solar collectors should be located on the building or site so that unwanted shading of the collectors by adjacent structures, landscaping, or building elements does not occur. In addition, considerations for avoiding shading of the collector by other collectors should also be made. Collector shading by elements surrounding the site may be addressed by considering the "solar window" concept. See Figures 3.23 and 3.24.

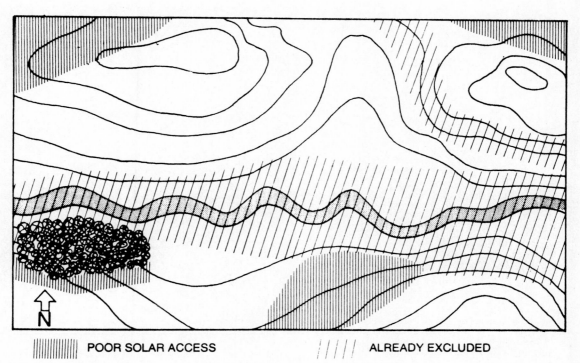

POOR SOLAR ACCESS   ////// ALREADY EXCLUDED

Figure 3.19   Areas of poor solar access should be marked on the base map.

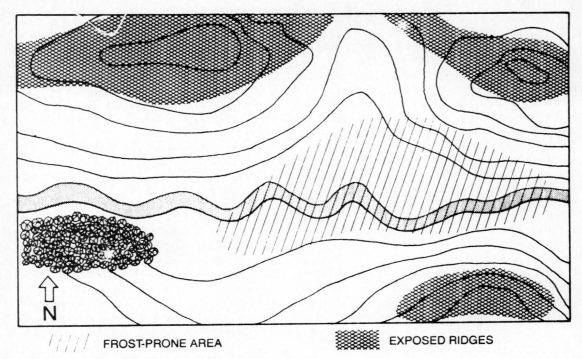

/ / / / FROST-PRONE AREA   EXPOSED RIDGES

Figure 3.20 Areas with poor energy-conservation features.

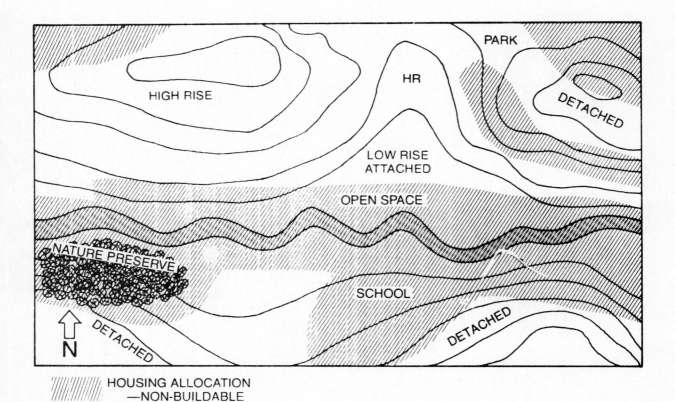

HOUSING ALLOCATION
—NON-BUILDABLE

Figure 3.21 Land use as allocated on the site.

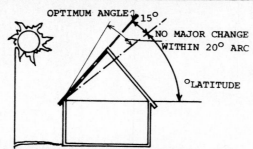

1. COLLECTOR TILT FOR HEATING The optimum collector tilt for heating is usually equal to the site latitude plus 10 to 15 degrees. Variations of 10 degrees on either side of this optimum are acceptable.

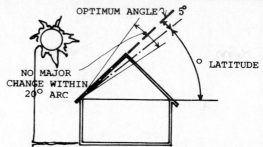

2. COLLECTOR TILT FOR HEATING AND COOLING The optimum collector tilt for heating and cooling is usually equal to site latitude plus 5 degrees. Variations of 10 degrees on either side of the optimum are acceptable.

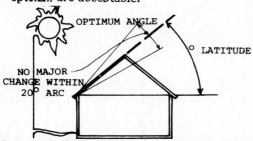

3. COLLECTOR TILT FOR DOMESTIC HOT WATER The optimum collector tilt for domestic water heating alone is usually equal to the site latitude. Again, variations of 10 degrees on either side of the optimum are acceptable.

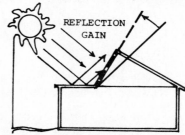

4. MODIFICATION OF OPTIMUM COLLECTOR TILT A greater gain in solar radiation collection sometimes may be achieved by tilting the collector away from the optimum in order to capture radiation reflected from adjacent ground or building surfaces. The corresponding reduction of radiation directly striking the collector, due to non-optimum tilt, should be recognized when considering this option.

5. SNOWFALL CONSIDERATION The snowfall characteristics of an area may influence the appropriateness of these optimum collector tilts. Snow buildup on the collector, or drifting in front of the collector, should be avoided.

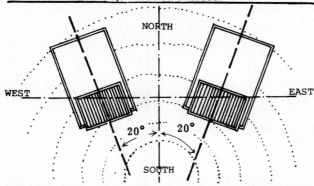

COLLECTOR ORIENTATION A collector orientation of 20 degrees to either side of true South is acceptable. However, local climate and collector type may influence the choice between East or West deviations.

Figure 3.22

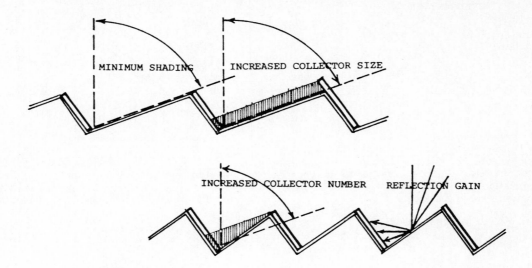

SELF-SHADING OF COLLECTOR   Avoiding all shelf-shading for a bank of parallel collectors during useful collection hours (9 AM and 3 PM) results in designing for the lowest angle of incidence with large spaces between collectors. It may be desirable therefore to allow some self-shading at the end of solar collection hours, in order to increase collector size or to design a closer spacing of collectors, thus increasing solar collection area. By making the collector's back slope reflective, one could increase the amount of solar radiation striking the adjacent collector, thus negating some of the shading loss.

Figure 3.23

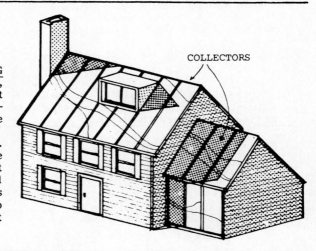

SHADING OF COLLECTOR BY BUILDING ELEMENTS   Chimneys, parapets, fire walls, dormers, and other building elements can cast shadows on adjacent roof-mounted solar collectors, as well as on vertical wall collectors. The drawing to the right shows a house with a 45° south-facing collector at latitude 40° North. By mid-afternoon portions of the collector are shaded by the chimney, dormer, and the offset between the collector on the garage. Careful attention to the placement of building elements and to floor plan arrangement is required to assure that unwanted collector shading does not occur.

Figure 3.24

3.3 Wind Orientation

Figures 3.25 and 3.26 show the effect of winds on various inland and shore homesites.

WINDBREAKS

Control of Wind by Plants   Basically, plants control wind by obstruction, guidance, deflection, and filtration. The dif-ferentiation is based not only on the degree of effectiveness of plants but on the techniques of placing them. There are books and articles that describe a number of ways in which plants control wind and their effectiveness in doing so. How-ever, plants, as natural elements, are not always predictable in their size, shape, and growth rate and thus in their absolute effectiveness.

Obstruction with trees, as with all other barriers, reduces wind speed by increasing the resistance to wind flow. Coni-

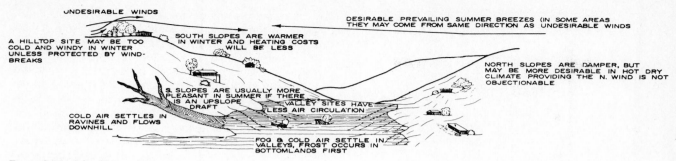

Figure 3.25 Inland homesites.

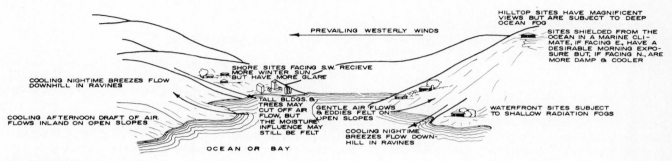

Figure 3.26 Shore homesites.

ferous and deciduous trees and shrubs used individually or in combination affect air movement. See Figure 3.27.

Plants may be used in conjunction with landforms and architectural materials to alter the airflow over the landscape and around or through buildings.

Deflection of wind over trees or shrubs is another method of wind control. Plants of varying heights, widths, species, and composition, planted either individually or in rows, have varying degrees of effect on wind deflection.

Coniferous evergreens that branch to the ground are generally the most effective year-round plants for wind control; and deciduous shrubs and trees, when in leaf, are most effective in summer. Wind velocity is cut from 15 to 25 percent of the open-field velocity directly leeward of a dense screen planting such as spruce or fir, while loose barrier of Lombardy poplars reduces leeward wind velocity to 60 percent of the open-field velocity. Wind velocity is cut from 12 to 3 miles per hour for a distance of 40 feet leeward of a 20-foot Austrian pine.

Filtration of wind under or through plants is still another method of control. It is sometimes desirable to speed up or slow down wind in this way.

Height The zones of wind reduction on the leeward and windward sides of a barrier depend largely upon the height of the barrier. The taller the trees, the greater the number of rows of trees required for protection. With an increase in the height of the trees, shelterbelts become more open. Instead of reducing the wind, avenues of trees open at the bottom increase wind speed as the airstream is forced beneath the tree canopy and around the tree trunks.

Width The field of effectiveness of a shelterbelt depends primarily on the shelterbelt's height and penetrability. The width of the planting is of secondary importance only insofar as it affects the degree of penetrability. The windbreak's width has a negligible influence on reducing wind velocity at its leeward edge, but it can cause a notable variation in the microclimate within the sheltered area. With a wide shelterbelt or forest, the maximum reduction in the velocity of the wind occurs within the shelterbelt or forest itself. Therefore, a wide shelterbelt or forest block actually consumes its own shelter to the extent that the wind velocity is reduced within the shelter itself.

An irregular windbreak, such as the tops of a picket fence, is more effective than a uniform one in breaking up a portion

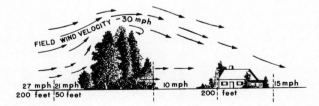

Figure 3.27 Windbreaks reduce wind currents. As shown here, part of the air current is diverted over the tops of the trees, and part filters through the trees.

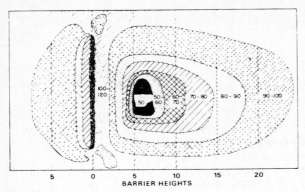

Figure 3.28

of the airstream deflected over it. A mixture of species and sizes of plants within the windbreak therefore produces a rough upper surface and is more effective in controlling wind.

Shelterbelts Studies have shown that shelterbelts and windbreaks are most effective when they are perpendicular to the prevailing winds. Wind velocity may be reduced by 50 percent for a distance of from 10 to 20 times the tree height downwind of a shelterbelt. The degree of protection and wind reduction depends upon the height, width, and penetrability of the plants used. See Figure 3.28.

Near a moderately dense shelterbelt, at the end of the windbreak wind speed is increased by more than 10 percent of the open-field velocity prior to its interception. The leeward sheltered zone is not confined within lines drawn perpendicular to the ends of the barrier but is broader than the length of the barrier.

Wind speed is also affected within or on the windward

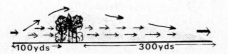

A 30ft. high shelterbelt affects wind speed for 100yds. in front of the trees and 300yds. down wind.

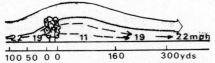

Effect of moderately penetrable windbreaks on wind.

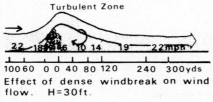

Effect of dense windbreak on wind flow. H=30ft.

Figure 3.29

side of a windbreak. For example, the speed is reduced for a distance of 100 yards on the windward or front side of a 30-foot-high shelterbelt and for a distance of 300 yards downwind or behind the shelterbelt. See Figure 3.29.

Partially penetrable windbreaks have effects on windflow that differ from those of dense windbreaks. Wind velocities immediately to the leeward of any windbreak are directly affected by the type of material or kinds of plants used.

The more penetrable the windbreak, the longer the distance that protection extends behind the windbreak. In passing through a penetrable windbreak, some wind retains some of its laminar flow characteristics at a reduced velocity, thus inhibiting turbulence behind it.

3.4 Topography: Slopes

To avoid costly construction and to make maximum use of grade variations, plans should always be studied in relation to accurately drawn topographical maps or sketches — if possible, directly over them.

Level Sites Even though project land may be so flat that topography does not control site planning, the grouping of buildings should be studied to devise a satisfactory system of drainage. Surfaces or recreation areas and yards require some pitch for discharging water to surface inlets. The locations of these facilities and the economical placing of areas for cutting and filling are important.

Steep or Broken Sites If there are marked differences in elevation, correlating the site plan with the topography will produce economies of first cost and maintenance, particularly in relation to sewer and drainage lines. A careful use of topographical variations may give a site plan individuality.

Very steep or broken sites may cause excessive development costs. On even moderate slopes, the practice of placing buildings parallel to contours will eliminate much costly construction, grading, and filling. This advice holds particularly true when rock is encountered in excavating. If buildings must be placed on comparatively steep portions of the site, the buildings themselves may serve as retaining walls. By a study of topographical sections, it is possible to determine whether to draw buildings closely together along the entrance side, leaving the greater portion of the slope to be taken up in yards and gardens, or to concentrate all garden areas on one side of each row, leaving only sufficient room for access walks on the other side. Adjustments in grades are preferably taken up between yards rather than close to buildings.

See Figure 3.30 for treatment of differing slopes.

Walks and Roads On sites having steeper slopes, say, greater than 5 percent, it is common practice, where conditions permit, to locate walkways close to buildings to reduce further costs of cutting and filling. Depending upon the project, roads may be similarly laid out. Roads should run parallel to contours to avoid grading expense. Since steps are considered undesirable, walkways also should parallel contours.

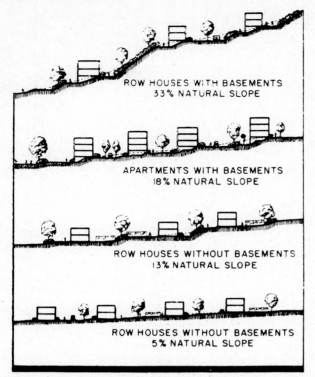

Figure 3.30 Cross sections showing treatment for varying degrees of slope. With a 33 percent slope, basement stories contain apartments, and buildings act as retaining walls. With an 18 percent slope, partially exposed basements contain apartments, laundries, and storage. With a 13 percent slope, buildings are fitted closely to the natural grade, and necessary adjustments are made between yards rather than near buildings. In all three of these cases walks are adjacent to buildings. In the bottom drawing (5 percent slope) buildings are also closely related to the natural grade, but walks and drives are midway between the buildings.

Abrupt changes in level may in some cases be overcome by the introduction of switchbacks.

Preservation of Trees and Buildings Topographical surveys should show existing trees. Efforts to preserve them may save time and expense in producing necessary shade besides adding charm. Slight variations in a plan to accommodate trees will help achieve a desirable informal appearance.

Desirable slope percentages are shown in the accompanying table, and required setbacks in Figure 3.31.

Figures 3.32 and 3.33 show suggested layouts for varying topography.

Relation of Buildings to Topography Few sites are entirely level; many are broken and steep, sometimes so much so as to be considered unbuildable. It is not good practice to fight against the land; yet many plans show evidence of this tendency. Level-land plans have been forced upon rugged sites, and occasionally planning that is characteristic

Desirable Slopes

	Percent of slope	
	Maximum	Minimum
Streets, service drives, and parking areas	8.0	0.5[a]
Collector and approach walks	10.0[b]	0.5
Entrance walks	4.0[c]	1.0
Ramps	15.0	. . .
Paved play and sitting areas	2.0	0.5
Paved gutters	. . .	0.5
Lawn areas	25.0[d]	1.0
Grassed playgrounds	4.0	0.5
Swales	10.0[e]	1.0[f]
Grassed banks	4:1 slope	
Planted banks	2:1 slope	
	(3:1 preferable)	

[a]0.75 percent for dished section.
[b]Less where icy conditions may occur frequently.
[c]Slopes up to 10 percent or more are satisfactory provided walks are long enough to employ a curved profile, so that a slope not exceeding 4 percent can be used adjoining the building platform.
[d]Steepest grade recommended for power mower.
[e]Less for drainage areas of more than approximately ½ acre.
[f]2 percent preferable in all cases, particularly if swales cross walks.

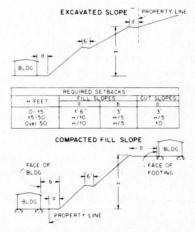

Figure 3.31 Required setbacks (after the Uniform Building Code of 1964).

of steep or rolling sites has been adapted artificially to level land. Although in some cases there may be justification for changing the fundamental character of the land, the basic idea is to follow the contours. However, this is a general principle only and not a rule to be accepted blindly. Single houses can be dotted along a slope without much change in natural grade, but houses in long rows and large apartments can follow only the general sweep of the surface. If slopes are steep, the following schemes should be used:

1. Buildings are placed on nearly level terraces cut into the

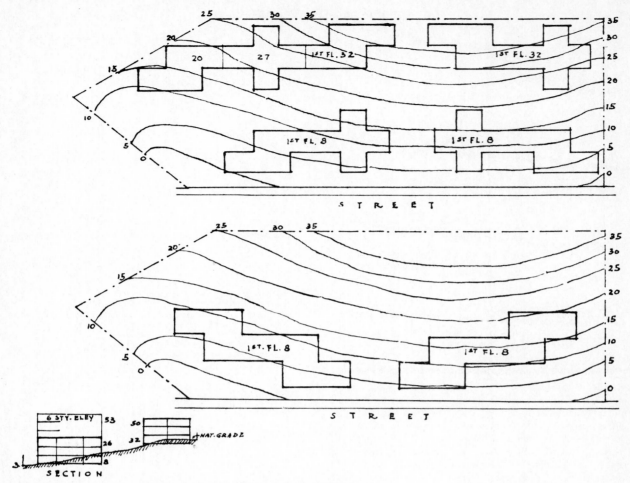

Figure 3.32   This site is too shallow and too steep for walk-up buildings.

hillsides. Streets are either parallel to the buildings and substantially parallel to the contours, or they are as nearly perpendicular to the contours as the maximum practicable gradient permits.

2. Buildings are built in a series of steps following streets that oppose the contours.

The choice of a site may not always be a part of the architect's job. However, all other factors being equal, level or gently sloping sites present few serious topographical difficulties as compared with those of hilly sites. See Figures 3.34, 3.35, and 3.36.

3.5 Gradients

Suggested gradient standards are shown in Figure 3.37 and in the accompanying table.

Slope Percentages   Several adaptations are given as a certain percentage slope or ratio of horizontal distance to vertical distance.

Slope is very important, as too steep an incline can be an effective barrier. In all cases the percentages of slope should be as small as possible; Figure 3.38 converts the percentage of slope to the horizontal and vertical distances. Distance is not measured in any particular unit as long as the horizontal and vertical distances are in the same units, i.e., inches to inches, feet to feet, meters to meters.

Calculating Slope   You may calculate the percentage of slope horizontal or vertical distance using the following formulas.

$$\text{Slope ratio} = H{:}V \text{ (horizontal to vertical)}$$

$$\%S = \frac{V}{H} \times 100$$

where H = horizontal distance
V = vertical distance
$\%S$ = percentage of slope

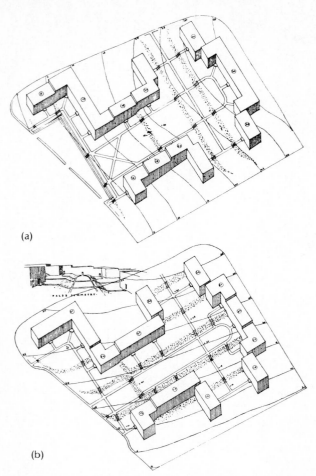

(a)

(b)

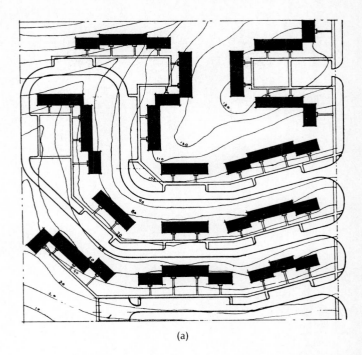

(a)

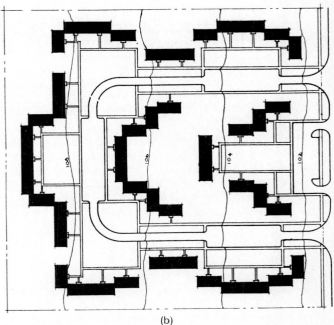

(b)

Figure 3.33 (*a*) A symmetrical layout of sloping land when the axis of symmetry is perpendicular to the contours; (*b*) an axis of symmetry parallel to the contours.

Figure 3.34 Contrasting access roads, cut and fill, and necessary surface drainage systems are shown for (*a*) a hilly site and (*b*) an almost level site.

Slopes Slopes should be designed at moderate grades to help ensure soil stability. In general longer and flatter slopes are more desirable than abrupt steep banks for purposes of maintenance and appearance (see Figure 3.39).

Slopes should be well rounded at the top and at the bottom to improve appearance and to reduce the possibility of erosion (see Figure 3.40).

Streets Street rights-of-way should be selected to allow provision of safe suitable finished grades for pavement, sidewalks, planting strips, drainage, and utilities and convenient access to adjoining properties and facilities.

Sharp vertical roadway curves should be avoided where sharp horizontal curves occur.

The top of sharp vertical curves should not be located where roads change direction just beyond the crest from an approaching vehicle.

To prevent contact between road surfaces and vehicle bumpers or undercarriages, crest or sag vertical curves should be used between tangents where the algebraic difference between gradients is greater than 9 percent. The minimum length of the curve should be 1 foot for each percentage point of difference between gradients (see Figure 3.41).

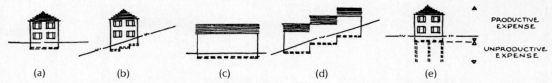

(a)   (b)   (c)   (d)   (e)

Figure 3.35 Construction problems and costs are compared for level and steep sites (*a* through *d*), and objections to sites containing soft, compacted fill are illustrated (*e*).

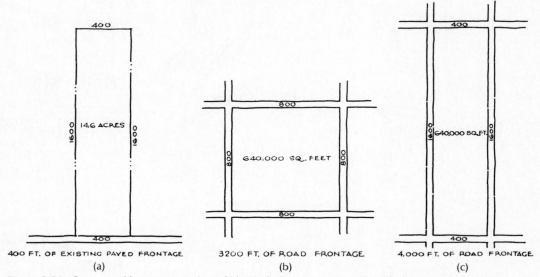

400 FT. OF EXISTING PAVED FRONTAGE   3200 FT. OF ROAD FRONTAGE   4,000 FT. OF ROAD FRONTAGE
(a)   (b)   (c)

Figure 3.36 Square and long, narrow sites of identical areas are compared. With the square site (*b*) the buildings may be serviced from peripheral streets, whereas the long, narrow site (a, c) requires interior roads. Where new peripheral roads, utilities, and so on must be provided at the project's expense, sites approaching square shapes have been found preferable.

The following criteria should be considered in street and driveway design. Under full rated load, guidelines for minimum approach, departure, and ramp breakover angles for passenger cars, station wagons, and ½-ton trucks designed since 1960 are as follows (see Figure 3.42):

A: approach angle, 16 degrees

B: ramp breakover angle, 10 degrees

C: departure angle, 10 degrees

Parking Lots   The maximum desirable grade of a parking lot in any direction is 5 percent. Steeper grades cause problems in opening and closing heavy car doors, increase the danger of unattended cars rolling, and make maneuvering difficult in slippery weather.

Parking wheel stops when used should be of durable material, firmly anchored and arranged to avoid trapping storm runoff.

In general steep walk grades (see grading requirements for limitations) or ramps are preferable to steps and steep banks that accompany them.

Avoid sharp changes in grade on long sections of walk, at intersections, or at building entrances. Where sharp breaks

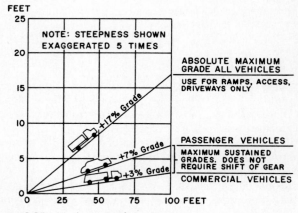

Figure 3.37 Maximum grades.

cannot be avoided, vertical curves should be used between intersecting tangents.

Grading   The primary objective when considering regrading is to retain as much of the existing topography as possible without affecting the use for which the site is intended.

In the regrading operation that *must* be done, it is impor-

Gradients (In Percentages[a])

Access and parking	Minimum		Maximum	
	Center line	Crown or cross slope	Center line	Crown or cross slope
Streets	0.5	1.0	14.0	5.0
Street intersections	0.5	1.0	5.0	5.0
Driveways[b]	0.5	1.0	14.0	5.0
Sidewalks[c]				
Concrete	0.5			
Bituminous	1.0			
Building entrances and				
short walks	1.0		12.0	5.0
Main walks	0.5		10.0	5.0
Adjoining steps			2.0	
Landings		1.0		2.0
Stepped ramp treads	1.0		2.0	
Parking		0.5	5.0	5.0

Slope gradients				
Slope away from				
foundations:				
Pervious surfaces	5.0[d]		21.0[e]	
Impervious surfaces	1.0[d]		21.0[e]	
Slope to upper end of a				
drainage swale	2.5[d,f]			
Pervious surfaces:				
Ground frost areas	2.0			
Non-ground frost areas	1.0[g]			
Impervious surfaces	0.5			
Usable open area			5.0	
Other areas[h]:			50.0	(2:1)
Slopes to be				
maintained by				
machine			33.0	(3:1)

[a]Approximate equivalents: 0.5 percent = 1/16 inch per foot, 1.0 percent = 1/8 inch per foot, 2.0 percent = 1/4 inch per foot, 5.0 percent = 5/8 inch per foot, 10.0 percent = 1 1/4 inches per foot, 12.0 percent = 1 1/2 inches per foot, and 21 percent = 2 5/8 inches per foot.
[b]Vertical transitions should prevent contact of car undercarriage or bumper with driveway surface.
[c]5.0 percent maximum for major use by elderly tenants.
[d]Minimum length, 10 feet or as limited by property lines.
[e]Minimum length, 4 feet.
[f]Can be used only where no steep adjacent slopes will contribute storm runoff.
[g]Areas having annual precipitation of more than 50 inches use a 2.0 percent gradient.
[h]After individual analysis the Department of Housing and Urban Development (HUD) may accept steeper slopes or require flatter slopes. See HUD technical study *Slope Protection for Residential Developments*.

tant that the amount of soil to be cut be equal (or as near as possible) to the required fill. This will avoid the expensive operational cost of either removing soil from the site or bringing it into the site.

Care should be taken to avoid cutting or filling slopes with the gradient over 33 percent because of the difficulty of establishing ground covers that will eventually hold these steep slopes.

Extensive regrading will require careful attention to existing topsoil, especially in areas where topsoil is expensive and unavailable. Usable topsoil must be stockpiled, screened, and respread evenly.

Site Considerations for Handicapped

Grading Access by means of grading, approach ramps, or walks should be provided for all buildings except privately owned residential projects.

Walkways should be provided to eliminate the need for the handicapped to walk or wheel behind parked vehicles.

Public walks end at buildings and connect the buildings with public transportation stops, handicapped parking areas, other walkways, and other buildings in the same complex.

Entrance Approach Building Level Types Entrance to buildings becomes difficult for handicapped people when the

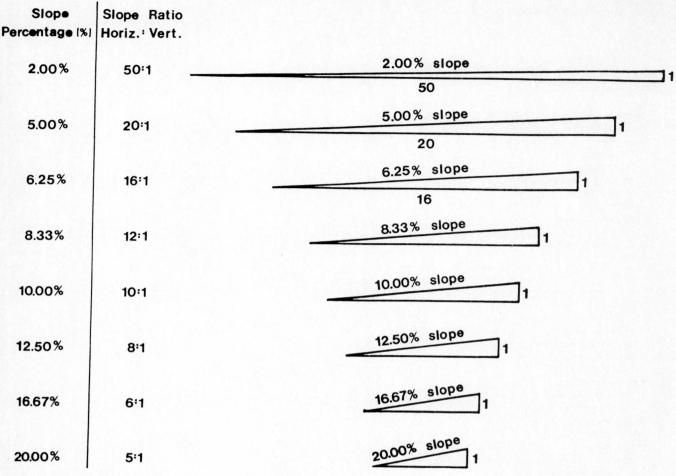

Slope Percentage (%)	Slope Ratio Horiz.: Vert.
2.00%	50:1
5.00%	20:1
6.25%	16:1
8.33%	12:1
10.00%	10:1
12.50%	8:1
16.67%	6:1
20.00%	5:1

Figure 3.38 Slope percentages.

entrance floor level is other than at or near ground level. See Figures 3.43, 3.44, and 3.45.

Most handicapped people require smooth, flat, or gradually sloping surfaces to reach entrance floor levels.

Existing buildings usually have entrance floor levels above or below ground level so that modifications for access will require installation of ramps, bridges, mechanical lifts, or changes in the shape of the land to provide a smooth, gradually sloped surface.

Building floor level types are diagramed in Figure 3.46. On the following pages, some typical entrance conditions for these building types are illustrated with recommended design solutions (see Figures 3.47 and 3.48).

Ramps Ramps must have 5 feet 0 inches by 5 feet 0 inches level platforms at top and bottom, must be at least 4 feet 0 inch wide, must not exceed 1 in 12 slope, must have handrails on both sides that are 2 feet 8 inches high and extend 1 foot 6 inches beyond the top and bottom of the ramp.

Surfaces of ramps must be slip-resistant materials.

Ramps can run parallel to exterior walls and turn corners if necessary.

When the distance from the main floor to the ground is greater than the distance from the ground level to the basement floor, it may be more economical to put ramps in below ground level to provide access to the lower level. The building must have means of vertical transportation such as elevators.

Ramps below ground can have exactly the same configurations as ramps above ground level.

Earthwork can be used to provide smooth, gradually sloping approaches to entrance floor levels.

Bridges On steeply sloping sites bridges might be used to connect floor levels to accessible site locations on the uphill side.

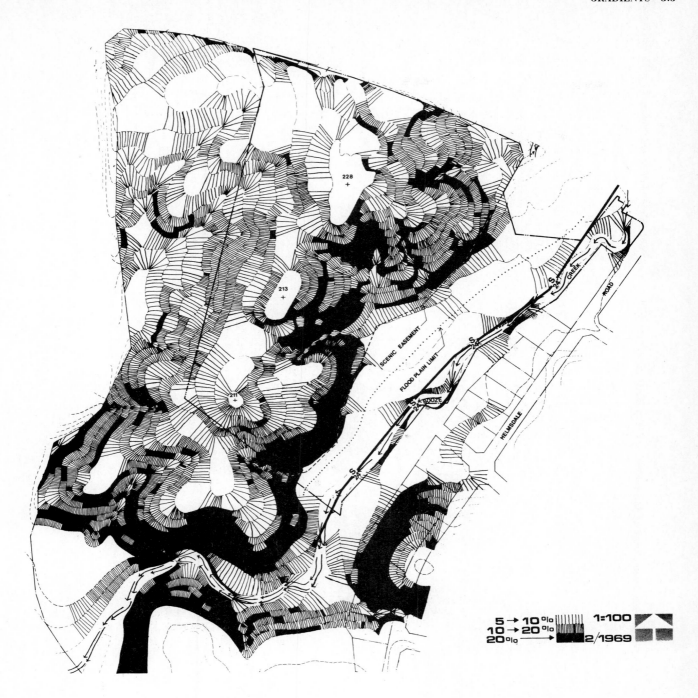

PLANNERS INCORPORATED

Figure 3.39 Slope analysis.

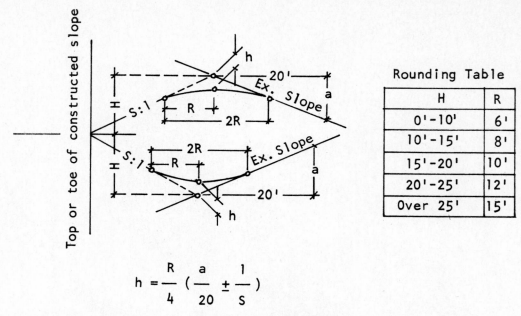

$$h = \frac{R}{4} \left(\frac{a}{20} \pm \frac{1}{S} \right)$$

Use + sign when original ground line slopes in opposite direction to constructed slope.

a = rise or fall of original ground in 20 ft
H = vertical height of constructed slope
h = vertical distance from intersection of
    slope angles to curve
R = 1/2 length of rounding curve
S = slope ratio = $\frac{horiz.}{vert.}$

Figure 3.40 Slope rounding.

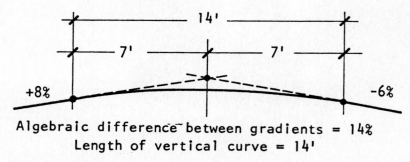

Algebraic difference between gradients = 14%
Length of vertical curve = 14'

Figure 3.41 Vertical curve length to prevent vehicle contact with road surface.

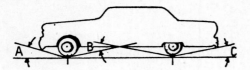

Figure 3.42 Grade change criteria.

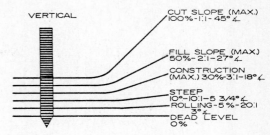

Figure 3.43 Acceptable grade slopes.

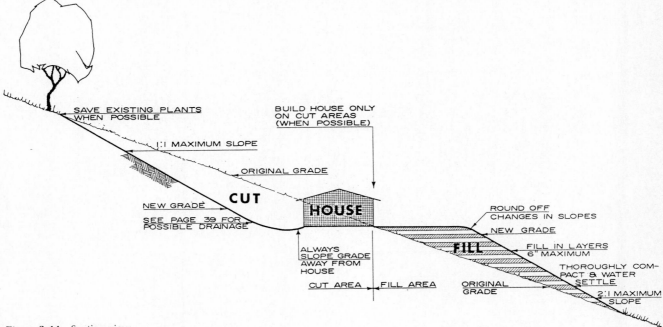

Figure 3.44 Section view.

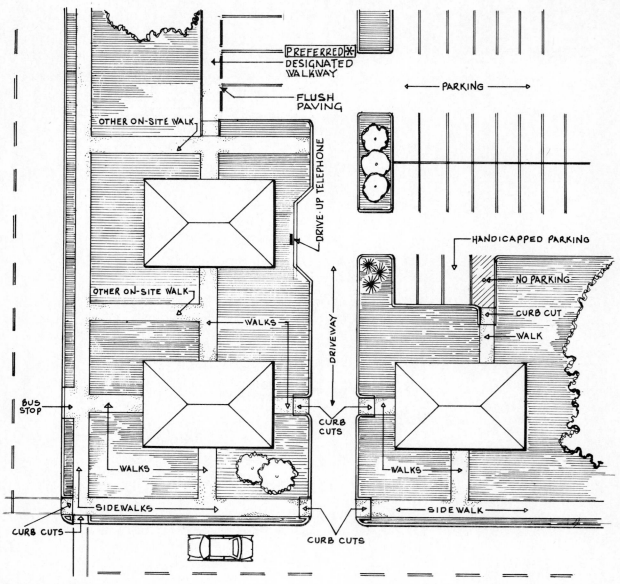

Figure 3.45 Site grading.

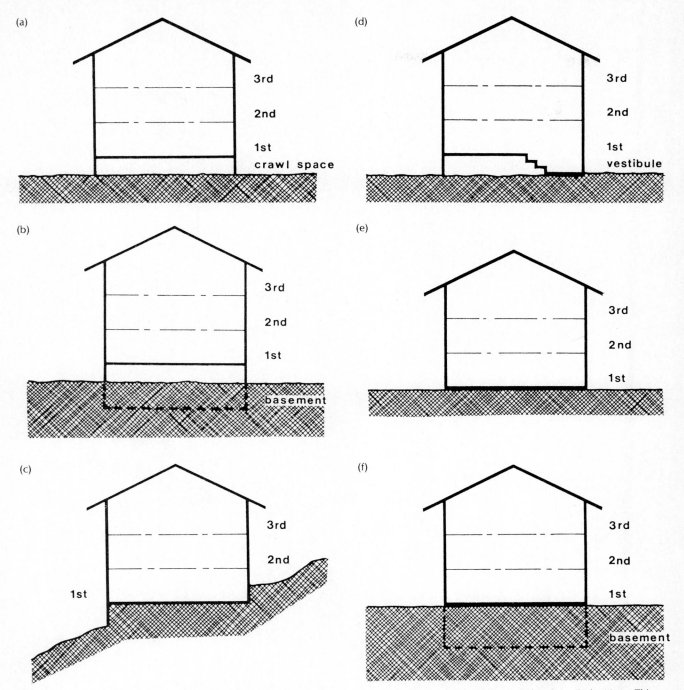

Figure 3.46 (*a*) Building with first floor and other floors above ground level. Crawl space, no basement. Will require installation of ramp or bridge or change in shape of land to provide access. (*b*) Building with first floor above ground and basement floor below ground level. Will require installation of ramp or bridge or change in shape of land to provide access to first floor or basement. (*c*) Building with first floor above and below grade on sloping site. Will require installation of ramp or change in shape of land for access to first floor, or bridge for access to second floor. (*d*) Building with first floor above grade with interior split-level vestibule. Note: This type of entrance is the most difficult to modify for accessibility. Will usually require raising portion of vestibule floor up to main floor level, raising door, and adding ramp, bridge, lift, or earthwork to provide accessibility. (*e*) Building with first floor at ground level. Other floors, if any: above. No basement. Will require minimal alteration to provide entrance access. (*f*) Building with first floor at grade level. Basement: below grade. Will require minimal alteration to provide entrance access.

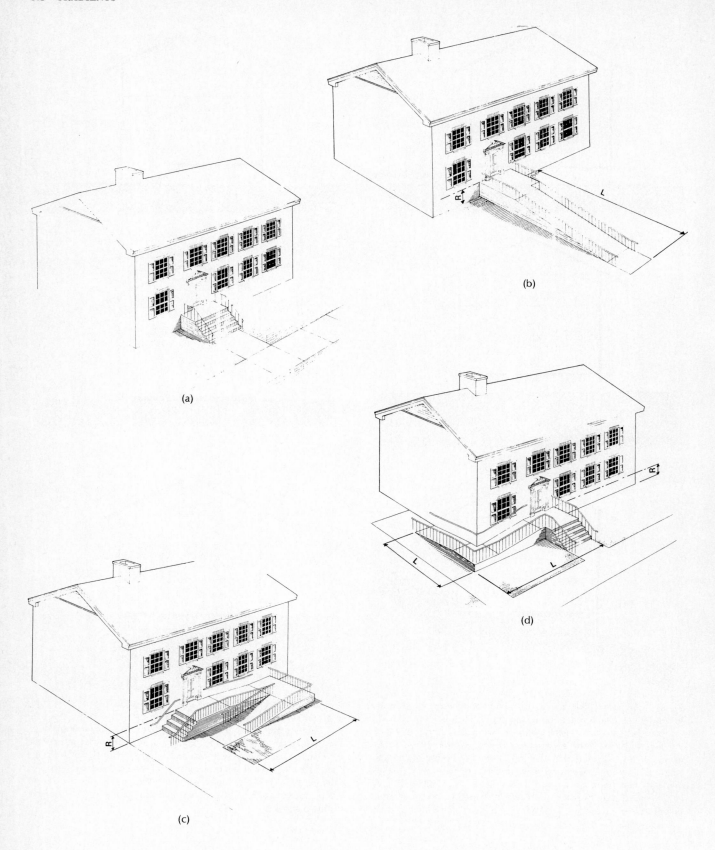

(a)

(b)

(c)

(d)

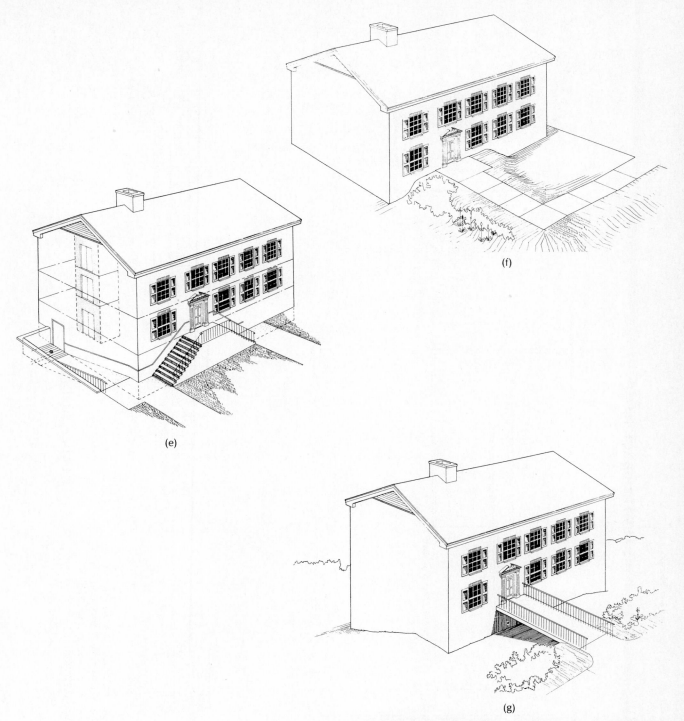

(f)

(e)

(g)

Figure 3.47 (*a*) Existing building. (*b*) Straight run. (*c*) Switch-back. (*d*) L-shape. (*e*) Straight run-down. (*f*) Earthwork. (*g*) Bridge.

Entrance Approach Space Requirements for Ramps

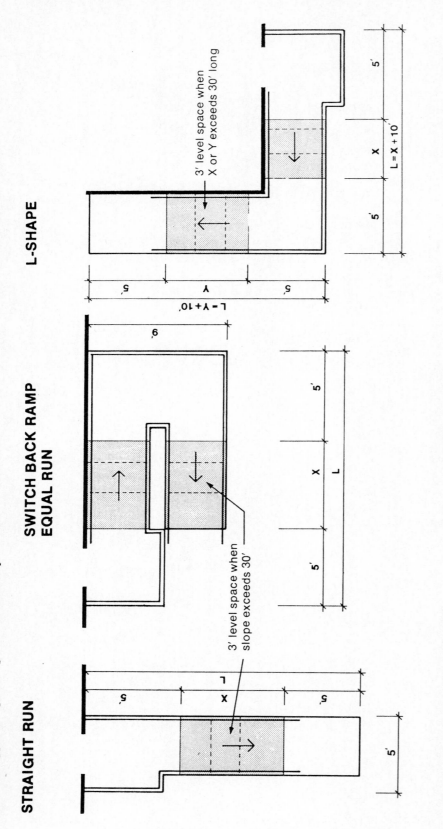

STRAIGHT RUN

SWITCH BACK RAMP EQUAL RUN

L-SHAPE

3' level space when slope exceeds 30'

3' level space when X or Y exceeds 30' long

L = X + 10'

L = Y + 10'

R	1'	2'	3'	4'	5'	6'
X	12'	24'	39'	51'	63'	75'
L	22'	34'	49'	61'	73'	85'

R	1'	2'	3'	4'	5'	6'
X	6'	12'	18'	24'	30'	39'
L	16'	22'	28'	34'	40'	49'

R	1'	2'	3'	4'	5'	6'
X + Y	12'	24'	36'	48'	60'	75'
			39'	51'	63'	78'

Whenever either X or Y exceeds 30' add 3' for rest area.

Wherever possible the length of the slope of ramps should be evenly divided.

L = total length of ramp
R = rise
X and Y = length of the slope

Note: Tables assume flat sites and 1 in 12 slopes. Sites which slope may require longer or shorter ramps depending on direction of ramp and slope of site. Ramps should be oriented to minimize their length.

Figure 3.48

3.6 Noise

SITE EXPOSURE TO AIRCRAFT NOISE

To evaluate a site's exposure to aircraft noise, you will need to consider all airports, both commercial and military, within 15 miles of the site. The most likely sources for the information required for this evaluation are listed below.

1. Federal Aviation Administration (FAA) area office or military agency in charge of the airport. Are the noise exposure forecast (NEF) or composite noise rating (CNR) contours available? (These contours have not yet been constructed for all airports. When they are available, they are superimposed on a map with a marked scale.) Is there any available information about approved plans for runway changes (extensions or new runways).

2. FAA control tower or airport operator if NEF or CNR contours are not available. What is the number of nighttime jet operations (10 P.M. to 7 A.M.)? What is the number of daytime jet operations (7 A.M. to 10 P.M.)? Are there any supersonic jet operations? What are the flight paths of the major runways? Is there any available information about expected changes in airport traffic; for example, will the number of operations increase or decrease in the next 10 or 15 years? Are there any plans for supersonic jet traffic?

In making your evaluation, use the data for the heaviest traffic condition, whether present or future.

If NEF or CNR contours are available, locate the site by referring to the marked scale. Also locate a point roughly in the center of the area covered by the principal runways. If the site lies outside the NEF 30 (CNR 100) contour, draw a straight line to connect these two points. Measure along this line the distances between (1) the NEF 40 (CNR 115) and NEF 30 (CNR 100) contours and (2) the NEF 30 (CNR 100) contour and the site. Now use the table "Site Exposure to Aircraft Noise" to evaluate the site's exposure.

If NEF or CNR contours are not available, determine the effective number of operations for the airport as follows. Multiply the number of nighttime jet operations by 17. Then add the number of daytime jet operations to obtain an effective total. Any supersonic jet operation automatically places an airport in the largest category of the table "Distances for Approximate NEF Contours," which governs noise acceptability.

On a map of the area that shows the principal runways, mark the locations of the site and of the center of the area covered by these runways (see Figure 3.49). Then, using the distances in the second table, you can construct approximate NEF 40 and NEF 30 contours for the major runways and flight paths most likely to affect the site. Again use the first table to evaluate the site's exposure to aircraft noise.

Example 1 Figure 3.50 shows two sites located on a map that has NEF contours. We draw a line from each of these sites to a point roughly in the center of the area covered by the principal runways. Measuring along these lines, we find that site 1 lies outside the NEF 30 contour at a distance greater than that between the NEF 30 and NEF 40 contours and that site 2 lies outside the NEF 30 contour at a distance less than that between the NEF 30 and NEF 40 contours. Therefore, the exposure of site 1 to aircraft noise is clearly acceptable, and the exposure of site 2 is normally acceptable.

Example 2 Figure 3.51 shows an airport for which NEF or CNR contours are not available. The airport has 20 nighttime and 125 daytime jet operations. There are no supersonic

Site Exposure to Aircraft Noise

Distance from site to center of area covered by principal runways	Acceptability category
Outside NEF 30 (CNR 100) contour, at a distance greater than or equal to distance between NEF 30 and NEF 40 (CNR 100, CNR 115) contours	Clearly acceptable
Outside NEF 30 (CNR 100) contour, at a distance less than distance between NEF 30 and NEF 40 (CNR 100, CNR 115) contours	Normally acceptable
Between NEF 30 and NEF 40 (CNR 100, CNR 115) contours	Normally unacceptable
Within NEF 40 (CNR 115) contour	Clearly unacceptable

Distances for Approximate NEF Contours

Effective number of operations	Distances to NEF 30 contour		Distances to NEF 40 contour	
	Distance 1	Distance 2	Distance 1	Distance 2
0–50	1000 feet	1 mile	0	0
51–500	½ mile	3 miles	1000 feet	1 mile
501–1300	1½ miles	6 miles	2000 feet	2½ miles
More than 1300 or any supersonic jet operations	2 miles	10 miles	3000 feet	4 miles

flights, and so we determine the effective number of operations as follows:

$$20(\text{nighttime}) \times 17 = 340$$

Add to this the actual number of daytime operations:

$$340 + 125(\text{daytime}) = 465$$

Using the distances in the table "Distances for Approximate NEF Contours," we construct approximate contours and then draw a line from the site to a point roughly in the center of the area covered by the principal runways. Measuring along this line, we find that the site lies outside the NEF 30 contour at a distance greater than that between the NEF 30 and NEF 40 contours. Therefore, the site's exposure to aircraft noise is clearly acceptable.

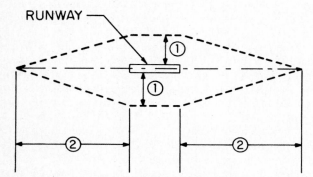

Figure 3.49 Construction of approximate NEF contours with the use of the table "Distances for Approximate NEF Contours."

SITE EXPOSURE TO ROADWAY NOISE

To evaluate a site's exposure to roadway noise, you will need to consider all major roads within 1000 feet of the site. The most likely sources for information required for this evaluation are listed below. Before beginning the evaluation, you should try to obtain any available information about approved plans for roadway changes, such as widening existing roads or building new roads, and about expected changes in road traffic. For example, will the traffic on this road increase significantly in the next 10 or 15 years?

1. Area map, city (county) engineer, or both. What are the distances from the site to the centerlines of the nearest and farthest lanes of traffic?

2. City (county) director of traffic. What is the peak hourly automobile traffic flow in both directions combined? What is the peak number of trucks and buses (buses count as trucks) per hour in each direction? If the road has a gradient of 3 percent or more, record uphill and downhill numbers separately because these figures will be necessary later. If not, simply record the total number of trucks.

You may also need to make adjustments for the following circumstances:

A road gradient of 3 percent or more

Stop-and-go traffic

Mean speed

A barrier

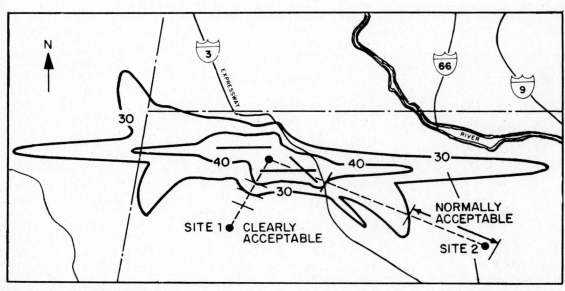

Figure 3.50 Example of NEF contours.

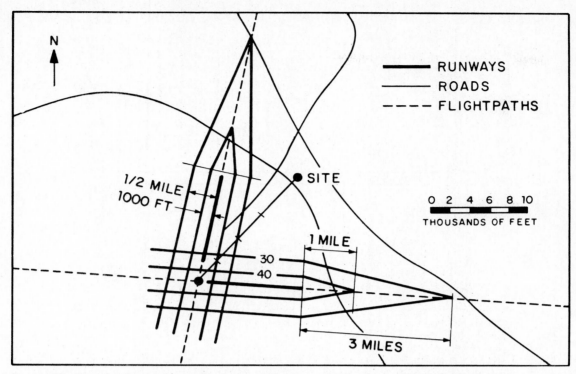

Figure 3.51 Example of approximate NEF contours drawn for an airport with an effective number of operations between 51 and 400.

The information required for these adjustments can be obtained from the city (county) director of traffic.

Traffic surveys show that the level of roadway noise depends on the percentage of trucks in the total traffic volume. To account for this effect, the guidelines discussed below provide for the separate evaluation of automobile and truck traffic.

Before proceeding with these separate evaluations, however, determine the effective distance from the site to each road by locating on Figure 3.53 the distances from the site to the centerlines of the nearest and farthest lanes of traffic. (See also Figure 3.52; the distances in this figure are used for all roadway examples.)

Now lay a straightedge to connect these two distances and read off the value at the point where the straightedge crosses the middle scale. This value is the effective distance to the road.

Automobile Traffic The numbers in Figure 3.54, which is used to evaluate the site's exposure to automobile noise, were arrived at with the following assumptions:

1. There is no traffic signal or stop sign within 800 feet of the site.

2. The mean automobile traffic speed is 60 miles per hour.

3. There is line-of-sight exposure from the site to the road; that is, there is no barrier that effectively shields the site from the road.

If a road meets these three conditions, proceed to Figure 3.54 for an immediate evaluation of the site's exposure to the automobile noise from that road. However, if any of these conditions are different, make the necessary adjustments and then use Figure 3.54 for the evaluation.

Adjustments for Stop-and-Go Traffic If there is a traffic signal or a stop sign within 800 feet of the site, multiply the total number of automobiles per hour by 0.1.

Adjustments for Mean Traffic Speed If there is no traffic signal or stop sign within 800 feet of the site and the mean automobile speed is other than 60 miles per hour, multiply the total number of automobiles by the adjustment factor shown in the table on page 235.

Barrier Adjustment This adjustment affects distance and applies equally to automobiles and to trucks on the same road. Therefore, instructions for this adjustment appear below after those for truck traffic.

Example 1 For road 1, the distance from the site to the centerline of the nearest lane of traffic is 300 feet. The distance to the centerline of the farthest lane of traffic is 366 feet. Figure 3.52 shows that the effective distance from the site to this road is 330 feet. For road 2, the distance to the centerline of the nearest lane of traffic is 150 feet. The distance to the centerline of the farthest lane of traffic is 186 feet. Figure 3.52 shows that the effective distance from the site to this road is 166 feet. For road 3, the distance to the centerline of the nearest lane of traffic is 210 feet. The distance to the centerline of the farthest lane of traffic is 320

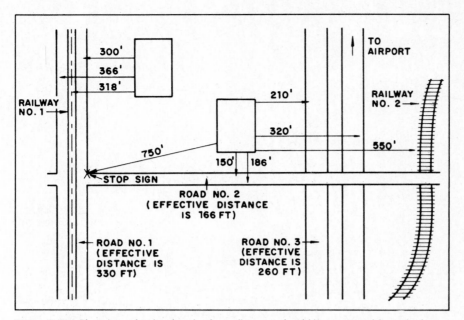

Figure 3.52 Plan view of a site showing how distances should be measured from the location of the dwelling nearest to the source. The site is exposed to noise from three major roads. Road 1 has four lanes, each 12 feet wide, and a 30-foot median strip that accommodates a rapid-transit line. Road 2 has four lanes, each 12 feet wide. Road 3 has six lanes, each 15 feet wide, and a median strip 35 feet wide.

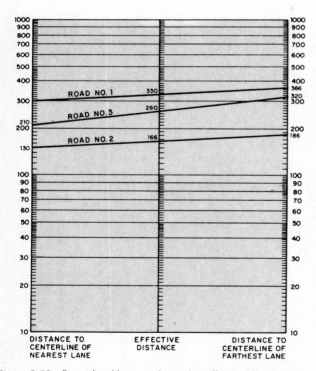

Figure 3.53 Example of how to determine effective distances.

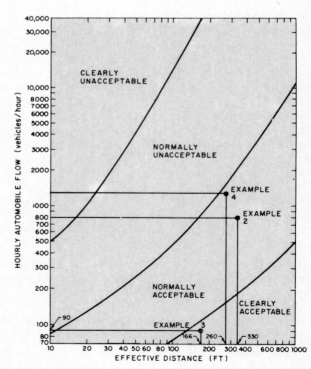

Figure 3.54 Example of how to evaluate site exposure to automobile noise.

Mean traffic speed (miles per hour)	Adjustment factor
20	0.12
25	0.18
30	0.25
35	0.32
40	0.40
45	0.55
50	0.70
55	0.85
60	1.00
65	1.20
70	1.40

feet. Figure 3.52 shows that the effective distance from the site to this road is 260 feet.

Example 2 Road 1 meets the three conditions that allow for an immediate evaluation. In obtaining the information necessary for this evaluation, we find that the hourly automobile flow is 800 vehicles. On Figure 3.54, we locate on the vertical scale the point representing 800 vehicles per hour and on the horizontal scale the point representing 330 feet (note that we must estimate the location of this point). Using a straightedge, we draw lines to connect these two values and find that the site's exposure to automobile noise from this road is normally acceptable.

Example 3 Road 2 has a stop sign at 750 feet from the site. The hourly automobile flow is reported as 900 vehicles. We adjust for stop-and-go traffic:

$$900 \times 0.1 = 90 \text{ vehicles}$$

We find from Figure 3.54 that the exposure to automobile noise is clearly acceptable.

Example 4 Road 3 is a depressed highway. There is no traffic signal or stop sign and the mean speed is 60 miles per hour. The hourly automobile flow is 1200 vehicles. The road profile shields all residential levels of the housing from line-of-sight exposure to the traffic. The only adjustment that can be made is the barrier adjustment. This adjustment is necessary, however, only when the site's exposure to noise has been found to be clearly or normally unacceptable. Figure 3.54 shows that the exposure to automobile noise is normally acceptable. Therefore, no adjustment for barrier is necessary.

Truck Traffic The numbers in Figure 3.55 which is used to evaluate the site's exposure to truck noise, were arrived at with the following assumptions:

1. There is a road gradient of less than 3 percent.
2. There is no traffic signal or stop sign within 800 feet of the site.
3. The mean truck traffic speed is 30 miles per hour.
4. There is line-of-sight exposure from the site to the road;

that is, there is no barrier that effectively shields the site from the road.

If a road meets these four conditions, proceed to Figure 3.55 for an immediate evaluation of the site's exposure to truck noise from that road. However, if any of the conditions are different, make the necessary adjustments and then use Figure 3.55 for the evaluation.

Adjustments for Road Gradient If there is a gradient of 3 percent or more, multiply the number of trucks per hour in the uphill direction by the appropriate adjustment factor.

Add to this adjusted figure the number of trucks per hour in the downhill direction.

Percent of gradient	Adjustment factor
3–4	1.4
5–6	1.7
More than 6	2.5

Adjustments for Stop-and-Go Traffic If there is a traffic signal or a stop sign within 800 feet of the site, multiply by 5 the total number of trucks.

Adjustments for Mean Traffic Speed Make this adjustment only if there is no traffic signal or stop sign within 800 feet of the site and the mean speed is not 30 miles per hour. If the mean truck speed differs with direction, treat the uphill

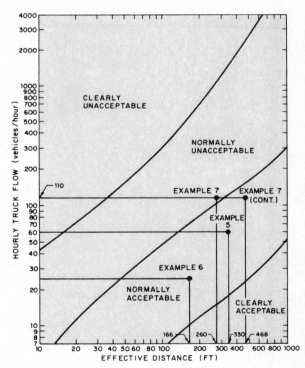

Figure 3.55 Example of how to evaluate the site's exposure to truck noise.

and downhill traffic separately. Multiply each by the appropriate adjustment factor below.

Example 5 Road 1 meets the four conditions that allow for an immediate evaluation. The hourly truck flow is 60 vehicles. Figure 3.55 shows that the site's exposure to truck noise from this road is normally acceptable.

Example 6 Road 2 has a stop sign at 750 feet from the site. There is also a road gradient of 4 percent. No trucks are allowed on this road, but four buses per hour are scheduled (two in each direction).

We adjust first for gradient:

Uphill $2 \times 1.4 = 2.8$ vehicles

Downhill   2.0 vehicles

Total flow   4.8 vehicles

We then adjust for stop-and-go traffic:

$$4.8 \times 5 = 24 \text{ vehicles per hour}$$

Figure 3.55 shows that the exposure to truck or bus noise from this road is normally acceptable.

Mean traffic speed (miles per hour)	Adjustment factor
20	1.60
25	1.20
30	1.00
35	0.88
40	0.75
45	0.69
50	0.63
55	0.57
60	0.50
65	0.46
70	0.43

Example 7 The profile of road 3 shields all residential levels of the housing from line-of-sight exposure to the traffic. The mean truck speed is 50 miles per hour. The hourly truck flow is 175 vehicles. We adjust for mean speed:

$$175 \times 0.63 = 110.24$$

$$= 110 \text{ vehicles}$$

We find from Figure 3.55 that exposure to truck noise is normally unacceptable. Therefore, we proceed with the barrier adjustment.

Road 3 has been depressed 25 feet from the 150-foot elevation of the natural terrain. The actual road elevation therefore is 125 feet. We find the effective road elevation to be

$$125 + 5 = 130 \text{ feet}$$

Six stories are planned for the housing, which is located at an elevation of 130 feet. The effective site elevation for the highest story is

$$6 \times 10 = 60 + 130 - 5 = 185 \text{ feet}$$

Barrier Adjustment A barrier may be formed by the road profile, by a solid wall or embankment, by a continuous row of buildings, or by the terrain itself. To be an effective shield, however, the barrier must block all residential levels of all buildings from line-of-sight exposure to the road, and it must not have any gaps that would allow noise to leak through. This adjustment is necessary only when the site's exposure to noise from a road has been found to be normally or clearly unacceptable.

Step 1: From the city (county) engineer obtain the elevation of the road. (Roads may be elevated above the natural terrain or may be depressed, as in our example. Make certain, therefore, that the figure you obtain for road elevation takes any such change into account.) Add 5 feet to this figure to obtain the effective road elevation.

Step 2: From the applicant obtain the ground elevation of the site and the number of stories in the proposed housing. Multiply the number of stories by 10 feet. Add the site elevation, and then subtract 5 feet from this total to obtain the effective site elevation.

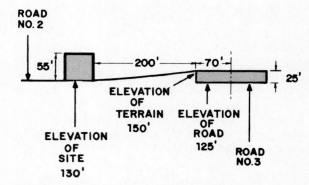

Figure 3.56 Detail of site showing the measurements necessary for a barrier adjustment.

Step 3: From the city (county) engineer or a contour map obtain the elevation of the terrain where the barrier is located. Add the actual height of the barrier to obtain the effective barrier elevation. (Note that in some cases, as in our example, the barrier is formed by the road profile, and the elevation of the terrain is the effective barrier elevation.) To find the barrier adjustment factor, you will need Figure 3.57 and a straightedge. The example of barrier adjustment below explains how to use Figure 3.57.

When you have determined the barrier adjustment factor, multiply the effective distance by the factor to obtain the adjusted distance from the site to the road.

Example 7 (Continued) The barrier, which is formed by the road profile, has no height other than the 150-foot elevation of the natural terrain. Thus, the effective barrier elevation is 150 feet. The difference in effective elevation between the site and the road is 55 feet, and between the barrier and the road is 20 feet. We now use Figure 3.57 to find the barrier adjustment factor.

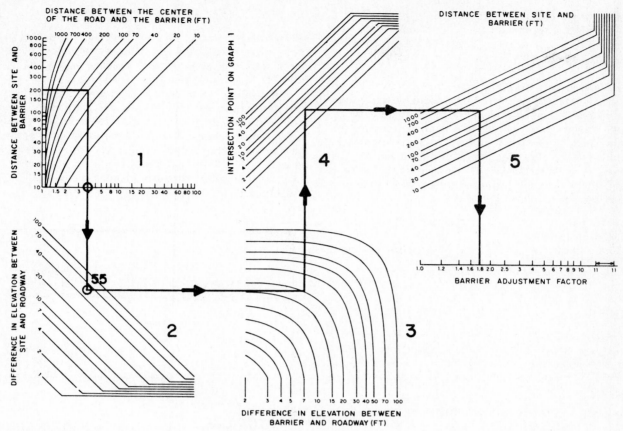

Figure 3.57 Example of how to find the adjustment factor.

Example of Barrier Adjustment

1. The distance from the site to the barrier is 200 feet.

2. The distance from the center of the road to the barrier is 70 feet.

3. The difference in effective elevation between the site and the road is 55 feet.

4. The difference in effective elevation between the barrier and the road is 20 feet.

On the vertical scale of graph 1 in Figure 3.57, we mark 200 feet and draw a straight horizontal line to meet the curve marked 70 feet. Then we draw a vertical line down to graph 2 to meet the point that represents 55 feet (note that we must guess the location) and a horizontal line over to graph 3 to meet the curve marked 20 feet. (If the line from graph 2 does not meet the appropriate curve on graph 3, the barrier is not an effective shield, and there is no adjustment.)

Next we draw a vertical line up to graph 4 to meet the curve marked 4 (the number intersected by the line going from graph 1 to graph 2) and a horizontal line over to graph 5 to meet the curve marked 200 feet. From graph 5 we draw

a vertical line down to the adjustment scale and find that our multiplier is 1.8.

Using this multiplier, we adjust the effective distance

$$260 \times 1.8 = 468 \text{ feet}$$

We find from Figure 3.55 that the site's exposure to truck noise from this road is normally acceptable.

SITE EXPOSURE TO RAILWAY NOISE

To evaluate a site's exposure to railway noise, you will need to consider all aboveground rapid-transit lines and railroads within 3000 feet of the site. The most likely sources for information required for this evaluation are listed below. Before beginning the evaluation, you should record the information on a work sheet.

1. Area map, county engineer, or both. What is the distance from the site to the railway right-of-way? Does a barrier effectively shield the site of the railway? Remember that an effective barrier blocks all residential levels of all buildings from line-of-sight exposure to the railway and has no gaps that would allow noise to leak through.

2. Supervisor of customer relations for the railway. What is the number of nighttime railway operations (10 P.M. to 7 A.M.)? Is there any available information about approved plans for changing the number of nighttime operations?

Distances in the table "Site Exposure to Railway Noise" were arrived at on the assumption that there are ten or more nighttime railway operations (10 P.M. to 7 A.M.). If a railway has ten or more nighttime operations, proceed to the table for an immediate evaluation of the site's exposure to noise from that railway. However, if a railway has fewer than ten nighttime operations, multiply the distance from the site to that railway by the appropriate adjustment factor; then proceed to the table.

Number of nighttime railway operations	Adjustment factor
1–2	3.3
3–5	1.7
6–9	1.2

Example 1   The distance from the site to railway 1 is 318 feet. There are two nighttime operations, and there is direct line-of-sight exposure to the right-of-way. Since there are fewer then ten nighttime operations, we adjust the distance as follows:

$$318 \text{ feet} \times 3.3 = 1049 \text{ feet}$$

We then proceed to the table "Site Exposure to Railway Noise," where we find that exposure to noise from this railway is normally acceptable.

Example 2   The distance from the site to railway 2 is 550 feet. There are twenty nighttime railway operations, and this site is completely shielded from the right-of-way. Since there are more than ten nighttime operations, we proceed immediately to the table and find that the site's exposure to noise from this railway is clearly acceptable.

Site Exposure to Railway Noise

Line-of-sight exposure (feet)	Shielded exposure (feet)	Acceptability category
More than 3000	More than 500	Clearly acceptable
601–3000	101–500	Normally acceptable
101–600	51–100	Normally unacceptable
Less than 100	Less than 50	Clearly unacceptable

Distance from site to right-of-way (possibly adjusted for number of nighttime operations)

LANDSCAPING FOR NOISE CONTROL

When external noise cannot be muffled at the source, landscape barriers can provide some control within the site. These barriers generally involve shielding or absorption, or both.

The combination of trees, low foliage, and ground cover provides noise attenuation when significant masses of such absorbing vegetation are involved. Generally, such foliage should be from 500 to 1000 feet deep to diminish properly the intensity of normal traffic noises. While relatively thin barriers serve effectively as a visual barrier or a sunscreen, a sonic barrier must be of much greater dimensions. See Figures 3.58 and 3.59.

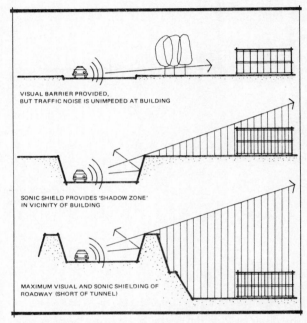

VISUAL BARRIER PROVIDED, BUT TRAFFIC NOISE IS UNIMPEDED AT BUILDING

SONIC SHIELD PROVIDES 'SHADOW ZONE' IN VICINITY OF BUILDING

MAXIMUM VISUAL AND SONIC SHIELDING OF ROADWAY (SHORT OF TUNNEL)

Figure 3.58   Development of topography for noise control.

3.7 Spatial Structure

Spatial structure is defined as the configuration of physical open space of a given site. Spatial structure is generally the result of topographic characteristics, vegetation massing, and topographic characteristics in conjunction with vegetation massing. See Figure 3.60. Because these three elements determine the size and, to a great extent, the quality of a space, they may be referred to as spatial determinants.

As the spatial configuration of a landscape is documented, the information thus obtained can be transferred to a site drawing (see Figure 3.61).

Spatial Structure of Landscape   After determining the spatial structure of a given landscape, one can establish the qualitative characteristics of a given space. An understanding

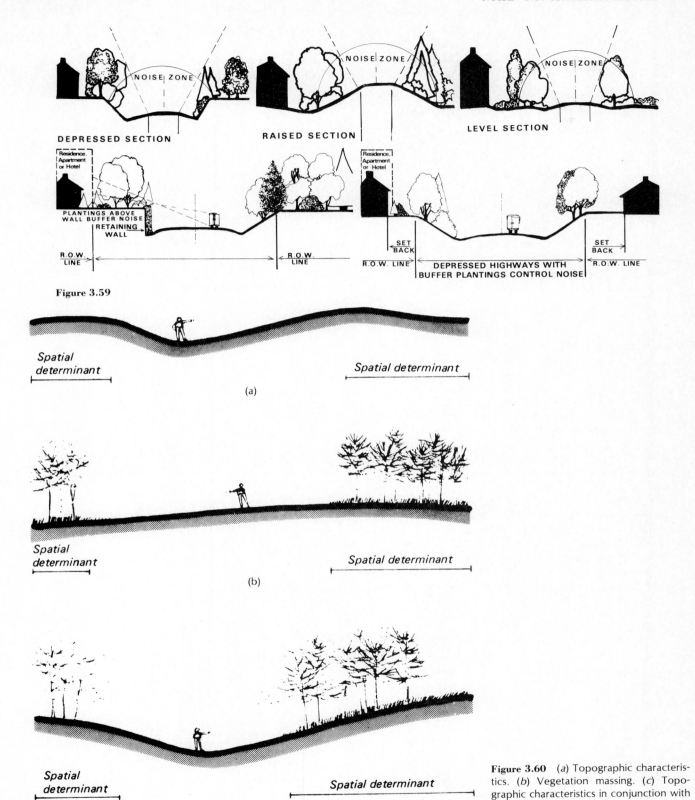

DEPRESSED SECTION

RAISED SECTION

LEVEL SECTION

NOISE ZONE

Residence, Apartment or Hotel

PLANTINGS ABOVE WALL BUFFER NOISE

RETAINING WALL

R.O.W. LINE

R.O.W. LINE

R.O.W. LINE

SET BACK

DEPRESSED HIGHWAYS WITH BUFFER PLANTINGS CONTROL NOISE

SET BACK

R.O.W. LINE

Figure 3.59

Spatial determinant

Spatial determinant

(a)

Spatial determinant

Spatial determinant

(b)

Spatial determinant

Spatial determinant

(c)

Figure 3.60 (*a*) Topographic characteristics. (*b*) Vegetation massing. (*c*) Topographic characteristics in conjunction with vegetation massing.

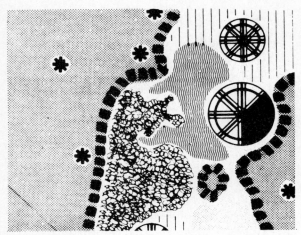

Figure 3.61 The aesthetic resources resulting from landform diversity, vegetation patterns, surface waters, and visual quality establish the major spatial definitions, high points, views and vistas, and site image.

of the overall spatial structure of a landscape, together with an understanding of the qualitative characteristics of the individual spaces, is extremely important in locating functions that are greatly influenced by visual factors, such as roadways, visitor centers, and walking and hiking trails.

Spatial Characteristics The spatial characteristics of a landscape generally depend upon three issues: size of space, degree of visual enclosure, and visual character.

Size of Space The size of a space is important in determining its total visual impact as well as its potential to absorb a given function. Size may be evaluated both in terms of acreage and in the relationship of its size to that of all other on-site spaces.

Degree of Visual Enclosure The degree of visual enclosure of a space is an important spatial factor, especially in locating functions that are greatly influenced by the need for circulation linkages (trails or roadways) or scenic views or vistas (scenic overlooks, visitor centers, or trails). Although the definition of space suggests enclosure, the spatial structure may be such as to evoke a feeling or image of one of the three diagrams in Figure 3.62.

Degree of enclosure is an important planning consideration not only in terms of spatial access but also in terms of visual form. For instance, a person near a topographic or vegetative mass will tend to look away from it (see Figure 3.63). This tendency may be used advantageously by the planner in

directing the visitor to more promising views or to other important visual phenomena.

Yet another important consideration of spatial enclosure relates to the "harbor quality" of a space. Harbor quality is the ability of a space to invite or attract use. Many spaces offer such a strong attraction. As shown in Figure 3.63, these spaces can best be employed as entry points wherever a strong sense of invitation and arrival is important.

Perhaps the single most important aspect of spatial structure consists in locating and developing sites that can support various land uses. A knowledge of on-site spatial enclosures arms the planner with greater opportunities for placing aesthetically unappealing activities (storage, service, water tanks) in less readily observable sites (see Figure 3.64).

Visual Character In determining the visual characteristics of a space, one should carefully interpret the space in terms of the inherent visual images it presents. For instance, a certain space may have a heavy background of conifers and a strong view of distant mountains; yet a centrally located lake may be so visually dominant that the entire scene evokes an image of a waterscape. Here, the particular space might be noted as follows:

Space X
  Major image: Lake and associated
                  water edge
  Subordinate
    images: Dense conifer forest
                  View of distant
                  mountains

A general listing of the natural or cultural features that may influence the visual characteristics of a space (example only) follows:

Natural features

1. Vegetation
    a. Dense conifer forest
    b. Grassland image
    c. Wetland image

2. Topography and geology
    a. Rock outcropping
    b. Scattered boulders
    c. Sand dunes

3. Hydrology
    a. Open lake
    b. Slow, meandering stream

Cultural features:

1. Land uses
    a. Agriculture
    b. Service and storage

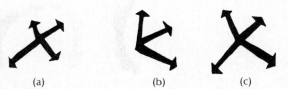

| (a) | (b) | (c) |

Figure 3.62 (*a*) Total enclosure; (*b*) semiopen enclosure; (*c*) no enclosure.

Figure 3.63   Spatial structure.

2. Utilities

   a. Telephone utilities

   b. Gas pipes

3. Barriers

   a. Corrals

   b. Fences

The inherent visual quality of a site will greatly influence the type of activity that occurs on it. Dense and heavily screened spaces will naturally promote an entirely different feeling from that produced by open, gently rolling spaces. If the final plan is to be successful, the activities planned for various sites should reflect the inherent quality of the site, that is, running, jumping, or passive recreation such as meditation (see Figure 3.65).

Figure 3.64

Figure 3.65

3.8 Site Accessibility

General Site Accessibility   (Figures 3.66 and 3.67.)

Waiting Areas Preferably located within 300 feet of building entry; area located between roadway and sidewalk to avoid traffic congestion; an overhead shelter is recommended for protection from weather; adequate seating and lighting should also be provided.

Signage   Should be provided to direct pedestrians to various destinations or areas of the site.

Site Entrance Well identified; obvious relationship to building and site it serves; signage to direct vehicular and pedestrian traffic to destinations on the site.

Walkways Should provide clear direct route through site; surfaces should be firm and level; curb cuts and ramps provided where necessary.

Rest Areas Provided where pedestrians must walk long distances; keep rest areas off walkway thoroughfares.

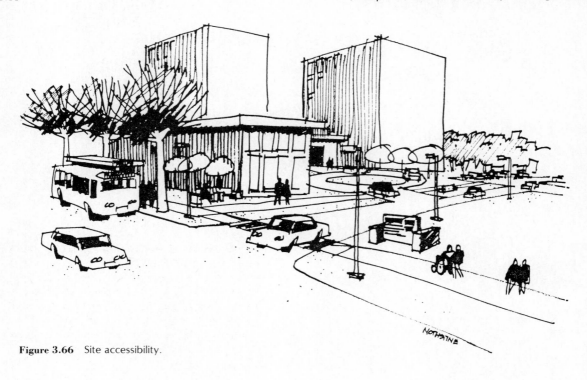

Figure 3.66 Site accessibility.

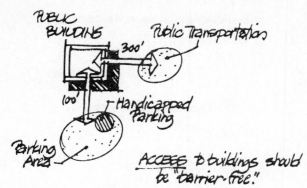

Figure 3.67   Access to buildings should be barrier-free.

Parking   Related directly to buildings which they serve; "handicapped" stalls no more than 100 feet from building entry.

Drop-Off Zones   Located as close to building entry as possible; no grade change between road surface and adjacent walkway. Direct vehicular connections between drop-off, site entrance, and parking areas; signage should be provided to direct both vehicles and pedestrians to destinations on the site.

Building Entry   Clearly identified; alternative means of entry provided for handicapped individuals (i.e., both ramps and stairs); public facilities located immediately off entry in lobby (lavatories, phones, drinking fountains, etc.); no grade change between entrance and facilities. See Figure 3.68.

Relating Site Components   Barrier-free site design is not some simplistic goal that can be achieved through the thoughtful handling of one or two problem areas on a site. The accessibility of any public or private outdoor area hinges on the physical relationships between design elements both inside and outside of the space. Unless there is a relationship of continuous accessibility between forms of transportation, site elements, and building entries, the value in making any one of these components more accessible is lost. Consequently, it is imperative that all elements of circulation be made as easily accessible as possible.

The following items should be considered to ensure a good interface between transportation, site, and building entry elements.

1. Special transportation facilities should be provided for people who are restricted in their use of the exterior environment. Care should be taken to separate varying types

Doorways & Entrances

PUBLIC LOBBIES SHOULD CONTAIN THE FOLLOWING ITEMS IN AREAS ACCESSIBLE TO THE HANDICAPPED AS CLOSE TO DOORWAYS AS POSSIBLE *:

- PUBLIC TELEPHONES
- REST ROOM FACILITIES
- DRINKING FOUNTAINS
- WAITING AREA WITH APPROPRIATE SEATING
- INFORMATION AND DIRECTIONAL SIGNAGE
- ELEVATORS, ESCALATORS, ETC.

* NOTE: ALL FACILITIES SHOULD BE FUNCTIONALLY USEABLE BY HANDICAPPED INDIVIDUALS.

- ENTRANCE PLAZAS REQUIRE 10'-0" LENGTH FROM DOORWAY TO CHANGE IN GRADE (STAIRS)
- PROVIDE 5'-0" MIN. CLEAR SURFACE AT ALL LANDINGS FOR BOTH STAIRS AND RAMPS.
- RAMPS: MAX. % SLOPE = 8.33 %
   MAX. LENGTH/RAMP = 30'-0"
   MIN. ONE-WAY WIDTH = 3'-0"
   MIN. TWO-WAY WIDTH = 6'-0"
- PROVIDE ADEQUATE RAILINGS, HANDLES, CURBS, AT ALL STAIR AND RAMP LOCATIONS, SEE "STAIRS, RAMPS, AND HANDRAILS."
- PROVIDE SIGNAGE SHOWING POINTS OF ACCESS FOR HANDICAPPED PEOPLE.

Doorways at Entrances

- MIN. 32" CLEAR OPENING.
- NO GRADE CHANGE AT THRESHOLD.
- HORIZONTAL THROW-BARS ARE RECOMMENDED OVER KNOBS, LATCHES, VERTICAL HANDLES, ETC.
- RECOMMENDED FORCE REQUIRED TO OPEN IS 5 lbs. TO 8 lbs.
- PROVIDE 18" SET BACK FROM NEAREST OBSTACLE (WALL, EDGE OF PAVEMENT, ETC.) SEE "GATES & DOORWAYS."
- PROVIDE AUTOMATIC DOOR AT HEAVILY USED LOCATIONS.

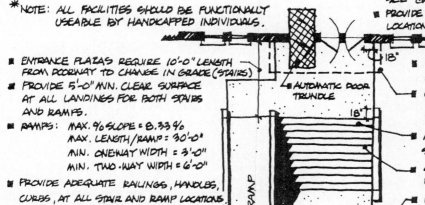

- PROVIDE 5 FOOTCANDLES LIGHTING AT ALL ENTRANCES.
- OVERHEAD CANOPY PROTECTS PEDESTRIANS DURING INCLEMENT WEATHER.
- MAX. GRADE CHANGE BETWEEN STAIRWAY LANDINGS = 6'-0"
- SEE "RAMPS, STAIRS, AND HANDRAILS," FOR DETAILS OF TREADS AND RISERS.
- PROVIDE MIN. 5 FOOTCANDLES OF ALL WALKWAYS, RAMPS, STAIRWAYS, SEE "LIGHTING CONSIDERATIONS."

Figure 3.68   Doorways and entrances.

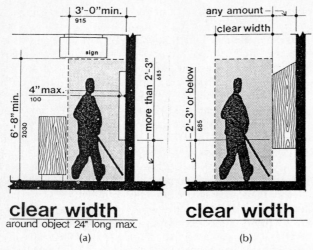

clear width
around object 24" long max.

(a)

clear width

(b)

Figure 3.69 (*a*) Objects 2 feet (610 millimeters) long or less that are fixed to wall surfaces should not project into accessible routes more than 4 inches (100 millimeters) if mounted with their leading edges between 2 feet 3 inches and 6 feet 8 inches (685 and 2030 millimeters) (nominal dimension) above the finished floor. (*b*) Objects fixed to wall surfaces may project more than 4 inches (100 millimeters) if mounted with the lower extreme of their leading edge at or below 2 feet 3 inches (685 millimeters) above the finished floor. These objects should not project into the required minimum clear width.

of transportation where practical, since their point of intersection is usually confusing, dangerous, and delaying. Vehicular traffic should be separated as much as possible from bicycle traffic, and both should be held apart from pedestrian traffic.

2. In general, access to transportation facilities, through the site, and to buildings should be smooth and free of barriers which may prove impossible for physically restricted people to negotiate. Paving surfaces should be hard and relatively smooth, curbs should have ramped cuts, walks should be sufficiently wide to accommodate two-way traffic, and entrance walks to buildings should slope gently to the platform before the doors. If situations are present in which stairs are normally required, then at least one major entrance should be served by a ramp as well.

3. Doors into public buildings should preferably be activated by automatic opening devices. When these items are prohibited by costs, horizontal levers or through bars should be installed on the doors.

4. Public conveniences such as restroom facilities, drinking fountains, telephones, elevators, and waiting areas should be well organized and located in close proximity to building entrances. This allows people with physical limitations to gain access to necessary facilities with a minimal amount of hardship or embarrassment. See Figures 3.69, 3.70, and 3.71.

clear width

(a)

fixed obstruction
plan view

(b)

fixed obstruction
elevation

(c)

Figure 3.70 (*a*) No protruding object should reduce the clear width of an accessible route or maneuvering space below the minimum required. (*b*) and (*c*) Objects mounted with their leading edge at or below 2 feet 3 inches (685 millimeters) above the finished floor may protrude any amount.

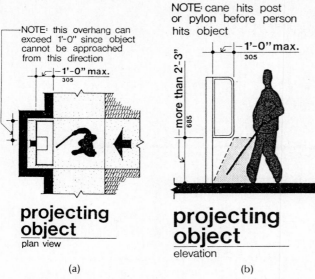

NOTE: this overhang can exceed 1'-0" since object cannot be approached from this direction

1'-0" max.
305

projecting object
plan view

(a)

NOTE: cane hits post or pylon before person hits object

—more than 2'-3"
685

1'-0" max.
305

projecting object
elevation

(b)

Figure 3.71 Free-standing objects mounted on posts or pylons may overhang the circulation path in the direction(s) of approach a maximum of 1 foot (305 millimeters) from 2 feet 3 inches to 6 feet 8 inches (685 to 2030 millimeters) above ground or finished floor surface.

SELECTING PLANT MATERIALS

GROUND COVER 1" TO 10"

KNEE HEIGHT 1.5 FT.

WAIST HEIGHT 3 FT.

EYE LEVEL 6 FT.

SCREEN 8 FT.

ABOVE 8 FT. INTO THE SKY

PEOPLE

PEOPLE–THEIR LINE OF VISION DETERMINES WHETHER A FENCE WILL PROVIDE PRIVACY OR MERELY SEPARATION. THEIR HEIGHT MEASURES FENCES, SHRUBS, TREES AND ALL VERTICAL AND OVERHEAD ELEMENTS.

3.9 Landscaping Elements

PLANT MATERIALS, DENSITY, AND VOLUME

The factors that determine the selection of plant materials, their density, and their volume are illustrated in Figures 3.72, 3.73, and 3.74.

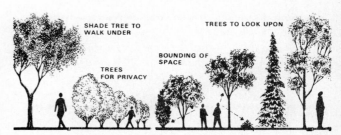

SHADE TREE TO WALK UNDER

TREES TO LOOK UPON

BOUNDING OF SPACE

TREES FOR PRIVACY

Figure 3.72

DENSITY

PETIOLE SIZE, SHAPE AND LENGTH

SOME PETIOLE SHAPES HAVE MORE STRUCTURAL STABILITY THAN DO OTHERS. ON PLANTS WITH MORE RIGID PETIOLES THERE IS LESS LEAF MOVEMENT, AND THE PLANT APPEARS MORE SOLID AND DENSE.

◈ SQUARE
◉ CIRCULAR
○ OVAL
▲ TRIANGULAR
⬡ OCTANGULAR

ARRANGEMENT OF LEAVES

LEAVES ARE ARRANGED IN GENERAL ORDER OF INCREASING COMPLEXITY OF FORM AND/OR MARGIN FROM LEFT TO RIGHT THIS INCREASING COMPLEXITY COMBINED WITH VARIOUS BRANCHING FORMS WILL PRODUCE VARYING DEGREES OF DENSITY.

EVERGREEN

DECIDUOUS

Figure 3.73

–245–

VOLUME

BRANCHING PATTERN — BRANCHING SPACING — BRANCHING HEIGHT

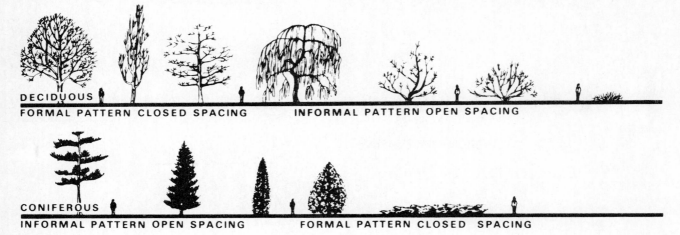

HEIGHT

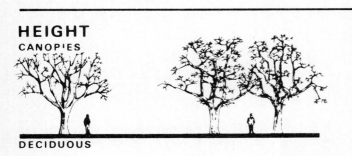

WIDTH

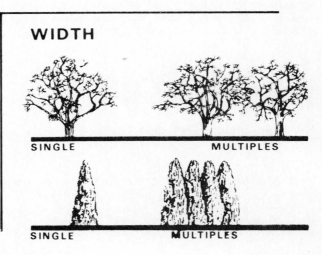

Figure 3.74

WALLS, CEILINGS, AND FLOORS

Factors that determine the use of plant materials as walls, ceilings, and floors are illustrated in Figure 3.75.

FLOORSCAPE AND CEILINGSCAPE

Factors affecting the role of plant materials in the role of floorscape and ceilingscape are shown in Figure 3.76.

SPACE ARTICULATORS

The various ways in which plants can be used to articulate space are shown in Figure 3.77.

WALLS
REPRESENTATIVE PLANTS INDICATING THE RELATIVE DEGREES OF VISUAL DENSITY.
(AT LEAST 6' HIGH)

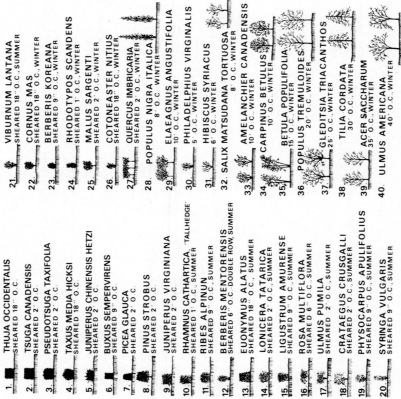

1. THUJA OCCIDENTALIS SHEARED 18" O.C.
2. TSUGA CANADENSIS SHEARED 2' O.C.
3. PSEUDOTSUGA TAXIFOLIA SHEARED 2' O.C.
4. TAXUS MEDIA HICKSI SHEARED 18" O.C.
5. JUNIPERUS CHINENSIS HETZI SHEARED 18" O.C.
6. BUXUS SEMPERVIRENS SHEARED 9" O.C.
7. PICEA GLAUCA SHEARED 2' O.C.
8. PINUS STROBUS SHEARED 2' O.C.
9. JUNIPERUS VIRGINIANA SHEARED 2' O.C.
10. RHAMNUS CATHARTICA, 'TALLHEDGE' SHEARED 1' O.C., SUMMER
11. RIBES ALPINUN SHEARED 9" O.C., SUMMER
12. BERBERIS MENTORENSIS SHEARED 6" O.C. DOUBLE ROW, SUMMER
13. EUONYMUS ALATUS SHEARED 18" O.C., SUMMER
14. LONICERA TATARICA SHEARED 2' O.C. SUMMER
15. LIGUSTRUM AMURENSE SHEARED 9" O.C., SUMMER
16. ROSA MULTIFLORA SHEARED 6" O.C. SUMMER
17. ULMUS PUMILA SHEARED 2' O.C., SUMMER
18. CRATAEGUS CRUSGALLI SHEARED 2' O.C. SUMMER
19. PHYSOCARPUS APULIFOLIUS SHEARED 9" O.C. SUMMER
20. SYRINGA VULGARIS SHEARED 9" O.C. SUMMER

21. VIBURNUM LANTANA SHEARED 18" O.C., SUMMER
22. CORNUS MAS SHEARED 18" O.C., WINTER
23. BERBERIS KOREANA SHEARED 9" O.C., WINTER
24. RHODOTYPOS SCANDENS SHEARED 1' O.C., WINTER
25. MALUS SARGENTI SHEARED 2' O.C. WINTER
26. COTONEASTER NITIUS SHEARED 18" O.C., WINTER
27. QUERCUS IMBRICARIA SHEARED 2' O.C. WINTER
28. POPULUS NIGRA ITALICA 8' O.C. WINTER
29. ELAEAGNUS ANGUSTIFOLIA 10' O.C. WINTER
30. PHILADELPHIUS VIRGINALIS 5' O.C. WINTER
31. HIBISCUS SYRIACUS 6' O.C. WINTER
32. SALIX MATSUDANA TORTUOSA 8' O.C. WINTER
33. AMELANCHIER CANADENSIS 10' O.C. WINTER
34. CARPINUS BETULUS 10' O.C. WINTER
35. BETULA POPULIFOLIA 15' O.C. WINTER
36. POPULUS TREMULOIDES 20' O.C. WINTER
37. GLEDITSIA TRIACANTHOS 25' O.C. WINTER
38. TILIA CORDATA 30' O.C. WINTER
39. ACER SACCHARUM 35' O.C. WINTER
40. ULMUS AMERICANA 40' O.C. WINTER

CEILINGS
REPRESENTATIVE PLANTS INDICATING THE RELATIVE DEGREES OF VISUAL DENSITY.
(SPACED TO GIVE UNIFORM CEILING DENSITY.)

1. THUJA OCCIDENTALIS AMERICAN ARBOR-VITAE
2. JUNIPERUS VIRGINIANA EASTERN RED-CEDAR
3. PSEUDOTSUGA TAXIFOLIA DOUGLAS-FIR
4. ACER PLATANOIDES NORWAY MAPLE, SUMMER
5. LARIX DECIDUA EUROPEAN LARCH, SUMMER
6. PICEA PUNGENS COLORADO BLUE SPRUCE
7. QUERCUS ALBA WHITE OAK, SUMMER
8. CELTIS OCCIDENTALIS AMERICAN HACKBERRY, SUMMER
9. ULMUS AMERICANA AMERICAN ELM, SUMMER
10. FRAXINUS PENNSYLVANICA LANCEOLATA GREEN ASH, SUMMER
11. QUERCUS ALBA WHITE OAK, WINTER
12. BETULA POPULIFOLIA GREY BIRCH, WINTER
13. MACLURA POMIFERA OSAGE ORANGE, WINTER
14. QUERCUS PALUSTRIS PIN OAK, WINTER
15. SORBUS AUCUPARIA EUROPEAN MOUNTAIN ASH, WINTER
16. CLADRASTIS LUTEA AMERICAN YELLOW-WOOD, WINTER
17. GINKGO BILOBA GINKGO, WINTER
18. GLEDITSIA TRIACANTHOS COMMON HONEY-LOCUST, WINTER
19. RHUS TYPHINA STAGHORN SUMAC, WINTER
20. ROBINIA PSEUDOACACIA BLACK LOCUST, WINTER

FLOORS
OF EXTERIOR SPACES CREATED BY PLANT MATERIALS RATED BY THEIR ABILITY TO WITHSTAND FOOT TRAFFIC AND THE EASE WITH WHICH THEY ARE ABLE TO BE WALKED UPON.

1. POA PRATENSIS KENTUCKY BLUEGRASS
2. THYMUS SERPYLLUM THYME
3. ANENARIA VERNA CAESPITOSA MOSS SANDWORT
4. ANTHEMIS NOBILIS CAMOMILE
5. MATRICARIA TCHIHATCHEWII TURFING DAISY
6. CERASTIUM TOMENTOSUM SNOW-IN-SUMMER
7. HEDERA HELIX ENGLISH IVY
8. AJUGA REPTANS BUGLEWEED
9. COTONEASTER ADPRESSA PRAECOX CREEPING COTONEASTER
10. ROSA WICHURAIANA MEMORIAL ROSE

Figure 3.75

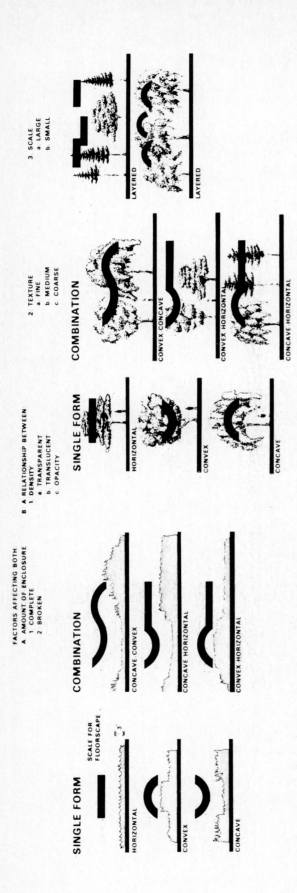

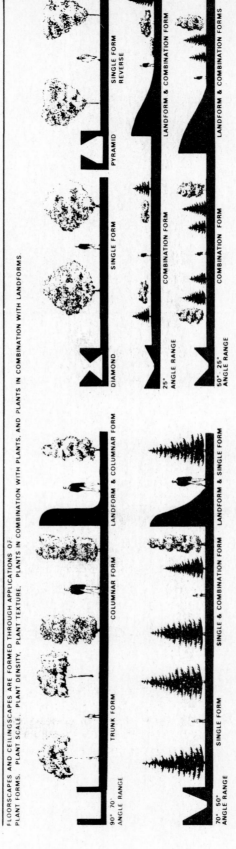

FLOORSCAPES AND CEILINGSCAPES ARE FORMED THROUGH APPLICATIONS OF: PLANT FORMS, PLANT SCALE, PLANT DENSITY, PLANT TEXTURE, PLANTS IN COMBINATION WITH PLANTS, AND PLANTS IN COMBINATION WITH LANDFORMS.

Figure 3.76 Floorscape and ceilingscape.

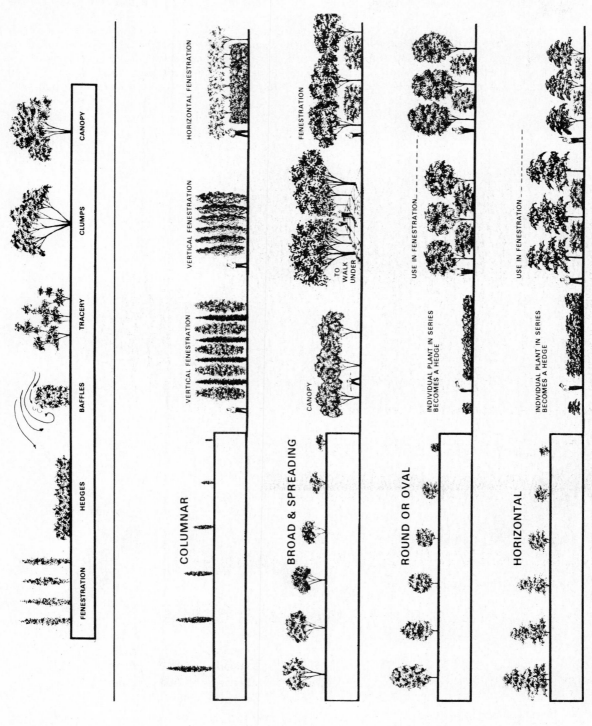

| FENESTRATION | HEDGES | BAFFLES | TRACERY | CLUMPS | CANOPY |

COLUMNAR

VERTICAL FENESTRATION

VERTICAL FENESTRATION

HORIZONTAL FENESTRATION

BROAD & SPREADING

CANOPY

TO WALK UNDER

FENESTRATION

ROUND OR OVAL

INDIVIDUAL PLANT IN SERIES BECOMES A HEDGE

USE IN FENESTRATION

HORIZONTAL

INDIVIDUAL PLANT IN SERIES BECOMES A HEDGE

USE IN FENESTRATION

Figure 3.77 Space articulators. The individual plant is a specimen that, through spacing, becomes fenestration, hedges, baffles, tracery, clumps, or canopy.

3.10 Regional Trees and Shrubs

Plant materials for each of the nine temperature regions of the United States are listed in the following pages. The correct list from which to select plants for a particular geographic location can be determined by examining the map in Figure 3.78, which shows the division of the United States into nine regions of average annual minimum temperature. Since extreme variations in altitude within the major regions or proximity to bodies of water may create somewhat colder or warmer areas within the particular regions, the experience of a local landscape architect familiar with the locality should be relied upon for a final judgment of a plant's hardiness in the area.

It should also be pointed out that although the plant materials listed for each region thrive best within that region, such plants are frequently usable in another region, thereby vastly increasing the number of plants from which a landscape architect can make a selection for a particular purpose.

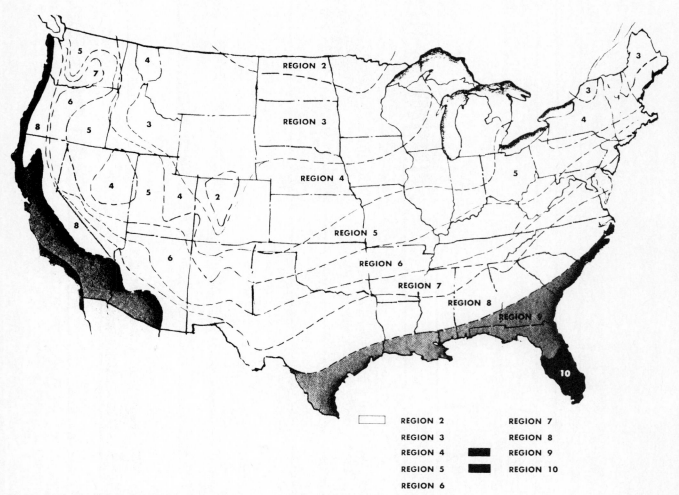

Figure 3.78 The U.S. Department of Agriculture has divided the country into twenty-three provinces. A simplification of these provinces dividing the United States into nine regions, would appear as on this map. Boundaries are determined by means of the average annual minimum temperatures of these regions.

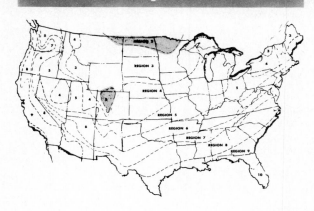

Region 2

Botanical Name	Common Name
Picea engelmanni	Engelmann spruce
Picea pungens	Colorado spruce

Medium Evergreen Trees

Botanical Name	Common Name
Picea glauca	Black Hills spruce
Pinus banksiana	Jack pine
Pinus cembra	Swiss stone pine
Pinus flexilis	Limber pine
Pinus sylvestris	Scotch pine
Thuja occidentalis	American arborvitae

Tall Deciduous Trees

Botanical Name	Common Name
Betula papyrifera	Canoe birch
Populus deltoides	Cottonwood
Populus nigra italica	Lombardy poplar
Populus tremuloides	Quaking aspen

Medium Deciduous Trees

Botanical Name	Common Name
Acer negundo	Box elder
Ainus incana	Speckled alder
Betula pendula	European birch
Fraxinus pennsylvanica	Red ash
Malus baccata	Siberian crab apple
Populus simoni	Simon poplar
Salix alba	White willow
Sorbus aucuparia	European mountain ash

Low Deciduous Trees

Botanical Name	Common Name
Acer spicatum	Mountain maple
Carpinus caroliniana	American hornbeam
Elaeagnus angustifolia	Russian olive
Malus ioensis plena	Bechtel crab apple
Prunus pennsylvanica	Pin cherry

Tall Evergreen Trees

Botanical Name	Common Name
Juniperus virginiana	Eastern red cedar
Larix decidua	European larch
Larix laricina	Eastern larch

Tall Deciduous Shrubs

Botanical Name	Common Name
Acer ginnala	Amur maple
Alnus rugosa	Smooth alder
Caragana arborescens	Siberian pea tree
Lonicera maacki	Amur honeysuckle
Rhamnus frangula	Alder buckthorn
Rhus glabra	Smooth sumac
Salix lucida	Shining willow
Tamarix pentandra	Five-stamen tamarix
Viburnum dentatum	Arrowwood
Viburnum lentago	Nannyberry

Medium-Tall Deciduous Shrubs

Botanical Name	Common Name
Syringa josikaea	Hungarian lilac
Viburnum trilobum	American cranberry bush

Medium Deciduous Shrubs

Botanical Name	Common Name
Ainus viridis	European green alder
Cornus alba sibirica	Siberian dogwood
Cornus stolonifera	Red osier dogwood
Myrica pennsylvanica	Bayberry
Physocarpus opulifolius	Eastern ninebark
Prunus tomentosa	Manchu cherry
Ribes alpinum	Alpine currant
Rosa rugosa	Rugosa rose
Shepherdia canadensis	Buffalo berry
Syringa villosa	Late lilac

Low-Medium Deciduous Shrubs

Botanical Name	Common Name
Amorpha canescens	Leadplant
Comptonia peregrina	Sweet fern
Symphoricarpus orbiculatus	Indiana currant

Low Deciduous Shrubs

Botanical Name	Common Name
Genista tinctoria	Dyer's greenweed

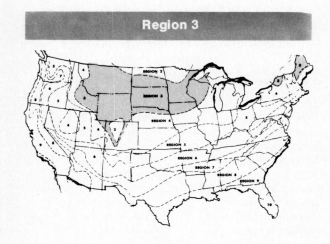

Region 3

Tall Deciduous Trees

Botanical Name	Common Name
Acer platanoides	Norway maple
Acer rubrum	Red maple
Acer saccharum	Sugar maple
Aesculus hippocastanum	Horse chestnut
Alnus glutinosa	European alder
Fagus grandifolia	American beech
Fraxinus americana	White ash
Fraxinus excelsior	European ash
Populus alba	White poplar
Prunus serotina	Black cherry
Tilia cordata	Littleleaf linden

Medium Deciduous Trees

Botanical Name	Common Name
Acer pennsylvanicum	Moosewood
Betula lenta	Sweet birch
Cladrastis lutea	American Yellowwood
Phellodendron amurense	Amur cork tree

Prunus avium	Sweet cherry
Quercus bicolor	Swamp white oak
Robinia pseudoacacia	Black locust

Low Deciduous Trees

Botanical Name	Common Name
Aesculus glabra	Ohio buckeye
Prunus cerasifera atropurpurea	Pissard plum
Prunus cerasus	Sour cherry

Tall Evergreen Trees

Botanical Name	Common Name
Abies veitchi	Veitchi fir
Chamaecyparis obtusa	Hinoki false cypress
Chamaecyparis pisifera	Sawara false cypress
Tsuga canadensis	Canada hemlock

Tall Deciduous Shrubs

Botanical Name	Common Name
Corylus avellana	European hazel
Euonymus europeus	Spindle tree
Hippophaë rhamnoides	Sea buckthorn
Ligustrum amurense	Amur privet
Rhus typhina	Staghorn sumac
Syringa vulgaris	Common lilac
Viburnum opulus	European cranberry bush
Viburnum prunifolium	Black haw

Medium-Tall Deciduous Shrubs

Botanical Name	Common Name
Elaeagnus umbellata	Autumn elaeagnus
Sambucus canadensis	American elder
Vaccinium corymbosum	Highbush blueberry

Medium Deciduous Shrubs

Botanical Name	Common Name
Clethra ainfolia	Summer sweet
Cotoneaster racemiflora	Coin-leaf cotoneaster
Euonymus alatus	Winged burning bush
Lonicera tatarica	Tatarian honeysuckle
Symphoricarpus albus	Snowberry
Viburnum acerifolium	Mapleleaf viburnum
Viburnum cassinoides	Withe rod

Region 4

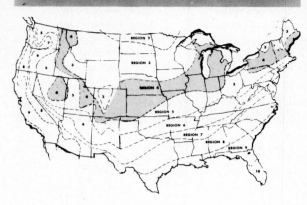

Tall Deciduous Trees

Botanical Name	Common Name
Betula nigra	River birch
Carya cordiformis	Bitternut
Carya glabra	Pignut
Carya ovata	Shagbark hickory
Catalpa speciosa	Northern catalpa
Cercidiphyllum japonicum	Catsura tree
Fagus sylvatica	European beech
Ginkgo biloba	Ginkgo
Gleditsia triacanthos	Common honey locust
Gymnocladus dioicus	Kentucky coffee tree
Juglans nigra	Eastern black walnut
Kalopanax pictus	Castor aralia
Liquidamber styraciflua	Sweet gum
Liriodendron tulipifera	Tulip tree
Magnolia acuminata	Cucumber tree
Nyssa sylvatica	Black tupelo
Quercus alba	White oak
Quercus velutina	Black oak
Tilia tomentosa	Silver linden

Medium Deciduous Trees

Botanical Name	Common Name
Ailanthus altissima	Tree of heaven
Amelanchier canadensis	Shadblow serviceberry
Castanea mollissima	Chinese chestnut
Catalpa bignonioides	Southern catalpa
Diospyros virginiana	Common persimmon
Juglans sieboldiana	Heartnut
Morus alba	White mulberry
Ostrya virginiana	Hop hornbeam
Oxydendrum arboreum	Sorrel tree
Quercus borealis	Red oak
Quercus coccinea	Scarlet oak
Quercus palustris	Pin oak

Sassafras albidum officinale	Sassafras
Sophora japonica	Japanese pagoda tree

Low Deciduous Trees

Botanical Name	Common Name
Carya tomentosa	Mockernut
Cercis canadensis	Eastern redbud
Cornus florida	Flowering dogwood
Cornus mas	Cornelian cherry
Crataegus crus-galli	Cockspur thorn
Crataegus phaenopyrum	Washington hawthorn
Halesia carolina	Carolina silver bell
Malus arnoldiana	Arnold crab apple
Malus coronaria charlottae	Charlotte crab apple
Malus floribunda	Japanese flowering crab apple
Malus hopa	Red flowering crab apple
Malus scheideckeri	Scheidecker crab apple
Malus spectabilis riversi	River's crab apple
Salix caprea	Goat willow
Syringa amurensis	Japanese tree lilac

Tall Evergreen Trees

Botanical Name	Common Name
Abies concolor	White fir
Abies homolepis	Nikko fir
Charmaecyparis nootkatensis	Nootka false cypress
Picea omorika	Serbian spruce
Pinus nigra	Austrian pine
Pseudotsuga taxifolia	Douglas fir
Taxodium distichum	Common bald cypress

Medium Evergreen Trees

Botanical Name	Common Name
Juniperus chinensis	Chinese juniper
Pinus bungeana	Lacebark pine
Pinus rigida	Pitch pine
Pinus virginiana	Scrub pine
Tsuga caroliniana	Carolina hemlock

Tall Deciduous Shrubs

Botanical Name	Common Name
Cephalanthus occidentalis	Buttonbush
Chionanthus virginicus	Fringe tree
Chornus mas	Cornelian cherry
Cornus racemosa	Gray dogwood
Euonymus yedoensis	Yeddo euonymus

Hamamelis virginiana	Common witch hazel
Hydrangea paniculata	Plumed hydrangea
Ligustrum vulgare	Common privet
Lindera benzoin	Spicebush
Photinia villosa	Oriental photinia
Rhus copallina	Shining sumac
Rosa setigera	Prairie rose
Syringa amurensis japonica	Japanese tree lilac
Tamarix parviflora	Small-flowered tamarix
Viburnum sieboldi	Siebold viburnum

Medium-Tall Deciduous Shrubs

Botanical Name	Common Name
Aesculus parviflora	Bottlebrush buckeye
Kolkwitzia amabilis	Beauty bush
Ligustrum ibolium	Ibolium privet
Sambucus racemosa	European red elder

Medium Deciduous Shrubs

Botanical Name	Common Name
Acanthopanax sieboldianus	Five-fingered aralia
Calycanthus floridus	Carolina allspice
Chaenomeles lagenaria	Flowering quince
Deutzia lemoinei	Lemoine deutzia
Elaeagnus multiflora	Cherry elaeagnus
Lonicera bella	Belle honeysuckle
Lonicera morrowi	Morrow honeysuckle
Philadelphus coronarius	Sweet mock orange
Philadelphus grandiflorus	Big scentless mock orange
Ribes adoratum	Clove currant
Rosa eglanteria	Sweetbrier
Spiraea prunifolia	Bridal wreath spirea
Spiraea vanhouttei	Vanhoutte spirea
Stephanandra incisa	Cut-leaf stephanandra
Viburnum tomentosum	Double-file virurnum

Low-Medium Deciduous Shrubs

Botanical Name	Common Name
Deutzia gracilis	Slender deutzia
Kerria Japonica	Kerria
Spiraea arguta	Garland spirea
Spiraea thunbergi	Thunberg spirea
Viburnum carlesi	Fragrant viburnum

Low Deciduous Shrubs

Botanical Name	Common Name
Aronia melanocarpa	Black chokeberry
Ceanothus americanus	New Jersey tea

Chaenomeles japonica	Japanese quince
Cotoneaster horizontalis	Rock spray
Daphne mezereum	February daphne
Hydrangea arborescens	Smooth hydrangea
Hypericum prolificum	Shrubby Saint-John's-wort
Rosa carolina	Carolina rose
Spiraea tomentosa	Hardhack spirea

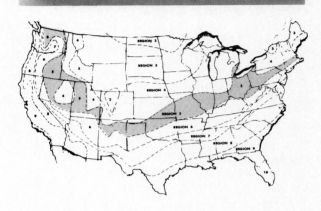

Region 5

Tall Deciduous Trees

Botanical Name	Common Name
Acer pseudoplatanus	Sycamore maple
Platanus acerifolia	London plane tree
Quercus imbricaria	Shingle oak
Quercus robur	English oak
Tilia petiolaris	Pendent silver linden
Zelkova serrata	Japanese zelkova

Medium Deciduous Trees

Botanical Name	Common Name
Asimina triloba	Pawpaw
Carpinus betulus	European hornbeam
Maclura pomifera	Osage orange
Paulownia tomentosa	Royal paulownia
Quercus phellos	Willow oak
Tilla eucheora	Crimean linden

Low Deciduous Trees

Botanical Name	Common Name
Acer campestre	Hedge maple
Acer palmatum	Japanese maple
Betula populifolia	Gray birch
Franklinia alatamaha	Franklinia
Koelreuteria paniculata	Goldenrain tree

Botanical Name	Common Name
Magnolia soulangeana	Saucer magnolia
Prunus persica	Peach
Prunus subhirtella	Higan cherry

Tall Evergreen Trees

Botanical Name	Common Name
Cedrus libani	Cedar of Lebanon
Chamaecyparis lawsoniana	Lawson false cypress
Cryptomeria japonica	Cryptomeria
Picea polita	Tiger's-tail
Sciadopitys verticillata	Umbrella tree
Thuja plicata	Giant arborvitae

Tall Deciduous Shrubs

Botanical Name	Common Name
Buddleia davidi	Butterfly bush
Cotinus coggygria	Smoke bush
Hibiscus syriacus	Shrub althea
Ligustrum ovalifolium	California privet
Magnolia stellata	Star magnolia
Prunus triloba	Flowering almond
Rhus chinensis	Chinese sumac
Spyringa chinensis	Chinese lilac
Xanthoceras sorbifolium	Shiny-leaf yellow horn

Medium-Tall Deciduous Shrubs

Botanical Name	Common Name
Colutea arborescens	Bladder senna
Hamamelis vernalis	Vernal witchhazel
Lonicera korolkowi	Blue-leaf honeysuckle
Ribes sangiuneum	Winter currant
Rosa multiflora	Japanese rose

Medium Deciduous Shrubs

Botanical Name	Common Name
Aronia arbutfilia	Red chokeberry
Berberis thunbergi	Japanese barberry
Cornus amomum	Silky dogwood
Cotoneaster divaricata	Spreading cotoneaster
Cytisus scoparius	Scotch broom
Deutzia scabra	Snowflake deutzia
Forsythia intermedia	Golden bell
Forsythia suspensa	Weeping forsythia
Forsythia viridissima	Green-stem forsythia
Hydrangea quercifolia	Oak-leaved hydrangea
Lanicera fragrantissima	Winter honeysuckle
Rhodotypos scandens	Jetbead
Rosa hugonis	Father Hugo rose

Botanical Name	Common Name
Syringa persica	Persian lilac
Syringa sweginzowi	Chengtu lilac

Low-Medium Deciduous Scrubs

Botanical Name	Common Name
Abeliophyllum distichum	Korean abelia leaf
Caryopteris incana	Bluebeard
Philadelphus lemoinei	Lemoine mock orange

Low Deciduous Shrubs

Botanical Name	Common Name
Berberis buxifolia	Dwarf Magellan barberry
Robinia hispida	Rose acacia
Weigela florida	Rose weigela

Region 6

Tall Deciduous Trees

Botanical Name	Common Name
Acer macrophyllum	Big-leaf maple
Celtis australis	European hackberry
Fraxinus oregona	Oregon ash
Juglans regia	English walnut
Platanus orientalis	Oriental plane tree
Quercus cerris	Turkey oak
Quercus garryana	Oregon white oak

Medium Deciduous Trees

Botanical Name	Common Name
Brupoussonetia papyrifera	Common paper mulberry
Cercis chinensis	Chinese redbud
Davidia involucrata	Dove tree
Quercus acutissima	Sawtooth oak
Quercus nigra	Water oak

Low Deciduous Trees

Botanical Name	Common Name
Cercis siliquastrum	Judas tree
Ficus carica	Common fig
Magnolia sieboldi	Oyama magnolia
Prunus amygdalus	Almond
Prunus conradinae semiplena	Double conradina cherry
Prunus mume	Japanese apricot
Quercus marilandica	Blackjack oak
Salix babylonica	Babylon weeping willow

Tall Evergreen Trees

Botanical Name	Common Name
Abies pinsapo	Spanish fir
Cedrus atlantica	Atlas cedar
Sequoia gigantea	Giant sequoia
Tsuga heterophylla	Western hemlock

Medium Evergreen Trees

Botanical Name	Common Name
Ilex aquifolium	English holly
Taxus baccata	English yew
Thuja orientalis	Oriental arbovitae

Low Evergreen Trees

Botanical Name	Common Name
Ilex pernyi	Perny holly
Laurus nobilis	Sweet bay
Myrica cerifera	Southern wax myrtle

Tall Deciduous Shrubs

Botanical Name	Common Name
Cercis chinensis	Chinese Judas tree
Clerodendron trichotomum	Harlequin glory-bower
Magnolia wilsoni	Wilson magnolia
Viburnum nudum	Smooth withe rod

Medium-Tall Deciduous Shrubs

Botanical Name	Common Name
Aesculus splendens	Flame buckeye

Medium Deciduous Schrubs

Botanical Name	Common Name
Berberis beaniana	Bean's barberry
Cytisus dallimorei	Dallimore broom
Cytisus multiflorus	White Spanish broom
Euonymus americana	Strawberry bush
Garrya wrighti	Wright silk tassel
Ligustrum quihoui	Quihou privet
Magnolia liliflora nigra	Purple lily magnolia
Pyracantha coccinea	Scarlet fire thorn
Sorbaria aitchisoni	Kashmir false spirea
Vitex agnus-castus	Chaste tree

Low Deciduous Shrubs

Botanical Name	Common Name
Berberis concinna	Dainty barberry
Spiraea cantoniensis	Reeve's spirea

Tall Evergreen Shrubs

Botanical Name	Common Name
Ilex crenata	Japanese holly
Laurus nobilis	Laurel
Myrica cerifera	Wax myrtle
Osmanthus ilicifolius	Holly osmanthus
Prunus laurocerasus	Cherry laurel
Taxus baccata	English yew

Medium-Tall Evergreen Shrubs

Botanical Name	Common Name
Mahonia beali	Leatherleaf mahonia

Medium Evergreen Shrubs

Botanical Name	Common Name
Erica darleyensis	Darley heath

Low Bamboos

Botanical Name	Common Name
Pseudosasa Japonica	Arrow bamboo
Sasa Pumila	Ground bamboo
Sasa veitchi	Veitchi bamboo
Shibataea kumasaca	Kumasaca bamboo

Region 7

Tall Deciduous Trees

Botanical Name	Common Name
Keteleeria fortune	Fortune keteleeria
Platanus racemosa	California plane tree
Populus fremonti	Fremont cottonwood
Quercus kelloggi	California black oak

Medium Deciduous Trees

Botanical Name	Common Name
Clethra delavayi	Dalavay clethra
Cornus nuttalli	Pacific dogwood
Diospyros kaki	Kaki persimmon
Fraxinus velutina	Velvet ash
Magnolia veitchi	Veitchi magnolia
Melia azedarach	Chinaberry
Podocarpus macrophyllus	Yew podocarpus
Quercus chrysolepis	Canyon live oak
Quercus laurifolia	Laurel oak
Quercus suber	Cork oak
Quercus virginiana	Live oak

Low Deciduous Trees

Botanical Name	Common Name
Albizzia julibrissin	Silk tree
Cercis racemosa	Raceme redbud
Prunus campanulata	Taiwan cherry
Prunus lusitanica	Portugal laurel

Tall Evergreen Trees

Botanical Name	Common Name
Araucaria araucana	Monkey-puzzle tree
Arbutus menziesi	Pacific madrone
Castanopsis chrysophylla	Giant evergreen chinquapin
Cedrus deodora	Deodar cedar
Cupressus macrocarpa	Monterey cypress
Cupressus sempervirens	Italian cypress
Magnolia grandiflora	Southern magnolia
Pinus coulteri	Big-cone pine
Pinus pinaster	Cluster pine
Sequoia sempervirens	Redwood

Medium Evergreen Trees

Botanical Name	Common Name
Camellia japonica	Common camellia
Cunninghamia lanceolata	Common China fir
Ilex latfolia	Luster-leaf holly
Juniperus drupacea	Syrian juniper
Juniperus excelsa	Greek juniper
Juniperus pachyphoea	Alligator juniper
Pinus radiata	Monterey pine
Umbellularia california	California laurel

Low Evergreen Trees

Botanical Name	Common Name
Cupressus arizonica bonita	Smooth Arizona cypress
Ilex cassine	Dahoon
Myrica californica	California bayberry
Photinia serrulata	Chinese photinia

Tall Deciduous Shrubs

Botanical Name	Common Name
Chilopsis linearis	Desert willow
Cotoneaster frigida	Himalayan cotoneaster
Cudrania tricuspidata	Silkworm tree
Lagarstroemia indica	Crape myrtle
Paliurus spina-christi	Christ's-thorn
Punica granatum	Pomegranate
Stewartia malacodendron	Virginia stewartia

Medium-Tall Deciduous Shrubs

Botanical Name	Common Name
Corylopsis griffithi	Griffith winter hazel
Ligustrum sinesse	Chinese privet
Spartium junceum	Spanish broom

Medium Deciduous Shrubs

Botanical Name	Common Name
Berberis potanini	Long-spine barberry
Chimonanthus praecox	Winter sweet
Hydrangea sargentiana	Sargent hydrangea
Leycesteria formosa	Formosa honeysuckle
Spiraea canescens	Hoary spirea
Styrax wilsoni	Wilson snowbell

Low-Medium Deciduous Shrubs

Botanical Name	Common Name
Abelia schumanni	Schumann abelia

Low Deciduous Shrubs

Botanical Name	Common Name
Genista cinerea	Ashy woadwaxen

Tall Evergreen Shrubs

Botanical Name	Common Name
Aucuba japonica	Japanese aucuba
Camellia japonica	Common camellia
Camellia sasanqua	Sasanqua camellia
Ilex cassine	Dahoon
Ligustrum Lucidum	Glossy privet
Michelia fuscata	Banana shrub
Myrica californica	California bayberry
Nerium oleander	Oleander
Nothopanax davidi	David false panax
Photinia serrulata	Chinese photinia
Rosa odorata	Tea rose

Medium-Tall Evergreen Shrubs

Botanical Name	Common Name
Berberis darwini	Darwin barberry
Cotoneaster henryana	Henry cotoneaster
Elaeagnus pungens	Thorny elaeagnus
Ilex yunnanensis	Yunnan holly
Ligustrum henryi	Henry privet
Ligustrum japonicum	Japanese privet
Osmanthus forunei	Fortune's osmanthus
Pieris formosa	Himalayan andromeda
Prunus lusitanica	Portugal laurel
Pyracantha crenulata rogersiana	Rogers fire thorn
Vaccinium ovatum	Box blueberry
Viburnum tinus	Laurustine

Medium Evergreen Shrubs

Botanical Name	Common Name
Arctostaphylos stanfordiana	Standord manzanita
Ceanothus delilianus	Delisle ceanothus
Choisya ternata	Mexican orange
Cistus albidus	White-leaf rockrose
Cistus cyprius	Spotted rockrose
Cistus laurifolius	Laurel rockrose
Cistus purpureus	Purple rockrose
Cotoneaster pannosa	Silverleaf cotoneaster
Hebe traversi	Travers hebe
Hypericum hookerianum	Hooker's Saint-John's-wort
Ilex cornuta	Chinese holly
Illicium floridanum	Florida anise tree
Lonicera nitida	Box honeysuckle
Nandina domestica	Nandina
Pieris taiwanensis	Formosa andromeda
Raphiolepsis umbellata	Yeddo hawthorn
Sarcococca ruscifolia	Fragrant sarcococca
Viburnum henryi	Henry viburnum
Viburnum japonicum	Japanese viburnum

Low-Medium Evergreen Shrubs

Botanical Name	Common Name
Daphne odora	Winter daphne
Erica mediterranea	Mediterranean heath
Hebe buxifolia	Box-leaf hebe
Skimmia japonica	Japanese skimmia
Sophora secundiflora	Mescal bean

Low Evergreen Shrubs

Botanical Name	Common Name
Danae racemosa	Alexandrian laurel
Gaultheria veitchiana	Veitchi wintergreen
Ruscus aculeatus	Butcher's broom
Viburnum davidi	David viburnum

Region 8

Medium Deciduous Trees

Botanical Name	Common Name
Juglans hindsi	Hinds black walnut
Prosopis glandulosa	Honey mesquite

Tall Evergreen Trees

Botanical Name	Common Name
Pinus canariensis	Canary pine

Medium Evergreen Trees

Botanical Name	Common Name
Gordonia lasianthus	Loblolly bay gordonia
Pinus torreyana	Torrey pine

Low Evergreen Trees

Botanical Name	Common Name
Arbustus unedo	Strawberry tree

Medium-Tall Deciduous Shrubs

Botanical Name	Common Name
Acacia farnesiana	Opopanax
Cassia corymbosa	Flowery senna

Medium Deciduous Shrubs

Botanical Name	Common Name
Severinia buxifolia	Chinese box orange

Low-Medium Deciduous Shrubs

Botanical Name	Common Name
Ceratostigma willmottianum	Willmott blue leadwort

Tall Evergreen Shrubs

Botanical Name	Common Name
Arbutus unedo	Strawberry tree
Euonymus japonica	Evergreen euonymus

Medium-Tall Evergreen Shrubs

Botanical Name	Common Name
Pittosporum tobira	Japanese pittosporum

Medium Evergreen Shrubs

Botanical Name	Common Name
Abelia floribunda	Mexican abelia
Carpenteria californica	Evergreen mock orange
Garrya elliptica	Silk tassel
Myrtus communis	Myrtle
Olearia haasti	New Zealand daisybush

Low-Medium Evergreen Shrubs

Botanical Name	Common Name
Gardenia jasminoides	Gardenia

Low Evergreen Shrubs

Botanical Name	Common Name
Philesia magellanica	Magellan box lily

Tall Bamboos

Botanical Name	Common Name
Arundinaria simoni	Simon bamboo
Phyllostachys aurea	Golden bamboo
Phyllostachys bambusoides	Timber bamboo

Region 9

Medium Deciduous Trees

Botanical Name	Common Name
Acacia decurrens dealbata	Silver wattle
Acer floridanum	Florida maple
Firmiana simplex	Chinese parasol tree
Pistacia chinensis	Chinese pistachio
Sapium sebiferum	Chinese tallow tree
Schinus molle	California pepper tree

Low Deciduous Trees

Botanical Name	Common Name
Parkinsonia aculeata	Jerusalem thorn

Tall Evergreen Trees

Botanical Name	Common Name
Pinus pinea	Italian stone pine
Quercus agrifolia	California live oak

Medium Evergreen Trees

Botanical Name	Common Name
Cinnamomum camphora	Camphor tree
Eugenia paniculata	Brush cherry eugenia
Juniperus lucayana	West Indies juniper
Lagunaria patersoni	Paterson sugarplum tree
Pinus halepenis	Aleppo pine
Quercus ilex	Holly oak
Schinus terebinthifolius	Brazil pepper tree

Low Evergreen Trees

Botanical Name	Common Name
Leptospermum leavigatum	Australian tea tree
Maytenus boaria	Chile mayten tree
Olea europaea	Common olive

Tall Deciduous Shrubs

Botanical Name	Common Name
Hibiscus rosa-sinensis	Chinese hibiscus

Tall Evergreen Shrubs

Botanical Name	Common Name
Callistemon lanceolatus	Lemon bottlebrush
Carissa grandiflora	Natal plum

Medium-Tall Evergreen Shrubs

Botanical Name	Common Name
Viburnum odloratissimum	Sweet viburnum

Medium Evergreen Shrubs

Botanical Name	Common Name
Heteromeles arbutifolia	Christmas berry
Viburnum suspensum	Sandankwa viburnum

Low Palms

Botanical Name	Common Name
Sabal minor	Dwarf palmetto
Serenoa repens	Saw palmetto

Region 10

Tall Deciduous Trees

Botanical Name	Common Name
Ceiba pentandra	Silk cotton tree

Medium Deciduous Trees

Botanical Name	Common Name
Acacia decurrens mollis	Black wattle
Brachychiton acerifolium	Flame bottle tree
Delonix regia	Flame tree
Jacaranda acutifolia	Sharp-leaf jacaranda

Low Deciduous Trees

Botanical Name	Common Name
Acacia baileyana	Cootamundra wattle
Acacia longifolia floribunda	Gossamer Sydney acacia
Acacia pendula	Weeping boree acacia
Bauhinia variegata	Buddist bauhinia
Cassia fistula	Golden shower Senna

Tall Evergreen Trees

Botanical Name	Common Name
Araucaria excelsa	Norfolk Island pine
Grevillea robusta	Silk oak grevillea
Pittosporum rhombifolium	Diamond-leaf pittosporum
Podocarpus elongatus	Fern polocarpus

Medium Evergreen Trees

Botanical Name	Common Name
Castanospermum australe	Moreton Bay chestnut
Casuarina equisetifolia	Horsetail beefwood
Ceratonia siliqua	Carob
Ficus macrophylla	Moreton Bay fig

Hymenosporum flavum	Sweet shade
Melaleuca leucadendron	Cajeput tree
Pittosporum eugenioides	Tarata pittosporum
Quillaja saponaria	Soapbark tree
Sparthodea campanulata	Bell flambeau tree

Low Evergreen Trees

Botanical Name	Common Name
Casuarina stricta	Coast beefwood
Leucadendron argenteum	Silver leucadendron
Macadamia ternifolia	Queensland nut

Tall Palms

Botanical Name	Common Name
Cocos nucifera	Coconut
Sabal palmetto	Palmetto
Washingtonia robusta	Mexican Washington palm

Medium Palms

Botanical Name	Common Name
Livistona australis	Australian pan palm
Roystonea regia	Royal palm

Low Palms

Botanical Name	Common Name
Erythea armata	Blue erythea
Phoenix reclinata	Senegal date palm
Ravenala madagascariensis	Madagascar traveler's-tree

DECIDUOUS TREES

NORWAY MAPLE
Height: 60-90 feet
Climate Zone IV

SUGAR MAPLE
Height: 70-125 feet
Climate Zone III

WHITE OAK
Height: 80-125 feet
Climate Zone IV

NORTHERN RED OAK
Height: 60-80 feet
Climate Zone IV

LINDEN
Height: 70-90 feet
Climate Zone IV

ENGLISH ELM
Height: 75-100 feet
Climate Zone V

BLACK LOCUST
Height: 40-80 feet
Climate Zone V

HORSECHESTNUT
Height: 60-80 feet
Climate Zone V

EUROPEAN BEECH
Height: 50-100 feet
Climate Zone IV

CAROLINA POPLAR
Height: 75-100 feet
Climate Zones IV, V

NORTHERN CATALPA
Height: 80-100 feet
Climate Zone V

MAIDENHAIR TREE
(Ginkgo Tree)
Height: 60-80 feet
Climate Zone IV

AILANTHUS
Height: 50-75 feet
Climate Zone V

**EUROPEAN
WHITE BIRCH**
Height: 50-75 feet
Climate Zones II, III

LOMBARDY POPLAR
Height: 75-100 feet
Climate Zone II

SWEETGUM
Height: 80-120 feet
Climate Zone IV

Figure 3.79 Profiles of deciduous trees.

RED MAPLE
Height: 50-125 feet
Climate Zone III

ORIENTAL PLANE TREE
(Sycamore, Buttonwood)
Height: 70-80 feet
Climate Zone IV

WEEPING WILLOW
Height: 30-40 feet
Climate Zone V

MAGNOLIA
(Cucumber Tree)
Height: 70-100 feet
Climate Zone V

AMERICAN ELM
Height: 80-100 feet
Climate Zone II

PIN OAK
Height: 60-125 feet
Climate Zone V

WHITE ASH
Height: 70-80 feet
Climate Zone III

AMERICAN BEECH
Height: 50-100 feet
Climate Zone IV

HONEY LOCUST
Height: 40-60 feet
Climate Zone IV

TULIP-TREE
Height: 100-120 feet
Climate Zone IV

BLACK WALNUT
Height: 75-150 feet
Climate Zone IV

APPLE
Height: 20-40 feet
Climate Zone III

Figure 3.79   (*Continued*)

EVERGREEN TREES

IRISH YEW
Height: 35 feet
Climate Zone VI

AMERICAN HOLLY
Height: 40-60 feet
Climate Zone V

DOUGLAS FIR
(young and mature)
Height: 100-300 feet
Climate Zone VI

BALD CYPRESS
(young and mature)
Height: 100-150 feet
Climate Zone V

MAGNOLIA, SOUTHERN
(Bull Bay)
Height: 70-80 feet
Climate Zone VII

WHITE FIR
Height: 100-150 feet
Climate Zone IV

BOX TREE
Height: 20-30 feet
Climate Zones VI, VII

NORWAY SPRUCE
Height: 40-100 feet
Climate Zone II

EUROPEAN LARCH
Height: 50-60 feet
Climate Zone II

ARBORVITAE
Height: 25-50 feet
Climate Zone II

COLORADO SPRUCE
Height: 70-100 feet
Climate Zone II

MONTEREY PINE
Height: 50-100 feet
Climate Zone X

JUNIPER
(Red Cedar)
Height: 25-50 feet
Climate Zone II

WHITE PINE
Height: 50-100 feet
Climate Zone II

AUSTRIAN PINE
Height: 60-100 feet
Climate Zone IV

CANADA HEMLOCK
Height: 60-100 feet
Climate Zone IV

RED PINE
Height: 60-80 feet
Climate Zone II

LIVE OAK
Height: 50-60 feet
Climate Zone VII

SAWARA CYPRESS
Height: 20-50 feet
Climate Zone III

Figure 3.80   Profiles of evergreen trees.

EVERGREEN SHRUBS

RHODODENDRON
(various)
Height: 6-36 feet
Subacid soil
Climate Zones IV, V

JAPANESE HOLLY
Height: 10-20 feet
Peaty (acid) soil
Climate Zone VI

OLEANDER
Height: 7-15 feet

DWARF BOX
Height: 10-12 feet
Soil rich in humus
Climate Zone VI

MUGHO PINE
Height: 6-8 feet
Climate Zone III

JAPANESE YEW
Height: 12-15 feet
Climate Zone IV

PFITZER'S JUNIPER
Height: 6-8 feet
Dry, sandy soil
Climate Zone IV

MOUNTAIN LAUREL
Height: 4-10 feet
Sandy, acid soil
Climate Zones III, IV

PITTOSPORUM TOBIRA
Height: 6-15 feet
Climate Zone IX

Figure 3.81 Profiles of evergreen shrubs.

DECIDUOUS SHRUBS

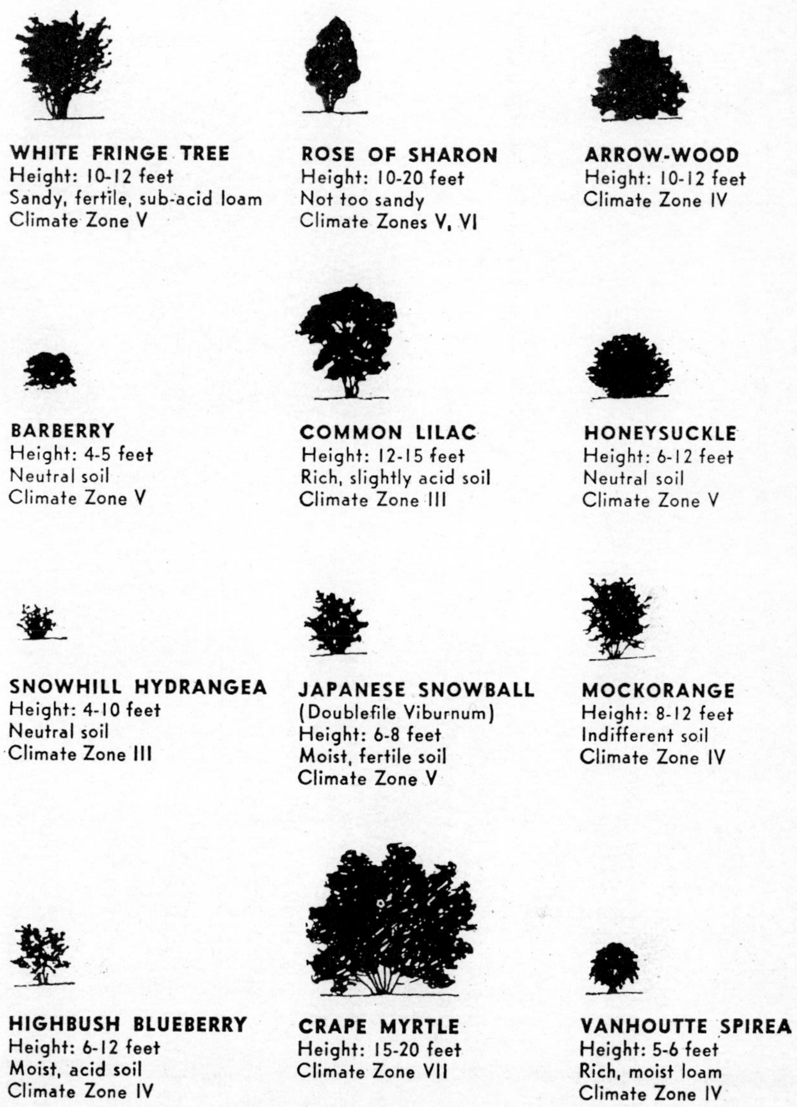

WHITE FRINGE TREE
Height: 10-12 feet
Sandy, fertile, sub-acid loam
Climate Zone V

ROSE OF SHARON
Height: 10-20 feet
Not too sandy
Climate Zones V, VI

ARROW-WOOD
Height: 10-12 feet
Climate Zone IV

BARBERRY
Height: 4-5 feet
Neutral soil
Climate Zone V

COMMON LILAC
Height: 12-15 feet
Rich, slightly acid soil
Climate Zone III

HONEYSUCKLE
Height: 6-12 feet
Neutral soil
Climate Zone V

SNOWHILL HYDRANGEA
Height: 4-10 feet
Neutral soil
Climate Zone III

JAPANESE SNOWBALL
(Doublefile Viburnum)
Height: 6-8 feet
Moist, fertile soil
Climate Zone V

MOCKORANGE
Height: 8-12 feet
Indifferent soil
Climate Zone IV

HIGHBUSH BLUEBERRY
Height: 6-12 feet
Moist, acid soil
Climate Zone IV

CRAPE MYRTLE
Height: 15-20 feet
Climate Zone VII

VANHOUTTE SPIREA
Height: 5-6 feet
Rich, moist loam
Climate Zone IV

Figure 3.82 Profiles of deciduous shrubs.

MATERIALS KEY

GROUP I—VINES 12-15″

Trailing and adaptable for slope planting and use as ground cover, also for planters. (medium to rapid growth).

TYPE A (Sun)

NE Rosa wichuriana (Memorial Rose)
SE Rosa banksiae (Lady Bank's Rose)
NCentral Hedera helix (English Ivy)
SCentral Rosa banksiae (Lady Bank's Rose)
SW Carpobrotus edulis (Ice plant)
Western Hedera helix (English Ivy)

GROUP I—VINES 12-15″

TYPE B (Shade Tolerant)

NE Akebia quinata, (Fineleaf Akebia)
SE Bignonia capreolata (Crossvine)
NC Vinca minor (Little leaf Periwinkle)
SC Trachelospermum jasminoides (Confederate jasmine)
SW Hedera canariensis (Algerian ivy)
 W Vinca major or V. minor (Big-leaf Periwinkle and Little-leaf Periwinkle)

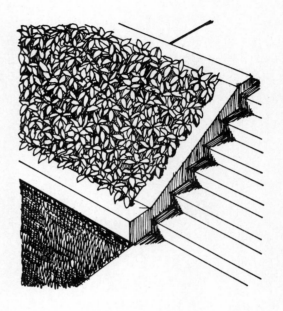

Figure 3.83

GROUP II—GROUNDCOVERS 12-15″

Low materials, medium to fast growth. Could also be vine-like in character.

TYPE A—Mounded form, adaptable to mass planting. (Sun)

NE Hypericum moserianum (Gold Flower)
SE Yucca filamentosa (Adams Needle Yucca)
NC Cotoneaster adpressa (Creeping cotoneaster)
SC Yucca filamentosa — (Adams Needle Yucca)
SW Juniperus chinensis (San Jose Juniper)
 W Cotoneaster microphylla (Rock Spray cotoneaster)

Figure 3.84

GROUP II—GROUNDCOVERS 12-15″

TYPE B—Mounded form, shade tolerant.

NE Juniperus horizontalis v. plumosa
SE Rhododrendron (many miniature azaleas)
NC Spiraea bumalda (Anthony waterer spiraea)
SC Rhododendron (many miniature azaleas)
SW Carissa grandiflora (Natal plum, "Boxwood Beauty")

Figure 3.85

GROUP II—GROUNDCOVERS 12-15″

TYPE C—Spreading or adaptable to mass planting.

NE Juniperus horizontalis (Creeping juniper)
NE Aegopodium podograria (Bishop's goutweed)
NE Pachysandra terminalis (Japanese spurge)

SE Juniperus conferta (Shore juniper)
SE Sedum acre (stone crop)
SE Liriope muscari (Liriope)

NC Pachysandra terminalis (Japanese spurge)
NC Lysimachia nummularia (Money wart or creeping Charlie)

SC Juniperus Chinensis 'Prostrata' (Creeping Chinese juniper)
SC Vinca major and v. minor (periwinkle)
SC Ajuga reptans (carpet bugle)

SW Carissa grandiflora (Natal plum "green carpet")
SW Rosmarinus officinalis 'Prostrata' — Prostrate Rosemary)

W Juniperus conferta (Shore juniper)
W Herdera helix (English ivy)

Figure 3.86

Guide to Selecting Ground Covers

Common name	Height (inches)	Hardiness (zone)	Type	Soil and light	Comments
Barrenwort *Epimedium alpinum; E. grandiflorum; E. pinnatum*	12	4–8	Woody herb	Tolerates almost any soil	Dense foliage; lasts into winter; white, yellow, lavender flowers
Bearberry *Arctostaphylos uvursi*	6–10	2–9	Evergreen shrub	Excellent in stony, sandy, acid soils	Low; hard to transplant; bright red fruit
Bergenia, heartleaf *Bergenia cordifolia*	12	5–10	Creeping, clumpy perennial; thick rootstocks	Sun or partial shade	Pink flowers; thick, heavy foliage
Broom *Genista pilosa; G. sagittalis*	6–12	5–9	Deciduous shrub	Well-drained soil; sun	Flowers are pea-shaped
Bugleweed *Ajuga reptans*	4–8	5–9	Perennial herb	Tolerates most soils	Densely packed plants; blue-purple flowers; rapid grower
Capeweed *Phyla nodiflora*	2–4	9–10	Creeping perennial herb	Sand and waste areas	Low-growing; spreads rapidly; cut like grass; light pink flowers
Coralberry *Symphoricarpos orbiculatus*	To 36	3–9	Deciduous shrub	Thrives in poor soils	Rapid growth by underground stems; requires yearly pruning
Cotoneaster *Cotoneaster adpressa* *C. apiculata* *C. dammeri* *C. horizontalis* *C. microphylla*	6–30	5–10 5–9 6–10 7–10	Semievergreen herb	Full sun, reseed	Stems will layer subject to fire blight
Cowberry *Vaccinium vitis-idaea*	To 12	5–9	Small evergreen shrub	Acid soil	Small pink flowers, dark-red berries
Creeping lilyturf *Liriope spicata*	To 12	5–10	Matted herb	Extreme heat, dry soil, stands salt spray	Dense mat, dark green leaves, purple flowers
Creeping lippia *Phyla nodiflora* var. *canescens*	2–4	5–10	Creeping perennial	Any soil, sun	White, lilac flowers
Creeping thyme *Thymus serpyllum*	To 3	5–10	Subshrubby with creeping stems	Tolerates dry soils, sun	Substitute for grass, extremely variable
Crownvetch *Coronilla varia*	12–24	3–7	Herb	Dry, steep banks, sun	Small pink flowers
Daylily *Hemerocallis*	18–60	3–10	Root, fleshy and tuberous parts	Sun, dry to boggy soils	Few problems; summer flowers
Dichondra *Dichondra repens*	1–2	9 and 10	Evergreen perennial	Sunny or shady locations	Poor drought resistance; rarely needs clipping; spreads rapidly
Dwarf bamboo *Sasa pumila, S. veitchii,* and *Shibataea kumasaca*	6–10		Low shrub	Sun; sandy soil	Foliage brown in winter; fire hazard; grass substitute

Plant	Height	Zone	Type	Requirements	Remarks
English ivy / *Hedera helix*	6–8	5–9	Evergreen vine	Sun or shade	Clip leaves to control leaf spots
Forsythia species / *Forsythia spp.*	Trim to 18	5–9	Deciduous shrub	Sun, well-drained soil	Stems root easily; yellow flowers in spring
Galax / *Galax aphylla*	6	5–7	Evergreen, stemless perennial herb	Moist, rich, acid soil, shade	White flowers in spring; leaves turn bronze in fall
Germander / *Teucrium chamaedrys*	To 10	6–10	Small woody perennial	Sun or partial shade	Winter damage without protection
Ground-ivy / *Glechoma hederacea*	3	3–9	Trailing perennial	Sun or shade; any soil	Becomes a pest in lawn if not trimmed; forms a low mat
Heath / *Erica carnea*	6–12	5–8	Evergreen shrub	Poor, acid soils, sun	Pink, purple, red, white varieties
Heather / *Calluna vulgaris*	6–24	4–7	Evergreen shrub	Acid soil, well-drained, low fertility, sun	Shear plants each spring
Holly, Japanese / *Ilex crenata*	Keep to 24	6–10	Evergreen shrub	Sun or semishade	Slow-growing; small bank plantings
Hollygrape, dwarf / *Mahonia repens*	To 10	6–9	Evergreen shrub	Sun or shade, any type soil	Yellow flowers
Honeysuckle, Japanese / *Lonicera japonica*	To 10	5–9	Twisting, trailing vine	Sun or partial shade	Prune yearly to keep in bounds; a semievergreen with white turning to yellow flowers
Iceplant / *Cephalophyllum, Carpobrotus, Delesperma, Drosanthemum, Malephora, Lampranthus*	4–6	10	Low succulent	Sun; well-drained soil	Temporary ground cover in cold climates; brilliant colored flowers open in full sunlight
Japanese spurge / *Pachysandra terminalis*	To 6	5–8	Evergreen herb	Semishade under tree	Spreads by underground stems
Juniper / *Juniperus horizontalis; J. sabina; J. procumbens; J. chinensis; J. conferta*	12–18; Trim to 36	3–10	Evergreen conifer	Sun; dry areas	Yearly pruning of upright forms; wide range of foliage colors; some turn purple in winter
Lantana / *Lantana sellowiana; L. montevidensis*	6–10	8–10	Trailing shrub	Sun, high salt tolerance	Wide range of flower colors
Lily-of-the-valley / *Convallaria majalis*	6–10	4–9	Rootstock	Rich, moist, high organic soil; partial shade	Fragrant white bell-shaped flowers
Lilyturf dwarf (Mondograss) / *Ophiopogon japonicus*	To 10	7–10	Matted herb	Any soil, sun or shade	Spikes of pale lilac flowers

Guide to Selecting Ground Covers (Continued)

Common name	Height (inches)	Hardiness (zone)	Type	Soil and light	Comments
Moss, pink *Phlox subulata*	6	4–10	Evergreen perennial	Porous soil; sun	Flowers are shade of pink and white
Moss sandwort *Arenaria verna*	3	2–9	Perennial herb	Fertile soil, moist; partial shade	Requires some winter protection
Periwinkle *Vinca minor* (small leaves) *V. major* (large leaves)	6–8	5–10	Trailing herb	Avoid high-nitrogen fertilizer, poorly drained soils	Purple, blue, and white flowers
Plantain lily *Hosta* spp.	12–16	4–10	Tufted plant with broad leaves	Moist, well-drained soils; shade	Needs frequent division
Polygonum, dwarf *Polygonum cuspidatum* var. *compactum*	12–24	4–10	Stout perennial	Rocky or gravelly soil; sun	Foliage turns red in fall
Rose, memorial *Rosa withuraiana*	6–12	5–9	Semievergreen low-growing shrub	Banks and sand dunes	2-inch white flowers
Saint-John's-wort *Hypericum calycinum*	9–12	6–10	Semievergreen shrub	Semishade; sandy soil	Yellow flowers in summer; red foliage in autumn
Sand strawberry *Fragaria chiloensis*	10–12	6–10	Perennial herb	Suitable for most soils	Spreads rapidly
Sarcococca *Sarcococca hookeriana*	To 72	6–10	Evergreen shrub	Shade	Shear for height control; small white flowers, large leaves
South African daisy *Gazania rigens*	6–9	9–10	Evergreen perennial	Avoid high-nitrogen fertilizers, poorly drained soils	Light green foliage; orange flowers
Stonecrop, goldmoss *Sedum acre*	4	4–10	Evergreen perennial	Dry areas	Forms mats of tiny foliage
Strawberry geranium *Saxifraga sarmentosa*	15	7–9	Perennial herb	Partial shade; rock gardens, heavy clay soils	Spreads by runners
Thrift *Armeria maritima*	6	5–9	Perennial herb	Sandy soil; full sun	Small pink flowers in spring
Wandering-Jew *Zebrina pendula*	6–9	10	Tender herb	Shade, acid or alkaline soils	Roots easily
Wintercreeper *Euonymus fortunei*	2–4	5–10	Clinging evergreen vine	Sun; shade; ordinary soil	Rapid, flat growth; subject to scale insects
Wintergreen *Gaultheria procumbens*	4	5–7	Creeping evergreen	Acid soil; moist shady areas	Creeps over area
Yarrow *Achillea millefolium*	2–3	5–9	Fernlike perennial herb	Adapted to poor, dry soil; full sun	Remains green even during drought

GROUP III DWARF SHRUBS 1-4″

TYPE A — rounded or upright oval, suitable for mass plantings.

NE Buxus microphylla v. Koreana (Korean boxwood)
SE Aucuba japonica 'Nana' (Dwarf Japanese Aucuba)
NC Mahonia aquifolium (Mahonia holly or Oregon grape)
SC Aucuba japonica 'Nana' (Dwarf Japanese Aucuba)
SW Thuja occidentalis v. "Tom Thumb" (Dwarf arborvitae)
 W Abelia grandiflora (glossy abelia)

Figure 3.87

GROUP III DWARF SHRUBS 1-4″

TYPE B — Upright — suitable for trimming into hedge.

NE Leucothoe catesbaei (Drooping leucothoe)
SE Ilex vomitoria 'Nana' (Dwarf Yaupon holly)
NC Viburnum opulus 'Nanum' (Dwarf cranberry bush)
SC Ilex cornuta rotunda (Dwarf Chinese Holly)
SW Myrtus communis "Microphylla compacta" (Dwarf compact myrtle)
 W Myrtus communis "Microphylla compacta" (Dwarf compact myrtle)
    or Myrsine Africana (African Boxwood)
    or Buxus microphylla var. Japonica (Japanese Little-leaf Boxwood)

Figure 3.88

GROUP IV — SMALL SHRUBS 4-6′

TYPE A — Upright oval — or upright irregular.

NE Cotoneaster divaricata (spreading cotoneaster)
SE Mahonia bealei (Leatherleaf mahonia)
NC Hydrangea aeborescens grandiflora (Snowhill hydrazea)
SC Nandina domestica (nandina)
SW Nandina domestica (nandina or heavenly bamboo)
 W Nandina domestica (heavenly bamboo) or Viburnum burkwoodi — (Burkwood viburnum)

Figure 3.89

GROUP IV — SMALL SHRUBS 4-6'

TYPE B — Rounded suitable for screening or trimming into hedge.

NE Thuja occidentalis v. woodwardi (Woodward dwarf arborvitae)
SE Berberis julianae (Wintergreen barberry)
NC Cotoneaster acutifolia (Peking cotoneaster)
SC Ilex crenata 'Bullota' (Japanese holly)
SW Ligustrum japonicum (Japanese privet — or Waxleaf privet)
 W Viburnum suspensum (Sandankwa Viburnum)

Figure 3.90

GROUP V — MEDIUM SHRUBS 6-10'

TYPE A — Upright, conical or rounded, suitable for screening.

NE Juniperus virginiana (red cedar)
SE Viburnum tinus (Laurestinus viburnum)
NC Taxus media "Hicksii (Hick's Anglojap Yew)
SC Nerium oleander (Common oleander)
SW Heteromeles arbutifolia (California holly)
 W Prunus carolinianum v. "Bright & Tight" ('Bright & Tight') Carolina cherry

Figure 3.91

GROUP V — MEDIUM SHRUBS 6-10'

TYPE B — Upright, irregular, broadleaf evergreen.
Suitable for screening, or specimen plant.

NE Rhododendron carolinianum (Carolina rhododendron)
SE Prunus caroliniana (Cherry laurel)
NC Spiraea vanhouttei — (Bridal wreath spirea)
SC Rhododendron carolinianum (Carolina rhododendron)
SW Citrus auranticum v. "Bouqet" (Sour orange)
 W Photinia fraseri (Photinia)

Figure 3.92

GROUP VI—LARGE SHRUBS 10′ AND OVER

TYPE A—Upright oval.

NE Syringa vulgaris (Common lilac)
SE Pittosporum tobira (Tobira pittosporum)
NC Syringa vulgaris (Common lilac)
SC Ligustrum japonicum (Japanese privet or Waxleaf privet)
SW Nerium oleander (Oleander)
 W Prunus laurocerasus (English Cherry Laurel)
 W Arbutus unedo (strawberry madrone)

Figure 3.93

GROUP VI—LARGE SHRUBS 10′ AND OVER

TYPE B—Upright conical evergreen.

NE Juniperus virginiana "Hilli" (Hill Dundee juniper)
SE Osmanthus fragrans (Sweet olive; fragrant olive)
NC Taxus cuspidata 'capitata' (Japanese Yew)
SC Osmanthus fragrans (Sweet olive; fragrant olive)
SW Juniperus canaerti (Canaert juniper)
 W Chamaecyparis lawsoniana "Fletcheri" (Fletcher chamaecyparis or cypress)

Figure 3.94

GROUP VII — PLANTS FOR SPECIAL EFFECTS

Types: May vary. Should be interesting in texture or branching. Should be suitable for specimen planting, and be **hardy** in the area to be used.

NE Pyracantha coccinea lalandi (Laland firethorn)
NE Tsuga canadensis pendula (Sargents weeping hemlock)

SE Fatshedera lizei (Fatshedera; Tree ivy)
SE Cycas revoluta (Sago palm)

NC Pinus mugho mughus (Dwarf mugho pine)
NC Rhus glabra (Smooth sumac)

SC Yucca (Common Yucca)
SC Cupressus sempervirens (Italian Cypress)

SW Caesalpinia Pulcherrina (Mexican Bird of Paradise)
SW Vitex agnus-castus (Vitex, monks pepper)

W Fatsia japonica (Japanese Aralia)
W Leptosperum laevigatum (Australian Tea Tree)

Figure 3.95

-274-

GROUP VIII — SMALL TREES 10-40'
(two examples are shown).

All types: Suitable for shade, screening or background. If for screening or background — medium to fast growth.

TYPE A — Single stem, broad umbrella canopy.

NE Malus hupehensis (Tea crabapple) 20'
NE Amelanchier canadensis (Downy shadblow) 30'

SE Cercis canadensis (Red bud) 30'
SE Cornus Florida (Flowering dogwood) 30'

NC Crataegus phaenopyrum (Washington Thorn) 25'
NC Syringa amurensis japonica (Japanese tree lilac) 20'

SC Sapium sebiferum (Chinese tallow)
SC Parkinsonia aculeata (Jerusalem thorn)

SW Albizzia julibrissin (Pink mimosa or Silk tree) 25'
SW Prosopis chilensis (Chilean mesquite) 30'

W Albizzia julibrissin (silk tree) 25'
W Maytenus boaria (Chile mayten tree) 35'

Palms:
Butia capitata (Cocos australis) (Butia palm or Pindo palm)
Chamaerops humilis (European fan palm or Mediterranean palm)
Trachycarpus fortunei (Windmill palm)

Figure 3.96

GROUP VIII—SMALL TREES 10-40'
(two examples are shown).

All types: Suitable for shade, screening or background. If for screening or background — medium to fast growth.

TYPE B—Multi-stem, upright, irregular.

NE Elaeagnus angustifolia (Russian olive) 20'
NE Betula pendula (white birch) (B. verrucosa)

SE Lagerstroemia indica (crape myrtle)
SE Myrica species (Common wax myrtle)

NC Betula pendula (laciniata)–cutleaf weeping white birch 30'
NC Amelanchier canadensis (shadblow service berry) 25'

SC Lagerstroemia indica (crape myrtle)
SC Eriobotrya japonica (Loquat; Japanese plum)
SC Myrica species (Common wax myrtle)

SW Rhus lancea (African sumac)
SW Olea Europea (Common olive var. 'Mission')

W Lagerstroemia indica (crape myrtle)
W Olea Europea (common olive)

Figure 3.97

GROUP VIII — SMALL TREES 10-40'

All types: Suitable for shade, screening or background. If for screening or background — medium to fast growth.

TYPE C — Conical or columnar conifer (two examples are shown)

NE Cryptomeria japonica (Cryptomeria or Japanese redwood) 40'
NE Thuja occidentalis (American arborvitae)

SE Juniperus virginiana (Eastern red cedar)
SE Cupressus sempervirens (Italian cypress)

NC Juniperus virginiana 'Glauca' (Silver red cedar)
NC Pinus nigra — (Austrian pine)

SC Juniperus virginiana (Eastern red cedar)
SC Cupressus sempervirens (Italian cypress)

SW Juniperus chinensis var. columnaris
SW Cupressus Arizonica (Arizona cypress or Cupressus glabra)

W Metasequoia glyptostroboides (Dawn redwood)
W Podocarpus macrophylla (Yew pine)

Figure 3.98

GROUP VIII — SMALL TREES 10-40'

All types: Suitable for shade, screening or background. If for screening or background — medium to fast growth.

TYPE D — Columnar broadleaf.

NE Populus nigra 'Italica' (Lombardy poplar)

NE Acer rubrum 'Columnare' (Columnar red maple)

SE Ilex opaca (American holly)
SE Prunus carolina (Carolina cherry laurel)

NC Pinus nigra 'Italica' (Lombardy poplar)
NC Populus alba "Bolleana" (Bolles poplar)

SC Ilex vomitoria (Yaupon holly)
SC Ilex opaca (American holly)

SW Brachychiton populneum (Bottle tree)
SW Grevillea robusta (Silk oak tree)

W Pittosporum eugenioides — (Tarata pittosporum)
W Popular alba Bolleana (Bolles poplar)
W. Calocedrus decurrens (Libocedrus decurrens) (Incense cedar)

Figure 3.99

GROUP IX—LARGE TREES—OVER 40'

All Types: Suitable for shade or background. Growth should be medium to fast growth.

TYPE A: Single stem, broad leaf, broad umbrella canopy.

NE Acer saccharum (Sugar maple)
NE Quercus rubrum (Q. borealis) (Red oak)

SE Acer rubrum (Red maple or swamp maple)
SE Acer saccharinum (Silver maple)

NC Aesculus hippocastanum (Common horsechestnut)
NC Gleditsia triancanthos inermis "Moraine" (Moraine locust)

SC Acer saccharinum (Acer dasycardum) (Silver maple)
SC Quercus phellos (Willow oak)

SW Pistacia chinensis (Chinese pistache)
SW Melia azedarach (Chinaberry)

 W Quercus rubra (Q. borealis) (Red oak)
 W Magnolia grandiflora (southern magnolia)

Figure 3.100

-279-

GROUP IX—LARGE TREES—OVER 40'

TYPE B: Single stem, broad leaf upright or columnar form

NE Tilia Americana (American linden basswood)
NE Liriodendron tulipifera (Tulip tree)

NC Liriodendron tulipifera (Tulip tree)
NC Quercus palustris (Pin oak)

SC Liriodendron tulipifera (Tulip tree)
SC Magnolia grandiflora (Southern magnolia)

SE Liriodendron tulipifera (Tulip tree)
SE Magnolia grandiflora (Southern magnolia.

SW Fraxinus velutina (Arizona ash)
SW Fraxinus Modesto ash (Modesto ash)

W Persea indica (Indian avocado)
W Liquid ambar styraciflua (American sweet gum)

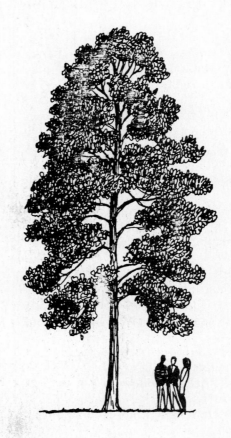

Figure 3.101

GROUP IX—LARGE TREES—OVER 40'

TYPE C: Columnar or conical conifer

NE Abies concolor (White fir)
NE Pseudotsuga menziesi (Douglas fir)

SE Cedrus deodara (Deodara cedar)
SE Taxoiun districhum (Common ball cypress)
SE Pinus sylvestris (Scotch pine)

NC Tsuga canadensis (Canadian hemlock)
NC Pinus resinosa (Red or Norway pine)

SW Taxodiun mucronatum (Montezuma cypress)

W Sequoia sempervirens (Sequoia)
W Pinus radiata (pine)

Figure 3.102

GROUP IX—LARGE TREES—OVER 40'

TYPE D: Columnar or irregular conifer

NE Pinus strobus (Eastern white pine)
NE Pinus resinosa (Red pine)

SE Pinus echinata (Short leaf pine)
SE Pinus polustris (Long leaf pine)

SC Pinus echinata (Short leaf pine)
SC Pinus polustris (Long leaf pine)

SW Cedrus deodara (Deodara cedar)
SW Pinus halepensis (Aleppo pine)

W Cupressus macrocarpa (Monterey cypress)
W Pinus halepensis (Aleppo pine)

Figure 3.103

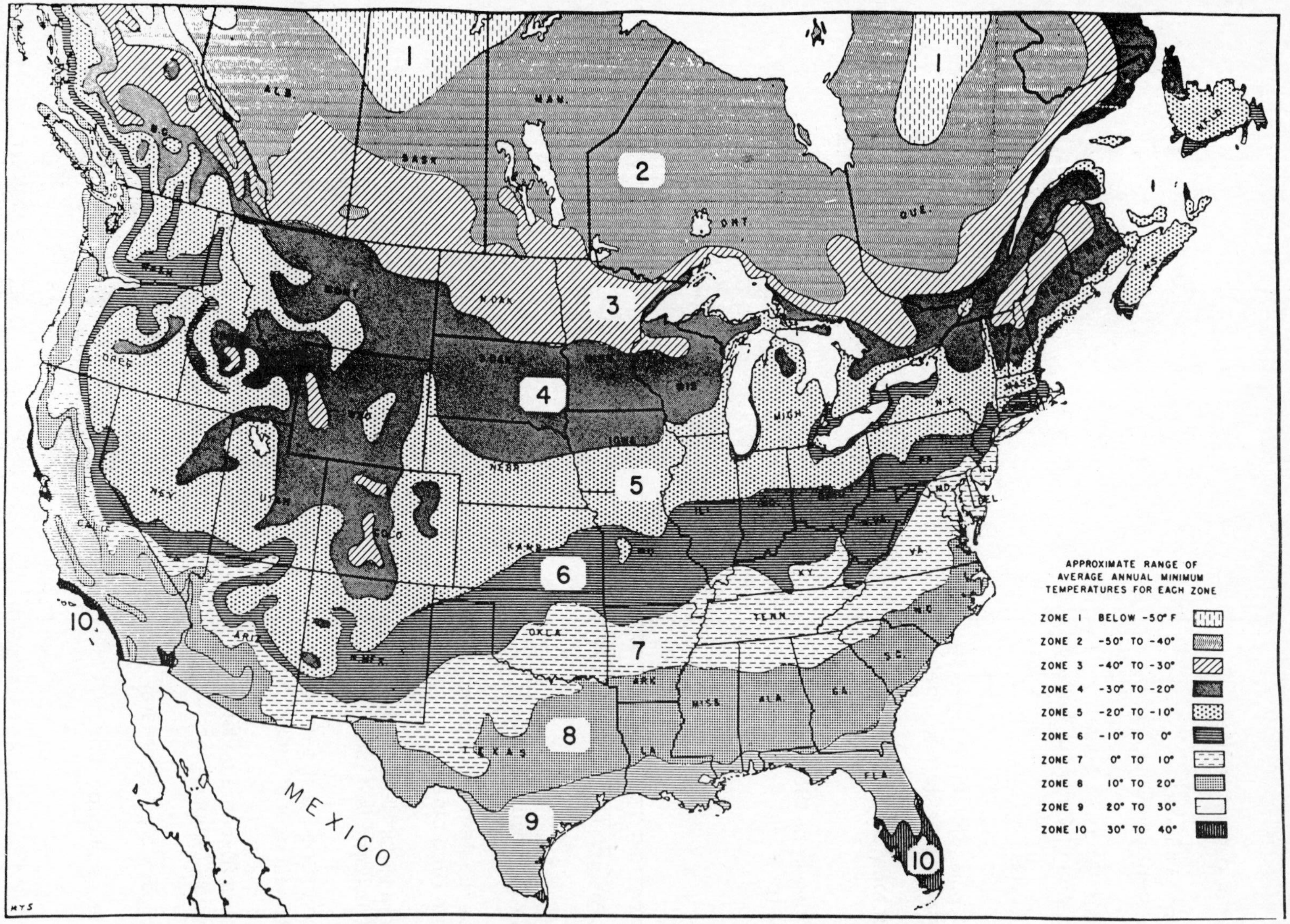

The Zones of Plant Hardiness. An adapted version for small scale reproduction.

This map is a general guide for locating your hardiness zone. However, mountains, bodies of water, alkaline conditions and other geographic and climatic factors also affect plant hardiness. These create zonal exceptions which are difficult to outline on this map. If you have questions, check with your local landscape or garden center operator or other plant authority.

Figure 3.104 Hardiness zone.

The Zones of Plant Hardiness

Zone 1 is located in Canada and is not shown here.

Hardiness Zone 2

Botanical and common names	Average ultimate height in cultivation (feet)	Average ultimate spread in cultivation (feet)	Evergreen semievergreen deciduous	Trim or informal	Spacing (feet) Single row	Spacing (feet) Multiple row	Recommended minimum size to plant (feet)	Plants also useful in following zones	Remarks
Betula fontinalis (fountain birch)	25–30	15	Deciduous	Informal	10–12	14	3–4	3,4,5	Medium-sized tree with brown bark and smallish leaves. Branches are spreading and arching. Native to the plains
Betula papyrifera (paper birch)	40–50	20	Deciduous	Informal	12–15	15–17	3–4	1,3,4,5	This native paper birch with its white bark and fine reddish brown branchlets is an excellent hardy tree. Adaptable to a wide variety of soils and moisture conditions
Lonicera maackii (Amur honeysuckle)	8–10	8	Deciduous	Either	6	8	2–3	3,4	A densely spreading tall shrub with grayish-green leaves and profuse white flowers followed by red berries. Adaptable to a wide range of growing conditions
Malus baccata (Siberian honeysuckle)	25–30	15	Deciduous	Informal	10–12	12–14	3–4	1,3,4,5	Has small fruits and profuse white or pink flowers. Some of the most popular ornamental crabapple varieties come from this plant
Picea glauca (white spruce)	40–50	20	Evergreen	Either	8–12	12–14	2–3	1,3,4,5	This spruce is one of the best for this zone. Used extensively in shelterbelts in colder areas. Needles bright green, trees conical and adaptable to wide variety of soils
Pinus sylvestris (Scotch pine)	40–50	20	Evergreen	Informal	10–12	12–14	2–3	1,3,4,5	Two-needle pine highly adaptable if northern seed sources are used. Makes a good specimen tree and is widely used in shelterbelts on the plains
Populus tremula "Erecta" (upright European aspen)	30–40	5	Deciduous	Informal	6–8	8–10	3–4	3,4,5	This upright aspen makes an excellent screen, especially when space is limited. Trees have good green foliage and clean trunks
Prunus virginiana (Shubert chokecherry)	15–20	10	Deciduous	Either	8–10	10–12	3–4	1,3,4,5	Young leaves are green, turning purplish-red with age. Tall shrub or small tree. Adaptable to wide range of soils
Quercus macrocarpa (burr oak)	30–60	20	Deciduous	Informal	12–20	20–25	3–4	3,4,5	This oak will grow into a magnificent tree on good soil. On poor soil it takes a shrublike multiple-stemmed form
Salix pentandra (laurel willow)	30–40	15	Deciduous	Informal	8–12	12–14	3–4	3,4	Fast-growing dense tree with glossy deep green leaves. Can be planted in both dry and wet locations
Syringa amurensis "Japonica" (Japanese tree lilac)	20–30	15	Deciduous	Either	10–20	20–25	2–3	3,4,5	Tall shrub or small tree. Probably best grown as multiple-stemmed small tree. Large masses of creamy white flowers are present after other lilacs have finished blooming
Syringa hyacinthiflora	6–8	6	Deciduous	Informal	8	10	2–3	3,4,5	Plants hardier than cultivars of S. vulgaris. Good range of colors; leaves have a tendency to turn purple in autumn
Syringa prestoniae (Preston hybrid lilac)	8–10	8	Deciduous	Informal	8–10	10–12	2–3	3,4,5	Colors range from light pink to dark purplish red
Viburnum lentago (nannyberry)	8–10	8	Deciduous	Either	8	10	2–3	3,4	Large shrub or small tree with white flowers and black fruit. Leaves shiny green turning to red in autumn. Adaptable to wide range of soil and moisture conditions
Virburnum trilobum (American cranberrybush)	4–10	4–6	Deciduous	Either	6	8–10	2–3	3,4,5	Handsome shrub with white flowers followed by red fruit (edible). Good fall cover and attracts birds

The Zones of Plant Hardiness (*Continued*)

Botanical and common names	Average ultimate height in cultivation (feet)	Average ultimate spread in cultivation (feet)	Evergreen semievergreen deciduous	Trim or informal	Spacing (feet) Single row	Spacing (feet) Multiple row	Recommended minimum size to plant (feet)	Plants also useful in following zones	Remarks
Hardiness Zone 3									
Acer ginnala (Amur maple)	20	15	Deciduous	Informal	6-10	12	4	2,4,5,6	Bushy, dense branching. Scarlet autumn color. Relatively pest-free. Grows in dry sandy soil
Acer platanoides "Columnare" (columnar Norway maple)	50	15	Deciduous	Informal	10	12	8	4,5,6	Narrow rugged crown of dark green leaves. Grows well in city conditions
Acer saccharinum "Skinner" (Skinner cutleaf maple)	70	35	Deciduous	Informal	15	20	5-6	4,5,6	Fast growth. Wide soil tolerance with pH preference neutral to slightly acid
Caragana arborescens (Siberian pea-tree)	15	12	Deciduous	Informal	5	8	3	2,4	Dense upright habit of growth. Very drought-resistant. Yellow pealike flowers, tolerates alkaline soil
Cornus alba "Siberica" (Siberian dogwood)	10	10	Deciduous	Informal	6	8	3	2,4,5,6	Dense spreading habit. Red twigs are very colorful in winter. Tolerates wet soil
Cornus racemosa (gray dogwood)	8	6	Deciduous	Either	5	7	2	4,5	Medium growth rate. Soil tolerant except dry, heavy soil. PH preference neutral to slightly acid. Does well in full sun or light shade
Cotoneaster lucida (= C. acutifolia) (hedge cotoneaster)	8	5	Deciduous	Either	4	8	2	4,5	Upright with purplish-black fruit in September. Deep glossy green foliage changing to bronze-red in fall
Elaeagnus angustifolia (Russian olive)	20	15	Deciduous	Informal	6	8	3	2,4,5	Upright habit. Attractive silver foliage. Rapid growth. Tolerates adverse conditions and very dry, sandy soil
Euonymus alatus (winged euonymus)	10	10	Deciduous	Either	4	6	2	4,5,6	Very dense habit. Corky twigs make a good screen in winter. Red autumn color, brilliant red in the variety Compacta
Fraxinus pennsylvanica "Lanceolata" (green ash)	50	30	Deciduous	Informal	25	35	8	2,4,5,6	Rapid vigorous growth. Tolerates a wide range of soil types, including alkaline soil
Juniperus virginiana (eastern red cedar)	50	10	Evergreen	Either	7	10	4	2,4,5,6,7,8	Tall, narrow growth. Tolerates dry soil and coastal conditions. Mostly available in selected grafted varieties
Ligustrum amurense (amur privet)	12	6	Deciduous	Either	2	4	3	4,5	Narrow, dense growth. Excellent as a formal hedge. Retains leaves late into the fall
Lonicera tatarica (tatarian honeysuckle)	10	10	Deciduous	Either	3	5	3	4,5,6	Dense rounded habit. Red fruit. Several varieties with red, pink, or white flowers. "Zabelli" is most popular. Rapid growth
Myrica pensylvanica (northern bayberry)	8	8	Semievergreen	Informal	5	7	15	2,4,5,6	Dense spreading growth. Aromatic gray berries. Thrives in poor, sandy soil. Excellent plant for seashore. Tolerates deicing salts along highways
Physocarpus opulifolius (eastern ninebark)	10	7	Deciduous	Informal	6	8	2	2,4,5	Rapid growth. Wide soil tolerance. Likes full sun and a pH slightly acid to neutral
Picea abies (= P. excelsa) (Norway spruce)	60	25	Evergreen	Informal	14	20	2	2,4,5	Fast growth. Wide soil tolerance except in soils with poor percolation. Prefers sun or light shade and a neutral to slightly acid pH
Picea glauca "Densata" (Black Hills spruce)	40	15	Evergreen	Informal	10	15	3	2,4	Thick pyramidal growth. Tolerates poor soil. Considered the hardiest spruce
Pinus strobus (eastern white pine)	70	40	Evergreen	Informal	12	16	3	4,5,6	Broad pyramidal growth. Tolerates a wide range of soil conditions, but not wet soil. Open branching when mature
Populus alba "Pyramidalis" (Bollena poplar)	50	12	Deciduous	Either	10	12	5-6	4,5	Fast growth. Wide soil tolerance. Prefers sun and pH neutral to slightly acid. Short lived. 12-20 years
Quercus borealis (= Q. rubra) (northern red oak)	75	20-30	Deciduous	Informal	10-15	15-25	5-6	4,5,6	One of the fastest-growing oaks with dense deep-green foliage turning red in fall. Withstands city conditions
Rhamnus frangula "Columnaris" (tallhedge buckthorn)	12	4	Deciduous	Either	2	4	3	4,5	Narrow growing, with dense shiny foliage. Plant in full sun. Numerous twigs make a good winter screen. Rapid growth

Plant									Remarks
Rhus canadensis (aromatic sumac)	6–8	6–8	Deciduous	Informal	4–6	8–12	18	4,5	A very dense growing shrub with heavy foliage. Turns bronze-red in fall and retains foliage late. Grows in all soils, except very wet ones
Syringa vulgaris (common purple lilac)	9	7	Deciduous	Informal	6	8	2	4,5	Medium to slow growth rate. Tolerant of well-percolated moist soil. Prefers slightly acid to neutral loam, fragrant purple flowers
Thuja occidentalis (American arborvitae)	30	10	Evergreen	Either	4	6	3	4,5,6	Tall narrow growth. Very adaptable to shearing for formal effects. Tolerates moist and acid soil
Thuja occidentalis "Nigra" (dark green arborvitae)	35	15	Evergreen	Either	10	14	2	2,4,5	Fast growth: wide soil tolerance except dry situations. Prefers sun or light shade with neutral to slightly acid soil
Viburnum dentatum (arrowwood)	10	6	Deciduous	Either	6	8	2	2,4,5,6	Dense shrub, holding branches well to ground line. White flowers, blue fruit, and bronze leaves in fall. Grows well in moist soil

Hardiness Zone 4

Plant									Remarks
Acer platanoides "Columnare" (columnar Norway maple)	40–50	6–12	Deciduous	Informal	8–12	10–20	5–6	3,5	Narrow rugged crown of dark green leaves. Grows well in city conditions
Cornus mas (Cornelian cherry)	20	18	Deciduous	Informal	8	10	4	5,6	Rounded oval habit. Glossy foliage. Yellow flowers in early spring. Red fruits in summer
Cotoneaster lucida (= C. acutifolia) (hedge cotoneaster)	8	5	Deciduous	Either	4	6	2	3,5	Upright with purplish-black fruit in September. Deep glossy green foliage changing to bronze-red in fall
Crataegus crus-galli (cockspur thorn)	20	15	Deciduous	Informal	10	15	4	5,6	Glossy foliage. The long thorns make it an excellent, vandal-proof barrier. Tolerates city conditions. Showy red fruit
Crataegus phaenopyrum (Washington hawthorn)	30	15–20	Deciduous	Informal	15–20	25–30	4–5	5,6	Very dense upright plant with thorny, twiggy growth. Blooms in June followed by small red fruits that persist all winter. Shiny deep green leaves change to orange and scarlet in fall
Juniperus chinensis "Hetzi" (Hetz juniper)	9	12	Evergreen	Either	6	10	2	5,6	Upright but wide-spreading growth. Silver blue foliage. Fast growing. Tolerant of a wide range of soils, except wet soil
Ligustrum amurense (Amur privet)	12	6	Deciduous	Either	2	4	3	5	Narrow, dense growth. Excellent as a formal hedge. Retains leaves late into the fall
Ligustrum ibolium (ibolium privet)	12	8	Deciduous	Either	2	4	3	5,6	Upright oval habit. Glossy leaves are retained late in the fall. Stands hard shearing and city conditions
Lonicera morrowii (Morrow honeysuckle)	8	12	Deciduous	Informal	4	8	3	5,6	Very dense and wide spreading. White flowers and red berries. Good wildlife plant
Lonicera tatarica (Tatarian honeysuckle)	10	10	Deciduous	Either	3	5	3	3,5,6	Dense rounded habit. Red fruit. Several varieties with red, pink, or white flowers. "Zabelli" is most popular. Rapid growth
Malus floribunda (Japanese flowering crab)	20	18	Deciduous	Informal	15	20	5	5,6	Wide spreading, with branches extending down to the ground. Dense, healthy foliage. Masses of fragrant pink flowers in spring
Malus "Spring Snow" (spring snow flowering crab)	15–18	12–15	Deciduous	Informal	10–15	25	3–4	5,6	White flowers followed by deep green dense foliage and absolutely no fruit. Hardy. Sterile form of well-known Dolgo crab
Myrica pensylvanica (northern bayberry)	5–10	8	Semievergreen	Informal	5	10	2	2,3,5,6	Dense spreading growth. Aromatic gray berries. Thrives in poor, sandy soil. Excellent plant for seashore. Tolerates deicing salts along highways
Philadelphus coronarius (sweet mock-orange)	9	9	Deciduous	Informal	6	8	3	5,6,7	Broad, rounded shrub with dense branches. Abundant fragrant white flowers in June. Will stand dry soil
Pinus sylvestris (Scotch pine)	60	40	Evergreen	Informal	8–12	20	2	3,5,6	Very picturesque in the landscape with its bright green foliage and cinnamon-colored bark. Fast-growing and tolerates dry and wet soils
Quercus palustris (pin oak)	80	50	Deciduous	Informal	30	40	8	5,6,7	Low-spreading branches retaining leaves in winter. Red fall color. Needs acid soil but will grow in wet or dry areas
Rhamnus frangula "Columnaris" (tallhedge buckthorn)	12–15	4	Deciduous	Informal	2	4	3	2,3,5,6	Narrow growing, with dense shiny foliage. Plant in full sun. Numerous twigs make a good winter screen. Rapid growth
Salix elegantissima (Thurlow weeping willow)	40	35	Deciduous	Informal	30	40	8	5,6	Rapid-growing tree with weeping branches. Thrives in very wet soil as well as drier sites
Salix pentandra (laurel willow)	20–25	15–18	Deciduous	Informal	10–20	20–30	2	5,6	Fast-growing dense tree with glossy deep green leaves. Can be planted in both dry and wet locations

The Zones of Plant Hardiness (Continued)

Botanical and common names	Average ultimate height in cultivation (feet)	Average ultimate spread in cultivation (feet)	Evergreen semievergreen deciduous	Trim or informal	Spacing (feet) Single row	Spacing (feet) Multiple row	Recommended minimum size to plant (feet)	Plants also useful in following zones	Remarks
Hardiness Zone 4 (Continued)									
Spiraea vanhouttei (Van Houtte spiraea)	6	5	Deciduous	Informal	4	5	2	5,6,7,8	Dense-growing shrub with gracefully arching branches covered with white flowers in May. Grows in a wide range of soils
Tsuga canadensis (Canadian hemlock)	60	30	Evergreen	Either	8	15	2	3,5,6,7	Thick foliage, pendulous branches. Tolerates shade, and hence uniquely useful for screening in shady areas
Viburnum dentatum (arrowwood)	10	6	Deciduous	Either	6	8	2	2,3,5,6	Dense shrub, holding branches well to ground line. White flowers, blue fruit, and bronze leaves in fall. Grows well in moist soil
Viburnum prunifolium (black haw)	15	15	Deciduous	Either	5	8	2	5,6,7	Tall rounded shrub or small tree. Flat clusters of white flowers, black berries, and red fall color. Tolerates acid, poor soil
Viburnum trilobum (American cranberrybush)	8–12	6–8	Deciduous	Informal	6	8–10	2–3	3,5,6	Handsome shrub with white flowers followed by red fruit (edible). Good fall cover and attracts birds
Hardiness Zone 5									
Amelanchier canadensis (shadblow serviceberry)	20	10	Deciduous	Informal	6	15–20	3	4,6,7	Tall, narrow shrub or small tree. Abundant white flowers in early spring and orange fall color. Excellent at the seashore. Tolerates wet acid soil
Carpinus betulus (European hornbeam)	30	20	Deciduous	Either	6	10	3	6	Pyramidal when young, later rounded. Retains leaves late in the winter. The variety "fastigeata" is an even more useful screening form. Withstands close shearing
Cornus mas (Cornelian cherry)	20	18	Deciduous	Informal	8	10	4	4,6	Rounded oval habit. Glossy foliage, yellow flowers in early spring. Red fruits in summer
Cotoneaster divaricata (spreading cotoneaster)	6	8	Deciduous	Either	4	6	2	6,7	Spreading rounded shrub with bright red fruit in fall. Excellent screen where space is limited
Crataegus phaenopyrum (Washington hawthorn)	30	15–20	Deciduous	Informal	15–20	25–30	4–5	4,6	Very dense upright plant with thorny, twiggy growth. Blooms in June followed by small red fruits that persist all winter. Shiny deep green leaves change to orange and scarlet in fall
Deutzia scabra ("Pride of Rochester")	8–9	4–6	Deciduous	Informal	4–6	6–8	2	6,7	Neat upright shrub with many branches. Pale pink to white flowers
Forsythia intermedia (border forsythia)	8–10	10	Deciduous	Informal	4–6	8	2	4,6,7	Upright dense shrub of very rapid growth. Dense masses of golden flowers in spring. "Spectabilis," "Spring Glory," and "Lynwood" are good varieties. Excellent for city planting
Juniperus scopulorum (Rocky Mountain juniper)	20–30	8–12	Evergreen	Either	5–8	10–15	2	4,6,7	Compact-growing evergreen of broad pyramidal form. Foliage has shades of purple to green and silver. Thrives in dry areas, not good in the Eastern states
Kolkwitzia amabilis (Beauty Bush)	9	7	Deciduous	Informal	6	12	2–3	6,7	Dense upright shrub with arching branches. Abundant pink flowers in June. Pest-free
Ligustrum ovalifolium ibolium (California privet)	15	6	Deciduous	Either	2	4	3	6,7	Narrow dense growth. Excellent as a formal hedge. Leaves retained late in the fall and semievergreen in warm areas. Thrives in the city. L. ibolium is somewhat hardier
Malus sargentii (Sargent crab)	8	10	Deciduous	Informal	8	12	5	6	Broad dense shrub. Fragrant white flowers in spring. Small red fruits held until late fall
Pseudotsuga menziesii (= P. taxifolia) (Douglas fir)	60	25	Evergreen	Informal	12	20	2–3	4,6	Tall pyramidal evergreen, dense and rapid growing in good soil. Only the Rocky Mountain strains are hardy in the east
Quercus palustris (pin oak)	80	50	Deciduous	Informal	30	40	6–8	4,6,7	Low, spreading branches retaining leaves in winter. Red fall color. Needs acid soil but will grow in wet or dry areas

Quercus robur "Fastigiata" (pyramidal English oak)	60	20	Deciduous	Informal	12	20	6	6,7	Densely columnar tree, holds foliage late into the fall. Makes a fine tall screen. Tolerates city conditions
Syringa chinensis (= S. rothomagensis) (Chinese lilac)	12	8	Deciduous	Informal	4	8	2–3	6,7	Upright shrub with small dense foliage and fragrant purple flowers. Pest-resistant. Requires good soil. Medium to slow rate of growth
Taxus media "Hicksii" (Hick's yew)	12	8	Evergreen	Either	4	8	2	4,6,7	Columnar in habit and will make a splendid narrow hedge. Thrives in many soil types but not in wet areas
Viburnum sieboldii (Siebold viburnum)	30	15	Deciduous	Informal	6	10	3	4,6,7	Forms a tall and rounded shrub or small tree with lustrous dark green foliage. Fruits are red in summer, turning black when mature
Weigela florida (pink weigela)	10	10	Deciduous	Informal	4	8	3	6,7	A rounded shrub with arching branches covered with pink flowers in late spring. W. vanicekii is an equally hardy, bright red variety
Hardiness Zone 6									
Abelia grandiflora (glossy abelia)	6	5	Semievergreen	Either	4	6	2	5,7,8	An arching shrub with glossy foliage and continuously in bloom from midsummer until frost. Flowers are white with a pleasant scent
Berberis julianae (wintergreen barberry)	6	5	Evergreen	Either	3	5	2	7,8	Dense evergreen shrub with attractive yellow flowers in the spring. Very thorny and highly resistant to human vandalism. An excellent city plant
Cryptomeria japonica (Japanese cryptomeria)	70	20	Evergreen	Informal	10	20	5	7,8	Forms a narrow pyramid of handsome, dark green foliage. Rapid growing in rich soil. The variety "Lobbi" is distinctly hardier than trees of seeding origin
Euonymus kiautschovica (= E. patens) (spreading euonymus)	6–9	6–8	Semievergreen	Either	4	6	2	7,8	Fully evergreen from zone 7 south. Thrives in city conditions. The variety "Manhattan" has more glossy and persistent foliage
Hamamelis vernalis (vernal witch-hazel)	6–10	5–8	Deciduous	Informal	5	8	2	5,7,8	A rounded, dense shrub with attractive foliage and small very fragrant flowers opening in late winter. The leaves are retained long after they turn brown in winter
Hibiscus syriacus (Rose-of-Sharon, shrub althea)	12–15	8–12	Deciduous	Informal	6	10	2	5,7	Upright in growth with many branches. Blooms, late in summer, have many colors, in both single and double flowers
Ilex opaca (American holly)	20	8	Evergreen	Either	8	16	2	7,8,9	Dense small tree with spiny leaves. Female plants bear abundant red berries
Ligustrum ovalifolium (California privet)	15	6	Semievergreen	Either	2	4	3	5,7,8,9	Narrow dense growth. Excellent as a formal hedge. Leaves retained late in the fall and semievergreen in warm areas. Thrives in the city. L. ibolium is somewhat hardier
Lonicera fragrantissima (winter honeysuckle)	6–8	4–7	Semievergreen	Either	4	6	2	5,7,8	Stiff half evergreen leaves. Bears small intensely fragrant white flowers in late winter. Grows well in city locations
Mahonia aquifolium (Oregon holly-grape)	6	5	Evergreen	Informal	4	6	2	5,7	Lustrous dark green foliage turning red-purple in the fall. Bears showy yellow flowers and blue berries. An excellent screening plant for shady areas
Osmanthus ilicifolius (holly osmanthus)	12	8	Evergreen	Either	4	8	2	7,8	Upright, thick shrub resembles a holly. Bears intensely fragrant flowers in July, and has shiny dark green foliage
Pyracantha coccinea "Lalandii" (Lalande firethorn)	10	10	Semievergreen	Either	5	8	2	7,8	Vigorous spiny shrub, evergreen in zones 7 and 8. Showy sprays of orange-red berries borne in the fall. Tolerates poor soil and city sites. Very drought-resistant
Quercus robur "Fastigiata" (pyramidal English oak)	60	20	Deciduous	Informal	12	20	6	5,7	Densely columnar tree, holds foliage late into the fall. Makes a fine tall screen. Tolerates city conditions
Salix babylonica (Babylon weeping willow)	30	30	Deciduous	Informal	30	50	4	7,8	A broad, fast-growing tree with very pendulous branches. Leaves are retained late into the fall. Thrives in wet soil

The Zones of Plant Hardiness (*Continued*)

Botanical and common names	Average ultimate height in cultivation (feet)	Average ultimate spread in cultivation (feet)	Evergreen semievergreen deciduous	Trim or informal	Spacing (feet) Single row	Spacing (feet) Multiple row	Recommended minimum size to plant (feet)	Plants also useful in following zones	Remarks
Hardiness Zone 6 (*Continued*)									
Thuja orientalis "Bakeri" (Baker oriental arborvitae)	25	10	Evergreen	Informal	6	12	3	7,8	Compact pyramidal evergreen with pale green foliage. Can withstand hot, dry locations. Rapid-growing
Viburnum rhytidophyllum (leatherleaf viburnum)	10	8	Evergreen	Informal	6	12	3	5,7,8	Upright shrub with dark green, wrinkled leaves. Grows best in rich, well-drained soil and thrives in a shady area
Vitex agnus-castus (chaste-tree)	9	8	Deciduous	Informal	4	6	2	7,8	Dense shrub with gray aromatic foliage and blue flowers borne from July to September. Withstands hot, dry conditions
Weigela florida (pink weigela)	10	10	Deciduous	Informal	4	8	3	5,7	A rounded shrub with arching branches covered with pink flowers in late spring. *W. vanicekii* is an equally hardy, bright red variety
Hardiness Zone 7									
Abelia grandiflora (glossy abelia)	6	5	Semievergreen	Either	4	6	2	5,6,8	An arching shrub with glossy foliage, continuously in bloom midsummer until frost. Flowers are white with a pleasant scent
Berberis julianae (wintergreen barberry)	6	5	Evergreen	Either	3	5	2	6,8	Dense evergreen shrub with attractive yellow flowers in the spring. Very thorny and highly resistant to human vandalism. An excellent city plant
Berberis mentorensis (Mentor barberry)	4-6	3-4	Semievergreen	Trim	2	3	2	5,6,8	Tolerates hot summer weather extremely well. This upright-growing shrub with thorny twigs has yellow flowers followed by dark red berries and colorful autumn foliage
Berberis thunbergii (Japanese barberry)	4-5	24 in	Deciduous	Informal	2	2	2	5,6,8	Considered most adaptable barberry. Will grow in most soil types and tolerates dry soils well. Will grow in semishade. Variety "Atropurpurea" recommended for sunny areas only
Cotoneaster parneyi (Parney cotoneaster)	6-8	4-6	Semievergreen	Either	4	6	5	8,9,10	Vigorous shrub thriving with little care. Suitable for any well-drained soil
Cupressocyparis leylandii (Leyland cypress)	25-30	10-15	Evergreen	Trim	6	8	10-15	5,6,8,9,10	Extremely rapid growing. Tolerates wide range of soils and pH
Elaeagnus angustifolia (Russian olive)	6-8	4	Semievergreen	Informal	2	3	2-3	3,4,5,6	Upright habit. Attractive silver foliage tolerates adverse conditions and dry, sandy soil
Elaeagnus x ebbinger	10-12	5-6	Evergreen	Either	5-6	8-9	8	8,9,10	Thornless and rapid growing. Suitable for any well-drained sunny location. Soil pH neutral to alkaline. *E. fruitlandii* also good. Has slightly larger rounded leaves with wavy margins. White and brown scales on underside of leaves
Elaeagnus umbellata (autumn olive)	8-12	6-8	Deciduous	Informal	4	8-10	2	4,5,6,8	Silver-green foliage holding color well into winter. Excellent winter bird food
Euonymus alatus "Compactus" (compact winged euonymus)	4-6	3-4	Deciduous	Trim	2	4	2	8	Sturdy plant with brilliant scarlet autumn color. This variety used as a hedge requires practically no clipping
Euonymus japonica (Japanese euonymus)	12	7	Evergreen	Informal	3-4	3-6	2	8	Highly attractive lustrous foliage. Will perform well in most soil conditions. Plant in full sun
Hibiscus syriacus (Rose-of-Sharon, shrub althea)	12	6	Deciduous	Informal	3-4	4-6	2	6,8	Upright in growth with many branches. Blooms, late in summer, have many colors, in both single and double flowers
Ilex altaclarensis "Wilsonii" (Wilson holly)	10-15	5-10	Evergreen	Either	6	7	6	6,8,9,10	Vigorous plant. Grows well in a wide range of soil and pH. Thick dark green leaves and bright red berries make this an attractive plant
Ilex opaca (American holly)	20	10	Evergreen	Informal	3-4	4-6	2	6,8	Prefers moist slightly acidic soil. Plant in full sun or partial shade. Foliage of this plant widely used in Christmas decorations

Plant									Remarks
Ilex x opaca "Foster #2"	12-14	4-5	Evergreen	Trim	3	4	2-3	8	Makes a beautiful hedge trimmed to 6 feet. Fast grower, likes loamy soil
Juniperus chinensis (Hetz juniper)	9	12	Evergreen	Either	6	10	2	5,6	Upright but wide-spreading growth. Silver blue foliage. Fast-growing. Tolerant of a wide range of soils, except wet soil
Juniperus chinensis "Pfitzeriana Glauca" (blue Pfitzer juniper)	10-12	15-20	Evergreen	Informal	8	10	2-3	4,5,6,8,9,10	Rapid-growing shrub. Likes well-drained garden soil. Tolerates wide range of soil pH. Plant noted for its blue-gray foliage
Juniperus scopulorum (Rocky Mountain juniper)	20	12	Evergreen	Informal	3-5	4-8	2	6,8	Vigorous grower. Adapted to most soil conditions. Many cultivars available with dark green to bright silver foliage
Juniperus virginiana (eastern red cedar)	50	10	Evergreen	Either	7	10	4	2,4,5,6,8	Tall, narrow growth. Tolerates dry soil and coastal conditions. Mostly available in selected grafted varieties
Ligustrum japonicum (= L. texanum) (Japanese privet)	6-9	4-5	Evergreen	Trim	5	7	4	8,9,10	Rapid-growing plant preferring loamy soil. Tolerates wide pH range
Ligustrum ovalifolium (California privet)	15	6	Semievergreen	Either	2	4	3	5,6,8,9	Narrow dense growth. Excellent as a formal hedge. Leaves retained late in fall and semievergreen in warm areas. Thrives in the city. L. ibolium is somewhat hardier
Ligustrum sinense (Chinese privet)	8-12	24 in	Deciduous	Trim	2	2	2	8	Flowers are white and in large panicles followed by black berries. Fast-growing plant preferring loamy soil. L. sinense "Pendulum" noted for its "weeping" branches
Lonicera fragrantissima (winter honeysuckle)	6-8	3-4	Semievergreen	Informal	2	3	2-3	6,8	Stiff, leathery leaves with fragrant flowers are the main attraction of this shrub. Fast grower. This plant prefers loamy soil
Photinia "Fraseri" (Fraser's photinia)	10	5-6	Evergreen	Either	5	7	5-6	6,8,9,10	New foliage a glistening coppery-red on bright red stems. Dark green mature leaves enhance this upright plant. Prefers sandy loam, neutral to acid
Podocarpus macrophylla (yew podocarpus)	15-20	6-8	Evergreen	Trim	6	6	6	8,9,10	Slow-growing plant. Can be trained by slight pruning to make a tall, full column of narrow deep green leathery leaves. Prefers garden loam soil
Populus alba "Bolleana" (Bolleana silver poplar)	40	10	Deciduous	Informal	3-4	4-6	4	6,8	Fast-growing columnar tree having wide range of soil tolerance. Leaves dark green above, wooly white on underside
Populus nigra "Italica" (Lombardy poplar)	40	6	Deciduous	Informal	3-4	4-6	3	6,8	Fast-growing tree. Erect, columnar habit of growth. Does well in all soil conditions
Prunus caroliniana ("bright 'n tight")	15-20	6-8	Evergreen	Trim	6	8	6	8,9,10	This selection has a showy, compact habit of growth. Has good deep green foliage on well-branched upright form. Moderately fast growing plant preferring sandy loam soil
Prunus laurocerasus (cherry-laurel)	15-20	10-15	Evergreen	Either	8	10	8	6,8,9,10	Rapid-growing plant preferring loamy soil, neutral to acid. Flowers are white and followed by purple to black cherries
Pyracantha coccinea ("Lalande Monrovia")	8-10	5-6	Evergreen	Either	5	8	4	5,6,8,9,10	Hardy vigorous plant. Does well in a wide range of soils. This strain has a superior upright habit and lustrous green foliage with scarlet-red fruits
Spiraea vanhouteii (Van Houtte spirea)	6	5	Deciduous	Informal	4	5	2	4,5,6,8	Dense-growing shrub with gracefully arching branches covered with white flowers in May. Grows in a wide range of soils
Tamarix chinesis (Chinese tamarix)	16	8	Deciduous	Informal	3-4	4-6	2-3	6,8	Fast-growing. Hardy, well adapted for a wide range of growing conditions. Bluish foliage with pink flowers
Thuja occidentalis (eastern arborvitae)	25-30	10-15	Evergreen	Trim	8	10	3	6,8,9,10	Moderately fast grower preferring neutral loamy soil. Plant has attractive reddish brown bark. The variety "Nigra" has dark green leaves
Thuja orientalis (oriental arborvitae)	16	6	Evergreen	Informal	3-4	4-6	2-3	8,9	Fast growing in wide range of soils. Numerous varieties exist
Tsuga canadensis (Canadian hemlock)	60	30	Evergreen	Either	8	15	2	5,6	Thick-foliaged plant with pendulous branches. Tolerates shade, and hence uniquely useful for screening in shady areas

The Zones of Plant Hardiness (*Continued*)

Botanical and common names	Average ultimate height in cultivation (feet)	Average ultimate spread in cultivation (feet)	Evergreen semievergreen deciduous	Trim or informal	Spacing (feet) Single row	Spacing (feet) Multiple row	Recommended minimum size to plant (feet)	Plants also useful in following zones	Remarks
Hardiness Zone 7 (*Continued*)									
Viburnum japonicum (Japanese viburnum)	10–20	5–10	Evergreen	Trim	5	8	5	8,9,10	Moderately fast growing plant preferring neutral to slightly acidic soil. Has lustrous green leaves and fragrant white flowers followed by red berries
Vitex agnus-castus (chaste-tree)	20	10	Deciduous	Informal	3–5	4–8	3	6	Fast growing in a wide range of soil conditions. Has dark green foliage and blue flowers in terminal spikes
Hardiness Zone 8									
Abelia grandiflora (glossy abelia)	6	5	Semievergreen	Either	4	6	2	5,6,7	An arching shrub with glossy foliage, continuously in bloom from midsummer until frost. Flowers are white with a pleasant scent
Callistemon citrinus (= *C. lanceolatus*) (lemon bottlebrush)	10–15	5–6	Evergreen	Either	5	7	8	9,10	Leaves lance-shaped, flowers with bright red stamens. Fast-growing plant preferring sandy loam and neutral to alkaline soil
Chilopsis linearis (flowing-willow)	16	8	Deciduous	Informal	3–5	5–6	3–4	7,9	Moderately fast grower. Prefers slightly acid to alkaline soils of any type. Plant in full sun; requires little care. Flower lilac-colored, trumpet-shaped
Cocculus laurifolius (laurel-leaf shail-seed)	20	15–20	Evergreen	Either	6	8	8	9,10	Leaves are shiny and leathery. Plant is a moderate to fast grower preferring neutral to acidic loam soil
Cortaderia selloana (= *C. argentea*) (pampas grass)	10	5–6	Evergreen	Informal	6	8	2	9,10	Rapid growing in most soil types. This perennial ornamental grass has long stalks towering above leaves bearing masses of decorative white flower plumes.
Cotoneaster parneyi (Parney cotoneaster)	6–8	4–6	Semievergreen	Either	4	6	5	7,9,10	Vigorous shrub thriving with little care. Suitable for any well-drained soil
Elaeagnus fruitlandi (fruitland silverberry)	10	8	Evergreen	Either	3–4	4–6	2	9,10	Handsome silver foliage; pruning enhances overall plant appearance. Moderately fast grower; adaptable to a wide range of soil conditions
Eucalyptus cinerea (silver dollar tree)	20–40	15–30	Evergreen	Trim	8	10	10	9,10	Whitish bark; flowers in clusters; fast-growing. Prefers light soil with a neutral to alkaline pH
Ilex altaclarensis "Wilsoni" (Wilson holly)	10–15	5–10	Evergreen	Either	6	7	6	6,7,9,10	Vigorous grower adaptable to a wide range of soil and pH. Thick dark green leaves and bright red berries make this an attractive plant
Ilex fosteri (Foster's holly)	12	8	Evergreen	Either	3–4	4–6	2	9	Handsome shrub producing bright red berries in profusion. Plant in full sun or partial shade. Prefers light to loamy acid soils. Makes a superb hedge with trimming
Ilex vomitoria (Yaupon holly)	20	8	Evergreen	Either	3–4	4–6	2	9	Bright red berries from late fall into winter. Moderately fast grower. Prefers well drained soil from acid to slightly alkaline
Juniperus chinensis "Pfitzeriana Glauca" (blue Pfitzer juniper)	10–12	15–20	Evergreen	Informal	8	10	2–3	4,5,6,7,9,10	Rapid-growing shrub. Likes well-drained garden soil. Tolerates wide range of soil pH. Plant noted for its blue-gray foliage
Lagerstroemia indica (crape myrtle)	20	8	Deciduous	Informal	3–4	4–8	2–4	9,10	Fast grower. Tolerant of a wide range of soil conditions. Plant in full sun. Many colors available. Produces masses of flower clusters all summer
Ligustrum japonicum (= *L. texanum*) (Japanese privet)	6–9	4–5	Evergreen	Trim	5	7	4	7,9,10	Rapid-growing plant preferring loam soil. Tolerates wide pH range
Ligustrum lucidum "Compacta" (compact glossy privet)	10	8	Evergreen	Either	3–4	4–6	2–4	9,10	Moderately fast grower. Adaptable to a wide range of soil conditions. Dark green foliage responds well to shearing
Magnolia grandiflora (southern magnolia)	30	10	Evergreen	Informal	4–5	6–8	4	9	Beautiful large white flowers and dark green foliage make this an attractive plant. Prefers loamy soils. For a screen plants should be left bushy from ground level

Nerium oleander (oleander)	8–12	8–10	Evergreen	Either	4	6	5	9,10	Flowers in variety of colors. Requires little attention. Fast-growing plant adaptable to most soils and pH
Photinia serrulata (Chinese photinia)	16	8	Evergreen	Informal	3–4	4–6	2	7,9	Moderately fast grower. Prefers well-drained soil having an acidic to moderately alkaline condition. Performs suitably in a partially shady situation
Photinia "Fraseri" (Fraser's photinia)	10	5–6	Evergreen	Either	5	7	5–6	6,7,8,10	New foliage a glistening coppery-red on bright red stems. Dark green mature leaves enhance this upright plant. Prefers sandy loam soil, neutral to acidic
Prunus caroliniana ("bright 'n tight")	15–20	6–8	Evergreen	Trim	6	8	6	7,9,10	This selection has a showy, compact habit of growth. Has good deep green foliage and well-branched upright form. Moderately fast-growing, prefers sandy loam soil
Prunus laurocerasus (cherry-laurel)	15–20	10–15	Evergreen	Either	8	10	8	6,7,9,10	Rapid-growing plant preferring loamy soil, neutral to acidic. Flowers are white and followed by purple to black cherries
Punica granatum "Wonderful" (wonderful pomegranate)	15	6	Deciduous	Informal	3–4	4–6	3	9,10	Has erect habit of growth, thornless and produces large double orange-red blossoms, edible fruit
Pyracantha koidzumii ("Victory")	10	8	Evergreen	Trim	5	7	5	7,9,10	Rapid-growing plant adaptable to most soils and pH. One of the showiest varieties of pyracantha
Thuja orientalis (oriental arborvitae)	16	6	Evergreen	Informal	3–4	406	2–3	7,9	Fast-growing plant in a wide range of soils. Numerous varieties exist
Xylosma congestum (= X. senticosum) (shiny xylosma)	8–10	8–10	Evergreen	Trim	4	6	5	9,10	Grows in most soils. Rate of growth is moderately fast. Has glossy light-green foliage on arching branches. Has an upright habit
Hardiness Zone 9									
Acacia longifolia (= A. latifolia) (bush acacia)	20	18–20	Evergreen	Either	8	10	10	10	Rapid-growing plant. Likes alkaline sandy soil. Masses of yellow flowers along branches in spring
Bambusa falcata nana	8	6	Evergreen	Either	3–4	4–6	2	10	Medium fast grower in any soil moderately acidic to slightly alkaline. Extremely dense canes and foliage; nonrunning type
Calistemon citrinus (= C. lanceolatus) (lemon bottlebrush)	10–15	5–6	Evergreen	Either	5	7	8	8,10	Leaves lance-shaped, flowers with bright red stamens. Fast-growing plant preferring sandy loam and neutral to alkaline soil
Cortaderia selloana (= C. argentea) (pampas grass)	10	5–6	Evergreen	Informal	6	8	2	8,10	Rapid growing in most soil types. This perennial ornamental grass has long stalks towering above leaves bearing masses of decorative white flower plumes
Cupressocyparis leylandii (Leyland cypress)	25–30	10–15	Evergreen	Trim	6	8	10–15	5,6,7,8,10	Extremely rapid growing. Tolerates a wide range of soils and pH
Dondonaea viscosa "Purpurea" (purple-leafed dondonaea)	12–15	6–7	Evergreen	Either	6	8	8	8,10	Attractive, fast growing, densely branched plant. Bronze green, willow-like leaves turn to reddish purple with cool weather
Elaeagnus fruitlandi (Fruitland silverberry)	10	8	Evergreen	Either	3–4	4–6	2	8,10	Handsome silver foliage; pruning enhances overall plant appearance. Moderately fast grower; adaptable to a wide range of soil conditions
Eriobotrya japonica (loquat)	18	8	Evergreen	Informal	3–4	4–6	2–3	8,10	Fast-growing with dark green leaves. Grows well in all soils moderately acidic to moderately alkaline. Produces edible fruit called Japanese plum
Eucalyptus cinerea (silver dollar tree)	20–40	15–30	Evergreen	Trim	8	10	10	8,10	Whitish bark, flowers in clusters; fast growing. Prefers light soil with a neutral to alkaline pH
Eucalyptus globulus "Compacta" (bushy blue gum)	30	20	Evergreen	Trim	8	10	10	10	Rapid growing in sandy loam soil. Has rich blue-green foliage
Ficus retusa nitida (Indian laurel)	20–30	10–15	Evergreen	Trim	8	10	8	10	Moderately fast growing in loamy soil. Prefers neutral pH. Attractive plant with erect branches and small rubbery leaves
Grevillea banksii	15–20	15	Evergreen	Either	8	10	10	10	Adapts to any well-drained soil neutral to alkaline. Moderately fast grower. Flowers red
Ilex altaclarensis "Wilsoni" (Wilson holly)	10–15	5–10	Evergreen	Either	6	7	6	6,7,8,10	Vigorous grower adaptable to a wide range of soil and pH. Thick dark green leaves and bright red berries make this an attractive plant

The Zones of Plant Hardiness (*Continued*)

Botanical and common names	Average ultimate height in cultivation (feet)	Average ultimate spread in cultivation (feet)	Evergreen semievergreen deciduous	Trim or informal	Spacing (feet) Single row	Spacing (feet) Multiple row	Recommended minimum size to plant (feet)	Plants also useful in following zones	Remarks
Hardiness Zone 9 (Continued)									
Ilex vomitoria (Yaupon holly)	20	8	Evergreen	Either	3-4	4-6	2	8	Bright red berries from late fall into winter. Moderately fast grower. Prefers well-drained soil from acid to slightly alkaline
Juniperus chinensis "Blue Vase" (Texas star juniper)	8	8	Evergreen	Either	4	6	2	8	Medium grower. Tolerant of all soil conditions from moderately acid to moderately alkaline. Slightly vase-shaped; blue foliage
Lagerstroemia indica (crape myrtle)	20	8	Deciduous	Informal	3-4	4-8	2-4	8,10	Fast grower. Tolerant of a wide range of soil conditions. Plant in full sun. Many colors available. Produces masses of flower clusters all summer
Ligustrum japonicum (= L. texanum) (Japanese privet)	6-9	4-5	Evergreen	Trim	5	7	4	7,8,10	Rapid-growing plant preferring loamy soil, tolerates wide pH range
Ligustrum lucidum "Compacta" (compact glossy privet)	10	8	Evergreen	Either	3-4	4-6	2-4	8,10	Moderately fast grower. Adaptable to a wide range of soil conditions. Dark green foliage responds well to shearing
Nerium oleander (oleander)	8-12	8-10	Evergreen	Either	4	6	5	8,10	Flowers in a variety of colors. Requires little attention. Fast-growing plant adaptable to most soils and pH
Pinus halepensis (Aleppo pine)	25	10	Evergreen	Informal	3-4	4-6	3	8,10	Moderately fast growing pine. Adaptable to most soils from slightly acidic to slightly alkaline. Erect habit of growth
Podocarpus gracilior (fern pine)	50	40-50	Evergreen	Trim	10	12	10	10	Slow to moderately fast grower. Prefers well-drained loam soil having neutral pH
Podocarpus macrophylla maki	16	8	Evergreen	Either	3-4	4-6	3	10	Relatively fast growing and adaptable to most soil types. Erect habit of growth
Prunus caroliniana ("bright 'n tight")	15-20	6-8	Evergreen	Trim	6	8	6	7,8,10	This selection has a showy, compact habit of growth. Has good deep green foliage and well-branched upright form. Moderately fast growing, prefers sandy loam soil
Prunus laurocerasus (cherry-laurel)	15-20	10-15	Evergreen	Either	8	10	8	6,7,8,10	Rapid-growing plant preferring loamy soil, neutral to acidic. Flowers are white followed by purple to black cherries
Pyracantha fortuneana "Graberi"	12	8	Evergreen	Informal	3-4	4-6	3	8,10	Fast growing. Any soil type from acid to alkaline. Has erect habit of growth. Outstanding for its huge large clusters of red berries
Viburnum odoratissimuu (sweet viburnum)	12	6	Evergreen	Informal	3-4	4-6	3	10	Moderately fast grower. Prefers well-drained soil slightly alkaline to moderately acidic. Has dark green foliage, white flowers. Useful in partially shady areas
Xylosma congestum (= X. senticosum) (shiny xylosma)	8-10	8-10	Evergreen	Trim	4	6	5	8,10	Grows in most soils. Rate of growth is moderately fast. Has glossy light-green foliage on arching branches. Has an upright habit
Hardiness Zone 10									
Aralia spinosa (devil's walking-stick)	10	3	Evergreen	Informal	2	3	2		Plant does extremely well in wet areas. Prefers acidic soil. Fast-growing plant with spines
Bambusa falcata nana	8	6	Evergreen	Either	3-4	4-6	2	9	Medium fast grower in any soil moderately acidic to slightly alkaline. Extremely dense canes and foliage; nonrunning type
Calliandra guildingii	6-8	5-8	Evergreen	Either	5	7	8	8,9	Grows in most soils. Moderate to rapid grower. Flowers in clusters having long conspicuous stamens
Calistemon citrinus (= C. lanceolatus) (lemon bottlebrush)	10-15	5-6	Evergreen	Either	5	7	8	8,9	Leaves lance-shaped; flowers with bright red stamens. Fast-growing plant preferring sandy loam and neutral to alkaline soil
Cassia splendida (golden shower cassia)	12	8	Evergreen	Informal	3-4	4-6	3	9	Fast-growing; not particular as to soil conditions. Becomes a solid mass of golden yellow blooms in late fall or early winter
Casuarina equisetifolia (horsetail-tree)	20-40	5-15	Evergreen	Either	5-10	7-15	2		Fast-growing plant. Prefers light loamy soil

Plant									Description
Chrysobalanus icaco (coco-plum)	6–10	4–8	Evergreen	Trim	3–4	5–6	2–3	9	Relatively slow grower. Has numerous white flowers. Prefers alkaline, heavy soil
Cinnamomum camphora (camphor tree)	20	10–20	Evergreen	Informal	5–8	8–10	3	9	Fast-growing. Any soil from acid to slightly alkaline. Dense plant with shiny bright green foliage; flowers yellow
Coccoloba uvifera (sea grape)	15–20	5–10	Evergreen	Either	3–4	5–6	2–3		Moderately fast grower. Considered very decorative
Cocculus laurifolius (laurel-leaf snail-seed)	20	10–20	Evergreen	Either	6	8	8	8,9	Moderate to fast-growing plant. Prefers neutral to acidic pH and loamy soil. Tolerates shade
Cordia boissieri	12–15	8	Evergreen	Informal	5	7–8	3	9	Moderately fast grower in most soils from slightly acidic to alkaline. Produces attractive white flowers in great numbers
Cortaderia selloana (= C. argentea) (pampas grass)	10	5–6	Evergreen	Informal	6	8	2	8,9	Rapid-growing in most soil types. This perennial ornamental grass has long stalks towering above leaves bearing masses of decorative white flower plumes
Cupressocyparis leylandii (Leyland cypress)	25–30	10–15	Evergreen	Trim	6	8	10–15	5,6,7,8,9	Extremely rapid growing. Tolerates a wide range of soils and pH
Dondonaea viscosa "Purpurea" (purple-leafed dondonaea)	12–15	6–7	Evergreen	Either	6	8	10–15	8,9	Attractive, fast-growing, densely branched plant. Bronze green, willow-like leaves turn to reddish purple with cool weather
Eriobotrya japonica (loquat)	18	8	Evergreen	Informal	3–4	4–6	2–3	8,9	Fast-growing with dark green leaves. Grows well in all soils moderately acidic to moderately alkaline. Produces edible fruit called Japanese plum
Eucalyptus globulus "Compacta" (bushy blue gum)	30	20	Evergreen	Trim	8	10	10	9	Rapid growing in sandy loam soil. Has rich blue-green foliage
Eucalyptus rostrata (red gum)	80	20	Evergreen	Informal	8	10–12	3–4	9	Fast-growing; tolerant of poor soil including high alkalinity and salinity. Pink flowers appear in late spring
Eugenia uniflora (Surinam-cherry)	10	5–6	Evergreen	Trim	2–3	4–5	3		Fast-growing; prefers loamy soil. Compact habit of growth, leaves glossy, flowers fragrant and white
Fejioa sellowiana (pineapple guava)	10	8	Evergreen	Informal	4–5	6–7	3	9	Fast-growing. Does well in most soils form acid to slightly alkaline. Foliage dark green above, silvery white below. Very attractive
Ficus retusa nitida (Indian laurel)	20–30	10–15	Evergreen	Trim	8	10	8	9	Moderately fast growing in loamy soil. Prefers neutral pH. Attractive plant with erect branches and small rubbery leaves
Hibiscus rosa-sinensis (Chinese hibiscus)	10–15	8–13	Evergreen	Either	3–5	5–7	4		Moderate to rapid grower. Does well in a wide variety of soils. Plant in full sun. Numerous varieties and colors available, flowers single or double
Ilex altaclarensis "Wilsoni" (Wilson holly)	10–15	5–10	Evergreen	Either	6	7	6	6,7,8,9	Vigorous grower adaptable to a wide range of soil and pH. Thick dark green leaves and bright red berries make this an attractive plant
Ixora coccinea (Ixora)	5–10	3–5	Evergreen	Trim	2–3	3–4	2–3		Moderately fast grower preferring acidic soil. Beautiful red flowers
Ligustrum lucidum "Compacta" (compact glossy privet)	10	8	Evergreen	Either	3–4	4–6	2–4	8,9	Moderately fast grower. Adaptable to a wide range of soil conditions. Dark green foliage responds well to shearing
Malvaviscus arboreus (Turk's cap)	6–8	4	Evergreen	Either	2	3	2–3		Considered a fast-growing plant which prefers light acid soil. Flowers red
Melaleuca leucodendron (Cajeput-tree)	20–30	4–6	Evergreen	Informal	2	3–4	3		Flowers creamy white in spikes 6 inches long. Considered a fast-growing plant preferring heavy acidic soils; does well in wet areas
Murraya paniculata (= Chalcas paniculata) (orange jessamine)	8–10	3–5	Evergreen	Either	3	4	3–4		This attractive small plant is a relatively slow grower. Plant bears clusters of white fragrant flowers and glossy foliage. Prefers acidic loamy soil
Musa paradisiaca sapientum (common banana)	12	6	Evergreen	Informal	3–4	4–6	2–3	9	Fast-growing. Does well in most soils if kept moist. Large dark green foliage
Nerium oleander (oleander)	8–12	8–10	Evergreen	Either	4	6	5	8,9	Flowers in a variety of colors. Requires little attention. Fast-growing plant adaptable to most soils and pH
Photinia "Fraseri" (Fraser's photinia)	10	5–6	Evergreen	Either	5	7	5–6	6,7,8,9	Moderately fast grower preferring sandy loam soil with a neutral to acidic pH
Pittosporum tobira (Japanese pittosporum)	6–8	6	Evergreen	Informal	2	3–4	3		Good seaside plant. Relatively slow grower. Prefers alkaline soil

The Zones of Plant Hardiness (Continued)

Botanical and common names	Average ultimate height in cultivation (feet)	Average ultimate spread in cultivation (feet)	Evergreen semievergreen deciduous	Trim or informal	Spacing (feet) Single row	Spacing (feet) Multiple row	Recommended minimum size to plant (feet)	Plants also useful in following zones	Remarks
Hardiness Zone 10 (Continued)									
Podocarpus macrophylla (yew podocarpus)	8–15	4	Evergreen	Trim	3	4	3		Slow-growing plant preferring heavy acidic soil. Foliage extremely dense
Prunus caroliniana ("bright 'n tight")	15–20	6–8	Evergreen	Trim	6	8	6	7,8,9	This selection has a showy, compact habit of growth. Has good deep green foliage and well-branched upright form. Moderately fast-growing, prefers sandy loam soil
Prunus laurocerasus (cherry-laurel)	15–20	10–15	Evergreen	Either	8	10	8	6,7,8,9	Rapid-growing plant preferring loamy soil, neutral to acidic. Flowers are white, followed by purple to black cherries
Syzygium paniculata (= Eugenia paniculata) (Australian bush-cherry)	30–40	10–15	Evergreen	Trim	5	7	6–8		Relatively slow growing plant preferring good loam soil with neutral to acidic pH
Tamarix aphylla (Athel tamarisk)	20	10	Evergreen	Informal	6–8	8–10	3	9	Fast-growing plant tolerant of poor soil conditions including alkalinity and salinity. Pink flowers
Thevetia nerifolia (yellow oleander)	14	8	Evergreen	Either	3–4	4–6	3	9	Fast-growing. Tolerant of wide range of soil conditions. Produces dark yellow blossoms profusely
Viburnum odoratissium (sweet viburnum)	12	6	Evergreen	Informal	3–4	4–6	3	9	Moderately fast grower. Prefers well-drained soil slightly alkaline to moderately acidic. Has dark green foliage, white flowers. Useful in partially shady areas
Viburnum suspensum (sandankwa viburnum)	4–6	4	Evergreen	Either	2–3	3	2		Slow-growing plant preferring loamy soil. Shiny leaves, pinkish flowers numerous
Xylosma congestum (= X. senticosum) (shiny xylosma)	8–10	8–10	Evergreen	Trim	4	6	5	8,9	Grows in most soils. Rate of growth is moderately fast. Has glossy light-green foliage on arching branches. Has an upright habit

Climate, soil, and altitude within a zone may vary widely. Check with your local American Association of Nurserymen member or other plant experts to determine whether your specific location has special plant requirements.

Vines for Screening

Zone	Botanical name	Common name	Type
3	Lonicera sempervirens	Trumpet honeysuckle	Semievergreen
4	Akebia quinata	Five-leaf akebia	Semievergreen
	Lonicera henryi	Silver vein creeper	Semievergreen
	L. japonica "Halliana"	Hall's Japanese honeysuckle	Semievergreen
5	Clematis paniculata	Sweet autumn clematis	Semievergreen
	Euonymus fortunei and varieties	Winter creeper	Evergreen
	Hedera helix	English ivy	Evergreen
	Muehlenbeckia complexa	Wirevine	Evergreen
6	Bignonia capreolata	Crossvine	Semievergreen
7	Kadsura japonica	Scarlet Kadsura	Evergreen
	Lonicera etrusca	Etruscan honeysuckle	Semievergreen
8	Trachelospermum asiaticum	Yellow star jasmine	Evergreen
	Ficus pumila	Creeping fig	Evergreen
9	Lonicera hildebrandiana	Giant honeysuckle	Evergreen
	Phaedranthus buccinatorius	Blood trumpet vine	Evergreen
	Trachelospermum asiaticum	Yellow star jasmine	Evergreen
	Trachelospermum jasminoides	Star jasmine	Evergreen

3.11 Regional Grasses

Listed below by four climatic zones are grasses suitable for planting within the United States. Since this is a rather general grouping of grasses and since climate and soil vary within these zones, the final selection of the best grass should be made by someone familiar with local soil and climatic conditions. The map in Figure 3.105 graphically displays the four zones into which the grasses have been tabulated.

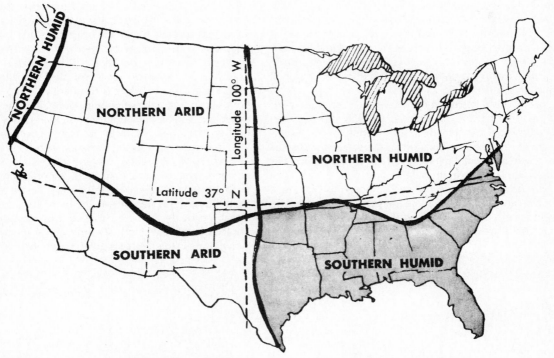

Figure 3.105 As shown on this map, grasses flourish differently in the four major climatic zones of the United States. Local soil and temperature conditions can cause some variation in growth.

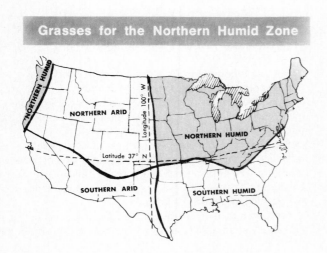

Grasses for the Northern Humid Zone

Canada bluegrass (Poa pratensis). Canada bluegrass has a rather shallow root system but can thrive on soils low in fertility. It is satisfactory for use where a dense turf is not essential.

Chewings fescue (Festuca rubra commutata). Chewings fescue is a fine-leaf grass that has a dense matted root system. It is best suited to well-drained soils such as sandy loams. A clay soil, which has poor drainage characteristics, will not allow vigorous growth of this grass. Although chewings fescue can be used alone, it is frequently sown in combination with other grasses such as bluegrass or redtop.

Kentucky bluegrass (Poa pratensis). Kentucky bluegrass is the most widely used grass for lawns and other turfing within the northern humid region. Soil and climatic requirements for this grass are more exacting than for other grasses, for it does not grow well on sandy soils or on soils that are low in fertil-

ity, and it must have an abundance of moisture. Because Kentucky bluegrass germinates rather slowly and requires time to become completely established, it is normally sown in combination with other grasses such as chewings fescue.

Red fescue (Festuca rubra). Red fescue is similar to chewings fescue in its requirements for growth and use. It tends to creep more than does chewings fescue.

Redtop (Agrostis alba). Redtop is a temporary grass that will usually persist for 2 or 3 years. A fast-growing, vigorous strain, it is best used as a temporary cover crop in combination with a permanent grass such as Kentucky bluegrass. Redtop is frequently preferred over ryegrass as a cover crop in that it is not as strong a competitor of the permanent strain with which it is sown.

Ryegrass (Lolium multiflorum). Ryegrass is another temporary grass that normally will persist for only 3 or 4 years. Because of its initial fast growth, it is frequently used as a temporary cover crop in combination with permanent grasses that are slower in establishing themselves. Ryegrass will make the best growth on fertile, moist soils.

Smooth brome (Bromus inermis). Smooth brome is a rather coarse grass with strong, creeping rhizomes that make it excellent for erosion and dust control. Fertile soil is required for its successful growth.

root system, bluestem is primarily useful for erosion control. A perennial grass, bluestem is adaptable to a variety of soils.

Common carpet grass (Axonopus affinis). This grass is more adaptable to clay than it is to sandy soils. The growth of common carpet grass is generally limited to the southern half of the southern humid zone. Since it requires more moisture for survival than does Bermuda grass, it is less useful in most parts of the southern humid zone than Bermuda is.

Rhodes grass (Chloris gayana). Rhodes grass is a vigorous grass that spreads by means of runners. It is not completely adaptable to all parts of the southern humid zone.

Ryegrass (Lolium multiflorum). Ryegrass is a temporary grass that normally will persist for only 3 or 4 years. Because of its initial fast growth, it is frequently used as a temporary cover crop in combination with permanent grasses that are slower in establishing themselves. Ryegrass will make the best growth on fertile, moist soils and hence may not be suitable in sections of the southern humid zone.

Sudan grass (Sorghum vulgare sudanense). Sudan grass is an annual that is useful only as a temporary cover crop. It is probably one of the best temporary dust control grasses. Even though it is killed by the first hard frost, the stubble that remains provides some erosion control and furnishes a temporary cover in which to seed perennial species of grasses. Sudan grass is tolerant of a great variety of soil situations.

Grasses for the Southern Humid Zone

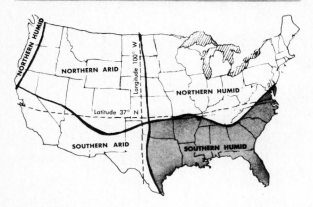

Grasses for the Northern Arid Zone

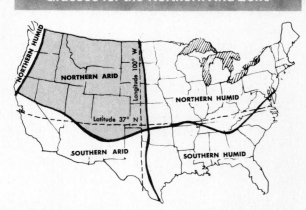

Bahia grass (Paspalum notatum). This grass is adaptable to sandy soils of the southern humid zone. Since it can also be grown on soils of low fertility and in climates with minimum rainfall, Bahia grass is quite useful. The stout rootstalks of this grass form a firm mat that is useful in dust and erosion control.

Bermuda grass (Cynodon dactylon). Bermuda grass is perhaps the best grass for use in the southern humid zone because of its tolerance of high temperatures and minimum rainfalls and its adaptability to a wide range of soils. It is sometimes considered a pest when used for individual lawns because of its invasive characteristic, which is difficult to restrain. Another disadvantage is that this grass will not grow satisfactorily in shaded areas.

Bluestem (Andropogon scoparius). Because of its deep

Alkali sacaton (Sporobolus airoides). This grass is best suited to heavy-textured soils such as clays. Alkali sacaton grows to a medium height and is a deep-rooted grass.

Blue grama (Bouteloua gracilis). Blue grama is a low-growing grass that adapts itself to a wide range of soil conditions. Thick stands of turf can be developed if the soil is of adequate fertility and the area is heavily seeded.

Buffalo grass (Buchloe dactyloides). Buffalo grass is not particularly adaptable to sandy soils, but it does thrive well on heavier, well-drained soils. It spreads rapidly by means of stolons and forms a dense, matted turf. Buffalo grass can withstand long periods of drought.

Sand dropseed (Sporobolus cryptandrus). Sand dropseed is generally adaptable in areas where blue grama and buffalo

grass can be successfully grown. Its primary usefulness is for erosion and dust control.

Smooth brome (Bromus inermis). Smooth brome is a rather coarse grass with strong, creeping rhizomes that makes it excellent for erosion and dust control. Fertile soil is required for its successful growth.

Sudan grass (Sorghum vulgare sudanense). Sudan grass is an annual that is useful only as a temporary cover crop. It is probably one of the best temporary dust-control grasses. Even though it is killed by the first hard frost, the stubble that remains provides some erosion control and furnishes a temporary cover in which to seed perennial species of grasses. Sudan grass is tolerant of a great variety of soil situations.

Western wheatgrass (Agropyron smithii). Wheatgrass is a perennial grass with sod-forming fibrous roots and strong, creeping root stalks. It can grow on a variety of soil types.

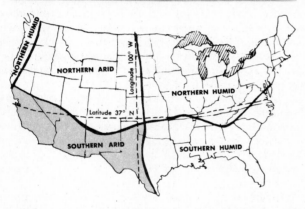

Grasses for the Southern Arid Zone

Alkali sacaton (Sporobolus airoides). This grass is best suited to heavy-textured soils such as clays. Alkali sacaton grows to a medium height and is a deep-rooted grass.

Bermuda grass (Cynodon dactylon). With irrigation Bermuda grass is an ideal grass for the southern arid zone because of its tolerance of high temperatures and its adaptability to a wide range of soils. It is sometimes considered a pest when used for individual lawns because of its invasive characteristic, which is difficult to restrain. Another disadvantage is that this grass will not grow well in shaded areas.

Blue grama (Bouteloua gracilis). Blue grama is a low-growing grass that adapts itself to a wide range of soil conditions. Thick stands of turf can be developed if the soil is of adequate fertility and the area is heavily seeded.

Buffalo grass (Buchloe dactyloides). Buffalo grass is not particularly adaptable to sandy soils, but it does thrive well on heavier, well-drained soils. It spreads rapidly by means of stolons and forms a dense, matted turf. Buffalo grass can withstand long periods of drought.

Rhodes grass (Chloris gayana). Rhodes grass is a vigorous grass that spreads by means of runners. In areas receiving less than 25 inches of annual rainfall, it is usually preferred to Bermuda grass.

Sand dropseed (Sporobolus cryptandrus). Sand dropseed is generally adaptable in areas where blue grama and buffalo grass can be successfully grown. Its primary usefulness is for erosion and dust control.

Sudan grass (Sorghum vulgare sudanense). Sudan grass is an annual that is useful only as a temporary cover crop. It is probably one of the best temporary dust-control grasses. Even though it is killed by the first hard frost, the stubble that remains provides some erosion control and furnishes a temporary cover in which to seed perennial species of grasses. Sudan grass is tolerant of a great variety of soil situations.

Weeping love grass (Eragrostis curvula). Weeping love grass is a quick-growing perennial grass of chief value in controlling dust and erosion on infertile areas. It grows well on sandy soils if the area receives from 15 to 20 inches of rainfall annually.

3.12 Site Security: Defensible Space

"Defensible space" is a term used to describe a series of physical design characteristics that maximize resident control of behavior, particularly crime, within a residential community. A residential environment designed under defensible-space guidelines clearly defines all areas as public, semiprivate, or private. In so doing, it determines who has the right to be in each space and allows residents to be confident in responding to any questionable activity or persons within their complex. The same design concepts improve the ability of police to monitor activities within the community.

Implementation of defensible space utilizes various elements of physical planning and architectural design such as site planning and the grouping and positioning of units, paths, windows, stairwells, doors, and elevators. Provision of defensible-space mechanisms is best achieved in a project's inception, as it involves major decisions with respect to the project.

However, a series of small-scale physical design techniques can be used to create defensible space and consequently to reduce crime in existing residential areas. These techniques consist of subdividing a project or building to limit access and improve neighbors' recognition, thus symbolically defining an area as coming under the sphere of influence of a particular group of inhabitants and improving the inhabitants' surveillance capacity.

The term "limiting access" refers to the use of physical design to prevent a potential criminal from entering certain spaces. Although no barrier is impregnable, physical barriers of this type are real and are relatively difficult to overcome.

In contrast, it is possible to use psychological or "symbolic" barriers that, while presenting no physical restriction, discourage criminal penetration by making an obvious distinction between stranger and intruder and bringing all activity under more intense surveillance. An intruder invading the space defined by such symbolic barriers becomes conspicuous to both residents and police.

Improved neighbor recognition plays a key role in the

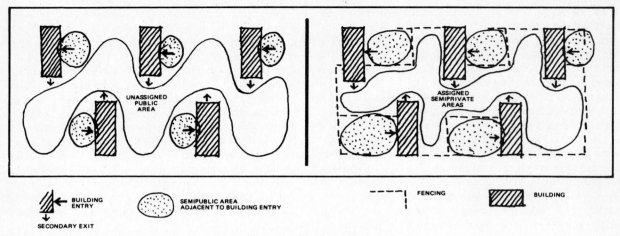

Figure 3.106 Alternative site plans with unassigned and assigned areas.

functional workings of psychological barriers. If, by newly defining areas, neighbors can be made to recognize one another, the potential criminal then can not only be seen but also be perceived as an intruder. This subdivision of space also reinforces the feeling in residents that they have the right to intervene on their own behalf.

Creating Territorial Areas Residential developments consisting of large superblocks devoid of interior streets have been found to suffer higher crime rates than projects of comparable size and density in which existing city streets have been allowed to continue through the sites.

Housing sites larger than a city block are best subdivided by through streets. The small scale of neighboring city blocks should be maintained where possible. This directive runs contrary to site planning principles aimed at removing vehicular traffic from the interior of large projects to free areas for recreation. However, large areas of low- and moderate-income projects that have closed off city streets but permitted public access have been considered dangerous by inhabitants and have consequently received minimal use. Through streets bring safety for three reasons:

1. They facilitate direct access to all buildings in the project by car and bus.

2. They bring vehicular and pedestrian traffic into the project and so provide the important measure of safety that comes with the presence of people.

3. They facilitate patrolling by police, provide easy access, and are a means of identifying building locations. Much of the crime deterence provided by police occurs while they pass through an area in a patrol car.

A project's site should be subdivided so that all of its areas are related to particular buildings or clusters of buildings. No area should be unassigned or simply left "public" (see Figure 3.106). Zones of influence should embrace all areas of a project, and the site plan should be so conceived. A "zone of

influence" is an area surrounding a building, or preferably an area surrounded by a building, that is perceived by residents as an outdoor extension of their dwellings. As such, it comes under their continued use and surveillance. Residents using these areas should feel that they are under natural observation by other project residents. A potential criminal should equally feel that any suspicious behavior will come under immediate scrutiny.

Grounds should be allocated to specific buildings or building clusters. This practice assigns responsibility and primary claim to certain residents. It also sets up an association between a building resident in his or her apartment and the grounds below.

Residents in projects that are subdivided have the opportunity of viewing a particular segment of the project as their own turf. When an incident occurs there, they are able to determine whether their area or another area is involved. When divisions do not exist within a project plan, an incident in one area is related to the complex and can create the impression of lack of safety in the entire project.

Defining Zones of Transition Boundaries can be defined by either real or symbolic barriers. Real barriers require entrants to possess a mechanical opening device, a familiar face or voice, or some other means of identification to indicate their belonging prior to entry. That is, access to a residence through a real barrier is by the approval of its occupants only, whether through the issuance of a key or through acceptance by their agents or by electronic signal.

Symbolic barriers define areas or relate them to particular buildings without physically preventing intrusion (see Figure 3.107). The success of symbolic versus real barriers in restricting entry rests on four conditions:

1. The capacity of the intruder to read the symbols

2. The capacity of the inhabitants or their agents to maintain controls and reinforce the space definition as symbolically defined

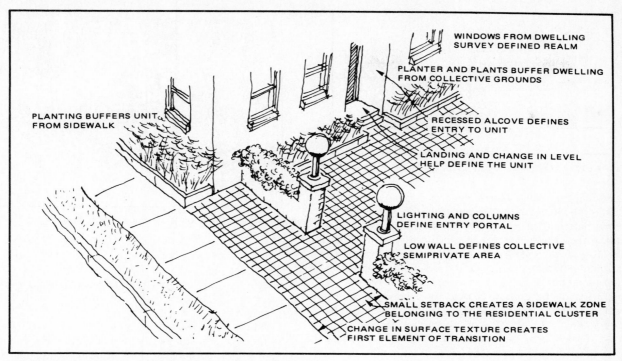

WINDOWS FROM DWELLING
SURVEY DEFINED REALM

PLANTER AND PLANTS BUFFER DWELLING
FROM COLLECTIVE GROUNDS

PLANTING BUFFERS UNIT
FROM SIDEWALK

RECESSED ALCOVE DEFINES
ENTRY TO UNIT

LANDING AND CHANGE IN LEVEL
HELP DEFINE THE UNIT

LIGHTING AND COLUMNS
DEFINE ENTRY PORTAL

LOW WALL DEFINES COLLECTIVE
SEMIPRIVATE AREA

SMALL SETBACK CREATES A SIDEWALK ZONE
BELONGING TO THE RESIDENTIAL CLUSTER

CHANGE IN SURFACE TEXTURE CREATES
FIRST ELEMENT OF TRANSITION

Figure 3.107 Symbolic barriers defining zones of transition.

3. The capacity of the defined space to require the intruder to make obvious his or her intentions

4. The capacity of the inhabitants or their agents to challenge the presence of an intruder and to take subsequent action

Since many of these components work in concept, a successful symbolic barrier is one that provides the greatest likelihood of the presence of all these conditions. By employing a combination of symbolic barriers, it is possible to indicate to entrants that they are crossing a series of boundaries without employing literal barriers to define the spaces along the route.

These symbolic tools for restricting space usage assume particular importance in existing projects that cannot be subdivided into territorial areas. When it is still the intent to make space obey semiprivate rules and fall under the influence and control of inhabitants, the introduction of symbolic elements along paths of access can serve this function.

Opportunities for the use of symbolic barriers to define zones of transition are many. As illustrated in Figure 3.108, the barriers can occur in moving from the public street to the semipublic grounds of the project, in the transition from outdoors to indoors, and in the transition from the semipublic space of a building lobby to the corridors of each floor.

Symbolic barriers can also be used by residents as boundary lines to define areas of comparative safety. Parents may use symbolic barriers to delimit the areas where young chil-

dren may play. Similarly, because symbolic barriers force outsiders to realize that they are intruding into a semiprivate domain, they can effectively restrict behavior to that which residents find acceptable.

Figure 3.108 Zones of transition between public street, project grounds, and building interior.

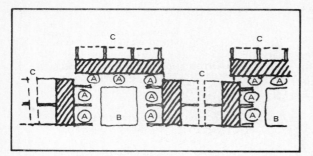

Figure 3.109 Ground areas assigned for particular uses.

Figure 3.111 Common entry to a cluster of buildings.

Locating Amenities Recreational and open-space areas should serve the needs of different groups. An understanding of what different age groups desire of open-space and recreational facilities is essential to the successful use of such areas. The design and location of these areas within the residential environment should follow the demands, capabilities, and expectations of their eventual users.

All areas of the grounds should be defined for specific uses and designed to suit those uses. Figure 3.109 illustrates different uses and users. The areas adjacent to each entry, labeled *A,* have been allocated for the use of 1- to 5-year-olds, with seating for adults. The larger areas in the center of each entry compound, labeled *B,* are provided with play facilities serving 6- to 12-year-olds. The areas labeled C are intended for more passive activity and as decorative green areas. The *C* areas, accessible from the building interiors only, are provided with barbecuing facilities and some seating.

Well-designed recreation facilities improve the security of an area if they provide for activities of a particular group of residents and are adjacent to the residents' interior environs. So designed, the facilities create outdoor zones that are effective extensions of the dwellings. By providing for out-

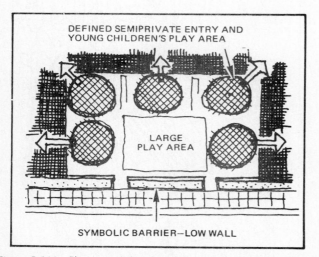

Figure 3.110 Play area defining a buffer to a multifamily building entry.

door activities adjacent to homes, these areas allow residents to assume a further realm of territory and further responsibility.

Children 1 to 5 years in age are most comfortable playing in an outdoor area immediately adjacent to their dwellings, preferably just outside the door in both single-family units and multiple dwellings. Figure 3.110 illustrates such an area. The location of these facilities adjacent to the entry door to the unit and the inclusion of benches for adults further create a semiprivate buffer zone separating the private zone of the residential interior from the more public zones.

In the design of a multifamily residential complex serving many groups of families, each with its own entry to its own building, the buffers that demarcate each entry zone can also define a larger subcluster within the project.

Figure 3.111, which extends the concept shown in Figure 3.110, illustrates the common entry area of a cluster of buildings. Each entry zone is provided with its own tot play area and surrounding seating. The five entries share a common central play facility for the 5- to 12-year-olds that is large enough to accommodate more active play and sufficiently separated from the dwelling units to reduce noise penetration. The large play area is, however, still very much in view of every dwelling.

Play areas for 12- to 18-year-olds should not be located immediately adjacent to home, but neither should they be too far away. They should be large enough to house activities of interest to this age group: basketball, football, handball, dancing.

These teen play areas should not be located in an isolated area of a development, disassociated from dwelling units. This is a common practice (see Figure 3.112) that results in the area's neglect, vandalization, or underuse. Rather, teen play areas should be bordered on three or four sides by the dwellings of residents, as illustrated in Figures 3.113 and 3.114.

The teen area should be provided with occasional

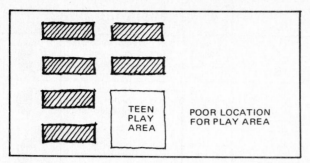

Figure 3.112   Teen play area located at the periphery of a project.

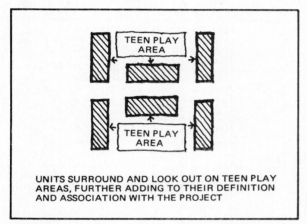

UNITS SURROUND AND LOOK OUT ON TEEN PLAY AREAS, FURTHER ADDING TO THEIR DEFINITION AND ASSOCIATION WITH THE PROJECT

Figure 3.113   Teen play area surrounded by buildings and their entries.

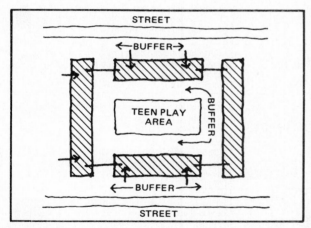

Figure 3.114   Teen play area located with a semiprivate zone.

benches bordering play areas. Benches allow children to gather and watch while only a few play. Children also use the benches for piling extra clothing and for resting after strenuous exercise. Benches give the play area a feeling of stability

and containment. When such areas are defined in this way, they frequently are adopted for social uses in the evening.

Green areas unencumbered by play facilities are the pride of the elderly and usually the thorn in the side of 7- to 15-year-olds, who are prevented from using these areas for playfields. It is therefore important to provide such green areas with protection by judicious placement and use of shrubs and fences. However, as Figure 3.115 illustrates, the best guarantee that these green areas will be respected for their decorative purpose is the provision of adjacent and separate play areas and equipment.

Creating Surveillance Opportunities   Surveillance is a major crime deterrent and a major contributor to the image of a safe environment. By allowing tenants to monitor activities in the areas adjacent to their apartment buildings, tenants in areas outside their homes feel that they are observed by other project residents. Surveillance also makes obvious to potential criminals that any overt act or suspicious behavior will come under the scrutiny of project occupants.

The ability to observe criminal activity may not, however, impel an observer to respond with assistance to the person or defense of the property being victimized. The decision to act will depend on the presence of the following conditions:

1. The extent to which the observer has developed a sense of his or her personal and proprietary rights and is accustomed to defending them

2. The extent to which the activity observed is understood to be occurring in an area within the influence of the observer

3. Identification of the observed behavior as being abnormal to the area

4. Identification on the part of the observer with either the victim or the property being vandalized or stolen

5. The extent to which the observer feels that he or she can effectively alter the course of events being observed

Linking opportunities for surveillance to territorially defined areas will go a long way toward ensuring that many of these required conditions will be satisfied. Figure 3.116 illustrates a territorially defined site plan that is supported by surveillance opportunities.

Designers should position all public paths so that access from public streets to units is as direct as possible. Access arteries should be limited in number to ensure that they are well peopled. They should also be evenly lit. The paths through a project should be designed to allow prescanning before use. There should be no (or few) turns on any artery, and all points along access routes should be observed from the point of origin to the point of destination. When a building is located for the particular use of the elderly, front entrances should face the street and be within 50 feet of the street.

Site Elements Security Capabilities Matrix

Valuation Key

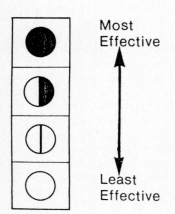

Most Effective

Least Effective

Figure 3.115

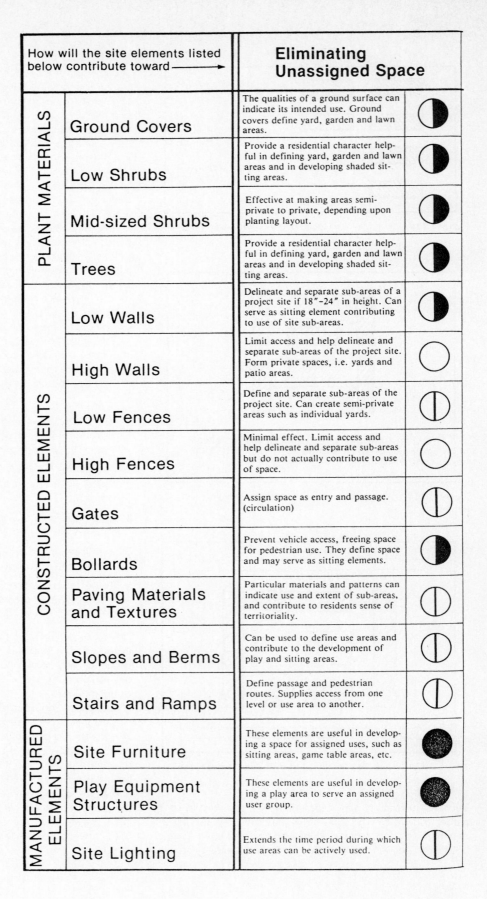

How will the site elements listed below contribute toward ➔		Eliminating Unassigned Space	
PLANT MATERIALS	Ground Covers	The qualities of a ground surface can indicate its intended use. Ground covers define yard, garden and lawn areas.	◐
	Low Shrubs	Provide a residential character helpful in defining yard, garden and lawn areas and in developing shaded sitting areas.	◐
	Mid-sized Shrubs	Effective at making areas semi-private to private, depending upon planting layout.	◐
	Trees	Provide a residential character helpful in defining yard, garden and lawn areas and in developing shaded sitting areas.	◐
CONSTRUCTED ELEMENTS	Low Walls	Delineate and separate sub-areas of a project site if 18″–24″ in height. Can serve as sitting element contributing to use of site sub-areas.	◐
	High Walls	Limit access and help delineate and separate sub-areas of the project site. Form private spaces, i.e. yards and patio areas.	○
	Low Fences	Define and separate sub-areas of the project site. Can create semi-private areas such as individual yards.	⊘
	High Fences	Minimal effect. Limit access and help delineate and separate sub-areas but do not actually contribute to use of space.	○
	Gates	Assign space as entry and passage. (circulation)	⊘
	Bollards	Prevent vehicle access, freeing space for pedestrian use. They define space and may serve as sitting elements.	◐
	Paving Materials and Textures	Particular materials and patterns can indicate use and extent of sub-areas, and contribute to residents sense of territoriality.	⊘
	Slopes and Berms	Can be used to define use areas and contribute to the development of play and sitting areas.	⊘
	Stairs and Ramps	Define passage and pedestrian routes. Supplies access from one level or use area to another.	⊘
MANUFACTURED ELEMENTS	Site Furniture	These elements are useful in developing a space for assigned uses, such as sitting areas, game table areas, etc.	●
	Play Equipment Structures	These elements are useful in developing a play area to serve an assigned user group.	●
	Site Lighting	Extends the time period during which use areas can be actively used.	⊘

Minimizing Penetrability		Maximizing Surveillance		Minimizing Design Conflicts	
Form subtle symbolic barriers when mass planted in a planting bed or panel.	◎	N.A.	○	When mass planted, they subtly define areas, but do not effectively separate conflicting uses or groups.	○
Capable of forming a symbolic barrier (dependent upon planting layout).	◎	Excellent as a means to define areas that require visual surveillance.	○	May provide adequate separation of use areas, however, mid-sized shrubs are more effective.	◎
Form symbolic barriers that may develop into real barriers depending upon plants and layout.	◐	May substantially block surveillance of adjacent areas, depending upon planting layout.	○	Excellent buffer/barrier to separate conflicting areas.	◐
Depending upon plant layout, can form symbolic barriers.	◎	Most large trees will not hinder surveillance, though smaller flowering trees may.	○	Mass plantings of small flowering trees can effectively separate use areas by forming screen barriers.	◐
Can define the project perimeter, and as symbolic barriers limit access to controllable points.	◐	If of sitting element height, they encourage use and activity in adjacent areas, thereby, contributing to surveillance.	◎	Provide good symbolic separation of use areas.	◐
Effective impenetrable barrier.	●	Effectively block visual surveillance.	○	Physically separate conflicting use areas. Can buffer noisy unattractive areas.	●
Symbolic barriers that can be breached, but do minimize penetrability of the project site.	◎	Do not hinder surveillance — if of metal picket or woven wire mesh fence types.	○	Symbolically separate uses, such as an active play area and an adjacent walkway.	◎
Excellent physical barrier, stopping penetrability of the project site.	●	If of chain link or picket construction, will only minimally interfere with surveillance.	○	Separate conflicting uses and activity areas.	●
Effective means to control access while stopping penetrability.	◎	As an access point, gates concentrate traffic, thereby increasing surveillance possibilities.	◎	Do not contribute toward separating conflicting uses of areas.	○
Prevent vehicular access, but permit free access for pedestrians and cyclists.	◎	Do not hinder surveillance.	○	Effectively separate auto and pedestrian traffic.	◎
Form subtle symbolic barriers. Contrasting patterns and materials can define transition zones.	◐	N.A.	○	Subtly define areas, but may not separate conflicts.	○
Form symbolic barriers that discourage penetration of the project site.	◐	Should be sized, shaped and located so as not to block surveillance.	○	Can separate conflicting uses as well as buffer noisy or unattractive areas.	◐
Can form symbolic barriers at the project perimeter or at on-site use areas and at building entrances.	◎	Due to change in grade, may limit surveillance from lower area.	○	May separate conflicts, while not disrupting pedestrian traffic. At certain locations could be extra wide for sitting.	◎
Capable of limiting penetrability of site sub-areas and encouraging outdoor activity of residents.	◎	These elements encourage outdoor activity and residential use, thereby increasing surveillance.	◐	Should be grouped and located to identify with a particular group and function. Adequate facilities must be provided for other groups to avoid conflicts over use. (Applies to both site furniture and play equipment structures.)	○
May function as symbolic barriers minimizing penetrability, particularly when actively used by residents.	◎	Encourages increased surveillance of area if play elements are actively used.	◐		
May limit, at least initially, penetrability of site areas.	◎	Effective and safe levels of lighting greatly aid night surveillance and residents' sense of security.	●	N.A.	○

Figure 3.116

Figure 3.117 Play areas adjacent to a decorative green.

Figure 3.118 Project designed to face surrounding streets and define interior areas as semiprivate.

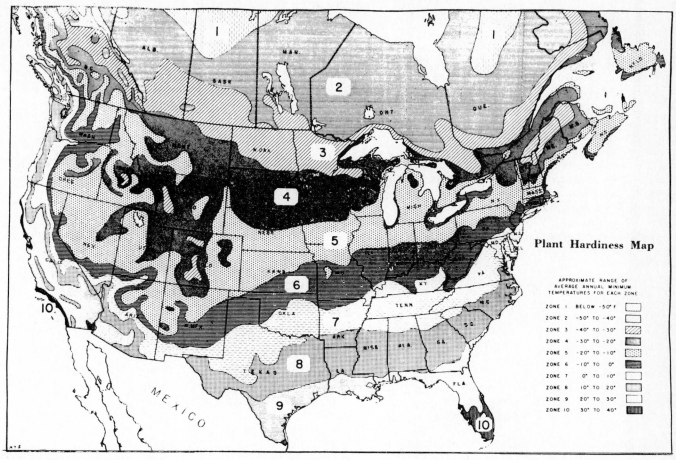

Figure 3.119 Plant hardiness map.

Plant Hardiness Map The plant hardiness zone map in Figure 3.119 will be helpful in determining which of the recommended plants will grow well in a particular area. The map divides the country into a series of climatic zones based on the average minimum winter temperature. The basic hardiness zones were developed by a committee of the American Horticultural Society, and the map itself was developed and published by the U.S. Department of Agriculture.

To use the map, one locates the climatic zone in which the particular project is located. If the climatic zone number is within the range of zone numbers cited in the list of recommended plants, the plant will be suitable for the particular location.

Especially in northern zones, the variety of hardy plants becomes limited. Only a few species can be included in the concise lists of recommended plants. A knowledgeable landscape nurseryman will be able to suggest additional suitable hardy plants.

Ground Covers

Description Ground covers are surface-growing plants that seldom achieve a height of more than 12 inches. They include such plants as ivy, pachysandra, vinca, or myrtle.

Security Aspects These low-growing plants can be used to define a separate specific area on a site as well as to help establish the boundaries of a site. When planted in a bed or panel, they present an attractive symbolic barrier, or buffer area. When planted in mass, these plants can also be used to "fill in" large, vacant, and anonymous areas for which no specific use is practical. Used in this manner, they neutralize the space, leaving it available as a visual experience but clearly indicating it is not to be walked on.

Examples

Baltic Ivy

- A tough, reliable, and attractive ground cover
- Has shiny dark green leaves (evergreen)
- Establishes itself readily — in either sunny or shady locations
- Excellent on banks or slopes

Japanese Spurge

- A very attractive, medium-textured evergreen ground cover
- Well suited to partial shade to full shade locations (but not full sun)
- Flowers in small white spikes in early May

Periwinkle

- An excellent low evergreen ground cover
- Has fine-textured (small), dark green lustrous foliage
- Well suited for either sunny or shady locations
- Excellent for holding soil on banks and slopes
- Small lilac-blue flowers in late April

Recommended Ground Covers*

Proper name	Common name	Type	Remarks	Hardiness zone*
Hedera helix "Baltica"	Baltic ivy	Evergreen	Shade or sun, mows well	4–9
Euonymus fortunei	Wintercreeper	Semievergreen	Spray to prevent scale attack	5–10
Pachysandra terminalis	Japanese spurge	Evergreen	Shaded areas only	5–8
Vinca minor	Periwinkle, myrtle	Evergreen	Sun or partial shade, mows well	5–10

*There are also a number of low junipers (*Juniperus horizontalis, J. subina, J. chinensis,* and *J. conferta*) which make excellent ground covers in the height range of 1 to 2 feet (12 to 15 inch plants should be planted 18 inches on centers).

Low Shrubs

Description   Low shrubs are bushy plants that do not grow beyond 3 feet in height. They consist of several woody stems rather than a single trunk, may be spreading, and are covered with either evergreen or deciduous foliage.

Security Aspects   Low shrubs, when used properly, can be an important element in security planning as well as add warmth and texture to a site. They can assist in reducing a site's penetrability, in removing design conflicts, and in assigning space for particular uses. When planted closely together, they form a tightly knit symbolic barrier that does not limit surveillance. This barrier can be used to define garden and lawn areas and to buffer sitting areas from more active uses, thus minimizing conflicts over use. Low shrubs planted along a site's perimeter at points where access is not desired can structure access to the site. They are subtle in their intent as opposed to constructed architectural elements.

Examples

Andorra Juniper

- A hardy and reliable shrub compact and horizontal in growth
- Its needlelike evergreen foliage is feathery in texture and medium green in color
- Turns an attractive purple during fall and winter
- An excellent shrub for holding soil on slopes and banks

Three-Spine Barberry

- A particularly hardy and attractive evergreen shrub
- Light to medium green foliage with scatterings of reds and yellows in fall and winter
- ⅜-inch-long spines appear in groups of three at regular intervals on stems and branches
- Can form an excellent low hedge

Japanese Holly

- A rugged and easily grown broadleaf evergreen shrub
- A dense shrub with lustrous dark green foliage
- This plant responds well to pruning and makes an excellent hedge approximately 2½ to 3 feet in height

Recommended Low Shrubs

Proper name	Common name	Type	Remarks	Hardiness zone
Berberis triacanthophora	Three-spine barberry	Evergreen	May be sheared	6
Euonymus alatus "compactus"	Dwarf-winged euonymus	Deciduous	May be sheared	5–8
Hex crenata	Japanese holly	Evergreen	Sheared periodically to keep within this height range	6–9
Juniperus horizontalis	Plumosa Andorra juniper	Evergreen		4–11
Taxus "repandens" baccata	Spreading English yew	Evergreen		4–7

Mid-Sized Shrubs

Description   Mid-sized shrubs can reach a height of 6 to 10 feet within 5 to 10 years. Their foliage can be either deciduous or evergreen, and their woody stems may be thorny. Some of these shrubs may be used in their natural form in small groupings, or closely spaced and sheared into a hedge.

Security Aspects   Shrubs form a substantial symbolic barrier with the potential of developing into a formal, impenetrable barrier. A few of these plants have thorns or spines which aid the plants in their own defense and establish a convincing barrier. Since shrubs with thorns can be a hazard, they should be planted at least 3 to 4 feet from walkways and other locations which are heavily used by residents.

These shrubs may grow tall enough to substantially block visual surveillance of site areas, but can be kept lower with pruning if they present a security hazard. Where space is at a premium, these plants may be grown as a hedge, effective at separating use areas over which there may be conflicts. Where more space is available, these shrubs may be used in their natural unsheared state to form an effective screen between use areas. Shrubs of this height may also be used to form screens defining semiprivate to private front and rear yard areas. They are also effectively used in small groupings to landscape lawn, garden, and yard areas.

Examples

Burford Holly

- Hardy and attractive shrub
- Has glossy dark green foliage (evergreen)
- Large, bright red berries in fall and winter
- Excellent in specimen and hedge plantings

Ibolian Privet

- Vigorous, quick-growing, and hardy
- Attractive evergreen foliage
- Small balck berries in fall and winter
- Forms excellent hedge and screen plantings

Firethorn

- Particularly hardy and rugged shrubs (with thorns)
- Numerous clusters of orange berries in fall and winter
- Foliage deciduous to evergreen depending upon region in the United States
- Can be planted as specimens, against walls, or to form a barrier hedge

Recommended Mid-Sized Shrubs

Proper name	Common name	Type	Remarks*	Hardiness zone
Berberis julianae	Wintergreen barberry	Evergreen	(H) spines	5–8
Euonymus alatus	Winged spindle tree	Deciduous	(H)	3–8
Hex cornuta Burfordii	Burford Chinese holly	Evergreen	Berries, spines	6–7
Hex crenata microphylla	Little-leaf Japanese holly	Evergreen	(H)	
Ligustrum ibolium	Ibolium privet	Deciduous	(H)	4–9
Prunus laruocerasus "schipkaensis"	Cherry-laurel	Evergreen	(H)	6–9
Pyracantha coccinea lalandei	Scarlet firethorn	Deciduous	(H), thorns, flowering berries	5
Viburnum dentatum	Arrowwood	Deciduous	(H)	2

*(H) indicates that the shrub is suitable for use in forming hedges.

Trees

Description   Characterized by a single stem or trunk, trees are woody plants which grow to a height of at least 10 feet or taller. They may be divided into two categories — small trees, those under 25 feet in height (many of which have conspicuous flowers), and large trees, most of which mature to a height of 40 to 60 feet and are noted for their overhead canopy and shade-projecting capabilities.

Security Aspects   Trees, with their heavy trunks and large canopies, can form an effective symbolic barrier when spaced 20 to 25 feet apart along a project property line. When the trees' vegetative canopies are above eye level, they do not hinder surveillance.

Trees also enhance sitting areas since their shade during the warm months of the year encourage people to use outdoor seating. This use provides neighbors with an opportunity to get to know and recognize one another and promotes informal surveillance on the site.

Large trees are particularly well suited for defining the limits of areas such as on-site playfields and adjacent walkways. When so used, they should be kept 10 feet or so away from play equipment or play courts so that lower branches cannot impede the play.

The shrubbier, smaller trees can also be used to define areas. They are particularly useful when a barrier or visual screen is desired, such as around a services area. Some species can even be sheared into a hedge form.

Examples of Large Trees

Bradford Pear

- A particularly hardy and low-maintenance shade tree
- Rounded pyramidal in form with dense foliage, turning red to glossy scarlet in the fall
- White flowers 1 inch in diameter appear in early May
- Highly regarded as an urban street tree

Japanese Pagoda Tree

- A shade tree which is wide-branching in habit with a rounded canopy

- Fine-textured, bright green foliage
- Valued for its large clusters of fragrant yellow-white flowers which appear in August
- The flowers become yellow pods which remain through the winter
- A street or ornamental shade tree adapted to city conditions

Examples of Small Trees

Crabapple

- The canopy of this tree is rounded in form with dense, vigorous foliage
- Flowers are 1½ inches in diameter, rose in color, and fragrant, appearing in early May
- The colorful fruit is orange to red in color, ¾ inch in diameter, and appears in late August to early October

Lavalle Hawthorn

- This tree is broadly upright in habit, its dense branches are twiggy and horizontal with spines to 2 inches long
- Foliage is a gloss green, turning bronzy red in fall
- White flowers (¾ inch in diameter) appear in late May
- Orangy-red fruits ⅝ inch in diameter are borne in the fall
- Forms excellent hedges and barrier plantings

Saucer Magnolia

- Graceful, light gray trunks and branches with lush, bright green foliage
- White or pink flowers (5 to 10 inch diameter) appear in May
- Cucumber-like pods with red seeds appear in early fall
- Withstands city conditions well, but will not withstand heavy abuse
- Plant as a flowering specimen in somewhat protected locations

Recommended Large Trees

Proper name	Common name	Remarks	Hardiness zone
Acer platanoides "Summershade"	Norway maple	Difficult to grow grass underneath	3–8
Liquidambar styraciflua	Sweet gum	Rather difficult to transplant in large sizes	4–9
Pyrus calleryana Bradford	Bradford pear		5–9
Sophora japonica (Regant)	Japanese pagoda tree	Flowers late summer. Pods remain on tree*	4–9
Tilia cordata	Little-leaf linden (XP110)	Very hardy	3–9
Zelkova serrata	Japanese Zelkova	Grows fast, close in shape to elm	5–9

*Pagoda trees are used extensively in urban areas, although the pods are considered poisonous. However, before falling, they deteriorate to the extent that they are not attractive to children as playthings.

Recommended Small Trees*

Proper name	Common name	Remarks	Hardiness zone
Crataegus lavallei	Lavalle hawthorn	Flowers late May, red fruits remain on tree through winter	4–7
Crataegus phaenopyrum	Washington hawthorn	Fruits remain all winter	4–7
Magnolia soulangiana	Saucer magnolia	Large flowers appear before leaves	5–10
Magnolia stellata	Star magnolia	Large flowers appear before leaves	5–9
Malus baccata	Siberian crab apple	White, early May, fruit red or yellow	2–9
Malus dorathea	Dorathea crab apple	Round shaped	4–9
Malus "Snowdrift"	Snowdrift crab apple	Vase-shaped	4–9
Prunus cerasifera "Atropurpurea"	Purpleleaf plum		4–9

*The above trees are deciduous. Needle-leaf evergreen trees are difficult to grow under city conditions. (*Pinus nigra* and *Tsuga caroliniana* are the best choices).

Planting Procedures

1. Trees should be well-formed nursery-grown stock of good health, without evidence of disease or insects. Trunks should be wrapped with a quality tree-wrapping paper. The root ball should be unbroken and of good quality soil, encompassing the entire root system and wrapped in burlap secured with cord (not wire).

2. Planting should be scheduled from March 15 to June 1 and from September 15 to December 1. Transplanting during these seasonal periods greatly increases the chances of success, especially when performed by relatively inexperienced project maintenance staff.

3. Large trees should be planted no closer than 15 to 20 feet to a building or another large tree. Small trees should be planted no closer than 10 to 15 feet.

4. Planting holes should be dug 2 times as wide and 1½ times as deep as the root ball, leaving a small mound of undisturbed earth on which to rest the root ball. Planting hole sides should be etched with a pick or shovel to promote root growth and drainage.

5. The tree should be positioned within the planting hole so that the trunk is straight and so that its base is at the finished grade at its new location.

6. Soil to be backfilled should be well mixed with soil conditioners, fertilizers, and possibly lime. Check with the local agricultural extension agent for specific recommendations.

7. Backfilled soil should be tamped in place in layers, filling the hole level with the top of the root ball. Tamp the backfilled soil firmly again, and form a shallow saucer around the tree to aid in watering.

8. One-third of the leaf area should be pruned in such a manner as to retain the natural form of the tree, then sprayed with a waterproofing spray.

9. The tree should be staked in accordance with good nursery practices. In most urban locations it is recommended that the tree trunk be protected by a wire mesh tree guard topped by a collar of rubber hosing, particularly for the first few years.

10. The area within the watering saucer should be mulched to a depth of 3 inches with pine bark, wood chips, or other organic mulch.

11. Immediately after planting, and again 2 days later, the tree should be watered thoroughly—slowly and gently—to avoid damage to the soil structure. The use of sprinklers is recommended. Trees should be watered for the next 3 months. Every watering should last from 2 to 3 hours, depending upon environmental conditions.

12. Finally, all excess soil and debris should be removed from the planted area, so that it is left neat and orderly.

Low Walls

Description   Generally between 18 and 36 inches in height, low walls are upright structures of masonry or wood construction. While they are not impenetrable, they do form excellent symbolic barriers.

Security Aspects   Low walls can be used to define the project perimeter, channeling pedestrian movement and limiting access to predetermined, controllable points. As a barrier they may define and separate on-site use areas such as private yards, entry and sitting courts, and playgrounds, and they are especially appropriate for this function when space is limited. If constructed between 18 and 36 inches in height, low walls can be casual sitting elements contributing to the use, activity, and surveillance of adjacent areas.

Examples

Concrete Wall (Poured in Place)

- Requires minimal maintenance
- Has the longest potential useful life—and the ability to withstand abuse
- Initial construction costs are relatively high
- Can be quite handsome in appearance

Brick Wall

- Attractive in appearance
- With moderate care will serve a long useful life
- A good value in terms of initial cost, appearance, and maintenance requirements

Concrete Block Wall (Stucco Surfaced or Painted)

- Low initial cost compared with other masonry walls
- Appearance can be quite utilitarian (especially painted walls)
- Requires more frequent maintenance

High Walls

Description   Similar in construction and form to low walls, these upright structures are a minimum of 4 feet in height, and are most frequently built at a height of 6 feet.

Security Aspects   At a height of 4 feet, walls can form a substantial physical separation, while at 6 feet they form a relatively impenetrable barrier as well as offering visual privacy and a buffering of adjacent areas.

With these characteristics, high walls can define the project permimeter and control penetrability into the project. They most effectively buffer off-site areas, such as noisy expressways or industry.

Used less often within the project grounds, high walls can buffer and separate areas of conflicting use, such as a maintenance yard and a quiet sitting area for elderly residents. Walls can also provide needed separation between two intensively used play areas, such as one for elementary-school-age children and one for teenagers. In examples such as these, the buffering or separating wall must be located with care so that it does not create a security hazard by eliminating surveillance opportunities from adjacent residential buildings and pedestrian walkways.

Examples

Concrete Wall (Poured in Place)

- Minimal maintenance required
- Will withstand abuse and has a long useful life potential
- Construction costs are only slightly higher than for brick construction
- Attractive appearance depending upon wall design

Brick Wall

- Attractive in appearance
- Moderate maintenance requirements, usually performed on a 10- to 15-year cycle
- With this moderate care, brick walls will serve a long useful life

Concrete Block Wall (Painted Finish)

- Lowest initial cost of the masonry wall types

- Requires more frequent maintenance
- Utilitarian in appearance (can be improved by stucco surfacing)
- Where walls are background elements, block walls may be the most economical and appropriate choice

Low Fences

Description

Usually between 36 and 42 inches in height, low fences are upright structures of wood or metal posts and rails with pickets or woven wire mesh. Traditionally, they have been used to indicate and maintain a property boundary, as well as serve as a barrier offering protection and/or confinement. Many types are now being standardized at 36 inches in height.

Security Aspects

Low fences can contribute to a site's security in several ways. They can be used to define a project's perimeter and to guide entry to fixed, observable points. They can also define and separate use areas on a site. Because of their low height, observation between separated areas is still possible.

High Fences

Description

Identical in construction and form to low fences, these structures are a minimum of 4 feet in height, though they are usually constructed to a height of 6 to 8 feet, increasing their effectiveness as a relatively impenetrable barrier.

Security Aspects

High fences are most frequently used to control access along a project perimeter by forcing those who wish to enter to pass to a controllable entranceway. High fences are used less frequently within the project grounds, but may form partial screens or complete enclosures, to contain activity on basketball and handball courts, minimizing conflict with adjacent use areas.

High fences also frequently enclose maintenance yards, keeping small children away from the hazards of the area and minimizing vandalism.

For the above uses, metal picket or chain link fences are most often used because they do not obstruct visibility and permit surveillance of adjacent areas. Where used to create private patios or terraces adjacent to low-rise residential buildings, they should be augmented to create a visual barrier as well.

Examples

Wrought-Iron Fence

- Attractive in appearance
- Difficult to vandalize

- Moderate maintenance requirements
- High initial cost is offset by potentially long useful life

Tubular Steel Fence

- Reasonably attractive in appearance
- A good value in terms of initial cost, low maintenance, and a reasonable resistance to vandalism
- With periodic maintenance, useful life approximates that of wrought iron

Chain-Link Fence

- Lowest initial cost
- Least resistance to vandalism
- Least attractive in appearance

Bollards

Description   Bollards are small posts constructed of wood, metal, or concrete. They are usually 12 inches or less in diameter and range from 24 to 30 inches in height.

Security Aspects   Bollards are used to separate and control vehicular and pedestrian traffic. Spaced as much as 5 feet apart, but never closer than 3 feet apart, they can bar vehicular access while permitting unhindered access for pedestrians and bicyclists.

In addition to controlling vehicular access, they can be useful for defining on-site areas such as playgrounds and as casual sitting elements. Both uses should contribute to resident use and activity within an area.

Examples

Wood Bollard

- Economical when 8 by 8 inches square or under
- Reasonably long useful life if pressure-treated
- Can be quite attractive

Concrete Bollard

- Attractive in appearance
- Minimal maintenance
- Long useful life

Pipe Bollard

- Utilitarian in appearance
- Durable and able to withstand abuse
- Well suited for service and delivery area

Construction

These rather basic structures can be constructed in a number of ways. Wooden bollards can be made up of lengths of new

or used railroad ties, telephone poles, or — for long life and minimal maintenance — pressure-treated 8- by 8-inch or 12- by 12-inch timber.

For satisfactory and uniform appearance, timber bollards should be made up in a shop rather than on site. Appearance is important because bollards are usually closely spaced and used in conspicuous places, such as the main entry to a community. Once cut to the proper length and shape (tops of bollards should be sloped so water does not puddle), all edges exposed above grade should be radiused (sanded round and smooth). If untreated timber is used, it should be thoroughly soaked in a wood preservative.

Bollards can also be constructed of heavy-gauge steel or iron pipe. The pipe should be filled with concrete and capped with a formed mortar or threaded metal cap. Concrete bollards can be made up by a concrete precasting plant, or cast on site, using wood or fiber tube formwork. They should be steel-reinforced.

Procedures for installing bollards are much the same as for fence posts. The foundation should extend a minimum of 2 feet 6 inches below grade. Earth at the bottom of the foundation hole should be compacted. Pipe bollards should be set in a concrete foundation with 6 inches of concrete surrounding the pipe. Within the foundation hole, wood and concrete bollards should be set on a compacted earth base and the hole backfilled with compacted gravel, capped with a concrete collar footing a minimum of 12 inches in depth and 12 inches greater in diameter than the bollard at its center. The top surface should be finished grade and formed so that water drains away from the bollard.

All bollards must be set plumb or truly vertical, and each in a group must be set to a consistent height above finished grade for satisfactory appearance. If one is out of line with the others, it is easily noticeable, especially when they are spaced as closely as 3 feet apart.

Bollards should contrast with the surrounding area and be easily seen both day and night. This is possible by selecting a bollard material of contrasting color or texture or by the application of paint. The area should be well lit at night.

Paving Materials and Textures

Description   A variety of paving materials are suitable for use in projects. They fall, in general, into two categories: those that are poured in a liquid state and harden into place, such as concrete or bituminous asphalt, and pavings of small blocks or units such as brick, granite sets, and precast asphalt pavers.

Security Aspects   Variations in paving material, texture, and color can be utilized to establish zones and use areas on a site. At the project perimeter and along entry and walkways, distinctive paving can indicate transitional areas between public streets and sidewalks and semiprivate and private residential areas. Within the project site, selection of a paving material and pattern for a particular use area can clearly indicate the boundary of that area and, by contrasting it with an adjacent area, minimize conflicts over use.

Distinctive paving is one effective way to "assign" or associate site areas with a particular group of residential units or buildings. For example, a paved entry court of a particular material and pattern will give identity and unity to the buildings which front on the court, and contribute to the residents' sense of territoriality. Introducing a selected variety of paving materials and patterns, corresponding to and identifying the use areas and entry zones on the project site, will contribute to the residents' sense of security, as well as add visual interest and variety.

Examples

Asphalt Paving

- Lowest in construction cost
- Requires the most frequent maintenance in the form of periodic resurfacing
- In most applications it is aesthetically unappealing and monotonous

Concrete Paving (Broom-Finished, Scored on 3-ft Grid)

- Very moderate in cost with a long useful life expectancy
- Minimal maintenance required
- Offers a reasonably attractive surface
- Offers a nonslip textured surface

Concrete Paving (Exposed Aggregate Finish)

- Presents a very attractive "pebbled" surface
- Usually costs approximately twice as much as ordinary concrete work
- Minimal maintenance required
- Has a reasonably long useful life expectancy

Precast Pavers (Concrete)

- Cost is only slightly higher than ordinary concrete paving
- Presents a very attractive textured surface similar to brick paving
- Has a long useful life expectancy and requires only minimal maintenance

Hexagonal Pavers (Asphalt)

- Attractive pavers with a slightly resilient surface — pleasant for walking
- Initial cost is somewhat more expensive than concrete but less than brick paving
- Will withstand heavy usage and traffic

Bomanite Paving (Concrete)

- Similar in appearance to brick or stone paving
- Less costly than brick or stone paving
- May deteriorate over time and is somewhat difficult to clean "joints" and to patch surface

Brick Paving

- A very attractive paving material
- It ages gracefully and requires minimal maintenance
- Its relatively high initial cost is offset by its potentially long useful life
- An especially appropriate material in older urban areas

Brick Grid (in Concrete Paving)

- An economical and attractive way to incorporate brick into paving surfaces
- Effective way to pave entrance plazas or courts at community or residential buildings
- If properly constructed will require minimal maintenance and serve a long useful life

Slopes and Berms

Description Slopes and berms are earthen barriers. A slope is an inclined surface which may be gentle, moderate, or steep, depending on its purpose and the site conditions. A berm is a mound of earth with sloping sides, located between areas of approximately the same elevation.

Security Aspects Berms and slopes are most effective on a project site in separating and buffering potentially conflicting use areas such as a playground and an intensively used roadway.

At the perimeter of a project, these earthen barriers can form an effective symbolic barrier that discourages penetration and directs pedestrians to proper entrances to the site.

Within the project site, earthen slopes and berms can define areas set aside for quiet sitting, as well as active play fields. Berms can be incorporated into the design of a lot or elementary-school-age playground (for example, slides can be incorporated into the slope of a berm). When located adjacent to a teen play or athletic court or field, berms and slopes form excellent casual spectator seating, while buffering adjacent residential buildings from noise.

Slopes and berms should be sized, shaped, and located with care so as not to block on-site surveillance.

Examples

3-ft height berm—2:1 slope with groundcover vegetation

- Serves as an excellent means of buffering sitting area from adjacent street and walkway

- With ivy ground cover the berm forms an attractive setting for the seating area
- Ground cover vegetation reduces the frequency of maintenance

3-ft height berm—3:1 slope with grass vegetation

- Low, gentle berms give visual relief to uniformly "flat" urban areas
- Provides an attractive parklike setting
- Reduces the visual and noise impact of adjacent roadways
- Gentle slope permits normal mowing operations

6-ft height slope—2:1 slope with ground cover vegetation

- Slope separates semiprivate residential areas from the public street and sidewalk
- Ground cover vegetation reduces periodic maintenance.

Design and Construction A certain amount of site evaluation and planning should be done before construction begins. There must be enough horizontal distance on the ground to accommodate the slope or berm. Maintenance is a factor in the amount of space required, as well as in considering how steep the slope or berm will be. If the barrier is to be surfaced with grass, the sides of the slope or berm should not be steeper than a 3:1 slope (1 foot of vertical drop for each 3 feet of horizontal distance) since it is difficult to handle mowing equipment on slopes steeper than 3:1. If the slope or berm is to be planted with ground cover plants, the sides of the berm or slope can be at a ratio fo 2:1.*

In certain locations, it may be impossible to introduce a slope because of existing buildings, drainage patterns, and roadways. In these locations, provided there is sufficient horizontal distance, a berm can be constructed as the barrier. In locations with inadequate space, a wall, fence, or hedge would be the logical choice for a barrier.

In constructing berms at most urban housing projects, soil fill will have to be purchased and trucked in. To gather soil material on-site by regrading open areas would be nearly impossible, or very expensive.

Where it is possible to construct slopes, this can generally be accomplished by "cutting and filling" soil in the vicinity. This may also result in more level and usable activity areas adjacent to the newly graded slope.

---

*A slope 3 feet in height at a 3:1 slope would be 9 feet in width; graded to a 2:1 slope would be 6 feet in width. A berm 3 feet in height and graded with 3:1 side slopes would be minimum of 18 feet in width, although an area approximately 25 feet in width would be required to grade the berm into the surrounding area. With 2:1 side slopes, the 3-foot-high berm would be a minimum of 12 feet in width and require an area approximately 17 to 19 feet in width to accommodate the berm.

Generally, if extensive regrading of a site area is contemplated, a grading plan detailing the changes to be made should be prepared by state-registered landscape architects or civil engineers. Among other considerations, they can ensure that underground utilities will not be adversely affected, surface drainage not obstructed, and that mature trees will not be harmed. They can also evaluate, and in many cases eliminate, a need for costly grading structures as retaining walls and outdoor stairs.

3.13 Sewage Disposal

GENERAL DESIGN

This subsection on sewage disposal will enable one to design private (self-contained) sewage disposal systems for residences, camps, summer cottages, schools, factories, hospitals, institutions, and the like for any number of occupants up to the equivalent of fifty persons in residence. For larger systems, a sanitation engineer should be consulted.

Past experience, engineering practice, and bacteriological research have proved that the old-time sewage cesspool is a menace to health and a nuisance. Sanitation engineers agree that all sewage disposal systems must include a septic tank, wherein sewage is changed by the action of anaerobic bacteria into gases and an effluent liquid, which is then rendered harmless by earth leaching, where aerobic bacteria oxidize all obnoxious components.

Influencing Factors A complete sewage disposal system, with all essential and optional elements, is presented below (see Figure 3.121). The final design of a specific installation is influenced by (1) the amount of sewage handled, which is based on the equivalent occupancy, and (2) the character of the soil as expressed by its relative absorption. Both of these factors can be determined for any project by methods described here.

Equivalent Occupancy The amount of sewage to be handled is related to the type of building and its occupancy. The base is the normal amount of sewage obtained under residential conditions per person per 24 hours (see table "Equivalent Occupancy"). That is, in residential service 50 gallons of sewage per person must be treated each 24 hours. Other types of buildings are related to this base by means of a conversion factor. To find the equivalent occupancy in any project, multiply the number of persons occupying the type of building by the conversion factor given in the table.

Equivalent occupancy governs the size of the septic tank and also influences the capacity required in the effluent disposal system.

Relative Absoprtion The porosity or absorption of the soil is a vital factor in design. A simple field method of determining the characteristics of any soil in relation to effluent disposal consists of digging a pit of fixed dimensions and measuring the rate of outflow of water from the pit.

The depth below grade of this test pit varies according to the unit under consideration, as developed in detail under the headings "Leaching Cesspools," "Subsoil Disposal Beds," and "Sand Filters." The size of the pit in which the water test

Equivalent Occupancy*

Type of building	Gallons per person	Conversion factor
Residence	50	1
Camp	25	1/2
Summer cottages	40	4/5
Day schools without showers or kitchens	15	3/10
Factories without showers or kitchens	15	3/10
Day schools with showers and kitchens	30	3/5
Institutions except hospitals	100	2
Hospitals	200	4

* To find equivalent occupancy, multiply the number of persons occupying the type of building by the conversion factor.

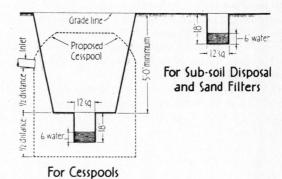

Figure 3.120 Test pits for relative absorption.

is made is always 12 inches square and 18 inches deep, as illustrated in Figure 3.120. Water to a depth of 6 inches (about 3¾ gallons) is poured quickly into this square test pit. The time required for the water to disappear is measured, and one-sixth of this time is taken as the average time for the water level to fall 1 inch. The latter time is the relative absorption factor for the soil, but for convenience it is expressed as rapid, medium, slow, semi-impervious, and impervious, as indicated in the table "Relative Absorption."

Relative Absorption

Time (1-inch drop in minutes)	Relative absorption
0–3	Rapid
3–5	Medium
5–30	Slow
30–60	Semi-impervious
60 and up	Impervious

Selection of Effluent Sewage Disposal System

Conditions	Type of disposal system		
	Leaching cesspool	Subsoil drainage	Sand filter
Relative absorption:			
Rapid	Yes	Yes	No
Medium	Yes	Yes	No
Slow	Yes	Yes	No
Semi-impervious	No	Yes	No
Impervious	No	No	Yes
Available area:			
Large	Yes	Yes	Yes
Moderate	Yes	Yes	Yes
Small	Yes	No	No
Groundwater:			
Below grade	8 feet minimum	2 feet minimum	2 feet minimum
Final disposal of effluent	Not necessary	Required only for semi-impervious soils	Always necessary
Relative initial cost	Low	Medium	High

Design Procedure A tentative layout of the proposed sewage disposal system similar to the diagrammatic plan in Figure 3.121 should be made over a topographic plot plan of the property. Test pits should be dug at the sites of any proposed leaching cesspool or other effluent disposal area and the relative absorption determined.

With these data on the equivalent occupancy known, reference should be made to the table "Selection of Effluent Sewage Disposal System" to select the type of system best adapted to project conditions.

Elements of Sewage Disposal Systems

House Sewer The house sewer extends from the house main to the septic tank. The house main is a continuation of the cast-iron soil line to a minimum of 5 feet outside the foundation. No trap or fresh-air inlet is required in the house main. The house sewer may be solid, glazed clay tile, cement bell-and-spigot pipe, or preferably, cast-iron pipe, laid with filled joints. Always use cast iron within 100 feet of any potable water supply and near trees. Never connect surface drainage lines to the sewage disposal system.

Requirements. Size: 6-inch preferable; minimum, 4-inch. Pitch: 1 inch in 8 feet for 6-inch pipe; 1 inch in 4 feet for 4-inch pipe. Grade: northern latitudes, 1 foot 6 inches minimum below surface; southern latitudes, sufficient depth to cover.

Grease-bearing waste and trap. These are optional elements, used to separate grease and oil from waste. When they are installed, run from the grease-carrying waste in building through a trap to the house sewer.

Septic Tank This is the essential element of a sanitary disposal system. Locate it as far to leeward of the building as possible. Its function is to retain the raw sewage out of con-

tact with air until anaerobic bacteria can break down the solids into gases that escape through vents and an effluent liquid that is subsequently purified by oxidation. Some solids settle as sludge. Construction and operating details are given under the heading "Septic and Siphon Tanks."

Siphon tank. This tank is required in large installations and when a sand filter is used; it is desirable but not essential with small septic tanks. It functions to collect effluent from the septic tank and periodically to discharge it to the effluent disposal system.

Sludge drain and pit. These elements are optional. They serve to draw sludge from the septic tank without interrupting its operation for cleaning. The drain is similar in construction to the house sewer. For details, see the section "Distribution Boxes."

Effluent Disposal There are three principal types of effluent disposal systems, the choice being governed by soil conditions and topography. All are designed to permit the effluent to come in contact with air and soil where it may be oxidized and rendered harmless by aerobic bacteria.

Septic Tanks and Dry Wells

ONE-COMPARTMENT SEPTIC TANK

Design for a typical 500-gallon concrete septic tank should provide adequate volume for settling and for sludge and scum storage and for access for cleaning. The structural design and materials used should be in accordance with generally accepted good engineering practice, providing a sound, durable tank that will safely sustain all dead and live loads and liquid and earth pressure involved in each case. The tank must be located so that it will achieve the minimum distances shown in the accompanying table. See also Figure 3.122c, which illustrates a 750-gallon tank.

Minimum Distances (in feet)

From	To			
	Septic tank	Absorption field	Seepage pit	Absorption bed
Well	50	100	100	100
Property line	10	5	10	10
Foundation wall	5	5	20	5
Water lines	10	10	10	10
Seepage pit	6	6		
Dry well	6	20	20	20

TWO-COMPARTMENT SEPTIC TANK

Liquid capacity of the tank should be based on the number of bedrooms proposed or that can be reasonably anticipated in the dwelling and should be at least as shown in the accompanying table. The design of such a tank is shown in Figure 3.122b.

Minimum Capacities for Septic Tanks*

Number of bedrooms	Minimum liquid capacity below outlet invert (gallons)
Two or less	750
Three	900
Four	1000
For each additional bedroom add	250

*These capacities provide for the plumbing fixtures and appliances commonly used in a single-family residence (automatic sequence washer, mechanical garbage grinder, and dishwasher are included).

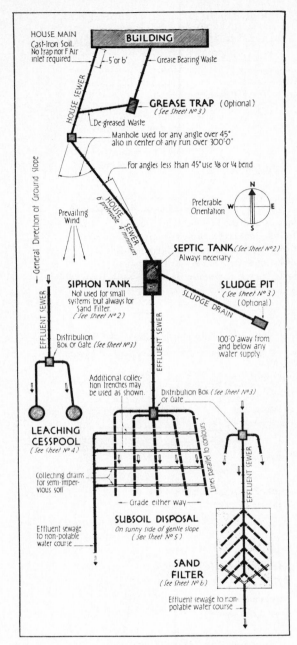

Figure 3.121 A key diagram of a sewage disposal system.

Effluent sewer. This element is common to all systems and is a closed sewerage line similar in construction and size to the house sewer, extending from the septic or siphon tank through a distribution box or gate to the chosen type of effluent disposal element. The minimum pitch may be 1 inch in 16 feet.

Distribution box or gate. This device serves to distribute the effluent to one part or another of the effluent disposal system in order to "rest" the part not in use. Details are given in the section "Distribution Boxes."

The choice of effluent disposal system is governed by factors included in the tables "Equivalent Occupancy" and "Relative Absorption." Selection is determined from the table "Selection of Effluent Sewage Disposal System" and from the sections "Leaching Cesspools." "Subsoil Disposal Beds," and "Sand Filters."

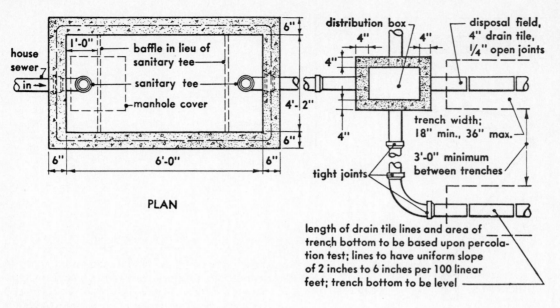

PLAN

house sewer in

6"
1'-0"
baffle in lieu of sanitary tee
sanitary tee
manhole cover
6"
4'-2"
6"
6'-0"
6"

distribution box
4"
4"
disposal field, 4" drain tile, 1/4" open joints

4"

trench width; 18" min., 36" max.

tight joints

3'-0" minimum between trenches

length of drain tile lines and area of trench bottom to be based upon percolation test; lines to have uniform slope of 2 inches to 6 inches per 100 linear feet; trench bottom to be level

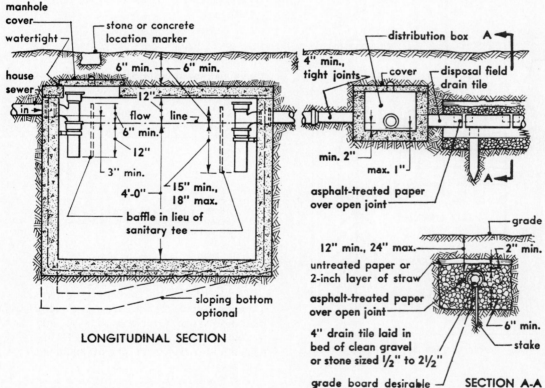

manhole cover watertight
stone or concrete location marker

house sewer in

6" min. — 6" min.
12"
flow   line
6" min.
12"
3" min.
4'-0"
15" min., 18" max.
baffle in lieu of sanitary tee

sloping bottom optional

LONGITUDINAL SECTION

distribution box   A

4" min., tight joints
cover
disposal field drain tile

min. 2"
max. 1"

asphalt-treated paper over open joint

A

grade
2" min.

12" min., 24" max.
untreated paper or 2-inch layer of straw
asphalt-treated paper over open joint
4" drain tile laid in bed of clean gravel or stone sized 1/2" to 2 1/2"
grade board desirable

6" min.
stake

SECTION A-A

Figure 1.22*a*   Concrete 750-gallon septic tank.

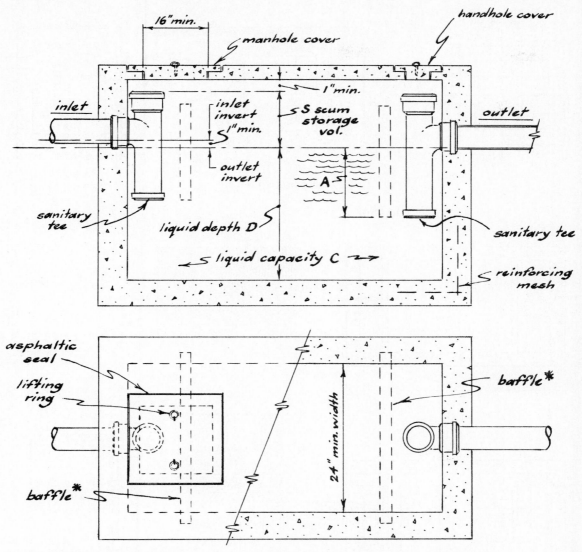

Figure 3.122b Two-compartment septic tank. Section *A* is approximately 40 percent of liquid depth *D*. *D* not less than 3-inch depth greater than 6 feet should not be considered in tank capacity. *S* is not less than 15 percent of liquid capacity *C*. In plan, below, baffles are optional to submerged inlet and outlet sanitary tee.

SEPTIC AND SIPHON TANKS

The septic and siphon tanks described here are of reinforced concrete and can be constructed by any competent contractor without requiring the use of any patented or manufactured element other than the automatic siphon, which is an essential part of a siphon tank. Septic tanks are made in commercial units and are available in all parts of the United States. They are built of steel, precast concrete, and other materials. The use of commercial septic and siphon tanks eliminates the need for detailed design of these units as presented here, but

the selection of the proper size of a commercial unit is indicated by the data presented.

Operation   Raw sewage from the house sewer enters the septic tank, where by a submerged intake it reaches the liquid in the tank below the overflow level. The liquid in the septic tank quickly forms three distinct layers: solid matter or sludge settles to the bottom, effluent sewage forms the main liquid content in the middle, and the upper stratum is a scum that serves to keep air out of contact with the effluent sewage and permits anaerobic bacterial action or septicization to take place. Most of the suspended solid matter is changed by

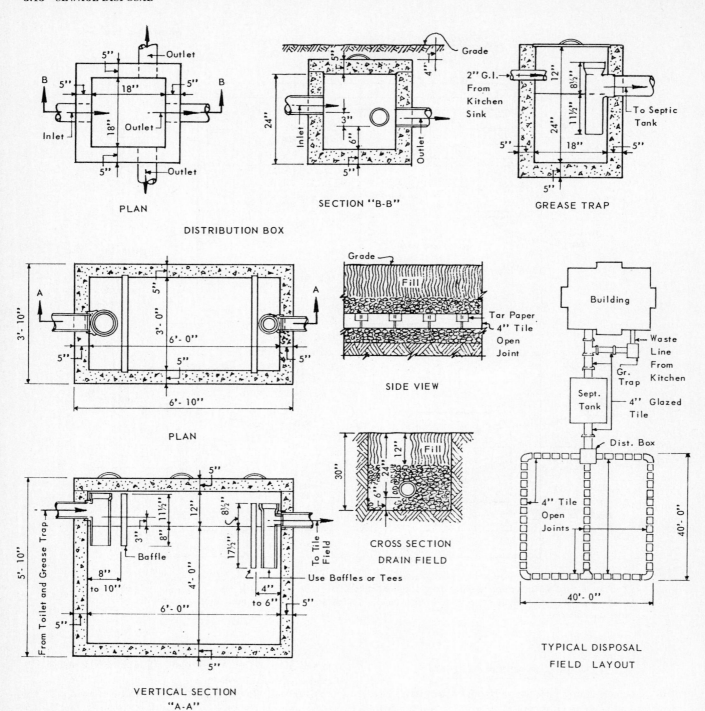

PLAN

SECTION "B-B"

GREASE TRAP

DISTRIBUTION BOX

PLAN

SIDE VIEW

VERTICAL SECTION
"A-A"

CROSS SECTION
DRAIN FIELD

TYPICAL DISPOSAL
FIELD LAYOUT

Figure 3.122c

Equivalent Occupancy

Type of building	Gallons per person	Conversion factor
Residence	50	1
Camp	25	1/2
Summer cottages	40	4/5
Day schools without showers or kitchens	15	3/10
Factories without showers or kitchens	15	3/10
Day schools with showers and kitchens	30	3/5
Institutions except hospitals	100	2
Hospitals	200	4

this action into (1) gases that escape through vents provided for the purpose and (2) effluent sewage that overflows either directly or through the siphon tank into the effluent sewer and then to the effluent disposal system.

The sludge that forms at the bottom of the septic tank must be removed periodically to avoid filling the tank with solid matter. In large installations, where interruption of the operation of the septic tank for cleaning purposes is undesirable, a sludge drain and sludge pit should be provided to permit removal of sludge while the tank is in continuous operation.

When an effluent disposal system of the sand filter type is used, when the system is designed for 1000 gallons or more daily capacity, and preferably in all residences, the septic tank should be equipped with a siphon tank. However, this unit is not actually required in small installations using leaching cesspools or subsoil disposal beds.

The siphon tank functions to collect overflow from the septic tank and to discharge it periodically through the action of the automatic siphon into the effluent sewer and disposal system. This permits the disposal units to absorb the effluent intermittently and prevents saturation of the disposal beds.

Design The size of a septic tank, and therefore the size of its related siphon tank, is governed wholly by the number of gallons of sewage to be treated per 24 hours. This can be determined from the data relating to equivalent occupancy given in the section "General Design," which is repeated here for convenience.

Rule 1 To find the equivalent occupancy in any project, multiply the number of persons occupying the type of building by the conversion factor given in the table "Equivalent Occupancy."

Rule 2 To find the dimensions and construction details for any reinforced-concrete septic tank and siphon tank as detailed in Figure 3.123, refer to the table "Selection and Design of Septic and Siphon Tanks," and find in the first column the equivalent occupancy nearest to that calculated for the project. Read horizontally to the right for all dimensions not given directly on the drawings.

Rule 3 To ascertain the capacity of any commercial septic tank, proceed as in Rule 2 but find in the table the capacity in gallons that corresponds to the equivalent occupancy of the project and select a unit guaranteed by the manufacturer to treat that quantity of sewage per 24 hours. The siphon tank adapted to the manufactured septic tank will be indicated by the manufacturer's own data.

Location When a septic tank is equipped with a sludge drain and pit, it can be buried and its manhole cover identified merely by the position of the protruding vent or vents.

Selection and Design of Septic and Siphon Tanks

Equivalent occupancy	Capacity (gallons)	Septic tank Length A	Septic tank Width B	Septic tank Air space C	Septic tank Liquid depth D	Siphon tank Length E	Siphon tank Width F	Siphon tank Depth G	Siphon Size L	Siphon Drawing depth M	Concrete thickness Walls J	Concrete thickness Top I	Concrete thickness Bottom K
1–4	325*	5'0"	2'6"	1'0"	3'6"	†							
5–9	450	6'0"	2'6"	1'0"	4'0"	3'0"†	2'6"†	3'0"†	3"	1'6"	6"	4"	6"
10–14	720	7'0"	3'6"	1'0"	4'0"	3'6"†	3'6"†	3'0"†	3'	1'6"	6"	4"	6"
15–20	1000	8'0"	4'0"	1'0"	4'0"	4'0"	4'0"	3'0"	4"	1'8"	6"	4"	6"
21–25	1250	9'0"	4'6"	1'0"	4'3"	4'6"	4'6"	3'0"	4"	1'8"	7"	5"	6"
26–30	1480	9'6"	4'8"	1'3"	4'6"	4'8"	4'8"	3'6"	4"	2'2"	8"	5"	6"
31–35	1720	10'0"	5'0"	1'3"	4'8"	5'0"	5'0"	3'6"	4"	2'2"	8"	5"	6"
36–40	1950	10'6"	5'3"	1'3"	4'9"	5'3"	5'3"	3'6"	4"	2'2"	9"	5"	6"
41–45	2174	11'0"	5'6"	1'3"	4'10"	5'6"	5'6"	3'6"	5"	2'2"	9"	5"	6"
46–50	2400	11'6"	5'9"	1'3"	5'0"	5'9"	5'9"	3'6"	5"	2'2"	9"	5"	6"

*Smallest size recommended. (Capacity of tanks is based on 50 gallons per equivalent occupancy per 24 hours)
†Siphon tank not essential for septic tanks under 1000-gallon capacity; rarely used on the smallest size.

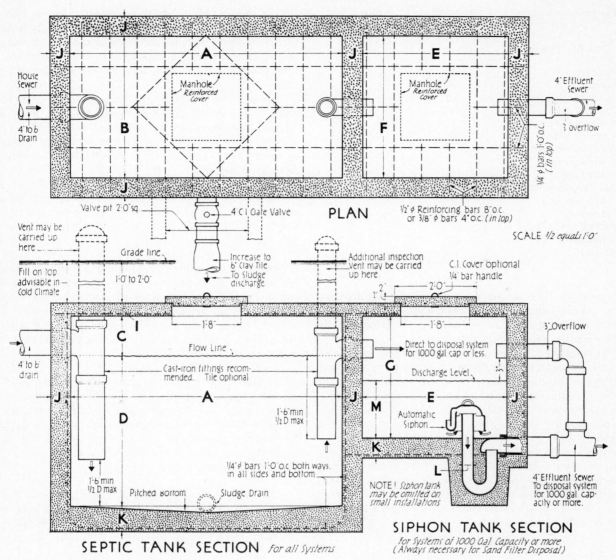

Figure 3.123 Septic tank section.

However, the manhole for access to the sludge drain gate valve should be carried near the surface so that it can be exposed conveniently for operating the valve. When no sludge drain and pit are provided, the septic tank should be so located that the covering earth may be removed periodically without disfiguring the property. The same precautions also pertain to the manhole cover for the siphon tank.

Maintenance Data   The owner of a septic tank should be provided by the designer with a written memorandum containing the following data: (1) A plan indicating the exact location of the septic and siphon tank manholes and sludge drain gate valve manhole, when used. (2) Advice for inspection of the septic tank each spring and fall by removing the vent caps and testing the depth of the sludge by means of a rod or a plumb bob. During severe weather the vents should be examined periodically to see that excess flowing on the interior has not obstructed their operation. (3) Instructions that whenever the sludge level appears to reach the low end of the intake or discharge pipes or there are any signs of flooding, the septic tank should be immediately cleaned or the sludge drawn off to the sludge pit. (4) Advice that whenever the siphon tank requires cleaning, the manhole cover of the tank should be removed and inspected and the automatic siphon cleaned.

DISTRIBUTION BOXES

This section provides data for the design of three types of units: grease traps, sludge pits, and distribution boxes. Of these, grease traps and sludge pits are optional units that may

or may not be required by the system, according to the conditions defined in the preceding sections. Distribution boxes are required in almost all systems.

Grease Traps   The function of a grease trap is to separate grease and oil from kitchen, laundry, and other specialized wastes and to prevent them from entering the sewage disposal system. Grease and oil may interfere with the formation of a proper scum in the septic tank and may clog or reduce the porosity of leaching cesspools, subsoil disposal beds, and sand filters. The use of a grease trap is therefore recommended in the majority of installations, but it is not a mandatory requirement in small installations where no great quantity of grease or oil occurs.

The grease traps described here are of concrete construction for use outside the house. Such a unit is not required if a metal grease trap has been installed indoors in waste lines carrying grease or oil. Indoor traps offer greater convenience for cleaning and may be used in small- or medium-sized projects if the odor arising during their cleaning operation is not a serious objection.

Complete design data are contained in Figure 3.124. The size of the grease trap does not vary materially with the size of the building it serves. However, when the quantity of waste causes rapid flow, it is advisable to use a retangular trap with a baffle.

Owners should be advised to clean grease traps frequently. Therefore, the trap should be located at a point where the loose earth over the cover may be removed and replaced without unduly impairing the appearance of the property.

Within reason, the grease trap should be located as far as possible from the building and to leeward to minimize objections to the odor that always follows a grease trap cleaning operation.

Sludge Pits   As indicated in the section "General Design," the use of a sludge pit depends upon the need for cleaning septic tanks without interrupting their operation. The location of a sludge pit is indicated in Figure 3.121.

Since a sludge pit must be of such a size that it has a capacity equivalent to the septic tank it serves, refer to tbe section "General Design" for methods of determining the size required and to the table "Sludge Pit Dimensions."

Sludge Pit Dimensions

Equivalent occupancy	Capacity (gallons)*	Length A	Width B	Air space C	Liquid depth D
1–4	325	5'0"	2'6"	1'0"	3'6"
5–9	450	6'0"	2'6"	1'0"	4'0"
10–14	720	7'0"	3'6"	1'0"	4'0"
15–20	1000	8'0"	4'0"	1'0"	4'0"
21–25	1250	9'0"	4'6"	1'0"	4'3"
26–30	1480	9'6"	4'8"	1'3"	4'6"
31–35	1720	10'0"	5'0"	1'3"	4'8"
36–40	1950	10'6"	5'3"	1'3"	4'9"
41–45	2175	11'0"	5'6"	1'3"	4'10"
46–50	2400	11'6"	5'9"	1'3"	5'0"

*Sludge pits should be of the same capacity as the septic tanks they serve.

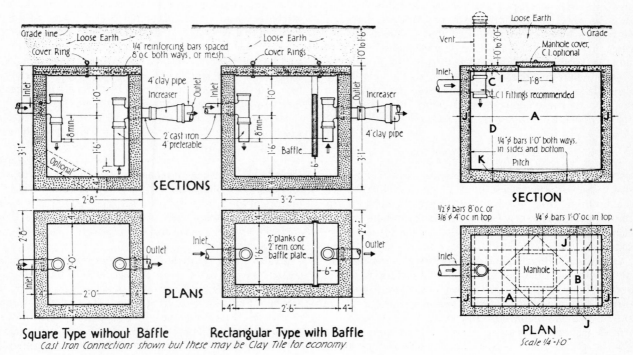

Figure 3.124   Grease traps and sludge pit.

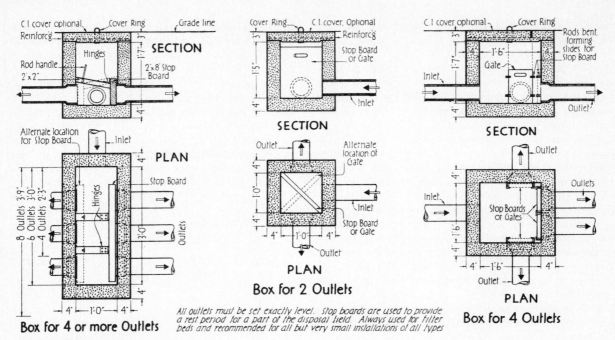

Figure 3.125 Distribution boxes.

Distribution Boxes The location and general use of distribution boxes is indicated in Figure 3.121 and in the section "General Design," as well as in each of the following sections relating to effluent disposal methods.

Distribution boxes function to control and direct the flow of effluent sewage from the effluent sewage main to various parts of the effluent disposal system, permitting part of that system to enter while another part or parts is functioning. The type of box varies with the number of outlets and the manner in which the flow must be controlled. In every installation the distribution box should be designed to provide one or more outlets in addition to those contemplated in the initial installation in order to facilitate the extension, removal, or relocation of the effluent disposal units. Complete design data for concrete distribution boxes are contained in Figure 3.125.

LEACHING CESSPOOLS

This section gives complete design data on leaching cesspools, which constitute one of three types of effluent disposal methods from which the designer may choose. The choice is governed largely by soil conditions and the amount of land area available, as defined in detail in the section "General Design."

Application The advantages of the leaching cesspool are that it requires a minimum of land area, it can be used on a site of any slope, its initial cost is low, and it seldom requires cleaning at more frequent intervals than about 2 years. It can be used in all reasonably absorptive soils.

The limitations on the use of the leaching cesspool are that it can never be used in a soil rated as semi-impervious or impervious, it requires a location where the normal ground-water level is at least 8 feet below grade or 2 feet below the bottom of the cesspool, it should never be located within 100 feet or more of a potable water supply, and it should be situated at least 15 feet from the building it serves.

Leaching cesspools are limited in capacity; hence several units may be required to handle the effluent from large septic tanks. The spacing and the land area required by multiple leaching cesspools are indicated in Figure 3.126. It is recommended that when two or more cesspools are used, at least the first pair be connected through a distribution box for alternate operation rather than be installed in tandem. When two or more are employed, tandem operation is permissible for the first pair because the more remote cesspool takes the overflow of the nearer unit when loads are heavy.

Operation Leaching cesspools receive the effluent sewage from the septic tank or siphon tank and allow the liquid to be absorbed by the surrounding porous earth. The walls of the pool are laid up below the inlet with open seepage joints to allow the liquid to pass through these joints to a surrounding layer of broken stone and thence to the earth. The bottom of the pool is also an absorptive surface. All masonry above the inlet should be laid with tight mortar joints to minimize the entrance of surface water as well to achieve structural strength.

Design To determine the size and number of leaching cesspools required by any project, it is first necessary to determine the equivalent occupancy (which governs the amount of effluent to be treated) and the relative absorption of the soil (which influences the capacity of the individual units). Methods of determining these two factors are given in the section "General Design" in general terms, but they are repeated here for the reader's convenience.

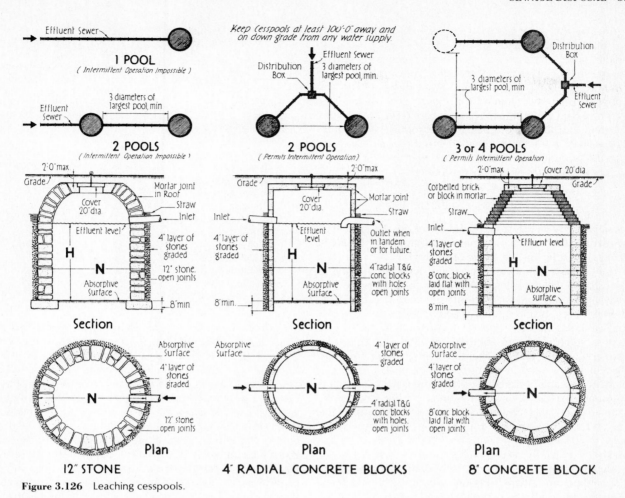

Figure 3.126 Leaching cesspools.

Rule 1   To find the equivalent occupancy in any project, multiply the number of persons occupying the type of building by the conversion factor given in the table "Equivalent Occupancy."

Equivalent Occupancy

Type of building	Gallons per person	Conversion factor
Residence	50	1
Camp	25	1/2
Summer cottages	40	4/5
Day schools without showers or kitchens	15	3/10
Factories without showers or kitchens	15	3/10
Day schools with showers and kitchens	30	3/5
Institutions except hospitals	100	2
Hospitals	200	4

Rule 2   To find the relative absorption of the soil at the site of any leaching cesspool, excavate a test pit at the site selected for the cesspool to a depth approximately half the distance from the inlet level to the bottom of the proposed cesspool but never less than 5 feet below grade. Make this pit large enough to work in conveniently. At the bottom of the pit carefully excavate a rectangular pit 12 inches square and 18 inches deep. Pour water into this small pit as quickly as possible, to a depth of 6 inches (requiring approximately 3¾ gallons). Note the time required for this 6 inches of water to be absorbed, and take one-sixth of this time as the aver-

Relative Absorption

Time (1-inch drop in minutes)	Relative absorption
0–3	Rapid
3–5	Medium
5–30	Slow
30–60	Semi-impervious (do not use leaching cesspools)
60 and up	Impervious (do not use leaching cesspools)

age time for the water to fall 1 inch. Refer to the table "Relative Absorption" to find whether the absorption is rapid, medium, or slow. If the rate of absorption exceeds 30 minutes, the site is not suitable for leaching cesspools.

Rule 3 To find the dimensions of any cesspool and the number of cesspools required for a given equivalent occupancy and determined relative absorption, refer to the table "Selection and Design of Leaching Cesspools" and Figure 3.126. Note that the table is divided into three parts according to the type of soil and that in each part the first column indicates the number of cesspools required.

SUBSOIL DISPOSAL BEDS

This section covers the design of subsoil disposal beds, which represent one of three methods for disposing of liquid effluent after it leaves the septic or siphon tank.

Application The advantages of the subsoil disposal bed as compared with leaching cesspools or sand filters are that it may be used in any soil except that rated as impervious. When it is used in soils rated as rapid, medium, or slow, distribution drains only are required; but when it is used in soils rated as semi-impervious, both distribution and collection drains are needed, and the filtered effluent sewage from the collection drains must be disposed of to more absorptive soil or to a nonpotable watercourse. These beds may be located on ground that is level or slightly sloping or, occasionally, on relatively steep slopes by a proper arrangement of drainage lines, and they require little or no cleaning if the septic tank is kept in good operating condition. When possible, the disposal beds should be placed on a southern slope.

Limitations on the use of this method are that groundwater should be more than 2 feet below grade. The initial cost of subsoil disposal beds is usually greater than the cost of leaching cesspools though less than that of sand filters. The land area required is greater than that for either cesspools or sand filters.

Operation Subsoil disposal beds consist of a series of drain lines laid with tight joints where slopes are relatively steep, leading to continuations of these lines laid with open joints through which the effluent sewage filters into the surrounding soil. The open-joint lines are laid at slopes ranging from 1 inch in 24 feet to 1 inch in 32 feet, and therefore usually follow the contour lines. The arrangement of lines shown in Figure 3.127 is purely diagrammatic.

Design The capacity of a subsoil disposal bed is governed by the number of lineal feet of 4-inch drainage lines laid with open joints. Drain lines laid with tight joints to effect proper separation of the seepage lines are not counted in computing the capacity of the bed. Capacity is related to both the equivalent occupancy upon which the entire system is designed and to the relative absorption of the soil. Methods of determining these factors are covered in the section "General Design" but are repeated here for convenience.

Rule 1 To find the equivalent occupancy in any project, multiply the number of persons occupying the type of building by the conversion factor given in the table "Equivalent Occupancy."

Equivalent Occupancy

Type of building	Gallons per person	Conversion factor
Residence	50	1
Camp	25	1/2
Summer cottages	40	4/5
Day schools without showers or kitchens	15	3/10
Factories without showers or kitchens	15	3/10
Day schools with showers and kitchens	30	3/5
Institutions except hospitals	100	2
Hospitals	200	4

Rule 2 To determine the relative absorption, proceed as follows. At the site of the proposed bed excavate a rectangular pit 12 inches square and 18 inches deep below the finished surface grade of that area, shaping the sides for accurate measurements as carefully as possible. Into this pit quickly pour water to the depth of 6 inches (requiring approximately 3¾ gallons), and measure the time needed for the water to be completely absorbed by the soil. Take one-sixth of this time as the average time required for the soil to absorb 1 inch. Refer to the table "Relative Absorption" and find the relative absorption as rapid, medium, slow, or semi-impervious. If the time for a 1-inch drop exceeds 60 minutes, subsoil disposal beds should not be used. If the time for a 1-inch drop is from 30 to 60 minutes, design the bed with collection drains as indicated in Figure 3.127 and carry the collection line to a nonpotable watercourse or to a cesspool or second disposal bed in a more absorptive soil.

Relative Absorption

Time (1-inch drop in minutes)	Relative absorption
0–3	Rapid
3–5	Medium
5–30	Slow
30–60	Semi-impervious (use collection drains)
60 and up	Impervious (do not use subsoil disposal)

Rule 3 To determine the lineal feet of 4-inch open-joint tile drain required for any equivalent occupancy and for any relative absorption up to impervious soils, refer to the table "Length of Subsoil Drainage Lines" and find the equivalent occupancy figure nearest to that determined for the project. Read to the right for the lineal feet of 4-inch open-joint drain tile in the column representing the relative absorption of the

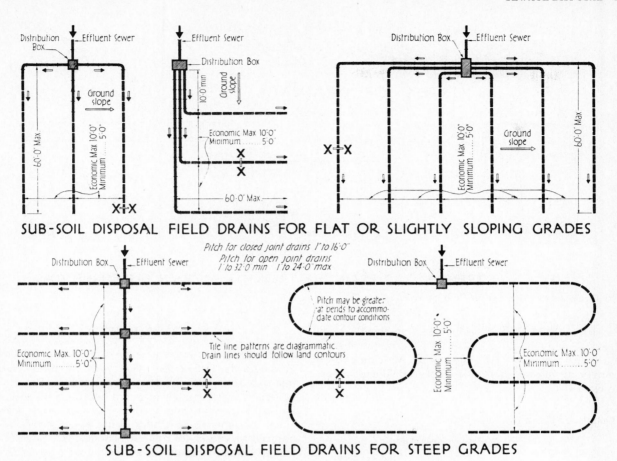

Figure 3.127 Subsoil disposal field drains.

Selection and Design of Leaching Cesspools

	Relative absorption											
	Rapid absorption; coarse sand or gravel				Medium absorption; fine sand or sandy loam				Slow absorption; clay with sand or loam			
Equivalent occupancy	Number of cesspools	Diameter N	Depth H	Absorptive area per person (square feet)	Number of cesspools	Diameter N	Depth H	Absorptive area per person (square feet)	Number of cesspools	Diameter N	Depth H	Absorptive area per person (square feet)
1–4	1	5′	5′	24.5	1	6′	6′	35.0	2	5′	5′	49.0
5–9	1	6′	6′	15.7	2	6′	6′	31.3	2	8′	7′	48.0
10–14	1	8′	6′	14.4	2	8′	6′	28.7	2	10′	8′	46.7
15–20	2	6′	6′	14.1	2	9′	7′	26.14	3	10′	8′	49.5
21–25	2	7′	6′	13.6	2	10′	8′	27.1	4	9′	8′	46.4
26–30	2	8′	6′	13.4	3	9′	7′	26.14	4	10′	8′	43.6
31–35	1 / 1	8′ / 9′	7′ / 7′	13.6	1 / 2	9′ / 10′	7′ / 8′	26.1	5	10′	8′	46.7
36–40	1 / 1	9′ / 9′	7′ / 8′	13.7	4	9′	7′	26.1	4	12′	10′	48.9
41–45	3	8′	6′	13.4	4	9′	8′	25.7	5	12′	10′	54.3
46–50	2	10′	8′	13.0	4	10′	8′	26.1	5	12′	10′	48.9

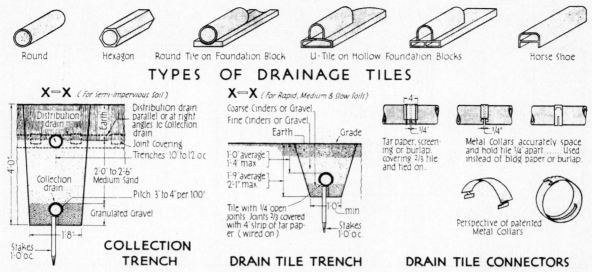

TYPES OF DRAINAGE TILES

Figure 3.128 Drainage tiles and trenches.

soil as determined by the preceding test. Note that the same number of lineal feet of tile is used for soils of slow absorption and those rated as semi-impervious, the difference in systems being reflected in the use of collection drains in semi-impervious soils.

Length of Subsoil Drainage Lines

	Relative absorption		
	Rapid; coarse sand or gravel	Medium; fine sand or sandy loam	Slow or semi-impervious; clay with sand or loam
Equivalent occupancy	Lineal feet of 4-inch open-joint tile drain required		
1–4	100	150	250
5–9	200	350	700
10–14	340	500	1000
15–20	475	650	1250
21–25	600	800	1500
26–30	725	1025	1800
31–35	850	1150	2100
36–40	975	1300	2400
41–45	1100	1450	2700
46–50	1200	1600	3000

Method of Laying Tile   Complete data on the layout of subsoil disposal drains are contained in Figures 3.127 and 3.128, which also show accepted methods for protecting the open joints between tiles. It is suggested that stakes and boards be used for accurately aligning the slope of drainage lines. The choice of various types of drainage lines is governed largely by their local availability and cost and the ease of laying them under project conditions.

SAND FILTERS

This section gives complete design data on sand filters, which constitute the last of the three types of effluent disposal methods from which the designer may choose. The choice is governed largely by soil conditions because this type of system is the only method adaptable to soils rated as impervious; the other two methods, being less expensive, would normally be chosen for other soil conditions.

Application   The sole advantage of sand filters lies in their adaptability to impervious soils. The limitations and disadvantages of this effluent disposal method are that collection drains must be used and the collected effluent be carried to a nonpotable watercourse or to leaching cesspools or subsoil disposal beds in more absorptive soils. The cost is relatively high because the entire area of the filter bed must be excavated and refilled with suitable filtering material, usually clean, coarse sand. The total area, however, is considerably less than the area of land required for subsoil disposal beds.

There are two types of sand filter. The closed type carries both the distribution and the collection drains underground in the filter bed, the upper layer of drains being covered with earth. These closed sand filters may be laid out in approximately rectangular or round patterns as indicated in Figure 3.129 or, when circumstances of site and capacity both permit, in the form of a long filter bed with a single pair of distribution and collection drains.

The open type is far less desirable because it exposes the effluent sewage and requires a filter bed free of any covering over the sand. In some instances it is less expensive to construct and may be adapted to institutions or large estates where the filter bed is removed from the building. The effluent sewage is conveyed in closed-joint drainage lines above the surface of the bed, with outlets discharging into wood troughs that serve as splashboards; they are laid out in

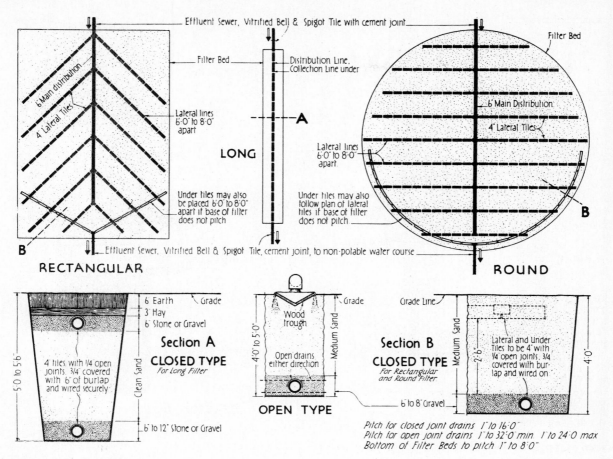

Figure 3.129   Sand filters.

the same manner as the lateral branches of the drain tile system.

Design   The capacity of a sand filter is expressed in its surface area in square feet and is related to the equivalent occupancy of the building it serves. Since this system is normally used only in impervious soils, it is advisable also to determine the relative absorption of the soil on the site. Methods of determining these two factors are given in the section "General Design" but are repeated here for convenience.

Rule 1   To find the equivalent occupancy in any project, multiply the number of persons occupying the type of building by the conversion factor given in the table "Equivalent Occupancy," on page 330.

Rule 2   To determine the relative absorption, proceed as follows. At the site of the proposed bed excavate a rectangular pit 12 inches square and 18 inches deep below the finished surface grade of that area, shaping the sides for accurate measurements as carefully as possible. Into this pit quickly pour water to the depth of 6 inches (requiring approximately 3¾ gallons), and measure the time required for the water to be completely absorbed by the soil. Take one-sixth of this time as the average time required for the soil to absorb 1 inch. Refer to the table "Relative Absorption," on

page 330, and find the relative absorption as rapid, medium, slow, or semi-impervious. If the time for a 1-inch drop exceeds 60 minutes, sand filters should be used.

Rule 3   To find the surface area of a sand filter bed required for a given equivalent occupancy refer to the table "Area of Filter Beds," on page 330, and find the equivalent occupancy nearest to that computed for the project. Read to the right for the area in square feet of earth for closed or open types of sand filters.

The detailed design for sand filters is clearly indicated in Figure 3.129.

DRY WELLS

Where sewers are not available for storm-water disposal and where the soil and the finished surface grade are suitable, the following means of disposal may be used:

1. Dry well (size is dependent on the area to be drained and the soil absorption).

2. Dry well with additional subsurface drains where soil absorption is such that the dry well alone cannot handle the load.

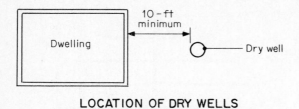

LOCATION OF DRY WELLS

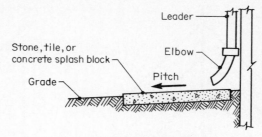

SPLASH BLOCK

Stone or concrete location marker — Grade

Concrete or stone slab cover

Outlet to other dry well or disposal field if required

6 in

Joints above inlet should be sealed with cement mortar

Inlet

Stone or concrete block or brick, laid up dry with open joints

3-in layer of coarse gravel backing may be placed to level of top of inlet

12-in coarse gravel

Ground water level

2-ft minimum

SECTION: LARGE DRY WELL

Figure 3.130 Dry wells.

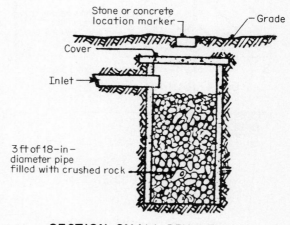

Stone or concrete location marker — Grade

Cover

Inlet

3 ft of 18-in-diameter pipe filled with crushed rock

SECTION: SMALL DRY WELL

Figure 3.131

3. Drainage for roof combined with drainage for surface and subsurface where the dry well is of limited capacity. The dry well is used for surface and subsurface drainage; roof drains lead to a splash block and are directed away from the dry well.

4. Splash block. When roof drains are discharged at grade, a splash block should be used to minimize soil erosion.

Figures 3.130 and 3.131 show the location and design of dry wells and splash blocks.

Equivalent Occupancy

Type of building	Gallons per person	Conversion factor
Residence	50	1
Camp	25	1/2
Summer cottages	40	4/5
Day schools without showers or kitchens	15	3/10
Factories without showers or kitchens	15	3/10
Day schools with showers and kitchens	30	3/5
Institutions except hospitals	100	2
Hospitals	200	4

Relative Absorption

Time (1-inch drop in minutes)	Relative absorption
0–3	Rapid (do not use sand filter)
3–5	Medium (do not use sand filter)
5–30	Slow (do not use sand filter)
30–60	Semi-impervious (do not use sand filter)
60 and up	Impervious (use sand filter)

Area of Filter Beds

Equivalent occupancy	Area (square feet) Closed type	Open type
1–4	200	100
5–9	900	450
10–14	1400	700
15–20	2000	1000
21–25	2500	1250
26–30	3000	1500
31–35	3500	1750
36–40	4000	2000
41–45	4500	2250
46–50	5000	2500

Design

Size In general 8-inch sewer pipe will be the minimum acceptable diameter. The use of 6-inch-diameter pipe will be given consideration where the design computations demonstrate that the properties connected thereto will be adequately served and the velocity when flowing full will not be less than 2 feet per second. Six-inch sewers will not be acceptable where future extensions are possible.

Depth In general street sewers must be designed deep enough to drain basements. Where for specific reasons shallow depths are necessary and can be justified, the sewer must be protected to prevent its being damaged. All sewers should be designed and bedded or encased to prevent cracking due to superimposed loads and weight of backfill material. Proper allowance for loads on the sewer should be made because of the width and depth of trench.

Capacity Sewer systems should be designed for the maximum hourly sanitary flow using an average daily per capita flow of sanitary sewage of not less than 75 gallons per day unless otherwise justified by sound engineering data. On this basis lateral and submain sewers should be designed with capacities, when running full, of not less than 4 times the average daily sewage flow. In similar manner, outfall sewers should be designed for 2.5 times the average sewage flow. Additional allowance should be made in each case for any wastes from shopping centers, etc., and for infiltration taking into account soil and groundwater conditions, type of pipe joint, and possibilities of future extension.

Velocity of Flow All sewers should be designed and constructed with hydraulic slopes sufficient to give mean velocities, when flowing full, of not less than 2.0 feet per second based on Kutter's or Manning's formula using an *n* value of 0.013. The following are the minimum slopes which should be provided especially where the depth of flow may be small and are desirable minima in all parts of the system:

Sewer size (inches)	Minimum slope in feet per 100 feet
6	0.65
8	0.40
10	0.28
12	0.22
14	0.17
15	0.15
16	0.14
18	0.12
21	0.10
24	0.08

Under special conditions, if full and justifiable reasons are given, slopes slightly less than those required for the 2 feet per second velocity when full may be permitted. Such decreased slopes will be considered only where the depth of flow will be 0.3 of the diameter or greater for design average flow.

Increasing Size When sewers are increased in size, or when a smaller sewer joins a larger one, the invert of the larger sewer should be lowered sufficiently to maintain the same energy gradient. An approximate method for securing these results is to place the 0.8 depth point of both sewers at the same elevations.

Alignment Sewers, up to and including those with a 24 inch diameter, must be laid with uniform slope and alignment between manholes

High-Velocity Protection In the case of sewers where the slope and volume are such that velocities of 15 feet per second are realized at average flow, special provisions should be made to protect against erosion and shock.

Manholes

Location Manholes should be installed at the end of each line; at all changes in grade, size, or alignment; at all intersections and at distances not greater than 400 feet for sewers 15 inches or less and 500 feet for sewers 18 to 30 inches. Greater spacing may be permitted in larger lines and those carrying a settled effluent. Lampholes may be used only for special conditions and should not be substituted for manholes or installed at the end of laterals greater than 150 feet in length.

Drop-Type Pipe A drop pipe should be provided for a sewer entering a manhole at an elevation of 24 inches or more above the manhole invert. Where the difference in elevation between the incoming sewer and the manhole invert is less than 24 inches the invert should be filleted to prevent solids deposition.

Diameter The minimum diameter of manholes should be 42 inches, with larger diameters being preferable.

Flow Channel The flow channel through manholes should be made to conform in shape and slope to that of the sewers.

Watertightness Solid manhole covers are to be used wherever the manhole tops may be flooded to an appreciable depth by street runoff or high water. Manholes of brick or segmental block should be waterproofed on the exterior with plaster coatings supplemented by a bituminous waterproofing coating where groundwater conditions are unfavorable.

Inverted Siphons Inverted siphons should have not less than two barrels with a minimum pipe size not less than 6 inches and should be provided with necessary appurtenances for convenient flushing and maintenance; the manholes should have adequate clearances for rodding; and in general sufficient head should be provided and pipe sizes selected to secure velocities of at least 3.0 feet per second for average flows. The inlet and outlet details must be arranged so that the normal flow is diverted to one barrel and so that either barrel may be cut out of service for cleaning.

Protection of Water Supplies

Water Supply Interconnections There should be no physical connection between a public or private potable water supply system and a sewer, or appurtenance thereto which would permit the passage of any sewage or polluted water into the potable supply.

Relation to Water Works Structures While no general statement can be made to cover all conditions, it is generally recognized that sewers should be kept remote from public water supply wells or other water supply sources and structures.

Relation to Water Mains Horizontal separation: Whenever possible, sewers should be laid at least 10 feet, horizontally, from any existing or proposed water main. Should local conditions prevent a lateral separation of 10 feet, a sewer may be laid closer than 10 feet to a water main:

1. If it is laid in a separate trench

2. If it is laid in the same trench with the water mains located at one side on a bench of undisturbed earth

3. If in either case the elevation of the top (crown) of the sewer is at least 18 inches below the bottom (invert) of the water main

Vertical separation: Whenever sewers must cross under water mains, the sewer should be laid at such an elevation that the top of the sewer is at least 18 inches below the bottom of the water main. When the elevation of the sewer cannot be varied to meet the above requirement, the water main should be relocated to provide this separation or reconstructed with mechanical-joint pipe for a distance of 10 feet on each side of the sewer. One full length of water main should be centered over the sewer so that both joints will be as far from the sewer as possible.

When it is impossible to obtain proper horizontal and vertical separation as stipulated above, both the water main and sewer should be constructed of mechanical-joint cast-iron pipe and should be pressure-tested to assure watertightness.

FOUR

Residential Development

Residential use is the largest single use of land in any community. Its successful design and development are essential for it to function efficiently and to be aesthetically pleasing. Almost all new residential development comes about through the subdivision of raw land. Such subdivision is the process whereby a relatively large tract of land is divided into blocks by streets that provide access. The blocks themselves are divided into lots for individual ownership. In the case of rental or cooperative development this division may not occur. Portions of the tract of land to be subdivided are reserved for parks or for schools. Adjacent areas normally contain shopping and other community facilities.

When this process occurs in a relatively undeveloped region, the result is generally referred to as a "new town" or a "new city." When it occurs adjacent to an existing built-up area, the development is generally considered simply as a subdivision or as an addition to the existing community. In both cases the process is basically the same.

This section deals with this residential development process and its various phases. The emphasis is on the physical development rather than on the administrative or legal aspects. The land use intensity standards provide a method of determining the maximum floor area of buildings on a given tract of land. They also determine the required amount of open space that must be provided for the specific floor area.

Types of streets and parking are important elements of any subdivision design. The street system must be laid out functionally and be appropriate for its particular use. It requires a separation, or classification, of local, collector, and major streets. Directly related to the street layout is the type of parking to be utilized. Open space and pedestrian circulation are other significant aspects of the subdivision process. Playgrounds and other recreational facilities are extensively covered in Section 5.

4.1 Land Use Intensity

Meaning of Land Use Intensity In the *Minimum Property Standards for Multifamily Housing* (1965) of the Federal Housing Administration (FHA), land use intensity means the overall structural-mass and open-space relationship in a developed property. It correlates the amount of floor area, open space, livability space, recreation space, and car storage space of a property with the size of its site, or land area.

Land use intensity is somewhat similar to density: living units or people per acre. But its approach is different; it covers a broader field of planning factors and correlates them.

Why and When the FHA Makes an Intensity Rating In analyzing certain housing developments, the FHA determines a land use intensity rating (LIR) of the site in order to apply the site planning standards of the *Minimum Property Standards for Multifamily Housing* (MPS). The rating ascribes to the site a set of interrelated standards for land area, floor area, open space, livability space, recreation space, and car storage capacity.

The rating of land use intensity is one of the most crucial steps in the preliminary consideration of a housing proposal. Correctly done, it establishes a workable basis for the planning, construction, and operation of a successful housing project—a project that is successful both as to market absorption and as to long-term values, successful whether the project is for rental or for home sales in a planned-unit development.

An unwarranted low rating tends to affect the project adversely through underuse of the land. An unwarranted high rating tends to affect the project adversely by lowering its livability beneath the level appropriate for its location in the community and thereby lowering its rentability or marketability. Of the two, the latter has the most detrimental effect, for an overly high intensity lowers the comfort, convenience, and appeal of the entire project, and this is reflected in the lower market appeal or lower rentals of the project during its life.

For the purpose of determining the FHA land use intensity rating of a site, data are gathered and analyzed. In some cases the market analyst, architect, site engineer, and sanitary engineer participate in the determining. They consider not only current conditions but also the future insofar as it can reasonably be anticipated.

Community Patterns The basic intent of the land use intensity ratios provided in the multifamily MPS is to establish the intensity for the housing development so that it will be appropriate to the characteristics of the site and its location in the anticipated community pattern. Although the characteristics of the site (steepness, shape, and so on) may affect the site intensity, the principal determinant of intensity is the location of the site in the anticipated community pattern. It is therefore necessary for the rating of site intensity to consider community patterns thoroughly.

The land use patterns of a community have the following primary dimensions:

1. Width and length, including the space arrangement of various land uses located near to each other, such as the dividing line between a town house area and a detached-home area
2. The intensity range of land uses in the community, ranging upward from very low urban residential uses, such as small houses on 2-acre lots
3. The time factor of the land use, the growth stages through which the community progresses through time, such as the rapidity of suburban development in a booming metropolis

Space Arrangement of Land Uses Communitty patterns have a great variety of space arrangements. Common arrangements are as follows:

1. A concentric pattern of development, typical of the medieval town and often found in small urban areas in the Plains states and elsewhere where topographic features do not shape community land usage
2. A string type of development where intensive uses are located in linear form, as along a main highway in a small town
3. A radial pattern, like the spokes of a wheel, which is often seen in modern cities, with the hub as the city center or highly intensive land use area and other intensive land uses ranging out along transportation lines, radial boulevards, and other arteries

As an urban area grows, these and other patterns frequently develop numerous subcenters extending from the center of the city or metropolitan area. These minor centers of growth and development combine to form a many-centered pattern of land use that is very common in large metropolitan areas.

The land use intensity rating of a site for housing varies substantially with the location of the site in the community's space pattern. This is true even in a relatively small community. The intensity is highest, of course, in the center of the community and lowest in the general outlying area. In the transitional areas between the two are intermediate intensities. The site might also be located along a corridor connecting the center to another urban area. In rating the land use intensity of a proposed site, it is therefore necessary to understand the present and prospective space arrangement of the land use pattern of the community and to identify the position of the subject site in that arrangement of land uses. See Figure 4.1.

Intensity Range of Land Use Intensity of land use is markedly affected by the relationship of the size of the total long-term demand for land for urban use and the size of the supply of usable land. In the proper rating of a site in a specific community, it is therefore necessary to reach an understanding regarding the immediate and potential demand and sup-

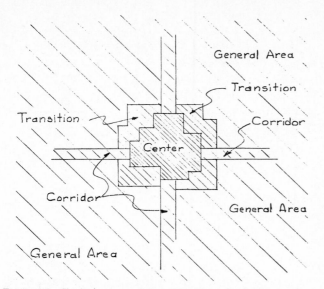

Figure 4.1 Typical community pattern. There is no intent to force this pattern or any other pattern on the growth of communities.

ply factors of land use. The amount of land available to meet the present and prospective demand, as well as the opportunities and limitations of its physical development, can usually be determined by examination of the area. Long-term economic factors are not so easily determined, and they are of major importance.

Data related to land economics are usually available from local planning commissions. Other statistics necessary to intelligent land use rating are available from surveys and forecasts of industrial and other economic activity, population size, and anticipated growth.

It is possible to reach general conclusions on the total range of intensities appropriate to the land in the community currently and for the foreseeable future from the data obtained and from consideration of the present size, rate of growth, and probable ultimate size of the community. A very small community with a very slow growth rate and prospective small ultimate size will have a much narrower and lower range of land use intensity ratings than a larger community will. For instance, an isolated town of less than 2000 population with slow growth will have a much lower upper limit to its range of land use intensities than will a well-located, rapidly developing suburb in a metropolitan area.

Time Stages of Land Use Intensity   A community's growth through time is one of the most significant dimensions of its land use pattern. In some cases, the intensity pattern as related to time is static. In others, as in a rapidly expanding metropolitan area, it may be characterized as explosive. The land use intensity may be recessive in depressed communities or in depressed sections of a community. The intensity pattern may be regenerative in cases such as a successful redevelopment area that revitalizes the economic activity of an entire community area or a major portion of it.

When the time pattern is static, the rating of land use intensity follows very closely the intensities of existing development. This may be true either of the community as a whole or of substantial portions of it. An example is a conservation area, where properties are being conserved rather than rebuilt.

With an explosive community pattern, the proper rating of land use intensity in fringe areas and satellite developments presents very difficult problems. It is necessary to forecast the emerging and future intensity pattern. Many fast-developing residential areas follow a normally predictable pattern. New stimuli or changed conditions, however, may warrant more intensive land use in the future development pattern. Where this is the case, a higher land use intensity than that currently present in the area is appropriate for a project. To determine the presence of such a situation and the appropriate degree of higher land use intensity, the space arrangement and intensity range are carefully studied in conjunction with the following:

1. A thorough study of the probable effect of higher densities on existing development in the area and on adjacent undeveloped areas

2. Consultation with planning, zoning, and engineering officials having jurisdiction in the community

3. Consultation with realtors and mortgage lenders active in the area

Such determinations are preceded by a complete documentation of the file with all facts, opinions, and judgments justifying a higher land use intensity rating.

With a recessive community pattern, special care must be taken to determine not only the presence of a need for additional housing but also the proper intensity level. The proper level may well be lower than that of existing development, inasmuch as the demand for land is receding and the greater livability of lower intensity may be essential for successful marketing.

In the regenerative community pattern, which usually involves a redevelopment area, the existing land uses and intensities are given little consideration, the prime focus being on future needs and the anticipated pattern. In this regard the analysis resembles that of the explosive community pattern, and the same thorough analysis is made prior to determination of appropriate land use intensity.

Land Use Intensity Compared with Density   The term "density" as commonly used in land use planning, zoning, and site planning means the number of living units or the number of people per unit of land area. Because the size of living units and the number of occupants of units of any given size vary widely, density is a rather crude measure of the degree of land use. This is why the FHA no longer uses density directly as a land use measure in its multifamily MPS.

In place of density the FHA now uses land use intensity, which is less variable and thus more reliable. The concept of land use intensity as employed here starts with the overall

Land Use Intensity Related to Floor Area and Density

| Land use intensity (LIR) (1) | Floor area ratio (FAR) (2) | Floor area per gross acre (square feet) (3) | Density in living units per gross acre | |
			1089 square feet per living unit (4)	871.2 square feet per living unit (5)
0.0	0.0125	544.5	0.5	0.625
1.0	0.025	1,089	1.0	1.25
2.0	0.05	2,178	2.0	2.5
3.0	0.1	4,356	4.0	5.0
4.0	0.2	8,712	8.0	10.0
5.0	0.4	17,424	16.0	20.0
6.0	0.8	34,848	32.0	40.0
7.0	1.6	69,696	64.0	80.0
8.0	3.2	139,392	128.0	160.0

structural-mass and open-space relationships of a developed property, as explained below. It begins with and is directly related to the floor area ratio, which is the relation of the floor area to the land area. Thus, it is a measure of the total permitted floor area on a site of a given size.

In residential property, of course, total floor area roughly determines the number of living units and the number of people. Thus, land use intensity ratings based on floor and land areas have some general relationships to density ratings based on living units and land area. Because density is unresponsive to wide variables in living-unit size and household size, density ratings can be compared with land use intensity ratings only in general terms and at the risk of gross misinterpretation. A valid comparison can be made, however, if the living-unit size or household size (the variable in density) is kept as a constant at some specific size. Some comparisons of this kind are made in the following subsections.

Floor Area: A Base for the Rating Scale   To rate or measure, it is necessary to have a measurement scale. For land use intensity, the rating scale is based first and most directly on the relationship of total floor area to total land area, that is, on the floor area ratio as defined in the multifamily MPS.

The rating scale appears in column 1 of the table "Land Use Intensity Related to Floor Area and Density." It starts at a floor area ratio (FAR) of 0.0125. This is arbitrarily called 0.0 on the scale of land use intensity rating (LIR 0.0). Since the FAR multiplied by the land area (LA) equals the floor area (FA), FAR 0.0125 times the 43,560 square feet in an acre equals 544.5 square feet of floor area per acre at LIR 0.0. To visualize LIR 0.0, this floor area per acre can be converted to living units per gross acre. For example, 544.5 square feet per acre equals one modest home with a floor area of 1089 square feet (2 × 544.5) with 2 acres of land. This is shown on the first line of the table. Each full unit on the intensity scale is equal to a 100 percent increase in the FAR. This increases the FA by 100 percent as related to the LA, thus doubling the number of living units of any given size for each full unit on the intensity scale, as illustrated in the table.

This doubling on the intensity scale and its origin at LIR 0.0

LIR	Living units per acre for living units of 1089 square feet
1	1
2	2
3	4
4	8
5	16
6	32
7	64
8	128

with FAR 0.0125 makes it easy to remember and visualize the entire scale. Note in line 2 and columns 3 and 4 of the table that LIR 1.0 means 1089 square feet of floor area or one house of that modest size per acre of land. Remember that LIR 1.0 equals one living unit per 1 acre for a house of a little over 1000 square feet. Also remember the doubling of the intensity scale. Then it is easy to remember the table above that visualizes the scale.

Memorizing another point on the scale is also helpful. LIR 3.0 is FAR 0.1. By starting from this, it is easy to find any FAR value through doubling:

LIR	FAR
3.0	0.1
4.0	0.2
5.0	0.4
6.0	0.8
7.0	1.6
8.0	3.2

Gross Land Area: Another Base for the Rating Scale   We have noted that the FHA does not use density directly as a measure in its multifamily MPS. A comparison of the land base of density ratings will be helpful, however, as a preliminary to considering land as a base in the rating of intensity.

In density ratings, the number of living units is sometimes related to all the land that benefits or is used by the development, including on-site streets and half of bordering streets. This is the gross land area, or gross acres; the resultant is gross density. Sometimes street areas are not counted in the land area; the resultant is net density. At times public streets are not counted in the area while privately owned streets are counted; the resultant is another variety of net density, an ambivalent measure of doubtful meaning. In recent years the FHA has used the latter variety of net density. The confusion mounts when paved parking areas in bays along streets are not counted in the net area but similar areas in stub compounds are counted.

Wiping the slate clean of such ambivalence and confusion, the FHA multifamily MPS now takes the gross land area as the firm base for the land use intensity rating. The intensity rating scale refers to the gross acres of all land benefiting a project. Within certain limits for land area (LA) established in the multifamily MPS, one-half of any abutting street right-of-way is included in the gross acreage. Also included in the LA for the intensity rating is one-half of any abutting park, river, or other beneficial open space that has a reasonable expectancy of perpetuity. Because the gross acreage of the total LA (all the area benefiting the project) is used in the rating, the land use intensity measures are more realistic and more reliable than are data based on net acres.

If existing information based on net area is used in site analysis, it is necessary to convert it to gross area for comparison with land use intensity ratings on the FHA's MPS-M intensity rating scale. Net area customarily excludes all street rights-of-way. These usually range from 15 to 30 percent of the total LA.

Gross area is converted to net area by subtracting the percentage of street area from 100 percent and multiplying the gross area by the difference.

Example If we know that the gross area is 125 gross acres and the street area is 20 percent of the gross area, what is the net area?

(100 − 20 percent) × 125 gross acres = 100 net acres

Net area is converted to gross area by subtracting the percentage of street area from 100 percent and dividing the net area by the difference.

Example If we know that the net area is 100 net acres and the street area is 20 percent of the gross area, what is the gross area?

100 net acres ÷ (100 − 20 percent) = 125 gross acres

Gross density is converted to net density by dividing the number of living units per gross acre by the difference between 100 percent and the percentage of street area.

Example If we know that there are four living units per gross acre and the street area is 20 percent of the gross area, how many living units are there per net acre?

4 ÷ (100 − 20 percent) = 5 living units per net acre

Net density is converted to gross density by multiplying the number of living units per net acre by the difference between 100 percent and the percentage of street area.

Example If we know that there are five living units per net acre and the streets are 20 percent of the gross area, how many living units are there per gross acre?

5 × (100 − 20 percent) = 4 living units per gross acre

Other Elements of the Rating Scale As we have mentioned, the land use intensity rating not only relates floor area to land area but also correlates other planning elements. The other elements are open space, livability space, recreation space, occupant car storage, and total car storage including guest cars. These planning elements are described and defined in the multifamily MPS.

Open space, livability space, and recreation space are expressed in the intensity rating system in terms of the open-space ratio (OSR), livability space ratio (LSR), and recreation space ratio (RSR). The ratio is simply the total area of the open space, livability (outside nonvehicular) space, or large recreation space divided by the total floor area. In the rating system, car storage capacity is expressed in terms of the car ratio, which is the number of cars divided by the number of living units.

We have seen that the floor area ratio (FAR) as a base for the intensity rating scale has a very simple mathematical relationship to the scale: it doubles with each unit of the LIR. These other elements of the rating scale, however, have a more complex relationship to the rating scale of land use intensity. This is seen in the ratio chart in Figure 4.2, reproduced from the multifamily MPS. The established ratios for these elements represent appropriate relationships readily attainable by proper planning. Good design can surpass them substantially while attaining the total floor area at a given intensity rating.

These other elements in the rating system are of importance in determining a proper intensity rating for a proposed housing site. Their relationship to the intensity rating scale deserves careful study (see Figure 4.2).

How Many Living Units? In the table "Land Use Intensity Related to Floor Area and Density" and the preceding discussion, we have seen that the rating scale for land use intensity is related directly to the floor area ratio and to the floor area and that the number of living units per acre is not related directly to an intensity rating because the size of living units varies. Columns 4 and 5 of the table show the great magnitude of this variation for living units of 1089 and 871.2 square feet. We have also seen that the rating of land use intensity is related to the amount of open space of several kinds, including livability space and recreation space, as well as to total open space.

Thus it becomes clear that the land use intensity rating scale is a measure of overall relationships of structural mass and open space. In the intensity rating these complex relationships are distilled into a single numerical scale. Grossly oversimplified, the intensity scale may be said to run from LIR 0.0 for very low rural-type land use, through LIR 3.0 for sub-

	LIR	Living units	Land per living unit	Land difference per LIR unit	Total cost per living unit*	FHA 203b home loan†		
						Down payment	Monthly payment	Difference per LIR unit
A	4.0	8	$1000		$12,200	$400	$70.14	
				$500				$2.97
B	5.0	16	500		11,700	400	67.17	

*$11,200 per living unit is assumed as the cost for building, land improvements, and other costs except raw land costs.
†Based on a 30-year mortgage at 5¼ percent interest plus ½ percent FHA premium.

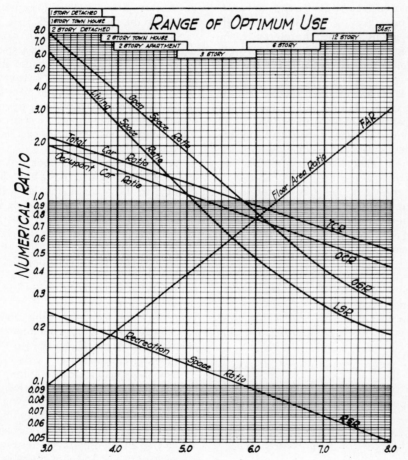

Figure 4.2 Land use intensity standards.

urban-type land use at about four modest houses per gross acre, to LIR 8.0 for very intensive urban land use for high-rise apartments. Actually, at each unit number of decimal subdivision of the intensity scale, there is a precise meaning of a definite degree of intensity of land use in terms of the relationships of maximum floor area for the land area and in terms of a minimum for open space, livability space, and recreation space for the floor area.

A natural question at this point is "How do the sponsors of a proposed development find out how many living units they can get on their site?" If it is an FHA development, the sponsors determine the number of living units acceptable to the FHA on the site by (1) obtaining the land use intensity rating of the site from the FHA and (2) then preparing a project planning program and project design properly related to the assigned site rating. An illustration of a simple approach to

developing a planning program for a project starting with the intensity rating is in the multifamily MPS. The FHA provides blank forms of the illustrated method for the use of sponsors and their design professionals.

A more important question is ''How would a higher land use intensity affect the financial soundness of the proposed project?'' If the livability offered by a property is lower than the level appropriate for its location in the community pattern, the property is at a disadvantage in competition with others in comparable locations. Consequently it will suffer in rentability and marketability. The ratio lines of Figure 4.2 show how livability decreases very rapidly as land use intensity increases.

Too high a land use intensity is a false economy. Let us consider the case of Developer A and Developer B, assuming that each developer has a land cost of $8000 per acre and $11,200 for all other costs to produce 1089-square-foot town houses on similar tracts. Developer A develops the property at a land use intensity of 4.0; Developer B develops the property at a land use intensity of 5.0, with results as shown in the table above Figure 4.2.

Note that Developer A, with a lower LIR of 4.0, offers a much more desirable property with only half the density of B's project, giving twice the amount of exterior livability space. A sells properties with an added cost of only $500 (4.3 percent), which the home buyer carries for an additional monthly payment of only $2.97.

Since raw land is a low-percentage component of the total housing cost, great reductions in total package costs cannot come from squeezing the land. We have seen that intensification of land use produces relatively small cost savings at the sacrifice of very great amounts of livability space. Because livability affects marketability and rentability, it is clear that an inappropriate intensification of land use is financially very hazardous. In determining the intensity rating of a specific site, an economic comparison of the type illustrated above can be prepared wtih applicable local data to aid in avoiding excessive intensity and its accompanying false economy.

To the extent practicable, FHA personnel participating in the intensity rating of a site inspect the site together, pool all collected data, analyze the data in consultation with each other, and coordinate their findings and recommendations. This is accomplished under the direction of the chief underwriter with the objective of reaching a concurrence of all participants. In any event, the intensity rating is made by the FHA chief underwriter after considering all data and recommendations.

How the Assigned Rating Is Used   The land use intensity rating assigned by the FHA to a proposed housing site tells the project sponsor and his or her design professionals the maximum land use intensity that the FHA will accept for a development of the site with FHA-insured financing. In essence, the FHA, by assigning its intensity rating to a site, ascribes to the site a selected set of land use standards consisting of the maximum floor area ratio and minimum ratios for open space, livability space, recreation space, and car storage. These are found in Figure 4.2 by locating the

assigned intensity rating on the scale at the bottom and following its vertical line to intersections with the ratio lines. The ratios and their application in the site planning of the project are defined and illustrated in the FHA *Minimum Property Standards for Multifamily Housing*.

4.2 Neighborhood Planning

The principles guiding development are as follows:

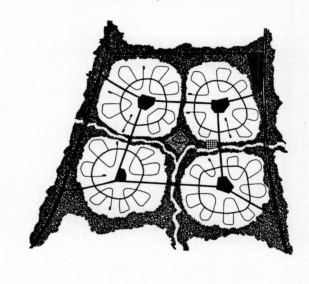

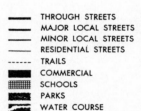

——— THROUGH STREETS
——— MAJOR LOCAL STREETS
——— MINOR LOCAL STREETS
——— RESIDENTIAL STREETS
------ TRAILS
▬ COMMERCIAL
▦ SCHOOLS
▨ PARKS
▧ WATER COURSE

Figure 4.3   Principles of neighborhood planning.

1. Residential areas should be planned on a neighborhood basis.
2. Major highways should pass around residential neighborhoods.
3. Primary street patterns should connect the centers of neighborhoods.
4. Secondary streets should provide access to residential streets of loop or cul-de-sac type designed to filter out traffic destined for the area.
5. Independent of vehicular routes, a system of pedestrian greenways, generally located in natural stream valleys, should be provided to connect homes, schools, and recreation areas.
6. Schools should be accessible by walkways. Church sites should be provided in each neighborhood.

7. Local and neighborhood shopping centers should be strategically located.

4.3 Street Classification

The overall street system for a housing development must conform to the circulation requirements of the master plan for the community. This will provide maximum accessibility to all parts of the community and ensure proper coordination with proposed circulation changes.

Direct access to a major arterial highway is essential. Such intersections must be adequately controlled with lights or other means. The practical minimum distance between intersections on the major arterial highway should be 800 to 1000 feet. No through streets should be provided. All circulation should be directed around the periphery of the development to the major arterial highway.

Each lane of traffic will carry from 600 to 800 cars per hour. Horizontal alignment of all collector, minor, loop, and access streets (see Figure 4.4) should provide for a minimum of 200 feet in clear sight distance. The vertical alignment should not exceed a grade differential of 6 to 8 percent. Sidewalks, when used, should be a minimum of 4 feet wide. When trees are planted between the curb and the sidewalk, the sidewalk should be set back approximately 8 feet. If no trees are used, the setback should be 4 feet.

Performance Objective The subdivision road system should permit the safe, efficient, and orderly movement of vehicles for transportation, garbage collection, snow removal and emergency services, pedestrains, and cyclists.

Design Considerations The road system should be related to other elements of subdivision design, such as land use planning, energy conservation, noise abatement, public transit, pedestrian circulation, and storm-water management. In addition, it should:

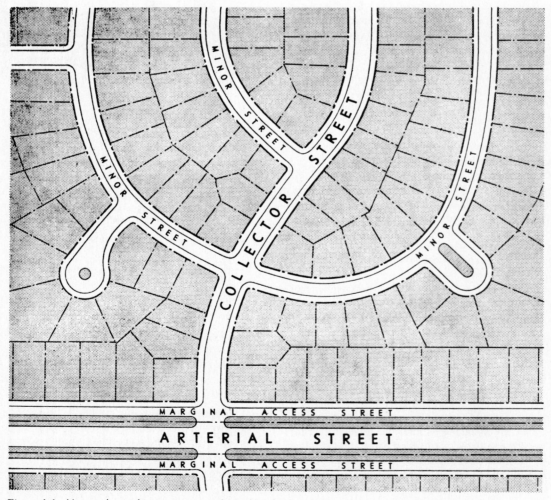

Figure 4.4 Nomenclature for street types adopted by the Association of State Highway Officials.

- Be cost-effective, so that length and construction specifications meet but do not exceed the needs of the population served
- Have a simple and logical pattern
- Respect natural features and topography
- Logically link into the existing street system
- Provide for visitor parking
- Accommodate future public transit needs
- Present an attractive streetscape
- Provide a system for draining surface water in major storms

Roads constructed in urban areas may be classified in four categories according to their service function:

- Local streets—minor and major
- Collector streets—minor and major
- Arterial roads
- Freeways

Figure 4.5 Street classification at the subdivision scale.

The first three are an integral part of a subdivision planning (see Figure 4.5). In most cases, the subdivision designer will be concerned mainly with local and collector streets and only occasionally with portions of an arterial road. The design of arterials and urban freeways is the responsibility of specialized engineers; but of course, these roads also are linked to subdivision development.

4.4 Types of Streets

Types of streets are described in the last table on page 343 and illustrated in Figure 4.4 and 4.5.

Collectors Collectors are the highest order of residential street. Designed to carry traffic volumes of 3000 ADT (aver-

age daily trips), a collector generally connects approximately 150 dwelling units each or connects a neighborhood and an activity area. No residential lots front directly onto these streets, and parking on them is usually prohibited. Generally, it is considered desirable for collectors to run north/south whenever possible to allow for maximum solar lot configurations off local east/west streets. However, local north/south streets can work efficiently for solar access under certain conditions.

Local streets carry limited traffic volume at low speeds; whereas arterial streets are designed to carry high volumes over greater distances at higher speeds. Design standards for collector streets fall between these two extremes. Four types of collector streets are discussed: residential, commercial, industrial, and neighborhood.

Residential Residential collectors are designed to handle traffic volumes of up to 2000 vehicles per day, while providing access to abutting property and on-street parking. Such streets are necessary adjacent to multifamily residential developments, schools, and local retail and public facilities. They are required also when more than fifty dwelling units or residential lots must utilize the street for access to the collector/arterial street system. Major entry streets to a residential community normally will be set up as residential collector streets. The right-of-way should be 60 feet wide and the pavement 44 feet wide.

Industrial/commercial A commercial or industrial collector is one serving as principal access to a commercial development or an industrial site. The length of such a street should not exceed 2 miles. Direct residential frontage should be discouraged to prevent a conflict between residential and commercial traffic. Multifamily development can front onto a commercial collector if ample off-street parking is provided and access is limited. For industrial streets, the right-of-way and pavement widths are 80 and 64 feet, respectively; and for commercial streets, 60 and 44 feet, respectively.

Neighborhood A neighborhood collector street is designed to traverse distances from ½ to 2 miles, to serve a variety of land uses, and to handle traffic volumes of up to 8000 vehicles per day. On-street parking usually is prohibited or restricted on neighborhood collectors; access to abutting property is limited, and uses may include multifamily dwelling units; schools; and retail, office, and community service facilities. Streets in a residential area serving more than 200 dwelling units should be designated as neighborhood collector streets. The right-of-way should be 70 feet wide and the pavement 44 feet wide. At major intersections, left-turn lanes may be required in addition to four through lanes. Where left-turn lanes are required, pavement must be 64 feet wide with an 80-foot-wide right-of-way.

Local Local streets provide access to abutting property, and parking on these streets is usually permitted. The main function of these streets is to link the collector/arterial system and the low-density residential development. Also, a few business and industrial streets can be considered in this class. However, because of the potential for increased development along these streets, most commercial streets should be designed as collectors.

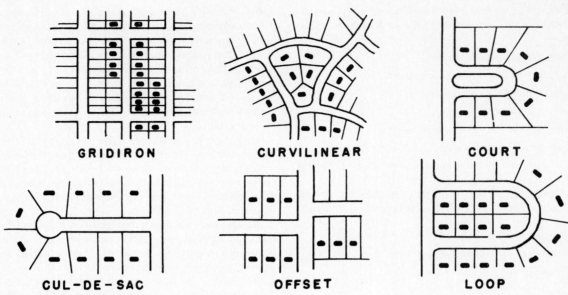

Figure 4.6 Street patterns.

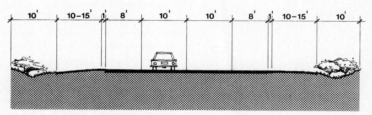

Figure 4.7 Standard road section.

Standard Criteria for Collectors*

	Ordinary terrain	Rolling terrain	Hilly terrain
Right-of-way width	60 ft (18 m)	60 ft (18 m)	60 ft (18 m)
Pavement width	36 ft (11 m)	36 ft (11 m)	36 ft (11 m)
Type of curb	Vertical face	Vertical face	Vertical face
Sidewalk width	5 ft (1.5 m)	5 ft (1.5 m)	5 ft (1.5 m)
Sidewalk distance from curb face	10 ft (3 m)	10 ft (3 m)	10 ft (3 m)
Minimum sight distance	250 ft (76 m)	200 ft (61 m)	150 ft (45 m)
Maximum grade	4%	8%	12%
Minimum spacing along major route	1300 ft (394 m)	1300 ft (394 m)	1300 ft (394 m)
Design speed	35 mph	35 mph	25 mph
Minimum centerline radius	350 ft (106 m)	230 ft (70 m)	150 ft (45 m)

*These figures are for medium-density development (2 to 6 units per acre).

Energy-Efficient Criteria for Collectors

	Alternative *a*	Alternative *b*
Right-of-way width	60 ft (18 m)	48 ft (14.5 m)
Pavement width:		
With curb	22–24 ft (6.7–7.3 m)	36 ft (11 m)
Without curb	20–22 ft + 4-ft shoulder (6–6.7 m + 1.2-m shoulder)	34 ft (10.3 m) (includes 12 ft–3.6 m—for bike path)
Sidewalk width		2 ft (0.6 m)
Sidewalk distance from curb		2 ft (0.6 m)
Minimum sight distance	250 ft (76 m)	
Maximum grade	8%	
Minimum spacing along major route	300 ft (91 m)	¾ mile (1200 m)
Design speed	35 mph	25 mph
Minimum centerline radius	350 ft (106 m)	
Maximum superelevation	0.08 ft/ft	
Maximum grade at intersection	3% for 100 ft (30 m)	

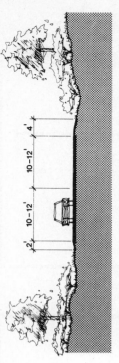

Figure 4.8 Collector street: road section—alternative *a*.

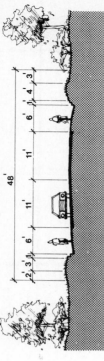

Figure 4.9 Collector street: road section—alternative *b*. (Earth Sheltered Community Design by the Underground Space Center, from Univesota of Minnesota, Copyright © 1981 by the University of Minnesota. Reprinted by permission of Van Nostrand Reinhold Company.)

Type of street	Purpose	Interval for intersections	Width of right-of-way	Paved width	Grade (percent)	traffic (miles per hour)	Sidewalks, etc.
Major roads (major arterials)	Provide unity throughout contiguous urban area; usually form boundaries for neighborhoods. Minor access control; channelized intersection; parking generally prohibited.	1½–2 miles	120–150 feet	84 feet maximum for four lanes, parking, and median strip	4	35–45	Require 5-foot-wide detached sidewalks in urban areas, planting strips (5 to 10 feet wide or wider), and adequate building setback lines (30 feet for buildings fronting on street and 60 feet for buildings backing on street).
Secondary roads (minor arterials)	Main feeder streets. Signals where needed; stop signs on side streets. Occasionally form boundaries for neighborhoods.	¾–1 mile	80 feet	60 feet	5	35–40	Require 5-foot-wide detached sidewalks, planting strips between sidewalks and curb (5 feet to 10 feet or wider), and adequate building setback lines (30 feet).
Collector streets	Main interior streets. Stop signs on side streets.	¼–½ mile	64 feet	44 feet (two 12-foot traffic lanes and two 10-foot parking lanes)	5	30	Require at least 4-foot-wide detached sidewalks; vertical curbs. Planting strips are desirable. Building setback lines are 30 feet from right-of-way.
Local streets	Local service streets; nonconducive to through traffic.	At blocks	50 feet	36 feet where street parking is permitted	6	25	Sidewalks at least 4 feet in width for densities greater than 1 dwelling unit per acre; curbs and gutters.
Cul-de-sac	Street open at only one end with provision for a practical turnaround at the other.	Only wherever practical	50 feet (90-foot-diameter turnaround)	30 to 36 feet (75-foot turnaround)	5		Should not have a length greater than 500 feet.

TYPICAL STREET CROSS-SECTIONS

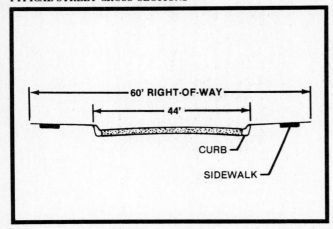

Figure 4.10

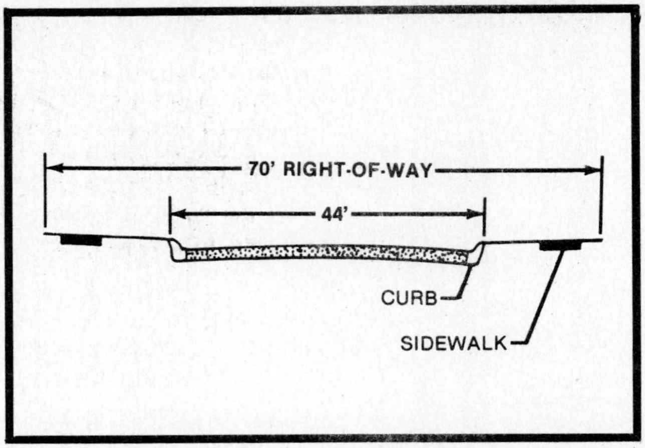

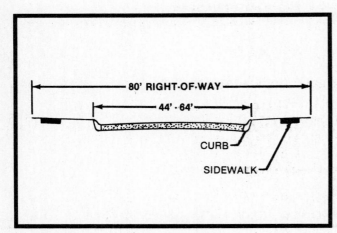

Figure 4.11

Figure 4.12

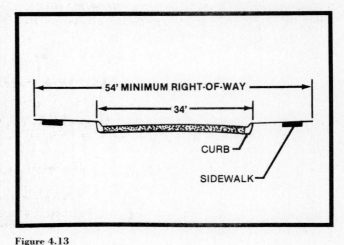

Figure 4.13

Traffic volumes on local streets should be less than 2000 vehicles per day and their length less than 3000 feet. The right-of-way requires a width of 54 feet, and the pavement should have a width of 34 feet.

Residential Streets Subcollectors are access streets that may provide frontage for residential lots. Designed to carry traffic volumes of 1000 ADT, subcollectors collect from smaller residential access streets while excluding external traffic. Sidewalks and parking may be included. Wider streets are usually required for multifamily development.

Residential streets usually have four classifications: place, lane, subcollector, or collector. Each is discussed below.

Place A place is a short street, cul-de-sac, or court, whose primary function is to conduct traffic to and from dwelling units to other streets within the neighborhood. Usually, a place is dead-end, with an ADT of less than 100 and with limited on-street parking.

Lane A lane is similar to a place in design and function, the primary difference being that a lane occasionally branches to connect two or three other lanes or places. Like a place, a lane does not serve through traffic, but its ADT range (75 to 350) is higher than that of a place.

Subcollector A subcollector with an ADT ranging between 200 and 1000, provides access to places and lanes and conducts traffic to an activity center or to a street of higher classification. The subcollector may be a loop connecting one collector or arterial street at two points, or it may be a fairly straight street conducting traffic between collector and/or arterial streets.

Collector A collector conducts traffic between arterial streets and/or activity centers. It is a principal traffic artery within residential areas and carries a relatively high ADT, ranging between 800 and 2000 vehicles.

Residential Access Roads Residential access roads are intended to carry the least amount of traffic at low speeds while providing immediate access to homes. Designed for 200 ADT, residential access roads may permit parking if twenty to twenty-five dwellings are served and front on the

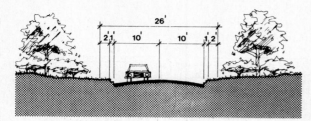

Figure 4.14 Subcollector street.

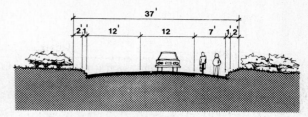

Figure 4.15 Subcollector street.

Standard Criteria for Subcollectors*

	Ordinary terrain	Rolling terrain	Hilly terrain
Right-of-way width	60 ft (18 m)	60 ft (18 m)	60 ft (18 m)
Pavement width	32 ft (9.7 m)	34 ft (10.3 m)	34 ft (10.3 m)
Type of curb	Vertical face	Vertical face	Vertical face
Sidewalk width	5 ft (1.5 m)	5 ft (1.5 m)	5 ft (1.5 m)
Sidewalk distance from curb	6 ft (1.8 m)	6 ft (1.8 m)	6 ft (1.8 m)
Minimum sight distance	200 ft (61 m)	150 ft (45 m)	110 ft (33 m)
Maximum grade	4%	8%	15%
Maximum cul-de-sac length	500 ft (151 m)	500 ft (151 m)	500 ft (151 m)
Minimum cul-de-sac radius (right-of-way)	50 ft (15 m)	50 ft (15 m)	50 ft (15 m)
Design speed	30 mph	25 mph	20 mph
Minimum centerline radius	250 ft (76 m)	175 ft (53 m)	100 ft (33 m)

*These figures are for medium-density development (2–6 units per acre).
(Source: Performance Streets.)

Energy-Efficient Criteria for Subcollectors

	Alternative *a*	Alternative *b*
Right-of-way width	50 ft (15 m) minimum	26–30 ft (7–9.9 m)
Pavement width:		
With curb	22–26 ft (6.7–7.9 m)	
Without curb	20 ft (36 ft with parking) (6 m—11 m with parking)	20–24 ft (6–7.3 m)
Sidewalk width		2 ft (0.6 m) on one side
Sidewalk distance from curb		0 ft (0 m)
Minimum sight distance	300 ft (91 m)	
Maximum grade	8%	
Maximum cul-de-sac length	1000 ft (303 m)	
Minimum spacing	125 ft (38 m)	
Design speed	30 mph	
Minimum centerline radius	140 ft (42 m)	
Maximum grade at intersection parking	5% for 50 ft (15 m), none permitted	
Driveways	permitted from 40–100 ft (12–30 m) apart	

(Source: Performance Streets.)

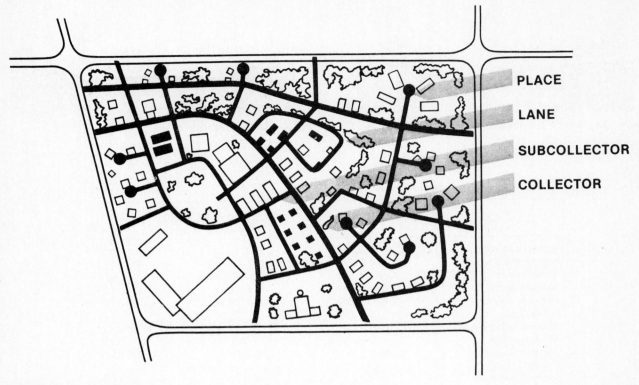

Figure 4.16 Street patterns.

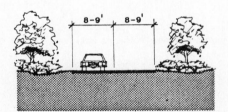

Figure 4.17 Access road section.

Standard Criteria for Residential Access Roads*

	Ordinary terrain	Rolling terrain	Hilly terrain
Right-of-way width	60 ft (18 m)	60 ft (18 m)	60 ft (18 m)
Pavement width	32 ft (9.7 m)	34 ft (10.3 m)	34 ft (10.3 m)
Type of curb	Vertical face	Vertical face	Vertical face
Sidewalk width	5 ft (1.5 m)	5 ft (1.5 m)	5 ft (1.5 m)
Sidewalk distance from curb	6 ft (1.8 m)	6 ft (1.8 m)	6 ft (1.8 m)
Minimum sight distance	200 ft (61 m)	150 ft (45 m)	110 ft (33 m)
Maximum grade	4%	8%	15%
Maximum cul-de-sac length	500 ft (151 m)	500 ft (151 m)	500 ft (151 m)
Minimum cul-de-sac radius (right-of-way)	50 ft (15 m)	50 ft (15 m)	50 ft (15 m)
Design speed	30 mph	25 mph	20 mph
Minimum centerline radius	250 ft (76 m)	175 ft (53 m)	110 ft (33 m)

*These figures are for medium-density development (2-6 units per acre).

(Source: Performance Streets.)

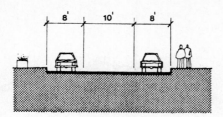

Figure 4.18 Road section — alternative *a*.

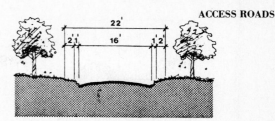

Figure 4.19 Road section — alternative *b*.

Energy-Efficient Criteria for Residential Access Roads

	Alternative *a*	Alternative *b*
Right-of-way width	50 ft (33 ft minimum) (15 m — 10 m minimum)	22 ft (6.7 m)
Pavement width:		
With curb	18 ft (5.4 m)	18 ft (5.4 m)
Without curb	16–18 ft (4.8–5.4 m)	16 ft (4.8 m)
With parking	26 ft (5.4 m)	
Sidewalk width		0–2 ft (0–.6 m) on one side
Sidewalk distance from curb		0 ft (0 m)
Minimum sight distance	250 ft (76 m)	
Maximum grade	10%	
Design speed	25 mph	10–20 mph
Minimum centerline radius	100 ft (30 m)	
Stopping distance	175 ft (53 m)	
Maximum grade at intersection	5% at 50 ft (15 m)	
Parking	8 ft (2.4 m)	None or 7 ft (2.1 m)

(Source: Performance Streets.)

street. The standard criteria for subcollectors and residential access roads do not differ. It is generally considered desirable for residential access roads to run east/west to allow for maximum solar lot configuration. North/south local roads can work efficiently for solar access in some cases.

Culs-de-sac

Physical Design Criteria Culs-de-sac, or turnarounds, are a type of access road designed to permit free turning for the largest service vehicles regularly servicing a neighborhood. The turning radius of trash collection vehicles is 28.5 to 35 feet (8.6 to 10.6 meters), suggesting that the minimum diameter of the cul-de-sac would be 70 feet (21 meters). Other standards for cul-de-sac diameters range from 70 to 80 feet (21 to 24 meters). Planted islands in the center reduce the total paved area of the cul-de-sac, as shown in Figure 4.20. The traffic lane is usually 14 feet (4 meters) wide. Recommended lengths for culs-de-sac range from 400 to 1000 feet (121 to 303 meters); a 500-foot (151-meter) length typically would servie 0.25-acre (0.1-hectare) lots. Shorter culs-de-sac help reduce traffic.

A number of alternatives to culs-de-sac are possible, depending on the lot sizes, configurations, and total number of lots served by the road. The T-type turnaround shown in Figure 4.21 requires less pavement area and space than a cul-

de-sac, but it also usually serves fewer lots. Another alternative is the loop road, which is used either when more lots are to be served or when a conventional cul-de-sac does not fit the topography. A smaller loop road can simply have a central green space with lots surrounding it; a larger loop street with lots in the central space is shown in Figure 4.22.

Energy-Efficient Strategies

- The cul-de-sac should use the smallest radius possible while still providing adequate space for service vehicles.

- The cul-de-sac should protect the easement space for use as an overhang area for occasional larger vehicles but should not have an excessively large radius.

- Mountable curbs should be used.

- The cul-de-sac should include a central planting strip to reduce the total amount of pavement.

- Designers should reconfigure the shape of the cul-de-sac to allow for service vehicles, such as snowplows, while reducing the total radius to 30 to 35 feet (9 to 10.6 meters), as shown in Figure 4.23.

- As shown in Figure 4.24, service raods for culs-de-sac should be extended by adding intermediate turnarounds rather than by using looped access roads.

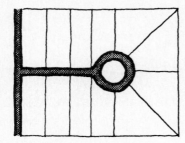

Figure 4.20 Cul-de-sac.

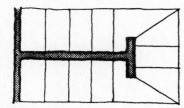

Figure 4.21 T-type turnaround.

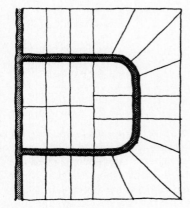

Figure 4.22 Loop road.

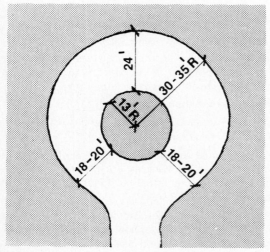

Figure 4.23 Alternative cul-de-sac dimensions.

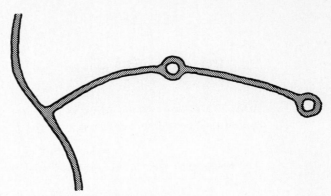

Figure 4.24 Intermediate turnaround.

Intersections

Physical Design Criteria Intersections should be carefully engineered according to design speeds, visibility, and terrain requirements.

In order to prevent dangerous jogs and turning movements, intersections are required to be either aligned directly opposite one another or offset by a minimum distance. The 125-foot (38-meter) offset shown in Figure 4.25 is considered a minimum by some sources; adjustments may be required to suit local conditions. It is also unacceptable to allow streets to intersect at a narrow angle. As shown in Figure 4.26 streets should intersect at a 90-degree angle for a minimum of 50 feet (15 meters) from the intersection.

Energy-Efficient Strategies

- The total number of gridded intersections should be reduced by using street layouts that are arranged in hierarchies.
- When appropriate, properly designed jogged intersections should be used to discourage through traffic.

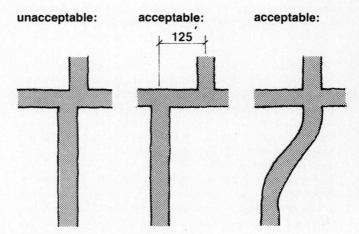

Figure 4.25 Intersection spacings.

unacceptable:

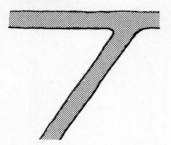

acceptable:

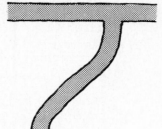

Figure 4.26   Intersection angles.

units per driveway has been recommended by the Bucks County Plan as shown in Figure 4.27, utilities could serve six to eight units adequately.

- Narrow driveways are recommended – 10 feet (3 meters) for one house, 16 feet (4.8 meters) for multiple-house service.

- Previous materials should be used wherever possible, and drainage should be designed to be collected in small collection service systems or surface drainage channels.

Driveways

Physical Design Criteria   Driveways are typically 10 to 20 feet (3 to 6 meters) wide. Usually a setback distance of 40 feet (12 meters) between an intersection and a driveway is established as a safety factor designed to avoid traffic conflicts at intersections.

Energy-Efficient Strategies

- Shared driveway access should be promoted where possible to reduce the total paved area and the numbers of intersections with subcollectors, and to provide less expensive common utility access. Although a maximum of four

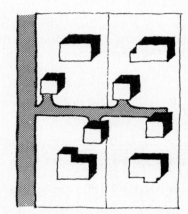

Figure 4.27   Common driveway.

Standard Criteria for Intersections

	Ordinary terrain, all densities	Rolling terrain, all densities	Hilly terrain, all densities
Approach speed	25 mph	25 mph	20 mph
Clear sight distance along each approach leg	90 ft (27 m)	90 ft (27 m)	70 ft (21 m)
Critical alignment within area	Flat	2%	4%
Minimum angle of intersection	75° (90° preferred)	75° (90° preferred)	75° (90° preferred)
Minimum curb radius:			
1. Local-local	20 ft (6 m)	20 ft (6 m)	20 ft (6 m)
2. Local-collector	25 ft (7.5 m)	25 ft (7.5 m)	25 ft (7.5 m)
Minimum centerline offset of adjacent intersections:			
1. Local-local	150 ft (45 m)	150 ft (45 m)	150 ft (45 m)
2. Local-collector	150 ft (45 m)	150 ft (45 m)	150 ft (45 m)
3. Collector-collector	200 ft (61 m)	200 ft (61 m)	200 ft (61 m)

Minimum Intersection Spacing

Major road type intersected	Spacing (ft)	(m)
Higher-order street	1000	303
Residential collector	300	91
Residential subcollector	125	38

4.5 Street Patterns

Modified Grids It is possible to avoid the monotony of the grid system (see Figure 4.28) by modifying the street pattern. A few simple street alterations can provide interesting grouping possibilities that not only eliminate the tediousness of the straight street but even provide opportunities for green areas within the groups.

The first arrangement (Figure 4.29) has a central loop that not only creates interesting grouping possibilities but also relieves the traffic load by setting up a resistance to through traffic. The problem of grouping is much less critical because there are no long, unarrested views. If the center section of the loop is retained as a green area, the grouping of surrounding houses should be governed by the fact that the park is the main focal feature. Every house should be sited to take advantage of the pleasant open space. The lot planting and house layout should stress the visual enclosure of the green.

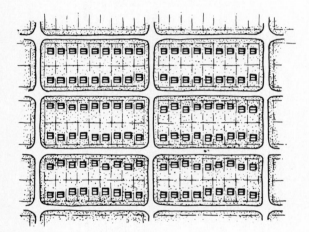

Figure 4.28 A typical grid pattern.

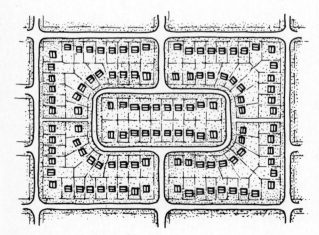

Figure 4.29 Modified grid 1.

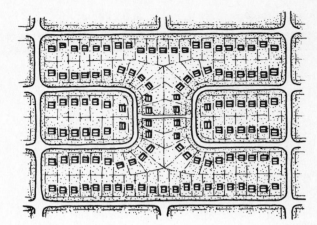

Figure 4.30 Modified grid 2.

The second arrangement (Figure 4.30) suggests another design for the same six-block area. In this case, instead of a center loop, two outside loops have been formed. The grouping problem is very similar to that in the first arrangement.

Cul-de-Sac With this type of street the grouping of houses presents a far less difficult problem than that of the straight street. A cul-de-sac is shaped so that the fronting houses automatically create an enclosed space.

A cul-de-sac that is longer than 500 feet may be too long to be effective for grouping purposes. It loses its compact shape and may have to be designed as a straight street with a distinct grouping of houses at the turnaround. The compactness of the shorter cul-de-sac usually permits a unified grouping of all the houses.

The more effective the enclosure of space at the circle, the more successful the house grouping. Some plans may look well on paper but in actuality may neither express the street shape nor enclose the space at the turnaround. This enclosure of space can be achieved by linking houses with screen walls or by carefully locating trees and hedges so that the space is defined.

Grouping 1 (Figure 4.31) uses two groups, one down the leg of the cul-de-sac and the other at the turning circle. The latter does not echo the street pattern; the houses are grouped to enclose a square terminating space.

In Grouping 2 (Figure 4.32) all the houses are illustrated as a single group. Their unbroken building line encloses a large symmetrical space that becomes the focal point of the group.

Grouping 3 (Figure 4.33) illustrates a recommended maximum-length cul-de-sac; the houses are treated as two groups. The space down the leg of the cul-de-sac has been emphasized by a definite setback, and to give the second group at the turning circle greater definition semidetached houses have been used to surround the space.

Loops Good opportunities exist for varied and interesting house groups on lots flanking looped streets. However, the typical loop arrangement (Figure 4.34) lacks design in both lot arrangement and house location. The narrow width

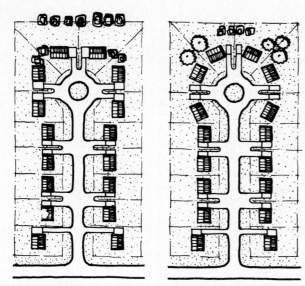

Figure 4.31   Grouping 1.   **Figure 4.32**   Grouping 2.

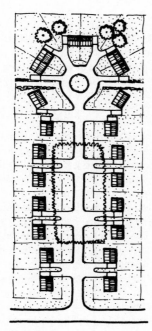

Figure 4.33   Grouping 3.

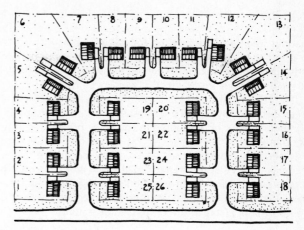

Figure 4.34   A typical loop.

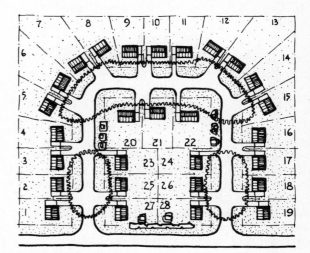

Figure 4.35   Layout 1.

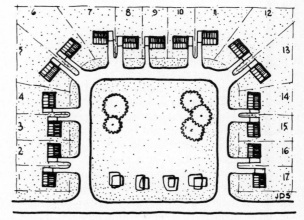

Figure 4.36   Layout 2.

at the building line of lots 5, 6, 7, 12, 13, and 14 is caused by the alignment of these lots on the radii of the looping road. The view from lots 8, 9, 10, and 11 is of backyard areas. The arrangement of houses is unsatisfactory since there is nothing to attract or arrest attention.

In Loop layout 1 (Figure 4.35) lots 5, 6, 7, 8, 12, 13, 14, and 15 have been enlarged. The interior arrangement of lots has been changed so that lots 20, 21, and 22 face the end of the

loop. This improves the outlook for all the houses at the head of the loop. A particularly pleasing feature is the openness that has been achieved by the recession of houses on lots 5 to 8 and 12 to 15. From either direction of the loop there are satisfactory focal points that can be further improved by skillful planting and landscaping.

The feature of the third layout (Figure 4.36) is the central green, which provides all the houses with a pleasant outlook. If a uniform setback is adhered to for all the houses surrounding the green, it will give strong definition to the space enclosed. Alternatively, the houses on the corner lots can be set back to give some variety to the whole group. All the houses surrounding the green should be related so that they present a unified grouping with the green as the focal point. Houses 1 and 17 act as the terminating elements.

4.6 Typical Sections of Streets and Roads

STREETS

Typical street profiles for four types of streets are shown in Figure 4.37.

ROADS

Road sections used on projects constructed within the National Park Service are designated as Types I, II, III, and IV. Type I sections are used on main circulatory roads carrying heavy volumes of traffic. The paved driving surface is 22 feet in width. Cut sections are to have a bituminous paved ditch 4 feet wide on a 5:1 slope and a uniform 1-foot paved or bituminous-treated shoulder. Fill sections should have a uniform 3-foot paved or bituminous-treated shoulder. Shoulders and paved ditch should be given a seal coat and chips, colored and textured to delineate the driving surface clearly. Transitions from the 1-foot shoulders in cuts to the 3-foot shoulders in fills should be carefully studied to provide for proper blending of the two shoulder widths into the roadway.

Type II sections are used on main circulatory roads carrying moderate traffic volumes. Design criteria conform to Type I sections except that the driving surface should be 20 feet. Typical sections for Types I and II are shown in Figure 4.38.

Type III sections may be used in lieu of Type II sections in dry and arid areas where drainage is a minor factor in design. They should be used on all secondary two-way circulatory roads such as campground and trailer access roads, picnic area access roads, residential and utility access and circulatory roads, access and circulatory roads in lodge and cabin areas, and loops and spurs serving overlooks, scenic viewpoints, interpretive features, and scenic overlooks. The driving surface of the Type III section is 20 feet wide. The shoulders may vary from 0 to 3 feet in width, depending upon the

intended use and facility to be served. It is desirable that shoulders be gravel or grass.

In general, the ditch sections are 3 feet wide and 1 foot deep, but this specification may be modified to meet special requirements such as curbs and gutters. Ditches are paved only when special drainage problems are encountered.

Type IV sections are used for all one-way circulatory roads (see Figure 4.39). The paved surface is to be 12 feet wide. One-foot grass or gravel shoulders are to be provided, unless the shoulders are used in conjunction with curbs, barriers, and so on, when the shoulder may be eliminated. In general, ditches should not be used, and the surface should be warped to provide adequate drainage. See also the accompanying slope chart.

Slope Chart

Fill (feet)	Slope	Cut (feet)	Slope	F
0–2	4:1	0–2	3:1	1
2–4	3:1	2–6	2:1	2
4–6	2:1	Over 6	1½:1	3
Over 6	1½:1	. . .	. . .	. . .

Elements of Residential Streets Curbs provide excellent control of drainage, furnish protection for pavement edge, and discourage drivers from encroaching beyond paved surfaces. For these reasons, curbs are recommended for most residential streets. However, when the sole purpose of the curb is to furnish pavement-edge protection, alternatives to curbs should be considered. (See Figures 4.40 and 4.41.)

Sidewalks along most residential collector streets are both desirable and necessary. However, for minor residential streets, sidewalks may not be justified on both sides, or even one side, of the street. In either case, before sidewalks are installed, the expected use of the sidewalk should be evaluated. For example, when residents will include children, and paved private driveways are not planned, sidewalks on at least one side of the street should be installed.

The sidewalk should be at least 4 feet wide, but wider in school areas. Also, the walks should be placed 3 to 10 feet from the curb—the greater distance being preferred along collector streets. In any case, the maximum separation of pedestrians and vehicles is desirable.

Parking for residential areas should be designed so that all residents park off street and only visitor parking overflows onto the street. For low- to medium-density housing areas, at least two off-street spaces should be provided per dwelling unit, and only parallel parking should be permitted on the street. (See Figure 4.42.)

Another item of increasing importance in residential parking is the recreational vehicle. These vehicles, as well as trailers and other special-purpose vehicles, should not be parked on residential streets, in front yards, or between residences.

Modern residential design has led to increased use of culs-de-sac and other dead-end roadways. Except on short drive-

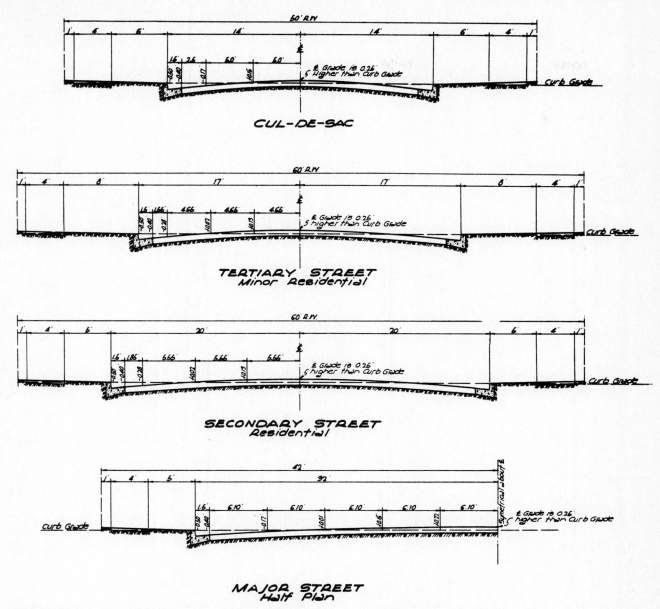

Figure 4.37   Street profiles: cul-de-sac; tertiary street/minor residential; secondary street/residential; major street/half plan.

ways serving individual dwelling units, every dead-end roadway that might be used by large vehicles should be provided with a turning facility. To accommodate a small truck and a single piece of fire equipment, a 40-foot curb radius is considered minimum. However, where parking is to be provided within the cul-de-sac, a 50-foot radius is recommended.

The usual length of street leading to a turnaround ranges from 400 to 600 feet. When using lengths longer than 500 feet, the maximum number of dwelling units that are provided access should not exceed 20.

For safety, security, and convenience, street lighting should be provided at every intersection. Energy savings cannot be justified when the trade-off involves pedestrian security. To add aesthetic value, underground wiring is recommended.

Low design speed on residential streets is desirable. A design speed of 20 to 25 miles per hour is recommended for places and lanes, and 25 to 35 miles per hour for collector streets. The lower speeds of each class should be used for hilly terrain, and the higher speeds for flat and rolling terrain.

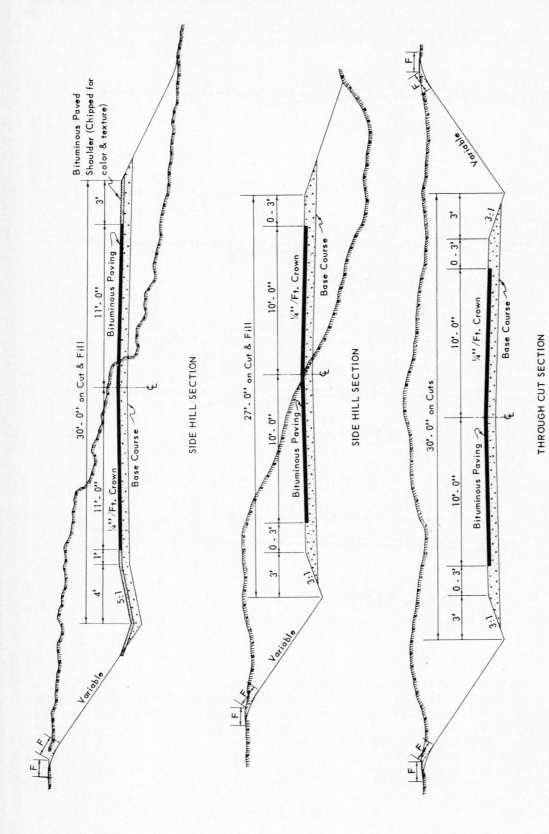

SIDE HILL SECTION

SIDE HILL SECTION

THROUGH CUT SECTION

Figure 4.38

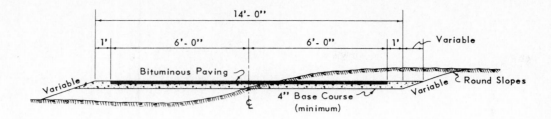

ONE-WAY ROAD
TYPICAL SECTION

NOTE: Surface to be Warped
to Provide Drainage

Figure 4.39

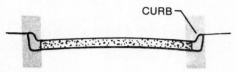

Figure 4.40

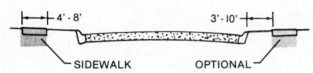

Figure 4.41

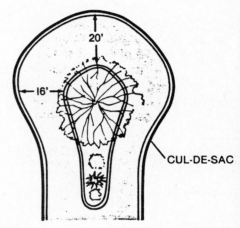

Figure 4.43

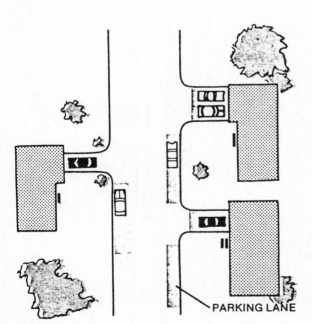

Figure 4.42

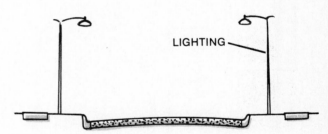

Figure 4.44

A residential roadway network should be designed to operate without any traffic control device, except along collector streets. The need for control devices can best be minimized by maintaining clear sight distance. (See Figure 4.46.) Clear sight distance, in turn, can best be provided by properly locating buildings, fences, shrubbery, or trees, and by restricting the height of any embankment. Sight distance can be controlled also by intersection location. When streets are laid out, the placing of intersections on a hilltop or slightly below a hilltop should be avoided. However, where one of these intersections must be used, the hillcrest is preferable because it offers two-directional visibility.

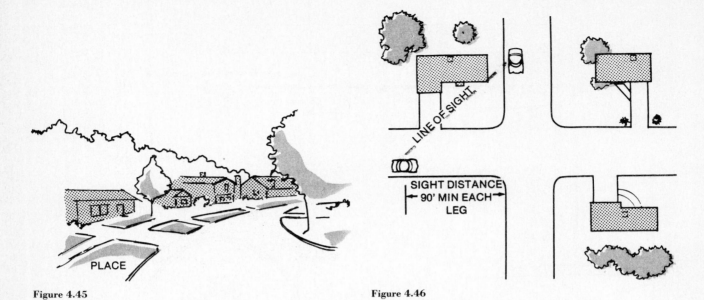

PLACE

Figure 4.45

SIGHT DISTANCE
90' MIN EACH
LEG

LINE OF SIGHT

Figure 4.46

Parking Lot Residential

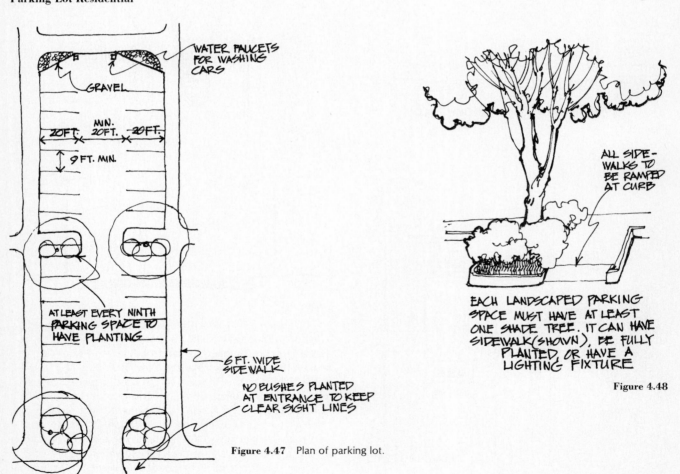

WATER FAUCETS FOR WASHING CARS

GRAVEL

20FT.   MIN. 20FT.   20FT.

9 FT. MIN.

AT LEAST EVERY NINTH PARKING SPACE TO HAVE PLANTING

6 FT. WIDE SIDEWALK

NO BUSHES PLANTED AT ENTRANCE TO KEEP CLEAR SIGHT LINES

Figure 4.47 Plan of parking lot.

ALL SIDE-WALKS TO BE RAMPED AT CURB

EACH LANDSCAPED PARKING SPACE MUST HAVE AT LEAST ONE SHADE TREE. IT CAN HAVE SIDEWALK (SHOWN), BE FULLY PLANTED, OR HAVE A LIGHTING FIXTURE

Figure 4.48

PLAN OF PARKING LOT

SHADE TREES & BUSHES IN PARKING LOTS PROVIDE SCREENS, SHADE & VISUAL RELIEF

Figure 4.49

4.7 Subdivisions

VICINITY MAP

Figure 4.50 shows a vicinity map of appropriate scale that covers sufficient adjoining territory to indicate clearly nearby street patterns, property lines, other adjacent properties in the subdivider's ownership, and other significant features that will have a bearing upon the subdivision.

CONTOUR MAP

Figure 4.51 shows a contour map of the parcel that has been made with suitable engineering accuracy and with contour intervals for determining general street and utility requirements. A contour map should show existing substantial buildings, significant trees that should be preserved, watercourses, drainage ditches, storm or sanitary sewers with size and flow-line elevation, manholes, culverts, water lines, gas lines, power lines, permanent easements, streets, and other features that will have a bearing upon the design of the subdivision or on the provision of utilities.

PRELIMINARY SKETCH

Figure 4.52 shows a sketch to designated scale of the proposed layout or alternative layouts of streets, proposed street names, lots and blocks with numbering, utility easements, storm and sanitary sewers, drainage courses, and water mains.

Streets and lots should be dimensioned to the nearest scaled foot. The sketch may be executed upon the contour map if the features required on the map are sparse enough to provide a legible and uncluttered result. The scale, north point, and total number of lots should be indicated.

GRADING PLAN

Figure 4.53 shows plans and specifications prepared for the grading plan, which should be sufficiently complete and of such engineering accuracy that they may be approved and used as final plans and specifications for rough grading.

STREET PROFILES

In selecting grades of proposed streets, consideration shall be given to topography with a view to securing safe and easy grades and avoiding unsightly and expensive cuts and fills. The subdivider shall furnish profiles of all proposed streets in the subdivisions (see Figure 4.54). The horizontal scale should be 40 feet to the inch and the vertical scale 4 feet to the inch for grades over 2 percent and 2 feet to the inch for grades of less than 2 percent.

SANITARY SEWERS

A profile of sanitary sewers is shown in Figure 4.55.

STORM SEWERS

A profile of storm sewers is shown in Figure 4.56.

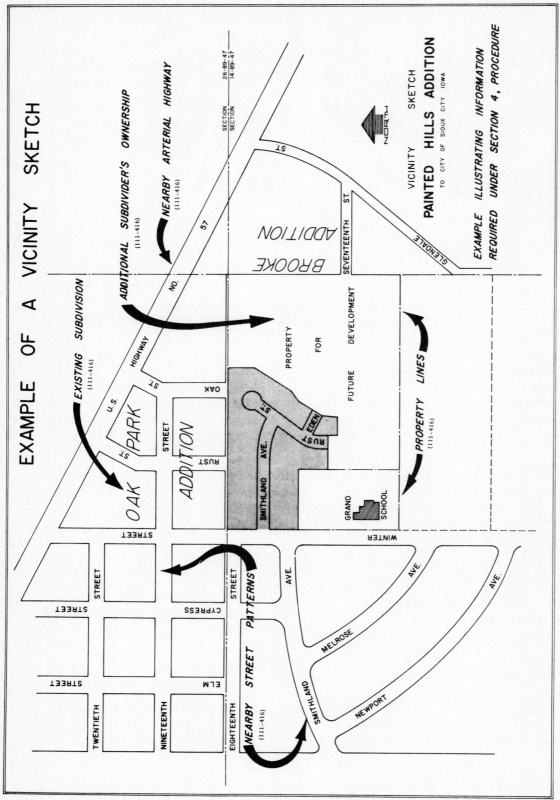

Figure 4.50 Example of a vicinity sketch.

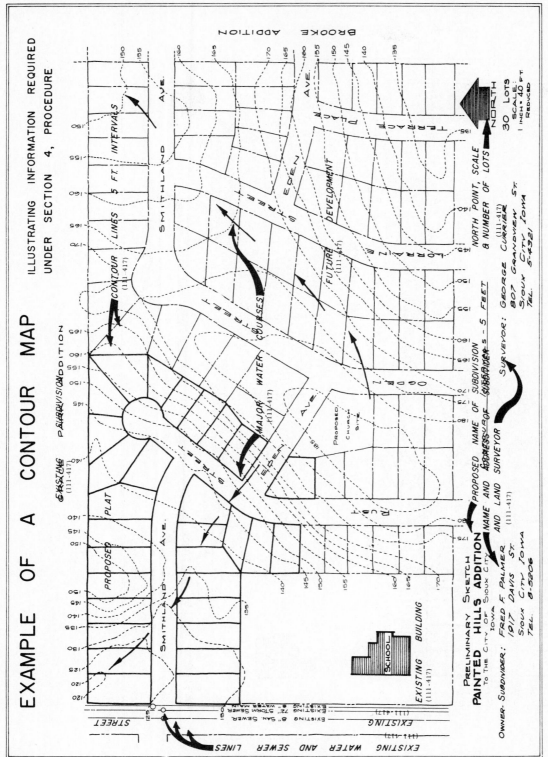

Figure 4.51 Example of a contour map.

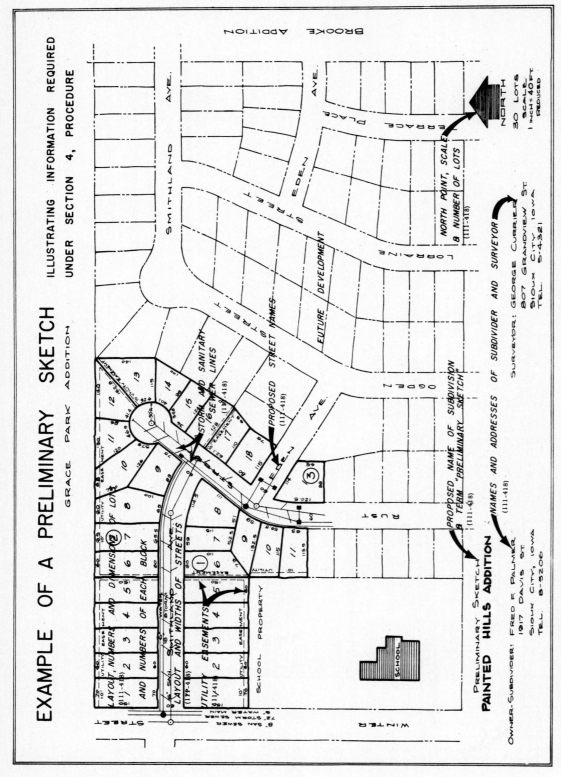

Figure 4.52 Example of a preliminary sketch.

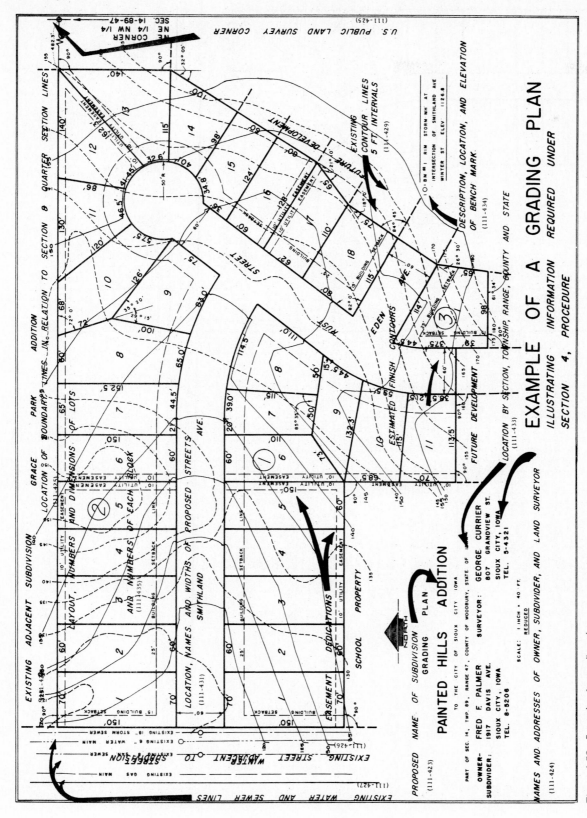

Figure 4.53 Example of a grading plan.

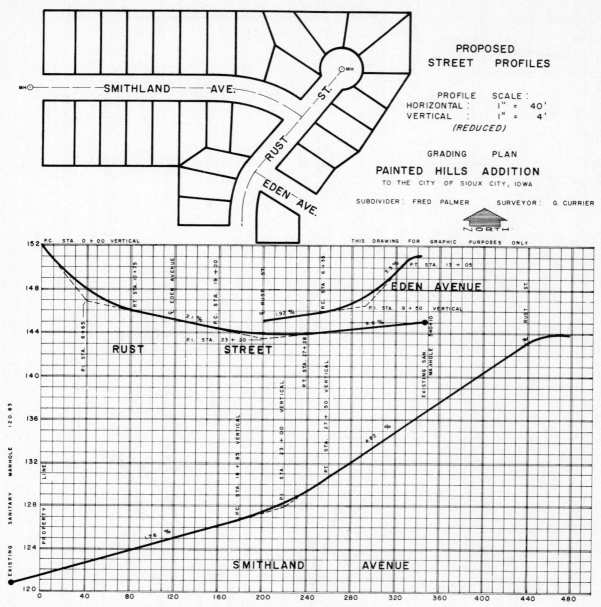

Figure 4.54 Example of a street profile to be submitted with the grading plan.

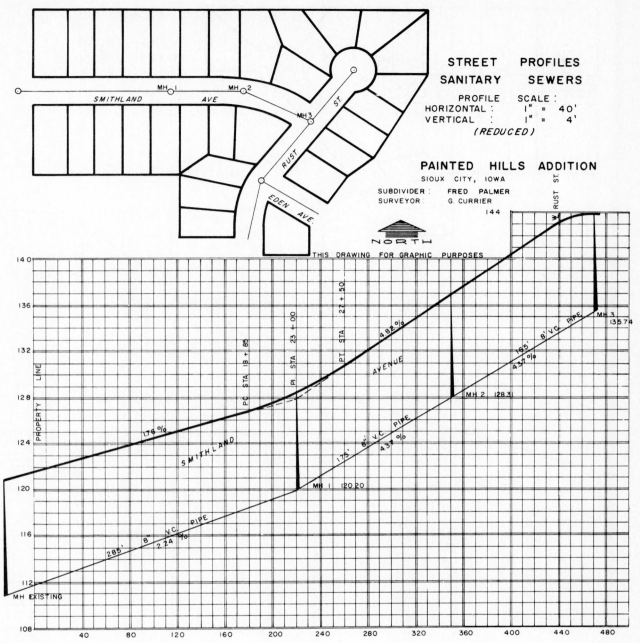

Figure 4.55 Example of sanitary sewers.

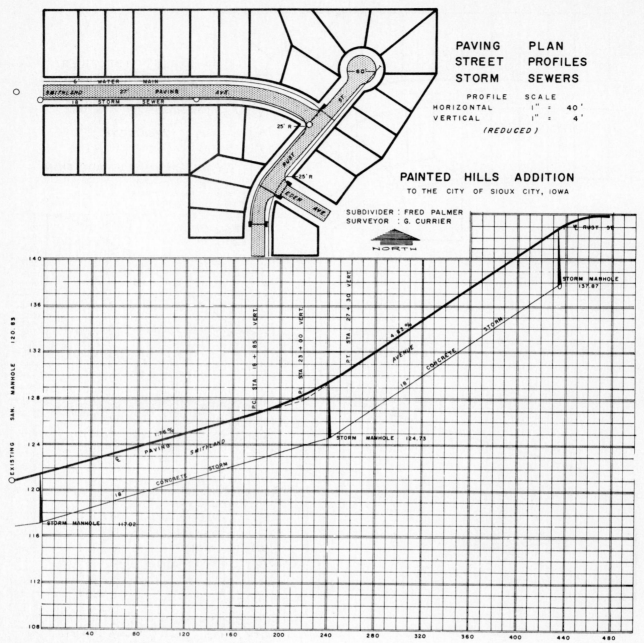

Figure 4.56 Example of storm sewers.

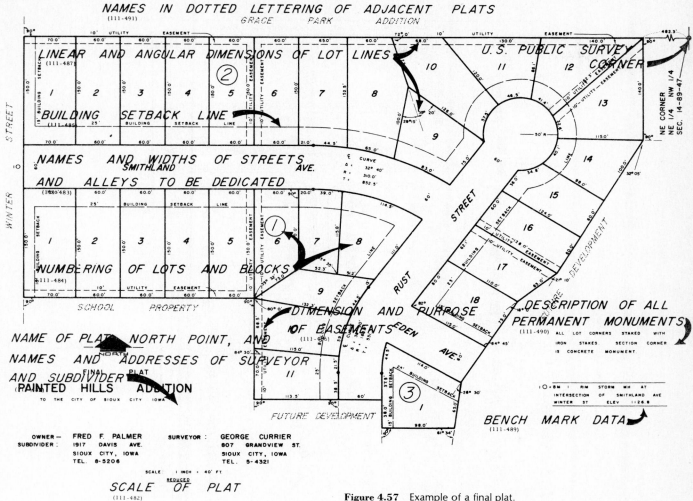

Figure 4.57 Example of a final plat.

FINAL PLAT

The final plat (Figure 4.57) should portray the following information:

1. The name of the subdivision, the points of the compass, the scale of the plat, the name of the subdivider, the date, and the name, address, and seal of the surveyor; also the location of boundary lines in relation to section and quarter section, all of which comprises a legal description of the property. All locations should be tied to a United States public-land survey corner.

2. The lines of all streets and alleys and other lands to be dedicated with their widths and names.

3. All lot lines and dimensions and the numbering of lots and blocks on a uniform system.

4. An indication of building lines with dimensions if desired.

5. Easements for any right-of-way provided for public use, drainage, services, or utilities, showing dimensions and purpose.

6. All dimensions, both linear and angular, necessary for locating the lines of lots, tracts, or parcels of land, streets, alleys, easements, and the boundaries of the subdivision. The linear dimensions are to be expressed in feet and decimals of a foot. The plat should show all curve data necessary to reconstruct on the ground all curvilinear boundaries and lines and the radii of all rounded corners.

DESIGN CONSIDERATIONS

In Figure 4.58 various design considerations for subdivisions are illustrated. Keyed to the illustration are twenty-six important points:

1. A 15-foot easement for a planting screen provides protection from nonresidential use.

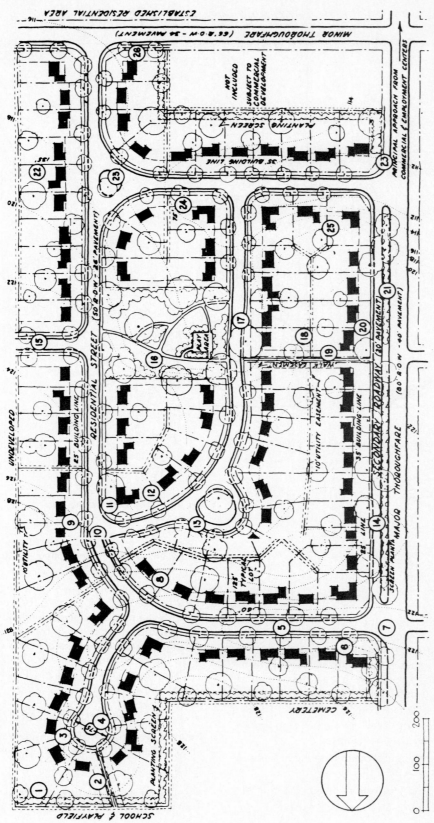

Figure 4.58 Design considerations for subdivisions.

2. A 10-foot walk easement gives access to the school.

3. A cul-de-sac utilizes an odd parcel of land to advantage.

4. A turnaround right-of-way is 100 feet in diameter.

5. Street trees are planted approximately 50 feet apart where no trees exist.

6. An additional building setback improves the subdivision entrance.

7. Street intersections at right angles reduce hazards.

8. The lot side line is centered on the street end to avoid car lights' shining into residences.

9. Residences opposite the street end are set back farther to reduce glare from car lights.

10. Three-way intersections reduce hazards.

11. Property lines are on 30-foot radii at corners.

12. Lot lines are perpendicular to street right-of-way lines.

13. An "eyebrow" provides frontage for additional lots in a deeper portion of the block.

14. A secondary roadway eliminates the hazard of entering a major thoroughfare from individual driveways.

15. There is provision for access to land now undeveloped.

16. A neighborhood park is located near the center of the tract. Adjacent lots are wider to allow for a 15-foot protective side-line setback.

17. The pavement is shifted within the right-of-way to preserve existing trees.

18. Aboveground utilities are in rear-line easements.

19. A 10-foot walk easement provides access to a park. Adjacent lots are wider to allow for a 15-foot protective side-line setback.

20. Variation of the building line along a straight street creates interest.

21. Screen planting gives protection from noise and lights on the thoroughfare.

22. Lots backing to uncontrolled land are given greater depth for additional protection.

23. Low planting at street intersections permits clear vision.

24. A wider corner lot permits equal building setbacks on each street.

25. Platting of the block end avoids siding properties to residences across the street.

26. Lots are sided to the boundary street where land use across the street is nonconforming.

4.8 Diagram of Gross Lots per Acre

When a site is studied for a possible subdivision, a critical determination is the number of lots per acre and the amount of open space for streets and park areas. The diagram in Figure 4.59 provides a quick means of determining the relationship of the number of lots per gross acre and percentages of open space. It is assumed that the site is level and entirely buildable, without steep slopes, marshy land, or other obstructions.

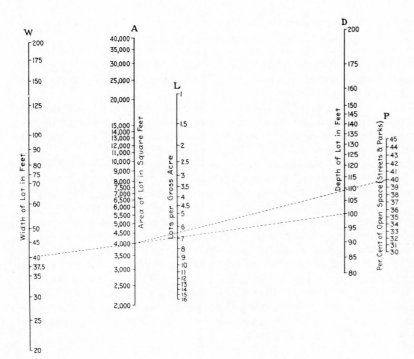

Figure 4.59 Diagram for determining lots per gross acre for varying lot sizes and percentage of open space in streets and parks. To use the diagram, start with values on the W and D scales. Lay a straightedge between them and read the area of the lot on the A scale; then choose a value on the P scale. Lay a straightedge between this value and the determined value on the A scale. Read the required answer on the L scale. In the example shown, W = 40 feet; hence A = 4000 square feet. With P = 40 percent, L = 6.5 lots per gross acre.

4.9 Types of Open Space

Orientation	Function	Space, design, and service area	Example
Home-oriented space	Should meet aesthetic qualities and accommodate informal activities of an active and passive nature, i.e., sitting, reading, gardening, sunning, children's play, and family activity	Varies according to housing type; immediately adjacent or within 500 feet of each dwelling unit	Front and back yards, driveways, sidewalks, porch, balconies, workshops, play rooms, recreation rooms
Home cluster or subneighborhood common space	Especially important in high-density areas, providing visual relief and aesthetic qualities for activities similar to those mentioned above, as well as meeting areas for small informal groups, walking, jogging, and dog walking	Must be visually accessible; varies from 500 square feet to 2 acres; designed to be as flexible and adaptable as possible; will serve an area of 100 yards to ¼ mile radius	Vacant lots, culs-de-sac, boulevards, green belts, walkways, trails, play lots, rest areas, vest-pocket parks, parkettes
Neighborhood space	Should accommodate neighborhood interest preferences; may include sports areas for minor leagues, outdoor skating rinks, water play, as well as special events and informal passive activities	Space should be associated with an elementary school; varies from 4 to 20 acres; will serve 5000 people within an area of ¼ to ½ mile radius	Neighborhood parks or park-school combinations; play fields for baseball, soccer, and football; adventure playgrounds, wading pools, neighborhood centers
Community space	Should accommodate social, cultural, educational, and physical activities of particular interest to the community; multipurpose, year-round, day/night activities; low-level competitive sports with limited spectator space	Space should be associated with a secondary school; varies from 15 to 20 acres; will serve several neighborhoods of 15,000 to 25,000 people within a radius of ½ to 1½ miles; accessible by walking, cycling, and public transit	Community park or park-school combinations; facilities for playgrounds, recreation center, meeting rooms, and library; track and field areas, sports fields, arena, and swimming pool

See Figures 4.61 through 4.66 for illustration of the text below.

Open-Space Alternatives   Seven alternative designs for the use of open space are shown in Figure 4.60.

Primary Space Types   There are three types of primary space, defined by their location on a typical block. These are the corner plaza, the mid-block plaza, and the through-block plaza. Appropriate size, proportion, and accessibility are determined by their location on a block.

Corner Plaza   A corner plaza is bounded by two intersecting streets. It is located at the intersection of major traffic circulation points and therefore is busy and noisy; yet it is highly visible to pedestrians approaching from several directions. This type of plaza would be an ideal location for a sidewalk café, a kiosk, or other uses which generate and are enhanced by pedestrian activity.

Mid-Block Plaza   A mid-block plaza is one which faces one street and is surrounded on three sides by buildings. The site is relatively free from noise, is not easily visible, and therefore lends itself for use as a quite area.

Through-Block Plaza   A through-block plaza is bounded by at least two parallel streets. It can serve as a pedestrian link between two side streets, as well as a resting place.

Residential Plaza   The residential plaza is an open space located at the base of a residential development which pro- vides amenities for both the public and the resident of the development. Residential plazas may contain two types of space, primary space and residual space.

Primary Space   A primary space is the larger portion of the residential plaza. It is where major recreational activity and public use occurs and occupies at least 60 percent of the plaza area.

Residual Space   A residual space is the remaining portion of a residential plaza. It usually surrounds a building and may be used as a visual landscaped amenity. It comprises not more than 40 percent of the total plaza area.

Mandatory Amenities   Mandatory amenities are those physical elements basic to all plazas. Regardless of their location, all primary spaces must provide seating, trees, bicycle racks, and a drinking fountain. Additional amenities are those elements that provide optional variety in the design of plazas.

Tree Planting   All primary spaces should provide a minimum of one tree per 1000 square feet of primary space. The tree should be 3½ to 4 inches in diameter. Each tree should be planted in at least 100 cubic feet of soil with a depth of soil not less than 3 feet 6 inches to ensure that it survives the urban environment. Trees should also be planted either flush to grade or in a planting bed with a minimum continuous area of 75 square feet.

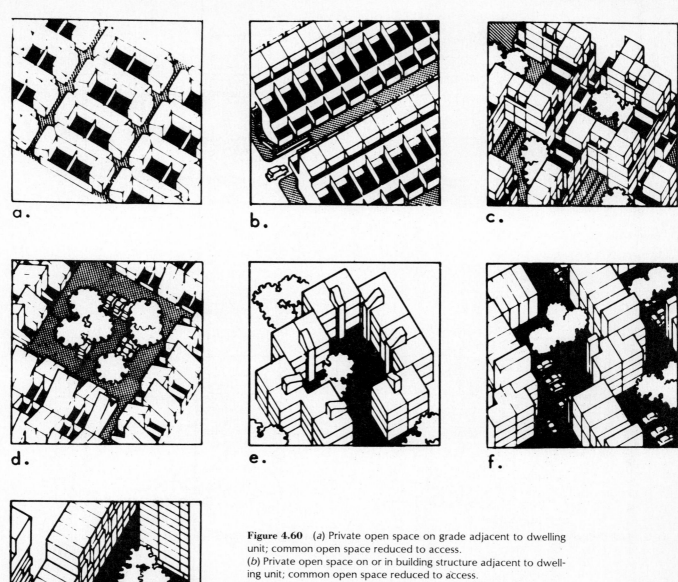

Figure 4.60   (*a*) Private open space on grade adjacent to dwelling unit; common open space reduced to access.

(*b*) Private open space on or in building structure adjacent to dwelling unit; common open space reduced to access.

(*c*) Private open space on grade or on or in building structure adjacent to dwelling unit; common open space shared by groups of dwelling units.

(*d*) Private open space on grade or on or in building structure adjacent to dwelling unit; common open space integrated with parking shared by groups of dwelling units.

(*e*) Common open space shared by groups of dwelling units.

(*f*) Common open space integrated with parking and shared by groups of dwelling units.

(*g*) Common open space shared by all dwelling units.

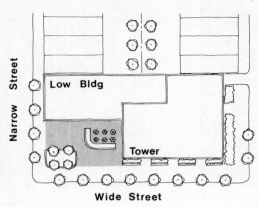

Figure 4.61 Corner plaza.

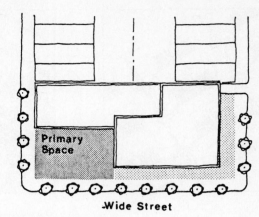

Figure 4.64 Primary space.

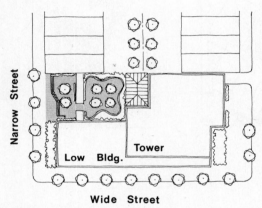

Figure 4.62 Mid-block plaza.

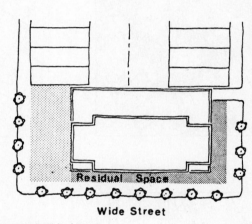

Figure 4.65 Residual space.

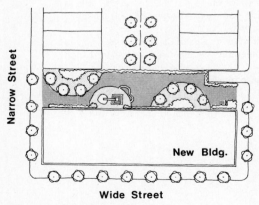

Figure 4.63 Through-block plaza.

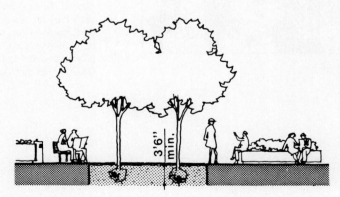

Figure 4.66 Tree planting.

Seating   One linear foot of seating must be provided for each 30 square feet of primary open space. As an example, on a 4000-square foot space area, at least 133 linear feet of seating should be provided. The seating should have a minimum depth of 1 foot 4 inches. Seating with backs at least 1 foot high should have a minimum depth of 1 foot 2 inches. Seating 2 feet 6 inches or more in depth should count double, provided there is access on both sides. Ten percent of the required seating should have backs, for the convenience of the disabled. Seating higher than 3 feet and lower than 1 foot above ground level, as well as steps, do not count toward the seating requirements.

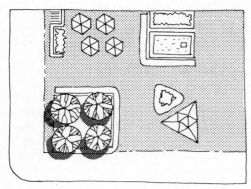

Figure 4.67   Seating area.

Bicycle Parking Facilities   Provisions are made for the inclusion of at least two bicycle parking spaces for every 1000 square feet of primary open space. A typical plaza area in the city will provide space for eight bicycles.

Figure 4.68   Bicycle parking facilities.

Drinking Fountain   Fresh drinking water is guaranteed to plaza users through the mandatory requirement of a drinking fountain for each plaza. If two drinking fountains are provided, one should be 2 feet 6 inches high, operated by hand and foot, and display the International Symbol of Access for the conveninece of the disabled.

Figure 4.69   Drinking fountain.

Planting   Planters, including hanging planters, or planting beds with seasonal flowers, shrubs, or ivy, or other plants are to occupy an area not less than 150 square feet for each 1000 square feet of primary space, when provided as an additional amenity. The area occupied by a single permanent planter or planting bed should be at least 30 square feet with a depth of soil of at least 1 foot 6 inches to guard against uprooting. Hanging planters should be exempt from these minimum size and placement requirements.

Figure 4.70   Planting.

Game Tables   Game tables for chess, backgammon, or checkers attract young and old. Enough game tables and seating to accommodate 16 people per primary space may be provided as an additional amenity.

Figure 4.71   Game tables.

Artwork A work of art such as a sculpture or mural lends visual excitement to the atmosphere of plaza activity and may be provided as an additional amentiy.

Figure 4.72 Artwork.

Fountains and Pools As moving water has the property of alleviating noise and cooling spaces, the provision of a fountain or a reflecting pool, occupying not less than 300 square feet, is an added attraction to a plaza and may be provided as an additional amenity.

Figure 4.73 Fountains and pools.

Play Equipment Play equipment is recommended in those areas with large populations of children. One play facility is recommended for each 1000 square feet of primary space, as an additional amenity. For safety, play equipment should not be located within 40 feet of any wide street frontage, and play equipment — climbers, swings, paddle pools, or similar facilities — are to meet the safety standards of the Federal Consumer Product Safety Council.

Figure 4.74 Play equipment.

Kiosk A kiosk is a one-story structure, made of predominantly light materials, such as glass, plastic, metal, or fabric. A kiosk may be a free-standing structure or may be attached to a wall of an adjoining building. The structure may house newsstands, flower stands, take-out food stands, and information booths. Kiosks in small primary spaces, of 2000 to 4000 square feet, should not exceed 60 square feet in area. Kiosks in spaces greater than 4000 square feet cannot exceed 100 square feet in area.

Open Air Café An open air café can promote the outdoor use of a residential plaza. Open air cafés should be permanently unenclosed, even though they may have a temporary fabric roof. No kitchen equipment should be installed within the open air café. A kiosk through which water or table service is provided may be joined to the café. Cafés should not occupy more than 20 percent of the total area of primary space.

Optional Amenities The primary space may also include additional numbers of the amenities mentioned above and other amenities such as arbors, trellises, litter receptacles, outdoor furniture, flagpoles, public telephones, awnings, canopies, bollards, subway station entrances, and drinking fountains which are operable by wheelchair users. The total area occupied by all amenities should not exceed 60 percent of the total primary space area to ensure ample pedestrian circulation space in the plaza.

Treatment of Adjoining Walls As plazas may often be enclosed by the party walls or adjacent buildings, treatment of these walls is recommended. The walls of a building adjoining a plaza should be covered with ivy, greenery, or artwork or otherwise decoratively treated. The wall surface should be treated up to a height of 20 feet, or the entire first floor.

Figure 4.75 Treatment of adjoining walls.

Lighting Provision of illumination during the hours of darkness increases plaza safety and usage. Lighting is important particularly for mid- and through-block plazas where the use is restricted and the visibility limited. The requirement that lighting intensity be not less than 2 horizontal foot-candles during hours of darkness will provide adequate illumi-

nation without creating glare visible in adjoining residential apartments.

Paving   Decorative paving is desirable for all residential plazas. Nonskid surface materials, such as brick and quarry tile, of varied dyes, aggregates, and textures are suggested. The materials used for paving must be extended to the curb line to render visual continuity along the sidewalk and to serve as a unifying element binding the plaza area to the street.

Access   Plazas should be designed to become an extension of the sidewalk, and to ensure openness and visibility. Accessibility to plazas is a function of their location and the design of the plaza entrance. For example, a plaza on the corner is more accessible and publicly visible that one at midblock. The elevation of the plaza is also crucial. Sunken plazas or raised plazas separate the spaces from street visibility and accessibility.

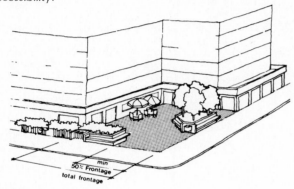

Figure 4.76   Corner access.

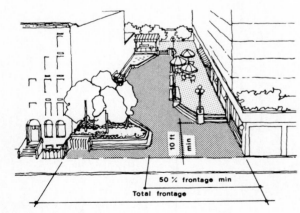

Figure 4.77   Interior access.

Access for the Disabled   As the disabled have been historically subjected to unnecessary barriers in the man-made environment, at least 60 percent of the primary plaza space should be designed for unobstructed use. Ramps should be provided alongside of steps which permit access to primary spaces. Ramps are to have a 3-foot minimum width and should be sloped no greater than 1 inch vertically for every 12 horizontal inches. Stairs should have no projecting nos-

ings, and both ramps and steps should provide lowered handrails. In addition, plazas should provide one major path at least 5 feet wide that traverses major portions of the plaza area. Nonskid surfaces are mandatory.

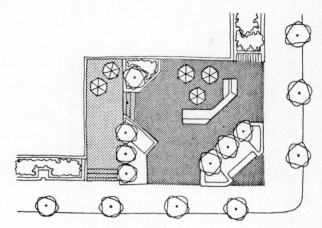

60% of Unobstructed primary space to be accessible to the disabled.

Figure 4.78   Barrier-free access.

Elevation   Plazas are encouraged to be at sidewalk grade. However, a 3-foot differential between the sidewalk and the plaza is permitted for physical definition of spaces as well as to encourage design flexibility. Mid- and through-block plazas must provide a clear 10-foot margin from the sidewalk before the base level of the plaza is raised or lowered.

Figure 4.79   Plaza elevations.

Residential Plazas

Figure 4.80

Figure 4.81

Shopping Plazas

Figure 4.82

Figure 4.83

Community Gardens

Site Development Community gardens can become major recreation resources in neighborhoods. A well-designed and maintained site provides opportunities for gardening, picnicking, outdoor games, and enjoyment of natural resource beauty.

Individual plot size is an important consideration and varies with the age and desires of the gardeners. For children's gardens, plots not to exceed 100 square feet are suggested. For adult gardeners, plots of 200 to 1000 square feet are realistic. In most operations the basic plot size made available to gardeners is 25 by 30 feet. Those desiring additional garden space then rent in multiples of the basic unit. Oversize plots sometimes result in discouragement of inexperienced gardeners and unsightly weed problems.

Access to the plots is provided by walkways or aisles. These aisles and walkways take different forms, depending on provisions by the management. Some projects provide a walk around each individual garden. Other projects provide access to each plot from one side only, except for the end plots which are accessible from two sides. Minimum width for walkways is 4 feet. Most groups either leave walkways in grass or tramp the dirt and cover with wood chips. Although wood chip paths are attractive and need little maintenance once in place, they are difficult to navigate with garden carts, wheelchairs, or baby strollers.

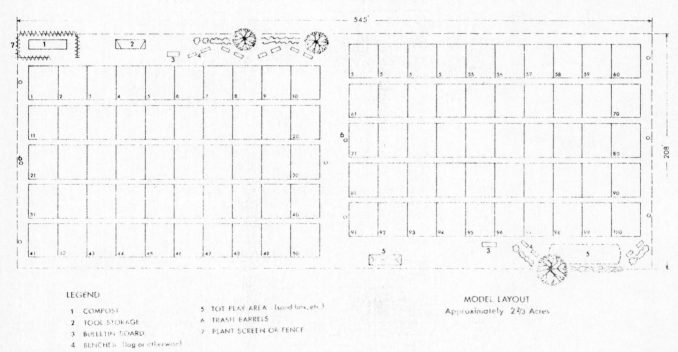

LEGEND

1 COMPOST
2 TOOL STORAGE
3 BULLETIN BOARD
4 BENCHES (log or otherwise)

5 TOT PLAY AREA (sand box, etc.)
6 TRASH BARRELS
7 PLANT SCREEN OR FENCE

MODEL LAYOUT
Approximately 2 2/3 Acres

Figure 4.84 Typical layout for community gardens.

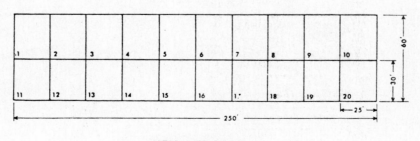

ALTERNATE PLAN
To Conserve Space

Figure 4.85 Alternate plan to conserve space.

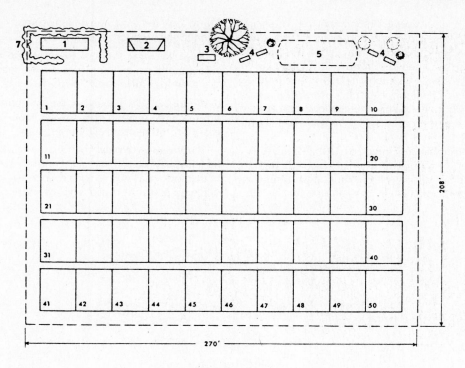

SITE PLAN
Approximately 1⅓ Acres

Figure 4.86 Site plan — approximately 1⅓ acres.

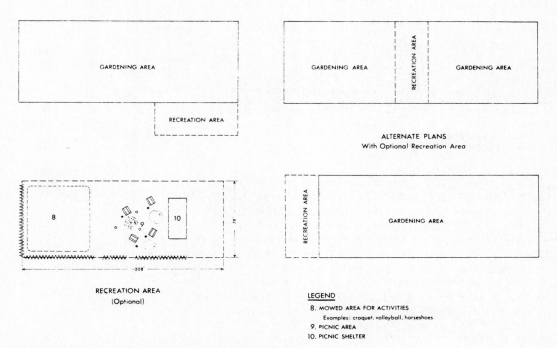

Figure 4.87 Layout of recreation area and alternate location.

Retreats   In addition to programmed facilities for social-izing, game playing, and gardening, people have a need for more quiet and secluded outdoor areas. They occasionally retreat to these areas when they are in a contemplative mood, want to take a walk without meeting others, or need a change of scenery. An important aspect of retreats is that they offer additional choices. While the goal of a retreat may be a quiet nice place, the process of getting there may be just as important.

Natural Features   Take advantage of natural features like tree groves, a river, or a pond to provide places to walk to on the site at some distance from housing, somewhat removed from sight, and with a natural vista if possible.

These can become convenient retreats for residents who want to get away from it all.

Built Retreats   Where there are no natural features to take advantage of, construct special features as retreats — like gazebos, duck ponds, or picnic areas.

Pathways   To maximize use of retreats, locate natural trails or create planned pathways from housing to retreats easily accessible from all housing units.

Challenge   Although barriers to mobility must generally be minimized along main paths, include some secondary trails with more challenging terrain such as steeper grades and unpaved paths for those who are more capable.

Figure 4.88

4.10 Lighting of Site

Lighting All access roads, paths, and parking areas must be adequately lighted at night. It is desirable to provide outdoor lighting around community facilities and some highly used recreation areas. Fixtures should not look institutional yet should be vandalproof. Lights should be of unbreakable plastic, recessed, or otherwise designed to reduce the damage and replacement problem.

The location, design of pole and fixture, and light intensity should be considered. The intensity, for example, may be varied — either increased or decreased — to change the mood and scale of a court or walkway from that of a parking area. Try to install fixtures of a scale and light intensity appropriate to the setting.

Walkway Lighting Walkways can be lighted with globe or hooded fixtures. Globe fixtures illuminate the walks as well as the trees. Hooded fixtures keep direct light away from windows of dwelling units close by. Both types can have varying intensities of light to match circumstances. Higher intensities are needed at street intersections and large open areas, and lower intensities along backyards, etc.

Dwelling Unit and Entrance Court Lighting These should be of an individual unit scale; low-intensity fixtures can be grouped together or installed at frequent intervals to provide enough light without being glaring to the pedestrian.

Fixtures for the above two lighting purposes should be mounted between 9 and 12 feet above finished grade and shielded to screen windows of dwelling from the direct light rays. For these purposes, only incandescent lamps should be used; mercury vapor is too bright.

Parking and Service Lighting Parking lots should be lit with enough intensity to discourage vandalism and create security. Parking lots should be lit by lights on poles at least 20 feet high and should not be installed on dwelling roofs. All lights should be hooded.

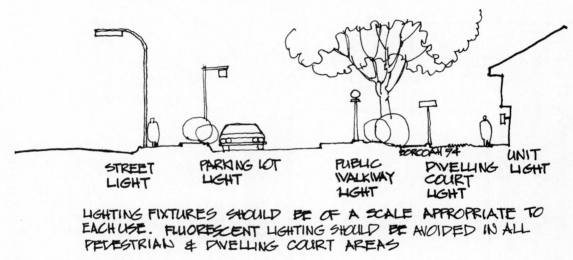

LIGHTING FIXTURES SHOULD BE OF A SCALE APPROPRIATE TO EACH USE. FLUORESCENT LIGHTING SHOULD BE AVOIDED IN ALL PEDESTRIAN & DWELLING COURT AREAS

Figure 4.89 Scale of lighting fixtures.

PUBLIC WALKWAY LIGHTING LOCATED AT FREQUENT INTERVALS TO AVOID DARK SPOTS

Figure 4.90 Placement of lighting fixtures.

TYPES OF PUBLIC WALKWAY LIGHTING FIXTURES

Figure 4.91 Types of public walkway lighting fixtures.

4.11 Fire Protection and Access

Site Access for Fire Apparatus The building site plan should provide adequate driveway widths, turning radii, and parking space on firm, level surfaces for fire apparatus. Avoid man-made and natural barriers that could interfere with movement of fire vehicles. Fire apparatus turning radii (R) typically vary from 28 to 40 feet and vehicle length (L) from 40 to 65 feet for ladder trucks and from 20 to 40 feet for pumpers.

Turning Clearances for Fire Apparatus in Driveway The table in Figure 4.93 gives the preferred minimum distance (d) in feet for equal legs of the right-triangle-shaped open area for "turning clearance." Table values are presented for private driveway width (W) and vehicle length (L).

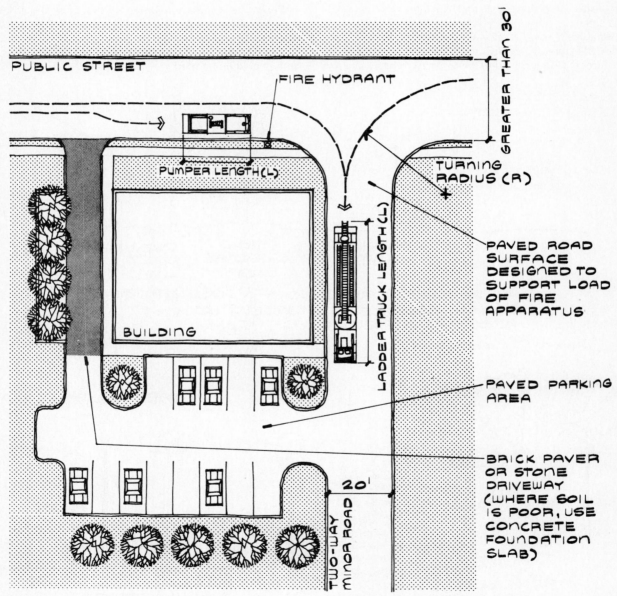

Figure 4.92 Site access for fire apparatus.

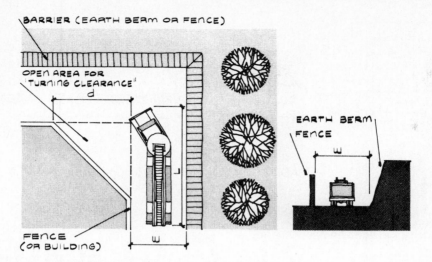

d Value Table

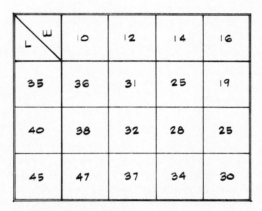

L\W	10	12	14	16
35	36	31	25	19
40	38	32	28	25
45	47	37	34	30

L AND W ARE IN FEET

Figure 4.93   Turning clearances for fire apparatus.

4.12 Block and Lot Grading

The raw land that forms the site for building houses should be converted to its new use in accordance with a design that employs natural topography to the best advantage. The levels of streets and buildings must be established so that the surface of the land may be shaped and stabilized to provide for the runoff of rain and melting snow. Permanent features such as driveways, retaining walls, trees, and hedges are part of the landscape design.

Drainage   The principal objective in shaping the ground is to provide satisfactory surface drainage so that water will flow away from buildings and be carried off by storm sewers or ditches. Areas that are not covered by buildings or paving must be provided with a permanent surface of grass or other material that will not be eroded by water and wind. For this reason there are great advantages in locating houses and streets so that the existing topsoil and trees are disturbed as little as possible. The replacement of this permanent groundcover is in itself an expensive landscape operation.

The most desirable topographical condition exists when the highest land forms a ridge along the rear lot lines of a block with a gentle slope toward the street on each side. Water will then flow toward the streets. Protective slopes are required to drain surface water away from the walls of buildings and from backfilled areas. To maintain the flow of surface water between buildings and to prevent pools from collecting in the rear of buildings, drainage swales are needed; these are formed by a slight dishing of the ground to provide natural channels. Grading layout 1 in Figure 4.94 illustrates such a grading plan by which water is moved from the rear yard and finally to the street for disposal.

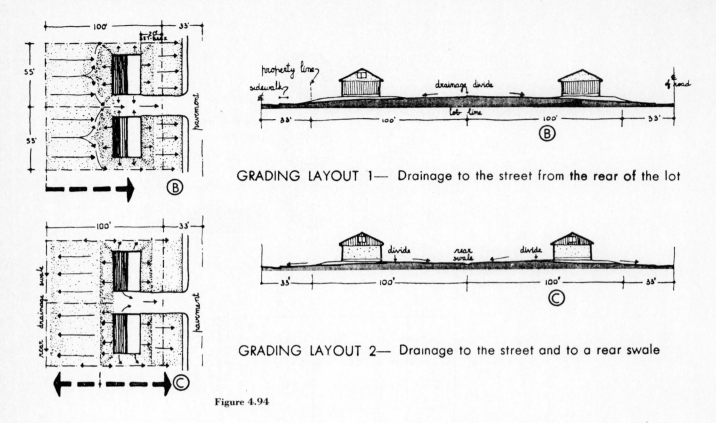

GRADING LAYOUT 1— Drainage to the street from **the rear of** the lot

GRADING LAYOUT 2— Drainage to the street and to a rear swale

Figure 4.94

For relatively flat land, as in grading layout 2, the surface water from the back-garden areas of lots can be drained to a rear swale, possibly designed as an easement, along the rear lot lines. In such an instance, the drainage divide occurs along the back edge of the landscaping plinth or protective slopes around the dwellings. All the remaining portion of the lot, including the front garden, is drained out to the street.

The slope across a block of houses may be such that lots on one side are higher than those on the other. In this instance, grading layout 3 (Figure 4.95), the drainage divide may occur along the front property line of the higher-level lots, with all drainage behind this line being directed between the dwellings and across the rear gardens. Intercepting drainage swales or easements must be provided along the rear lot lines or at intermediate locations leading out to the lower-level street. Such drainage easements must be permanently established by proper legal methods with continuous maintenance assured. Lots on the lower side of the block should be drained out to the street.

In another very common situation, a cross slope runs diagonally across a block and hence diagonally across each individual lot, either from back to front or from front to back. Grading layout 4 (Figure 4.95) shows such a situation. Drainage of these lots is more complicated, and special attention is needed to ensure that the width of individual lots is adequate to allow for protective slopes on either side of the dwelling, for the ground to fall away following the line of the

slope, and for side swales. Small retaining walls or steep banks may be necessary along side lot lines, and these should be carefully carried out so that the stepping up of the slope as seen from the lower end of the street does not have a jerky and disorganized appearance.

The most important factor in lot grading and drainage is the relationship of house floor elevation to street elevation. If the floor elevation is too low in relation to adjoining street grades, adequate protective slopes and drainage swales cannot be provided to drain the lot satisfactorily. If the floor elevation is too high, unnecessary terracing and outside stairs will be required, and the house will appear to be dissociated from the natural lie of the land.

Block Grading Patterns

LOT GRADING

Two types of lot grading, one in which all drainage is to the street and the other in which drainage is both to the rear and to the street, are shown in Figures 4.101 and 4.102.

Storm Drainage A valley may be created along rear lot lines to drain surface water from rear yards to a swale that terminates at a cross street or a rear-of-lot catch basin (Figure 4.103). The swale is usually located in an easement to facilitate access by municipal maintenance staff. The front yard

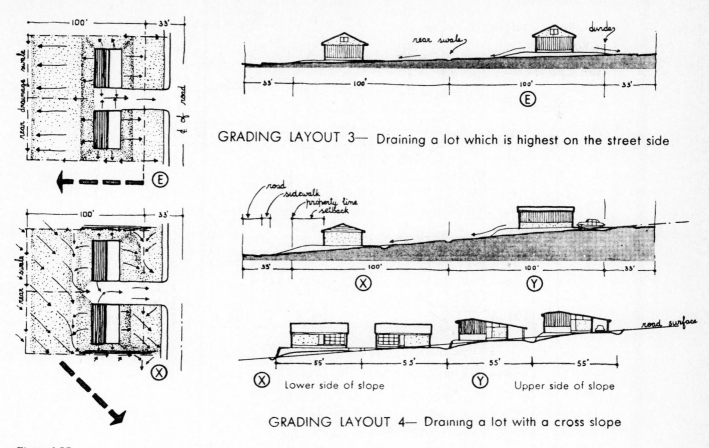

GRADING LAYOUT 3— Draining a lot which is highest on the street side

GRADING LAYOUT 4— Draining a lot with a cross slope

Figure 4.95

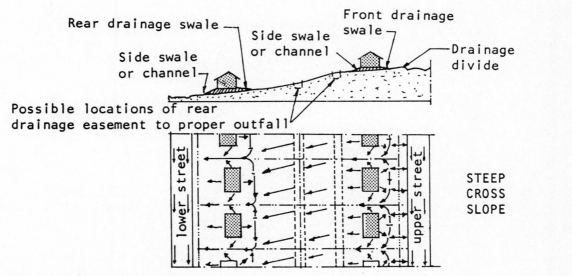

Figure 4.96   Steep cross slope.

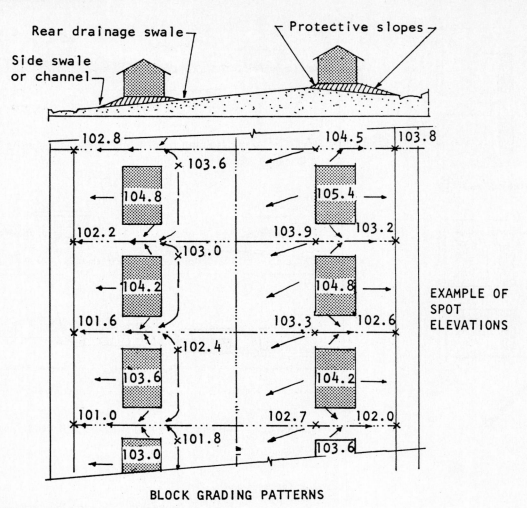

BLOCK GRADING PATTERNS

Figure 4.97 Example of spot elevations.

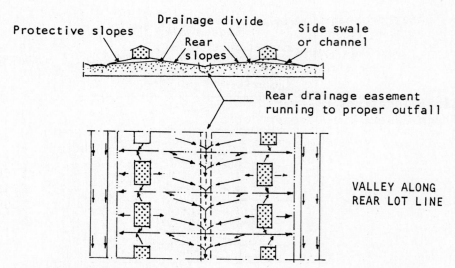

Figure 4.98 Valley along rear lot line.

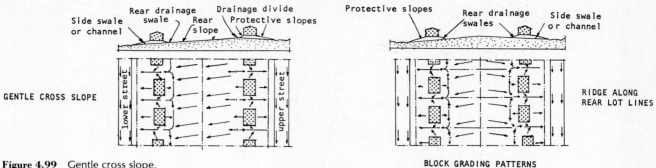

GENTLE CROSS SLOPE

BLOCK GRADING PATTERNS

RIDGE ALONG REAR LOT LINES

Figure 4.99 Gentle cross slope.

Figure 4.100 Ridge along rear lot lines.

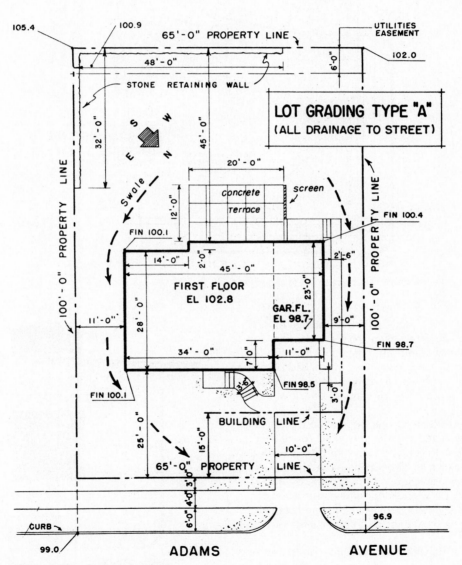

Figure 4.101 Drainage to street.

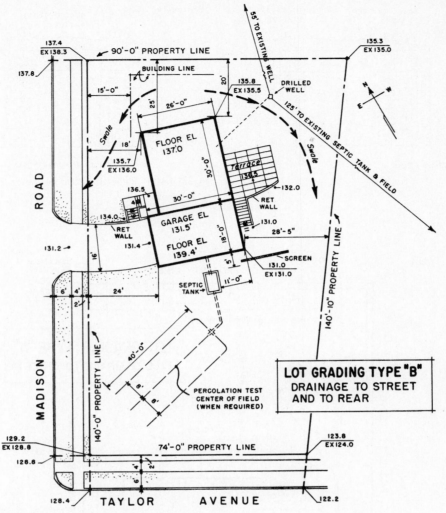

Figure 4.102 Drainage to street and to rear.

drains to the street. The system has the advantage of directing surface runoff completely away from the building, and it is the best method available for continuous row housing. Emergency overflows from rear-of-lot catch basins to the street are desirable to prevent flooding in the event that the basins become blocked with silt, debris, or snow (Figure 4.104).

A ridge may be built along rear lot lines so that each lot handles its own drainage without the need for rear-of-lot catch basins (Figure 4.105). A protective apron is provided at the rear of the building to direct roof water to a drainage swale, which carries water around the building by side-yard swales to the street. Care is needed in the construction of the swales between buildings, particularly in medium-density areas, where side yards are narrow, and in areas of heavy snowfall.

Across-the-block drainage is a variation of the second alternative. Instead of using a rear-yard ridge, the elevation of lots along one street may be raised, so that surface water

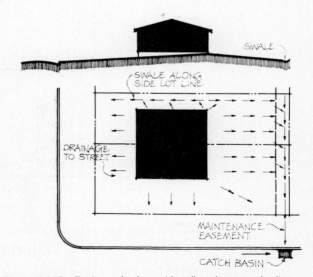

Figure 4.103 Drainage for lots with valley along rear lot line

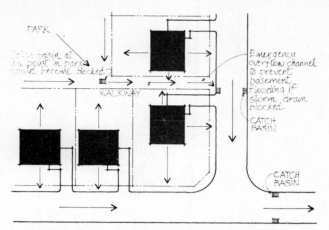

Figure 4.104 Emergency storm-water overflow channel.

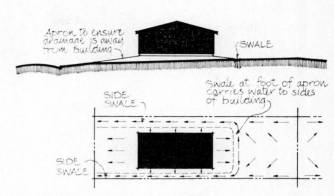

Figure 4.105 Drainage for lots with ridge at rear lot line.

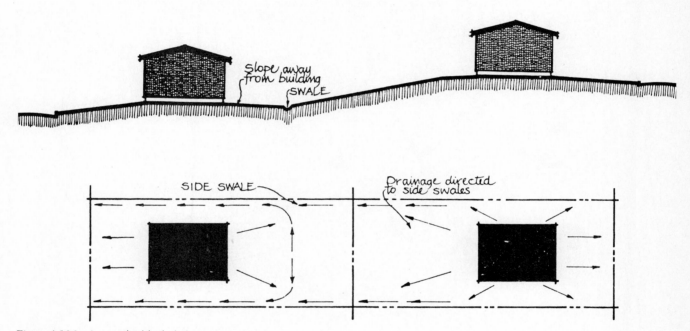

Figure 4.106 Across-the-block drainage.

drains from them, along side-yard swales to the lower street (Figure 4.106). Grading plans should be adjusted to protect buildings with a surrounding apron, to provide adequate swales between buildings, and to prevent erosion along the drainage path.

Site Grading

Performance Objective Site grading should produce a usable and easily maintained ground surface for the subdivision (Figure 4.108). Through initial rough grading and final site grading, it should:

- Preserve existing vegetation, trees, and topsoil wherever possible
- Provide final road and lot grades that ensure suitable pedestrian and vehicular access to dwellings and also permit adequate drainage of the site
- Balance cut and fill to localize the movement of earth, thus reducing both haulage costs and the risk of property damage during construction
- Achieve maximum cost-effectiveness with respect to the amount of earth moved and the lot yield of the site

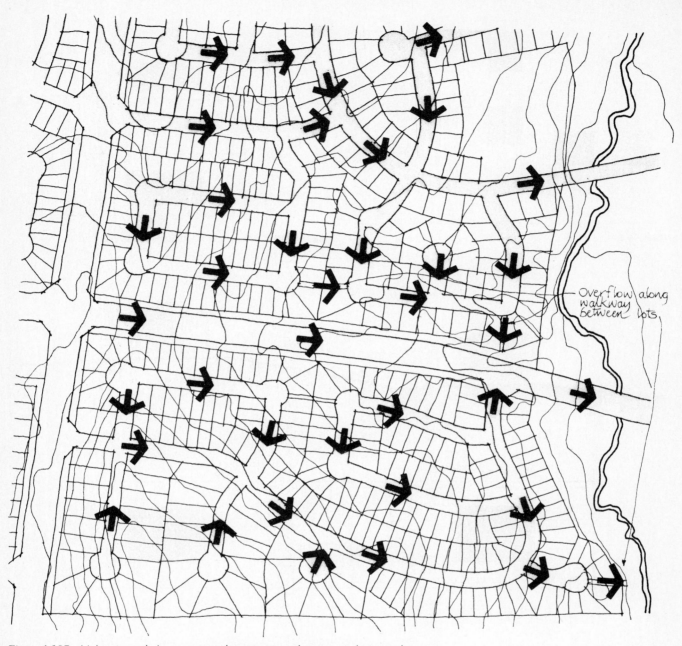

Overflow along walkway between lots.

Figure 4.107 Major storm drainage system along streets and easements between lots.

- Match the proposed building form to the existing topography in order to provide usable and easily maintained outdoor living areas at reasonable cost (Figure 4.108)
- Ensure that the elevation of the lowest floor of buildings will permit gravity connections into sanitary sewers
- Match existing grades at the perimeter of the subdivision.

Design Considerations Subdivision design should relate proposed dwellings and the cut-and-fill requirements of lots to the finished level of the road and adjacent homes. When lots are high in relation to the road, driveways may be inconveniently steep. When lots are low, large amounts of fill may be required to provide proper drainage away from buildings. In fact, inadequate drainage is one of the most common complaints associated with subdivision maintenance. In medium-density areas, problems can become acute because distances between buildings are often too small to accommodate drainage swales and changes in elevation, and the lot area is too small to accommodate excavated material.

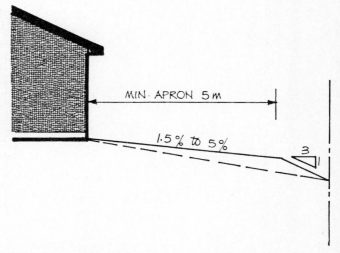

Figure 4.108 Building form designed to fit topography.

There are five basic design criteria for grading residential sites:

- Drainage for the total site should be positive — that is, away from the dwellings — with gradients of at least 2 percent for grass-covered areas and 1 percent for hard surfaces.

- Driveways should not have a slope greater than 12 percent. Steep slopes may be difficult to use in winter, and abrupt changes in grade between the driveway and curb or garage entrance create insufficient clearance between the underside of vehicles and the pavement.

- Outdoor living areas should have gradients between 1.5 and 5 percent for effective drainage away from the building and for practical use by residents (Figure 4.109).

- The balance of the change in grade between the rear lot line and the rear building wall should be taken up either by a grass slope of not more than 33 percent or, where slopes are greater than this, by a retaining wall, a slope covered with stone or other riprap material, or ground-cover plantings (Figure 4.110).

- The elevation of the house basement floor should normally be a minimum of 1 meter above the elevation of the street

sanitary sewer invert, in order to permit connection of rear house drains at a minimum gradient.

Figure 4.109 Gentle slope away from building for outdoor living area.

Figure 4.110 Slopes steeper than 33 percent can be planted with ground cover.

4.13 Sewage Disposal Systems

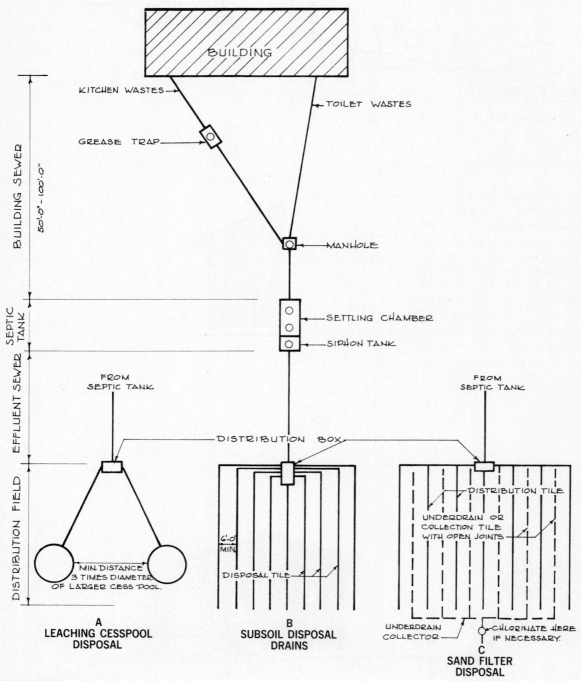

Figure 4.111 Types of sewage disposal systems.

METHOD	INSTALLATION	OPERATING COST	ADVANTAGES	DISADVANTAGES
UNDERGROUND DISPOSAL SYSTEM	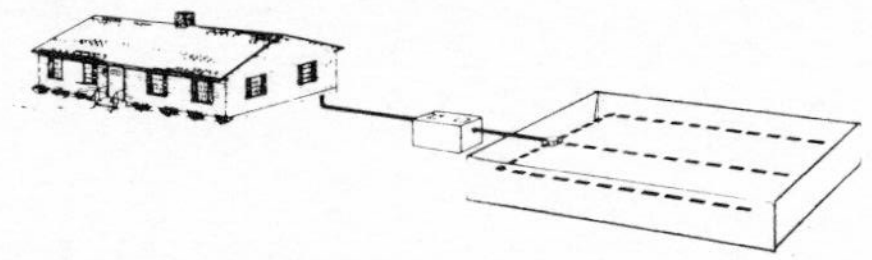 **SEPTIC TANK/SUBSURFACE DISPOSAL BEDS**	Very little. Septic tank requires pumping every 2-3 years.	1. Least costly of all disposal methods. 2. Requires little maintenance. 3. Provides highly effective and reliable treatment. 4. No power and no moving parts unless a pump is required. 5. Operationally stable.	1. Requires specific soil conditions with adequate space. 2. Cannot be used in shallow bedrock areas. 3. Cannot be used in areas of high water table. 4. Cannot be close to wells and surface waters. 5. Soil may become clogged if maintenance program is not followed.
	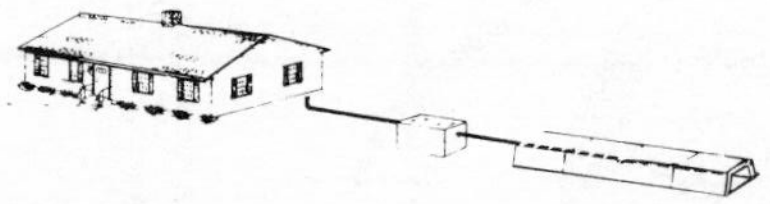 **SEPTIC TANK/LEACHING CHAMBERS**	Very little. Septic tank requires pumping every 2-3 years.	1. Next least costly of all disposal methods. 2. Requires little maintenance. 3. Provides highly effective and reliable treatment. 4. No power and no moving parts unless a pump is required. 5. Operationally stable.	1. Requires specific soil conditions with adequate space. 2. Cannot be used in shallow bedrock areas. 3. Cannot be used in areas of high water table. 4. Cannot be close to wells and surface waters. 5. Soil may become clogged if maintenance program is not followed.
	 **WASTE SEPARATION-NON-DISCHARGE TOILET/SUBSURFACE DISPOSAL**	Varies. Depends on type of non-discharge toilet used. Septic tank requires pumping every 2-3 years.	1. Grey water disposal bed may be used on sites which are marginally suitable for underground disposal. 2. Requires a smaller disposal bed area than for combined black and grey wastewaters.	1. Requires specific soil conditions with adequate space. 2. Cannot be used in shallow bedrock areas. 3. Cannot be used in areas of high water table. 4. Cannot be close to wells and surface waters. 5. Soil may become clogged if maintenance program is not followed. 6. Non-discharge toilets require regular maintenance and service.
OVERBOARD DISCHARGE METHOD	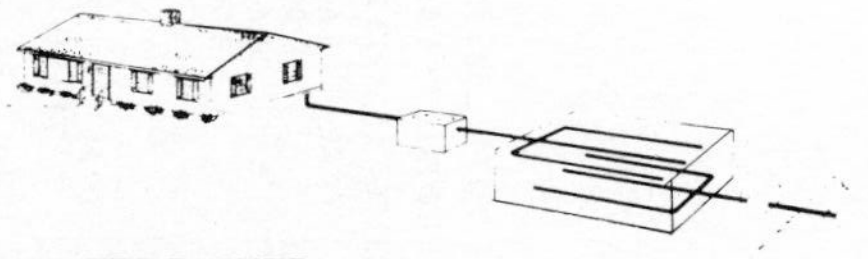 **UNDERDRAINED SAND FILTER/CHLORINATION**	Cost of chlorine tablets depend on amount of use—Septic tank requires pumping every 2-3 years.	1. No mechanical parts. 2. Little or no maintenance. 3. Can be used in areas where impervious soils or bedrock exist. 4. Most reliable overboard treatment method.	1. The filter may have to be replaced after several years if it is not constructed properly. 2. Requires more space than a mechanical unit.
	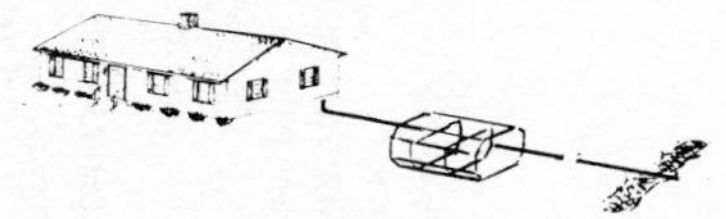 **MECHANICAL UNIT/CHLORINATION**	$30 - $60 a year plus chlorine costs. Service Contract: $100 - $175 a year.	1. Mechanical units can be used in areas unsuitable for underground disposal. 2. Mechanical units provide good quality treatment if properly operated and maintained. 3. They require a small area for installation.	1. There is a possibility of mechanical failure. 2. Mechanical units may be adversely affected by surge flows or nonregular use. 3. Mechanical units require regular maintenance and service.
ALTERNATIVE METHOD	 **HOLDING TANK**	$120 - $200 a month for a family of four with a pump-out once a week.	1. Holding tanks can be used for storage of wastes in areas where underground disposal is not suitable. 2. Can be used in areas where an overboard discharge is not possible.	1. Requires frequent pumping out. 2. May be expensive to purchase. 3. Cannot be used in inaccessible areas. 4. Very costly to operate. 5. Requires waiver from Human Services.

Figure 4.112 Types of sewage disposal systems.

4.14 Comparison of Site Development Proposals

Four types of site development are shown in Figures 4.113, 4.114, 4.115, and 4.116. Figure 4.117 shows a typical grid street pattern, and Figure 4.118 illustrates a curvilinear street pattern.

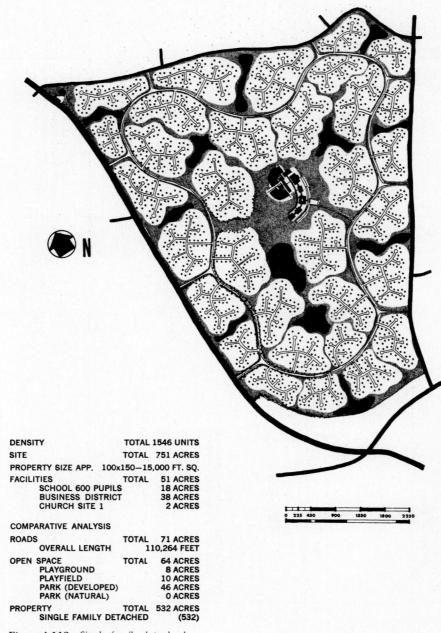

DENSITY	TOTAL 1546 UNITS	
SITE	TOTAL 751 ACRES	
PROPERTY SIZE APP.	100x150—15,000 FT. SQ.	
FACILITIES	TOTAL	51 ACRES
SCHOOL 600 PUPILS		18 ACRES
BUSINESS DISTRICT		38 ACRES
CHURCH SITE 1		2 ACRES

COMPARATIVE ANALYSIS

ROADS	TOTAL	71 ACRES
OVERALL LENGTH		110,264 FEET
OPEN SPACE	TOTAL	64 ACRES
PLAYGROUND		8 ACRES
PLAYFIELD		10 ACRES
PARK (DEVELOPED)		46 ACRES
PARK (NATURAL)		0 ACRES
PROPERTY	TOTAL	532 ACRES
SINGLE FAMILY DETACHED		(532)

Figure 4.113 Single-family detached.

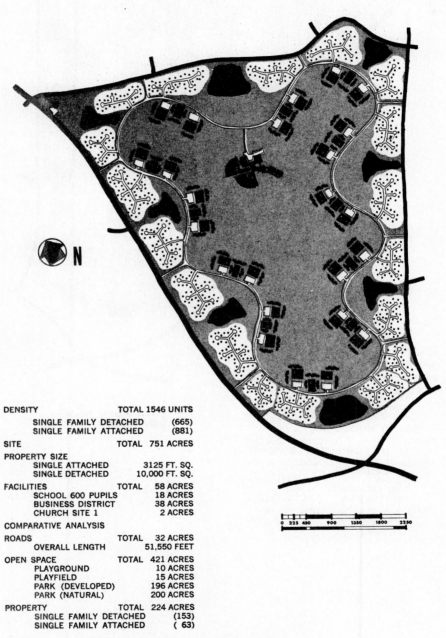

N

DENSITY	TOTAL 1546 UNITS	
SINGLE FAMILY DETACHED	(665)	
SINGLE FAMILY ATTACHED	(881)	
SITE	**TOTAL 751 ACRES**	
PROPERTY SIZE		
SINGLE ATTACHED	3125 FT. SQ.	
SINGLE DETACHED	10,000 FT. SQ.	
FACILITIES	TOTAL 58 ACRES	
SCHOOL 600 PUPILS	18 ACRES	
BUSINESS DISTRICT	38 ACRES	
CHURCH SITE 1	2 ACRES	
COMPARATIVE ANALYSIS		
ROADS	TOTAL 32 ACRES	
OVERALL LENGTH	51,550 FEET	
OPEN SPACE	TOTAL 421 ACRES	
PLAYGROUND	10 ACRES	
PLAYFIELD	15 ACRES	
PARK (DEVELOPED)	196 ACRES	
PARK (NATURAL)	200 ACRES	
PROPERTY	TOTAL 224 ACRES	
SINGLE FAMILY DETACHED	(153)	
SINGLE FAMILY ATTACHED	(63)	

Figure 4.114   Single-family detached/single-family attached.

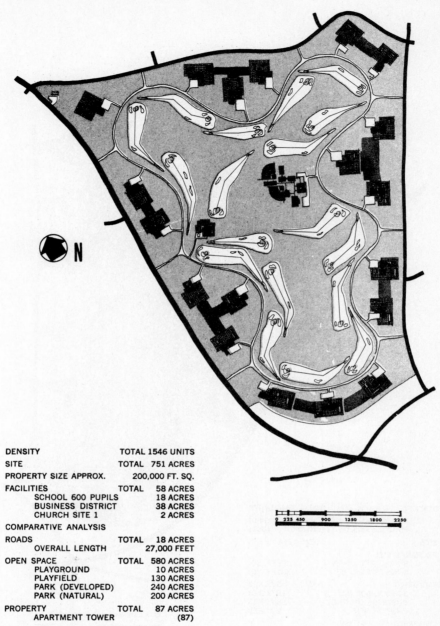

DENSITY	TOTAL 1546 UNITS	
SITE	TOTAL 751 ACRES	
PROPERTY SIZE APPROX.	200,000 FT. SQ.	
FACILITIES	TOTAL 58 ACRES	
SCHOOL 600 PUPILS	18 ACRES	
BUSINESS DISTRICT	38 ACRES	
CHURCH SITE 1	2 ACRES	
COMPARATIVE ANALYSIS		
ROADS	TOTAL 18 ACRES	
OVERALL LENGTH	27,000 FEET	
OPEN SPACE	TOTAL 580 ACRES	
PLAYGROUND	10 ACRES	
PLAYFIELD	130 ACRES	
PARK (DEVELOPED)	240 ACRES	
PARK (NATURAL)	200 ACRES	
PROPERTY	TOTAL 87 ACRES	
APARTMENT TOWER	(87)	

Figure 4.115   Apartment tower.

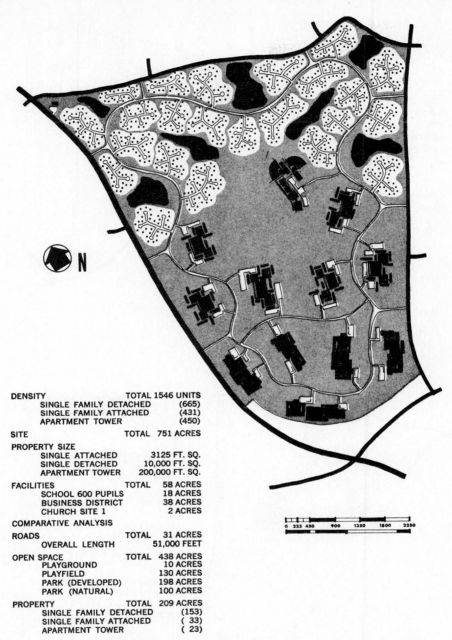

DENSITY   TOTAL 1546 UNITS
    SINGLE FAMILY DETACHED   (665)
    SINGLE FAMILY ATTACHED   (431)
    APARTMENT TOWER   (450)

SITE   TOTAL 751 ACRES

PROPERTY SIZE
    SINGLE ATTACHED   3125 FT. SQ.
    SINGLE DETACHED   10,000 FT. SQ.
    APARTMENT TOWER   200,000 FT. SQ.

FACILITIES   TOTAL   58 ACRES
    SCHOOL 600 PUPILS   18 ACRES
    BUSINESS DISTRICT   38 ACRES
    CHURCH SITE 1   2 ACRES

COMPARATIVE ANALYSIS

ROADS   TOTAL   31 ACRES
    OVERALL LENGTH   51,000 FEET

OPEN SPACE   TOTAL 438 ACRES
    PLAYGROUND   10 ACRES
    PLAYFIELD   130 ACRES
    PARK (DEVELOPED)   198 ACRES
    PARK (NATURAL)   100 ACRES

PROPERTY   TOTAL 209 ACRES
    SINGLE FAMILY DETACHED   (153)
    SINGLE FAMILY ATTACHED   (33)
    APARTMENT TOWER   (23)

Figure 4.116 Single-family detached, single-family attached, apartment tower.

SINGLE DETACHED

GRID

DENSITY	TOTAL	1546 UNITS
SITE	TOTAL	751 ACRES
PROPERTY SIZE APP.	100x150—15,000 FT. SQ.	
FACILITIES	TOTAL	51 ACRES
SCHOOL 600 PUPILS		18 ACRES
BUSINESS DISTRICT		38 ACRES
CHURCH SITE 1		2 ACRES

COMPARATIVE ANALYSIS

ROADS	TOTAL	71 ACRES
OVERALL LENGTH		110,264 FEET
OPEN SPACE	TOTAL	64 ACRES
PLAYGROUND		8 ACRES
PLAYFIELD		10 ACRES
PARK (DEVELOPED)		46 ACRES
PARK (NATURAL)		0 ACRES
PROPERTY	TOTAL	532 ACRES
SINGLE FAMILY DETACHED		(532)

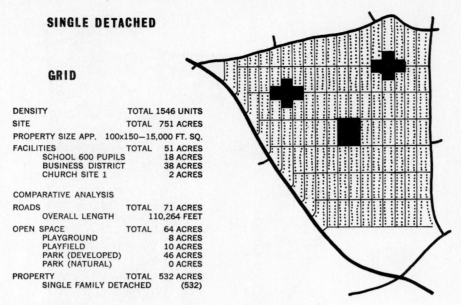

Figure 4.117   Grid pattern.

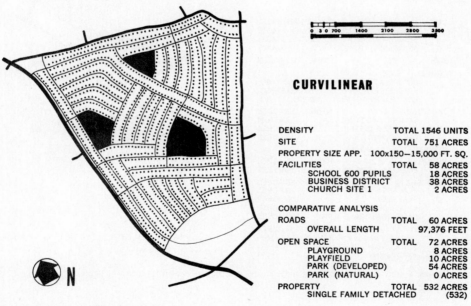

Figure 4.118   Curvilinear pattern.

CURVILINEAR

DENSITY	TOTAL	1546 UNITS
SITE	TOTAL	751 ACRES
PROPERTY SIZE APP.	100x150—15,000 FT. SQ.	
FACILITIES	TOTAL	58 ACRES
SCHOOL 600 PUPILS		18 ACRES
BUSINESS DISTRICT		38 ACRES
CHURCH SITE 1		2 ACRES

COMPARATIVE ANALYSIS

ROADS	TOTAL	60 ACRES
OVERALL LENGTH		97,376 FEET
OPEN SPACE	TOTAL	72 ACRES
PLAYGROUND		8 ACRES
PLAYFIELD		10 ACRES
PARK (DEVELOPED)		54 ACRES
PARK (NATURAL)		0 ACRES
PROPERTY	TOTAL	532 ACRES
SINGLE FAMILY DETACHED		(532)

4.15   Subdivision Plat for a Flood-Prone Area

Plat Features Related to Flooding

1. Clustering lots to avoid flood areas
2. Sewer and water protected against flooding
3. Drainage facilities
4. Common areas dedicated or reserved for park space (shaded in Figure 4.119)
5. Boundary of floodway and flood fringe shown on plat
6. Deed restrictions preventing development in floodway areas and requiring elevation of flood fringe uses
7. Bridge designed to pass flood flows without substantially increasing flood heights

8. Access road protected against flooding through elevation on fill

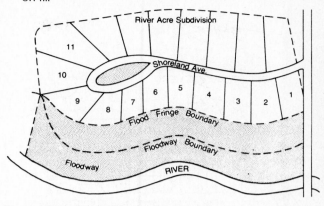

Figure 4.119   Subdivision of a flood-prone area.

Subdivision Ordinance Operative Provisions

Provision	Content	Purpose	Commentary
Standards and suitability	Subdivision regulations often require that plats be suitable for their intended uses. Subdivision approval may be refused because of flooding, inadequate drainage, unstable soils or geology, susceptibility to mudslides or earthslides, severe erosion potential, unfavorable topography, inadequate water supply or waste disposal, or other problems	Ensure that buyers will not be "victimized" by purchase of lands unsuitable for intended purposes. Prevent threats to safety, extraordinary costs for governmental services	Subdivision ordinances are typically adopted in combination with zoning, which identifies areas of flood or other hazards.
Building site improvements	No subdivision is permitted which would block floodway areas or increase flood damages. Subdivision approval is granted for sites outside of floodway areas if the site is elevated above the regulatory flood through fill or is otherwise protected. Commercial and industrial subdivisions are permitted below the flood protection elevation if structures will be adequately floodproofed. Deed restrictions may be required where a portion of a site is subject to flooding	Ensure that building sites will not block flood flows and will be protected against flooding	Subdivision regulations typically require not only sufficient area for building free of flood problems but for waste disposal and adequate access
Drainage ties	Subdividers are required to install storm drainage facilities sufficient to convey a specified frequency of flood. Below ground (for smaller floods) and above ground (for larger floods) may be required	Reduce community expense for drainage system. Ensure that storm runoff does not result in serious basement flooding or other drainage problems. Ensure that the subdivision facilities will mesh with community drainage plans	Some community drainage regulations require not only conveyance areas, but also ponding areas to prevent increases in the total amount or rate of runoff which may flood downstream lands
Roads	Subdividers are required to install roads at an elevation compatible with emergency access	Reduce community expense for roads. Ensure that subdivision residents can be safely evacuated	
Waste disposal facilities	Lots must be provided with adequate waste disposal (liquid, domestic). Soil absorption systems for on-site waste disposal may be prohibited in flood areas or other unsuitable soils	Protect public safety; prevent water pollution; reduce community expenses	Waste disposal is a major problem for rural subdivisions with on-site waste disposal.
Water supply	Water systems including individuals wells must be floodproofed	Protect public safety	

4.16 Comparison of Subdivision Plats

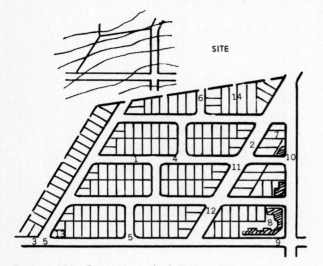

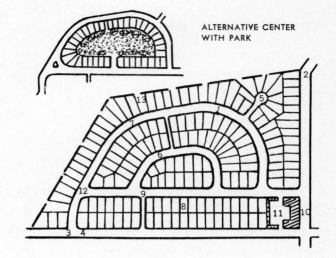

Figure 4.120 Comparison of subdivision plats.

Poor planning and what is wrong with it:

1. Gridiron street pattern without purpose
2. Heavy traffic within subdivision
3. Angular intersections
4. Nonabutting cross streets
5. Numerous subdivision entrance streets
6. Dead-end streets
7. Small, uneconomical blocks
8. Ribbon shopping district
9. No off-street parking space
10. Stores amid residences
11. Lots not perpendicular to streets
12. Angular lots
13. Small corner lots
14. Deep lots

Good planning and what is good about it:

1. Curved street pattern to add subdivision appeal
2. Heavy traffic diverted
3. Safe, perpendicular intersections
4. Few subdivision entrance streets
5. Quiet street
6. Local streets for local traffic only
7. Streets that fit topography
8. Long economical blocks
9. Cross walks in long blocks
10. Organized shopping center
11. Off-street parking space
12. Wide corner lots
13. Lots perpendicular to streets
14. Provision for interior park

Comparison

	Gridiron plan	Long-block plan (without park)	Long-block plan (with park)
Number of residential lots	156	169	144
Size of lots (average), feet	50 × 100	50 × 100	50 × 100
Length of major streets, feet	1050		
Length of minor streets, feet	4800	3850	3200
Length of cul-de-sac, feet		200	200
Total length of streets, feet	5850	4050	3400
Total street length per lot, feet	37.5	24.0	23.6

4.17 Revision of Subdivision Plat

Layout A: Poor Subdivision Planning and Why

1. Gridiron street pattern without purpose or appeal
2. Through traffic within subdivision (noisy streets)
3. Four-way intersections (additional traffic hazard)
4. Small uneconomical blocks. Long streets require maximum public improvements
5. *Uneconomical* (to both developer and municipality)

Layout B: Good Subdivision Planning and Why

1. Curved street pattern adds neighborhood appeal

2. Through traffic diverted (quiet streets local streets for local traffic only)
3. Four-way intersection eliminated (less traffic hazards)
4. Long economical blocks — minimum public improvements
5. *Economical* (to both developer and municipality)

Through revision of layouts	Layout A	Layout B
Number of plots at $1200 per plot	481	499
Linear feet of streets	21,740	19,315

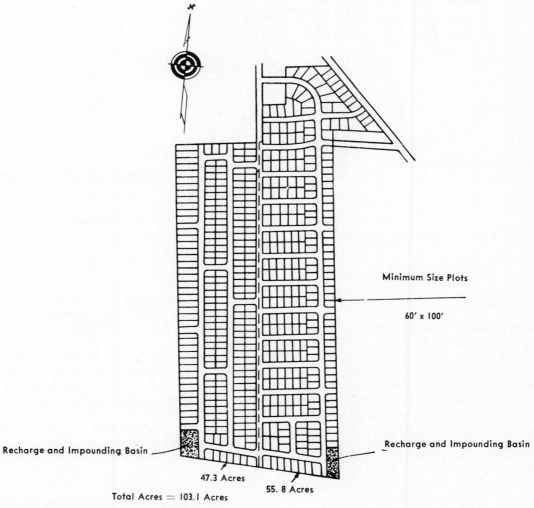

Minimum Size Plots

60' x 100'

Recharge and Impounding Basin

Recharge and Impounding Basin

47.3 Acres

55. 8 Acres

Total Acres = 103.1 Acres

Figure 4.121 Layout A: poor subdivision planning.

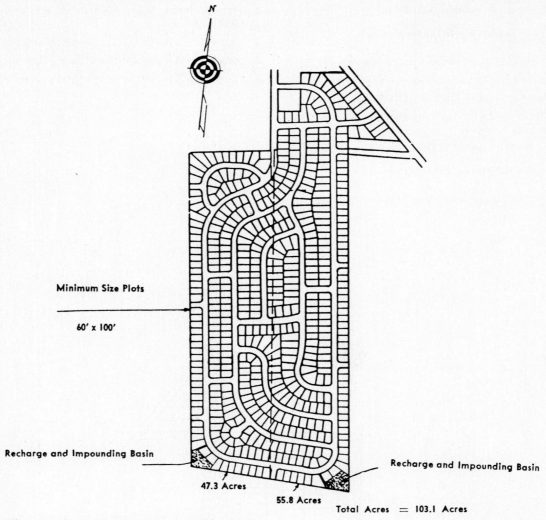

N

Minimum Size Plots

60' x 100'

Recharge and Impounding Basin

Recharge and Impounding Basin

47.3 Acres

55.8 Acres

Total Acres = 103.1 Acres

Figure 4.122 Layout B: good subdivision planning.

4.18 Cluster Housing

Development Program A site planner who has not been given a development program by the developers must participate in its preparation. The program must not be specific at the onset but must take form during the site planning process according to the given physical environment.

The planner works initially with the developers to establish a feeling for their expectations, keeping decisions flexible, covering: approximate number of units; approximate ratio of resident, guest, and recreational vehicle parking spaces; anticipated household mix; desired community feeling, and recreation and service facilities to be included.

Programs are usually based on a financial evaluation of all developmental costs measured against anticipated sales or rental receipts. However, since costs can be balanced in many ways, site planners are becoming involved in the cost/benefit process as a more effective way to justify certain site-related decisions.

A minimum unit count can often be determined from existing zoning (developers never build less than the allowable density); sewer or water capacity; what nearby developers have done; or what the developers have built in the past.

A complete range of household types is theoretically possible on most larger sites. However, since developers usually have experience with only a limited range of household types, they may have to be convinced of a more balanced community structure. Ideally, the mix should approximate the normal percentages in each group in the market area.

The United States breakdown of household types is:

15 percent singles

15 percent young couples

30 percent couples with young children

12 percent couples with teenagers

18 percent couples with grown children

5 percent elderly couples

5 percent elderly singles

Desired variations of income, education, or ethnicity within a cluster should be discussed and strategies for attracting residents implemented.

Parking count depends on accessibility between the site and shopping, jobs, and schools. If there is good public transportation, one car per family, with 0.5 guest parking per unit may do. If not, 1.5 to 2 cars plus 0.5 guest parking per unit may be required. Elderly need only one parking space per unit, while two or more singles living together may need one car per occupant. The planner should avoid committing to a precise number of parking spaces, instead look for a range which can later be refined.

Functional diagrams should be prepared of alternative ways to incorporate each household type within a cluster or development. Theoretically all household types can live together, but some separation often makes life more comfortable. Thus, which household types complement each other and which types may cause each other problems if they live side by side need to be considered. Generally, elderly people and young singles prefer to be separated from families with children. On the other hand, young couples or couples with grown children may serve as reasonable buffers between families with children and elderly or singles.

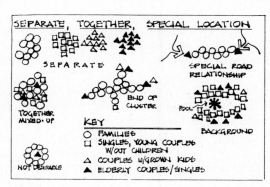

Figure 4.123

Subtractive Process—What Should Be Reserved Land inappropriate for housing may be "subtracted" from the site. This includes all steep land, using two slope categories: 10 to 25 percent as "maybe" and 25 percent as "too steep." The 10 to 25 percent category may be refined by eliminating

those areas with soil conditions subject to slippage. Also to be subtracted are major site amenities which convey an essential character or quality of the land, for instance, a meadow, a stand of trees, land bordering streams, steep cliffs, rock outcrops, old buildings, and prime vistas. Other land to be "subtracted" for preservation includes habitat areas likely to support diversity of wildlife: streams, edges of forests, bogs, or ponds. An overlap among drainage, wildlife, steep slopes, and amenity may be good because the amount of land to be reserved is minimized.

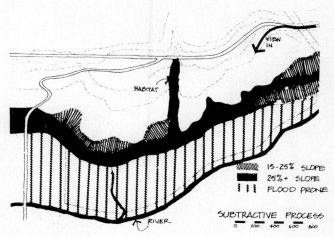

Figure 4.124

Each category to be subtracted (slope, amenity, natural process) should be outlined on the plan with a different color marker. If the diagram becomes confusing, separate plans may be drawn for each characteristic. Because the process of subtracting valued features is judgmental, each category should be accompanied with a list of implications and directions for development. Implications relate to increased construction and maintenance costs, amenity or sense of place values, or to intangible environmental factors associated with balance of life. Development directions describe actions, such as save, enhance, eliminate, relate to, use, protect, plant, extend, and so forth. Thus, each site feature becomes a guiding constraint and simplifies the decision-making process. Design becomes easier as the number of options decreases.

In simplest terms, good ecological site planning is *doing the least to the land.* This is opposite to most suburban development standards which require wide roads, tree removal, artificial drainage structures, and uniform lotting. The less clearing, grading, and disturbance you do the better; the less disturbance to natural drainage, the better.

Community open space is inextricably associated with the notion of amenity and natural factors and is the single most important environmental image maker. Determining its shape and location is the first step in concept development. At this

point, we as planners select the most outstanding natural or cultural site amenities and plan the open space to include and enhance them. We should plan to make generally available the site areas most suited for public or resident use, reserving and supplementing existing vegetation. Neglected areas of the site should be improved through landscape development programs.

Figure 4.126 Typical clusters.

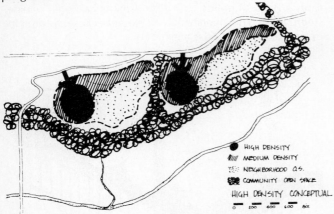

Figure 4.125

The *subtracted* elements discussed above should be linked together, attempting to create an identifiable, visible open space network. Two extreme solutions are possible:

1. An active, resident-use open space system for children's play, bike riding, walking, sports, and so on. These activities are enhanced by a continual, linear open space system, free from auto intrusion, with level areas for specific activities.

2. Visual open space for light, air, amenity, and views. Very steep land, areas rich in wildlife, or fragile land may be best looked at and not made easily accessible. This approach limits resident mix to young working couples, elderly, or those without children or suggests providing a centralized recreation facility elsewhere on the site for those needing active recreation.

Recreation Facilities Level open space should be examined for its potential as location for some community recreation facility. Although specific facilities will be discussed later, for now it is necessary to consider only *active* or *passive* and *centralized* or *decentralized*. Active implies specialization, requiring facilities such as tennis courts, basketball courts, swimming pools or ponds, and horsehoe courts, which may be disruptive to nearby residents. Passive recreation, such as picnicking or walking, may happen anywhere on the site, needs no facilities, and is not disruptive to nearby residents. Centralized means that most active facilities are located in one place and a community recreational focus developed. Decentralized means that the active facilities are spread out more evenly throughout the site.

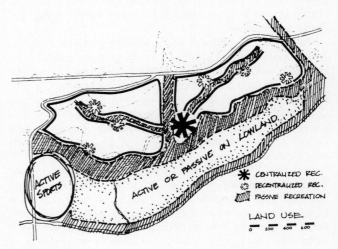

Figure 4.127

The decision to centralize or decentralize is based on a combination of site factors and program requirements which include the following:

1. Sites with several large, level areas may use a decentralized approach with small active recreational facilities located on a number of small level areas throughout the site.

2. Small developments often benefit from grouping all facilities together to make them appear larger and to reduce construction and maintenance costs.

3. Clusters may focus on special recreational facilities in a decentralized system. In this approach, each cluster would have one special facility, such as a swimming pool, tennis court, or children's play area.

4. Sites with a large, level area may benefit from a centralized facility, particularly if the open space is a valley overlooked by the units. In addition, each individual cluster would have some open space around and within it. The advantage of allocating all the funds to one complex is efficiency in both construction and operation. Further, the complex can be located far enough from units to ensure privacy and quiet for all.

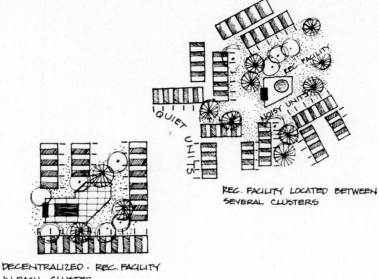

REC. FACILITY LOCATED BETWEEN
SEVERAL CLUSTERS

DECENTRALIZED - REC. FACILITY
IN EACH CLUSTER

Figure 4.128

Some typical locations for recreation facilities include:

1. At the entrance, where residents and guests can see them upon entry. Recreation facilities can offer a degree of status.

2. On a high point overlooking the development.

3. In a valley, overlooked by housing units.

4. At the apex of pedestrian or auto circulation.

5. Alongside or near a natural amenity. Open space enlarges and enhances the recreation experience. It is essential not to surround recreation facilities with roadways.

Community Structure   By this time the planner should understand the site and almost be able to see community structure emerging. Four community structures are possible.

Graduated has a center (or several centers) from which units move out in concentric rings. Typically density decreases from the center out, and unit mix changes depending on how urban or rural the site is.

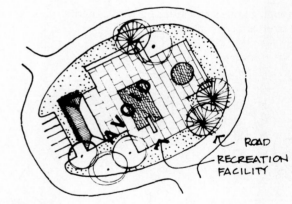

ROAD

RECREATION
FACILITY

Figure 4.129

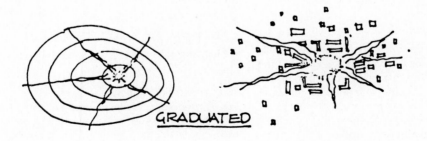

GRADUATED

Figure 4.130

Focus relates each cluster of level space to the open space which formerly separated it. Open space can either separate or bind clusters. In this case clusters are bound together by their focusing on open space.

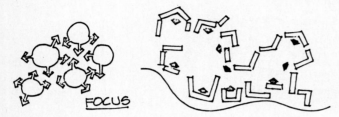

Figure 4.131

Linear unites the level areas with circulation in a linear pattern.

Figure 4.132

Nodal relates each cluster to itself using the open space for buffer and separation.

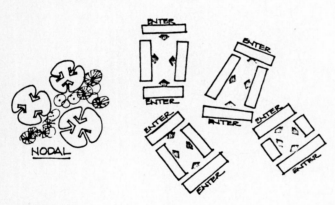

Figure 4.133

Sense of place, or fundamental essence of a site, usually implies how the site might be best developed and used. If the sense of place is strongly related to a valley, meadow, or forest and the site's character would be diminished by construction on it, development plans should preserve it. How closely houses may abut natural features is judgmental. However, if the feature is important enough to be shared by all residents, adequate room for a public walk should be provided and privacy in adjacent gardens maintained.

Figure 4.134

What to do:

- Build heavily in impermeable soils
- Use surface drainage for impermeable soils
- Collect roof water runoff for site distribution
- Use porous pavement or porous paving blocks
- Use swales and berms to direct runoff
- Use vegetation to slow runoff
- Select valuable trees for preservation
- Determine "preservation strip" width for streams (100 to 500 feet)
- Create storm-water retention ponds
- Keep roadways away from drainage swales or creeks

Design objectives:

- Minimizing disturbance to natural drainage patterns
- Reducing erosion
- Maintaining groundwater balance
- Preserving biologically productive areas
- Protecting habitats of fauna and birds
- Maintaining diversity of vegetation
- Maintaining strength through size
- Maintaining continuity of the natural features
- Preventing slippage/landslide
- Preventing flooding
- Avoiding stream degradation
- Maintaining water quality
- Maintaining the overall quality or character of the land

One word of caution at this point. The subtractive process leaves predominantly level areas with little natural attractiveness. The authors are not suggesting these are the only areas to build on. In fact, it may be best to build on more difficult areas, leaving level areas for active recreation; or to build at high densities on a small portion of sensitive land to take

advantage of a dramatic view; or any of a hundred other design alternatives. Therefore, the choice is to:

Avoid potentially hazardous or ecologically productive land, or

Build on these lands with an understanding of the risks, and design to minimize them.

Focusing Down Once site amenities and development constraints have been identified, site planning is a matter of simultaneously interrelating several levels of design, from the total development at the largest scale to specific concerns of each cluster and unit. Each unit must become a perfect environment in every way, while clusters of units must become a perfect neighborhood.

Auto Access External connections to the nearest municipality should be considered next. If several existing roads abut the development, the planner should determine the most effective route based on the following criteria:

- Which road connects most directly to the nearest town?
- Which road has sufficient capacity to handle additional traffic?
- Which road could be connected to the site at the lowest cost?
- Which is the most visually pleasant route?
- Which road has adequate sight distance for moving into or away from the site?

Main entrance roads meeting existing roads should: (1) Intersect at right angles, assuring visibility both ways and reducing the chance of "slipping" dangerously into traffic from an angled intersection. (2) Be long enough to allow stacking while cars wait to enter the main road. Judgment is required to anticipate the number of cars leaving the development during commuting hours and to gauge traffic levels on the existing road. Twenty feet of stacking distance should be allowed per car with no intersections crossing this space. In addition, main entrance roads should be located no closer than 200 feet from any existing intersection. (3) Be opened visually to alert drivers to their location and allow them time to slow down. (4) Serve developments up to 200 units in size per single entrance. (5) Be approached from the driver's right side, enabling a free right turn into the development. If left turns are necessary, the possibilities of a left turn lane should be explored.

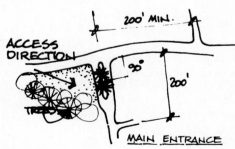

Figure 4.135

Roadway Location The road is the most difficult and disruptive site planning element. Minimizing or eliminating the road will usually improve the site plan. The planner should determine how extensive a road system is required for the site, using three road categories, major, local, and access. Large sites may require all three, while smaller sites may need only an extensive access road system. Avoiding or minimizing major roads will make the development a better place to live.

Conceptually there are only two road patterns:

1. Branching pattern—like a tree with wide trunk and main branches accommodating large volumes of traffic, and smaller branches where theoretically traffic diminishes.

2. Grid pattern—a network of roads forming squares, rectangles, or triangles, allowing uniform through traffic on all roads. In a sense, this pattern turns its back on the issue of controlling auto traffic by saying, "let the motorist decide where to go."

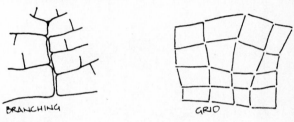

Figure 4.136

Small-scale developments should use the dead-end *branching pattern,* which by limiting unnecessary traffic enhances the development's environmental quality. Larger complexes of over 500 units probably require a combination of *grid and branching* patterns. The grid allows easy auto access between major living clusters, while the branching assures higher environmental quality immediately adjacent to internal clusters.

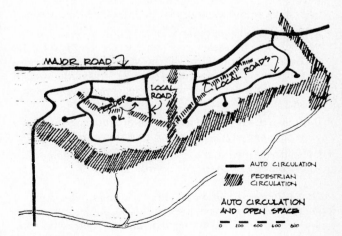

Figure 4.137

All elements of the design need to be tackled at the same time without worrying about detail, size, or exact scale. Broad felt-tipped pens and easy flowing lines may be used to create bubble diagrams of road and cluster locations.

Topography is the significant design determinant because it must be considered in all circulation, open space, and house layouts. A coordinated response to topography automatically and naturally brings each system into accord with other elements of the environment; i.e., all elements of the scheme flow together. Topography is also a structuring device which gives form to any community.

Roads may be located either parallel or perpendicular to the contours. Roads *parallel* to the contours require extensive cross-sectional grading but allow easy access to downhill units and open views for all units, while roads *perpendicular* to the contours minimize grading but cause difficult unit access problems and limit views. As with all extremes, there are compromise solutions, such as a diagonal street. Other aspects of road alignment the planner should consider are:

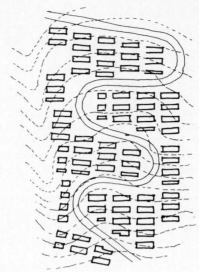

Figure 4.139

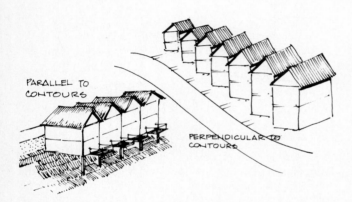

Figure 4.138

- Avoiding difficult topography where possible, and trying not to cross streams or steep valleys with roads. If they must be crossed, the feasibility of a bridge crossing, with pedestrians moving under the road, should be explored.
- Relating roads to topography. If land forms are hilly, a curvilinear route paralleling one contour may be appropriate. Following the contours with an easy gradient provides comfort for pedestrians. Steep roads or long grades require extra power for the auto to traverse, creating bothersome noise.
- Designing roads to take advantage of unfolding views and vistas and focusing attention on desirable visual elements.
- Determining the natural drainage pattern and siting roads so the drainage pattern can be maintained.

Road Patterns for Small Sites   A prime site planning objective is to combine the space allocated for housing, open space, and circulation into arrangements separating pedestrian traffic from automobile traffic. Although this is

rarely possible, conflicts between pedestrians and automobiles can be minimized and roads treated according to the severity of the problem. Separating auto from pedestrian traffic should occur when large numbers of pedestrians are forced regularly to cross a street or when large numbers of autos conflict with pedestrian movement.

Although many sites abutting an existing county road are easiest to develop by connecting several cul-de-sac access roads to the county road, development identity will be difficult to achieve because the access roads will not be connected to each other. If suburban clusters are to relate to each other, an internal local road is in order.

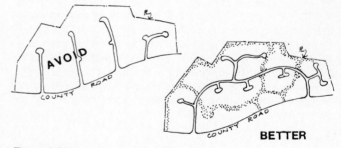

Figure 4.140

Long Culs-de-sac   Confining automobiles to the periphery of the site on long culs-de-sac is one way to create interconnected, separate pedestrian ways. This branching pattern is highly efficient and desirable. But, because it has only one exit, auto traffic can become congested.

How long is a long cul-de-sac? The FHA suggests 400 feet maximum, but the authors suggest 1000 feet, which allows deep penetration into the site, opens up pedestrian areas,

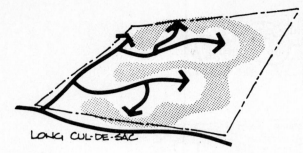

Figure 4.141

minimizes the need for local roads. However, 1000 feet requires careful planning to prevent confusion and to assure safety and convenience. A mid-way cul-de-sac would allow some people to turn around without traveling the entire length.

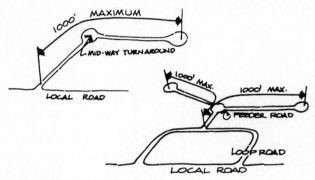

Figure 4.142

Fire trucks must have adequate passage in every site plan. (It should be pointed out that fire trucks are becoming smaller in size, and more flexible in turning, and that most firemen reach a fire even if they have to cross lawns or other landscaping.) Often the open space system can be used for emergency access.

At this point it is worth mentioning that road alignments should allow enough room for development, with the road either along the edge of the development serving one direction, or entering near the center and serving both directions. Often there is no choice, since topography forces roads along more level portions for cost efficiency and ease of use. Roads in the middle of the site divide it into two open space systems with half the units relating to each space, rather than to the whole. If the site is large, this is not a problem, but if it is small, the road should be aligned closer to one side and a single centralized open space developed.

The road may be single-loaded or double-loaded. Double-loaded roads, which should be used where possible, are more efficient and cheaper to build since costs are shared by units on both sides. The site planner must be prepared to defend the use of a single-loaded roadway.

The *Ring Road,* related to the grid, encircles the site as a large loop near its perimeter and serves interior units, or as a tighter loop closer to the center, it serves units on both sides. If the road is placed next to the property line, at least 15 feet of buffer should be left and the space carefully planted. Native plant materials are most appropriate, since property line edges are seldom adequately maintained. Locating units between a ring road and the property line is a tricky task, which can most easily be accomplished by placing large family units outside the ring road, with direct auto access and oversized private gardens. Large private gardens take the place of public open space for these units. A row of single-family units also softens the transition between two properties, and often eases neighborhood concerns about higher-density development. If the property abuts a park or other open space, at least 25 feet should be reserved for pedestrian circulation and units faced onto the adjoining space. Roads serving a small number of these units may weave around trees, be narrow in places, or be looped with one-

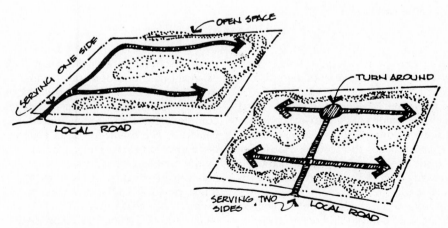

Figure 4.143

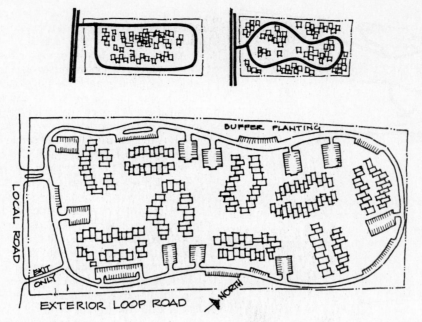

Figure 4.144

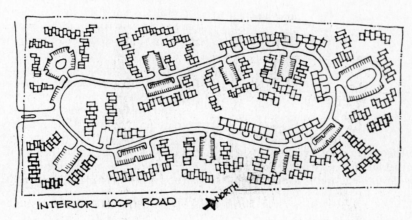

Figure 4.145

way traffic. All culs-de-sac must be two-way. A design speed of 10 to 15 miles per hour may improve abutting environmental conditions.

Local and Feeder Roads on Medium-Sized Sites Local roads usually connect two major roads with a curvilinear alignment that keeps speeds down, assures a level of automobile inconvenience, and reduces shortcutting. Houses or clusters of units may front directly on local roads, although the majority of units should be served by feeder roads which are connected to a local road.

As the road system becomes larger and more complex, the number of pedestrian/auto conflicts and noise problems also increases. A principal site planning goal is to develop patterns which minimize environmental nuisances to residents. Balanced neighborhood traffic plans, called "Environmental Areas" by the Buchanan report *Traffic in Towns,* are defined as "having no extraneous traffic," and as "areas within which considerations of environment predominate over the use of vehicles." The exact size of an environmental area is partly dependent on size, density, and auto use, and is governed by the need to prevent traffic from building up to a volume that causes irritation. British housing experts suggest that traffic on local roads in environmental areas should not exceed 120 vehicles per hour. Translated into houses, that means 200 to 300 houses comprise an identifiable environmental area.

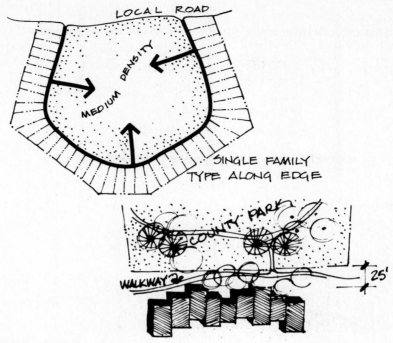

Figure 4.146

Major roads form the bulkiest but shortest element of our travel network, serving as efficient distributors of *auto* traffic. They provide optimum routes from any point within the site to any other point that is more than a short distance away and tend to reduce travel time for longer distances, but should not be freeways. The closest American road type is the boulevard, followed by the arterial, which in its truest form is supposed to serve similar purposes. Since suburban neighborhoods are likely to have existing networks of arterials, many developments may be able to connect one or two major roads running through the property, these arterials thus complementing or extending the existing road network.

Major roadway locations are determined by:

• Existing and potential connections
• Physiography
• Location of facilities and amenities
• Intensity of residential development

The *grid pattern*, whether slightly modified or bent around topography, is the basis for a major road system. Grid patterns distribute traffic uniformly, serve large areas, have flexibility since they are infinitely divisible into smaller units, and can be adjusted to avoid natural features or topography. The planner should first block out a basic grid, say at 2500-foot intervals, and evaluate its usefulness. Adjustments have to be made to avoid unique areas, forests, streams, and steep topography. Precise dimensions or details are not important at this stage because the design will be adjusted and refined.

The process is one of trial and error, combining the planner's best intuition about the site with standard roadway layout patterns. While planners are siting roads, they should remember to favor the pedestrian by developing large, auto-free areas for pedestrian activity.

Intersections of major roads should be a minimum ¼ mile apart, with no direct access to business or residential clusters. This last point is worth repeating; if driveways interrupt major roads, traffic is slowed down and drivers use local roads, causing congestion and noise closer to the neighborhoods.

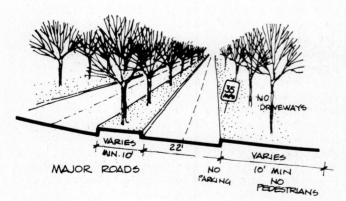

Figure 4.147

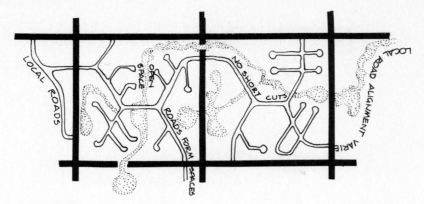

Figure 4.148

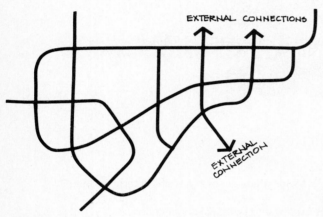

Figure 4.149

Travel speed should be maximum 35 miles per hour on major roads with no parking or sidewalks allowed alongside.

Linear road patterns may be used on narrow sites, serving development on one or two sides. A linear major road may be aligned alongside but at some distance from a stream or natural feature. If the road is to abut a natural feature directly, it can be relatively close, allowing room for pedestrian walks. But if units are to be placed between the road and the natural feature, the distance must be greater to allow local or feeder road penetration into the site. The approximate distance can be quickly calculated by adding up assigned widths for each unit.

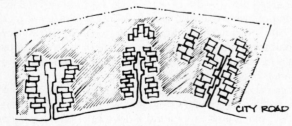

Figure 4.150

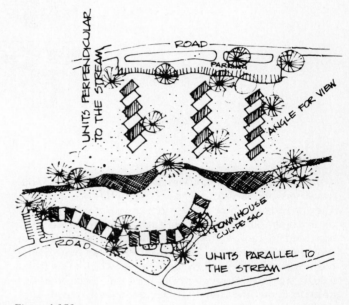

Figure 4.151

In summary, an ideal circulation diagram may consist of a large circular or grid loop for travel flexibility and reduction of congestion, with branching dead-end streets penetrating into the center to serve housing clusters. Open space fills internal spaces between the branches, completing the travel network via pedestrian movement. Minimizing the length of road reduces costs substantially, since utilities, sidewalks, lighting, and landscaping are also reduced. As many units as possible should be located off the road. Adequate room should be maintained for units not directly on the road. Internal circulation should be varied to match accessibility requirements. On nearing housing units, circulation should become *less* efficient and environmental conditions should improve. The farther from main auto access points, the fewer houses are served and the narrower roads can become.

Cluster Arrangements The total site should now appear as a number of individual clusters, categorized according to

Figure 4.152

the most appropriate user, anticipated densities, and desired environmental character. We are now ready to work within each cluster to determine the most appropriate distribution of units and the cluster arrangement. Three general arrangements related to topography, land size, amenity, orientation, and life style emerge.

Linear   Units in a row, either parallel or perpendicular to contours, work well on steeper sites. Since access from adjacent roads or parking courts is direct, the pattern suggests individuality and dependence on the car.

Focus   Court or cul-de-sac arrangements are organized around a shared access or interior space. This pattern creates the greatest sense of community.

Carpet   Internal individually focused units such as patio houses or bungalows stress private space and cover large areas, particularly land without natural amenities. The pattern is rolled out like a carpet, adjustable at the edges to conform to unusual boundaries.

The cluster arrangements we have been talking about are superblocks, layouts with auto-free internal pedestrian spaces and with auto penetration from the perimeter. In suburban communities the designer has more opportunity to determine the size and shape of superblocks than in urban areas where existing street patterns and other constraints force the superblock to be regular and small in size. How large can superblocks be? They can be large enough for pedestrian enjoyment — for long uninterrupted walks, for play facilities and socializing areas — and at the same time small enough for automobile flexibility.

Superblocks have two dimensions, length and width, the width being critical. To determine minimum widths, we should diagram several possible housing arrangements including:

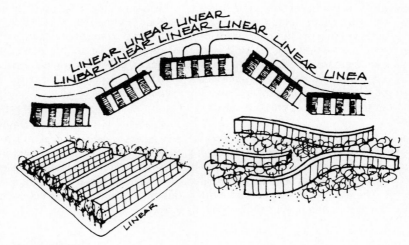

Figure 4.153

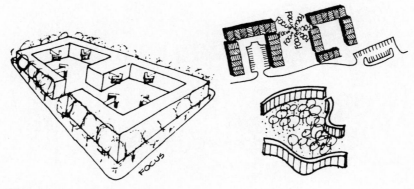

Figure 4.154

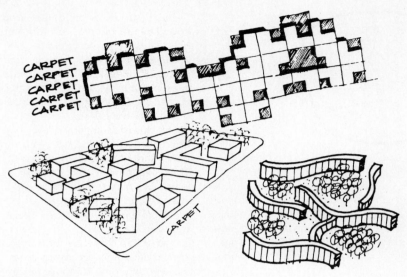

Figure 4.155

1. *Units parallel to the road.* To approximate the space needed, we should start with an adequate setback, parking if necessary, the width of the building, a small space for private gardens, and an adequate amount of common open space, and repeat the section on the other side with back yard, house, front yard or setback, parking, and roadway. Main disadvantages are that unit back yards face each other directly. One row receives insufficient sunshine if the open space is not aligned north and south. In addition, views from the road and pedestrian spaces are likely to be boring.

2. *Units perpendicular to the road.* This suggests small parking courts or rear entrance roads serving single or double rows of units. A full range of open spaces, including private gardens, space between buildings, and large internal open space, is possible. The width of the superblock must be wider than that with units aligned parallel to the road.

3. *Rows of units parallel and perpendicular to roads* can be adjusted to fit around desirable existing site factors. Wide sections should be used for units perpendicular to the road and narrow portions for units parallel to the road. Parking arrangements may vary: for row houses or units perpendicular to the road, parking in separate lots or under buildings; for units parallel to the road, internal parking. Parking lots at one end also work well for units parallel to the road.

Parking In suburban locations auto access for residents and guests is necessary, although it is important that pedestrian routes be clear and comfortable for those preferring public transit to the automobile. Auto arrival sequences necessitate a visual connection between parking and final destination, which can be either bold and direct or subtle and indirect.

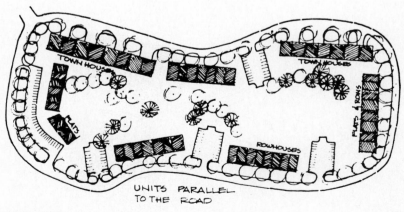

Figure 4.156

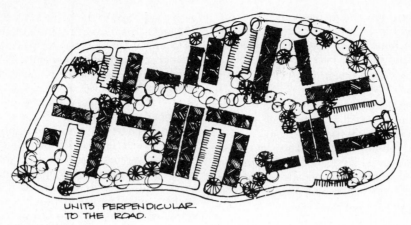

UNITS PERPENDICULAR
TO THE ROAD.

Figure 4.157

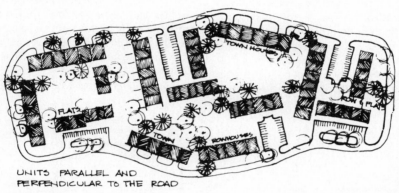

UNITS PARALLEL AND
PERPENDICULAR TO THE ROAD

Figure 4.158

The bold approach is one in which the final destination can be seen from the parking area, while with the subtle approach the driver can see the destination *before* parking and can orientate to it after parking.

The planner should experiment with different parking arrangements described below.

1. Parking between units and the roadway is convenient since each unit has its parking space immediately in front. The major disadvantage is the relatively poor visual appearance created by a large parking lot at the entrance.

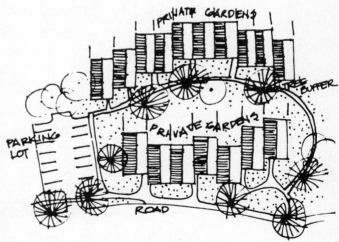

Figure 4.160

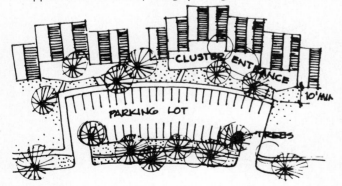

Figure 4.159

2. Parking at the ends of the buildings in small lots of ten or twelve cars works for units parallel or perpendicular to the road, is readily accessible, and minimizes walking distances. In addition, it looks better than parking immediately in front of houses.

3. Parking in individual or group garages under the units conserves open space and is convenient for residents. The driveways to these garages create visual and functional disruption and must be wide for back-out safety. Since there is no central entrance, a sense of community is difficult to achieve with this arrangement.

4. Parking in large, shared garages either detached from or under a medium-density building is another parking arrangement. In considering the detached garage, centrality and land-use efficiency advantages must be weighed against the disadvantages of higher construction costs, impersonal parking, disorientation, and lack of clarity between the garage and the housing units. Detached garages do not work well for guest parking.

Courtyard parking on suburban sites need not be as compact and efficient as in urban developments. Parking may be around an informal center island, or may serve town houses off an entrance court using an end entrance (I or U shapes). The court may be extended as a cul-de-sac serving three or four units in a hard-to-reach corner of the site.

Road/Unit Relationship   Now the planner should relate the road system to the housing program. How close should each road come to the house? Will people be willing to walk 100 feet? 100 yards? Upstairs? Along a covered path? How far people will walk varies with their age, health, family composition, and attitude. In general a 200-foot walk is maximum. A factor to consider is the distance garbagemen and tradesmen are willing to walk from their vehicles, usually 150 to 200 feet. If stairs are necessary, the distance may have to be shortened.

The cluster entrance sequence should be considered next. Where should cars leave the road and enter the cluster? This, of course, relates to parking arrangements. Long linear sites or steep sites suggest parking near the road rather than penetrating the site with a feeder road. Conversely, deep sites are best served by a feeder road leading to parking courts or individual drives. Town houses with internal parking require road access. Patio and row houses can be located off the road, with parking handled nearby in grouped arrangements.

The highest-density units should be located so they connect efficiently to the external road/transit system and lower density units located toward distant edges. It would be desirable to designate buffer zones to separate living areas from busy roads.

Topography   Sites with steep topography or unique nat-

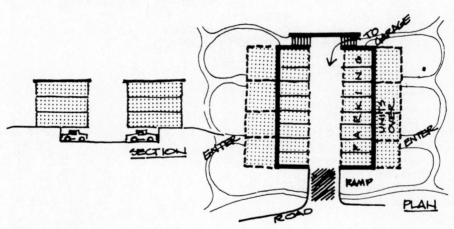

Figure 4.161

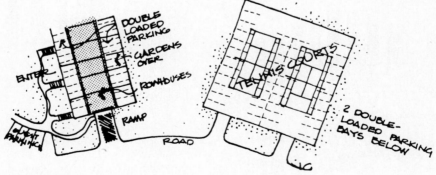

Figure 4.162

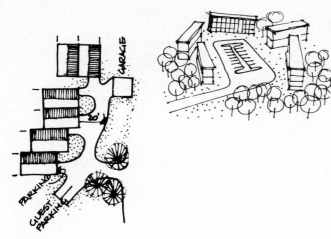

Figure 4.163

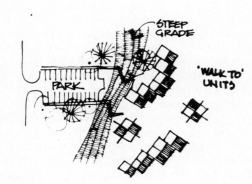

Figure 4.164

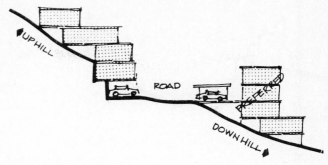

Figure 4.165

Views Units whose major living areas have a view are always the first units sold or rented. We, as planners, should consider the following questions regarding views. Is there a dramatic view which should be shared by all units? Is there a unique land form such as a meadow, lake, forest, hillside, or stream which should be reserved for all, and might determine the clustering arrangement? Are there scattered small features such as clumps of trees, outcrops, or small glens providing a pleasant close-up view around which housing units might be clustered? Is there a *best orientation* providing sunlight, winter wind protection, and view? We should study each slope with sections to find ways to use distant views without retaining walls or steep banks. We should note carefully the exact percent of slope so that building spacing can be determined to assure views to units behind. Figuring floor-to-floor heights at 9 feet, we may experiment with using half-levels created by living spaces above one-floor units.

• Territorial views suggest arranging units along the edge of the level land, giving each part of the view

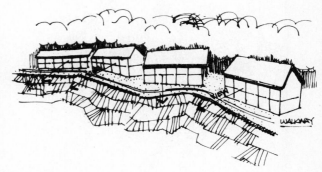

Figure 4.166

• Dramatic views suggest tall, compact units, giving all units an uninterrupted view
• Views into a valley are preferable to views out from a valley

ural features benefit from clustered units because density is concentrated so some areas are left free from development. Conversely, level sites or those without unique features may be developed more uniformly over the entire site, with open space allocated privately or for a group of units. Although steep land is more expensive to build on, it may be necessary if level land is scarce. If there is only enough level land for either recreational uses or housing, it is more acceptable to spend money to create level areas for houses than to create level areas for recreation and open space. With population mixes requiring large, level areas for open space use, the best land must be selected for that use first and slightly steeper slopes used for clusters of living units.

Uphill versus Downhill Roads paralleling contours have an uphill and downhill side, which can create difficulties for cluster access. Generally, downhill access is preferred since the cluster is overviewed from the car and is more imageable. Access is direct and change of grade can be accommodated within individual units instead of in public circulation routes.

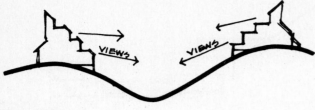

Figure 4.167

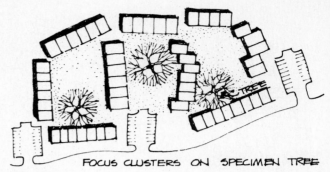

FOCUS CLUSTERS ON SPECIMEN TREE

Figure 4.170

• Panoramic views suggest tall buildings on the higher elevations, with lower units terracing down the slope

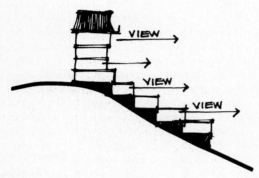

Figure 4.168

• V-shaped clusters can be oriented so each unit looks out on the view

Figure 4.169

• Natural attractions, such as trees, interesting topography, or rock outcrops, relate well as a focus for small clusters of units

Placing Buildings on Sloping Land We should avoid placing buildings exactly on top of a hill. Since the top of a hill is likely to be windy, it is best to build on the brow, just before the grade steepens, so that protection against the wind will be gained and the hill's intrinsic quality saved.

Siting houses on steep slopes is conceptually quite simple; it involves constructing a level area with foundations, framing, or grading and constructing a house on the level area. Five approaches are possible.

Landfill Grading soil out from a slope to create a level area. There may be problems: first, fill material may be expensive to acquire; second, sloping between level pads consumes land which is costly; third, the chance of erosion is increased; and last, fill may settle, causing structural problems within the building.

Cut Cut, the opposite of fill, is created by carving a level area from a slope. The level area, then, is on existing stable soil, erosion is minimized, and slopes can be steeper. The only problem is how to dispose of the dirt removed.

Cut and Fill This is the in-between or balanced solution. If structures are constructed on cut areas, and parking, roads, or other activity areas are relegated to fill area, a balanced grading scheme can be developed.

Stepped Foundations Working up a slope with stepped concrete foundations creates voids beneath the main floors. This method can be environmentally sound, though contractors tend to tackle the job with oversized equipment, causing erosion and displacing vegetation.

Pole Foundations Telephone or concrete poles are the least disruptive to the landscape if properly installed, and are relatively inexpensive.

Units running parallel to the contours have all entrances and gardens at approximately the same elevation. Entrances on steep slopes are on an upper floor, while living/eating rooms are on a lower floor with direct access to the garden. On the other hand, in units running perpendicular to the contours each entrance is at a different elevation. Units should be fit to the site, terracing down the hill to minimize cut and fill, and to add variety. (See Figure 4.172.)

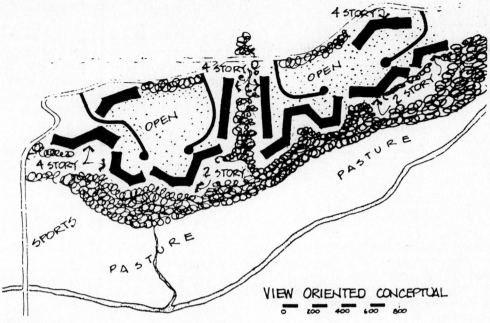

VIEW ORIENTED CONCEPTUAL

0   200   400   600   800

Figure 4.171

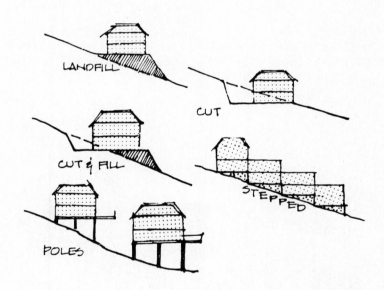

LANDFILL

CUT

CUT & FILL

STEPPED

POLES

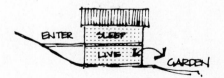

ENTER   SLEEP

LIVE

GARDEN

Figure 4.172

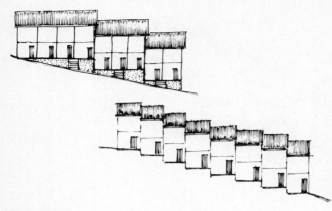

Figure 4.173

If soil and geologic conditions indicate no slippage problems, building on steep slopes can work. Since steep land suggests no direct auto access to units, walking may be required. Thus, steep developments may not be good for large families, elderly, or the handicapped, but work well for more agile resident types.

Look, Do Not Touch   At high densities, fragile landscapes may not recover from overuse and should be treated as visual rather than usable areas. The secret to preservation lies in limiting or eliminating access by:

• Planting with impenetrable vegetation.

• Dividing fragile lands into private ownership with individual responsibility for maintaining the land. Determining where people should and should not go is a prime site planning function.

• Aligning walkways and use areas away from fragile land. Most people are instinctively attracted to fragile areas; we are all intrigued by steep cliffs, marshy areas, small creeks, and unique vegetation. A walk might be constructed through a portion of the fragile areas and residents allowed some access in trade for no access elsewhere.

Access Roads   Access roads are laid out last—inserted after open space and house type have been determined. In principle, feeder roads are located in the worst location environmentally, to the north, and away from desirable views. This low priority does not mean we can ignore it as a site planning problem since adequate space for maneuvering, parking, and buffer must be provided, and a careful relationship created between the auto and each unit. Enough room for access roads must be reserved during the early site planning phases.

For most clusters, the road need not serve autos efficiently. In other words, roads may be narrow and turning radius small, so that overall pedestrian safety and use increase. Parking must be organized to minimize conflict, and defined with expanded sidewalk or planter areas. Major pedestrian walks may cross access roads as raised speed bumps or paved with different materials. Pavement widths for two-way access roads may be as little as 18 feet, certainly no wider than 20 feet. Parking bays should be 7 feet wide for parallel parking, and 20 feet for perpendicular. And, turning radius at corners may be as little as 2 feet, to limit travel speed. (See Figures 4.175 and 4.176.)

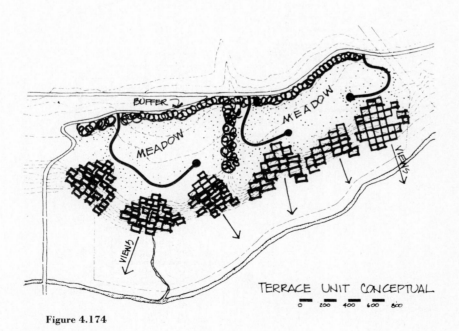

TERRACE UNIT CONCEPTUAL

Figure 4.174

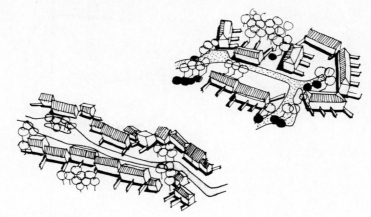

Figure 4.175

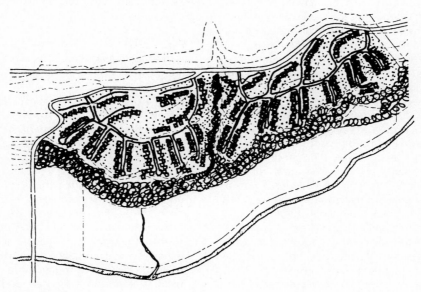

Figure 4.176

A loop cul-de-sac is different from the cul-de-sac discussed earlier in that the loop is larger and facilitates traffic flow. Internal parking courts for small clusters (10 to 12) may wrap around a lawn area.

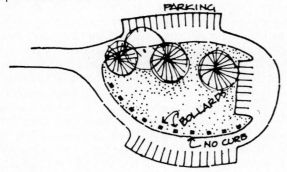

Figure 4.177

Design, and not just traffic considerations, plays a major role in local and access road layout. The site planner is orchestrating a sequence of open space views from the auto, eliminating the view in places and opening it up elsewhere. By simulating an auto trip through the site during design, we can sense where logical open space expansion is appropriate, e.g., at the development entrance, at bends in roads, at view spots, or at recreational facilities. Auto efficiency may be low as long as experience is high.

The right-of-way width may be varied to relate to abutting features. For instance, the road may be widened to include a grove of trees, stream, meadow, or view of an interesting natural feature. Portions of open space should be allocated to roadways, so that part of the feeling of an open space community is experienced from the automobile. Two opposite open space solutions are presented here, with numerous compromises possible.

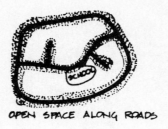

OPEN SPACE ALONG ROADS

Figure 4.178

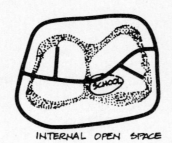

INTERNAL OPEN SPACE

First, pedestrian and auto circulation may be *combined* with the open space allocated along the roadways. This maximizes pleasant views from the auto, and minimizes construction and maintenance costs since only one pedestrian walk system has to be provided. However, resident recreational needs, as well as peace, quiet, and safety, are often not best served by open space adjacent to a road. Second, *separating* most of the pedestrian system from the roadway ensures privacy, quiet, and safety for residents. Open space may be linked to roads in key locations such as at a school, park, recreation center, or pedestrian drop-off space.

Safety trade-offs must be considered in determining the best location for pedestrian circulation. The road, our most public zone, is easy to survey and control, and therefore provides more safety for pedestrians from attack; yet it is not safe from autos and the dangers they imply. Pedestrian circulation adjacent to a road has many surveillance possibilities: from passing autos, nearby front doors, windows of adjacent buildings, or other pedestrians. Additionally, night walking is lit by street lights and emergency call boxes are nearby. Abutting residents can count on pedestrian traffic being on only one side of their homes.

On the other hand, separating pedestrians from vehicular traffic reduces surveillance potential, places pedestrians on both sides of houses, but provides more protection from automobile dangers.

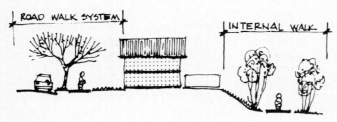

ROAD WALK SYSTEM

INTERNAL WALK

Figure 4.179

Open space serves three broad functions which when related to architecture create a total unified environment.

These functions may overlap for multiple use. The three functions are:

1. *Pleasurable open space.* This space is necessary to ensure a desirable quality of living and includes walks, malls, greenways, parks, and play areas.

2. *Service open space.* This functional open space includes streets, parking areas, roadways, and service space. These areas, often only marginally developed, must be totally landscaped to become an integral and pleasant part of the envronment.

3. *Open space to create form and image.* The appearance and feeling of the community is determined by a combination of buildings and landscape. In higher-density areas, buildings and building groupings dominate the landscape, forming the identifiable spaces. Here, open space plays a minor role as form giver but is vitally important for recreation, service, and continuity. In less dense areas, where buildings tend to be separate and lack identifiable form, open spaces provide the unifying element.

For interest, tight spaces should be contrasted with larger open areas. Contrast is a basic principle in medium-density design, that is, making each area identifiable. The open space system should be extended to all areas of the site: over land reserved for utilities, along peripheral barriers or screens, along roadways, and around (as screening) site services such as parking, garbage disposal, and laundry. Open space should integrate all portions of the site as part of a pedestrian or bicycle path, or be used for community gardens, or even for wildlife habitat. The hierarchy of open spaces should be developed in three categories: *community open space,* consisting of the largest-scale elements and amenities relating to the whole community; *neighborhood open space,* consisting of smaller-scaled elements oriented to the immediate residential community; and *development open space,* consisting of public components immediately around a cluster such as

streets, sidewalks, playgrounds, meeting places, and even private gardens.

Creating Open Space Amenity If a site has open space with no character, elements of interest and focus may be added. For instance, a lake might be created as the central focus if soil conditions and topography are right and there is a source of water. Public circulation must be provided along any amenity, even if it must pass near residential units. If this circulation is placed below units, on a hill, there is little loss of privacy for residents.

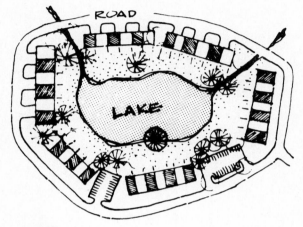

Figure 4.180

Focus may be on recreational spaces (pool, sauna, recreation building, and outdoor activity areas) or on natural elements (such as streams or drainage routes) which can be adapted to passive recreational purpose. For instance, a path system might be designed along and across a stream which might be dammed in places to create ponds; the stream might serve the dual purpose of providing views for units.

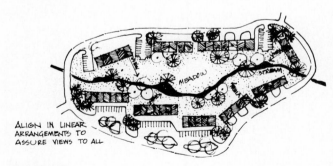

Figure 4.181

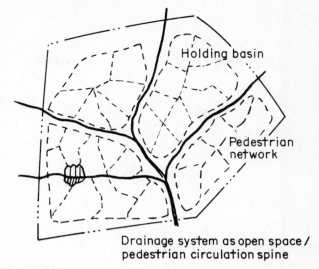

Drainage system as open space / pedestrian circulation spine

Figure 4.182

A golf course, though expensive to construct, serves as an open space focus and as a promotional feature. Paths can be worked in carefully for long walks (though golfers usually are not happy sharing their space with nongolfers), views opened from the road, and high-density units placed strategically around it, with low-density units in areas with less amenity. Since an 18-hole golf course needs approximately 140 acres, the total development size would have to be substantial.

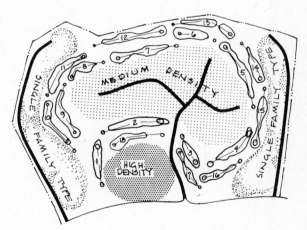

Figure 4.183

S E C T I O N

FIVE

Layout of Recreational Facilities

Recreational facilities are common to all kinds of site development. They constitute major elements in the complexes of all schools, from elementary schools to universities. A wide range of recreational facilities is an essential accessory use for most residential developments, especially cooperatives and condominiums. In fact, many newer communities are designed around specialized recreational activities such as golf courses, tennis courts, and marinas. Even industrial parks today provide selective recreational amenities within their boundaries.

Recreational areas are planned and designed to provide proper facilities for games and sports. Each activity requires a specific area and layout for its best utilization. This section includes the layouts of the most popular games and sports as recommended by the various athletic associations. Since each activity is specific as to layout, the site must be prepared to its requirements of area, grading, and orientation.

After selecting the individual activities to be included within a particular site development, the site designer should take great care

1. To avoid conflicts or dangerous conditions in the juxtapositioning of the various activities

2. To provide physical and visual barriers between the recreational facilities and adjacent uses, such as residential or community facilities

3. To provide adequate parking areas to accommodate both participants and spectators

4. To provide proper access to the facilities for service and maintenance

5.1 General Area Requirements for Games and Sports

Area requirements for popular games and sports are shown in the accompanying table.

Activity	Play area (feet)	Total area		Number of players
		Dimensions	Square feet	
Archery	Length; 90–300	50 (minimum) × 450 (maximum)		
Badminton	Single; 17 × 44	25 × 60	1,500	2
	Double; 20 × 44	30 × 60	1,800	4
Baseball	Diamond; 90 × 90	300 × 300 (minimum)	90,000	18
		350 × 350 (average)	122,500	
Basketball (men)	Minimum; 42 × 74	60 × 100 (average)	6,000	10
	Maximum; 50 × 94			
Basketball (women)	45 × 90	55 × 100 (average)	5,500	12–18
Bocci	18 × 62	30 × 80	2,400	2–4
Bowling (lawn)*	One alley; 14 × 100	120 × 120 (eight alleys)	14,400	32–64
Box hockey	4 × 10	16 × 20	320	2
Box lacrosse	Minimum; 60 × 160	85 × 185 (average)	15,725	14
	Maximum; 90 × 200			
Clock golf	Circle; 20–24 in diameter	30-foot circle	706	2–8
Cricket	Between wickets; 66	420 × 420	176,400	22
Croquet	30 × 60	30 × 60	1,800	2–8
Curling	Between hacks; 138	14 × 150	2,100	8
Deck tennis	Single; 12 × 40	20 × 50	1,000	2
	Double; 18 × 40	26 × 50	1,300	4
Field ball	180 × 300	210 × 340	71,400	22
Field hockey	Minimum; 150 × 270	200 × 350 (average)	70,000	22
	Maximum; 180 × 300			
Football	195 × 300; plus end zones	195 × 480	93,600	24
Handball	20 × 34	30 × 45	1,350	2 or 4
Hand tennis	16 × 40	25 × 60	1,500	2 or 4
Horseshoes (men)	Between stakes; 40	12 × 50	600	2 or 4
Horseshoes (women)	Between stakes; 30	12 × 40	480	2 or 4
Ice hockey	Minimum; 65 × 165	85 × 185 (average)	15,725	12
	Maximum; 85 × 200			
Lacrosse	Minimum; 210 × 450	260 × 500 (average)	130,000	24
Paddle tennis	Single; 16 × 44	28 × 70	1,960	2
	Double; 20 × 44	32 × 70	2,240	4
Polo	Maximum; 600 × 960	600 × 960	576,000	8
Quoits	Between stakes; 54	25 × 80	2,000	2 or 4
Roque	30 × 60	30 × 60	1,800	4
Shuffleboard	6 × 52	10 × 64	640	2 or 4
Soccer (men)	Minimum; 150 × 300	240 × 360 (average)	86,400	22
	Maximum; 300 × 390			
Soccer (women)	Minimum; 120 × 240	200 × 320	64,000	22
	Maximum; 180 × 300			
Softball	Diamond; 60 × 60	250 × 250 (minimum)	62,500	20

Activity	Play area (feet)	Total area		Number of players
		Dimensions	Square feet	
Speedball (men)	Minimum; 160 × 240	180 × 300	54,000	22
	Maximum; 160 × 360	180 × 420	75,600	
Swimming (pool)†	75 × 45		4–8 lanes	
Competitive	165 × 7		4–8 lanes	
Synchronized	50 meters × 45 feet			
Table tennis	5 × 9	12 × 20	240	2 or 4
Tennis	Single; 27 × 78	50 × 120	6,000	2
	Double; 36 × 78	60 × 120	7,200	2 or 4
Tether tennis	Circle; 6 in diameter	20 × 20	400	2
Touch football	160 × 300	175 × 330	57,750	22
Volley ball	30 × 60	50 × 80	4,000	12–16
Water polo	55 yards × 15 yards × 6 feet			

*Most bowling greens in public recreation areas measure 120 × 120 feet and provide for eight alleys. The amount of space required for a single alley would be 20 by 120 feet.
†A 1-meter diving board requires a 9-foot depth of water; a 3-meter board requires 12 feet of water; a 10-meter diving platform requires 10 feet of water.

5.2 Layout of Games and Sports Activities

Running Track and Football Field Layouts of a running track and a football field are shown in Figures 5.1 and 5.2.
Field and Track Layouts for the high jump, javelin throw, pole vault, combination long jump and triple jump, discus throw, and shot put and hammer throw are shown in Figures 5.3 through 5.8.

Baseball Suggested layouts for a baseball diamond and for home and pitcher's plates are shown in Figures 5.9 and 5.10.

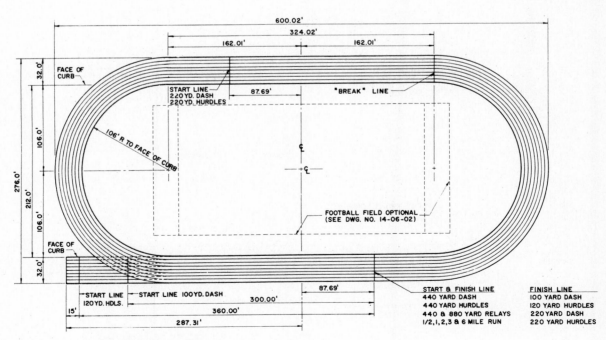

Figure 5.1 A 440-yard running track.

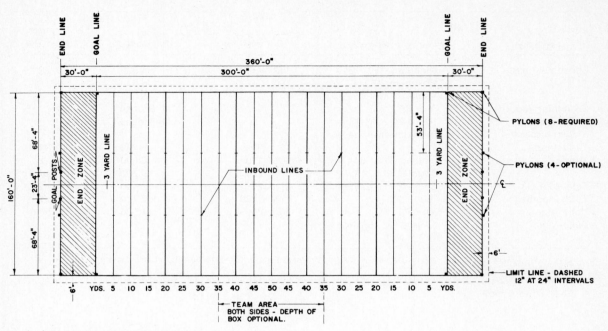

Figure 5.2 An eleven-man football field.

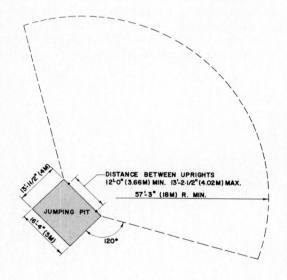

Figure 5.3 High jump.

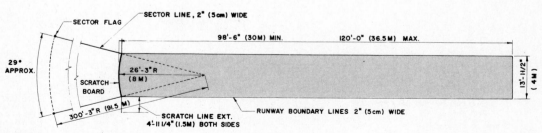

Figure 5.4 Javelin throw.

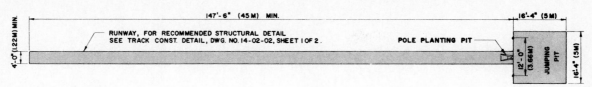

Figure 5.5   Pole vault.

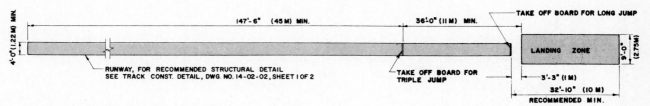

Figure 5.6   Combination long jump and triple jump.

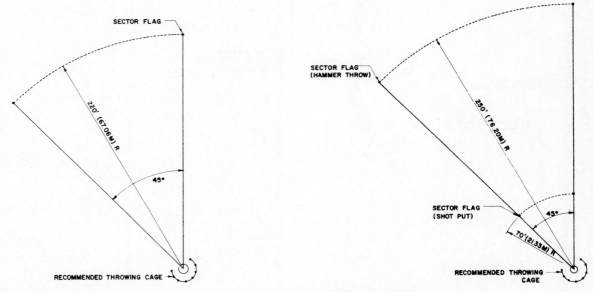

Figure 5.7   Discus throw.

Figure 5.8   Shot put and hammer throw.

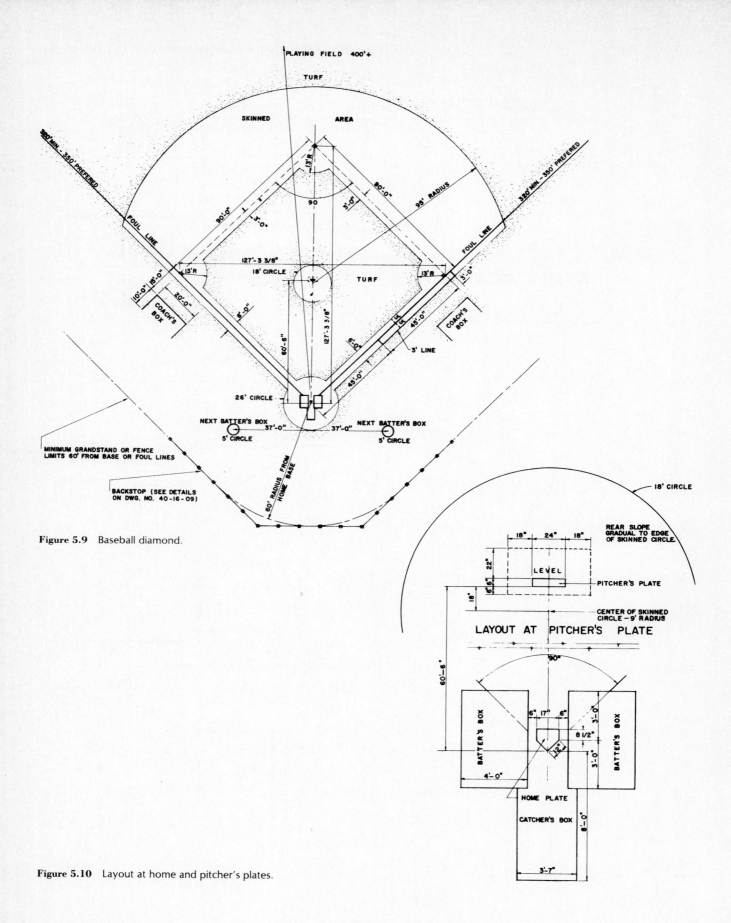

Figure 5.9 Baseball diamond.

Figure 5.10 Layout at home and pitcher's plates.

Pony League Baseball

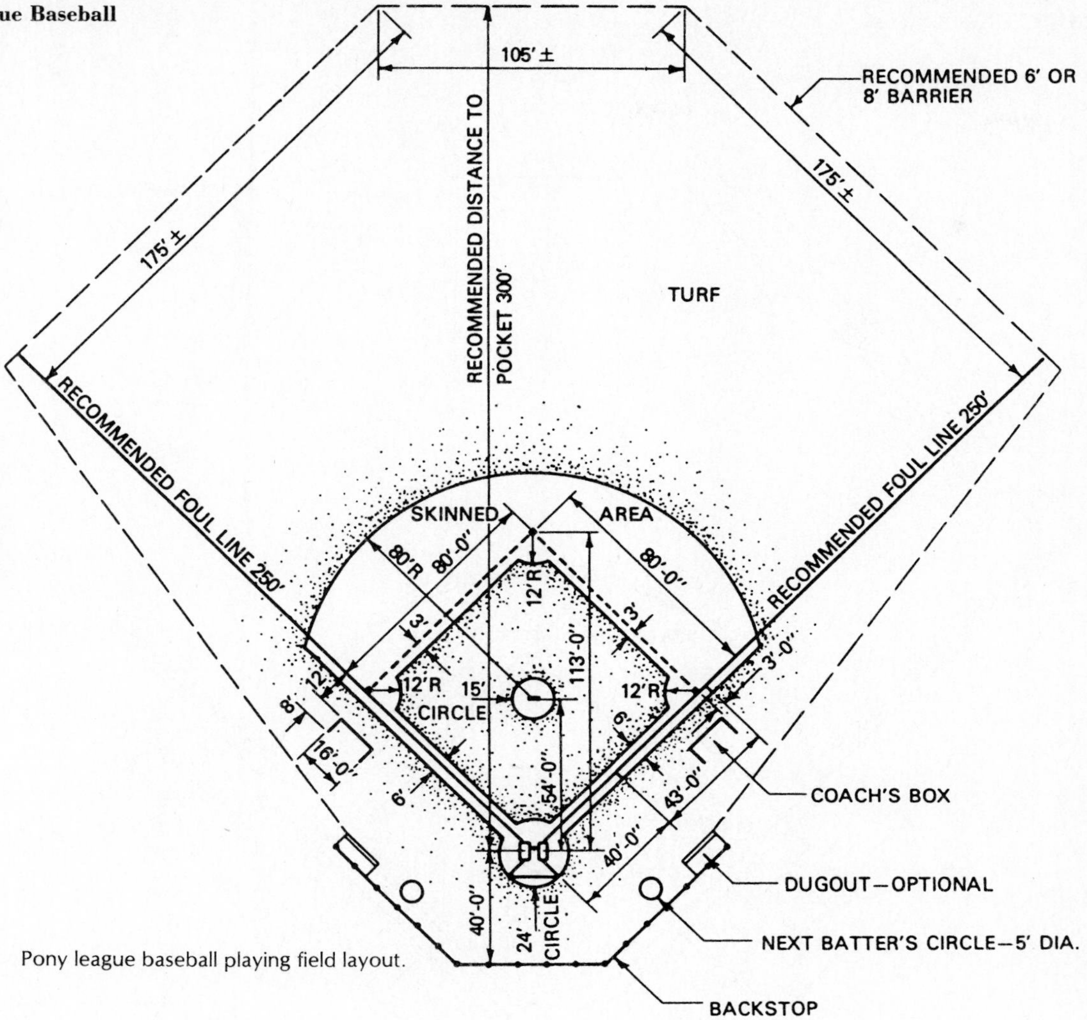

Figure 5.11   Pony league baseball playing field layout.

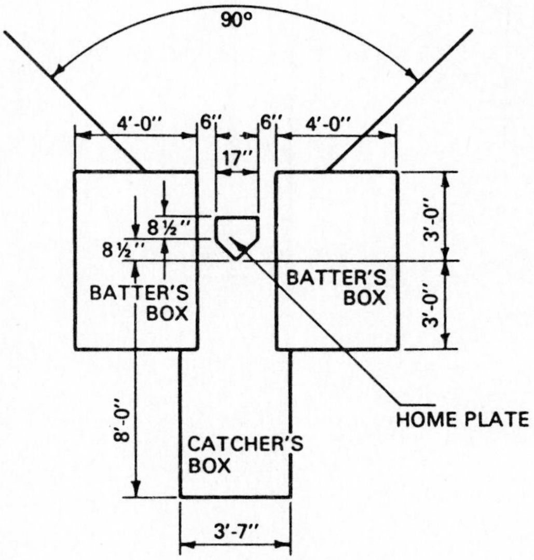

Figure 5.12   Optional layout at home plate.

Bronco and Colt League Baseball

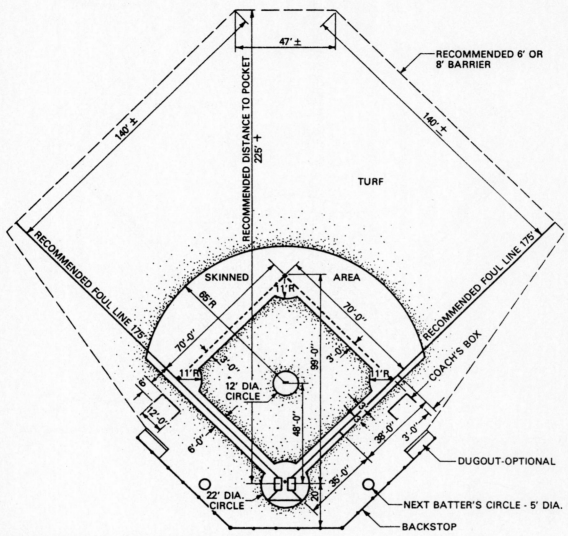

Figure 5.13 Bronco league baseball playing field layout.

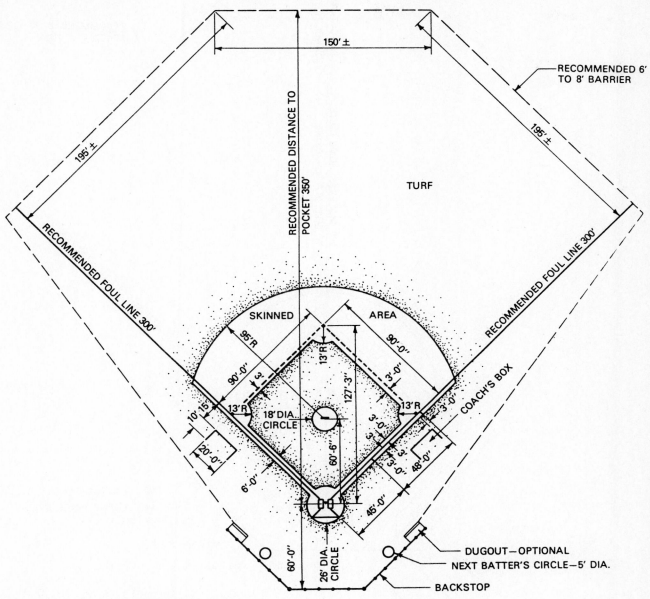

Figure 5.14   Colt league baseball playing field layout.

Little League Baseball

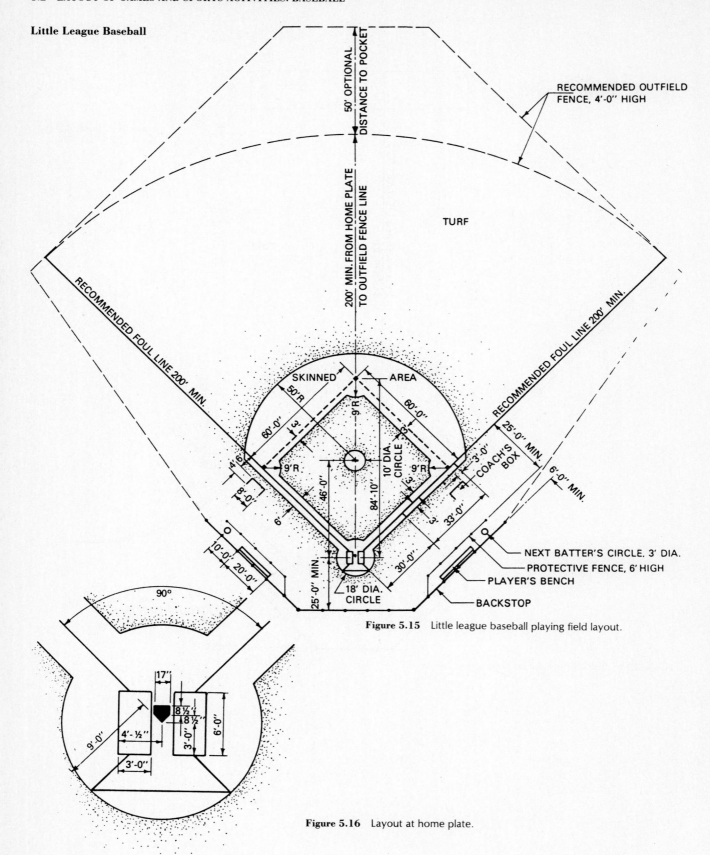

Figure 5.15 Little league baseball playing field layout.

Figure 5.16 Layout at home plate.

Softball

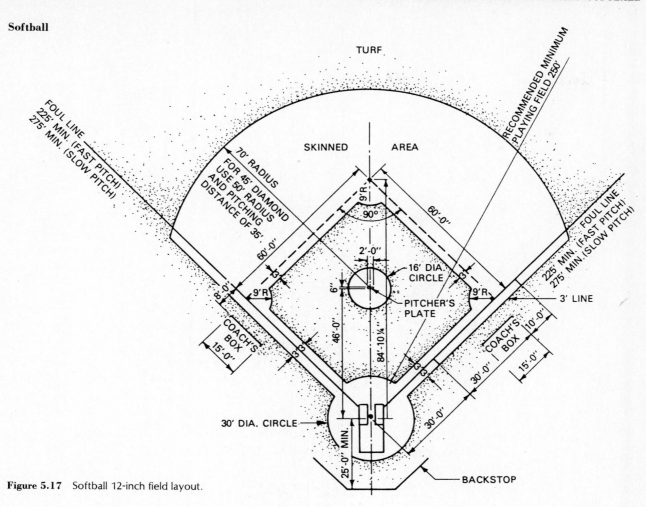

Figure 5.17   Softball 12-inch field layout.

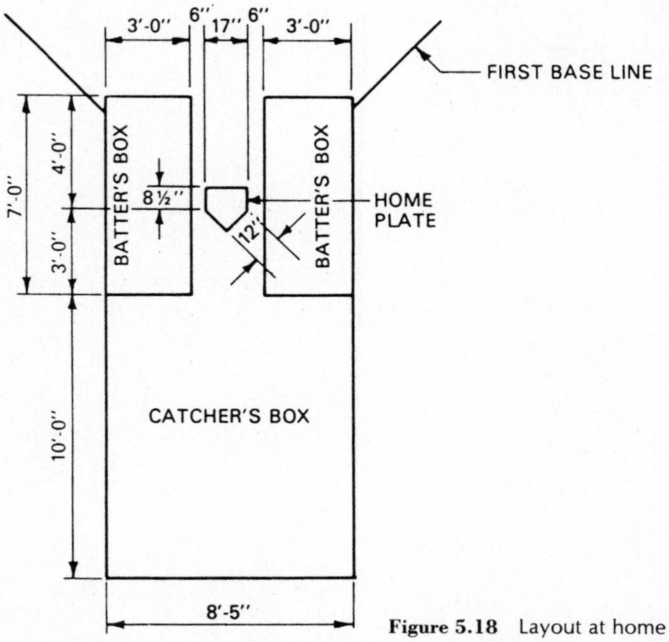

Figure 5.18   Layout at home plate.

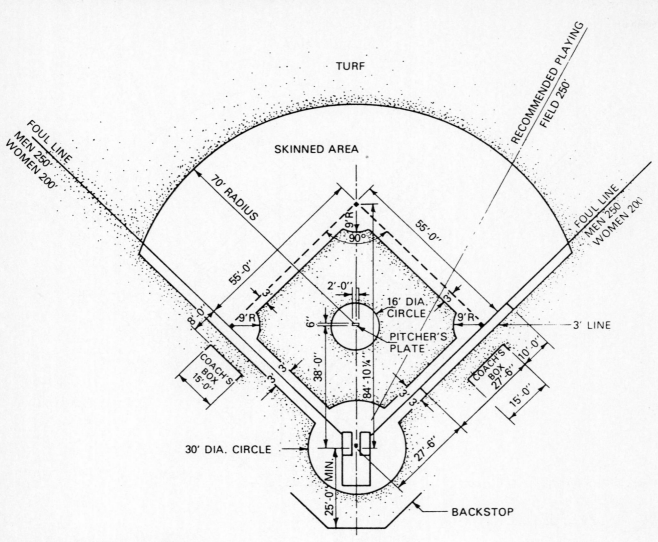

Figure 5.19   Softball 16-inch field layout.

Combination Ballfields

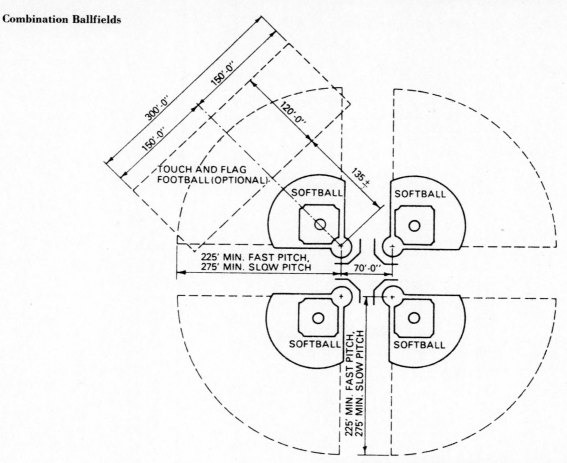

Figure 5.20 Multiple softball fields combined with optional touch and flag football.

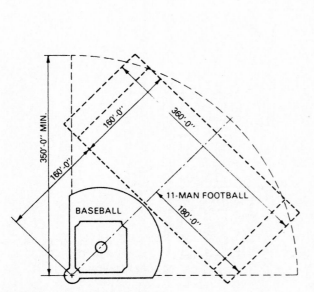

Figure 5.21 Baseball and football.

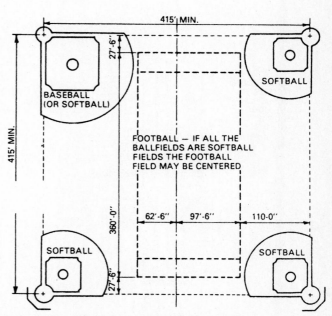

Figure 5.22 Multiple baseball, softball, and football.

Tennis   A layout for two tennis courts is shown in Figure 5.23.

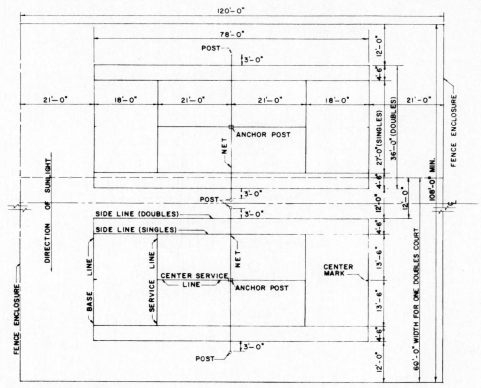

Figure 5.23   Tennis courts.

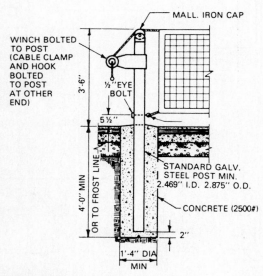

Figure 5.24   Tennis net and post details.

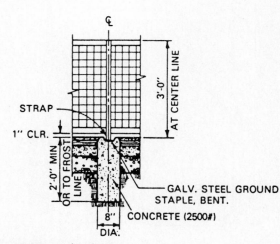

Figure 5.25   Anchor footing.

Badminton

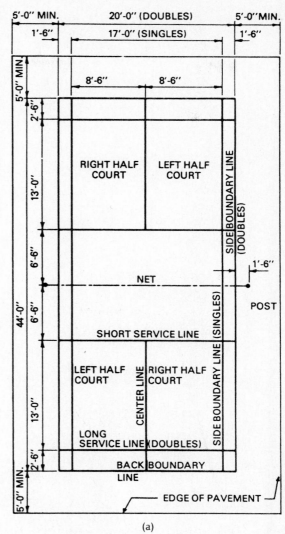

(a)

Figure 5.26 Badminton: (*a*) court layout; (*b*) isometric showing net.

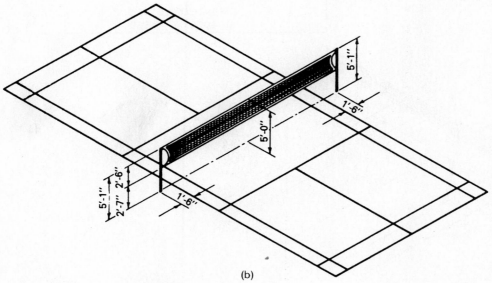

(b)

Deck Tennis

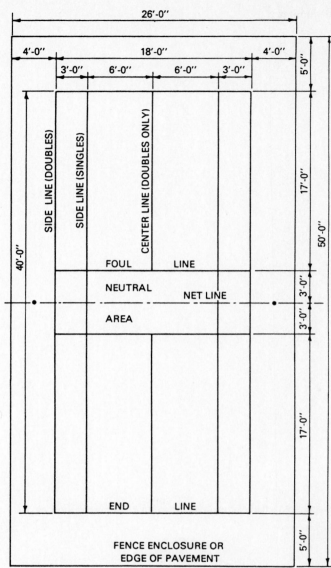

(a)

Figure 5.27 Deck tennis: (a) court layout; (b) isometric showing net.

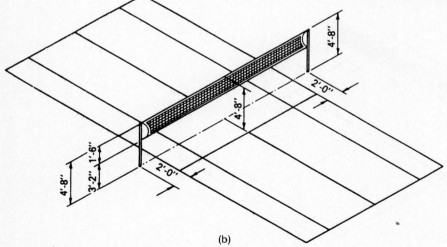

(b)

Paddle Tennis

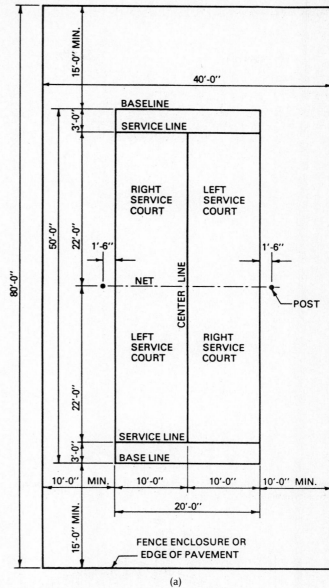

(a)

Figure 5.28 Paddle tennis: (*a*) court layout; (*b*) isometric showing net.

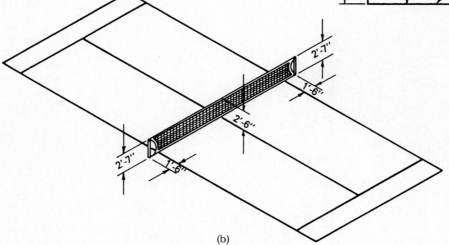

(b)

Platform Tennis

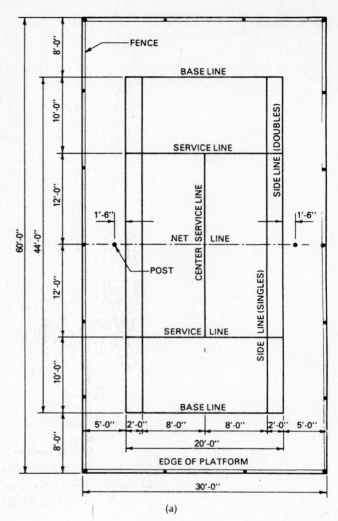

(a)

Figure 5.29 Platform tennis: (*a*) court layout; (b) isometric showing fence (typical wood construction).

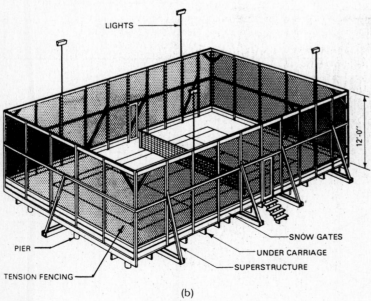

(b)

Volleyball and Polo

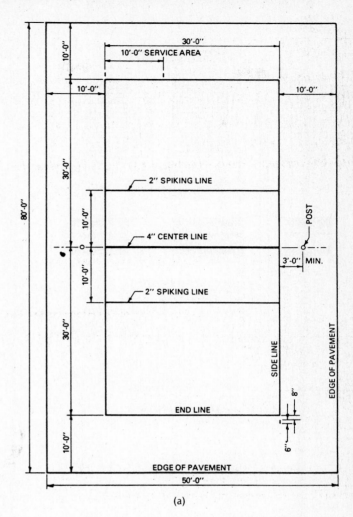

(a)

Figure 5.30 Volleyball: (*a*) court layout; (*b*) isometric showing net.

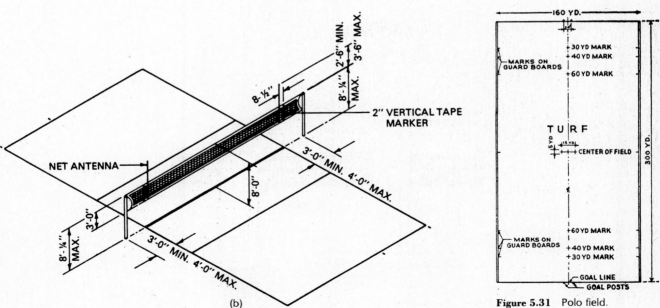

(b)

Figure 5.31 Polo field.

Fixed Net Posts

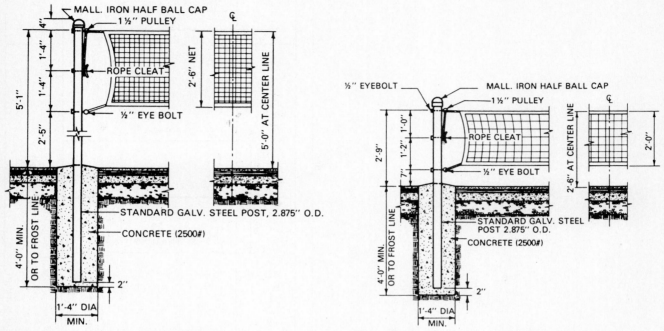

Figure 5.32 Badminton net and post details.

Figure 5.33 Paddle tennis net and post details.

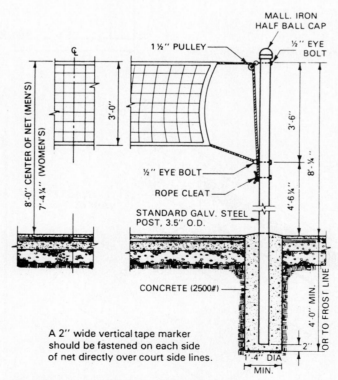

A 2" wide vertical tape marker should be fastened on each side of net directly over court side lines.

Figure 5.34 Volleyball net and post details.

Handball A layout and an elevation for a handball court are illustrated in Figure 5.35*a* and *b*.

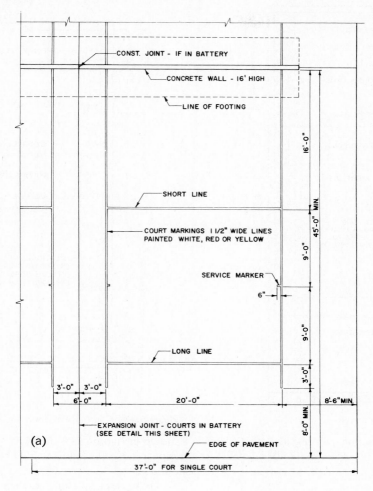

(a)

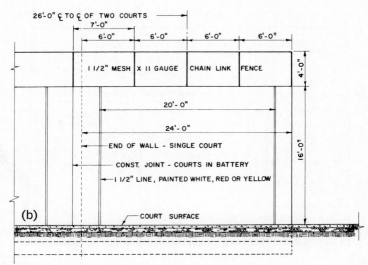

(b)

Figure 5.35 One-wall handball court: (*a*) plan; (*b*) elevation.

Three- and Four-Wall Handball

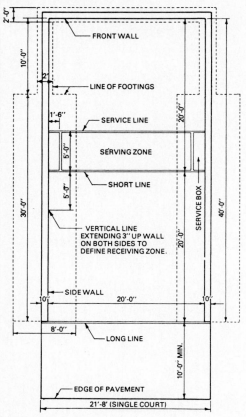

Figure 5.36 Court layout, three-wall.

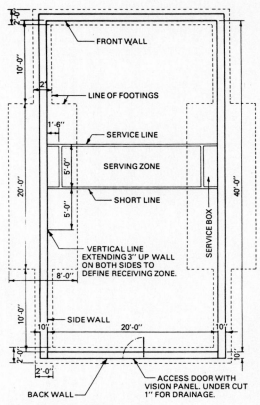

Figure 5.37 Court layout, four-wall.

Basketball Layouts of National Collegiate Athletic Association and Amateur Athletic Union basketball courts are shown in Figures 5.38 and 5.39. Two types of backboards appear in Figures 5.40 and 5.41.

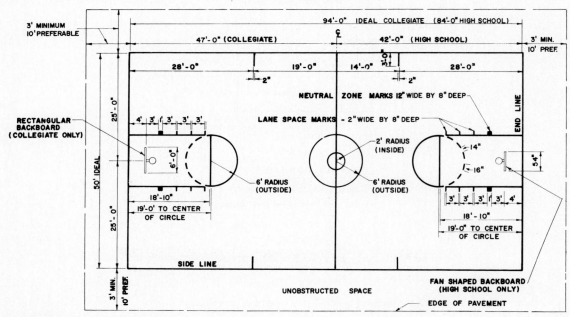

Figure 5.38 NCAA basketball court.

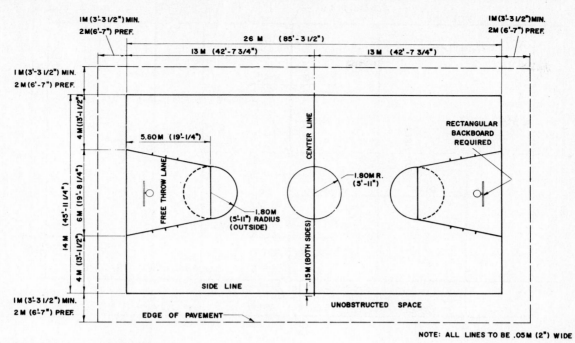

Figure 5.39   AAU basketball court.

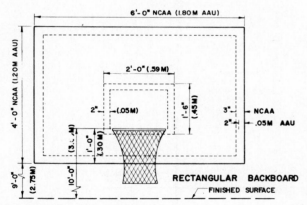

Figure 5.40   Rectangular backboard.

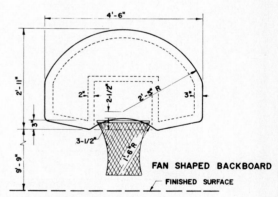

Figure 5.41   Fan-shaped backboard.

Lacrosse

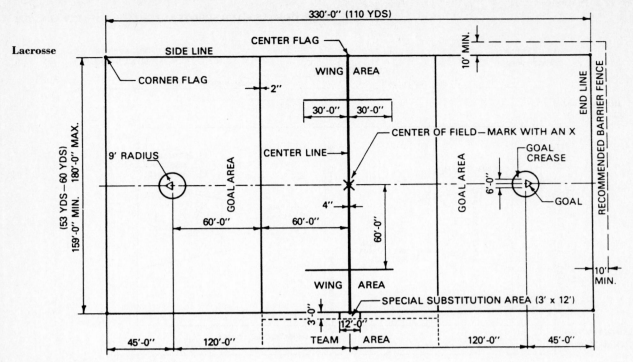

Figure 5.42 Lacrosse, men's.

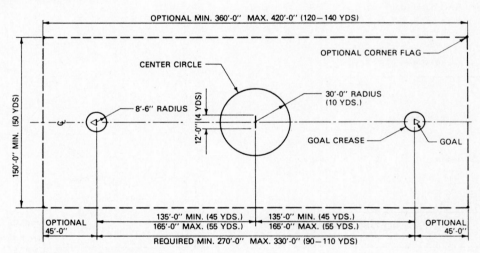

Figure 5.43 Lacrosse, women's.

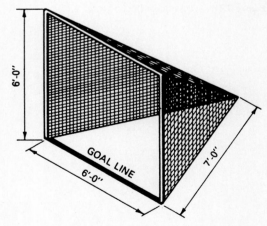

Figure 5.44 Detail of goal.

Soccer

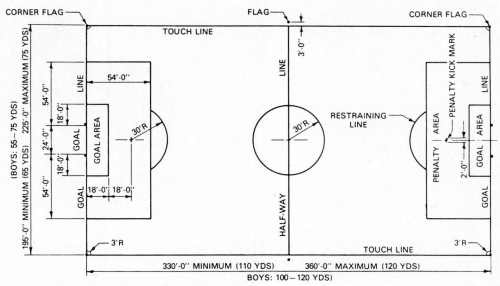

Figure 5.45 Soccer, men's and boys'.

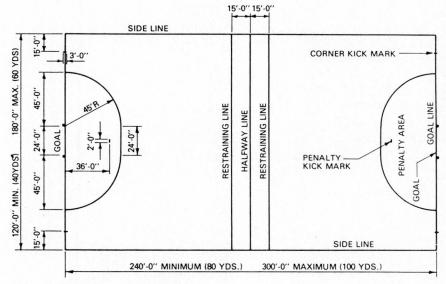

Figure 5.46 Soccer, women's and girls'.

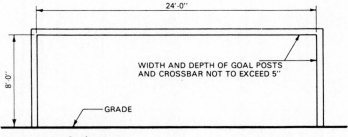

Figure 5.47 Goal posts.

Field Hockey and Rugby

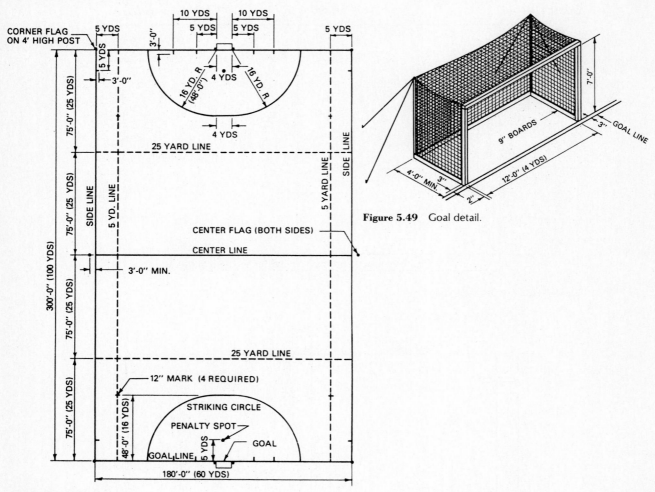

Figure 5.49 Goal detail.

Figure 5.48 Field hockey.

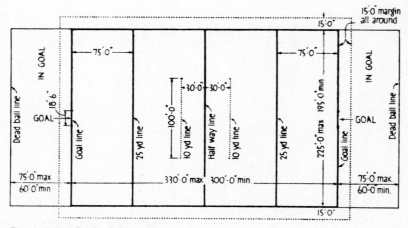

Figure 5.50 Rugby field.

Roque

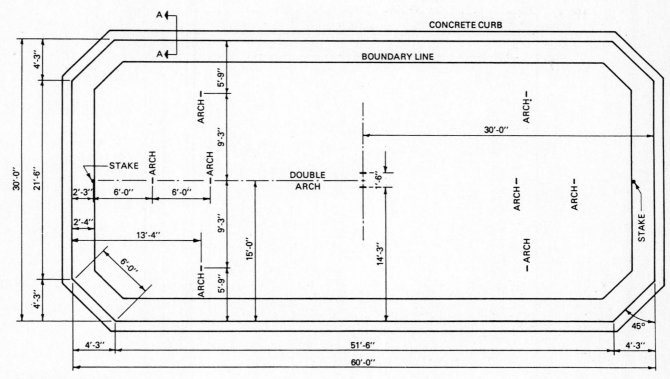

Figure 5.51   Roque.

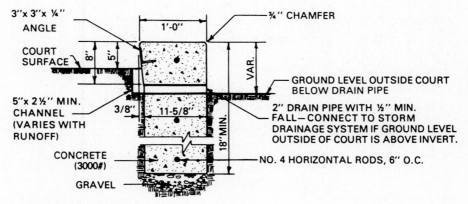

Figure 5.52   Curb section *A-A*.

Croquet

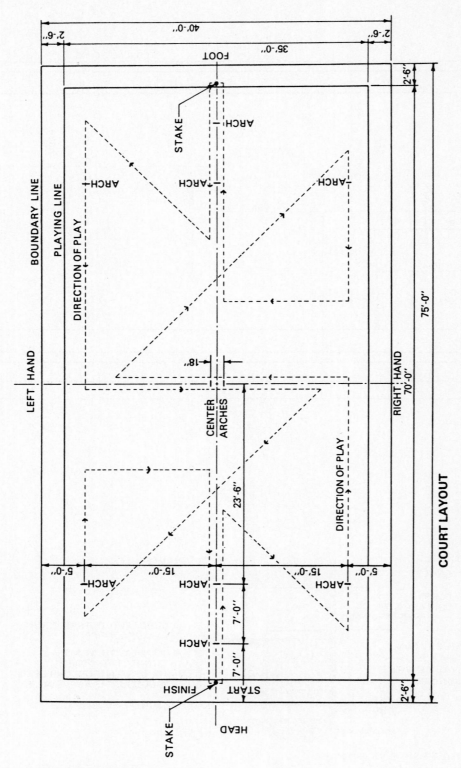

Figure 5.53 Croquet.

Archery   Figures 5.54 and 5.55 illustrate an archery range and the design for a target.

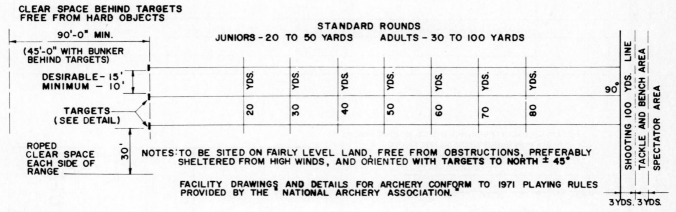

Figure 5.54   Archery range.

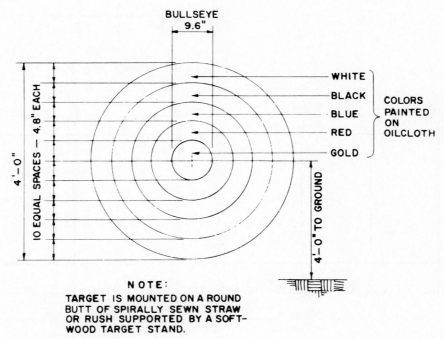

NOTE:
TARGET IS MOUNTED ON A ROUND
BUTT OF SPIRALLY SEWN STRAW
OR RUSH SUPPORTED BY A SOFT-
WOOD TARGET STAND.

Figure 5.55   Target detail.

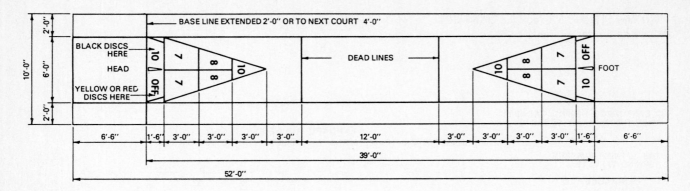

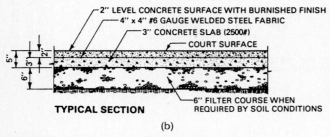

TYPICAL SECTION

(b)

Figure 5.56 Shuffleboard: (a) court layout; (b) typical section.

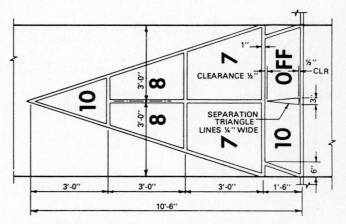

Figure 5.57 Court marking detail.

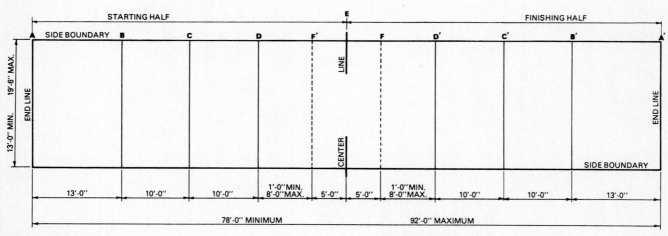

Figure 5.58 Boccie court.

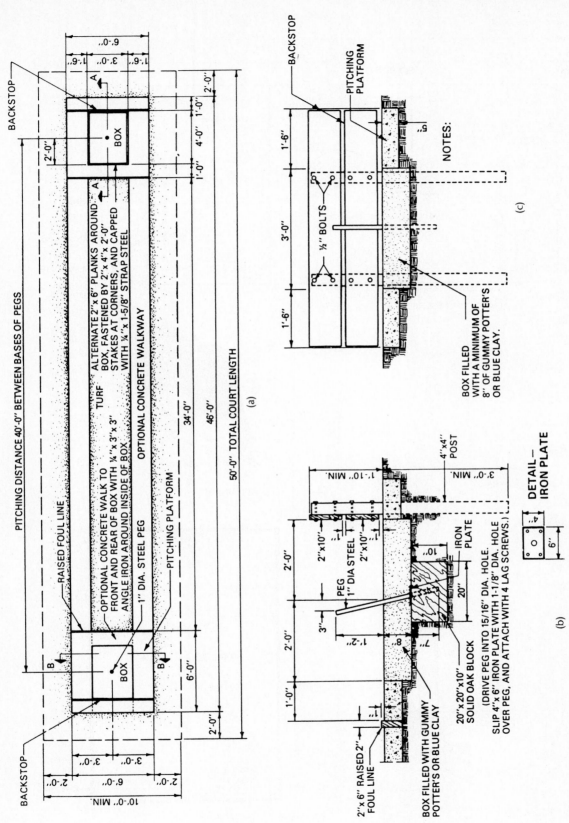

Figure 5.59 Horseshoes: (a) court layout; (b) section A-A; (c) section B-B.

Ice Sports: Hockey, Skating, Polo

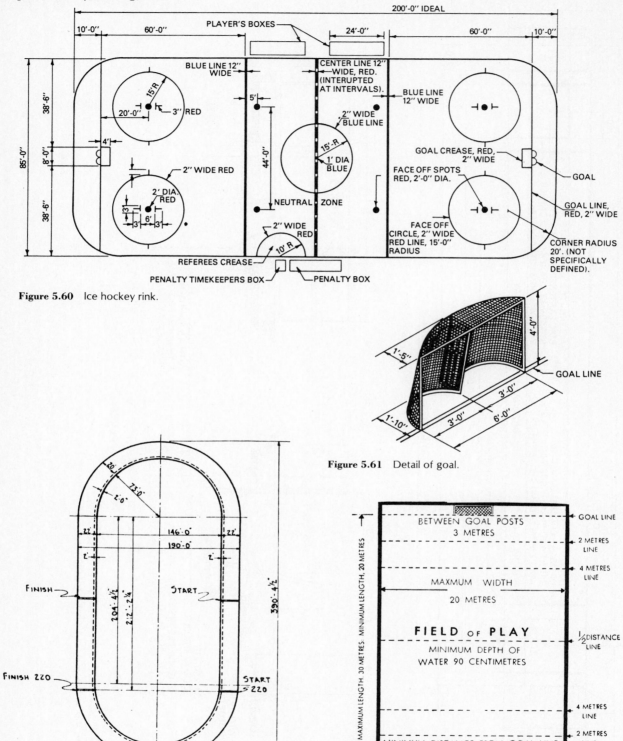

Figure 5.60 Ice hockey rink.

Figure 5.61 Detail of goal.

Figure 5.62 Six-lap ice skating rink.

Figure 5.63 Water polo field.

Multiple Recreation Courts

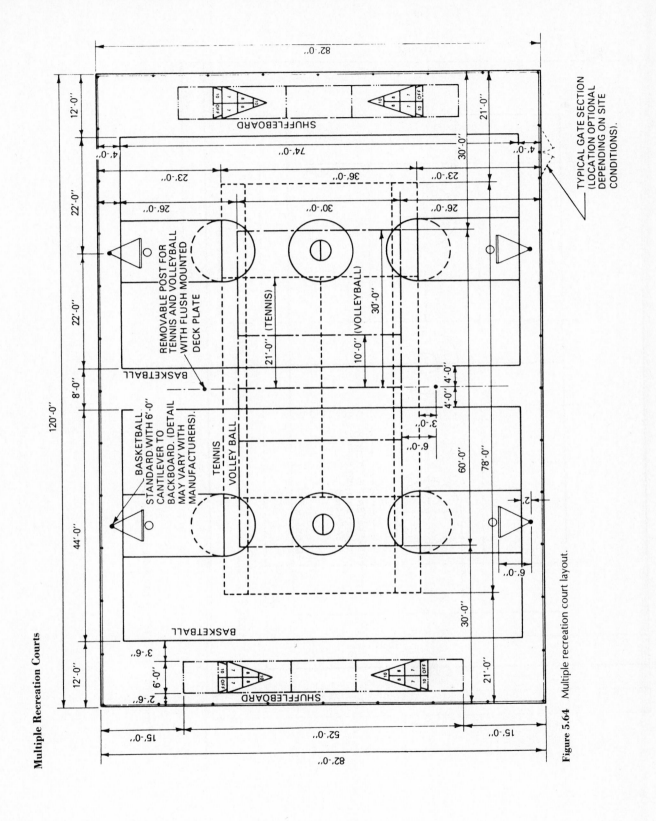

Figure 5.64 Multiple recreation court layout.

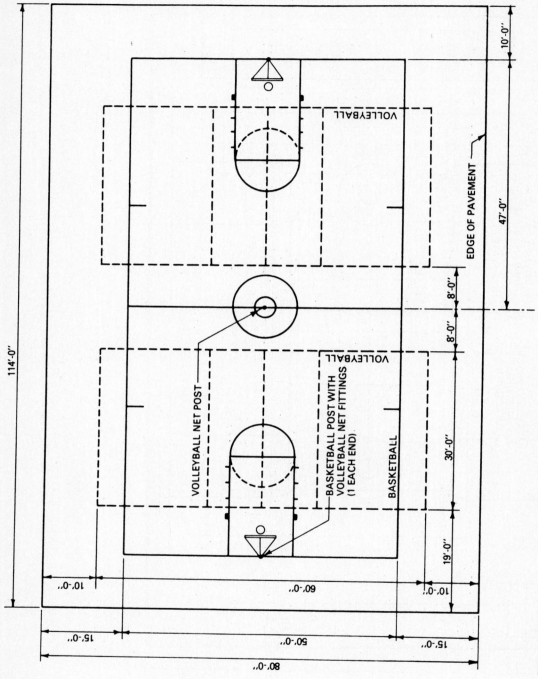

Figure 5.65 Layout for one basketball and two volleyball courts.

Bicycling: Track Layouts

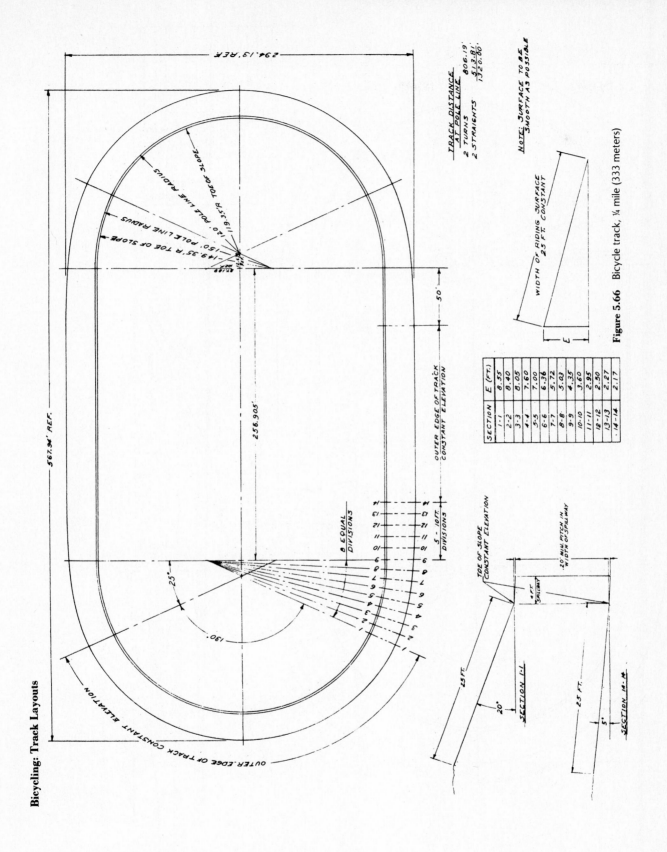

SECTION	E (FT.)
1-1	8.55
2-2	8.40
3-3	8.05
4-4	7.60
5-5	7.00
6-6	6.36
7-7	5.72
8-8	5.03
9-9	4.35
10-10	3.60
11-11	2.95
12-12	2.50
13-13	2.27
14-14	2.17

TRACK DISTANCE
AT POLE LINE
2 TURNS   806.19
2 STRAIGHTS 513.81
             1320.00

NOTE: SURFACE TO BE
SMOOTH AS POSSIBLE

WIDTH OF RIDING SURFACE
25 FT. CONSTANT

Figure 5.66 Bicycle track, ¼ mile (333 meters)

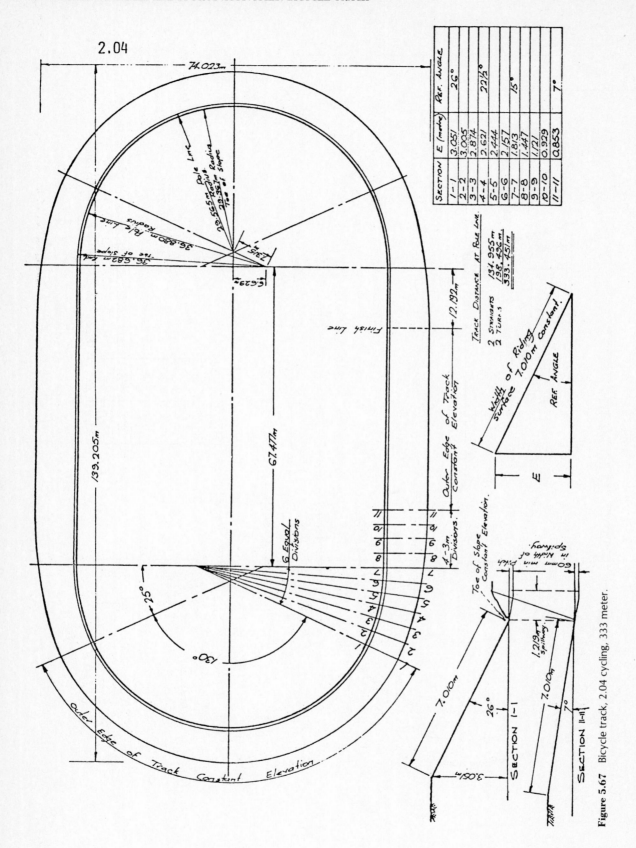

SECTION	E (meter)	REF. ANGLE
1–1	3.051	26°
2–2	3.005	
3–3	2.874	22½°
4–4	2.621	
5–5	2.444	
6–6	2.157	15°
7–7	1.813	
8–8	1.447	
9–9	1.721	
10–10	0.929	7°
11–11	0.853	

TRACK DISTANCE AT POLE LINE.

2 STRAIGHTS 134.955m
2 TURNS   198.496m
             333.451m

Figure 5.67 Bicycle track, 2.04 cycling, 333 meter.

Horse Shows

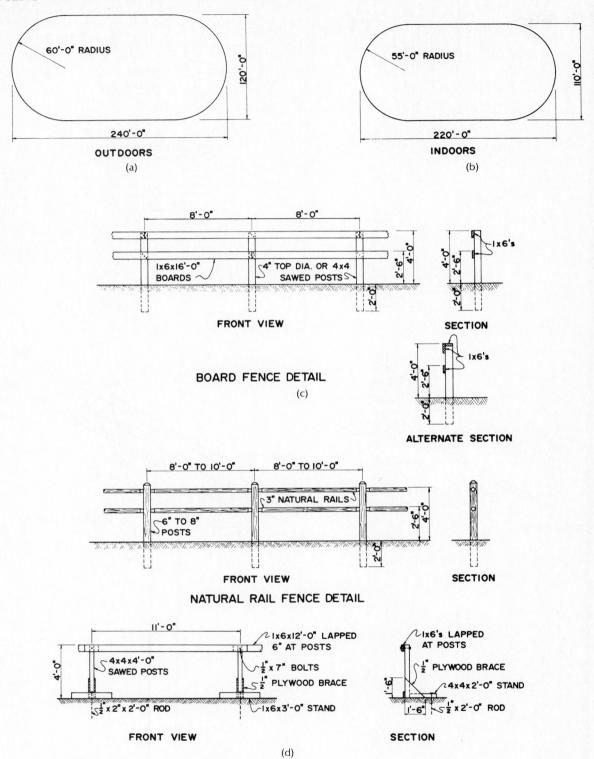

Figure 5.68 Horse show rings and details.

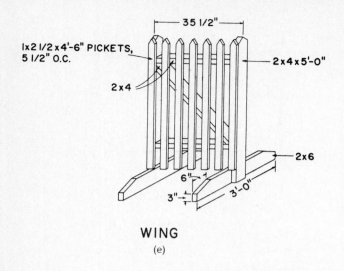

WING

(e)

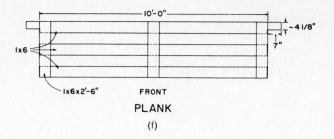

PLANK

(f)

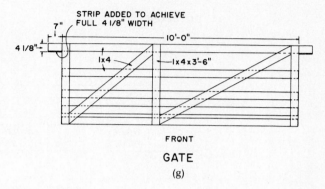

GATE

(g)

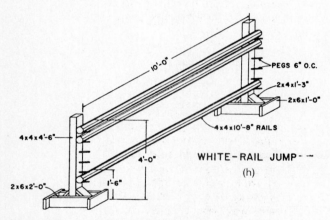

WHITE-RAIL JUMP

(h)

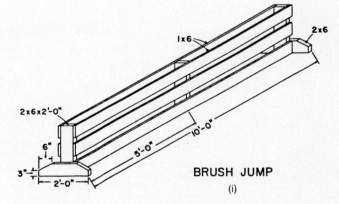

BRUSH JUMP

(i)

Figure 5.68 Horse show rings and details (*Continued*).

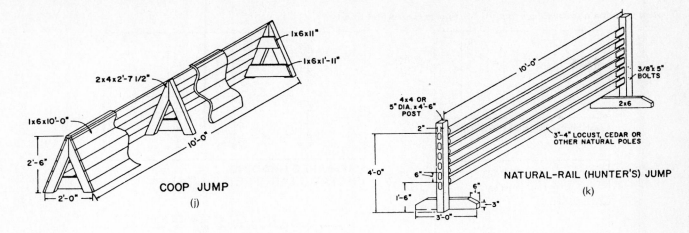

COOP JUMP
(j)

NATURAL-RAIL (HUNTER'S) JUMP
(k)

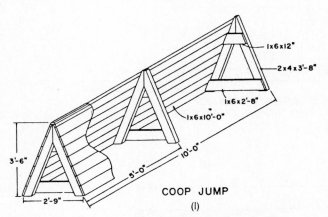

COOP JUMP
(l)

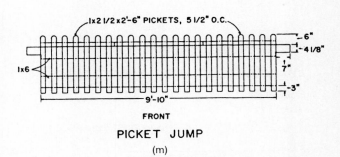

FRONT

PICKET JUMP
(m)

Figure 5.68 Horse show rings and details (*Continued*).

Typical Grading and Drainage Details for Sports Courts and Fields

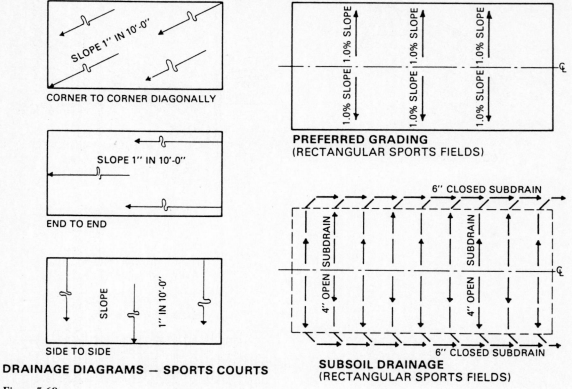

DRAINAGE DIAGRAMS — SPORTS COURTS

Figure 5.69

SUBSOIL DRAINAGE
(RECTANGULAR SPORTS FIELDS)

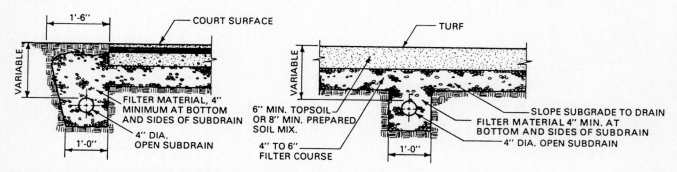

SECTION — PERIMETER DRAIN (SPORTS COURTS)

Figure 5.70

SECTION — SUBSOIL DRAIN

Court Surfaces Paved playing surfaces should be in one plane and pitched from side to side, end to end, or corner to corner diagonally, instead of in two planes pitched to or from the net. Minimum slope should be 1 inch in 10 feet 0 inches. Subgrade should slope in the same direction as the surface. Perimeter drains may be provided for paved areas. Underdrains are not recommended beneath paved areas.

Playing Fields Preferred grading for a rectangular field is a longitudinal crown with 1 percent slope from center to each side.

Grading may be from side to side or corner to corner diagonally if conditions do not permit the preferred grading.

Subsoil drainage is to slope in the same direction as the surface. Subdrains and filter course are to be used only when subsoil conditions require. Where subsoil drainage is necessary, the spacing of subdrains is dependent on local soil conditions and rainfall.

Subdrains are to have a minimum gradient of 0.15 percent.

Typical Playing Surfaces

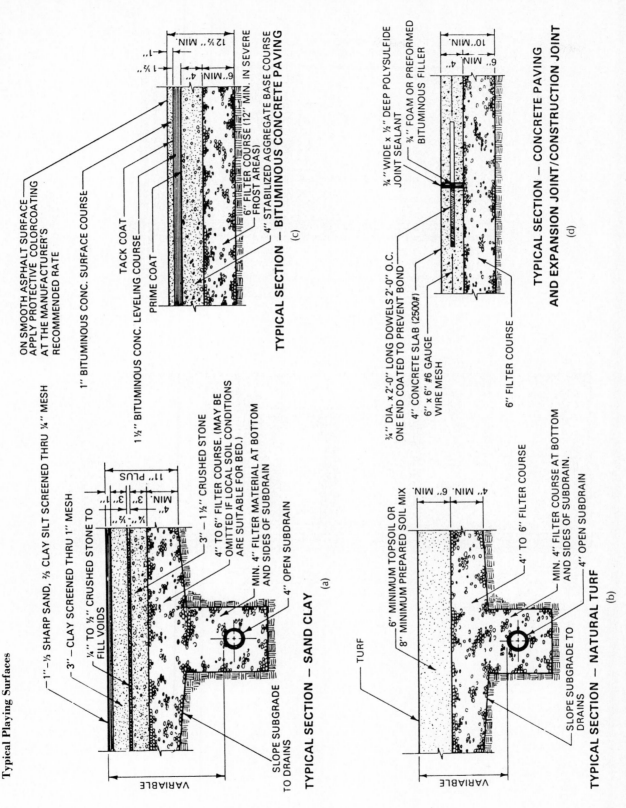

TYPICAL SECTION — SAND CLAY

(a)

TYPICAL SECTION — NATURAL TURF

(b)

TYPICAL SECTION — BITUMINOUS CONCRETE PAVING

(c)

TYPICAL SECTION — CONCRETE PAVING
AND EXPANSION JOINT/CONSTRUCTION JOINT

(d)

Figure 5.71 Typical playing surfaces.

Fence Enclosures for Sports Facilities

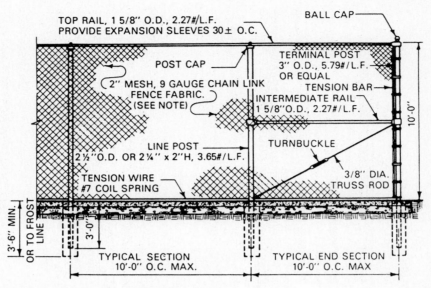

Figure 5.72 Elevation of typical fence.

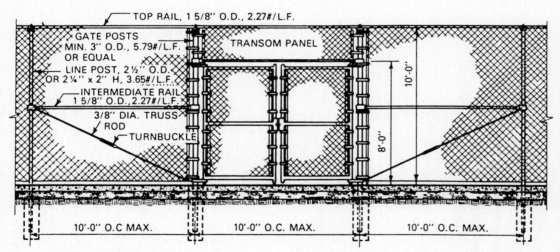

Figure 5.73 Elevation of double gate.

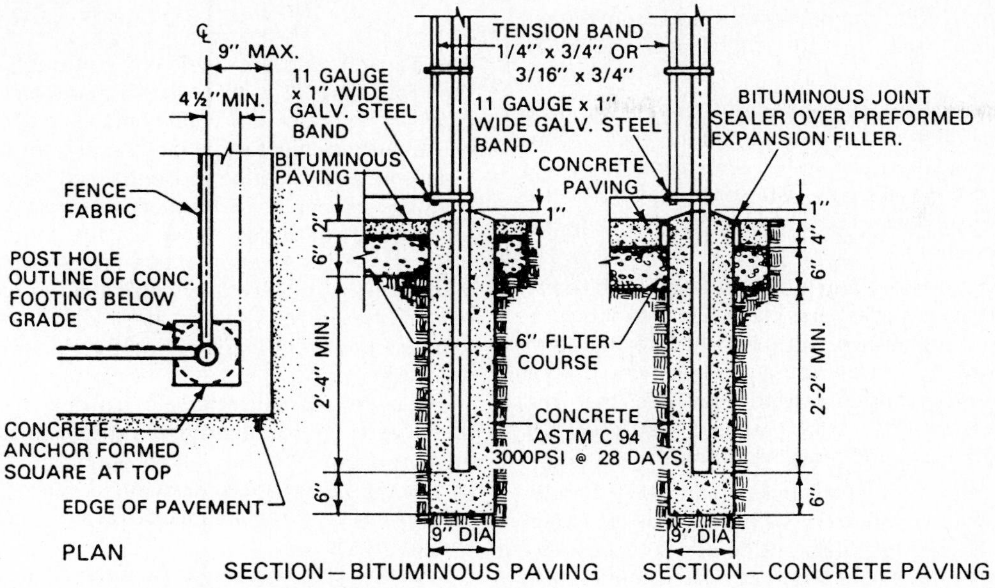

Figure 5.74 Corner post and footing details.

5.3 Play Lots

Play lots should be provided for preschool children up to 6 years of age primarily in conjunction with multifamily (town house and apartment) developments and, where desirable, in single-family neighborhoods remote from elementary schools. They are a necessary element of such developments to complement common open-space areas. Play lots may include (1) an enclosed area for play equipment and such special facilities as a sand area and a spray pool, (2) an open turfed area for active play, and (3) a shaded area for quiet activities.

Location   Play lots should be included as an integral part of the housing area design and be located within 300 to 400 feet of each living unit served. They should be accessible without crossing any street, and walkways leading to them should have easy gradients for pushing strollers and carriages. Play lots may be included in playgrounds close to housing areas to serve the preschool age group in the adjoining neighborhood.

Size   The enclosed area for play equipment and special facilities should be based on a minimum of 70 square feet per child, which is equivalent to 21 square feet per family on the average basis of 0.3 preschool child per family. A minimum enclosed area of approximately 2000 square feet will serve some thirty preschool children (about 100 families). Such a size will accommodate only a limited selection of play equipment. To accommodate a full range of equipment and special facilities, including a spray pool, the minimum enclosed area should be about 4000 square feet, which will serve up to fifty preschool children (about 165 families).

Additional space is required to accommodate the ele-

ments of the play lot outside the enclosed area. A turfed area at least 40 feet square should be provided for active games.

Activity Spaces and Elements   A play lot should comprise the following basic activity spaces and elements:

1. An enclosed area with play equipment and special facilities including
   a. Play equipment such as climbers, slides, swing sets, play walls and playhouses, and play sculpture
   b. A sand area
   c. A spray pool
2. An open turfed area for running and active play
3. A shaded area for quiet activities
4. Miscellaneous elements including benches for supervising parents; walks and other paved areas wide enough for strollers, carriages, tricycles, wagons, and so on; and play space dividers (fences, walks, trees, and shrubs), a step-up drinking fountain, trash containers, and landscape planting

Layout   The specific layout and shape of each play lot will be governed by existing site conditions and the facilities to be provided. The general layout principles are as follows:

1. The intensively used part of the play lot with play equipment and special facilities should be surrounded by a low enclosure with supplemental planting and provided with one entrance-exit. This design will discourage intrusion by animals or older children, provide adequate and safe con-

trol over the children, and prevent the area from becoming a thoroughfare. Adequate drainage should be provided.

2. Equipment should be selected and arranged with adequate surrounding space in small, natural play groups. Traffic flow should be planned to encourage movement throughout the play lot in a safe, orderly manner. This traffic flow may be facilitated by walks, plantings, low walls, and benches.

3. Equipment that enables large numbers of children to play without taking turns (climbers, play sculpture) should be located near the entrance, yet positioned so that it will not cause congestion. With such an arrangement, children will tend to move more slowly to equipment that limits participation and requires turns (swings, slides), thus modifying the load factor and reducing conflicts.

4. Sand areas, play walls, playhouses, and play sculpture should be located away from such pieces of equipment as swings and slides for safety and to promote a creative atmosphere for the child's world of make-believe. Artificial or natural shade is desirable over the sedentary play pieces, where children will play on hot days without immediate supervision. Play sculpture may be placed in the sand area to enhance its value by providing a greater variety of play opportunities. A portion of the area should be maintained free of equipment for general sand play that is not in conflict with traffic flow.

5. Swings or other moving equipment should be located near the outside of the equipment area and should be sufficiently separated by walls or fences to discourage children from walking into them while the equipment is moving. Swings should be oriented toward the best view and away from the sun. Sliding equipment should preferably face north, away from the summer sun. Equipment with metal surfaces should be located in available shade.

6. Spray pools should be centrally located, and step-up drinking fountains strategically placed for convenience and economy in relation to water supply and waste disposal lines.

7. The open turfed area for running and active play and the shaded area for such quiet activities as reading and storytelling should be closely related to the enclosed equipment area and serve as buffer space around it.

8. Nonmovable benches should be conveniently located to assure good visibility and protection of the children at play. Durable trash containers should be provided and conveniently located to maintain a neat, orderly appearance.

5.4 Playgrounds

ADVENTURE PLAYGROUND

Plans for an adventure playground in Freeport, New York, are shown in Figure 5.75.

PLAYGROUND REQUIREMENTS

The playground is the chief center of outdoor play for children from 5 to 12 years of age. It also offers some opportunities for recreation for young people and adults.

At every elementary school the playground should be of sufficient size and design and be properly maintained to serve both the elementary educational program and the recreational needs of all age groups in the neighborhood. Since education and recreation programs complement each other in many ways, unnecessary duplication of essential outdoor recreational facilities should be avoided. Only where this joint function is not feasible should a separate playground be developed.

A playground may include (1) a play lot for preschool children, (2) an enclosed playground equipment area for elementary school children, (3) an open turfed area for active games, (4) shaded areas for quiet activities, (5) a paved multipurpose area, (6) an area for field games, and (7) circulation and buffer space.

Location A playground is an integral part of a complete elementary school development. School playgrounds and other playgrounds should be readily accessible from and conveniently related to the housing area served. There should be a playground within ¼ to ½ mile for every family housing unit.

Size and Number The recommended size of a playground is a minimum of 6 to 8 acres, which will serve approximately 1000 to 1500 families. The smallest playground that will accommodate essential activity spaces is about 3 acres, serving approximately 250 families (about 110 elementary school children). This minimum area should be increased at the rate of 0.2 to 0.4 acre for each additional 50 families. More than one playground should be provided where (1) a complete school playground is not feasible, (2) the population to be served exceeds 1500 families, or (3) the distance from some housing units is too great.

Activity Spaces and Elements A playground should contain the following basic activity space and elements:

1. A play lot

2. An enclosed playground equipment area with supplemental planting for elementary school children

3. An open turfed area for informal active games for elementary school children

4. Shaded areas for quiet activities such as reading, storytelling, quiet games, handicrafts, picnicking, and horseshoe pitching for both children and adults

5. A paved and well-lighted multipurpose area large enough for

   a. Activities such as roller skating, dancing, hopscotch, foursquare, and captain ball

   b. Games requiring specific courts, such as basketball, volleyball, tennis, handball, badminton, paddle tennis, and shuffleboard

6. An area for field games (including softball, junior baseball,

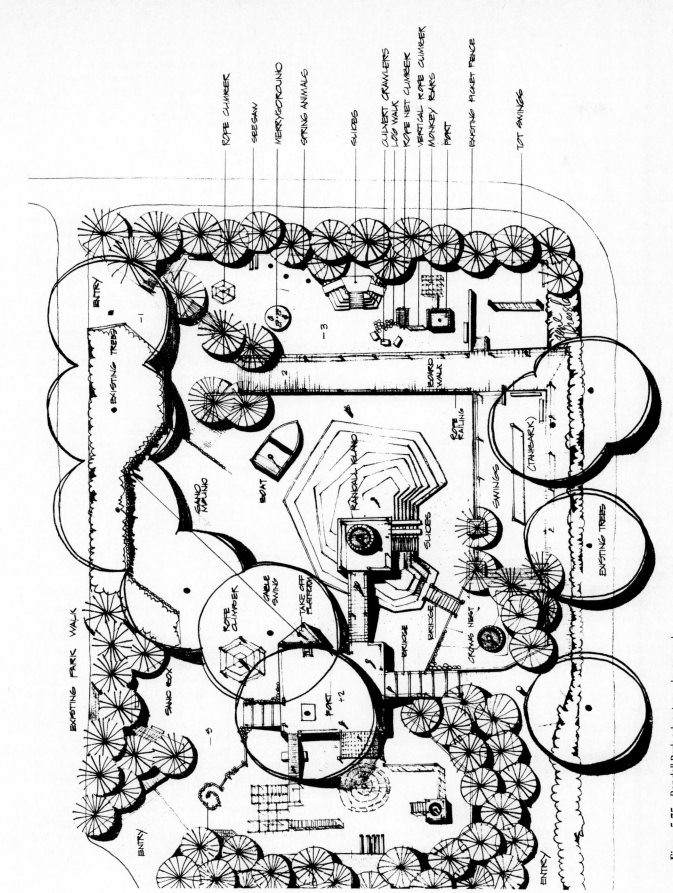

ROPE CLIMBER
SEE-SAW
MERRY-GO-ROUND
SPRING ANIMALS
SLIDES
CULVERT CRAWLERS
LOG WALK
ROPE NET CLIMBER
VERTICAL ROPE CLIMBER
MONKEY BARS
FORT
EXISTING PICKET FENCE
TOT SWINGS

ENTRY
EXISTING TREES
BOARD WALK
ROPE RAILING
SWINGS
(TANBARK)
EXISTING TREES
RANDALL ISLAND
SLIDES
BOAT
SAND MOUND
ROPE CLIMBER
CABLE SWING
TAKE OFF PLATFORM
BRIDGE
BRIDGE
CROWS NEST
FORT
SAND BOX
EXISTING PARK WALK
ENTRY
ENTRY

Figure 5.75 Randall Park adventure playground.

touch or flag football, soccer, track and field activities, and other games), preferably well lighted, that will also serve for informal play of field sports and kite flying and be used occasionally for pageants, field days, and other community activities

7. Miscellaneous elements such as public shelters, storage space, toilet facilities, drinking fountains, walks, benches, trash containers, and buffer zones with planting

Layout The layout of a playground will vary with the size of the available site and its topography and the specific activities desired. It should fit the site with maximum preservation of the existing terrain and such natural features as large shade trees, interesting forms, rock outcrops, and streams. These features should be integrated into the layout to the maximum extent feasible for appropriate activity spaces, as natural divisions of various use areas, and for landscape interest. Grading should be kept to a minimum consistent with activity needs, adequate drainage, and erosion control.

The general principles of layout are as follows:

1. The play lot and the playground equipment area should be located adjacent to the school and to each other.

2. An open turfed area for informal active play should be located close to the play lot and the playground equipment area for convenient use by all elementary school children.

3. Areas for quiet activities for children and adults should be somewhat removed from active play spaces and be close to tree-shaded areas and other natural features of the site.

4. The paved multipurpose area should be set off from other areas by planting and be located near the school gymnasium so that it may be used for physical education without disturbing other school classes. All posts or net supports required on the courts should be constructed with sleeves and caps that will permit removal of the posts and supports.

5. The area for field games should be located on fairly level, well-drained land with finished grades not in excess of 2.5 percent. A minimum grade of 1 percent is acceptable on pervious soils having good percolation for proper drainage.

6. In general, the area of a playground may be divided as follows:

   a. Approximately half of the area should be parklike, including the open turfed areas for active play, the shaded areas for quiet activities, and the miscellaneous elements as described in paragraph 7.

   b. The other half of the area should include ¾ to 1 acre for the play lot, the playground equipment area, and the paved multipurpose area and 1¾ acres (for softball) to 4 acres (for baseball) for the field games area.

7. The playground site should be fully developed with landscape planting for activity control and traffic control and

for attractiveness. This site also should have accessible public shelters, storage for maintenance and recreation equipment, toilet facilities, drinking fountains, walks wide enough for strollers and carriages, bicycle paths, benches for adults and children, and trash containers.

PLAYGROUND EQUIPMENT

Basic Play Equipment Play equipment may include swings, slides, and merry-go-rounds; various types of climbers; balancing equipment such as balance beams, conduits, leaping posts, and boxes; hanging equipment such as parallel bars, horizontal bars, and ladders; play walls and playhouses; and a variety of play sculpture forms. Different types of play equipment should be provided for preschool children and for elementary school children to meet the developmental and recreational needs of the two age groups.

Play Lot Equipment for Preschool Children The accompanying table indicates the requirements for various kinds of equipment totaling about 2800 square feet. This area, plus additional space for circulation and play space dividers, will accommodate a full range of play lot equipment serving a neighborhood containing approximately fifty preschool children (about 165 families).

Equipment	Number of pieces	Play space requirements (feet)
Climber	1	10 × 25
Junior swing set (four swings)	1	16 × 32
Play sculpture	1	10 × 10
Play wall or playhouse	1	15 × 15
Sand area	1	15 × 15
Slide	1	10 × 25
Spray pool (including deck)	1	36 × 36

Smaller play lots may be developed to serve a neighborhood containing about thirty children (about 100 families), using a limited selection of equipment with play space requirements totaling about 1200 square feet. This area, plus additional space for circulation and play space dividers, should be planned with the following desirable priorities: (1) a sand area, (2) a climbing device such as a climber, a play wall, or a piece of play sculpture, (3) a slide, and (4) a swing set. If several play lots are provided, the equipment selections should be complementary rather than of the same type. For example, one play lot may include play walls or a playhouse, while another may provide a piece of play sculpture. Also such a costly but popular item as a spray pool may be justified in only one out of every two or three play lots provided.

Playground Equipment for Elementary School Children The accompanying table indicates types, quantities, and minimum play space requirements totaling about 6600 square feet. This area, plus additional space for circulation, miscellaneous elements, and buffer zones, will accommodate

a full range of playground equipment serving approximately fifty children at one time.

Equipment	Number of pieces	Play space requirements (feet)
Balance beam	1	15 × 30
Climbers	3	21 × 50
Climbing poles	3	10 × 20
Horizontal bars	3	15 × 30
Horizontal ladder	1	15 × 30
Merry-go-round	1	40 × 40
Parallel bars	1	15 × 30
Senior swing set (six swings)	1	30 × 45
Slide	1	12 × 35

Surfacing The selection of suitable surfacing materials for each type of play area and for circulation paths or walks, roads, and parking areas should be based on the following considerations:

1. *Function.* The surface should suit the purpose and the specific function of the area, such as surfaces for court games or field games and surfaces under play equipment. Choice of the surface should also depend on whether the area is multipurpose or single-purpose and whether it is intended for seasonal or year-round usage.

2. *Economy.* The factors of economy are the initial cost, replacement cost, and maintenance cost. Often an initially more expensive surfacing is the least expensive in the long run because of reduced maintenance.

3. *Durability.* The durability of the surface should be evaluated in the light of its resistance to the general wear caused by the participants and of its resistance to extended periods of outdoor weathering such as sunlight, rain, freezing, sand, and dust.

4. *Cleanliness.* The surface should be clean and attractive to participants, it should not attract or harbor insects or rodents, and it should not track into adjacent buildings or discolor children's clothing.

5. *Maintenance.* Maintenance must be evaluated in the light not only for cost but also of the time when the facility is not available for use because of repair or upkeep.

6. *Safety.* The safety of participants is a primary consideration in selecting a play surface and should not be compromised for the sake of economy.

7. *Appearance.* A surface that has an attractive appearance and harmonizes with its surroundings is very desirable. Surfacing materials should encourage optimum use and enjoyment by all participants, and channel the activities in an orderly manner by providing visual contrasts.

Evaluation of Surfacing Materials The various types of surfacing material have their advantages and disadvantages.

1. *Turf.* This material is generally considered to be the best surface for many of the recreation activities carried on at play lots and playgrounds. Although turf is not feasible for heavily used play areas, most park and recreation authorities recommend using it wherever practicable. Underground irrigation sprinkler systems with rubber-top valves should be specified in areas with inadequate seasonal rainfall to maintain a turf cover. The major reasons for using turf are that it is relatively soft, providing greater safety than other surfaces, and that it has a pleasing, restful appearance with great appeal to participants. A turf surface is especially suitable for open and informal play areas for younger children and the large field game areas for sports and general recreation use.

2. *Bituminous concrete.* This flexible paving material is the most generally used material for paving play areas. The designer should note that various asphalt grades and mixes, as well as color coatings to improve appearance and maintenance, are available. A suitable mix and careful grade control should be used to obtain a smooth, even surface, economical construction, and little or no maintenance. Bituminous concrete pavement is especially useful for paved multipurpose areas, for tennis, basketball, and volleyball courts, for roller-skating and ice-skating rinks, and for walks, roads, and parking areas.

3. *Portland cement concrete.* This rigid paving material is the most favored type of surface for use in specialized areas where permanence is desired. It provides uniformity, maximum durability, and little or no maintenance. A portland cement concrete surface is especially useful for court games requiring a true, even surface, such as tennis and handball, for shuffleboard courts, for roller-skating and ice-skating rinks, and for walks, curbs, roads, and parking areas.

   Portland cement concrete and bituminous concrete surfaces are generally considered for many of the same uses. Choice of either one should include appropriateness for the purpose intended, initial cost, and long-term cost.

4. *Synthetic materials.* Synthetic materials that have a cushioning effect are being used by some school, park, and recreation departments, primarily for safety, under play equipment. Several companies have developed successful resilient materials that provide excellent safety surfaces; these have been more expensive than the other materials discussed here. A number of cities have conducted studies using these surfaces.

5. *Miscellaneous materials.* Materials used for specific areas include sand, sawdust, tanbark, or wood chips around and under play equipment, earth on baseball diamond infields, and brick, flagstone, or tile on walks and terraces.

See Figure 5.76 for a playground design and the accompanying table for a comparison of various surfacings.

Space requirements for various types of playground equipment are shown in Figures 5.77, 5.78, 5.79, 5.80, and 5.81.

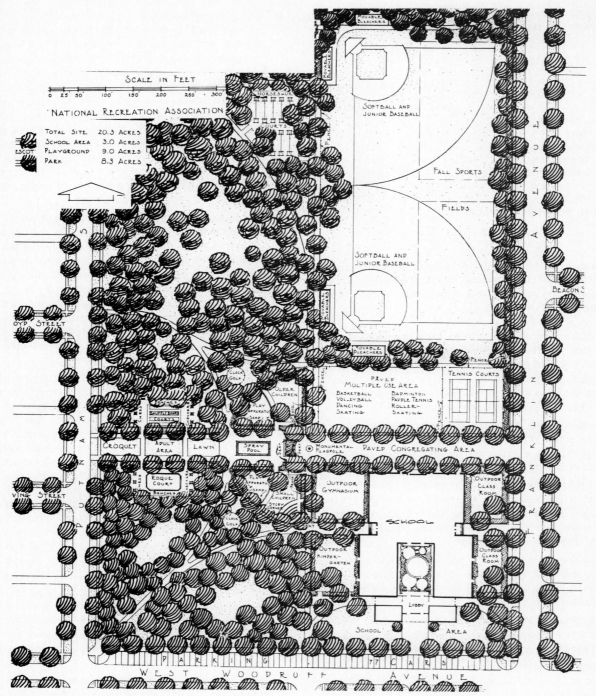

Figure 5.76 Playground layout.

Advantages and Disadvantages of Various Surfacing for Recreation Areas

Surfacing type	Advantages	Disadvantages
Turf	Soft surface, ideal for many play purposes; low first cost	Cannot be used in wet weather; difficult to maintain
Natural soil	Low first cost; soft surface	Muddy in wet weather; dusty in dry weather
Gravel	Low first cost; pleasing appearance	Thrown about by children to such extent that it is unsuitable for any use as surfacing in housing developments
Sand-clay and clay-gravel	Low cost when suitable material is available; reasonably soft surface	Difficult to get properly proportioned mixture
Brick on sand cushion	Attractive appearance	Initial cost relatively high
Stone paving blocks on sand cushion or natural soil	Low cost when salvaged from old pavements; satisfactory appearance; durability	Surface too rough for play use; maintenance cost relatively high
Precast concrete slabs on sand or natural soil	Year-round utility; satisfactory appearance	
Flagstones on sand or natural soil	Year-round utility; pleasing appearance; durability	
Bituminous concrete	Good surface for most play purposes when properly specified and laid; not so hard on feet as portland cement concrete; year-round utility	Rough and abrasive unless properly specified and constructed (competent inspection essential for good workmanship); hot for bare feet; possibility of becoming soft; unattractive in large areas
Cork asphalt	Resiliency; excellent surface for many play purposes; year-round utility; satisfactory appearance	Comparatively high cost (competent inspection essential for good workmanship); softening in very hot weather
Portland cement concrete	Year-round utility; minimum maintenance expense; good surface for wheeled toys, roller skating, and some court games	Lack of resiliency; initial cost relatively high; large areas requiring expansion joints; whiteness and glare of large areas unattractive

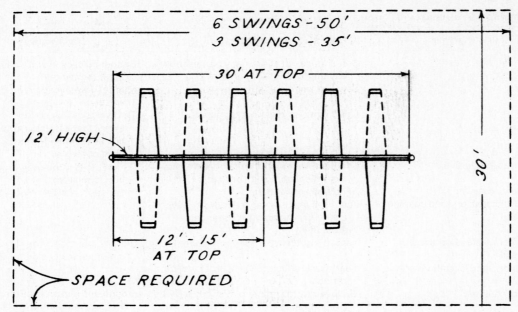

Figure 5.77 Swings.

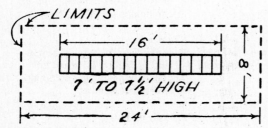

Figure 5.78 Horizontal ladder.

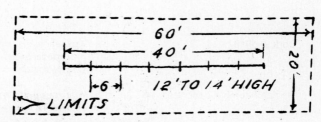

Figure 5.79 Traveling rings.

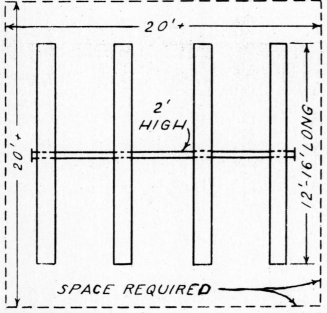

Figure 5.80 Seesaws.

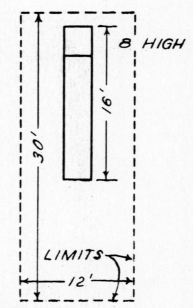

Figure 5.81 Slides.

AREAS FOR CHILDREN'S GAMES

Space requirements for seesaws and swings are shown in Figure 5.82, and areas for a number of games are given in the accompanying table and in Figures 5.83 and 5.84.

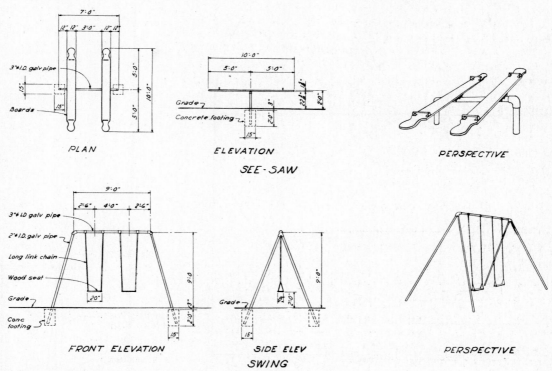

Figure 5.82 Seesaws and swings.

Children's Areas

Name	Dimensions of game areas (feet)	Use dimensions (feet)	Space required (square feet)	Number of players
Archery	60–150 in length; targets 15 feet apart	50 × 135 (minimum)	6,750	2 or more
		50 × 235 (maximum)	11,750	
Baseball	75-foot diamond	250 × 250	62,500	18
	82-foot diamond			
Basketball	40 × 60	50 × 70	3,500	10
Field hockey	120 × 200 (maximum)	150 × 250 (maximum)	37,500	22
Hopscotch	5 × 12½	10 × 20	200	2 or 4
Horseshoes	Stakes 25 feet apart	12 × 40	480	2 or 4
Marbles	10-foot diameter	18 × 18	324	2–6
Paddle tennis	13½ × 39 (singles)	25 × 60	1,500	2
	18 × 39 (doubles)	30 × 60	1,800	4
Soccer	100 × 200	125 × 240	30,000	22
Softball	45-foot diamond	175 × 175 (average)	30,625	18
Speedball	120 × 220	150 × 260	39,000	22
Team dodge ball:				
Boys	Circle 40 feet in diameter	60 × 60	3,600	20
Girls	Circle 35 feet in diameter	50 × 50	2,500	20
Touch football	120 × 240	140 × 280	39,200	18–22
Volleyball	25 × 50	40 × 70	2,800	12–16

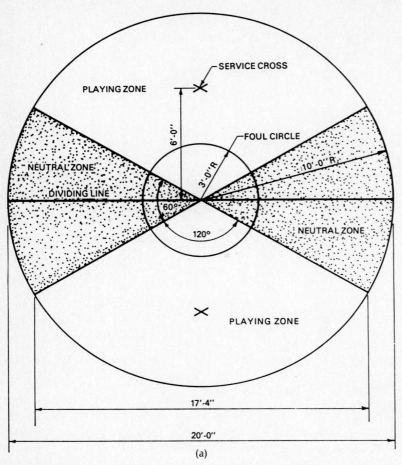

(a)

Figure 5.83 Tether ball: (*a*) court layout; (*b*) section; (*c*) detail, removable post.

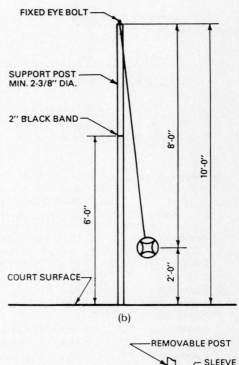

(b)

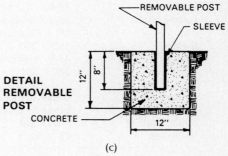

(c)

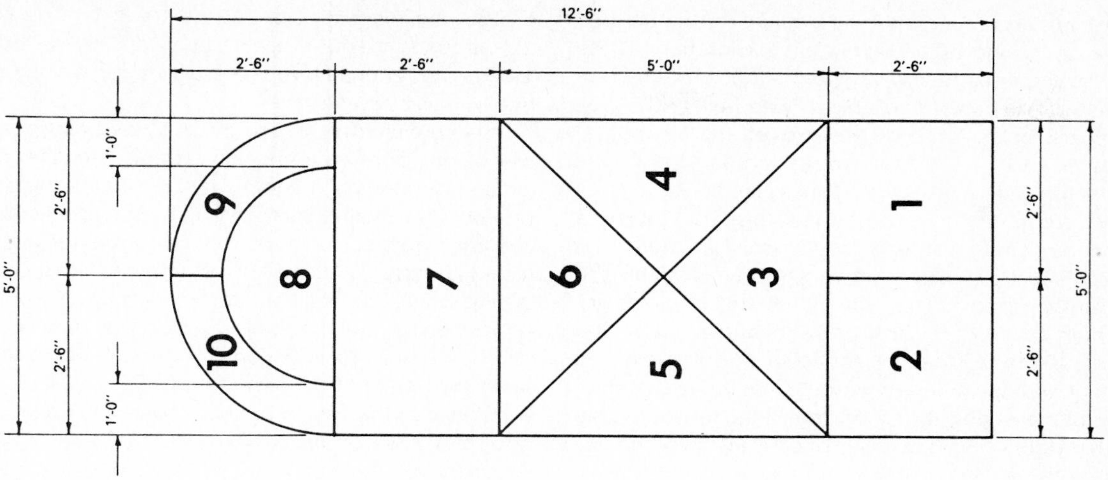

Figure 5.84 Hopscotch.

5.5 Outdoor Swimming Pools

PUBLIC SWIMMING POOLS

In the development of any outdoor swimming pool and related facilities, there should first be a master plan of the area in which the pool is to be located. A swimming pool by itself is never as successful as a pool planned in conjunction with other recreational facilities, such as baseball diamonds, tennis courts, volleyball courts, a craft center, and a community center building. One activity complements another, and a pool thus related to other recreational facilities will be much more successful and easier to administer.

Residential pools and pools designed for hotels and motor courts are used almost exclusively for recreational swimming and sunbathing. On the other hand, pools at private clubs, city parks, and recreation areas are normally used for both recreational and competitive swimming. Competitive swimming is the impetus for wide usage of the pools by youths and young adults.

Publicly owned and operated pools probably constitute the largest market today for outdoor pools. An increasing number of public pools are being built each year. The private swimming-club-pool idea is fast expanding. The club pool meets a great need and is an excellent idea, but it generally will not take care of the needs of a community. The vast majority of potential swimmers come from families that cannot afford a family membership in a private pool, or an individual membership, or a daily gate fee. Therefore, the building of swimming pools for the general public today is largely a governmental responsibility, whether it be municipal, county, state, or federal.

Classification of Outdoor Pools

Public Pools Public pools are ordinarily those owned and operated by municipalities, counties, schools, park dis-

tricts, states (as in state parks), and the federal government (as in a national park).

Privately Owned Pools The second category of sponsorship is the privately owned pool that serves the public. The YMCA, YWCA, boys' clubs, YMHA, and other organizations build outdoor pools or indoor-outdoor pools that serve both winter and summer needs.

At most country clubs today, an outdoor swimming pool is considered a necessity if the club is to be a family-type club rather than just a golf club. These pools must be attractively designed, well built, and expertly operated. The private swimming club is expanding and serves a real need, particularly in suburban areas.

Most apartment houses are including a swimming pool in their complexes because they realize that this is a necessity if the apartment units are to be rented at capacity. Rooftop pools are no longer unique.

Many public housing projects are including swimming pools. These are publicly owned but are considered private since use of a pool is restricted to the occupants of a particular housing project.

Hotels are fast coming to realize that they must have a pool, and the outdoor pool is the usual thing. In many northern communities, the pool has been enclosed by either a permanent or a semipermanent structure.

Pools are included as a necessary attraction at resorts and resort hotels and motels. Motels for transients must also include a pool, even though some of them seem to be satisfied with the smallest pool that will answer the purpose: in many instances a pool measuring only 20 by 40 feet is installed.

Some motels have become aware of the nonuse of the pools by their guests in the middle of the day, and the pools are being made available to community groups during this slack period. In several instances, the pools have been opened to underprivileged or low-income groups that lack other swimming facilities.

Commercial Pools Commercial pools are those operated for a profit by an individual owner or corporation. These pools are usually separate and independent of other recreational facilities. Many such pools have been built in the past 25 or 30 years, but experience has proved that a pool by itself is not as good a business proposition as a pool with other recreational facilities. Such facilities can include a par-3 golf course, a golf driving range, a miniature golf course, a bowling lane, a skating rink, or a group picnicking area.

Types of Pools by Design A rectangle was probably the original pool shape, and a great many rectangular pools are still being built. Some of the first pools measured 50 by 100 feet or even 100 by 200 feet. Many indoor pools measured 30 by 60 feet and, in some few cases, 35 by 75 feet. A rectangular pool measuring 45 by 75 feet is quite acceptable today both indoors and outdoors, since it includes the 25-yard short course for competitive swimming. In some cases, a pool 60 feet wide, or preferably 75 feet (25-yard short course), and 50 meters (164 feet ½ inch) long is also quite acceptable.

The rectangular pool has a comparatively low construction cost and is easy to supervise. However, in small pools there is a low percentage of shallow water, and only one competitive activity can take place at a time. The rectangular pool is a traditional type of pool, but it is not as dramatic or as interesting as some of the free-form pools, which can easily include the proper competitive swimming lengths, make possible larger deck areas, offer an interesting shape that fits into the landscape, and give variety and interest to the activities.

The T-shape pool was probably one of the first pools to break away from the rectangular shape, and it is today a very fine shape for a pool. The top of the T can be 45 feet by 164 feet ½ inch, and sometimes 75 feet by 164 feet ½ inch, thereby providing both short and long courses and an entire area of so-called shallow water that is 3 feet 3 inches to 5 feet in depth.

The lower part of the T should be reserved for the diving area and be separated from the top of the T by a float line. This is probably one of the most popular pool designs, although the L-shaped pool includes the same features, except that the deep area is placed at one end rather than in the middle.

The Z-shaped pool is also extremely functional, having the competitive swimming lengths and general swimming area in the center of the Z, the diving area offset from the main part on one side, and the extremely shallow area (extending to 2 feet) off the other side of the general swimming area. In this shape the swimming area could be the 25-yard or 50-meter distance. In some cases, the Z-shaped pool can include both the 25-yard and the 50-meter lengths. Although recreational swimming is the prime objective, the objectives of a good recreational area and competitive swimming lengths can be combined in such a pool. The most obvious advantages in the T-, L-, and Z-shaped pools are the separation of divers and swimmers, tending to eliminate the danger of collisions; and the provision of a larger shallow-water area. These shapes do, however, present some supervision problems.

The multiple-pool idea is an extremely good one, especially if funds are not available to build all the swimming facilities desired at one time. The program can be phased so that one pool is built at a time.

In a country club where there is need to spread out the use of the pool area and have pools for different groups (adults and juniors), a competitive pool, a diving pool, and a play pool for small children will be found most advantageous. The adult pool has been conceived to provide a special area for adults only. This is particularly desirable in a country club setting where alcoholic beverages are served around a pool.

The junior pool is a pool ranging from 2 to 3 feet in depth where swimming instruction can be conducted for children up to the age of 10.

The tots' play pool, more commonly known as the wading pool, has a water depth of from 0 to 15 or 18 inches. It is designed for the comfort and convenience of youngsters below the age of 6. The more completely equipped tots' play pool includes a 6- to 10-foot walk surrounding the pool; the entire area is enclosed with a barrier, such as a 3-foot chain link fence (knuckle-finish top and bottom) or a masonry sitting wall for parents or others accompanying the children.

The separate diving pool has many advantages. Since there is usually a conflict between swimmers and divers, the safety feature is a good justification for the separation of the main pool from the diving pool.

There are advantages, certainly, to separate pools, safety being the principal one. However, separate pools cost more, require a greater number of lifeguards, need more extensive mechanical and water treatment systems, and occupy more space.

Free-form pools are ordinarily restricted to use as resort pools and residential pools, This shape is not favored for the public or institutional types of pool.

The spray pool is popular in many communities. In some instances, the water goes to waste, and in other locations the spray pool is combined with a wading area. Spray pools measuring 30 to 40 feet in diameter have been designed to serve also as dance and roller-skating areas. In the majority of cases, the pool with a wading area has proved more popular than the spray pool.

Recommended Sizes for Outdoor Pools It has been customary for cities to construct large pools. A large pool may sometimes be desirable. Most authorities agree, however, that a pool 50 meters long and 45, 60, or 75 feet wide, with a diving area to one side (as is possible in the T or L shapes), is as large a pool as any community needs. If a pool of this size does not adequately serve the needs of the community, it would probably be advantageous to have a second pool in another location. A pool is usually more popular and more successful if children and adults do not have to travel out of their own neighborhood.

In cities where a pool with 25-yard and 50-meter competitive swimming lengths and a large recreational swimming area can be justified, this may be the most desirable method of providing swimming facilities. As this pool is outgrown or as the need for additional swimming facilities is evident, it may be that the next pool or next group of pools should be

what is ordinarily called the neighborhood pool. This is a rectangular or fan-shaped pool measuring approximately 45 by 75 feet. An L-shaped pool with a swimming area of 45 by 75 feet and a diving area of 40 by 42 feet would be even better. The neighborhood pool has the advantage of making it possible for persons to put on their bathing suits at home. It also enables them to swim with their friends.

In designing a pool for a neighborhood or a community, the size of the pool should be in direct proportion to the number of people to be served. In cities of 30,000 population or less, the daily average attendance may be expected to be from 5 to 6 percent of the population, about one-third of whom will be at the pool at any one time during the day. On peak days the pool may be used by as much as 10 percent of the population.

General Planning Considerations Adequate area is needed around a pool not only to dignify its setting and location but to serve as a buffer from nearby streets and residences. Space must also be provided for parking and for other recreational areas and facilities.

A pool should not be located in a low spot. Water from the surrounding area will drain into the pool, and unless precautions are taken for proper drainage around the edge of the pool, considerable water will penetrate the area under the deck and floor and be a source of constant annoyance and engineering problems.

A pool should not be located in a grove or trees. Leaves fall into the pool and keep it dirty, clog filters and the hair and lint catcher, and in many ways leave a pool in an unsatisfactory condition. Trees also keep the sun away from the pool, and to be successful a pool area must have sunshine. The pool should be located so that buildings and trees to the west are at such a distance that they will not shade the pool in the late afternoon.

Location in relation to streets is most important. A pool must be near main traffic arteries for good circulation and for accessibility, but the pool itself should not be too near a busy street. The dirt from the street will blow into the pool and give considerable trouble in the filtration system. If possible, the pool should be set back from 200 to 300 feet or more from the street.

Recreation facilities supplementing a swimming pool can consist of recreation buildings or community centers, softball and baseball diamonds, a football field, pitch-and-putt, par-3, or regulation golf courses, a multiple-use paved area, an area for playground equipment for both small children and older groups, and parking areas. Games and sports such as tennis, croquet, shuffleboard, badminton, handball, horseshoes, paddle tennis, table tennis, deck tennis, roller skating, and volleyball can also be included.

Pools are often built in an area just large enough for the pool, no provision being made for other facilities. A swimming pool should be a part of an overall recreational development whenever possible. If circumstances will permit, the bathhouse should be designed as a section of a community recreation center. This is especially true when the building includes a gymnasium, the same dressing rooms serving the

swimming pool during the summer and the gymnasium in the winter. It is quite possible that the dressing rooms will require a little more space, but this will certainly be more economical than designing multiple dressing areas. The bathhouse can be used the year around rather than just during the summer months. Relating the pool to the community center reduces the cost of administration and operation.

If the pool and a community recreation building are not related, a separate bathhouse is required. Of the many new ideas in bathhouse design, one of the most interesting is the bathhouse without a roof other than the roof covering the immediate dressing space around the wall and the toilet facilities. Showers are usually out in the middle in the sunshine, and, in some cases flowers and grass areas have been incorporated in the interior of the bathhouse. The clothes checkroom is, of course, completely covered and can be secured. This type of design is economical to build, and while it will probably require more space, it is the kind of bathhouse that is ventilated. It is, however, subject to vandalism to a greater extent than the closed bathhouse.

When space is at a premium, a roof is essential for either a summer or a year-round bathhouse. The best plan is a roof with sky domes to permit natural light. All windows can be omitted since adequate light will come in through the sky domes. Vent domes and exhaust fans can be used for circulating the air. This will provide a bathhouse that is vandalproof, economical to construct, and easily maintained. Rest rooms and dressing rooms should be kept to a minimum size, and construction should be simple so that the cost will be minimized.

The pool and deck area should be completely surrounded with a chain link fence at least 7 feet high with knuckle-finish top and bottom. This fence is a safety feature for those using the pool and contributes to maintaining proper control of the facility. Plantings on both sides of the fence are desirable. The wading pool or play pool and space for the small children, usually located immediately adjacent to the swimming pool, should be separated from it by a fence and a gate.

Prevailing winds are an important consideration. When the water is cool, swimmers do not want the wind to dry them off too fast, thereby causing them to become chilled. Therefore, swimmers should be protected from prevailing winds by a proper orientation of the pool and bathhouse so that the bathhouse is on the side of the prevailing wind. A canvas may be hung on the fence, a plastic or glass panel may be provided, or a masonry wall may be installed to shield bathers from the wind. In northern climates windbreaks are almost essential.

For a pool to receive full usage, overhead and underwater lights should be provided for night use. Overhead lighting should be a minimum of 1.2 to 3 watts per square foot of pool surface. Overhead lighting can be provided by overhead floodlights mounted on 30-foot-high steel poles.

The underwater lighting of a pool should be planned with great care. If possible the lights should be placed on the sidewalls rather than on the end walls. If they are placed on the end walls, they should be on a separate electrical circuit so

that they can be turned off during swimming meets. Otherwise, the swimmers will be looking directly into the lights. Underwater lights placed on the end walls also present difficulties in that swimmers, in making flip turns at the ends of the swimming lanes, might kick and break the glass lenses. The lights should be placed near the bottom of the pool and should be located directly under the float lines and away from the center of the swimming lanes so that swimmers will not kick the lights in turning. Underwater lights are valuable for safety and for aesthetics. It is recommended that 2 to 2.5 watts per square foot of water area be provided.

Construction and Design Factors

Water Depth Pools used for both recreational and competitive swimming often have a water depth in the shallow end of only 3 or 3½ feet, but provisions are made for flooding the gutters to secure extra depth during swimming meets. The water depth at the shallow end of a pool, as recommended by the Amateur Athletic Union (AAU), should be 4 feet. This depth will enable swimmers to make turns. Between 75 and 85 percent of the water area of a pool should have a depth of less than 5 feet. This is important because more than 85 percent of the swimmers will use the shallow water; only a small percentage will be in the deep water at one time.

Diving Facilities Nothing adds more to the attractiveness of a swimming pool, from the standpoint of both the swimmers and the spectators, than good diving facilities. While 1- and 3-meter boards have long been in use, many pools are now installing diving towers with 3-, 5-, 7½, and 10-meter diving platforms.

For the 10-meter platform, which is 32 feet 10 inches above the water surface, the minimum depth of the water must be 16 feet at a point 7 feet from the back wall of the pool. The hopper bottom should be at least 20 feet wide, and it should rise gradually to a water depth of 14 feet at a point 42 feet from the back wall of the pool. The overall diving area should extend a minimum length of 60 feet to a water depth of 6 feet, and the minimum width of the diving area should be 45 feet.

The 5-meter platform should be 16 feet 5 inches above the water surface, with a minimum water depth of 12½ feet, although a 14-foot depth is recommended. The 3-meter board should be 10 feet above the water surface, and the minimum depth of the water should be 12 feet. The board is installed so that the end of the board is 5 feet beyond the edge of the pool wall. The 1-meter board is similarly installed, the board being 39.37 inches above the surface of the water and the depth of the water being 9 feet.

The diving platforms will provide facilities for practice for various events and for official AAU and accredited Olympic tryouts and meets. The diving tower should preferably be constructed of reinforced concrete.

Pool Finishes It is the generally accepted practice that outdoor pool walls and floor be painted with a good pool paint. A pool should be painted pure white and not an off-color white. A good reason for this practice is that dirt or silt in the pool is immediately evident and attracts the attention of those responsible for keeping the pool clean. More important is the safety feature, since objects on the bottom of the pool are easily visible in a pure-white pool. Then, too, the appearance of the pool water is improved by using pure-white background, which brings out the natural blue color of the clear water. A pool should never be painted with blue or green paint.

A very fine pool finish is obtained by using plaster of white marble dust, which should be applied during the construction period; however, it may be applied to an old pool. This finish is more durable than paint and, if cared for properly, will last eight years or more. For the finish to be successful, it must be applied by qualified technicians.

The finish of the floor and walls of a pool should be neither too rough nor too smooth. If the floor is too smooth (for instance, glazed tile), the swimmer will slide down. If it is too rough, the swimmer's feet will be scratched and hurt, and pool will be difficult to keep clean.

Water Temperature Water temperature is important in a swimming pool. If the water is too cold, as is often the case when its source is a deep well or a spring, it is not comfortable for swimming. The most desirable temperature for water in an outdoor pool will vary with the region of the country and the atmospheric temperature. The colder the air temperature, the warmer the water, and vice versa. In colder climates, water temperatures of 78 to 80 degrees Fahrenheit should be maintained, and in air temperatures of 90 degrees Fahrenheit or more, the pool water should be maintained at 72 to 74 degrees to be comfortable for the bathers.

Deck Space It is important that the maximum amount of deck space be provided for the outdoor pool because this space greatly increases the pool's capacity. Surveys have shown that approximately one-third to one-half of the total number of swimmers in the pool area are in the water at one time. The cost per square foot of the walk or deck around a pool will usually be less than 10 percent of the cost per square foot of the water area. In addition, the deck area, if it is of sufficient width, can be used by the swimmers for sunbathing and lounging. If possible, the deck area should exceed by 100 percent the square footage of the water area of a pool and should be raised from 6 to 9 inches above the surface of the water.

SEMIPRIVATE SWIMMING POOLS

Location and Site The swimming pool, and all structures housing appurtenances thereto, should be not less than 5 feet from the nearest lot line.

The pool should be bounded on all sides by a deck or runway which has a uniform slope of ⅛ inch to the foot away from the pool; or if the pool is equipped with a continuous gutter, the deck surface may slope at this grade to the pool gutter. The surface should be smooth and easily cleaned but of nonslip construction insofar as possible. Decks which do not drain to a continuous pool gutter should be equipped at

the outer edge with drain receptacles every 20 feet along the deck periphery, or other device which will assure positive drainage. The deck or runway should not be less than 12 feet wide for a minimum of one-half of the perimeter of the pool and not less than 6 feet wide for the remainder.

The pool should be equipped with a substantial protection barrier which should be sufficient to protect persons or animals from trespassing to assure they are not subject to danger or harm. All openings in the barrier should be equipped with gates or doors which may be locked.

The pool or mechanical equipment should be located so as to minimize the noise and traffic and lessen the nuisance to nearby occupants of apartments and houses.

Walkways not less than 3 feet wide should be provided to all appurtenances such as filter buildings and toilet and shower facilities which are essential to the use and operation of the swimming pool.

No overhead obstructions less than 12 feet above the pool or deck will be permitted. In the diving area no obstruction less than 12 feet above the height of the highest diving board or platform will be permitted over the pool or deck within 12 feet of the diving board or platform.

Design Pools may be of any dimension or shape, provided satisfactory circulation of water can be obtained and undue hazards are absent.

Pools should provide a minimum of 250 square feet with at least 25 square feet of water surface per bather at time of maximum load. Where minimum area is proposed, not more than one-third of the pool area should have a water depth exceeding 5 feet. The loading design should be approved by the health authority having jurisdiction.

No diving surfaces higher than 3 meters above the water will be permitted. Platforms and diving boards which are over 1 meter high should be equipped with guard railings. Where multiple diving boards are used, the space between the centerlines shall be not less than 12 feet, and no board

should be closer than 10 feet to a sidewall. These dimensions apply at both the end wall and the point of maximum depth.

Pools should have a minimum water depth of 24 inches. The minimum depth of water in diving areas with board or platforms 1 meter or less above water should be 8 feet 6 inches. If the board or platform is over 1 meter but not over 3 meters above the water, the depth should be at least 10 feet. The minimum required depth should extend at least 12 feet beyond the end of the diving board or platform.

At depths of less than 5 feet, the slope of the swimming pool bottom should be uniform and not greater than 1:12. End and sidewalls should be vertical (having an incline of no more than 11 degrees from the vertical) except at depths of 6 feet or greater. At pool depths of 6 feet or more, the wall should be vertical for a depth of 3 feet below the operating water surface level. Below this point, the wall may be curved to the bottom with a radius of curvature equal to the water depth less 3 feet.

Two or more means of egress should be provided in all pools except that a ladder or stairway may be omitted at the shallow end of the pool if the depth from the bottom of the pool to the lip of the deck does not exceed 24 inches. At least one means of egress should be located at each portion of the pool where the water depth exceeds 5 feet and one at the shallow section except that such ladder or stairway may be omitted as provided above. No steps or stairways may protrude into the pool proper more than 6 inches unless steps are guarded by handrail or are located in a corner of the pool. Stairways should be equipped with at least one handrail. A recess in the pool wall for the stairway is acceptable.

SWIMMING POOL SIZES

Figure 5.85 shows a plan (*a*) and a typical section (*b*) for a residential swimming pool. Data in the illustrations are elaborated in the table below.

Recommended Swimming Pool Dimensions

Size	A	B	C	D	E	F	G	H	J	K	L	Length of springboard	Overhang	Height of diving board stand
12 × 28	1′ 6″	7′ 0″	7′ 0″	6′ 6″	6′ 0″		2′ 6″	3′ 0″	4′ 6″	6′ 6″	5′ 0″	0	0	None
12 × 30	1′ 6″	7′ 0″	9′ 0″	6′ 6″	6′ 0″		2′ 6″	3′ 0″	4′ 6″	6′ 6″	5′ 0″	0	0	None
12 × 32	1′ 6″	7′ 0″	9′ 0″	8′ 6″	6′ 0″		3′ 0″	3′ 6″	5′ 0″	7′ 6″	6′ 0″	8′ 0″	1′ 6″	Deck level
15 × 30	1′ 6″	7′ 0″	8′ 6″	7′ 0″	6′ 0″		3′ 0″	3′ 6″	5′ 0″	7′ 6″	6′ 0″	8′ 0″	1′ 6″	Deck level
15 × 32	1′ 6″	7′ 0″	9′ 0″	8′ 6″	6′ 0″		3′ 0″	3′ 6″	5′ 0″	7′ 6″	6′ 0″	8′ 0″	1′ 6″	Deck level
15 × 35	2′ 0″	8′ 0″	10′ 6″	8′ 6″	6′ 0″		3′ 0″	3′ 6″	5′ 0″	8′ 6″	7′ 0″	10′ 0″	2′ 0″	Deck level to 12″
16 × 35	2′ 0″	8′ 0″	10′ 6″	8′ 6″	6′ 0″		3′ 0″	3′ 6″	5′ 0″	8′ 6″	7′ 0″	10′ 0″	2′ 0″	Deck level to 12″
16 × 40	2′ 0″	8′ 0″	10′ 6″	13′ 6″	6′ 0″		3′ 0″	3′ 6″	5′ 0″	8′ 6″	7′ 0″	10′ 0″	2′ 6″	12″ to 18″
18 × 38	2′ 0″	8′ 0″	12′ 6″	9′ 6″	6′ 0″		3′ 0″	3′ 6″	5′ 0″	8′ 6″	7′ 0″	10′ 0″	3′ 0″	12″ to 18″
18 × 40	3′ 0″	9′ 0″	11′ 6″			16′ 6″	3′ 0″	See Note A	5′ 0″	9′ 0″	7′ 6″	12′ 0″	3′ 0″	12″ to 39″
20 × 40	3′ 0″	9′ 0″	11′ 6″			16′ 6″	3′ 0″	See Note A	5′ 0″	9′ 0″	7′ 6″	12′ 0″	3′ 0″	12″ to 39″

NOTE A: Floor is to slope from 3 feet deep to a uniform slope to the 5-foot depth.
NOTE B: Provide about 30 square feet of floor space for filtration equipment (preferably inside the building).
NOTE C: Slope walks away from pool at ¼ inch per foot; provide drains as required.
NOTE D: Provide self-closing, self-latching gates capable of being locked.

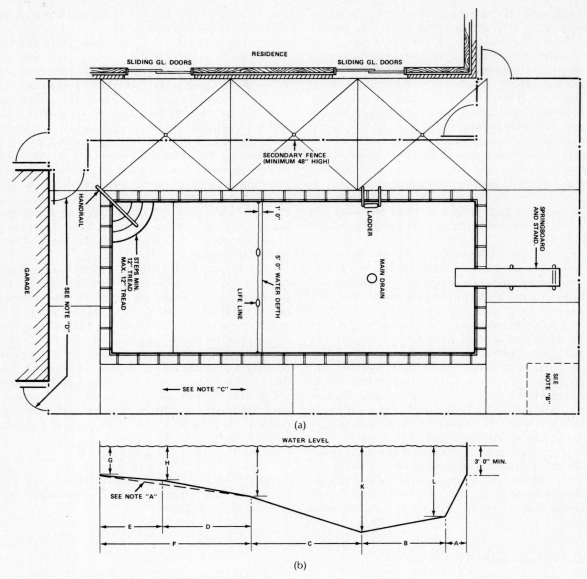

Figure 5.85 Residential pool: (a) plan; (b) typical section.

SWIMMING POOL ENCLOSURES

The following guidelines, recommendations, and suggestions for the design and construction of a protective enclosure will help to make a residential pool safer:

1. Substantial fence or protective barrier with a minimum height of 6 feet

2. Fence bracing and framing on the pool side

3. Fencing material with vertical segments or close mesh to minimize footholds and handholds

4. Fence flush with hard deck surface; 6 inches below ground level if installed on dirt surface

5. No solid fencing on living-area side

6. Secondary fence 48 inches high if the pool is contiguous with the home or another building

7. All openings equipped with self-closing, self-latching gates with provisions for locking

8. Gate latch located 3½ to 4 feet above the walking surface

9. Gate material and construction the same as that of fencing

A residential swimming pool and its enclosure are illustrated in Figure 5.86.

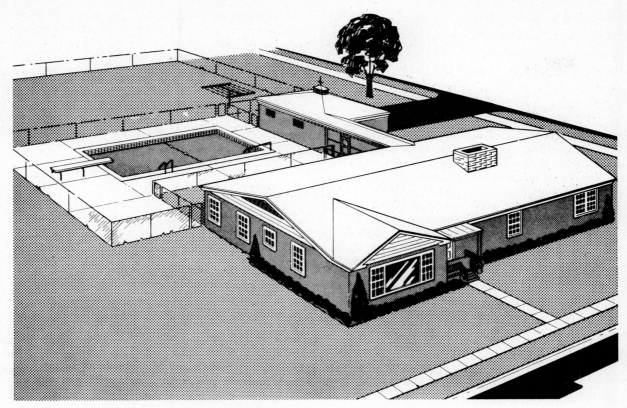

Figure 5.86 A residential swimming pool and its enclosure.

5.6 Waterfront Development

SWIMMING

In the following discussion of the location and construction of waterfront facilities, natural and artificial waterfronts are discussed separately.

Planning the Waterfront Location

Criteria for Natural Waterfronts The natural waterfront site should have certain characteristics to make it desirable for use in aquatic programs. The recommended criteria for the selection of waterfront or beach sites are discussed below. Helpful checklists have also been provided.

Water characteristics. The water content should be of a sanitary quality affording safe usage. The health conditions of a site are judged primarily by a careful examination of both its surrounding environs and its water content. The first is accomplished by a careful field analysis, the second by a laboratory analysis. Both examinations can indicate the bacterial quality and physical clarity of the water.

Checklist

1. Surrounding water source

2. Water quality (bacterial content)

3. Water clarity (visibility test)

Water condition characteristics. The circulation of the water through the potential waterfront site should be examined. Slow-moving water can produce swampy or built-up mud conditions, while fast-moving water can produce undercurrents and erosive conditions.

The ideal water temperature for swimming ranges from 72 to 78 degrees Fahrenheit, depending upon the air temperature. The American Public Health Association indicates that less than 500 gallons of additional water per bather per day is too small a diluting volume unless there is sufficient application of disinfection.

Checklist

1. Rate of water flow

2. Rate of water turnover

3. Water-level fluctuation

4. Water constancy

5. Availability of water

6. Types of currents and undertow

7. Outlet for water

8. Eddies, floods, waves, or wash
9. Weeds, fungi, mold, or slime
10. Parasites, fish, and animals
11. Debris, broken glass, and so on
12. Oil slick
13. Odor, color, and taste

Bottom characteristics. The waterfront bottom should be unobstructed and clear of debris, rock, muck, mulch, peat, and mud. The waterfront should not be in an area where the channel shifts or silt builds up. The most desirable bottom is white sand with a gradual pitch sloping from the shallow to the deep end. The bottom should not be precipitous or too shallow or have holes, pots, channels, bars, or islands.

The bottom should be of gravel, sand, or stable hard ground to afford firm and secure footing. Soundings should be taken in a boat, and an actual underwater survey should be undertaken before a final decision is made on the location of the waterfront.

Checklist

1. Bottom movement
2. Amount of holes and debris
3. Slope of subsurface
4. Amount of area
5. Condition of soil
6. Porosity of bottom
7. Average depth and various depths
8. Bottom color

Climatic characteristics. Continuous dry spells or numerous rainy seasons raise water retention problems. Dangerous storms, including tornadoes, lightning, hurricanes, and northeasters, create extremely dangerous waterfront conditions. The severity of the winter can also affect the waterfront. Ice and ice movement can cause damage to waterfront facilities and bottom. The ideal is a south-southeast exposure, in which maximum benefit is derived from the sun and there is the least exposure to the force of the wind.

Checklist

1. Number of storms and type
2. Prevailing winds
3. Amount of ice
4. Change of air temperature
5. Amount of precipitation
6. Fluctuation of temperature
7. Sun exposure

Environmental characteristics. The locale of the waterfront should be carefully examined for all influences on its construction and utilization. Zoning regulations, building codes, insurance restrictions, health ordinances, title covenants, and many other legal restrictions by the Coast Guard, conservation department, water resources commission, public works agencies, and fire department should be studied. The arrangement of land uses and their compatibility to the project, transportation, utilities, community facilities, population, and area economics should also be considered.

Checklist

1. Ownership and riparian rights
2. Availability of water supply
3. Zoning and deed restrictions
4. Local, state, and federal regulations
5. Adjacent ownerships
6. Water patrol and a control agency

Program characteristics. The waterfront should be so situated that it can be protected by a fence or other controlled access, particularly in a camp, marina, or other small area. It should also be internally segregated; that is, bathing should be separated from boating, boating from fishing, and so on. The site should also have storage room for waterfront equipment, a safety area near the lifeguard station or post, and ready access to a road.

Checklist

1. Distance of waterfront from other areas
2. Access road
3. Separation of waterfront activities
4. Area for unity of controls
5. Space available for adjunct activities

Access characteristics. The waterfront facility must be accessible by transportation available to the user. There should always be a means of vehicular access for emergency or maintenance use. The site around the waterfront and along its approach should be free of poison ivy, sumac, poison oak, burdock thistle, and other irritating plants.

Checklist

1. Location for access road
2. Poisonous plants
3. Area accessible yet controllable

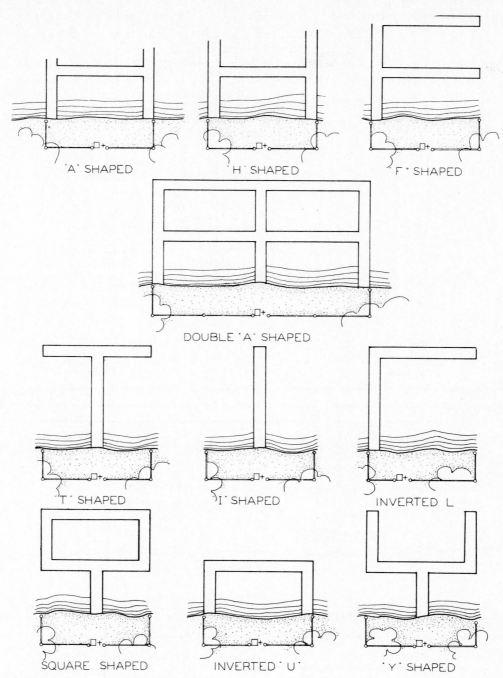

Figure 5.87   Typical waterfront shapes.

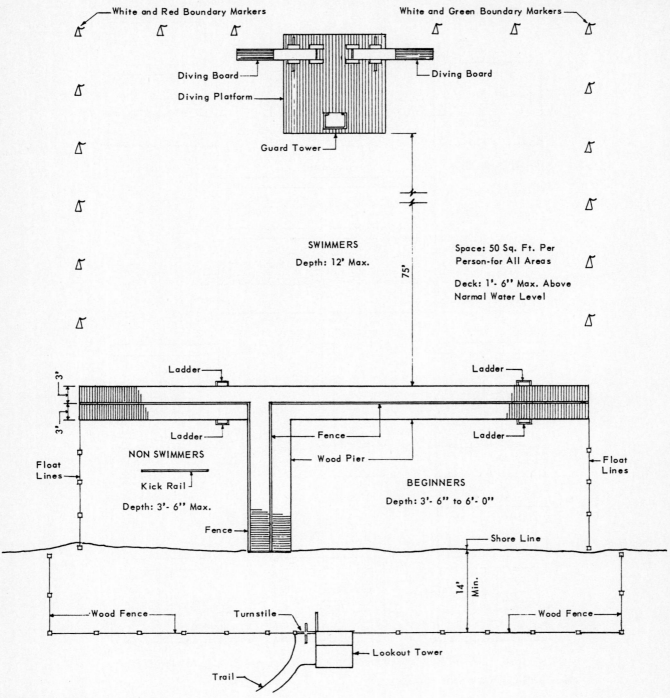

Figure 5.88 Waterfront layout.

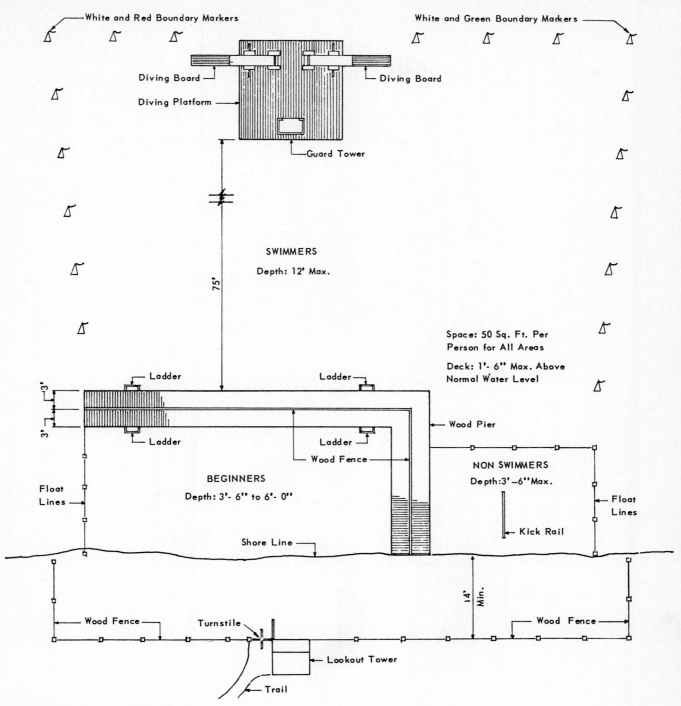

Figure 5.89   Waterfront layout.

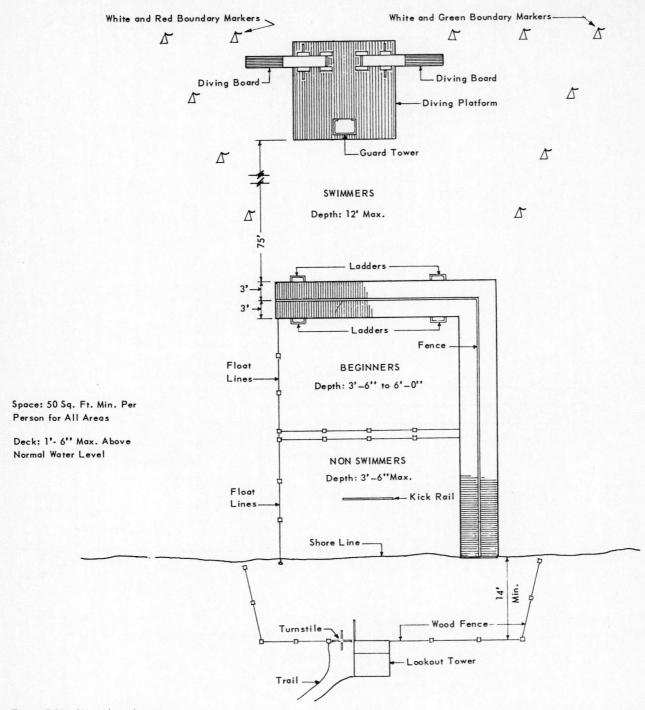

Figure 5.90 Waterfront layout.

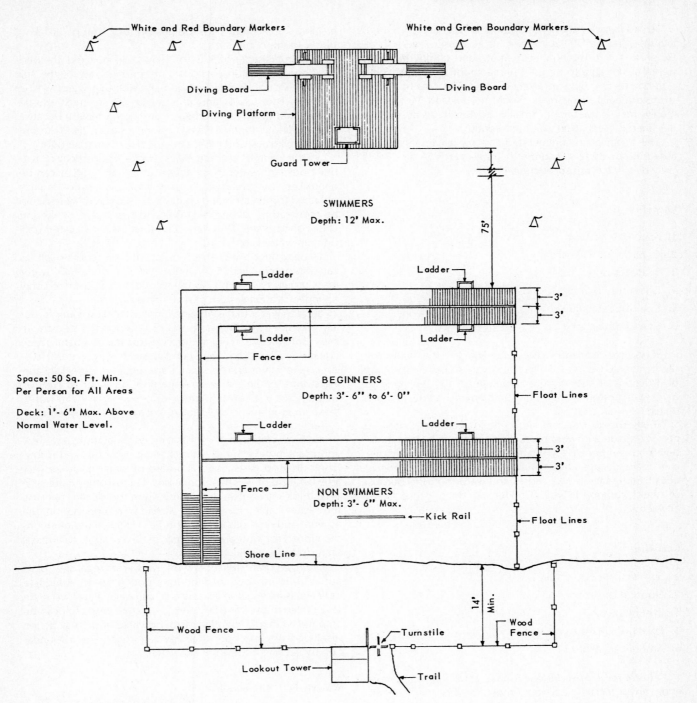

Figure 5.91 Waterfront layout.

Area characteristics. The waterfront bathing area should allow for at least 50 square feet for each user. There should be areas for instruction, recreation, and competition. The depth of the area to be used primarily for the instruction of nonswimmers should not exceed 3 feet. The area to be used for intermediate swimmers should not exceed 5½ feet (primarily for competition). Smaller or larger areas may be designed if users are divided differently.

The minimum recommended size for a camp swimming area is 60 by 30 feet, and the desirable size is 75 by 45 feet, providing a 25-yard short course.

Checklist

1. Space for bathing
2. Capacity of waterfront
3. Water depths
4. Division of bathing area into stations
5. Size of boating area
6. Size of fishing area

Shore characteristics. The shoreline for the waterfront facility should be free of irregular rocks, stumps, debris, and obstructions. It should be a minimum of 100 feet long for bathing in a camp area and can be many miles long in a park beach.

There should be trees adjacent to waterfront areas to provide shade and wind protection. Large, high trees should be eliminated because they attract lightning, and moldy trees have many decayed overhanging branches. Too many deciduous trees create mucky shores and water bottoms because of their autumn leaves. Coniferous trees cause fewer problems.

Checklist

1. Surrounding vegetation
2. Slope of the shore
3. Existing beach
4. Extent of clearing
5. Amount of debris

Criteria for Artificial Waterfronts In locating and considering an artificial waterfront, most of the characteristics described for natural areas should be examined. Additional criteria to be considered are outlined below.

Environmental characteristics. If all available bodies of water are being utilized, artificial-waterfront facilities must be developed. In some cases, waterfront locations are unsatisfactory or unavailable for new camps or resorts. Thus, consideration must be given to utilizing undeveloped sites with sufficient watershed (runoff water), water table (underground water), and water bodies (surface water) for lakes, pools, or impoundments.

Water characteristics. Before any site is selected, the percolation rate and, in particular, the permeability of the soil should be carefully checked to make sure that water will be retained. The stability and structure of the soil must also be determined (from test borings or test pits, or both) because of the various types of dams, pump houses, dikes, pools, berms, spillways, and other structures that must be built.

Water content characteristics. Unlike the content of natural bodies of water, the content of artificial bodies can be controlled by chlorination and filtration. Runoff water obtained from storms and contained in a pond or lake should be collected by diversion ditches and fed to a reservoir and chlorination plant. This water can then be recirculated until potable water is obtained.

Underground water that is obtained from wells or springs can also be contained in a pond. This type of artificial water body usually has a continuous flow and thus would need only a simple filtration system plus chlorination.

Surface water that is obtained from running streams is usually contained in a bypass pond or in a pond in the stream itself. Both methods require the construction of a dam. These artificial water bodies have continuous running water. However, gate valves and floodgates are required, especially during storms, when there is a large flow of water to control. Unless there is a constant turnover or supply of clean water, these impoundments will require a filtration and chlorination system.

Climatic characteristics. Climatic considerations are very important in developing artificial bodies of water and waterfronts. In most cases, natural bodies of water fluctuate very little because of weather conditions. On the other hand, artificial bodies are solely dependent upon the climate because the water table, runoff, and streamflow depend on the amount and time of rainfall. All other climatic considerations mentioned for natural waterfronts generally apply to artificial waterfronts as well.

Drainage characteristics. A low-lying area, regardless of its appeal, is not a good location for a pool or pond. Adequate drainage is essential so that surface and deck water will drain away from the water body and the water body itself can be emptied without pumping. Groundwater and frost action resulting from improper drainage can undermine a foundation by causing it to heave and settle.

Waterfront Construction

Criteria for Natural Waterfronts A natural waterfront facility should have features that make it both safe and usable. The following criteria are suggested as a basis for the construction of such a facility.

Bottom characteristics. Most swimming facilities around natural bodies of water require the dragging and grading of the bottom subsurfaces to eliminate hazards. In many cases where definite improvement of the bottom is required, mat,

mesh, or plastic sheets must be laid down on top of muck and staked down. Once these sheets have been laid, sand must be spread over the mat surface. When the bottom is firm, sand can be spread 6 inches thick on top of ice in the crib area during winter. As the ice melts, the sand will fall fairly evenly over the bottom. However, this can only be accomplished when the ice does not shift or break and float away.

Shore characteristics. When a beach is constructed, a gentle slope of from 6 to 12 feet in 100 feet should be maintained. If the waterfront requires a great deal of construction, a dock shoreline is recommended rather than trying to maintain an unstable beach. The ground above the water can then be developed with turf, terraces, decks, and boardwalks, depending upon the nature of the project. When the bottom drops off very quickly, the shore can be dug out to the grade desired underwater. This forms a crescent-shaped waterfront with an excellent beach.

Access characteristics. If possible, access roads and streets around waterfront areas should be acquired by the owner to keep the area buffered from conflicting uses. These roads should be made durable and be attractively maintained. Access roads should have clear horizontal and vertical vision so that pedestrian and vehicular conflict can be prevented.

Program characteristics. The waterfront in small recreation areas, such as camps or resorts, should be completely enclosed by planting or fencing. There should be a central control for ingress and exit. Many facilities require the use of boards for checking in and out, tickets, and similar devices for controlling the use of the area. The waterfront bathing, boating, and fishing facilities should be separated, each with its own control.

Criteria for Artificial Waterfronts Both artificial and natural waterfronts should have features that make them safe and usable. In improving and developing an artificial waterfront, most of the same points should be considered as illustrated for the natural waterfront. The following additional criteria should be carefully considered in providing an artificial waterfront.

Bottom characteristics. When an artificial beach is constructed, the grade should be the same as that recommended for natural shores: 6 to 12 feet in 100 feet. For reservoirs and ponds, there should be a minimum of 9 inches of large crushed stones, then 4 inches of well-graded, smooth gravel to fill in the voids, and finally 9 inches of washed medium sand. If the sand beach terminates at a depth of approximately 7 feet of water, it is recommended that riprapping be established to resist the tendency of the beach sand to move down the slope. The area above the beach should be ditched where the natural slope of the ground exceeds that of the beach. Thus, the slopes of the beach should be approximately 6 percent below the water and 10 percent above it. For areas in tidal waters, a maximum slope of 1 foot for 15 feet can be established for the bottom below the waterline.

For pools, bottoms are usually concrete, with the sides of concrete, welded steel, or aluminum plate. Such innovations as precast concrete slabs, plastic, and rubber are being tried.

Shore characteristics. In creating a shoreline for artificial bodies of water, there should be either a berm or a dike if the water is to be confined. A steep slope to eliminate shallow areas is usually required to prevent weeds and other plant materials from growing in the water.

If the soil conditions will not allow a steep slope underwater (3 feet deep to retard water plant growth), bulkheads or docks will be required. Otherwise only a limited beach can be provided.

Typical waterfront shapes are shown in Figure 5.87, and a typical waterfront layout in Figure 5.92.

DOCKS AND FLOATS

Permanent structures are usually set on concrete, wood, or steel foundations or on piers or piles. The decks should be made in sections of 10 to 20 feet for ease in removing for repairs or winter storage. The dock should be constructed with at least a 1-foot air space between deck and water. Underwater braces and other crossbeams should be limited to prevent swimmers from becoming entangled in them. When water levels change, allowances should be made for piers to be outside the deck limits so that the deck can move up and down on sleeves or brackets. Walkways or decks should be a minimum of 6 feet wide and preferably 8 to 10 feet wide. They should be cross-planked so that swimmers will avoid splinters. The planking should not be less than 2 inches thick by 4 to 6 inches wide. Boards should be spaced a maximum of ¼ inch apart to prevent toe stubbing. The deck should be treated with a non-creosote-based preservative, since creosote will burn feet, plus a plastic nonlead paint that is not heat-absorbing. The paint should be white with a blue or green tint to reduce the glare and aid in reflection.

Square Feet Needed per Person

Type of area	Water	Beach	Backup and buffer	Total
High density	30	45	400	475
Medium density	40	60	800	900
Low density	60	90	1200	1350

Recommended Density Limits for Boating

Type of boating	Maximum number of boats per acre of reservoir water
High-speed motorboats, unrestricted engine size	0.33
Low-speed motorboats ≤ 10 hp	1
Nonmotorized boats (no sail)	2
Fishing	1
(trolling)	0.5
Sailboats	1

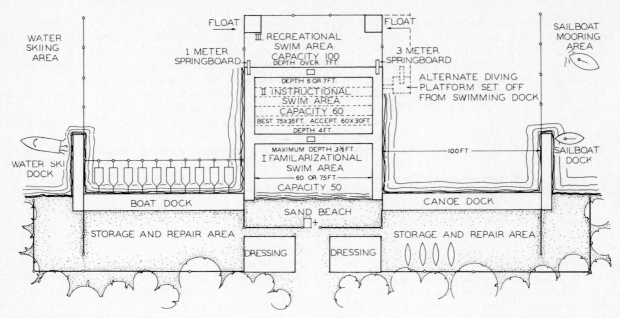

Figure 5.92 Waterfront layout.

Layout Criteria for Marinas

Layout Planning The proper siting of the various components of a small-craft harbor is a prerequisite for the functional soundness of the overall plan. The following general principles will provide guidance in determining the best allocation of land area within the confines of the harbor boundary (Figure 5.95).

For several reasons the larger craft should generally be berthed near the entrance. They are less influenced by residual wave action entering the harbor where action is greater near the entrance. Larger craft require greater maneuver space, which is usually provided in the harbor area near the entrance for the larger volume of traffic. More physical space is needed than for the smaller boats, and if the larger craft are berthed so they need not traverse the inner fairways of the harbor, those fairways may be proportionately narrower. The deeper drafts of the larger craft require a deeper channel and basin; hence the inner parts of the harbor can be shallower if they are not used by the larger craft. This allowance for a shallower inner harbor is also a factor in initial channel and berthing area excavation.

Commercial craft usually fall in the same category as large private recreational craft with regard to their water area requirements. The berthing areas of commercial and recreational craft should generally be separated because of different adjacent land use requirements. If possible, commercial boats should be located near the entrance in a separate basin or across a fairway from the recreation craft. A commercial fishing fleet will require special hoists and other equipment for moving the fish out of the holds onto perimeter docks and for sorting and preparing the catch for market.

Charter boats for sport fishing must have adjacent facilities for selling their services, for controlling the boarding and debarking of clients, and for parking cars. Fish-cleaning stations are often provided for clients to have their catch cleaned by professionals. A viewing area for prospective clients to watch the cleaning operations will help to advertise the charter boat service.

Rental boats should be berthed in the same area and not mixed with private recreational craft. This berthing area should be close to the office where the rentals are handled, with easy access to the harbor entrance. The car parking area for rental boat clients should be separate from the slip rental parking area, but it may be shared with visitors to other facilities in the harbor complex.

Sailboats without auxiliary power should be berthed in slips that open to leeward of the prevailing winds and that can be reached via wide fairways and channels or routes that allow for sailboat tacking with least interference to the powered craft.

Ramps or hoists for launching trailered craft should be separated from the berthing areas as far as possible. The boating habits of the owners of these craft are usually different from those of the berthed craft; conflicts may result if the same fairways are used. If possible, the trailered craft should have a separate entrance or be launched directly into the main water body without using the harbor. If trailered craft must use the same protected waters, the launching area should be as near the entrance as possible and physically separated from the berthing areas so that vehicle traffic to berthing areas and trailer traffic to the launching area do not merge. Occasionally, the launching area must be at the inner

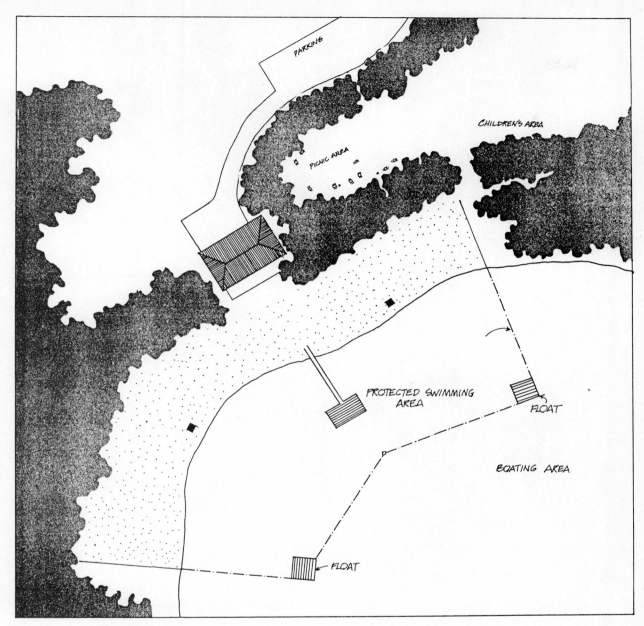

Figure 5.93 Beach layout (plan).

end of the harbor complex where more parking space is available. The launched craft may then have to share the main channel with the berthed craft, and the channel made wide enough to accommodate traffic from both sources without overcrowding.

The best location for a boat fueling dock is near the entrance in an area that is protected from waves in the entrance channel, thereby causing no interference with the entrance channel traffic. The adjacent land area must be suitable for buried fuel storage tanks and easily accessible for fuel distributing vehicles. The pumpout station should normally be located in the same area and is often on or along the fueling dock so that its operation can be supervised by the station manager. However, it should not be so close to the fuel pumps that a client has to wait for a pumpout operation before he can dock for fueling. It should not be necessary for any boat to go far out of its way for these services. Also, the station should not be in a location where it interferes with traffic flow or constitutes a fire hazard because of its proximity to other harbor facilities or berthed craft.

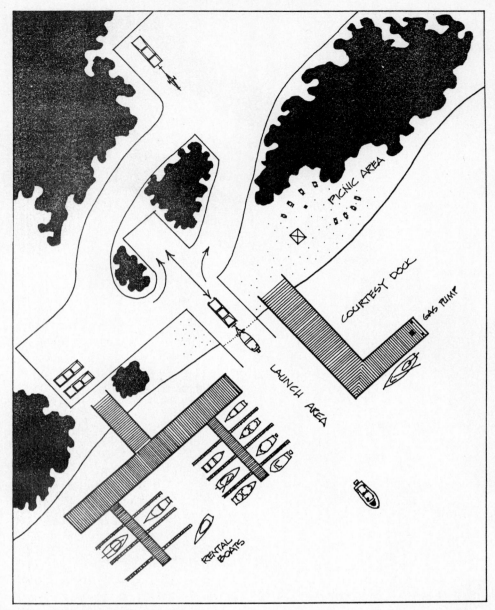

Figure 5.94   Plan of boating facilities.

The harbor administration area should also be located near the entrance and guest docks, where owners of visiting craft can easily come ashore to obtain information. The harbormaster's office should be either a part of or close to the administration complex to provide a good view of boats passing through the entrance. A good view of the berthing areas is recommended, but is not essential and may be impossible in a large harbor.

Vehicle parking lots for the berthing basins should be located so that no parking space in any lot is more than about 500 feet from the head of the pier for the particular lot it is intended to serve. Parking lots for ancillary facilities should be adjacent to parking lots for the berthing basins so that overflow from one lot can be absorbed by the other under time-staggered peak-use conditions. Lots should be well marked for the major area or function they serve.

The boat repair and servicing yards should be located in a remote part of the harbor that has adequate navigation access for the largest craft. If a marine railway, large hoist, drydock, or other device for launching or retrieving large craft is to be provided, it should be included as a part of this operation. The marine traffic generated by the repair yard

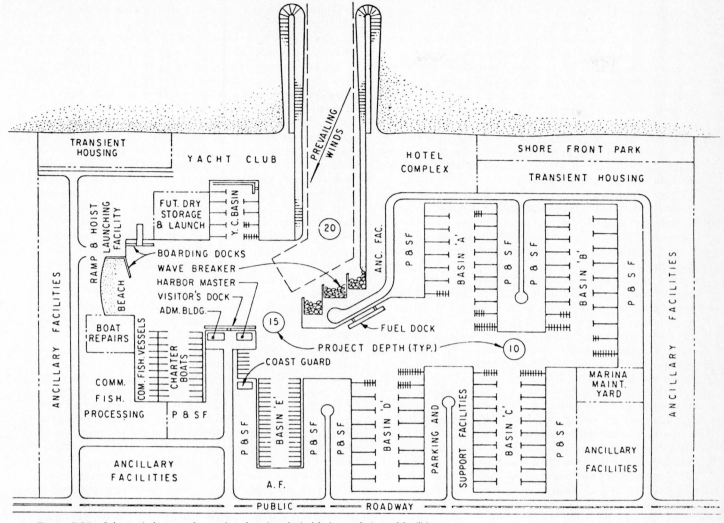

Figure 5.95   Schematic layout of a marina showing desirable interrelation of facilities.

will be minimal in comparison with the regular entrance channel traffic and need not be a consideration. However, the yard should be readily accessible to large tractor-trailers used for hauling new cruisers or boats to be launched.

If an operational dry storage facility is a part of the harbor complex, it should be located generally in accordance with the criteria that apply to launching ramps. The launchings and retrievals in such a facility are generally accomplished by hoist rather than by ramp, and a well-protected launching basin is necessary for this purpose. If the dry storage facility is for off-season layup only, it should be in a remote area not required for the more important facilities of the harbor; in fact, it need not be in the harbor complex at all. In many places, vehicle parking lots are used in the off-season for dry storage in lieu of providing a separate facility for this purpose.

If a U.S. Coast Guard vessel is to be docked in the harbor complex, it should be located near the entrance where it can

move out quickly with the least interference from regular harbor traffic. U.S. Coast Guard personnel should, of course, be consulted and their concurrence obtained as to the exact site.

Boat sales and chandlery facilities should be located along the main access route to the harbor where all vehicular traffic must pass. Restaurants should be located where they have a commanding view of harbor activities but do not interfere with or occupy areas that are needed for more vital harbor-support functions. A good restaurant site would be one overlooking the entrance or the main water body just outside the entrance. If transient housing facilities are included in the harbor complex, they should be near the restaurants or adjunct to them. Any recreational facility should be near or readily accessible from the housing facilities. General shopping areas for food supplies, clothing, and drugstore items should be near the harbor's land boundary and away from a water-oriented activity. Yacht clubs should generally be located away

from the main activities of the harbor for a more exclusive atmosphere. Higher ground that overlooks the harbor from a nearby viewpoint often makes a good yacht club site. A variation is a yacht club with its own separate basin.

Space Allocation   The total area available to the harbor development often places a restriction on the number of boats that the harbor can accommodate and on the size and scope of the ancillary activity it can support. Several general relationships, found valid for most harbors, may help the planner to make tentative allocations of space, which can later be adjusted to definite dimensions in the final planning stages. Such allocations are important in making adequate allowance for future expansion.

The average harbor with all-slip moorage can berth about 15 to 20 boats per acre of navigable water area, including main interior channel, fairways, and slip areas, but not the entrance channel. This general rule applies only when the average boat length is 30 to 35 feet and where good basin geometry can be obtained. Because of the wider fingers needed for two-boat slips, they will occupy almost the same area as that required for single-boat slips. When bow-and-stern moorings are used in lieu of slips, about 2 to 4 times as much water area (depending on the water depth) will be required, exclusive of fairways and channels. Single-point moorings require about 6 times the area occupied by the same number of bow-and-stern moorings if full-circle clearance is provided.

For the normal distribution of boats, a minimum of three vehicle spaces in the parking lot will be required for every four boats in the berthing area. About 90 cars can be parked in an acre, so that roughly one-sixth of an acre of parking lot is required for each acre of water area in the harbor. Where the average size of the berthed craft is large — and many are used for social occasions and multifamily cruising — the ratio may have to be increased to a maximum of about three spaces per berth.

An average launching ramp or hoist will launch and retrieve about 50 trailered boats on a peak day, and because of staggered usage, car-trailer parking spaces will be required for only 80 percent of the peak-day ramp or hoist traffic. About 30 car-trailer units can be parked in an acre if pull-through parking at 45 degrees is provided. This works out to 1.33 acres of parking lot per ramp lane or hoist.

Land area required for harbor service facilities, ancillary facilities, and road varies from one harbor to another. The minimum requirement is an area roughly equal to the parking area required for berths and operational launchings. This will generally provide enough space only for harbor support facilities and roads. To obtain a good revenue versus cost balance it is usually necessary to supplement slip rentals with leaseholds for ancillary facilities; with the additional parking area required, the minimum leasehold and supplemental parking area needed for the extra services that convert a simple small-craft harbor into a complete marina is about twice the area needed for boatowner parking alone. Thus, once the parking area requirement for slips and launching has been determined, it should be multiplied by 4 to obtain the total

minimum land area required for a complete marina. Any additional land that can be obtained may be put to beneficial use later, as a good marina will upgrade its surroundings and attract more revenue-producing ancillary development.

Slip arrangements vary, usually for best conformance to the shape of a basin. The most common is a series of piers or headwalks extending perpendicular to the bulkhead to a pierhead line, with finger piers extending at right angles from the headwalk on either side. For power craft, widths of fairways between finger ends are usually 1.75 to 2 times the length of the longest slips served, while for sailboats the width is 2 to 2.5 times the slip length. The average headwalk width is about 8 feet, with a range of about 5 to 16 feet. The wider headwalks usually have some width for bearing-pile risers, locker boxes, firefighting equipment, and utility lines. The narrower piers often have all obstructions moved to knees at the junctions of finger piers. Extra-wide headwalks are usually in fixed-pier installations because of the higher cost of floating construction. Long, fixed headwalks can also serve as roadways for service vehicles.

Boarding fingers for single-boat slips are usually about 3 feet wide, normally the minimum allowed for floating construction because of the instability of narrower floats. For this reason, floating fingers longer than 35 feet are usually 4 feet wide. In double-boat slip construction, a finger width of 4 feet is common for all slip lengths.

Launching Ramps   Launching ramps have essentially the same design criteria. The slope of the ramp ranges between 12 and 15 percent. Few trailered boats can be launched with a ramp slope flatter than 12 percent without submerging wheel hubs of the pulling vehicle. Slopes steeper than 15 percent are dangerous for all but the most skilled drivers. Many trailered sailboats cannot be launched without hub submergence, even on a 15 percent slope. The best alternatives are to use a trailer-tongue extension or to launch the craft with a hoist.

The surface of the ramp should be paved down to an elevation of about 5 feet below extreme low water level; the top should be rounded over on a 20-foot vertical curve until it becomes nearly level at about 2 feet above extreme high water. The bottom of the ramp should end in a level shelf of loose gravel so that a vehicle losing brakes or traction would be stopped before sliding deeper into the water.

The wetted part should be paved with portland cement concrete, as asphaltic or bituminous paving does not hold up well from traffic in submerged areas. Unpaved ramps will soon deteriorate under even moderate use. A single-lane ramp should not normally be narrower than 15 feet, and a multiple-lane ramp should not have raised divider strips. Lane marking is not necessary and may lead to less than optimum use during peak hours.

Ample maneuver room should be provided beyond the top of the ramp, usually 40 to 60 feet, on a gentle rampward slope (about 1:50) for proper surface drainage. About 50 pull-through parking spaces should be provided for each ramp lane, with a clearly marked traffic-circulation pattern between the parking area and ramp.

Boarding docks should be provided, preferably on each side of the ramp and extending out into or along the sides of the basin, with a total boarding length of at least 50 feet for each ramp. The ramp should adjoin fairly quiet water, although not necessarily as quiet as that needed for a berthing site. Ample protected *holding area* in the water just off the ramp and boarding dock location should also be available for boats awaiting their retrieval turn during peak hours.

Ramps leading into salt water or polluted waters should have a conveniently located washdown facility, just outside the maneuver area if possible. This area should be large enough to accommodate one car and trailered boat per ramp. A waiting area of about the same size for boats just retrieved should also be provided. Fresh water piped from the local supply main and washdown hoses of ample length should be available at this site, together with an adequate drainage system. Ramp users often beach their craft temporarily before retrieving them. Accordingly, users will appreciate having a small sandy beach reserved for this purpose near the launching area. A typical launching-ramp facility is shown in Figure 5.96.

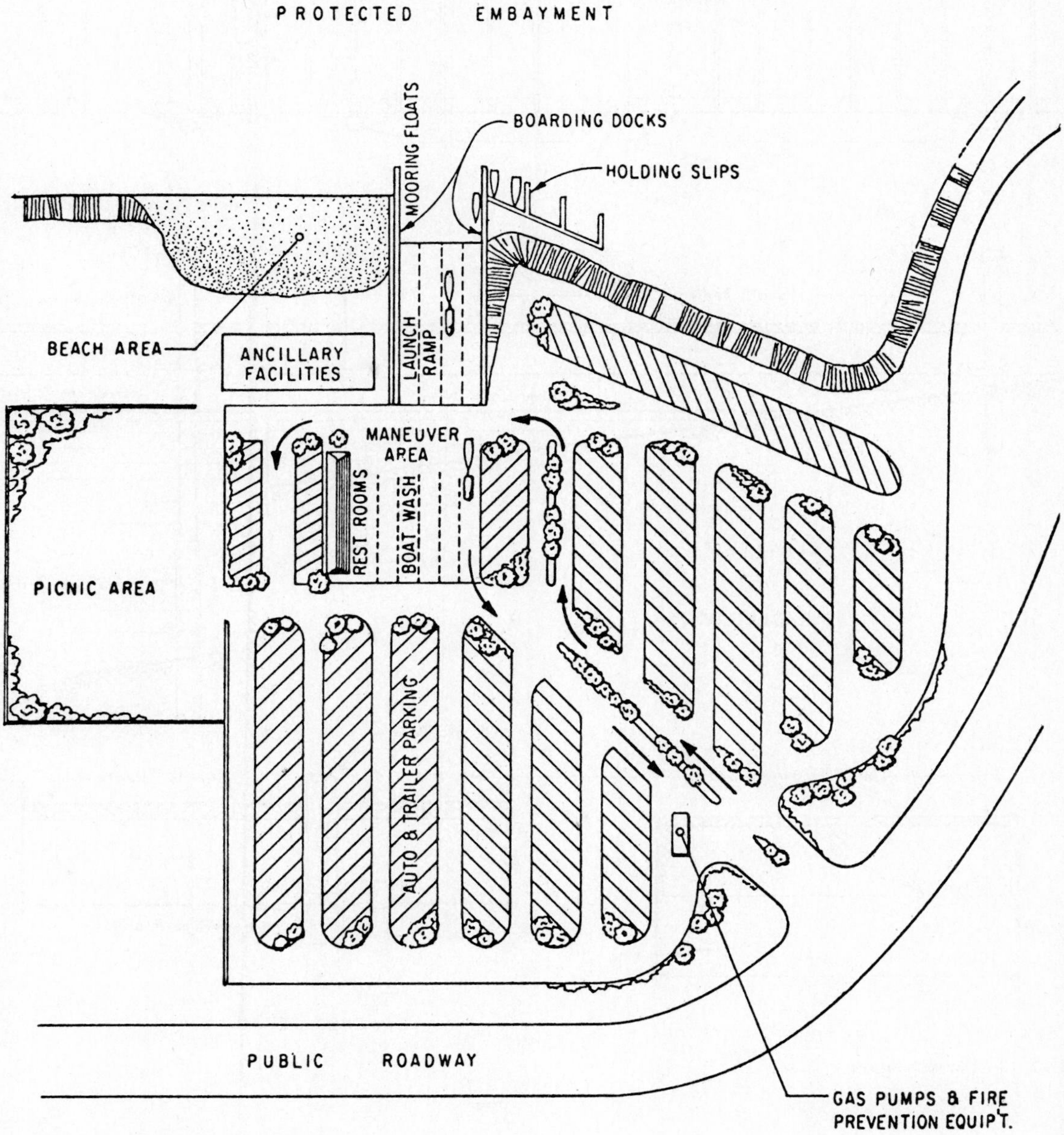

PROTECTED   EMBAYMENT

MOORING FLOATS

BOARDING DOCKS

HOLDING SLIPS

BEACH AREA

ANCILLARY FACILITIES

LAUNCH RAMP

MANEUVER AREA

REST ROOMS

BOAT WASH

PICNIC AREA

AUTO & TRAILER PARKING

PUBLIC   ROADWAY

GAS PUMPS & FIRE PREVENTION EQUIP'T.

Figure 5.96 Layout of a typical launching-ramp facility.

Boat Landing Dock

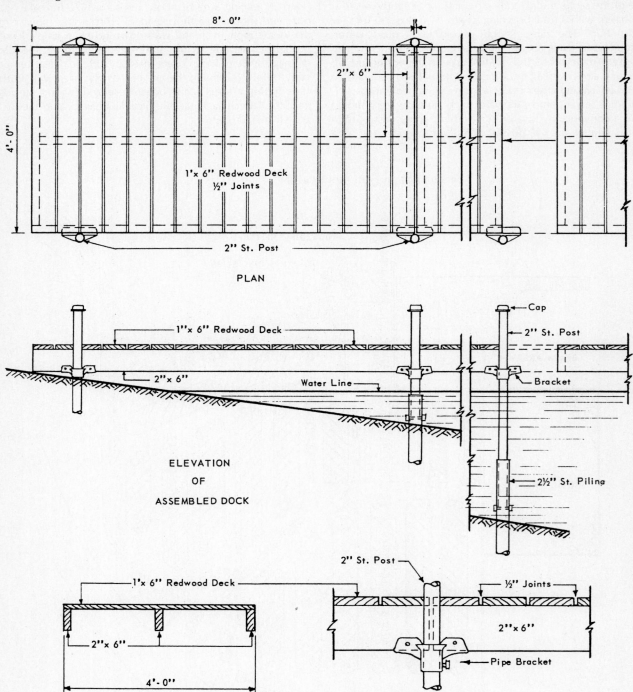

Figure 5.97 Boat launching dock.

Parking Lot

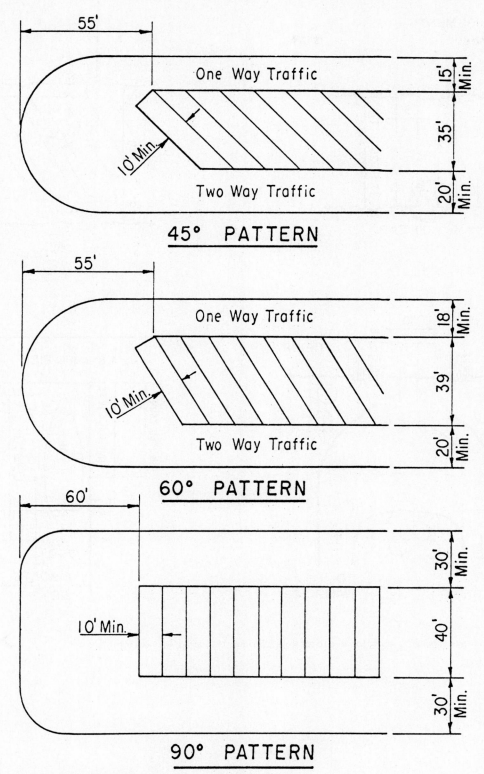

Figure 5.98 Standard parking lot layout criteria for marinas.

Auto/Boat Parking

SPACE REQUIREMENTS

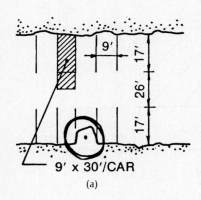

9' x 30'/CAR

(a)

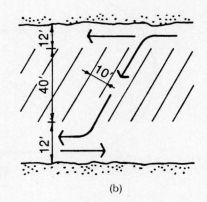

(b)

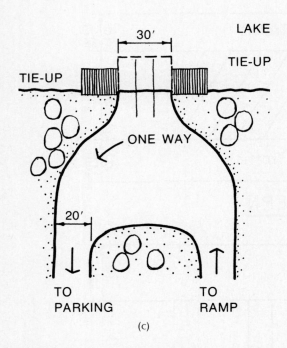

(c)

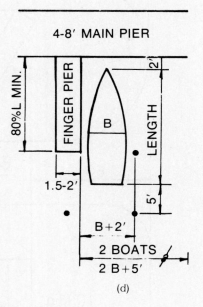

(d)

Figure 5.99 Auto/boat parking: (*a*) auto; (*b*) auto/boat trailer; (*c*) launching ramp; (*d*) boat slip.

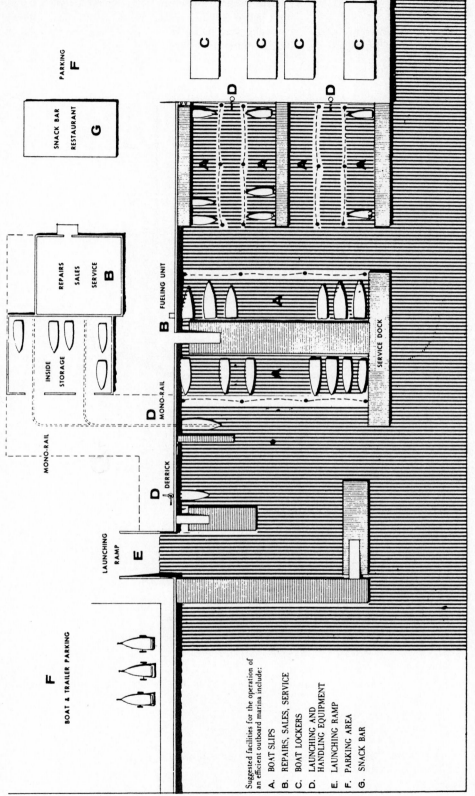

Typical Marina Layout

Suggested facilities for the operation of an efficient outboard marina include:

A. BOAT SLIPS
B. REPAIRS, SALES, SERVICE
C. BOAT LOCKERS
D. LAUNCHING AND HANDLING EQUIPMENT
E. LAUNCHING RAMP
F. PARKING AREA
G. SNACK BAR

Figure 5.100 Typical marina layout.

BOAT BASIN; MARINA

A design for a typical pleasure boat basin is shown in Figure 5.101.

The accompanying table should be used in conjunction with Figure 5.102 to obtain the widths of slips, the lengths of catwalks, and the locations of stern anchor piles. The table is based on the use of traveler irons.

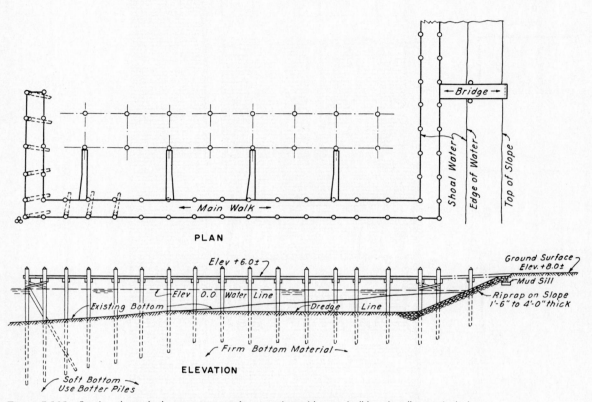

Figure 5.101   Section through slope revetment for a marina without a bulkhead wall; a typical pleasure boat basin.

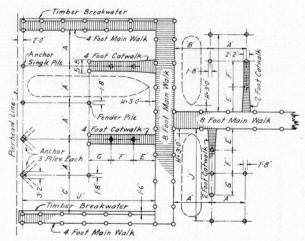

Figure 5.102   Dimension diagram for slips and catwalks for a typical pleasure boat basin.

Dimensions for Slips and Catwalks

Length group for yachts (feet)	Beam to be provided for	Minimum clearance for beam	Minimum clear width of slip	Allowance for half anchor pile	Allowance for half of catwalk	Gross slip width: Type A	Gross slip width: Type B	Gross slip width: Type C	Gross slip width: Type D	Usable width of catwalk	First catwalk span: length E	Second catwalk span: length F	Third catwalk span: length G	Total length of catwalk	Distance J to anchor pile
20–25	7' 6"	3' 0"	10' 6"	10"	1' 1"	12' 5"	12' 2"	12' 5"	...	2' 0"	10' 0"	8' 0"	...	18' 0"	28' 0"
	8' 6"	3' 0"	11' 6"	10"	1' 1"	13' 5"	13' 2"	13' 5"	...	2' 0"	10' 0"	8' 0"	...	18' 0"	28' 0"
25–30	7' 6"	3' 0"	10' 6"	10"	1' 1"	12' 5"	12' 2"	12' 5"	...	2' 0"	10' 0"	10' 0"	...	20' 0"	33' 0"
	9' 6"	3' 0"	12' 6"	10"	1' 1"	14' 5"	14' 2"	14' 5"	...	2' 0"	10' 0"	10' 0"	...	20' 0"	33' 0"
30–35	8' 6"	3' 0"	11' 6"	10"	1' 1"	13' 2"	13' 2"	13' 5"	.	2' 0"	12' 0"	10' 0"	...	22' 0"	38' 0"
	11' 6"	3' 0"	14' 6"	10"	1' 1"	16' 5"	16' 2"	16' 5"	...	2' 0"	12' 0"	10' 0"	...	22' 0"	38' 0"
35–40	9' 6"	3' 6"	13' 0"	10"	1' 1"	14' 11"	14' 8"	14' 11"	...	2' 0"	12' 0"	12' 0"	...	24' 0"	42' 0"
	12' 0"	3' 6"	15' 6"	10"	1' 1"	17' 5"	17' 2"	17' 5"	...	2' 0"	12' 0"	12' 0"	...	24' 0"	42' 0"
40–45	9' 6"	4' 0"	13' 6"	10"	1' 1"	15' 5"	15' 2"	15' 5"	...	2' 0"	14' 0"	12' 0"	...	26' 0"	47' 0"
	12' 6"	4' 0"	16' 6"	10"	1' 1"	18' 5"	18' 2"	18' 5"	...	2' 0"	14' 0"	12' 0"	...	26' 0"	47' 0"
45–50	10' 6"	4' 0"	14' 6"	10"	1' 1"	16' 5"	16' 2"	18' 5"	...	2' 0"	9' 0"	9' 0"	10' 0"	28' 0"	52' 0"
	13' 6"	4' 0"	17' 6"	10"	1' 1"	19' 5"	19' 2"	19' 5"	...	2' 0"	9' 0"	9' 0"	10' 0"	28' 0"	52' 0"
50–60	11' 6"	5' 0"	16' 6"	1' 7"	1' 1"	19' 2"	18' 11"	18' 5"	...	2' 0"	11' 0"	11' 0"	12' 0"	34' 0"	61' 0"
	14' 6"	5' 0"	19' 6"	1' 7"	1' 1"	22' 2"	21' 11"	21' 5"	...	2' 0"	11' 0"	11' 0"	12' 0"	34' 0"	61' 0"
60–70	12' 6"	5' 0"	17' 6"	1' 7"	2' 10"	21' 11"	19' 11"	21' 2"	...	4' 0"	11' 0"	11' 0"	12' 0"	34' 0"	72' 0"
	14' 6"	5' 0"	19' 6"	1' 7"	2' 10"	23' 11"	21' 11"	23' 2"	...	4' 0"	11' 0"	11' 0"	12' 0"	34' 0"	72' 0"
	16' 0"	5' 0"	21' 0"	1' 7"	2' 10"	25' 5"	23' 5"	24' 8"	...	4' 0"	11' 0"	11' 0"	12' 0"	34' 0"	72' 0"
70–80	13' 0"	5' 0"	18' 0"	1' 7"	2' 10"	22' 5"	20' 5"	21' 8"	...	4' 0"	11' 0"	11' 0"	12' 0"	34' 0"	82' 0"
	16' 6"	5' 0"	21' 6"	1' 7"	2' 10"	25' 11"	24' 11"	26' 2"	24' 7"	4' 0"	11' 0"	11' 0"	12' 0"	34' 0"	82' 0"

5.7 Ponds

Types of Ponds Farm ponds and reservoirs may be divided into two general types, embankment ponds and excavated ponds. An embankment pond is a body of water created by constructing a dam across a stream or a watercourse. It is usually built in an area where the land slope ranges from gentle to moderately steep and where the stream valley is sufficiently depressed to permit the storage of water to a considerable depth.

An excavated pond is a body of water created by excavating a pit or a dugout. It is usually constructed in a relatively level area. The fact that its capacity is obtained almost entirely by excavation limits its use to a location where only a small supply of water is required.

Ponds are also built in gentle to moderately sloping areas where capacity is obtained by both excavation and the construction of a dam. For the purpose of classification, these are considered to be embankment-type ponds if the depth of water impounded against the embankment exceeds 3 feet.

Selecting the Pond Site The selection of a suitable pond site should begin with preliminary studies of possible sites. If more than one site is available, each should be studied separately with a view of selecting the most practical and economical site.

From an economic point of view, a pond should be located where the largest storage volume can be obtained with the least amount of earth fill. This condition generally occurs at a site where the valley is narrow, side slopes are relatively steep, and the slope of the valley floor permits a large, deep basin. Such a site tends to minimize the area of shallow water, but it should be examined carefully for adverse geologic conditions. Unless the pond is to be used for wildlife, large areas of shallow water should be avoided because of excessive evaporation losses and the growth of noxious aquatic plants.

If water must be conveyed for use elsewhere, as for irrigation or fire protection, the pond should be located as close to the point of use as is practical.

Ponds to be used for fishing, boating, swimming, and other forms of recreation should be readily accessible by automobile. This is particularly true when the general public is charged a fee for use of the pond. The success of such an income-producing enterprise may well depend on the accessibility of the pond.

Pollution of farm pond water should be avoided by selecting a site where drainage from farmsteads, feeding lots, corrals, sewage lines, mine dumps, and similar areas will not reach the pond. Where this cannot be done practically, the drainage from such areas should be diverted from the pond.

The pond should not be located where a sudden release of the water because of failure of the dam would result in loss of life, injury to persons or livestock, damage to residences or industrial buildings, railroads, or highways, or interruption of use or service of public utilities.

Preliminary Site Studies In addition to the considerations mentioned for the selection of a pond location, other physical characteristics of the drainage area and the pond site should be investigated before the final selection is made.

Adequacy of the Drainage Area Where surface runoff is the main source of water supply, the contributing drainage area should be large enough to yield sufficient runoff to maintain the water supply in the pond during all periods of intended use. The drainage area should not be so large, however, as to require large and expensive overflow structures to bypass the runoff safely.

The amount of runoff that can be expected annually from a watershed of a given area depends on so many factors that no set rule can be given for its determination. The physical characteristics of the watershed that have a direct effect on the yield of water are land slopes, soil infiltration, vegetal

cover, and surface storage. Storm characteristics such as the amount, intensity, and duration of rainfall also affect water yield. All these characteristics vary widely throughout the United States. Figure 5.103 can be used as a general guide for estimating the size of a watershed required for each acre-foot of capacity in a pond or reservoir to maintain normal pool level if more precise local data are not available.

Minimum Pond Depth For a permanent water supply, it is necessary to provide sufficient water depth to meet the intended use and to offset seepage and evaporation losses. Such losses vary in different sections of the United States and also from year to year in a given section. Figure 5.104 shows recommended minimum depths of water for farm ponds if normal seepage and evaporation losses are assumed. Greater depths are desirable when a year-round water supply is essential or when seepage losses may exceed 3 inches per month. See state standards and specifications for local minimum depths.

Recommended Minimum Depths of Ponds and Reservoirs

Climate	Annual rainfall (inches)	Minimum water depth over 25 percent of the area (feet)
Superhumid	Over 60	6
Humid	40–60	8
Subhumid: moist	30–40	9
Subhumid: dry	20–30	10
Semiarid	10–20	12
Arid	Under 10	14

Drainage Area Protection To maintain the required depth and capacity of a farm pond, it is necessary that the inflow be reasonably free from sediment. The best protection is adequate erosion control on the contributing drainage area. Land under a cover of permanent vegetation, such as

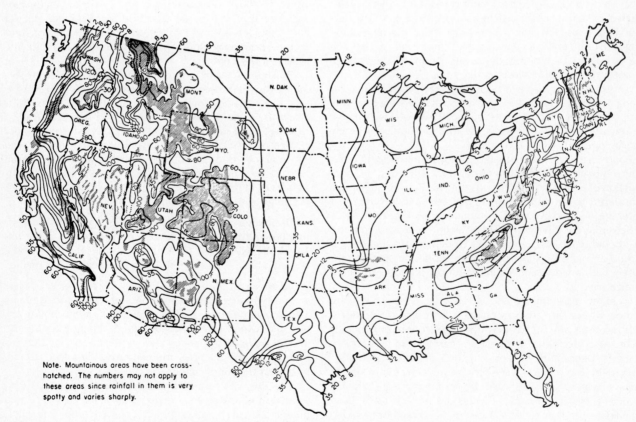

Note. Mountainous areas have been cross-hatched. The numbers may not apply to these areas since rainfall in them is very spotty and varies sharply.

Figure 5.103 A guide for estimating the approximate size of the drainage area (in acres) required for each acre-foot of storage in an embankment or excavated pond.

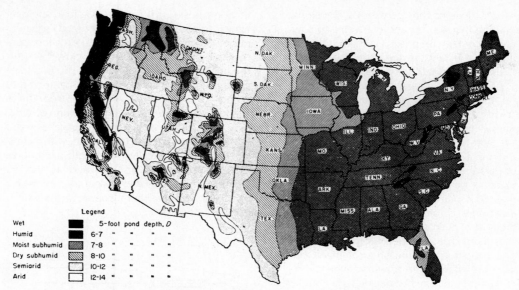

Figure 5.104 Recommended minimum depths of water for ponds in the United States.

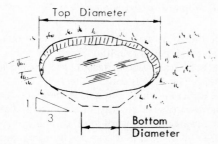

Figure 5.105 A circular pond.

trees or grasses, makes the most desirable drainage area. If such an area is not available, cultivated areas that are protected by necessary conservation practices, such as terracing, contour tillage, strip cropping, conservation cropping systems, vegetated desilting areas, and other soil improvement practices, may be utilized as a last resort. Allowance should be made for the expected sedimentation during the effective life of the structure.

Figure 5.105 shows the design of a circular pond. The capacity of ponds of this type is given in the accompanying table.

Capacity of Circular Ponds (Bank Slopes of 3:1)

Water depth (feet)	Top diameter (feet)	Bottom diameter (feet)	Capacity (gallons)
8	75	27	131,000
8	100	52	281,000
12	100	28	320,000
12	150	78	947,000
16	150	54	1,051,000

5.8 Golf Courses

Par The United States Golf Association has set a general standard for par in relation to the yardage of any given hole: "'Par' is the score that an expert golfer would be expected to make for a given hole. Par means errorless play without flukes and under ordinary weather conditions allowing two strokes on the putting green." The method for computing par on any hole is shown in the accompanying table.*

Distance (Yards)

Men	Women	Par
Up to 250	Up to 210	3
251–470	211–400	4
471 and over	401–575	5
	576 and over	6

Regulation Golf Course In most cases a regulation golf course has a par of 70, 71, or 72 (occasionally 69 or 73). Many older courses built in the United States play to a total par of 70. However, in recent years par 72 has become the standard of excellence in the minds of many developers and golfers. It should be stressed, however, that the size and natural characteristics of a site determine what the total par should be; therefore, many courses are built outside the standard par 72. In many cases the golf course architect will determine that a shorter par-70 course may be much better than a forced par 72 because it is more demanding and natural. Neither par nor total yardage should be the criterion of

*Golf Committee Manual and USGA Golf Handicap System, United States Golf Association, New York, 1969.

quality, for the objectives of the recreational development golf course should be that it is fair and enjoyable to play.

A regulation golf course comprises eighteen holes with a combination of par 3s, 4s, and 5s, the sum of which equals pars 70 to 73. The standard mix for a par-72 golf course is ten par 4s, four par 3s, and four par 5s. Par-71 courses generally drop a par 4 and replace it with a par 3 or drop a par 5 and replace it with a par 4. A par-70 golf course generally has either six par 3s, eight par 4s, and four par 5s or four par 3s, twelve par 4s, and two par 5s. A par-73 golf course generally has an additional par 5 in place of a par 4. It is these combinations of pars that comprise what is considered to be the norm to qualify a course as regulation in the minds of golfers. However, it is neither total yardage nor par that determines the amount of area used, the quantity of lot frontage, and the cost of maintenance and control of the golf facility once it has been built. The needs of the project, the shape of the total property, and the physical characteristics of the site all have an influence on how and where the golf course architect, planner, and owner decide to lay out a regulation golf course.

Basic Golf Course Designs There are five basic golf course design types, with several possible needs of an individual development. After a feasible location has been determined by studying the topography and the natural site characteristics, the developer and design team can determine which type or combination of types would be most appropriate for the project from every standpoint.

The five basic prototypal configurations for an eighteen-hole regulation golf course are (1) a single-fairway eighteen-hole course with returning nines, (2) a single-fairway continuous eighteen-hole course, (3) a double-fairway eighteen-hole course with returning nines, (4) a double-fairway continuous eighteen-hole course, and (5) an eighteen-hole core golf course. (See below, Figures 5.109, 5.110, 5.111, 5.112, and 5.113.)

In the following discussion, hypothetical diagrams illustrate the five basic golf course designs. For ease of comparison, each diagram is shown with a 10-acre clubhouse site. All five layouts are 72 par, and each includes four par-5 holes of 500 yards, four par-3 holes of 200 yards, and ten par-4 holes of 420 yards, totaling 7000 yards in length. The diagrams are laid out on the same scale to facilitate visual and mathematical comparison of golf course areas, real estate frontage, and function within a particular development.

It should be noted that any of these illustrative layouts, when applied to an existing site, will vary in area and configuration from the figures and plans shown to accommodate irregularities in property boundaries and topographic features. In addition, the architect may choose to combine the characteristics of several of the basic formats to serve best the requirements of the development and the dictates of the site. Site characteristics and topography must be accommodated before striving for good frontage figures.

The diagrams and dimensions of the five prototypal configurations are merely illustrations. The figures given for acreage and length of frontage were computed for comparative purposes, not for use as a rigid standard. The acreage figures for each prototype are slightly greater than would be necessary in most actual situations. Boundaries adjacent to tees can be from 75 to 90 feet closer to the centerline than they must be in the target areas. The boundary must then be gradually angled away from the centerline to the 150-foot distance at a point 450 to 500 feet from the back of the tee, as shown in Figure 5.106. If this is done on single- and double-row courses, the total acreage figure would decrease markedly. The acreage in a core course, with its fewer boundaries, will decrease minimally. The length of frontage for each prototype should also be regarded as the maximum figure because all features were separated as much as was feasible to achieve the maximum frontage possible.

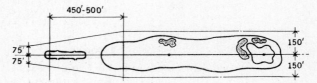

Figure 5.106 Conserving land in the tee areas.

Different types of golf holes and bunker locations are shown in Figures 5.107 and 5.108.

The five basic golf course designs are shown in Figures 5.109, 5.110, 5.111, 5.112, and 5.113.

Selection of Course Site The golf architect usually considers a number of prospective sites for a course and selects the one that, at a reasonable cost of land, can be converted into a good course at a minimum construction cost and can be maintained properly at minimum expense.

The size of the property is important, 50 acres for a nine-hole course and 110 acres for eighteen holes generally being considered the minimum. Even these areas involve risk of injury to players playing parallel holes. For better courses, 80 acres for a nine-hole course and 160 acres for eighteen holes are about right. Irregularly shaped plots often afford opportunities for most interesting course design.

The land should not be too rugged. A gently rolling area with some trees is preferable. Land that is too hilly is tiring on players, usually necessitates too many blind shots, and is more costly to keep well turfed.

The course should have a practice fairway area close to the clubhouse. Some public and daily-fee courses have installed practice ranges, lighted for night use, adjoining their courses and on highways. From these ranges they derive considerable income and develop golfers for day play on the courses.

The usual experience of a golf club that leases its land with an option to buy is that the installation of a course increases the value of the land enough to make the option an excellent investment when it is exercised. When finances permit and the community's prospects warrant, it also is well to tie up enough land so that property bordering on the course may be sold for residential sites and the proceeds used to pay off

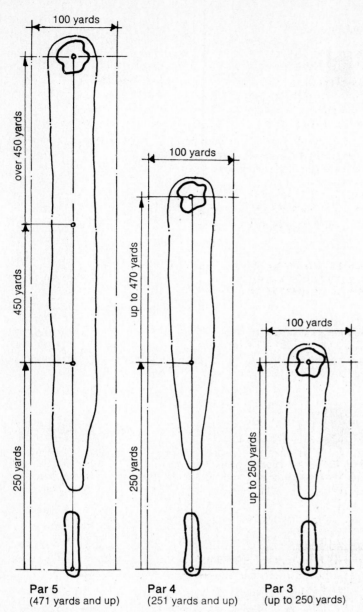

Figure 5.107 Different types of golf holes in relation to par.

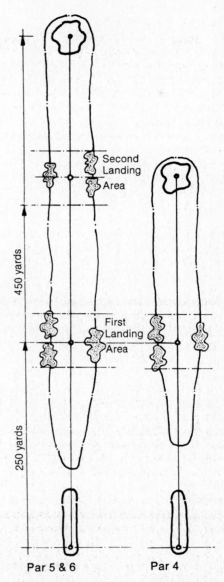

Figure 5.108 Strategic fairway bunker locations.

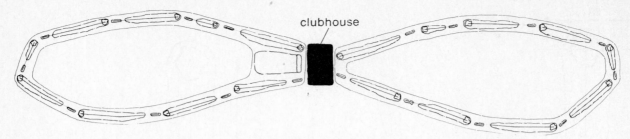

Figure 5.109 Single-fairway eighteen-hole course with returning nines. The course covers 175 acres, the minimum width between developed areas is 300 feet, and the length of the lot frontage is plus or minus 44,400 feet.

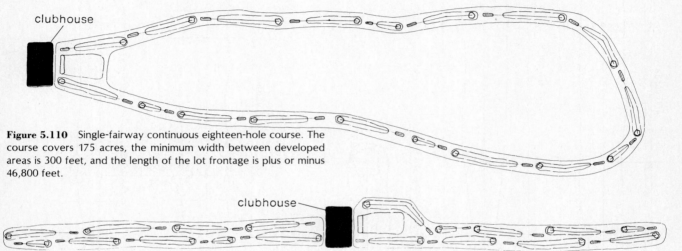

Figure 5.110 Single-fairway continuous eighteen-hole course. The course covers 175 acres, the minimum width between developed areas is 300 feet, and the length of the lot frontage is plus or minus 46,800 feet.

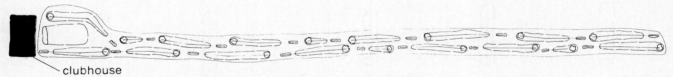

Figure 5.111 Double-fairway eighteen-hole course with returning nines. The course covers 150 acres, the minimum width between developed areas is 500 feet, and the length of the lot frontage is plus or minus 24,200 feet.

Figure 5.112 Double-fairway continuous eighteen-hole course. The course covers 150 acres, the minimum width between developed areas is 500 feet, and the length of the lot frontage is plus or minus 25,000 feet.

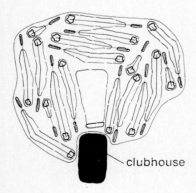

Figure 5.113 Core golf course. The course covers 140 acres, the minimum width between developed areas is zero feet, and the length of the lot frontage is plus or minus 10,000 feet.

the club's loans. Frequently reference to this increase in value of surrounding property due to establishment of a golf club is so attractive to landowners that enough property exclusively for course use becomes available at a bargain price. The owners correctly surmise that not only is the net value of their entire holdings increased by the location of the golf course but that their property surrounding the course is more readily salable.

Accessibility Unless absolutely unavoidable, a golf course should not be off the beaten track. This is especially important in the case of a small-town course for which greens fees from transients are counted on to help to meet maintenance costs. Locate your course along the main highway into town. All other things being equal, design the course so that one or two holes parallel the highway; it is good advertising.

Another reason for not locating the course in an out-of-the-way spot is that the club should have good transportation for the members. It should be as near to town as possible, the cost of land being taken into consideration, and the main highway from town to the club should be one that is kept in good condition and not merely a country lane, unpaved and liable to become impassable with every heavy rain.

Soil Factors The condition of the soil is extremely important because the better the stand of turf raised on fairways and greens, the more satisfactory and more popular the course will be. The ideal golf course soil is a sandy loam. It is not impossible but is expensive to grow a good stand of grass on a heavy clay. Be sure to take the character of the soil into consideration when choosing a site.

Soil analysis of areas of the golf course site will be made at low cost by state agricultural departments or county agents. Considerable helpful information can be supplied by state agricultural experiment stations and county agents in determining the most desirable site from the viewpoint of a good possibility of golf turf development and in recommending the grass seeding, growing, and maintenance program.

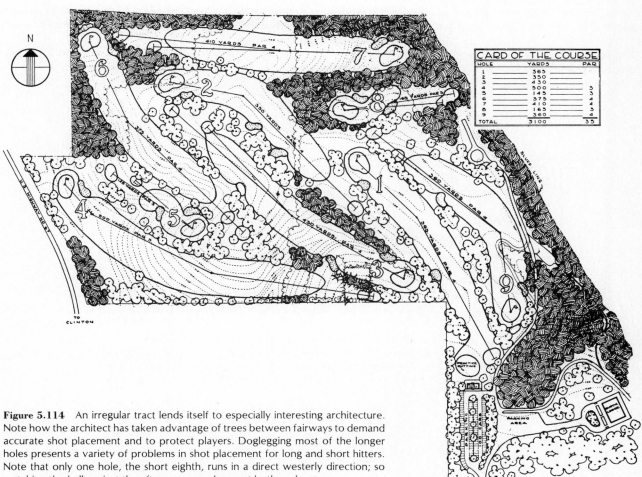

Figure 5.114 An irregular tract lends itself to especially interesting architecture. Note how the architect has taken advantage of trees between fairways to demand accurate shot placement and to protect players. Doglegging most of the longer holes presents a variety of problems in shot placement for long and short hitters. Note that only one hole, the short eighth, runs in a direct westerly direction; so watching the ball against the afternoon sun does not bother players.

Past Use Closely linked with soil factors is the use to which the land has been put in the past. Is the plot a run-down farm with soil from which a large part of the plant food has been removed, or is the soil rich in the elements that will be necessary for the successful cultivation of turf? Has the land lain idle for many years, or has it been intensively cultivated by its farmer owner without plant food being returned to the soil?

The selection of property that has been well kept up as pastureland is highly advisable. Much money is saved in putting the course into excellent condition. For instance, one southern Ohio club built its course on property that had been in bluegrass for about 30 years, and for that reason its fairways cost practically nothing. Frequently the scenic attractions of a site are such that to the susceptible and uninformed organizers of a golf club they totally outweigh soil conditions. A happy balance should be maintained between soil and scenic factors. Pick a site that will offer no serious handicaps to the attempts of the club to grow a stand of grass and maintain it thereafter.

Power and Water Availability Water and power are absolute necessities for any modern golf course. Even in the smallest communities, grass-green courses with a clubhouse are being built. To water only greens and tees or the whole course and to operate a clubhouse, you must have power and water.

The source of water should be close to the site and be reliable and pure enough to drink and to irrigate fine turf. It may be a city system, wells, a lake, a river, or some combination of these. The cost of connections to water and power supplies must be included in your plans.

How Much Clearing? Consider next the amount of clearing that must be done in building the course. Will it be necessary to move many trees or grub out many stumps? Will removing stones from the soil be expensive? Are there large swamp areas that must be filled in or drained? Do not misunderstand the question about clearing out trees. A golf course should, if possible, have patches of woodland, as trees offer one of the best natural hazards if properly placed with reference to the course. However, it is an expensive matter to remove large growing trees, and the site selected should not have too many of these in the portions that are intended to be fairways.

Natural Golf Features The last consideration in selecting the site is whether or not it possesses natural golf features. This may seem to the uninitiated the first and most important thing to look for, but natural golf features, while extremely desirable, are not nearly as important as the character of the soil and the site location. Rolling terrain, creek valleys, woodlands, ravines, ponds, and the like make the job of designing an interesting course much easier, but all these features or a substitute for them can be secured through artificial hazards. For this reason, the presence or absence of natural golf features is perhaps less important than any of the other factors mentioned above.

Figures 5.114, 5.115, 5.116, 5.117, 5.118, and 5.119 show desirable — and undesirable golf course designs.

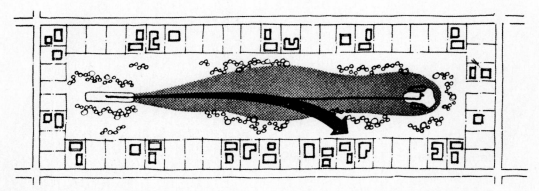

Figure 5.115 Single fairways are an expensive way to get frontage. Although they permit the maximum number of fairway lots, they present the greatest danger of golf balls flying into backyards. For safety's sake, single fairways should be at least 100 yards wide in the landing area, which extends from 150 to 250 yards from the tee. This width can be reduced if there are plenty of trees in the landing area.

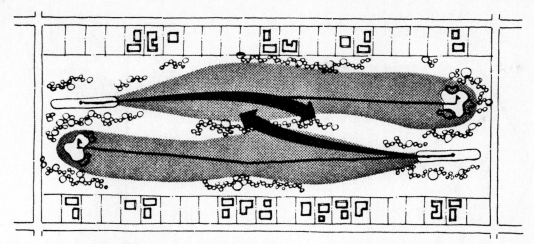

Figure 5.116   Double fairways require a smaller safety margin. They are laid out so that a sliced drive from either tee will go into the adjacent fairway. As a result, the combined width of the two landing areas need be only 150 yards. The double fairway thus requires 25 percent less maintenance. Since a ball sliced into another fairway is easier to find than one sliced into a border area or a house lot, play is speeded up.

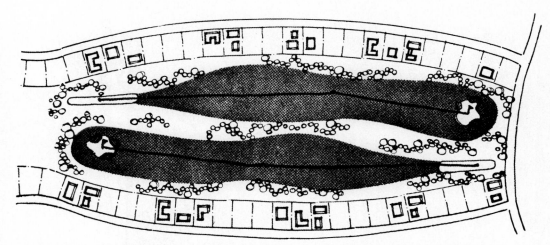

Figure 5.117   A curved building line saves even more land. Here is a good example of the need for coordinating course and housing. Tees and greens require less of a buffer zone around them than does the landing area, and the combined tee and green area at either end of this double fairway can be as narrow as 100 yards. If the adjacent building land and streets can be curved to fit this pattern, the result will be further saving of land, a deeper cut in maintenance costs, and from the housing point of view, a more interesting subdivision.

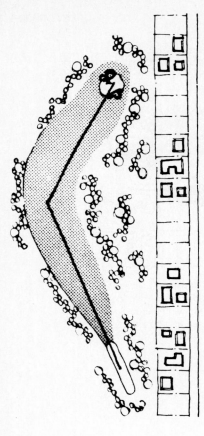

Figure 5.118b This dogleg layout wastes land. Houses near this hole are relatively safe from bombardment, and they have a very nice view. But the developer has lost a big piece of land – enough for half a dozen or so lots. And unless the wasted area is carefully manicured (which means extra maintenance), many golf balls could be lost in this area.

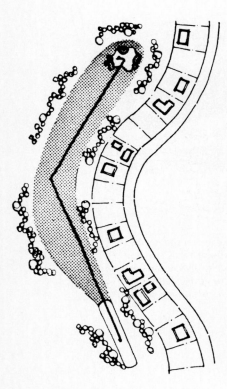

Figure 5.118a This dogleg layout is dangerous. Golfers always take the short route on a dogleg, and they will try to drive over a greenhouse if this will give them a shorter second shot. On this hole seven or eight lots are in the danger zone.

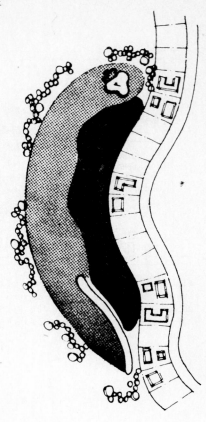

Figure 5.119b This dogleg is safe and beautiful. A lake has been built on the inside of the dogleg, protecting the houses. Golfers are forced to keep their drives away from them. The lake makes a challenging hole; the closer the drive to the water, the shorter the second shot. And it creates premium, high-priced lots for houses, town houses, or apartments that command a view of both fairway and water.

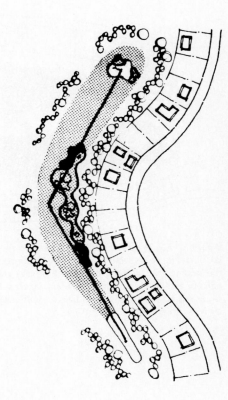

Figure 5.119a This dogleg plan is better. Golfers will think twice before they cut this corner. Trees are planted along the lot lines, including the area next to the tee. Traps and water hazards are placed on the right-hand side of the land area. As a result, the golfer who tries to go the short way is almost sure to get into trouble.

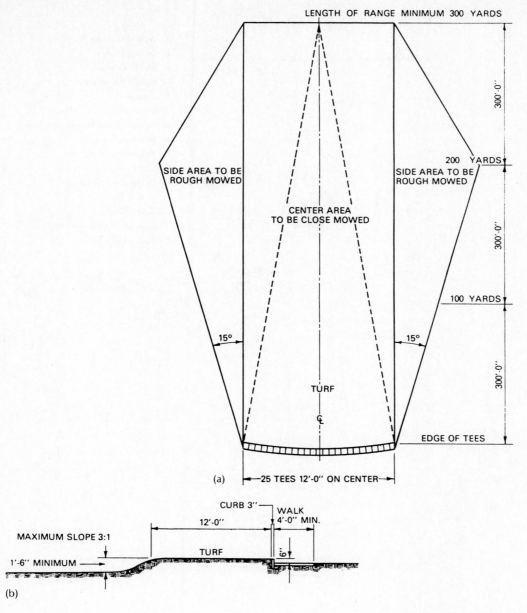

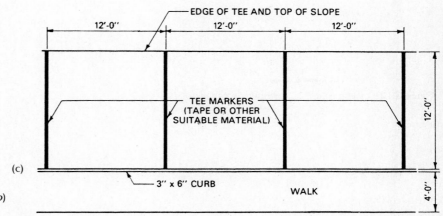

Figure 5.120 Golf driving range; (*a*) layout; (*b*) section of tees; (*c*) plan of tees.

5.9 Camp Sites

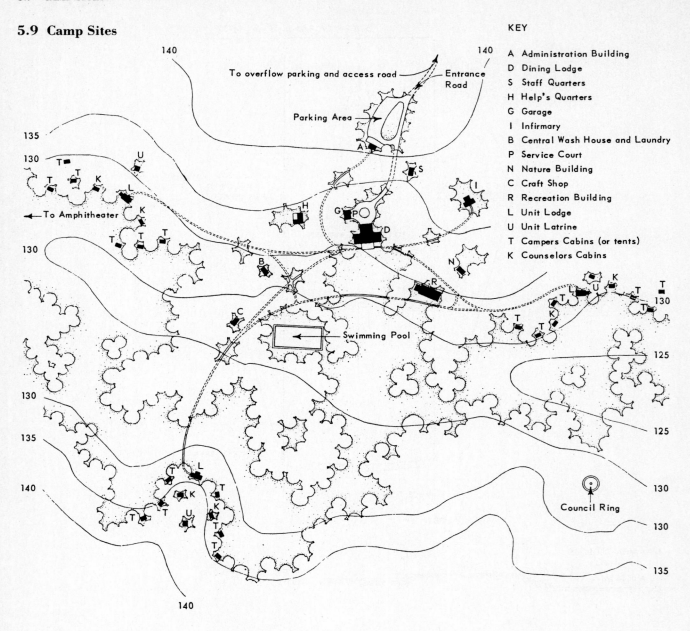

KEY

A Administration Building
D Dining Lodge
S Staff Quarters
H Help's Quarters
G Garage
I Infirmary
B Central Wash House and Laundry
P Service Court
N Nature Building
C Craft Shop
R Recreation Building
L Unit Lodge
U Unit Latrine
T Campers Cabins (or tents)
K Counselors Cabins

To overflow parking and access road — Entrance Road

Parking Area

To Amphitheater ←

Swimming Pool

Council Ring

Paths – Service Trails

0'   120'   240'

Figure 5.121 Camp site, 72- to 96-person capacity.

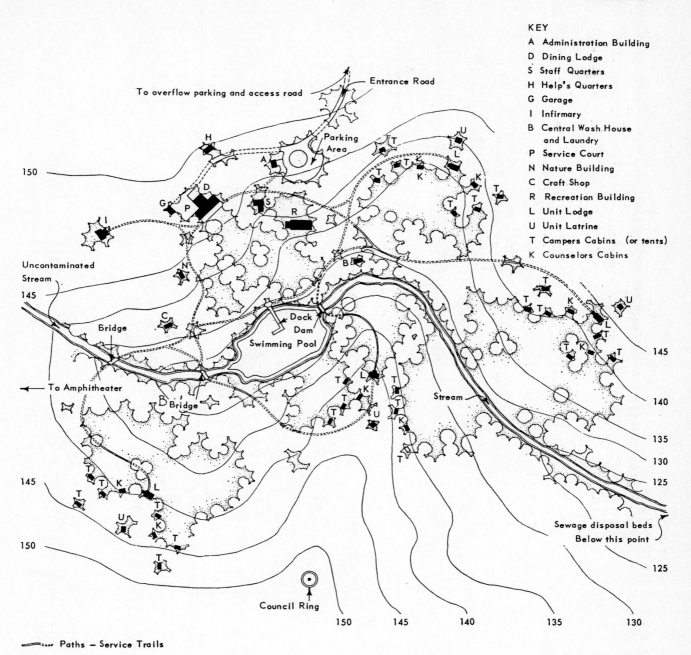

KEY

A Administration Building
D Dining Lodge
S Staff Quarters
H Help's Quarters
G Garage
I Infirmary
B Central Wash House
   and Laundry
P Service Court
N Nature Building
C Craft Shop
R Recreation Building
L Unit Lodge
U Unit Latrine
T Campers Cabins (or tents)
K Counselors Cabins

To overflow parking and access road

Entrance Road

Parking Area

150

Uncontaminated Stream

145

Bridge

To Amphitheater

Bridge

Dock
Dam
Swimming Pool

Stream

145

140

135

130

125

Sewage disposal beds
Below this point

125

145

150

Council Ring

150   145   140   135   130

Paths — Service Trails

0'   120'   240'

Figure 5.122 Camp site, 72- to 96-person capacity.

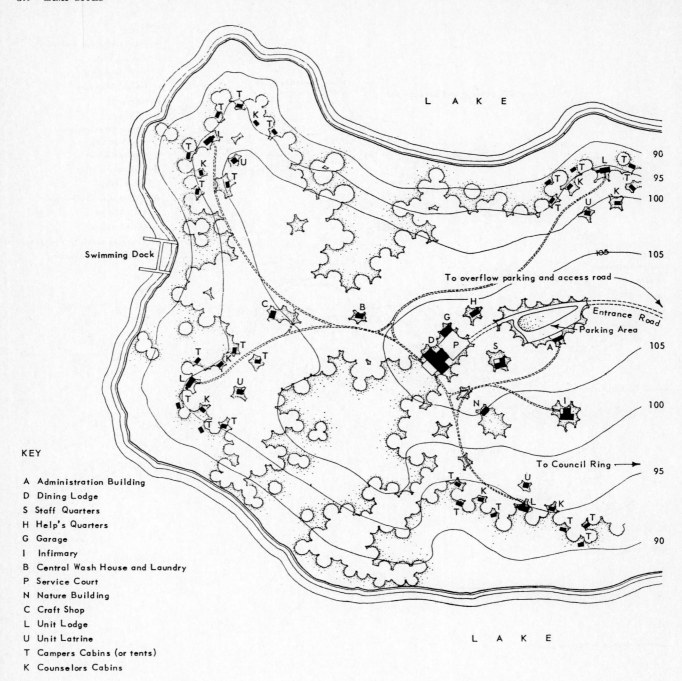

L A K E

90
95
100

105 — 105

To overflow parking and access road

Entrance Road

Parking Area

105

100

95

90

Swimming Dock

To Council Ring

KEY

A Administration Building
D Dining Lodge
S Staff Quarters
H Help's Quarters
G Garage
I Infirmary
B Central Wash House and Laundry
P Service Court
N Nature Building
C Craft Shop
L Unit Lodge
U Unit Latrine
T Campers Cabins (or tents)
K Counselors Cabins

⸗᠁ Paths — Service Trails

0'   120'   240'

Figure 5.123   Camp site, 72- to 96-person capacity.

L A K E

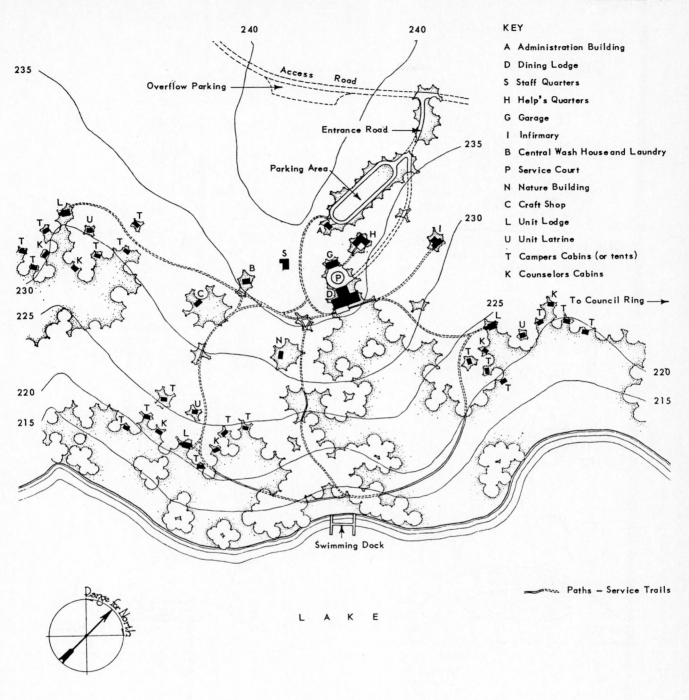

KEY

A Administration Building
D Dining Lodge
S Staff Quarters
H Help's Quarters
G Garage
I Infirmary
B Central Wash House and Laundry
P Service Court
N Nature Building
C Craft Shop
L Unit Lodge
U Unit Latrine
T Campers Cabins (or tents)
K Counselors Cabins

Paths — Service Trails

Figure 5.124 Camp site, 72- to 96-person capacity.

Space Requirements of Site Elements

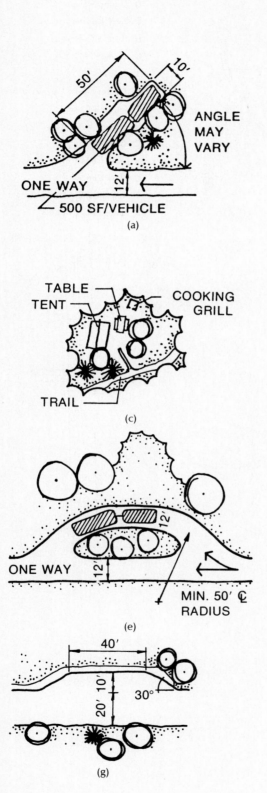

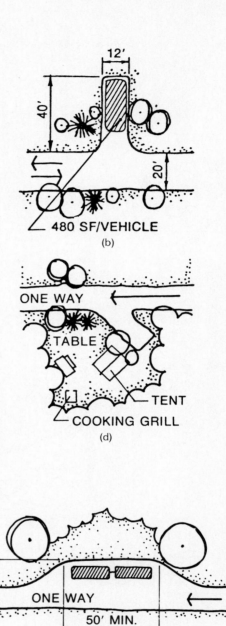

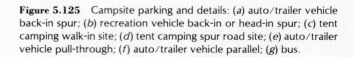

Figure 5.125 Campsite parking and details: (*a*) auto/trailer vehicle back-in spur; (*b*) recreation vehicle back-in or head-in spur; (*c*) tent camping walk-in site; (*d*) tent camping spur road site; (*e*) auto/trailer vehicle pull-through; (*f*) auto/trailer vehicle parallel; (*g*) bus.

5.10 Tables for Parking-Lot Sizes

The series of quick-reference tables presented here simplifies the design of parking layouts. These tables (Figures 5.126, 5.127, 5.128, 5.129, and 5.130) will immediately answer the two questions that most commonly arise when a parking layout is being planned: (1) What parking pattern can be most advantageously fitted to this particular site? (2) If a certain parking angle has been specified, how many parking stalls at this angle can be imposed upon this site, or, alternatively, how big a site will be needed to accommodate a given number of cars parked at a specified angle?

In each table figures are given for five different stall widths from the utmost (and, we believe, shortsighted) economy of 8 feet to the new luxury standard of 10 feet. To fit the site without waste and the maximum number of stalls, a different

a	b	c	d	e	f¹	f²	a	b	c	d	e	f¹	f²
parking angle	stall width	stall to curb [19' long stall]	aisle width	curb length per car	center-to-center width of two-row bin with access road between curb-to-curb	overlap c-c	parking angle	stall width	stall to curb [19' long stall]	aisle width	curb length per car	center-to-center width of two-row bin with access road between curb-to-curb	overlap c-c
0°	8'0"	8.0	12.0	23.0	28.0	—	50°	8'0"	19.7	14.0	10.5	53.4	48.3
	8'6"	8.5	12.0	23.0	29.0	—		8'6"	20.0	12.5	11.1	52.5	47.0
	9'0"	9.0	12.0	23.0	30.0	—		9'0"	20.4	12.0	11.7	52.8	47.0
	9'6"	9.5	12.0	23.0	31.0	—		9'6"	20.7	12.0	12.4	53.4	47.3
	10'0"	10.0	12.0	23.0	32.0	—		10'0"	21.0	12.0	13.1	54.0	47.6
20°	8'0"	14.0	11.0	23.4	39.0	31.5	60°	8'0"	20.4	19.0	9.2	59.8	55.8
	8'6"	14.5	11.0	24.9	40.0	32.0		8'6"	20.7	18.5	9.8	59.9	55.6
	9'0"	15.0	11.0	26.3	41.0	32.5		9'0"	21.0	18.0	10.4	60.0	55.5
	9'6"	15.5	11.0	27.8	42.0	33.1		9'6"	21.2	18.0	11.0	60.4	55.6
	10'0"	15.9	11.0	29.2	42.8	33.4		10'0"	21.5	18.0	11.5	61.0	56.0
30°	8'0"	16.5	11.0	16.0	44.0	37.1	70°	8'0"	20.6	20.0	8.5	61.2	58.5
	8'6"	16.9	11.0	17.0	44.8	37.4		8'6"	20.8	19.5	9.0	61.1	58.2
	9'0"	17.3	11.0	18.0	45.6	37.8		9'0"	21.0	19.0	9.6	61.0	57.9
	9'6"	17.8	11.0	19.0	46.6	38.4		9'6"	21.2	18.5	10.1	60.9	57.7
	10'0"	18.2	11.0	20.0	47.4	38.7		10'0"	21.2	18.0	10.6	60.4	57.0
40°	8'0"	18.3	13.0	12.4	49.6	43.5	80°	8'0"	20.1	25.0*	8.1	65.2	63.8
	8'6"	18.7	12.0	13.2	49.4	42.9		8'6"	20.2	24.0*	8.6	64.4	62.9
	9'0"	19.1	12.0	14.0	50.2	43.3		9'0"	20.3	24.0*	9.1	64.3	62.7
	9'6"	19.5	12.0	14.8	51.0	43.7		9'6"	20.4	24.0*	9.6	64.4	62.7
	10'0"	19.9	12.0	15.6	51.8	44.1		10'0"	20.5	24.0*	10.2	65.0	63.3
45°	8'0"	19.1	14.0	11.3	52.2	46.5	90°	8'0"	19.0	26.0*	8.0	64.0	—
	8'6"	19.4	13.5	12.0	52.3	46.3		8'6"	19.0	25.0*	8.5	63.0	—
	9'0"	19.8	13.0	12.7	52.6	46.2		9'0"	19.0	24.0*	9.0	62.0	—
	9'6"	20.1	13.0	13.4	53.2	46.5		9'6"	19.0	24.0*	9.5	62.0	—
	10'0"	20.5	13.0	14.1	54.0	46.9		10'0"	19.0	24.0*	10.0	62.0	—
							90° Back in†	8'0"	18.5	22.0*	8.0	59.0	—
								8'6"	18.5	21.0*	8.5	58.0	—
								9'0"	18.5	20.0*	9.0	57.0	—

* Two-way circulation.
† For attendant parking only. Two-way traffic in aisles possible, but not desirable.

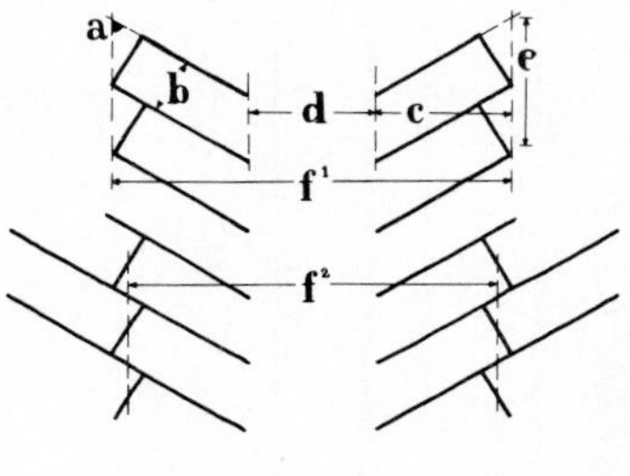

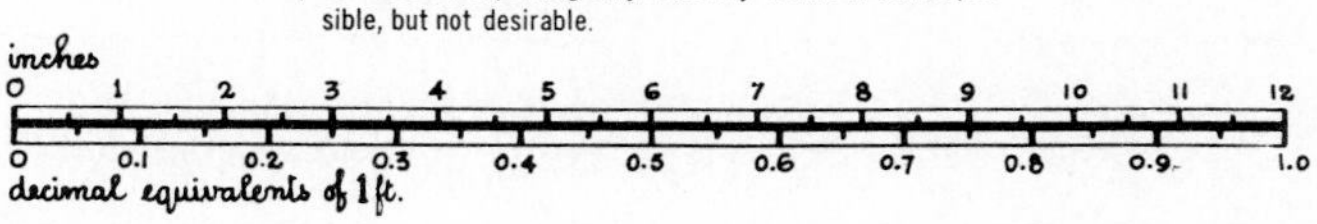

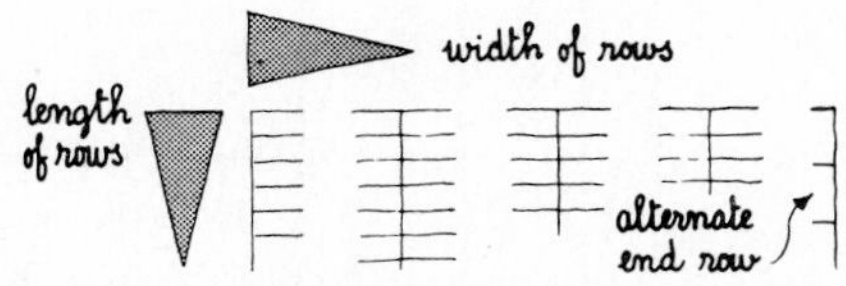

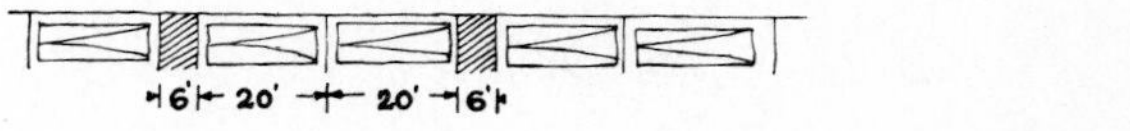

Figure 5.126 Criteria for parking lot.

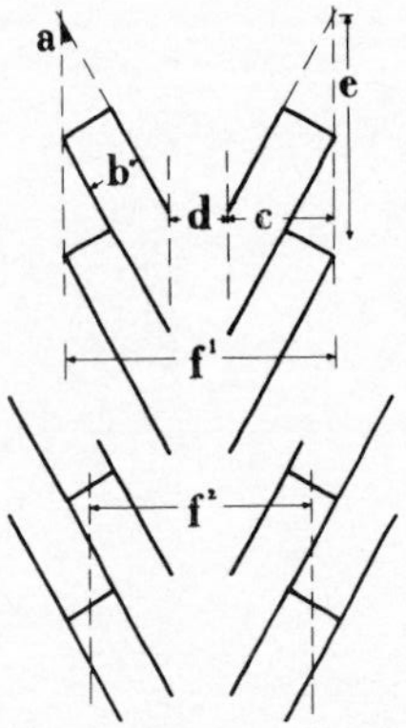

TOTAL LENGTH

NO. OF STALLS	SINGLE ROW — WIDTH OF STALL: 8'0"	8'6"	9'0"	9'6"	10'0"	OVERLAP — WIDTH OF STALL: 8'0"	8'6"	9'0"	9'6"	10'0"
1	20'6"	20'6"	21'0"	21'0"	21'6"	20'0"	21'6"	22'6"	24'0"	25'0"
2	36'6"	37'6"	39'0"	40'0"	41'6"	36'6"	38'0"	39'0"	40'6"	41'6"
3	52'6"	54'6"	57'0"	59'0"	61'6"	36'6"	38'6"	40'6"	43'0"	45'0"
4	68'6"	71'6"	75'0"	78'0"	81'6"	52'6"	55'0"	57'0"	59'6"	61'6"
5	84'6"	88'6"	93'0"	97'0"	101'6"	52'0"	55'6"	58'6"	62'0"	65'0"
6	100'6"	105'6"	111'0"	116'0"	121'6"	68'6"	72'0"	75'0"	78'6"	81'6"
7	116'6"	122'6"	129'0"	135'0"	141'6"	68'0"	72'6"	76'6"	81'0"	85'0"
8	132'6"	139'6"	147'0"	154'0"	161'6"	84'6"	89'0"	93'0"	97'6"	101'6"
9	148'6"	156'6"	165'0"	173'0"	181'6"	84'0"	89'6"	94'6"	100'0"	105'0"
10	164'6"	173'6"	183'0"	192'0"	201'6"	100'6"	106'0"	111'0"	116'6"	121'6"
11	180'6"	190'6"	201'0"	211'0"	221'6"	100'0"	106'6"	112'6"	119'0"	125'0"
12	196'6"	207'6"	219'0"	230'0"	241'6"	116'6"	123'0"	129'0"	135'6"	141'6"
13	212'6"	224'6"	237'0"	249'0"	261'6"	116'0"	123'6"	130'6"	138'0"	145'0"
14	228'6"	241'6"	255'0"	268'0"	281'6"	132'6"	140'0"	147'0"	154'6"	161'6"
15	244'6"	258'6"	273'0"	287'0"	301'6"	132'0"	140'6"	148'6"	157'0"	165'0"
16	260'6"	275'6"	291'0"	306'0"	321'6"	148'6"	157'0"	165'0"	173'6"	181'6"
17	276'6"	292'6"	309'0"	325'0"	341'6"	148'0"	157'6"	166'6"	176'0"	185'0"
18	292'6"	309'6"	327'0"	344'0"	361'6"	164'6"	174'0"	183'0"	192'6"	201'6"
19	308'6"	326'6"	345'0"	363'0"	381'6"	164'0"	174'6"	184'6"	195'0"	205'0"
20	324'6"	343'6"	363'0"	382'0"	401'6"					

a	b	c	d	e	f¹	f²
parking angle	stall width	stall to curb [19' long stall]	aisle width	curb length per car	center-to-center width of two-row bin with access road between — curb-to-curb	overlap c-c
30°	8'0"	16.5	11.0	16.0	44.0	37.1
	8'6"	16.9	11.0	17.0	44.8	37.4
	9'0"	17.3	11.0	18.0	45.6	37.8
	9'6"	17.8	11.0	19.0	46.6	38.4
	10'0"	18.2	11.0	20.0	47.4	38.7

ALTERNATE END ROWS FOR 30°
Change in overall width of parking:

With one end row of:	WIDTH OF STALL (same in end row as in rest of parking) 8'0"	8'6"	9'0"	9'6"	10'0"
0°	− 7'6"	− 7'6"	− 7'6"	− 7'6"	− 7'0"
45°	+ 5'6"	+ 5'0"	+ 4'6"	+ 4'6"	+ 4'6"
60°	+12'0"	+11'6"	+11'0"	+10'6"	+10'6"
90°	+17'6"	+16'0"	+16'0"	+14'6"	+14'0"

TOTAL WIDTH (including access roads)

STALL WIDTH	AISLE WIDTH	1	2	3	4	5	6	7	8	9	10	11	12	13	14	15	16	17	18	19	20
8'0"	11'0"	27'6"	44'0"	64'6"	81'0"	101'6"	118'0"	139'0"	155'6"	176'0"	192'6"	213'0"	229'6"	250'0"	266'6"	287'0"	303'6"	324'0"	341'0"	361'6"	378'0"
8'6"	11'0"	28'0"	45'0"	65'6"	82'0"	102'6"	119'6"	140'0"	157'0"	177'6"	194'6"	215'0"	232'0"	252'6"	269'0"	290'0"	306'6"	327'0"	344'0"	364'6"	381'6"
9'0"	11'0"	28'6"	45'6"	66'0"	83'6"	104'0"	121'0"	142'0"	159'0"	179'6"	197'0"	217'6"	234'6"	255'0"	272'6"	293'0"	310'6"	330'0"	348'0"	368'6"	385'0"
9'6"	11'0"	29'0"	46'6"	67'0"	85'0"	105'6"	123'6"	144'0"	162'0"	182'6"	200'0"	221'0"	238'6"	259'0"	277'0"	297'6"	315'6"	336'0"	354'0"	374'6"	392'0"
10'0"	11'0"	29'0"	47'6"	68'0"	86'0"	106'6"	125'0"	145'6"	163'6"	184'0"	202'0"	223'0"	241'0"	261'6"	279'6"	300'0"	318'6"	339'0"	357'0"	377'6"	395'6"

All dimensions in feet and inches are to nearest 6 inches

Figure 5.127 Tables for 30-degree parking.

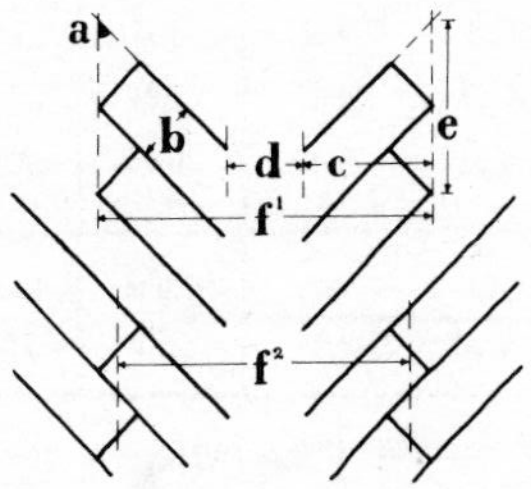

Figure 5.128 Tables for 45-degree parking.

All dimensions in feet and inches are to nearest 6 inches

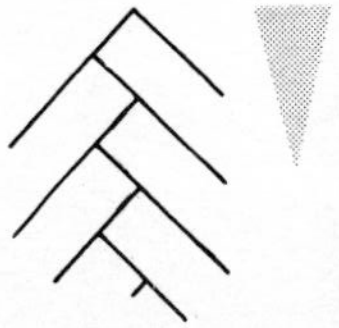

45°

a	b	c	d	e	f¹ curb-to-curb	f² overlap c-c
parking angle	stall width	stall to curb [19' long stall]	aisle width	curb length per car	center-to-center width of two-row bin with access road between	
45°	8'0"	19.1	14.0	11.3	52.2	46.5
	8'6"	19.4	13.5	12.0	52.3	46.3
	9'0"	19.8	13.0	12.7	52.6	46.2
	9'6"	20.1	13.0	13.4	53.2	46.5
	10'0"	20.5	13.0	14.1	54.0	46.9

ALTERNATE END ROWS FOR 45°.
Change in overall width of parking:

With one end row of:	WIDTH OF STALL (same in end row as in rest of parking)				
	8'0"	8'6"	9'0"	9'6"	10'0"
0°	−11'0"	−11'0"	−11'0"	−10'6"	−10'6"
30°	−2'6"	−2'6"	−2'6"	−2'6"	−2'6"
60°	+6'6"	+6'6"	+6'0"	+6'0"	+6'0"
90°	+12'0"	+11'0"	+10'0"	+10'0"	+9'6"

HERRINGBONE for odd number of stalls, opposite page

	WIDTH OF STALL				
	8'0"	8'6"	9'0"	9'6"	10'0"
for odd number of stalls = same as overlap table					
for even number of stalls add to overlap table:	+7'6"	+7'6"	+7'0"	+6'6"	+6'0"

TOTAL LENGTH

NO. OF STALLS	SINGLE ROW WIDTH OF STALL:					OVERLAP WIDTH OF STALL:				
	8'0"	8'6"	9'0"	9'6"	10'0"	8'0"	8'6"	9'0"	9'6"	10'0"
1	19'0"	19'6"	19'6"	20'0"	20'6"	17'0"	18'0"	19'0"	20'0"	21'0"
2	30'6"	31'6"	32'6"	33'6"	34'6"	30'6"	31'6"	32'6"	33'6"	34'6"
3	41'6"	43'6"	45'0"	47'0"	48'6"	28'6"	30'0"	32'0"	33'6"	35'6"
4	53'0"	55'6"	58'0"	60'6"	62'6"	41'6"	43'6"	45'0"	47'0"	48'6"
5	64'0"	67'6"	70'6"	73'6"	77'0"	39'6"	42'0"	44'6"	47'0"	49'6"
6	75'6"	79'6"	83'0"	87'0"	91'0"	53'0"	55'6"	58'0"	60'6"	63'0"
7	87'0"	91'6"	96'0"	100'6"	105'0"	51'0"	54'0"	57'0"	60'6"	63'6"
8	98'0"	103'6"	108'6"	114'0"	119'0"	64'6"	67'6"	70'6"	73'6"	77'0"
9	109'6"	115'6"	121'6"	127'6"	133'0"	62'0"	66'0"	70'0"	73'6"	77'6"
10	120'6"	127'6"	134'0"	140'6"	147'6"	75'6"	79'6"	83'6"	87'0"	91'0"
11	132'0"	139'6"	146'6"	154'0"	161'6"	73'6"	78'0"	82'6"	87'0"	91'6"
12	143'6"	151'6"	159'6"	167'6"	175'6"	87'0"	91'6"	96'0"	100'6"	105'0"
13	154'6"	163'6"	172'0"	181'0"	189'6"	85'0"	90'0"	95'6"	100'6"	106'0"
14	166'0"	175'6"	185'0"	194'6"	203'6"	98'0"	103'6"	108'6"	114'0"	119'0"
15	177'0"	187'6"	197'6"	207'6"	218'0"	96'0"	102'0"	108'0"	114'0"	120'0"
16	188'6"	199'6"	210'0"	221'0"	232'0"	109'6"	115'6"	121'6"	127'6"	133'6"
17	200'0"	211'6"	223'0"	234'6"	246'0"	107'6"	114'0"	120'6"	127'6"	134'0"
18	211'0"	223'6"	235'6"	248'0"	260'0"	121'0"	127'6"	134'0"	140'6"	147'6"
19	222'6"	235'6"	248'6"	261'6"	274'0"	118'6"	126'0"	133'6"	140'6"	148'0"
20	233'6"	247'6"	261'0"	274'6"	288'6"					

TOTAL WIDTH (including access roads)

STALL WIDTH	AISLE WIDTH	NUMBER OF ROWS																			
		1	2	3	4	5	6	7	8	9	10	11	12	13	14	15	16	17	18	19	20
8'0"	14'0"	33'0"	52'0"	79'6"	98'6"	126'0"	145'0"	172'6"	191'6"	219'0"	238'0"	265'6"	284'6"	312'0"	331'0"	358'6"	377'6"	405'0"	424'0"	451'6"	471'0"
8'6"	13'6"	33'0"	52'6"	79'0"	98'6"	125'6"	145'0"	172'0"	191'0"	218'0"	237'6"	264'6"	284'0"	310'6"	330'0"	357'0"	376'6"	403'6"	422'6"	449'6"	469'0"
9'0"	13'0"	33'0"	52'6"	79'0"	99'0"	125'0"	145'0"	171'6"	191'0"	217'6"	237'6"	264'0"	283'6"	310'0"	330'0"	356'0"	376'0"	402'6"	422'0"	448'6"	468'6"
9'6"	13'0"	33'0"	53'0"	79'6"	99'6"	126'0"	146'0"	172'6"	192'6"	219'0"	239'0"	265'6"	285'6"	312'0"	332'0"	358'6"	378'6"	405'0"	425'0"	451'6"	471'6"
10'0"	13'0"	33'6"	54'0"	80'6"	101'0"	127'6"	148'0"	174'0"	194'6"	221'0"	241'0"	268'0"	288'6"	315'0"	335'6"	362'0"	382'6"	408'6"	429'0"	455'6"	476'0"

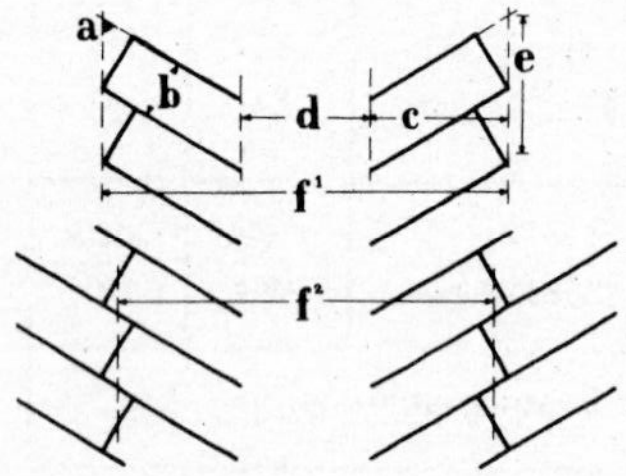

TOTAL LENGTH

NO. OF STALLS	SINGLE ROW — WIDTH OF STALL					OVERLAP				
	8'0"	8'6"	9'0"	9'6"	10'0"	8'0"	8'6"	9'0"	9'6"	10'0"
1	16'6"	17'0"	17'6"	17'6"	18'0"	16'0"	17'0"	18'0"	19'0"	20'0"
2	25'6"	26'6"	27'6"	28'6"	29'6"	25'6"	26'6"	27'6"	28'6"	29'6"
3	35'0"	36'6"	38'0"	39'6"	41'0"	25'6"	27'0"	28'6"	30'0"	31'6"
4	44'0"	46'6"	48'6"	50'6"	52'6"	35'0"	36'6"	38'0"	39'6"	41'0"
5	53'0"	56'0"	59'0"	61'6"	64'0"	34'6"	37'0"	39'0"	41'0"	43'0"
6	62'6"	66'0"	69'6"	72'6"	75'6"	44'0"	46'6"	48'6"	50'6"	52'6"
7	71'6"	75'6"	79'6"	83'6"	87'0"	43'6"	46'6"	49'6"	52'0"	54'6"
8	81'0"	85'6"	90'0"	94'6"	98'6"	53'6"	56'0"	59'0"	61'6"	64'0"
9	90'0"	95'6"	100'6"	105'6"	110'0"	53'0"	56'6"	60'0"	63'0"	66'0"
10	99'0"	105'0"	111'0"	116'6"	121'6"	62'6"	66'0"	69'6"	72'6"	75'6"
11	108'6"	115'0"	121'6"	127'6"	133'0"	62'0"	66'0"	70'0"	74'0"	77'6"
12	117'6"	124'6"	131'6"	138'6"	144'6"	71'6"	75'6"	79'6"	83'6"	87'0"
13	127'0"	134'6"	142'0"	149'6"	156'0"	71'6"	76'0"	80'6"	85'0"	89'0"
14	136'0"	144'6"	152'6"	160'6"	167'6"	81'0"	85'6"	90'0"	94'6"	98'6"
15	145'0"	154'0"	163'0"	171'6"	179'0"	81'0"	86'0"	91'0"	96'0"	100'6"
16	154'6"	164'0"	173'6"	182'6"	190'6"	80'6"	86'0"	91'0"	96'0"	100'6"
17	163'6"	173'6"	183'6"	193'6"	202'0"	90'0"	95'6"	100'6"	105'6"	110'0"
18	173'0"	183'6"	194'0"	204'6"	213'6"	89'6"	95'6"	101'6"	107'0"	112'6"
19	182'0"	193'6"	204'6"	215'6"	225'0"	99'0"	105'0"	111'0"	116'6"	121'6"
20	191'0"	203'0"	215'0"	226'6"	236'6"	99'0"	105'6"	112'0"	118'0"	123'6"

60°

parking angle — a: stall width	b: stall to curb [19' long stall]	c: aisle width	d: curb length per car	center-to-center width of two-row bin with access road — f¹: curb-to-curb	f²: overlap between
8'0"	20.4	19.0	9.2	59.8	55.8
8'6"	20.7	18.5	9.8	59.9	55.6
9'0"	21.0	18.0	10.4	60.0	55.5
9'6"	21.2	18.0	11.0	60.4	55.6
10'0"	21.5	18.0	11.5	61.0	56.0

ALTERNATE END ROWS FOR 60°

Change in overall width of parking:

With one end row of:	WIDTH OF STALL (same in end row as in rest of parking)				
	8'0"	8'6"	9'0"	9'6"	10'0"
0°	−12'6"	−12'0"	−12'0"	−11'6"	−11'6"
30°	−4'0"	−4'0"	−3'6"	−3'6"	−3'6"
45°	−1'6"	−1'6"	−1'0"	−1'0"	−1'0"
90°	+5'6"	+5'6"	+5'6"	+4'0"	+3'6"

TOTAL WIDTH (including access roads)

STALL WIDTH	AISLE WIDTH	NUMBER OF ROWS																			
		1	2	3	4	5	6	7	8	9	10	11	12	13	14	15	16	17	18	19	20
8'0"	19'0"	39'6"	60'0"	95'0"	115'6"	151'0"	171'6"	207'0"	227'0"	262'6"	283'0"	318'6"	339'0"	374'0"	394'6"	430'0"	450'6"	486'0"	506'0"	541'6"	562'0"
8'6"	18'6"	39'0"	60'0"	95'0"	115'6"	150'6"	171'6"	206'6"	227'0"	262'0"	282'6"	317'6"	338'6"	373'6"	394'0"	429'0"	450'0"	485'0"	505'6"	540'6"	561'0"
9'0"	18'0"	39'0"	60'0"	94'6"	115'6"	150'0"	171'0"	205'6"	226'6"	261'0"	282'0"	316'6"	337'6"	372'0"	393'0"	427'6"	448'6"	483'0"	504'0"	538'6"	559'6"
9'6"	18'0"	39'0"	60'6"	95'0"	116'0"	150'6"	172'0"	206'6"	227'6"	262'0"	283'0"	317'6"	339'0"	373'6"	394'6"	429'0"	450'6"	485'0"	506'0"	540'6"	560'6"
10'0"	18'0"	39'6"	61'0"	95'6"	117'0"	151'6"	173'0"	207'6"	229'0"	263'6"	285'0"	319'6"	341'0"	375'6"	397'0"	431'6"	453'0"	487'6"	509'0"	543'6"	565'0"

All dimensions in feet and inches are to nearest 6 inches

Figure 5.129 Tables for 50-degree parking.

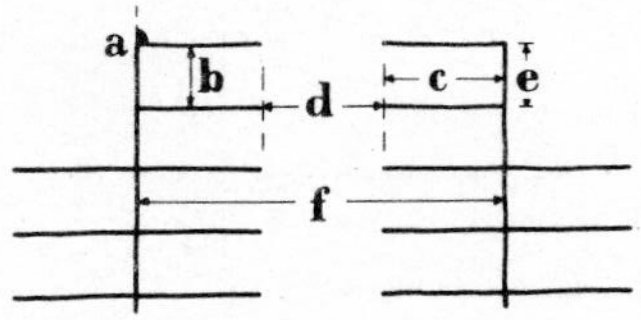

TOTAL LENGTH

NO. OF STALLS	WIDTH OF STALL: 8'0"	8'6"	9'0"	9'6"	10'0"
1	8'0"	8'6"	9'0"	9'6"	10'0"
2	16'0"	17'0"	18'0"	19'0"	20'0"
3	24'0"	25'6"	27'0"	28'6"	30'0"
4	32'0"	34'0"	36'0"	38'0"	40'0"
5	40'0"	42'6"	45'0"	47'6"	50'0"
6	48'0"	51'0"	54'0"	57'0"	60'0"
7	56'0"	59'6"	63'0"	66'6"	70'0"
8	64'0"	68'0"	72'0"	76'0"	80'0"
9	72'0"	76'6"	81'0"	85'6"	90'0"
10	80'0"	85'0"	90'0"	95'0"	100'0"
11	88'0"	93'6"	99'0"	104'6"	110'0"
12	96'0"	102'0"	108'0"	114'0"	120'0"
13	104'0"	110'6"	117'0"	123'6"	130'0"
14	112'0"	119'0"	126'0"	133'0"	140'0"
15	120'0"	127'6"	135'0"	142'6"	150'0"
16	128'0"	136'0"	144'0"	152'0"	160'0"
17	136'0"	144'6"	153'0"	161'6"	170'0"
18	144'0"	153'0"	162'0"	171'0"	180'0"
19	152'0"	161'6"	171'0"	180'6"	190'0"
20	160'0"	170'0"	180'0"	190'0"	200'0"

a parking angle	b stall width	c stall to curb [19' long stall]	d aisle width	e curb length per car	f¹ center-to-center width of two-row bin with access road between curb-to-curb	f² overlap c-c
90°	8'0"	19.0	26.0*	8.0	64.0	—
	8'6"	19.0	25.0*	8.5	63.0	—
	9'0"	19.0	24.0*	9.0	62.0	—
	9'6"	19.0	24.0*	9.5	62.0	—
	10'0"	19.0	24.0*	10.0	62.0	—

ALTERNATE END ROWS FOR 90°
Change in overall width of parking:

With one end row of:	WIDTH OF STALL (same in end row as in rest of parking) 8'0"	8'6"	9'0"	9'6"	10'0"
0°	—11'0"	—10'6"	—10'0"	—9'6"	—9'0"
30°	—2'6"	—2'0"	—1'6"	—1'0"	—1'0"
45° } 60° }	Fewer cars in more width, so not worth consideration.				

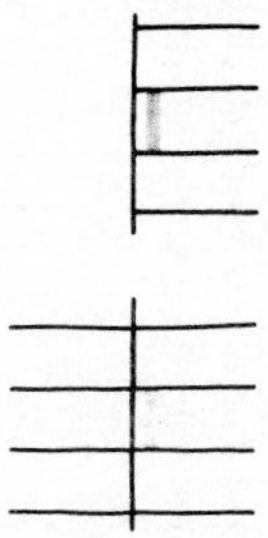

90° BACK IN*
Total length is same as ordinary

All dimensions in feet and inches are to nearest 6 inches

TOTAL WIDTH (including access roads)

STALL WIDTH	AISLE WIDTH	NUMBER OF ROWS 1	2	3	4	5	6	7	8	9	10	11	12	13	14	15	16	17	18	19	20
8'0"	26'0"	45'0"	64'0"	109'0"	128'0"	173'0"	192'0"	237'0"	256'0"	301'0"	320'0"	365'0"	384'0"	429'0"	448'0"	493'0"	512'0"	557'0"	576'0"	621'0"	640'0"
8'6"	25'0"	44'0"	63'0"	107'0"	126'0"	170'0"	189'0"	233'0"	252'0"	296'0"	315'0"	359'0"	378'0"	422'0"	441'0"	485'0"	504'0"	548'0"	567'0"	611'0"	630'0"
9'0"	24'0"	43'0"	62'0"	105'0"	124'0"	167'0"	186'0"	229'0"	248'0"	291'0"	310'0"	353'0"	372'0"	415'0"	434'0"	477'0"	496'0"	539'0"	558'0"	601'0"	620'0"
9'6"	24'0"	43'0"	62'0"	105'0"	124'0"	167'0"	186'0"	229'0"	248'0"	291'0"	310'0"	353'0"	372'0"	415'0"	434'0"	477'0"	496'0"	539'0"	558'0"	601'0"	620'0"
10'0"	24'0"	43'0"	62'0"	105'0"	124'0"	167'0"	186'0"	229'0"	248'0"	291'0"	310'0"	353'0"	372'0"	415'0"	434'0"	477'0"	496'0"	539'0"	558'0"	601'0"	620'0"
90° BACK IN* 8'''0	22'0"	40'6"	59'0"	99'6"	118'0"	158'6"	177'0"	217'6"	236'0"	276'6"	295'0"	335'6"	354'0"	394'6"	413'0"	453'6"	472'0"	512'6"	531'0"	571'6"	590'0"
8'6"	21'0"	39'6"	58'0"	97'6"	116'0"	155'6"	174'0"	213'6"	232'0"	271'6"	290'0"	329'6"	348'0"	387'6"	406'0"	445'6"	464'0"	503'6"	522'0"	561'6"	580'0"
9'0"	20'0"	38'6"	57'0"	95'6"	114'0"	152'6"	171'0"	209'6"	228'0"	266'6"	285'0"	323'6"	342'0"	380'6"	399'0"	437'6"	456'0"	494'6"	513'0"	551'6"	570'0"

* For attendant parking only

Figure 5.130 Tables for 90-degree parking.

parking angle may be used in one or both boundary rows. A table in each group shows the effect of these adjustments on the overall width.

Finally, for those who already know what they want and how to arrive at it, the tables can save a great deal of tiresome figuring. We have found them particularly useful for testing a number of alternative proposals in the early discussion stages of design.

The parking pattern is always considered to be viewed (as it normally will be by the designer) when looking along the length of the parking aisles, which should point toward the direction in which most parkers want to go when they leave their cars.

The large gray arrows on each page of tables refer to length and width as shown in the drawing on page 517. Figures in the width tables always include the necessary aisles. The length tables do not include the circulation road along the row ends. The minimum width of this road might be taken as 24 feet.

Width of aisle and width of stall are interrelated. With self-parking particularly, a narrower stall will often require a wider aisle if the drivers are to turn into a stall in a single sweep without damage or delay. For this reason, as the tables show, a narrow stall may actually take more space per unit than one more generously proportioned. Stall width has a comparatively slight effect on the overall area.

Complete tables are given for 30-, 45-, 60-, and 90-degree parking. For those who wish to investigate the subject more extensively, basic dimensions for every 10 degrees appear in Figure 5.126. One subtlety that cannot be shown in a table is an advantageous way of improving the 0-degree (parallel) parking pattern by pairing the stalls, as shown in the small drawing in the figure. The allowance per car, 23 feet in length, remains the same, but cars can park and unpark more quickly, and with less traffic delay on the street; and pedestrians can pass through the line of parked cars.

Figure 5.127 shows tables for 30-degree parking.
Figure 5.128 shows tables for 45-degree parking.
Figure 5.129 shows tables for 60-degree parking.
Figure 5.130 shows tables for 90-degree parking.

5.11 Facility Modifications for the Handicapped

Facility type	Specific standards	Design considerations	Locational criteria
Parking areas	Car space of min width 12 ft	Areas should be level, paved, and sheltered from wind, rain	Close to building entrances; adapted for special use
Walkways	Min width of 5 ft and gradient less than 5%	Surface material should be firm, not slick Level throughout	Every public area/building
Building entrances/doors	Min doorway opening 2 ft 6 in	All doors should open easily Automatically operated doors timed for slower movement of wheelchair/crutches	Accessible entrances should permit entry to main parts of building
Floor areas	All floor areas should be maintained in nonslip condition	Floor coverings should permit easy wheelchair movement All rugs secured to floor	
Stairs	Landing should be used to reduce length of stair flights	Stairs should be well lighted	An alternative to stairs should be provided nearby
Ramps	Min width of 3 ft Gradient 1:12 or less Level platform at top of ramp at least 5 × 5 ft	Nonslip surface Outside ramps protected from rain and snow	Within view of main entrance or well marked All public buildings and areas should be accessible by ramps
Railings	At least 1½ in required between rail and wall 32 in from floor	Solidly supported	On both sides of all stairs At least one side of ramps and docks
Drinking fountains and telephones	Low enough for use by children as well as adults, max 45 in from ground	Controls within reach and easily operable	All fountains and phones in public areas accessible
Seating/furniture: Outdoor park and picnic areas	Permanent outdoor cooking facilities should have variable height	Park benches should be easy to use and comfortable Picnic tables adapted for use by wheelchair-bound persons	Provide enough benches at convenient locations in areas heavily used by the elderly Locations of adapted picnic tables clearly marked

Facility type	Specific standards	Design considerations	Locational criteria
Indoor theaters, auditoriums, and arenas	At least two flat areas near exits, elevators, ramps with min size 50 × 32 in	Movable portable seats to allow wheelchair-bound persons to sit with companions	In all theaters, auditoriums, or public centers
Swimming pools/showers	At least 1 community pool adapted Shower stalls at least 3 × 4 ft deep	Adapted pools Ramp with side rails Swing lift from wheelchair to water, ramp, and stairs Showers with grab bars and folding seat Water wheelchairs	Towel rentals and lockers easily accessible No lips or coverings between deck, showers, drying room.
Washroom facilities	1 toilet stall large enough for wheelchair, min 3 ft wide by 4⅔ ft deep; 32 in swing-out doors with handrails 33 in high	Wash basins and accessories within reach	

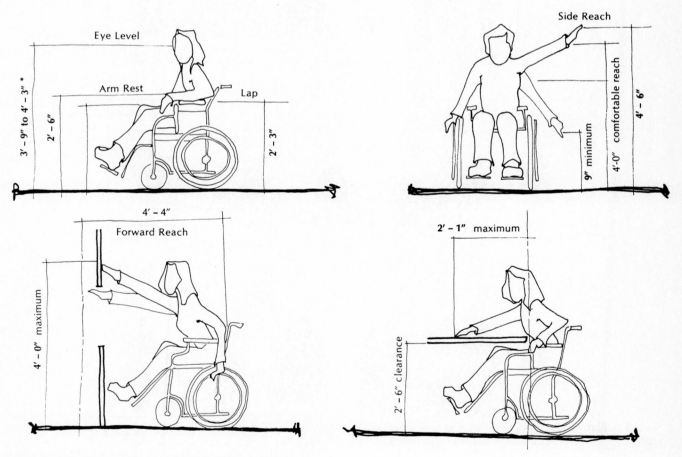

Figure 5.131 Human dimensions for person in wheelchair.

5.12 Dimensions and Clearances for People Involved in Outdoor Activities

The data in Figure 5.132 have been gathered and condensed from a myriad of reports. Although the dimensions have been determined by methods other than anatomical research, the information presented synthesizes the varying and sometimes contradictory recommendations published by the wide number of sources. These specific dimensions best represent the collective average of the recommendations found in different publications. Accordingly, they should be

used with the understanding that the dimensions are not finite or absolute but rather are general guidelines.

5.13 Pathways and Trails—Handicapped

Provision of recreation trails, like all recreation facilities, should reflect the range of people's preferences and abilities. A classification system designed to provide such a spectrum can be used by recreation planners and managers during initial park design or later reassessment of facilities.

Dimensions for People Outdoors

Figure 5.132 Dimensions for people at outdoor activities.

Trails are rated Class I through V based on criteria of width, slope, surface preparation, cross slope, and trail edge. Benefits to both users and providers of such a system include:

- Encouraging individual choice of a trail based on the degree of difficulty one can expect to encounter

- Eliminating stigmatizing labels, i.e., senior citizens/handicapped trail

- Providing an easy framework for inventory of the total system of recreational trails offered

This type of classification system can be a very flexible planning tool.

Trail Surfaces In large part, the surface of a trail determines its accessibility. Balancing topography with the need to provide a wide range of trail accessibility, one can choose from a variety of trail surfaces. In order of decreasing accessibility, possible choices include concrete; asphalt; wooden planking (over wet, fragile, or sandy areas); solidly packed, fine crushed rock; well-compacted pea gravel; bound wood chips; coarse gravel; rock; unbound woodchips; sand. By using surfaces in combination with other features such as slope, trail width, and rest areas, recreation providers can satisfy a large and extremely varied constituency.

In developed areas build either Class I or Class II trails depending on frequency and type of use.

If ramping or switchbacks are needed, provide stairs for alternate and more direct route.

Stairs should have rounded nosings. Treads should be a minimum of 11 inches from nosing to nosing. Risers should be no higher than 7 inches and no less than 4 inches.

Cross slope of pathways should be just enough to provide drainage (about 2 percent).

Expansion joints should be less than ½ inch.

Gratings for storm drainage should be placed off of pathways. Where they do occur, openings should be less than ½ inch.

Prune tree branches overhanging pathway to a height of 8 feet 6 inches.

If pruning will damage or detract from the tree, consider rerouting the path around it.

Provide firm, wide paths leading to amphitheater and support facilities (restrooms, parking, concessions).

Surface of pathways should be slip-resistant under wet and dry conditions.

Whenever possible, grade site to provide access at several different seating levels without the use of steps or ramps. Otherwise, provide ramps (in addition to stairs), at least 4 feet wide, with a slope of 1:12 or less.

Provide wheelchair seating in as many different viewing areas as possible.

Cross aisle may be used to seat wheelchairs if it provides 4 feet of clear aisle space behind the chair.

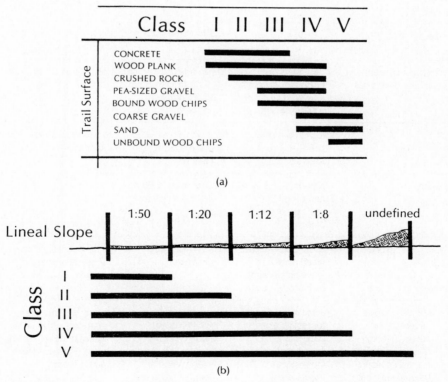

Figure 5.133 Lineal slopes and widths for different classes of pathways and trails are shown above and on page 526.

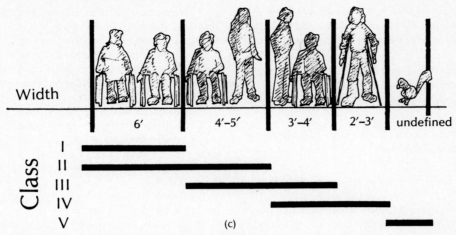

Figure 5.133 Lineal slopes and widths for different classes of pathways and trails are shown above and on page 525 (*Continued*).

Trail Planning Classifications*

Class of trail	I	II	III	IV	V
Approx. length of trail	0–¼ mile	¼–1 mile	1–3 miles	3–10 miles	Over 10 miles
Rest stop spacing and types (use natural materials whenever possible for benches, shelters, etc.)	100–150 ft, benches, shelter, interpretation	200–300 ft, benches,† shelter, interpretation	500–600 ft, natural benches occasionally, interpretation	Rest area or interpretation every 1 mile	None—unless extremely unique interpretation
Width of trail	1-way: 4 ft 2-way: 6 ft	1-way: 3–4 ft 2-way: 4–5 ft	3–4 ft	2–3 ft	Undefined
Shoulder of trail	1½ ft grass; slight slope toward trail	Clear understory brush to 1 ft from trail: gradual slope either direction	Clear understory brush to 1 ft from trail: no abrupt dropoffs adjacent	Clear understory brush to ½ ft from trail	Undefined
Slope of trail	1:50	1:20 with 5 ft level space at 100 ft intervals	1:12 with 5 ft level space at 30 ft intervals	1:8 with occasional level space when possible	Steps or natural
Cross slope	None	1:50 for max. of 30 ft and varied from one side of trail to other	1:25 for max. of 50 ft vary from side to side	1:20	Undefined
Surface of trail	Concrete, asphalt	Asphalt, perpendicular wood planking, very fine crushed rock solidly packed	Firm surface, well compacted	Bound woodchips, class V gravel mixture coarse	Sandy, rough unbound wood chips, rocks
Trail edge (rails, curbs, etc. Use natural materials whenever possible)	Curbs used where necessary for safety; 3 ft high rails for safety or for resting along lineal slope where necessary	Gradual ramping; rails for resting along lineal slope and to provide safety on cross slope or hazard area	Compacted earth level with trail edge; definite texture change. Rails for holding slope at steepest grade and for safety	Texture change with immediate drop to natural terrain from trail edge. Rails used to guard hazard	Nothing

*Courtesy of Minnesota Department of Natural Resources.
†Benches may mean commercial type or a big log or boulders suitable for sitting.

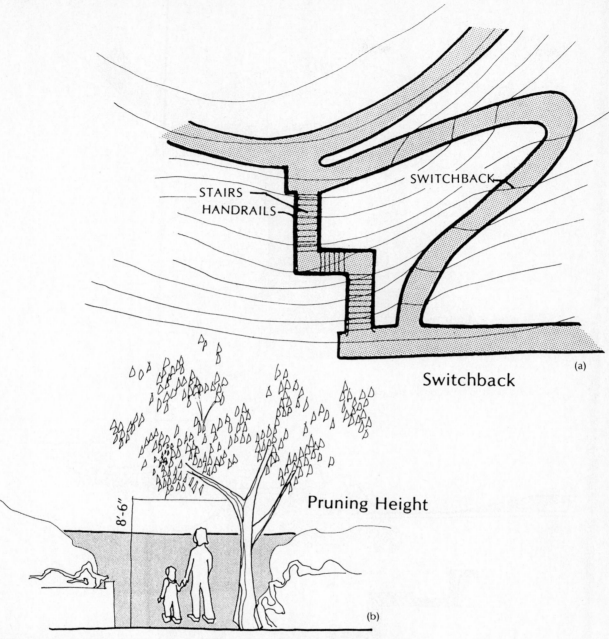

STAIRS
HANDRAILS
SWITCHBACK
Switchback
(a)

Pruning Height
8'-6"
(b)

Figure 5.134   Switchback path: (*a*) stairs and switchback; (*b*) pruning height for trees.

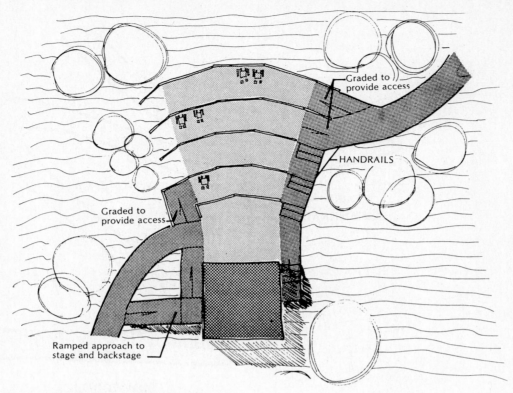

Figure 5.135 Typical amphitheater plan.

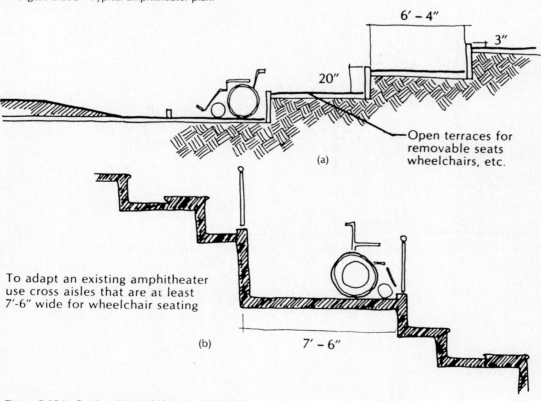

To adapt an existing amphitheater use cross aisles that are at least 7'-6" wide for wheelchair seating

Figure 5.136 Seating: (*a*) amphitheater; (*b*) bench-type.

Swimming

Pathway from Parking to Beach This pathway requires a firm stable path such as concrete, asphalt, or wood plank 4 to 5 feet wide.

Handrails 36 inches high assist path users but may restrict movement of maintenance vehicles.

Pathway from Beach to Water Stabilized sand or wood plank minimizes impact to beach while still providing good access.

A wooden boardwalk built in sections can be used seasonally on beaches with severe winter ice problems.

Repair and maintenance of the boardwalk is usually limited to periodically scraping algae and sand from the path and removing handrailings before winter.

Sand stabilized with hardened clay will need to be reconstructed each season.

Handrails may obstruct view or movement across beach and should be used with discretion.

Entering the Water Provide handrails 36 inches high and sloped curb along edge of platform to gently stop wheelchairs.

The handrailing gives added assurance to people in the water or on the platform and serves as a grab bar for beginning swimmers.

A rubber mat (laid directly on the sand) can also improve access to the water. Although it offers the swimmer less support, it is less subject to water erosion or displacement by waves.

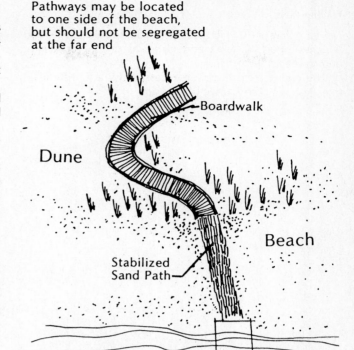

Pathways may be located to one side of the beach, but should not be segregated at the far end

Dune

Boardwalk

Beach

Stabilized Sand Path

Slip-Resistant Platform or Rubber Mat

Figure 5.137 Pathway from parking to beach.

Swimming Platform

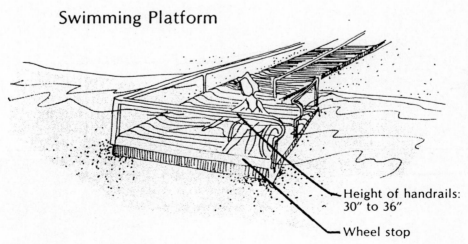

Height of handrails: 30" to 36"

Wheel stop

Figure 5.138 Swimming platform.

Fishing Piers

Design fishing piers which provide shade, space for gear, and benches (optional). Allow enough space for free movement behind seated fishermen.

The fishing area should be accessible by a firm-surfaced trail, flush with the surface of the pier.

An arm rest and bait shelf assists seated fishermen.

Provide a bait shelf, 8 to 12 inches wide, and an arm/pole rest inclined about 30 degrees.

Provide a 4-inch kickplate along the edge of the pier for safety.

Spacing between planks on the deck should be less than ½ inch.

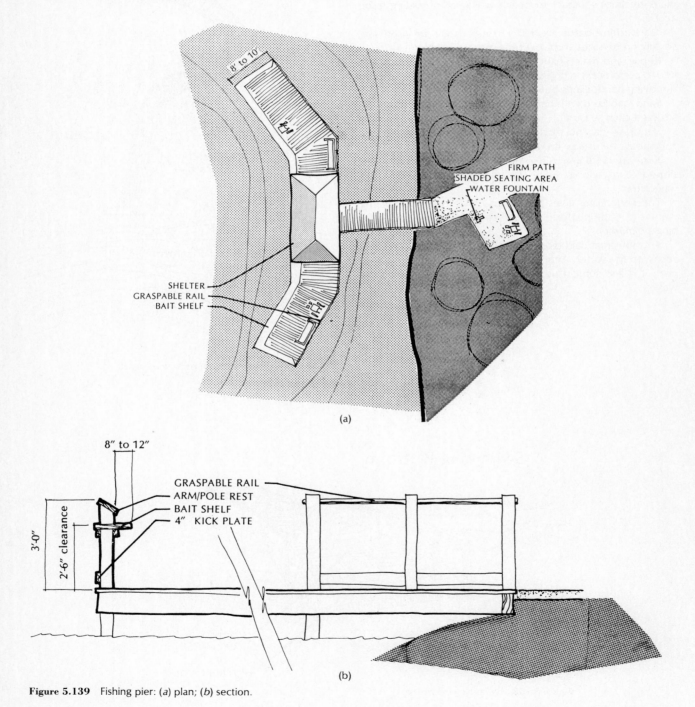

Figure 5.139 Fishing pier: (a) plan; (b) section.

Pools and Docks

Pools   Design considerations such as ramps, handrails, and textured surfaces can be easily incorporated into the modification of older pools and the design of new facilities.

Give people the option of entering the water from a ramp with handrails or from wide steps where they can also sit and relax.

Both stairs and ramps should have slip-resistant surface.

Provide handails 2 feet 6 inches to 3 feet 0 inch high along ramp.

The slope of the ramp should be 1:12 or less.

A ramp 3 feet 0 inch wide allows an individual to grasp both rails at once, whereas a wider ramp allows two people to pass freely.

Use color and textural cues to indicate edge and other high-risk areas.

Use a slip-resistant surface on walkways near the pool.

Dressing rooms, bathrooms, and services should be accessible by firm paths. These facilities should also conform to ANSI standards.

Boat Docks   A properly designed dock provides safety without restricting access to boats.

Access to the dock site should be across a hard-surfaced path.

Ramps should have a slope no greater than 1:12, although this may be difficult to achieve for floating docks in tidal waters.

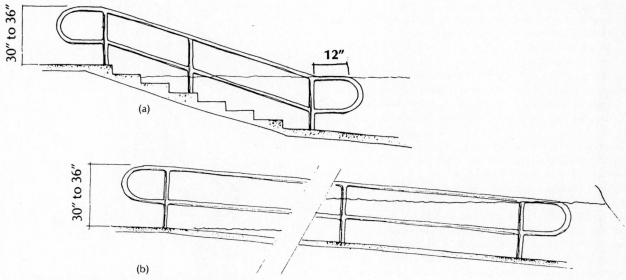

Figure 5.140   Entrance into pool: (*a*) steps; (*b*) ramp.

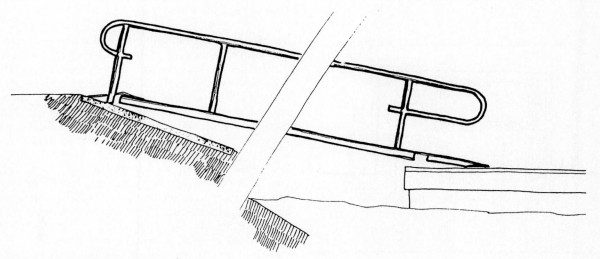

Figure 5.141   Floating dock section.

Ramps 3 to 5 feet wide allow unencumbered access to docks.

Easy-to-grasp safety railings 30 inches to 36 inches high should be provided in all areas where they will not interfere with boating activities.

Sitting Areas and Landscaping

Sitting Area Include a space beside the bench for a wheelchair or stroller (about 30 inches wide for the average wheelchair).

Seating areas should be located adjacent to (but not obstructing) pathways and developed trails, particularly along inclines.

Texture change in walkways adjacent to seating areas will cue the blind to the location of benches.

Benches that contrast in color from surroundings are more easily distinguished by visually impaired people.

Back and arm rests provide comfort and assistance for people who have difficulty standing or sitting.

Drinking Fountain Hand levers are easier to operate than buttons or knobs.

Fountains and faucets should be accessible by firm, level paths and should not protrude into pathways.

In order to accommodate the reach of the average wheelchair user, faucets should be 3 feet 0 inch to 3 feet 4 inches high.

Openings in drains should not exceed ½ inch.

Dual fountains accommodate children and people in wheelchairs as well as standing adults.

A lower fountain should be cantilevered 17 to 19 inches from an upright or wall.

The spout of the lower fountain should be no more than 36 inches high with 27 inch clearance underneath.

Keep in mind that many people with disabilities require water frequently.

Landscaping prior to Occupancy Obviously mature trees are needed to provide shade. Whenever possible, sites should be developed to retain all existing trees and to gain

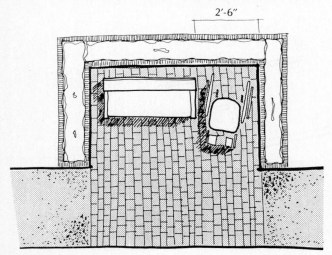

Figure 5.142 Sitting area.

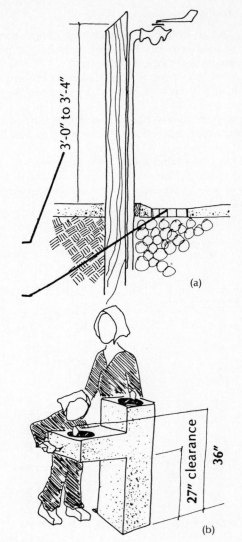

Figure 5.143 Drinking fountain.

the maximum benefit from them. Unfortunately, this is either not done or many project sites have no trees to begin with. If there are no trees on the site, they should be planted as soon as the site is developed and prior to occupancy. This is critical not only because this will permit the plant materials to become established and grow right from the start but, perhaps more importantly, trees and other plantings placed on the site at its inception can either encourage or inhibit certain activities. For example, if an interior court area is provided but left entirely open, it will encourage ball playing and other very active usage which might not be desirable in such a space. It is generally much more difficult to stop an activity already in progress than to prevent it from occurring in the first place through proper planning. In this case, the court area should be landscaped right from the start with trees, masses of shrubs, and possibly earthmounds to discourage ball playing.

Wind Protection

Fences Protection from wintry winds can be provided by fences, plantings massed as hedges or screens, or by earthmounds. In order to be effective, a wind screen must be dense enough to at least retard the force of the wind, high enough to direct the wind over the play area, and in the path of the prevailing winds so as to be effective most of the time. In order for fences to satisfy these requirements, they must be massive. Such a fence would become a dominant visual feature and would be quite expensive. And fences by their very presence invite climbing, battering, and other types of abuse. Therefore, a fence must be built solidly and maintained regularly. A fence in disrepair is a real eyesore.

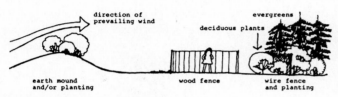

Figure 5.144 Types of windbreaks.

Plantings as Windbreaks Shrubs and trees, on the other hand, can serve as excellent windbreaks and can enhance the appearance of the project site as well. Shrubs do have some disadvantages, however. Unlike a fence, which provides instant protection, it takes some time for any plant material to mature and obtain its desired fullness and height. Unfortunately, when windbreaks are needed most (i.e., from late fall to early spring), many hedges have no leaves and therefore are less effective. Evergreen plants provide a solution to this problem, but they have one major drawback in that they are quite slow-growing. Although many deciduous shrubs such as tall hedge and privet grow several feet per season, the common evergreen shrubs may grow only a few inches each year. One way to provide a quickly established windbreak, as well as an effective year-round screen, is to plant a fast-growing deciduous hedge with a low fence and to back this up with an evergreen planting. The deciduous hedge can serve as a windbreak until the evergreens have matured and are effective.

Earthmounds Earthmounds can be completely effective as windbreaks if properly placed. They also provide many other benefits to the site. Children love to climb, slide, and roll up and down hills no matter how large or small. Earthmounds not only shelter play areas, they screen them from view. Therefore, the mounds can enhance the attractiveness of a site by adding changes in elevation while minimizing the visual impact of a play area.

Ideally mounds should be formed as the site is developed by using soil excavated from building foundations. This obviously saves the cost of transporting waste soil away from the site. On developed sites, good clean fill would have to be brought in, or open areas could be excavated and mounds

formed from the excavated soil. If this is done, careful consideration would have to be given to potential drainage problems; but the concept is quite feasible and worthy of consideration if open areas exist on the site.

For mounds to be effective as windbreaks, they should be at least as high as the people who will be using the play area. If the mounds are to be covered in turf, they should have no more than a 3:1 slope. On steeper slopes, retaining walls or terraces can be used to stabilize the soil. Terraces can be effectively used to provide seating areas on the sides of the mounds adjacent to the play area.

Figure 5.145 Earthmounds can serve many functions.

Planting Several important considerations must be made when planting trees and shrubs in and around play areas. No plant can be guaranteed to survive the onslaught of groups of children, but chances for survival may be greater if the following concepts are followed:

Plant the most mature trees and shrubs your budget will permit. Whips will seldom survive abuse by children and will take far too long to become effective.

Plant trees and shrubs in masses rather than scattered singly all over the site. This will maximize their visual impact in the site and lessen the chance of their being trampled or abused.

If plantings are to be placed across an obvious circulation route, do not try to make a shrub serve as a fence—it will certainly be trampled. Instead, provide an opening in the planting and lay a hard-surfaced pathway or walk through the area. Proper design of the planting scheme can accommodate circulation patterns without sacrificing the functional or aesthetic objectives of the planting.

The most important consideration is the use of plant materials which are native to the area or at least do best under the conditions to which they will be subjected. Climatic conditions, soil conditions, availability of water, amount of sun and wind, and air quality all are critical factors when choosing plant materials.

Trees and shrubs come in all sizes and shapes, and there is at least one variety to suit every landscaping need. By choosing the plant materials which will mature to the desired height and size and which will grow best under existing environmental conditions, a great deal of pruning, thinning, clipping,

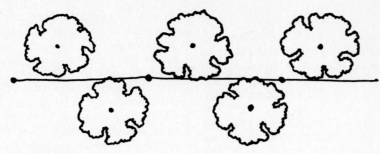

Figure 5.146

fertilizing, and spraying can be eliminated, thus saving considerable time and money.

The susceptibility of young plants to abuse can be minimized by following these planting procedures:

Use the most mature and fullest plantings your budget will permit.

Erect a simple but sturdy low (3 to 4 feet) wire fence along the line of the desired windbreak.

Place the plant materials in staggered double rows along both sides of the fence. (See Figure 5.146.)

Plant only in early spring and late fall, when the plants will have the best chance to take root and begin new growth quickly.

S E C T I O N

SIX

Typical Site Details

After the preparation of the overall site plan, many design details must be developed to show specific methods of construction. These details are an integral part of the design process and serve two important purposes. First, they stipulate the aesthetic and structural elements of the plan; and second, they provide the basis for costing the project.

This section offers a wide range of representative details of the various aspects of site development. Even though the scope of the details presented may seem extensive, they are only a very small segment of the available data. It is assumed that these details will act as a guide to assist site designers in solving their particular problems. The section is not intended to present aesthetic or design solutions but to indicate how others have handled similar technological difficulties. It should be emphasized that none of these details should be followed to the letter. Those that are more general in scope should be reviewed carefully against their anticipated use. Other details, which require structural or mechanical expertise, must be reviewed by appropriate professional engineers for their conformance with current codes and standards.

6.1 Walks

Walks should be designed to allow the greatest diversity of people to move safely, independently, and unhindered through the exterior environment (see Figure 6.1). Items to consider in the design or modification of walk systems are discussed in the following paragraphs.

Surfaces The surface of walks should be stable and firm, be relatively smooth in texture and have a nonslip quality. The use of expansion and contraction joints should be minimized, and they should be as small as possible, preferably under ½ inch in width.

Rest Areas Occasional rest areas off the traveled path are enjoyable and helpful for all pedestrians and especially for those with handicaps that make walking long distances exhausting.

Gradients Pedestrian paths with gradients under 5 percent are considered walks. Walks with gradients in excess of 5 percent are considered ramps and have special design requirements. Routes with gradients up to 5 percent can be negotiated independently by the average wheelchair user, but sustained grades of 4 and 5 percent should have short (5-foot) level areas approximately every 100 feet to allow a chair-bound person using the walk to stop and rest. Gradients up to 3 percent are preferable where their use is practical.

Lighting Lighting along walkways should vary from ½ to 5 footcandles, depending on the intensity of pedestrian use,

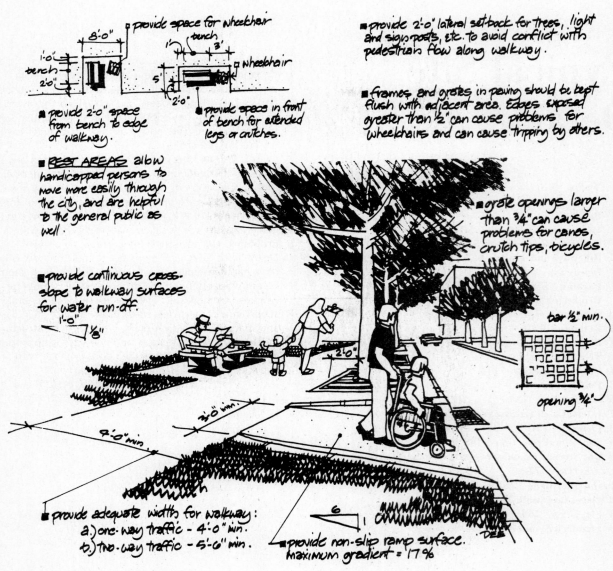

Figure 6.1 General dimensions of walks.

the hazards present, and the relative need for personal safety.

Maintenance Proper maintenance of walks is imperative. If walks are deteriorating, repairs should be made to eliminate any conditions that may cause injury.

Curb Ramps Changes in grade from street to sidewalk and from sidewalk to building entrances create the most numerous problems for people with physical handicaps. To facilitate movement over low barriers, a curb ramp should be installed. Surfaces should be nonslip but not corrugated, for grooves may fill with water, freeze, and cause the ramp to become slippery.

Drainage Structures Improperly designed, constructed, or installed drainage structures may be hazardous to people who must move over them. They should be placed flush with the surface on which they occur, and grates having narrow parallel bars or patterns with openings larger than ¾ inch should not be used. Grates should be kept clean so as not to lessen the efficiency of the overall storm system. Obviously, a surface buildup of water, especially in winter, may present a hazard. For this reason, drainage structures should not be located between a curb ramp and the corner of a street or immediately downgrade from a curb ramp.

Dimensions Walkways vary in width according to the amount and type of traffic using them. They should be a minimum of 4 feet wide, with 5 feet 6 inches (6 feet preferred) being the minimum for moderate two-way traffic.

Wheel Stops Wheel stops are necessary where wheeled vehicles may roll into a hazardous area. They should be 2 to 3 inches high and 6 inches wide and should have breaks in them every 5 to 10 feet to allow water drainage off the walk.

Walks

Provide walks on grade to connect drop-off zones, reserved parking, and other applicable facilities to usable entrances.

Make walks subject to use by the handicapped at least 6 feet wide and uninterrupted by any abrupt change in level.

Use slopes no greater than 1 in 24 (4.2 percent), blending to a common level with landings and other walks and with driveways and parking lots where curb ramps are not used. When slopes exceed 1 in 30 (3.3 percent), provide level landings at least 6 by 6 feet at 60-foot intervals for the purpose of rest and safety.

For walks terminating at doorways, provide a level landing at least 6 by 6 feet, extending a minimum of 18 inches beyond the strike jamb of the doorway.

Make surfaces fixed, firm, and nonslip. For water drainage, surfaces may be crowned at ⅛ inch per foot.

Avoid grates and manholes in walks. If grates must be used, have no opening between bearing bars greater than ⅝ inch in width with cross bars no more than 4 inches on center. Grates should be set so that bearing bars are perpendicular to the path of travel.

Grade ground surface up to walks and compact to avoid drop-off. Where grounds drop off or recede at slopes greater than 1 in 6 (16.5 percent), provide guards.

Set lighting elements, signposts, and street furniture back from walks at least 1 foot. Street furniture will be placed on surfaces that are fixed, firm, and nonslip.

New landscape elements should be planted to allow clearances of at least 1 foot on sides and at least 7 feet vertical from walk.

Where walks exceed 200 feet in length, it is desirable to provide rest areas adjacent to the walk at convenient intervals with space for bench seats and wheelchair parking. See Figure 6.2.

If the slope of the walk is 5 percent or less, no handrail is required (see Figure 6.3).

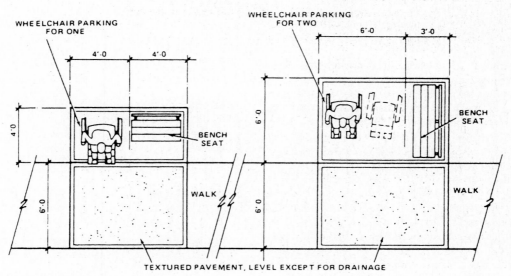

Figure 6.2 Rest areas.

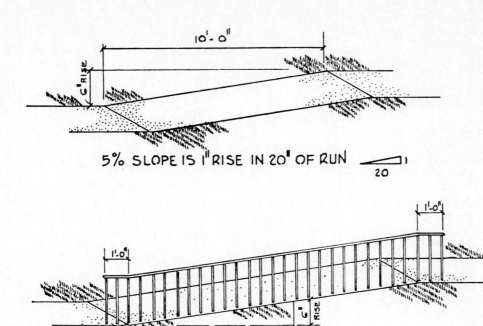

5% SLOPE IS 1" RISE IN 20' OF RUN

8.33 % SLOPE IS 1" RISE IN 12" RUN

Figure 6.3 Slopes.

If the slope of the walk is greater than 5 percent, a handrail is required on one side.

The slope of a walk may not exceed 8.33 percent.

Walks should have a continuous common surface not interrupted by steps or abrupt changes in level greater than ½ inch.

Vertical level changes greater than ½ inch obstruct the small wheels of the chair and may trip those with semiambulatory disabilities (see Figure 6.4).

Walks terminating at doors should have a 5 feet 0 inch by 5 feet 0 inch level platform extending 1 foot 6 inches beyond the strike jamb on the pull side of the door (Figure 6.6).

Geometric Design   Safety and volume of pedestrian traffic are the primary controls for geometric design of walks. The traffic volume used for design should be the average of the maximum hour for each day for a year. However, since sufficient data are rarely available to determine this value, a design pedestrian traffic volume (pedestrians per hour) must be estimated on the basis of available data, engineering judgment, and pedestrian traffic at existing similar installations.

Width   The minimum width for walks should be 3 ft for single-family residences and for low-volume traffic. Walks will normally be in increments of 2 feet (width of pedestrian traffic lane) as required to accommodate the anticipated volume of pedestrian traffic. An extra foot of width should be added to walks adjacent to curbs or where obstacles encroach on the walk. Width of walks will be determined on the basis of the capacities (pedestrians per hour) shown in the accompanying table.

Capacity of walks in pedestrians per hour	Location of walk	Minimum width (feet)
Less than 10	Any location	3
Up to 100	Any location	4
100 to 750	Shopping centers	6
100 to 1000	All other locations	6
Greater than 750	Shopping centers	$2\dfrac{P_T - C_1}{C_1} + 6$
Greater than 1000	All other locations	$2\dfrac{P_T - C_2}{C_2} + 6$

P_T = design pedestrian traffic volume in pedestrians per hour
C_1 = 750 pedestrians per hour
C_2 = 1000 pedestrians per hour

For instance, assume that the design pedestrian traffic for walks at a particular shopping center is 1700 pedestrians per hour, the width for these walks would be determined as follows:

$$W = 2\frac{P_T - C_1}{C_1} + 6$$

where W = width required by traffic
P_T = design traffic
C_1 = 750

Therefore

$$W = 2\frac{1700 - 750}{750} + 6$$

$$W = 8.5 \text{ feet}$$

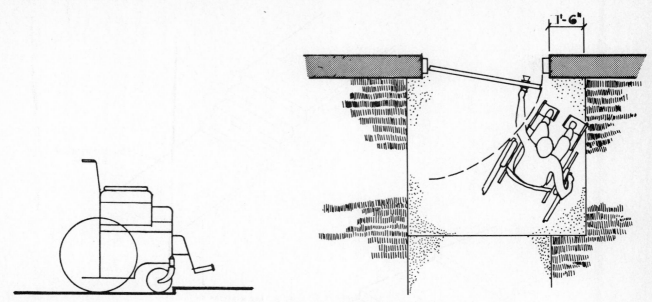

Figure 6.4 Level changes.

Figure 6.5 Entrance platform.

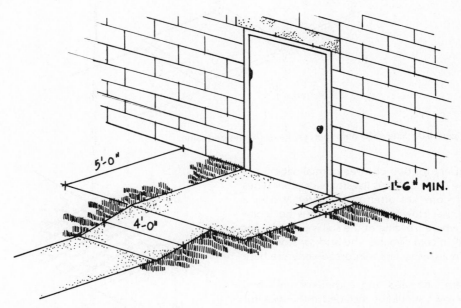

Figure 6.6 Entrance platform.

Walks in this shopping center would be 8 feet wide (nearest even-foot width).

Grade The grade of walks should follow the natural grade of the ground as nearly as possible. The transverse grade will not be less than ¼ inch per foot. The longitudinal grade should not be greater than about 15 percent. Steps will be used where the maximum longitudinal slope would otherwise be too great. Steps should be grouped together, rather than spaced as individual steps, and located so that they will be lighted by adjacent street or night lights. The sum of the depth of tread and height of riser should not be less than 18 inches, and risers should not be less than 5 inches or greater than 7 inches on any steps.

When slopes to the house are greater than a 5 percent grade, stairs or steps should be used. This may be accomplished with a ramp sidewalk, a flight of stairs at a terrace, or a continuing sidewalk (Figure 6.7*a*). These stairs have 11-inch treads and 7-inch risers when the stair is 30 inches or less in height. When the rise is more than 30 inches, the tread is 12 inches and the riser 6 inches. For a moderately uniform slope, a stepped ramp may be satisfactory (Figure 6.7*b*). Generally, the rise should be about 6 to 6½ inches and the length

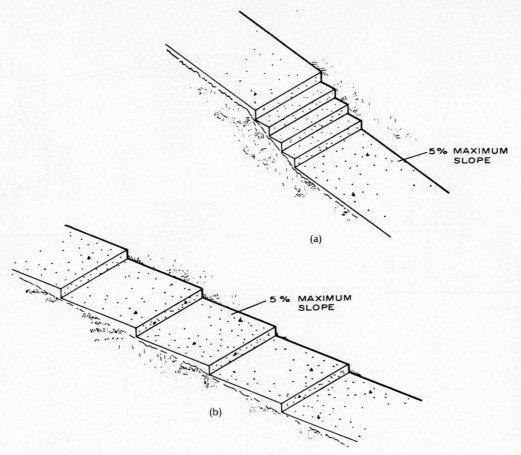

Figure 6.7 Sidewalks on slopes: (*a*) stairs; (*b*) stepped ramp.

between risers sufficient for two or three normal paces.

Walks can also be made of brick, flagstone, or other types of stone. Brick and stone are often placed directly over a well-tamped sand base. However, this system is not completely satisfactory where freezing of the soil is possible. For a more durable walk in cold climates, the brick or stone topping is embedded in a freshly laid reinforced-concrete base (Figure 6.8).

As in all concrete sidewalks and curbed or uncurbed driveways, a slight crown should be included in the walk for drainage. Joints between brick or stone may be filled with a cement mortar mix or with sand.

Sidewalks adjacent to streets should follow the general grade of the street. It is desirable from the standpoints of safety and appearance to maintain sidewalk grades at or above adjacent street and curb levels.

Recommended Grades

Paved Areas A minimum gradient of 9.5 percent is required to ensure fast runoff; this can be reduced to 0.35 percent for gutters or pavement next to curbs.

Maximum grades for paved areas:

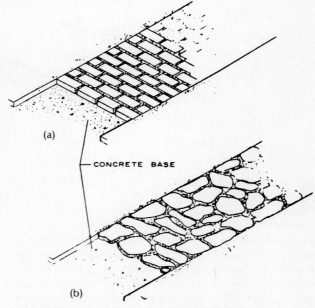

Figure 6.8 Masonry paved walks: (*a*) brick; (*b*) flagstones.

Roadways: Where the gradient is related to the design speed, avoid a sustained slope of 17 percent and over.

Under general conditions it would be advisable not to exceed 6 percent for roadways.

Parking should not be planned on slopes in excess of 5.75 percent.

Walkways must be kept within a gradient of 5.75 percent unless the construction of a ramp is required.

Ramps require a nonslip surface and should not be designed in excess of 12 percent.

Transition between grades by means of a ramp calls for special treatment. Sudden changes in grade between two level areas may cause under-car damage.

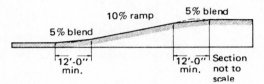

Figure 6.9 Transition of grades created by a ramp.

The transition grade, being approximately half the ramp grade, should be at least 10 to 12 feet in length for automobile traffic (Figure 6.9).

Make changes of grade in walkways clearly visible, using at least two risers where steps are required. The National Building Code requires handrails for any steps in excess of three risers.

Keep risers in steps between 4 and 6 inches with the product of run and riser not exceeding 84 inches. Low risers with long treads make for easy outdoor walking. Do not make treads longer than can be used by a person with a small stride.

Avoid drainage across walkways and introduce a minimum cross slope of 1 percent to prevent water accumulation.

Provide a 1 percent slope for paved areas and 2 percent for lawns for a distance of 10 feet away from any building wall.

Avoid warping of paved areas. Slight differences in grade can be camouflaged in lawns but become very pronounced in paved areas, especially where pavement slopes adjoin horizontal lines of structures. Avoid gradients in excess of 4 percent.

Lawn Areas Minimum gradient required to provide satisfactory drainage is 2 percent.

A 4 percent grade can be considered sufficiently level to be suitable for general activities.

Grades approaching 10 percent create a distinct sloping feature on the site.

Slopes in excess of 33 percent (3:1) are not only difficult to mow with ordinary mowing equipment, but increase maintenance costs.

Newly created slopes of 50 to 60 percent require special treatment to prevent erosion.

Treatment of Slopes Each grading situation must be judged on its own merits and its relation to the surrounding contours.

Slopes steeper than 3:1 are permitted only in granular materials. Clay embankments must be sloped at 3:1 or less.

The angle of repose of different soils is not related to the angle at which slopes can withstand the erosive effect of water.

Seepage behind slopes will increase the risk of slip (Figures 6.10 and 6.11)

Possible slope treatments:

- Cutting out terraces, intercepting surface water (Figure 6.12)
- Grading to direct surface water away from the slope
- Not allowing surface water to spill over the embankment but directing it toward flumes
- Installation of drain tile parallel to the top of the slope or along the side of the slope, preventing excessive seepage
- The use of flumes to ensure a fast runoff
- A dwarf wall at the toe of a slope to reduce the danger of slippage

When slopes consist of recently placed fill, stabilization can be obtained by using straw mats, wire mesh, wire-reinforced brush mats, etc. This material is placed in stepped manner in the slope during its construction.

Leave established slopes undisturbed wherever possible and preserve existing groundcover.

Slopes of less than 1:1 can be sodded if certain precautions are taken. For example, chicken wire, pegs, or straw embedded in the topsoil can be used for stabilization.

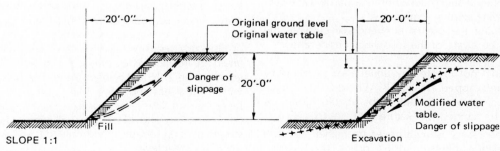

Figure 6.10 Slope 1:1.

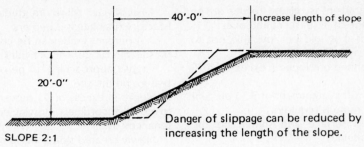

SLOPE 2:1

Danger of slippage can be reduced by increasing the length of the slope.

Figure 6.11 Slope 2:1.

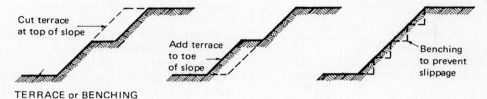

TERRACE or BENCHING

Figure 6.12 Terrace or benching.

Wire mesh is placed over the topsoil, the sod laid, and pegs are driven through the sod, mesh, and topsoil into the undisturbed subsoil.

Sod is laid parallel to the length of the slope, and the joints are staggered to slow down any water runoff.

Heavy jute mesh can be spread over the slope and fastened with long pegs into the soil to stabilize the slope after seeding.

Retaining walls, riprap, or the use of retaining cribs will generally be required for slopes of 1:1 or steeper.

CURB RAMPS

Curbing, though a commonly specified element on most sites, is one of the most neglected items with regard to the physical barriers it creates. The problem is twofold. It stems, first, from the attitude of most designers that 6-inch concrete curbs are an unavoidable necessity and, second, from municipalities that aggravate the problem by writing in curbing clauses to building ordinances for no reason other than that they have always been a requirement. While this subsection by no means advocates the retraction of municipal curbing requirements, it does seem that viable alternatives should be allowed where they would reduce potential barriers and hazards while satisfying existing requirements.

When specifying the use of conventional curbing, the designer should be aware of the following considerations:

1. Curbing should not create any unnecessary barriers to physically handicapped individuals. Where barriers have been created, previously laid curbs should either be removed or be ramped. See Figure 6.13.

2. Curbing, if necessary, should never be higher than the maximum height of one step, or 6½ inches. This is particularly important where there is any pedestrian traffic crossing over or vehicles parking adjacent to the curb.

3. Double, or stepped, curbs are difficult for the handicapped to negotiate and in darkness are hazardous to all pedestrians. Their use should be limited if not restricted.

Ideally, curb ramps should be installed at all intersections and wherever walks cross streets and drives. When installed at intersections, a consistent pattern of orientation should be observed areawide to provide repetitive "cuing" to blind people. Locate curb ramps to direct pedestrian traffic into crosswalk areas. See Figures 6.14, 6.15, 6.16, and 6.17.

Ramp surfaces themselves should be nonslip with a texture discernible by visually impaired individuals. Most importantly, the lower ramp edge should blend with street or gutter surfaces, free of "lips" or elevation changes.

When located adjacent to parking areas, care in placement must be observed to preclude blockage by automobiles.

Avoid ramps that project from a curb into traffic areas. Projecting ramps may create traffic, drainage, and maintenance problems.

Widths of ramps as shown represent practical minimums. Increase widths as required to suit snowplows and local and state standards. Additionally, verify the requirement for railings with the appropriate jurisdiction.

Use curb ramps with flared sides. Make ramp width a minimum of 40 inches, excluding flared sides. Use slopes no greater than 1:12 (8.3 percent), blending to a common level with both street and walk. Make transitions between surfaces smooth. Use slopes for flared sides approximating 1:6 (16.5 percent) and blend gradually into ramp slope. Run street curbs up to ramps location and blend with flared sides down to street level.

■ avoid "lip" greater than ½" wherever ramp meets adjacent paving at top or bottom.

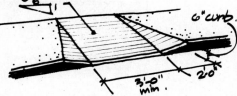

1. Flared Ramp

■ corrugated lines in ramps should be avoided since they can hold water in freezing weather and become icy.

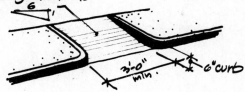

2. Ramp With Continuous Curb

■ use of this type often interferes with curb-side storm drainage & snow plowing.

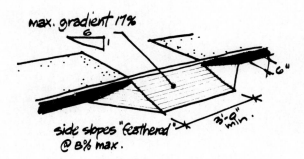

side slopes "feathered" @ 8% max.

3. Extended Ramp

Figure 6.13 Curb ramps.

■ locate handrail to avoid conflict with adjacent pedestrian walkway.

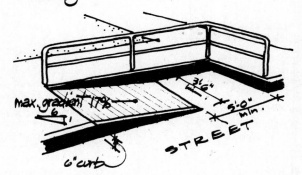

4. Parallel Ramp

Curb Ramps

1.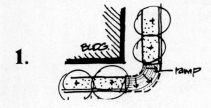

2.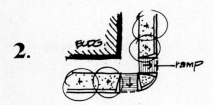

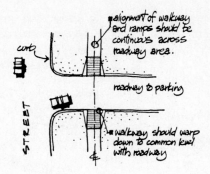

Alignment of Ramps

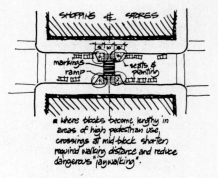

- Where blocks become lengthy in areas of high pedestrian use, crossings at mid-block shorten required walking distance and reduce dangerous "jaywalking".

Mid-Block Crossings

3.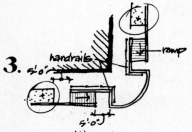

- Wherever possible, curb ramps should occur as a natural extension of the alignment of the walkway.

Curb Ramps at Corners

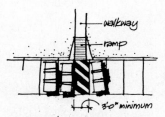

- Access to curb cut and walkway must be kept clear of parked cars. Area should be clearly marked for usability.

Access to Ramps

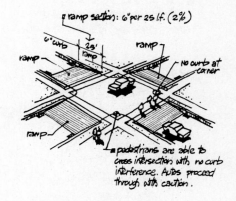

- Pedestrians are able to cross intersection with no curb interference. Autos proceed through with caution.

Ramped Intersection

Figure 6.14 Curb ramps.

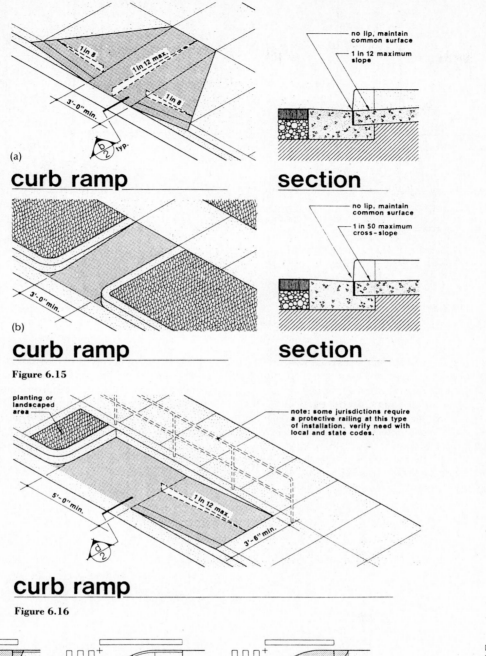

no lip, maintain
common surface

1 in 12 maximum
slope

1 in 8

1 in 12 max.

1 in 8

3'-0" min.

b/2 typ.

(a)

curb ramp

section

no lip, maintain
common surface

1 in 50 maximum
cross-slope

3'-0" min.

(b)

curb ramp

section

Figure 6.15

planting or
landscaped
area

note: some jurisdictions require
a protective railing at this type
of installation. verify need with
local and state codes.

1 in 12 max

5'-0" min.

3'-6" min.

d/2

curb ramp

Figure 6.16

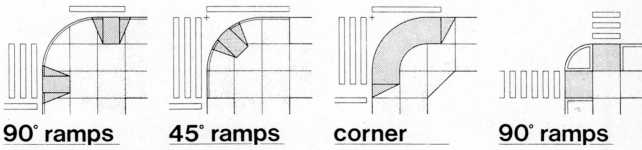

90° ramps   # 45° ramps   # corner   # 90° ramps

Figure 6.17   Arrangement of curb ramps.

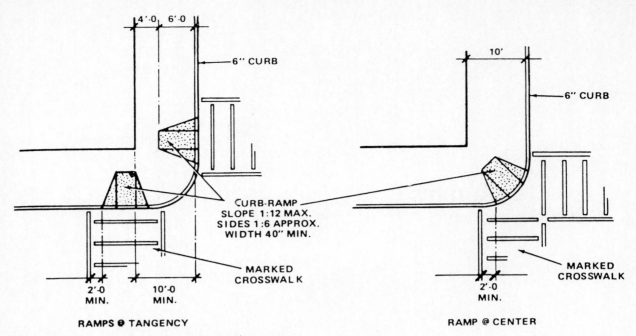

Figure 6.18 Location of curb ramp at intersection corner.

Provide minimum 4-foot clear walkway at head or curb ramp. Make sure curb ramps are clear of obstructions such as utility poles, hydrants, and street-sign poles.

At street intersection corners, preferred location of curb ramps is at points of tangency no less than 10 feet from the point where curbs would intersect. See Figures 6.18, 6.19, 6.20, and 6.21. Curb ramps located in the center of the corner may be used only where corners are clearly protected from corner-cutting vehicles and where cross walks can be laid out to assure safety.

At street crossings, provide marked cross walks and traffic warnings to ensure unobstructed passage of pedestrians. Place markings at least 2 feet from outer edges of curb-ramp flares. See Figure 6.18.

Make ramp surfaces firm and nonslip by use of gritted concrete, paint-on gritted epoxy, etc. Avoid corrugated lines running across slope of ramp, since they may hold water in cold weather and become icy.

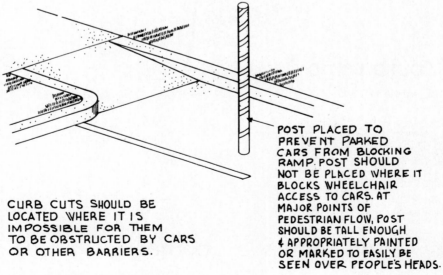

CURB CUTS SHOULD BE LOCATED WHERE IT IS IMPOSSIBLE FOR THEM TO BE OBSTRUCTED BY CARS OR OTHER BARRIERS.

POST PLACED TO PREVENT PARKED CARS FROM BLOCKING RAMP. POST SHOULD NOT BE PLACED WHERE IT BLOCKS WHEELCHAIR ACCESS TO CARS. AT MAJOR POINTS OF PEDESTRIAN FLOW, POST SHOULD BE TALL ENOUGH & APPROPRIATELY PAINTED OR MARKED TO EASILY BE SEEN OVER PEOPLE'S HEADS.

Figure 6.19 Posts at curb ramp.

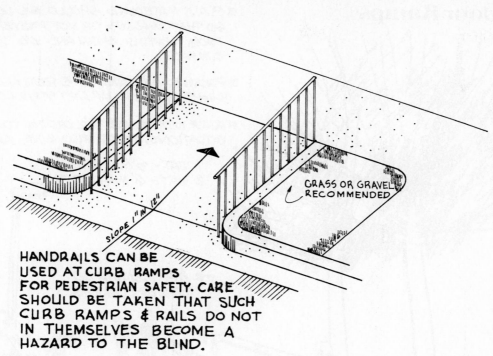

HANDRAILS CAN BE
USED AT CURB RAMPS
FOR PEDESTRIAN SAFETY. CARE
SHOULD BE TAKEN THAT SUCH
CURB RAMPS & RAILS DO NOT
IN THEMSELVES BECOME A
HAZARD TO THE BLIND.

Figure 6.20 Handrails at curb ramp.

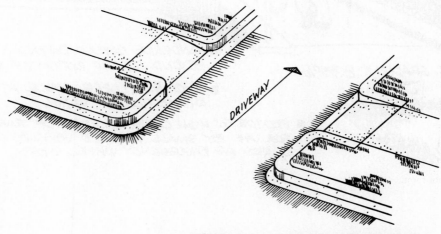

IIx3.2(b) ALL WALKS CROSSING
DRIVEWAYS MUST WARP
DOWN TO A COMMON
LEVEL WITH THE STREET.
CARE SHOULD BE TAKEN SO
THAT CURB CUT IS NOT IN
IT SELF A HAZARD TO THE
BLIND.

Figure 6.21 Curb ramps at driveway.

Outdoor Ramps

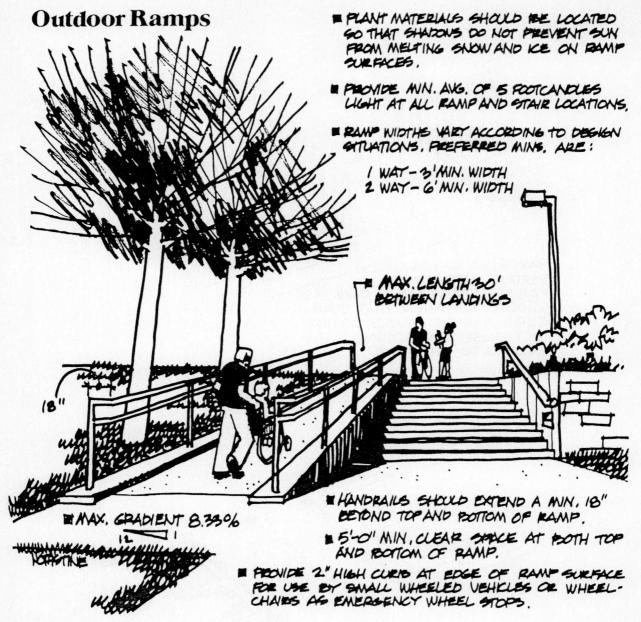

- PLANT MATERIALS SHOULD BE LOCATED SO THAT SHADOWS DO NOT PREVENT SUN FROM MELTING SNOW AND ICE ON RAMP SURFACES.

- PROVIDE MIN. AVG. OF 5 FOOTCANDLES LIGHT AT ALL RAMP AND STAIR LOCATIONS.

- RAMP WIDTHS VARY ACCORDING TO DESIGN SITUATIONS. PREFERRED MINS. ARE:

  1 WAY - 3' MIN. WIDTH
  2 WAY - 6' MIN. WIDTH

- MAX. LENGTH 30' BETWEEN LANDINGS

18"

- MAX. GRADIENT 8.33%

NORMLINE

12
1

- HANDRAILS SHOULD EXTEND A MIN. 18" BEYOND TOP AND BOTTOM OF RAMP.

- 5'-0" MIN. CLEAR SPACE AT BOTH TOP AND BOTTOM OF RAMP.

- PROVIDE 2" HIGH CURB AT EDGE OF RAMP SURFACE FOR USE BY SMALL WHEELED VEHICLES OR WHEEL-CHAIRS AS EMERGENCY WHEEL STOPS.

Figure 6.22 Outdoor ramp.

Outdoor Ramps. Ramps are alternate routes for people who are not able to use stairs; however, they do not take the place of stairs since certain portions of the population find ramps more difficult to use. Any surface pitched above 5 percent is considered a ramp. See Figures 6.22 through 6.26.

1. The maximum gradient for a ramp of any extended length should not exceed 1:12 (8.33 percent), not including curb ramps.

2. The maximum length for a single ramp at 1:12 should not exceed 30 feet 0 inch. Ramps of lesser grades can, of course, be lengthened.

3. The minimum clear width of any ramp is 3 feet 0 inch. Where ramps are heavily used by pedestrians and service deliveries, there should be sufficient width to accommodate both, or provisions should be made for alternate routes.

4. The bottom and top approach to a ramp should be clear and level for a distance of at least 5 feet 0 inch, allowing for turning maneuvers by strollers, dollies, wheelchairs, etc.

5. A textural signal prior to the ramp, at both top and bottom, may be used to warn the pedestrian of the upcoming obstacle. (See "Signage Considerations, Textural Paving" for details.)

6. Ramps should be designed to carry a minimum live load of 100 pounds per square foot.

7. Low curbs along the sides of ramps and landings should be provided as surfaces against which wheeled vehicles can turn their wheels in order to stop.

8. Ramps should be illuminated to an average maintained light level which ensures their safe use in darkness. It is important that the heel and toe of the ramp be particularly well illuminated.

9. Ramps should be maintained properly to keep them from being hazardous. Debris, snow, and ice should be kept off the surface. Handrails should, at all times, be properly secured.

Ramps for Outdoor Use

1. Straight-Run

2. Angled Landing

3. Intermediate/Switch-Back Landing

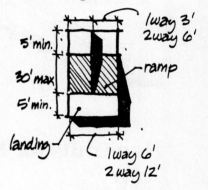

Figure 6.23 Configuration of outdoor ramps.

Conditions at Tops & Bases of Ramps

1. Traffic Goes Straight

2. Traffic Turns

3. Traffic Turns to Gate/Doorway

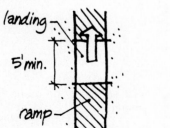
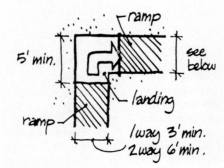
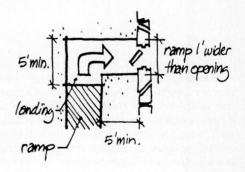

Figure 6.24 Minimum dimension of landings and turns.

–549–

Handrails for Ramps

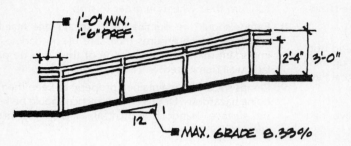

■ 3'-0" IS THE MOST COMFORTABLE HT. FOR HANDRAILS ON RAMPS.

■ A SECOND HANDRAIL, USEFUL TO PEOPLE IN WHEELCHAIRS AND CHILDREN SHOULD BE PLACED AT 2'-4".

■ HANDRAILS SHOULD EXTEND A MIN. 1'-0" BEYOND BOTH ENDS OF A RAMP.

Handrails for Stairways

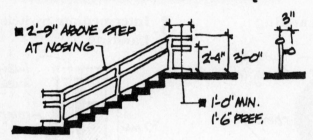

■ 3'-0" IS THE MOST COMFORTABLE HT. FOR RAILINGS AT BOTH ENDS OF STAIRWAYS. 2'-9" IS THE ACCEPTED HT. ON STAIRWAYS.

■ A SECOND HANDRAIL, USEFUL TO CHILDREN SHOULD BE PLACED AT 2'-4"

■ HANDRAILS SHOULD EXTEND A MIN. OF 1'-0" BEYOND STAIRWAYS.

Handrails for Extra-Wide Stairways

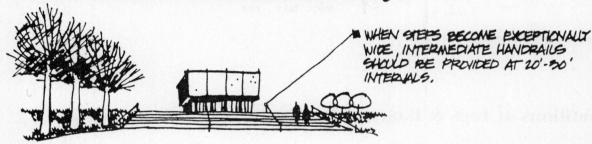

■ WHEN STEPS BECOME EXCEPTIONALLY WIDE, INTERMEDIATE HANDRAILS SHOULD BE PROVIDED AT 20'-30' INTERVALS.

Figure 6.25 Handrails.

Handrails for Ramps and Stairways Handrails serve the primary function of providing support for people who are in the process of climbing or descending stairs or ramps, whereas railings are placed more for reasons of preventing people from entering or falling into a dangerous area.

The designer should take into account the following items in regard to handrails and railings:

1. General

a. Handrails and railings should preferably be round or oval, 1½ to 2 inches in diameter.

b. There should be a minimum 3 inches spacing between handrails and adjacent walls, and wall surfaces should preferably be nonabrasive.

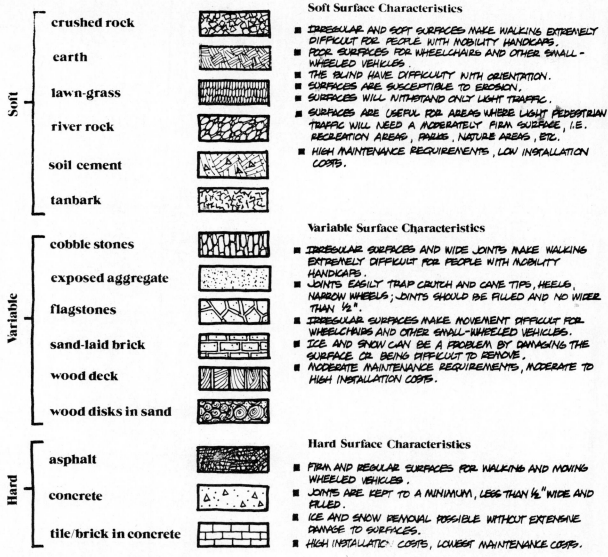

Soft Surface Characteristics

- *IRREGULAR AND SOFT SURFACES MAKE WALKING EXTREMELY DIFFICULT FOR PEOPLE WITH MOBILITY HANDICAPS.*
- *POOR SURFACES FOR WHEELCHAIRS AND OTHER SMALL-WHEELED VEHICLES.*
- *THE BLIND HAVE DIFFICULTY WITH ORIENTATION.*
- *SURFACES ARE SUSCEPTIBLE TO EROSION.*
- *SURFACES WILL WITHSTAND ONLY LIGHT TRAFFIC.*
- *SURFACES ARE USEFUL FOR AREAS WHERE LIGHT PEDESTRIAN TRAFFIC WILL NEED A MODERATELY FIRM SURFACE, I.E. RECREATION AREAS, PARKS, NATURE AREAS, ETC.*
- *HIGH MAINTENANCE REQUIREMENTS, LOW INSTALLATION COSTS.*

Variable Surface Characteristics

- *IRREGULAR SURFACES AND WIDE JOINTS MAKE WALKING EXTREMELY DIFFICULT FOR PEOPLE WITH MOBILITY HANDICAPS.*
- *JOINTS EASILY TRAP CRUTCH AND CANE TIPS, HEELS, NARROW WHEELS; JOINTS SHOULD BE FILLED AND NO WIDER THAN ½".*
- *IRREGULAR SURFACES MAKE MOVEMENT DIFFICULT FOR WHEELCHAIRS AND OTHER SMALL-WHEELED VEHICLES.*
- *ICE AND SNOW CAN BE A PROBLEM BY DAMAGING THE SURFACE OR BEING DIFFICULT TO REMOVE.*
- *MODERATE MAINTENANCE REQUIREMENTS, MODERATE TO HIGH INSTALLATION COSTS.*

Hard Surface Characteristics

- *FIRM AND REGULAR SURFACES FOR WALKING AND MOVING WHEELED VEHICLES.*
- *JOINTS ARE KEPT TO A MINIMUM, LESS THAN ½" WIDE AND FILLED.*
- *ICE AND SNOW REMOVAL POSSIBLE WITHOUT EXTENSIVE DAMAGE TO SURFACES.*
- *HIGH INSTALLATION COSTS, LOWEST MAINTENANCE COSTS.*

Figure 6.26 Surfaces for walkways.

c. Where handrails or railings are fully recessed into walls, a space of 6 inches should be allowed between the top of the rail and the top of the recess, and a space of 3 inches should be allowed between the bottom of the rail and the bottom of the recess.

d. The ends of handrails should be rounded off or turned into the wall so that they are not hazardous.

e. Handrails, railings, and their appurtenances should be maintained free of slivers, sharp protrusions, etc.

2. Handrails for Ramps

a. Handrails should be provided on both sides of every ramp. They should extend past the heel and toe, 1 foot 0 inch to 1 foot 6 inches, except in places where the extension in itself presents a hazard.

b. The vertical dimension from the ramp surface to the top of a single handrail should be between 2 feet 8 inches and 3 feet 0 inch.

c. A second rail is advantageous to children and wheelchair-dependent people. Where two rails are used, the top rail should be placed at 3 feet 0 inch to 3 feet 3 inches, and the lower rail should be placed at 2 feet 4 inches.

WALKWAYS

Various types of walkway construction are shown in Figure 6.27, and drainage is shown in Figure 6.28.

WALKS AND TERRACES

Methods of construction for walks and terraces of different types are shown in Figures 6.29 through 6.33.

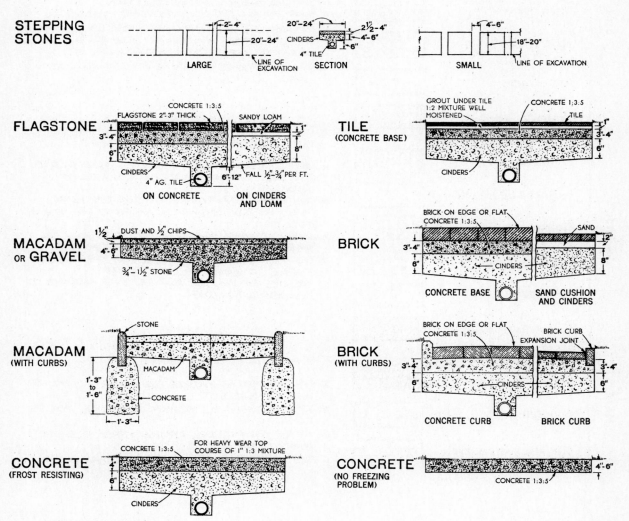

Figure 6.27 Walkway construction.

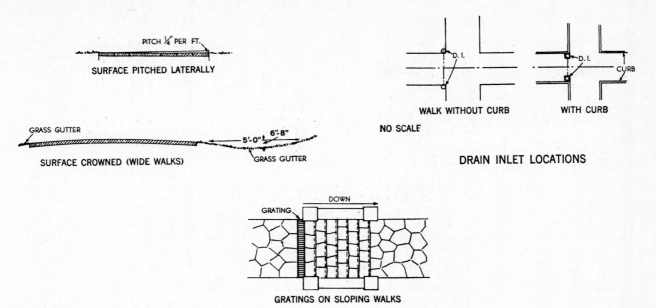

Figure 6.28 Walk drainage.

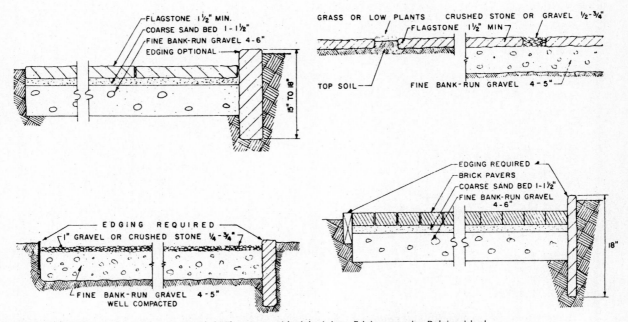

Figure 6.29 Dry construction. *(upper left)* Flagstone with tight joints. Edging may be Belgian block, granite, or flagstone. *(upper right)* Flagstone with gravel or grass joints. The type shown to the left will heave with frost. Grass or low plants in joints are hard to maintain. *(lower left)* Gravel. Edging may be 1-inch or thicker flagstone 18 inches deep or ³⁄₁₆- by 5-inch metal; or Belgian block, granite, or precast concrete curb may be used. In frost-free zones, bricks on edge, ⅛- by 4-inch metal, or 1- by 6-inch redwood with 1- by 3- by 18-inch stakes 4 feet on center may be used. Metal edging should be used for curved walks. *(lower right)* Brick with tight joints. Edging may be 1- by 8-inch redwood, 1½- by 18-inch flagstone, or ¼- by 5-inch metal.

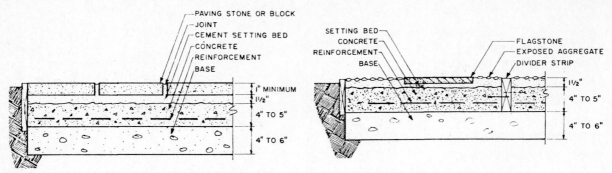

Figure 6.30 Frost-resisting construction. The base should be coarse bank-run gravel or slag or 1- to 3-inch crushed stone; it must be well drained. Concrete should have asphaltic expansion joints every 20 feet. The top surface should be rough for bond to the setting bed. Reinforcement should be 6- by 6-inch welded wire mesh. Joints should be ¾ inch or narrower and be grouted; care must be taken to keep grout from the face of the paving at all times. Edgings are required only during construction. In frost-free zones, the base and reinforcement may be omitted. *(left)* Flagstone or other paving units. Brick, quarry tile, slate, marble, Belgian block, or asphalt block may be set to the same details. There must be total contact, with no air pockets, between paving stone and cement setting bed. *(right)* Exposed aggregate concrete. Exposed aggregate ⅜ to 1 inch in size is set in natural or colored cement, with or without flagstones. If desired, dividers of 2- by 8-inch redwood may be used, with 4-inch galvanized spikes 12 inches on center to prevent heaving.

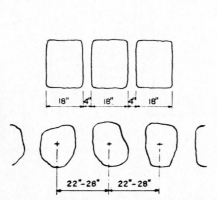

Figure 6.31 Stepping-stones: *(top)* rectangular; *(bottom)* irregular. Stones should be not less than 1½ inches thick. Common rectangular sizes are 12 by 15, 15 by 20, 18 by 24, 20 by 30, and 24 by 36 inches.

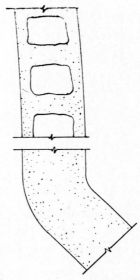

Figure 6.32 Gravel path with or without stepping-stones: width: single, 1 foot 6 inches to 3 feet; double, 3 feet 6 inches to 5 feet. Edging is required.

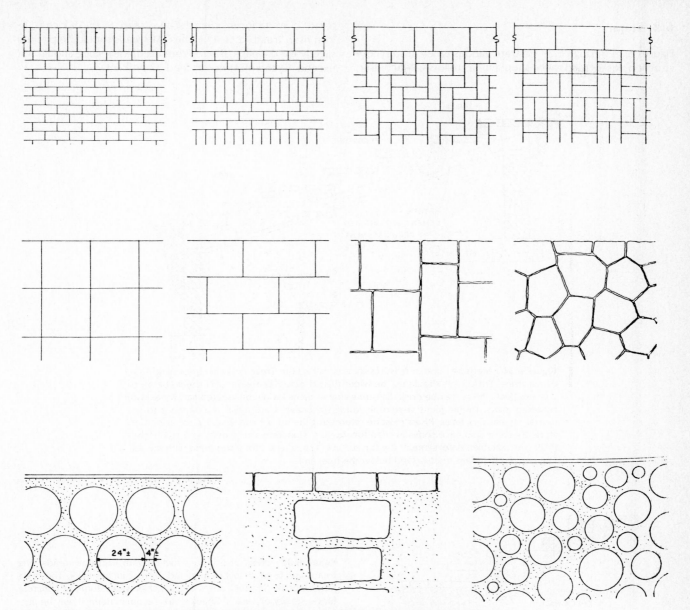

Figure 6.33 *(top row, left to right)* Brick pavement patterns: running bond; running-bond variation; herringbone; basket weave. Borders may be brick or stone. Edging is required with dry construction. *(center row, left to right)* Flagstone pavement patterns: squares; regular rectangular; random rectangular; irregular or polygonal. All joints should be tight. Some standard patterns are available in random irregular (for this pattern, joints should be broken every three stones or oftener). Square and rectangular shapes are also available in precast concrete. *(lower row, left to right)* Precast concrete in gravel or exposed aggregate concrete; flagstone or other paving blocks in exposed aggregate concrete; log sections in gravel. For logs, use cypress, redwood, chestnut, locust, or other durable wood, 6 inches thick; spacing is optional. Tanbark or low plants may be used instead of gravel.

6.2 Steps

Figures 6.34 through 6.39 illustrate various types of step construction. Riser-tread ratios should be as follows: 4-inch riser, tread 16 to 18 inches; 5-inch riser, tread 14 to 16 inches; 6-inch riser, tread 12 to 14 inches. All treads should be pitched slightly for drainage. If ramps are used, their slope should not exceed 10 and must not exceed 12 percent.

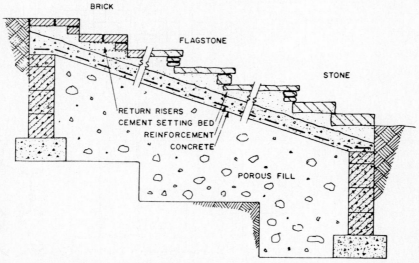

Figure 6.34 Steps of bonded frost-resistant construction. Treads may be brick, flagstone, or cut stone. Brick joints should not be wider than ⅜ inch. Flagstone treads should be 1½ to 2 inches thick. Risers may be brick, flagstone strips 4 inches wide, or selected flat stones with recessed joints. Visible joints should be rodded concave. Care must be taken not to get mortar on masonry faces. Risers must be returned at the sides if cheek wall is not used. Cut-stone treads should have some overlap for support. Concrete slab should be 5 to 6 inches thick with suitable reinforcement; the top surface is roughened for good bond with the setting bed. Foundations must extend below the frost line.

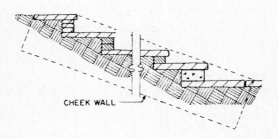

Figure 6.35 Steps of dry construction with flagstone treads. The treads are 16 to 20 inches long and 1½ to 2 inches thick. The risers may be brick, flagstone strips 4 to 5 inches wide, stone, or concrete block, set dry on treads. Cheek wall is recommended; it may be flagstone 2 to 3 inches thick by 18 inches deep. The steps are set dry on soil fill; no sand.

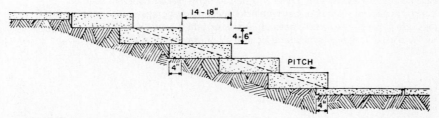

Figure 6.36 Cut-stone steps of dry construction. The treads should overlap 4 inches for support. The steps are set dry on soil fill; no sand.

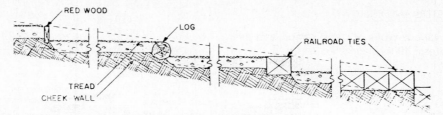

Figure 6.37 Variations of wood-and-gravel steps. Risers may be (1) redwood, 2 by 6 inches minimum, with cheek walls 2 by 8 inches and aluminum or stainless-steel nails; (2) logs, 6 inches or more in diameter, of cypress, juniper, arborvitae, black locust, or other decay-resistant wood; or (3) railroad ties. Treads may be (1) tanbark or ⅜-inch crushed stone 1 inch thick over a 4-inch base of bank-run gravel; or (2) railroad ties (also for cheek walls). Risers should have a maximum height of 4 inches; treads should measure 16, 40, or 64 inches.

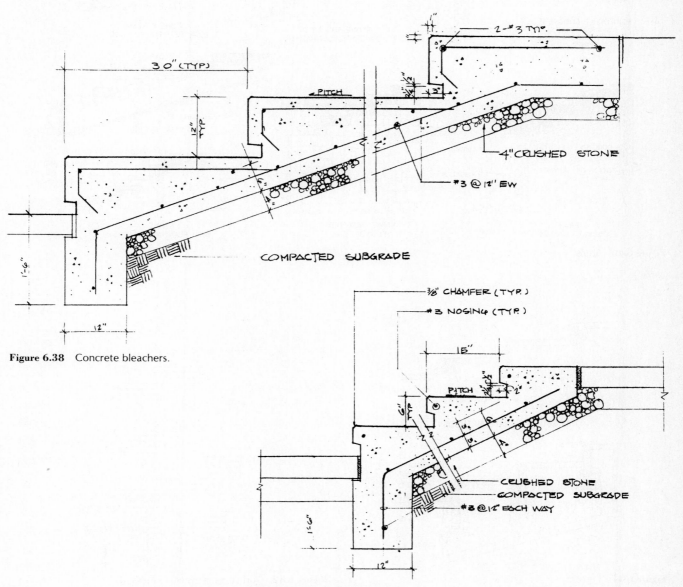

Figure 6.38 Concrete bleachers.

Figure 6.39 Concrete steps.

STEPS AND PERRONS

The construction fo various types of steps, ramps, and perrons is shown in Figures 6.40 through 6.56.

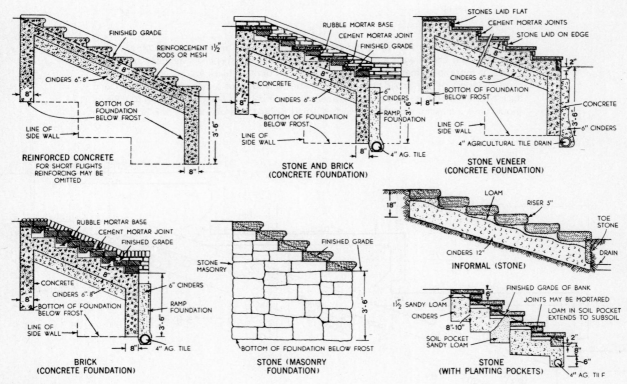

Figure 6.40 Steps.

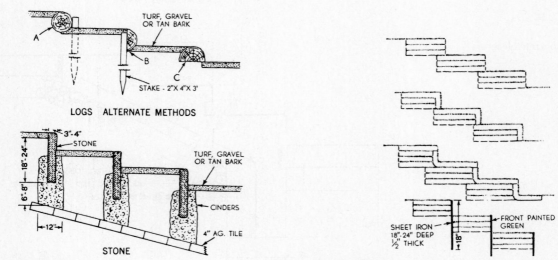

Figure 6.41 Perrons.

Figure 6.42 Turf steps: alternative methods.

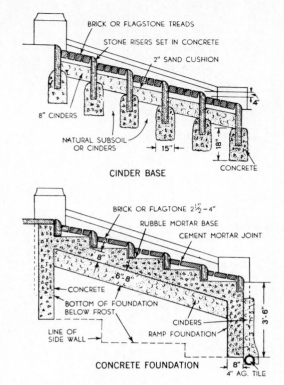

Figure 6.43 Ramps.

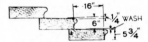

Figure 6.44 Wash, or pitch, on steps for drainage.

Walks The walk system should be designed to provide convenient and safe pedestrian access and circulation and should include ramps, not to exceed 7½ percent, for the physically handicapped. Width of walks should be based on pedestrian traffic volume. The minimum width for public walks should be 4 feet; this width should normally be increased in increments of 2 feet (width of a pedestrian traffic lane) as required to accommodate the anticipated volume of traffic, and two extra feet for walks adjacent to curbs or where obstacles encroach on the walk. The grade of walks should follow the natural pitch of the ground as nearly as possible. Walks should have a cross slope of ¼ inch to 1 foot in the direction of the ground slope with the top of the walk 1 inch above grade. The use of steps in walks should be avoided whenever possible; single risers in particular are extremely hazardous. In preference to using steps, grades of walks may be as steep as 15 percent in nonfreezing climates and 10 percent in freezing climates, if handrails are provided when gradients are 10 percent or greater. When steps are unavoidable, they should have at least three risers; handrails should be provided as a safety measure.

Exterior Steps Cheek walls for steps of three or more risers are needed for appearance and safety and to reduce maintenance. They should be constructed on both sides of the steps and extend a few inches above the top line of the outer edge of the treads. Cheek walls permit smooth grades adjacent to steps which are fitted into the ground without leaving difficult-to-maintain erosion-prone little slopes on both ends of each lower riser rear tread junction. Such slopes usually become eyesores and can result in safety hazards through the erosion of loose soil onto step treads.

The horizontal dimension of handrails should not exceed 2⅝ inches. Handrails should be 30 to 34 inches above the nose of step treads.

The rise of a single run of steps should be limited to 6 feet between landings for comfort and safety.

For user comfort step-tread riser design proportion should be based on the formula: twice the riser plus the tread equals 26 inches.

Outdoor Stairs Stairs should be designed to provide for the minimum amount of energy expenditure, a factor which is particularly important to elderly and semiambulant people. They should be wide enough for people to pass one another, be of safe design, and have proper appurtenances to ensure their safe use

1. The minimum clear width for any stairway should be 3 feet 0 inch. Where stairs are heavily used, widths should be increased to handle traffic requirements.

2. The maximum rise between landings for external unprotected stairs is 4 feet 0 inch. Where the stairs are protected, a 6 feet 0 inch rise is acceptable. Stairs should not be used where there are only a few in a series. These are dangerous and usually not necessary.

3. All steps in a series should have uniform tread width and riser height.

4. Stair treads should be deep enough to allow a man to place his whole foot on it. The preferred range is between 11 and 14½ inches.

5. Risers for exterior stairs should be between 4 and 6½ inches in height, with 5¾ inches being preferred.

6. Nosings should be rounded or chamfered. A 1-inch rounded nosing is most acceptable. It should be of a color contrasting that of the treads and risers to make identification easier. Abrupt, square nosings provide less frictional resistance and cause tripping.

7. Stairways should have an average maintained light level which ensures their safe use in darkness. Light should be cast down toward risers so that the treads will not be in shadow. (For recommended lighting levels, see "Lighting" section.)

Steps for Handicapped Wood handrails are popular, especially inside buildings. Choose handrails that are grooved, which are much easier to grip.

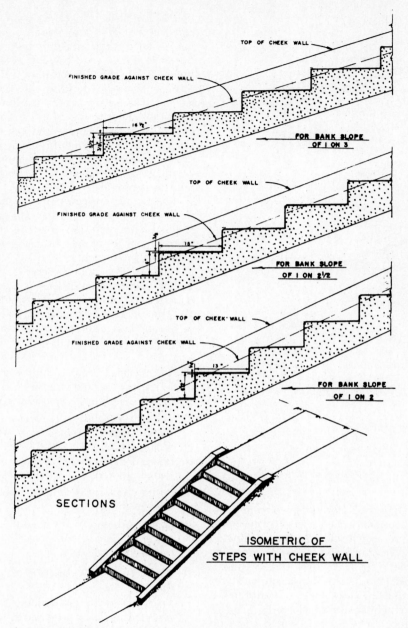

Figure 6.45 Steps to fit various slopes.

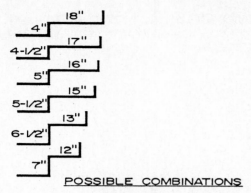

POSSIBLE COMBINATIONS

Figure 6.46   Riser and tread dimensions for outdoor steps.

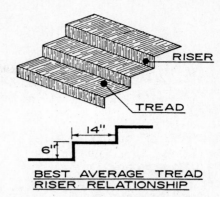

BEST AVERAGE TREAD
RISER RELATIONSHIP

Figure 6.47   Most common tread-riser relationship.

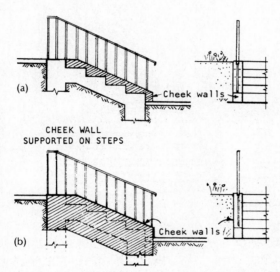

CHEEK WALL
SUPPORTED ON STEPS

Figure 6.48   Cheek walls: (*a*) supported on steps; (*b*) supported on subgrade.

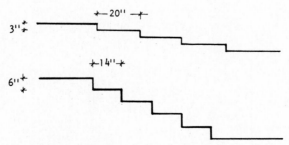

Figure 6.49   Step-tread riser proportions.

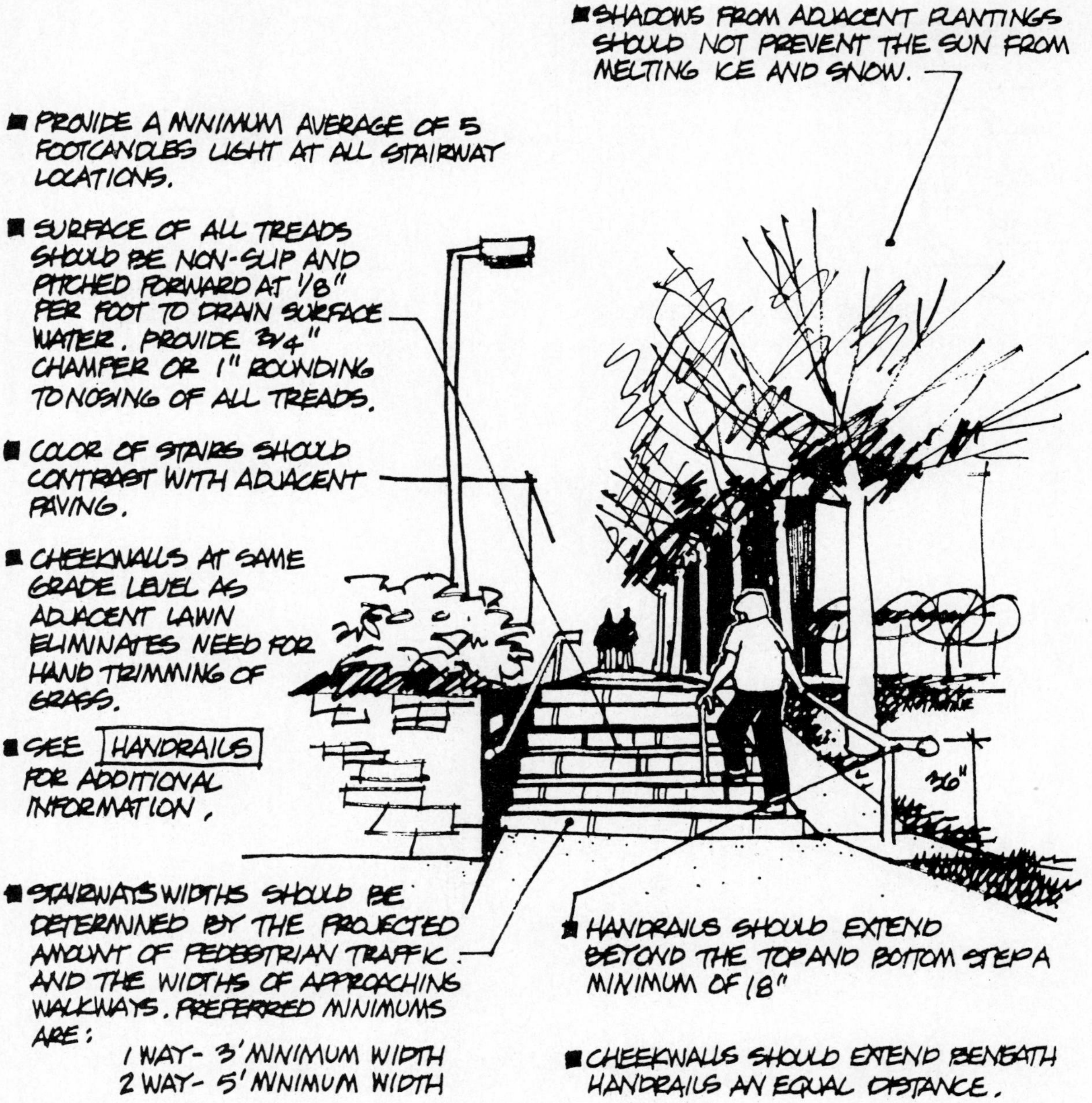

■ PROVIDE A MINIMUM AVERAGE OF 5 FOOTCANDLES LIGHT AT ALL STAIRWAY LOCATIONS.

■ SURFACE OF ALL TREADS SHOULD BE NON-SLIP AND PITCHED FORWARD AT 1/8" PER FOOT TO DRAIN SURFACE WATER. PROVIDE 3/4" CHAMFER OR 1" ROUNDING TO NOSING OF ALL TREADS.

■ COLOR OF STAIRS SHOULD CONTRAST WITH ADJACENT PAVING.

■ CHEEKWALLS AT SAME GRADE LEVEL AS ADJACENT LAWN ELIMINATES NEED FOR HAND TRIMMING OF GRASS.

■ SEE HANDRAILS FOR ADDITIONAL INFORMATION.

■ STAIRWAY WIDTHS SHOULD BE DETERMINED BY THE PROJECTED AMOUNT OF PEDESTRIAN TRAFFIC AND THE WIDTHS OF APPROACHING WALKWAYS. PREFERRED MINIMUMS ARE:

    1 WAY - 3' MINIMUM WIDTH
    2 WAY - 5' MINIMUM WIDTH

■ SHADOWS FROM ADJACENT PLANTINGS SHOULD NOT PREVENT THE SUN FROM MELTING ICE AND SNOW.

■ HANDRAILS SHOULD EXTEND BEYOND THE TOP AND BOTTOM STEP A MINIMUM OF 18"

■ CHEEKWALLS SHOULD EXTEND BENEATH HANDRAILS AN EQUAL DISTANCE.

Figure 6.50 Outdoor stairs.

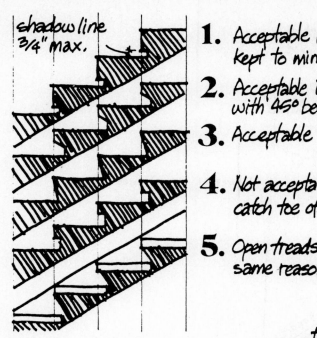

shadow line 3/4" max.

1. Acceptable if shadow line is kept to minimum.

2. Acceptable if nosing is provided with 45° bevel below.

3. Acceptable

4. Not acceptable; recesses can catch toe of shoes, braces, etc.

5. Open treads not acceptable for same reasons as above.

Figure 6.51 Outdoor step types.

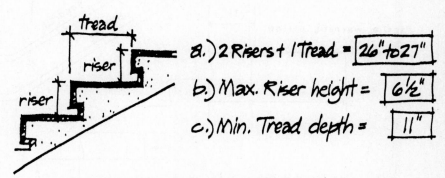

tread

riser

riser

a.) 2 Risers + 1 Tread = 26" to 27"

b.) Max. Riser height = 6½"

c.) Min. Tread depth = 11"

Figure 6.52 Outdoor steps rules of thumb.

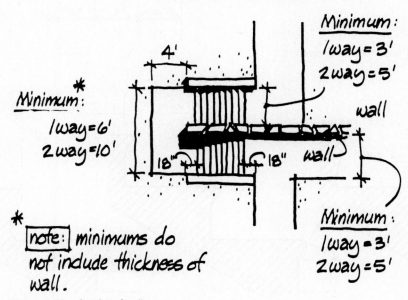

4'

Minimum:
1 way = 3'
2 way = 5'

wall

Minimum*:
1 way = 6'
2 way = 10'

18"   18"   wall

Minimum:
1 way = 3'
2 way = 5'

* note: minimums do not include thickness of wall.

Figure 6.53 Outdoor landings.

Provide 5 footcandles lighting on stair and landing areas.

■ Where total grade change exceeds 6'-0", intermediate landings are necessary.

■ Provide landings at 4'-0" intervals.

Figure 6.54 Height between landings.

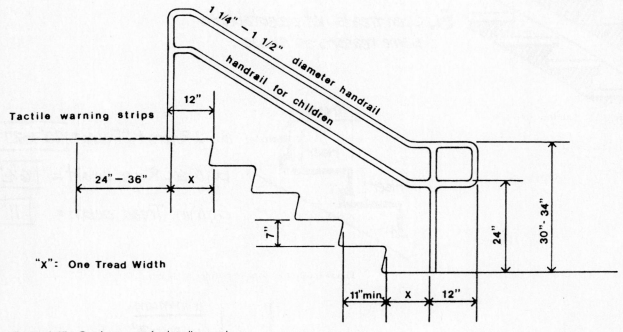

Tactile warning strips

1 1/4" - 1 1/2" diameter handrail
handrail for children

12"

24" - 36"   X

7"

24"

30" - 34"

11"min.   X   12"

"X": One Tread Width

Figure 6.55 Outdoor steps for handicapped.

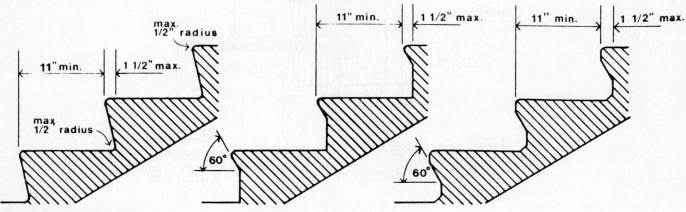

max. 1/2" radius

11" min.   1 1/2" max.

max. 1/2 radius

60°

11" min.   1 1/2" max.

11" min.   1 1/2" max.

60°

Figure 6.56 Detail of steps.

WOOD STEPS

The construction of wood steps and decks is illustrated in Figure 6.57. Redwood or cypress with fastenings of aluminum or stainless steel should be used for all exterior wood construction.

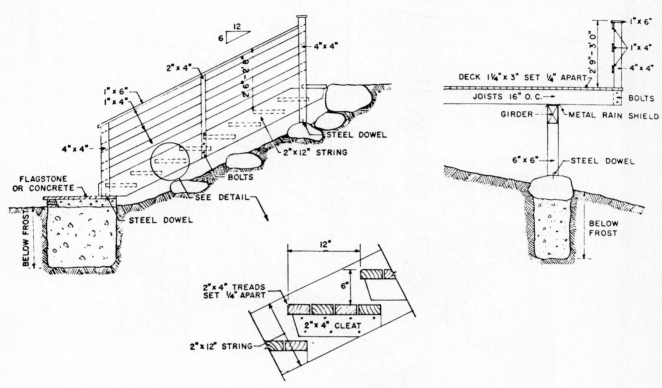

Figure 6.57 Wood decks and steps: *(left)* Wood steps. If steps are more than 30 inches wide, add a center carriage member. *(center)* Detail of stair treads. *(right)* Elevated wood deck.

RAMPS AND STEPPED RAMPS

Ramps and stepped ramps present a slipping hazard similar to that of stairs. Therefore, they should have a natural or applied slip-resistant surface. Broom-finished concrete, rough brick or stone, or the application of slip-resistant safety strips or cleats would reduce the slipping hazard.

For safety in negotiating changes of elevation in exterior applications, the design of ramps and stepped ramps should incorporate the following characteristics:

1. Ramps
    a. Preferable slope of 7 to 15 degrees, maximum, 20 degrees

2. Stepped ramps
    a. Maximum riser height, 5 inches
    b. Minimum tread width, 15 inches
    c. Ramp gradient maximum, 1 degree (¼ inch per foot, or 2 percent); minimum, ½ degree (⅛ inch per foot, or 1 percent)
    d. Ramp length; one or three easy strides (3 or 6 feet suggested)
    e. Overall gradient; 15 degrees (3¼ inches per foot, or 27.1 percent) or as low as 10 degrees (2⅛ inches per foot, or 17.7 percent) with 4-inch risers and 1 percent treads

Recommended designs are shown in Figures 6.58 and 6.59.

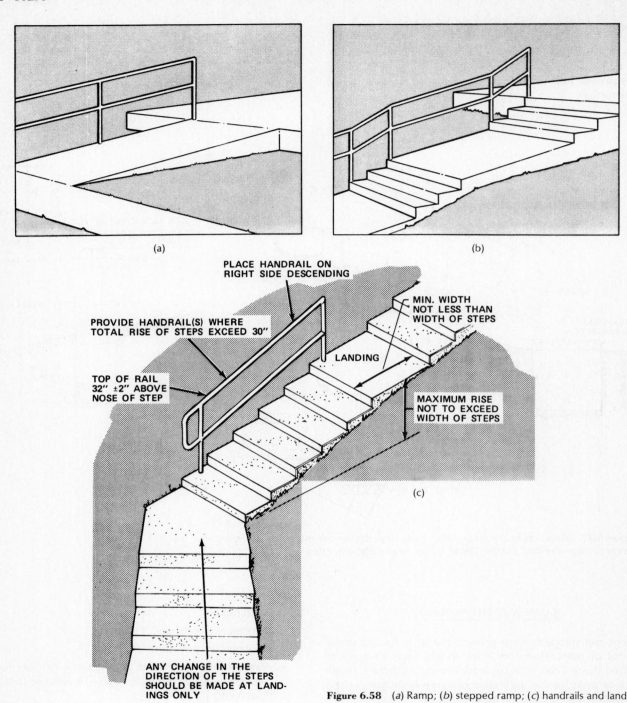

(a)

(b)

PLACE HANDRAIL ON
RIGHT SIDE DESCENDING

PROVIDE HANDRAIL(S) WHERE
TOTAL RISE OF STEPS EXCEED 30"

MIN. WIDTH
NOT LESS THAN
WIDTH OF STEPS

LANDING

TOP OF RAIL
32" ±2" ABOVE
NOSE OF STEP

MAXIMUM RISE
NOT TO EXCEED
WIDTH OF STEPS

(c)

ANY CHANGE IN THE
DIRECTION OF THE STEPS
SHOULD BE MADE AT LAND-
INGS ONLY

Figure 6.58 (*a*) Ramp; (*b*) stepped ramp; (*c*) handrails and landing.

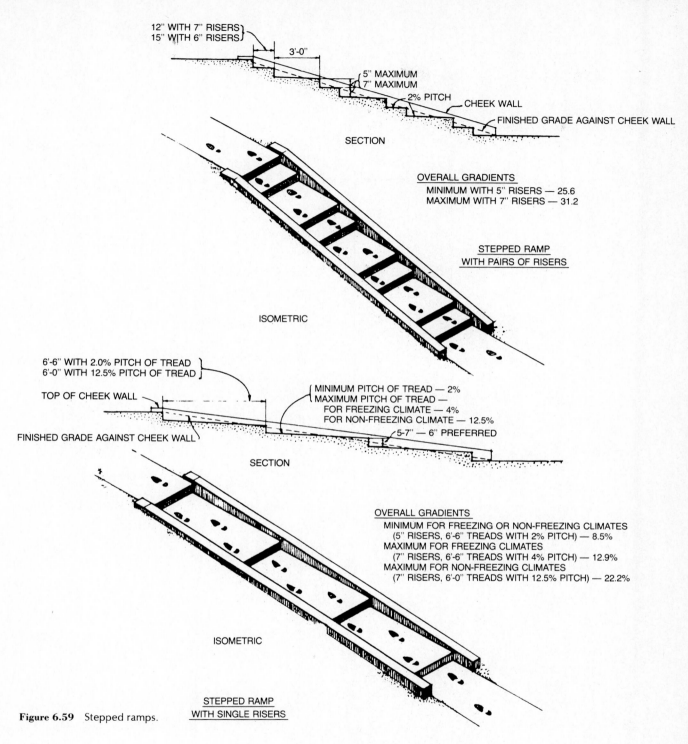

12" WITH 7" RISERS
15" WITH 6" RISERS

3'-0"

5" MAXIMUM
7" MAXIMUM

2% PITCH
CHEEK WALL

FINISHED GRADE AGAINST CHEEK WALL

SECTION

OVERALL GRADIENTS
MINIMUM WITH 5" RISERS — 25.6
MAXIMUM WITH 7" RISERS — 31.2

STEPPED RAMP
WITH PAIRS OF RISERS

ISOMETRIC

6'-6" WITH 2.0% PITCH OF TREAD
6'-0" WITH 12.5% PITCH OF TREAD

TOP OF CHEEK WALL

MINIMUM PITCH OF TREAD — 2%
MAXIMUM PITCH OF TREAD —
FOR FREEZING CLIMATE — 4%
FOR NON-FREEZING CLIMATE — 12.5%

FINISHED GRADE AGAINST CHEEK WALL

5-7" — 6" PREFERRED

SECTION

OVERALL GRADIENTS
MINIMUM FOR FREEZING OR NON-FREEZING CLIMATES
(5" RISERS, 6'-6" TREADS WITH 2% PITCH) — 8.5%
MAXIMUM FOR FREEZING CLIMATES
(7" RISERS, 6'-6" TREADS WITH 4% PITCH) — 12.9%
MAXIMUM FOR NON-FREEZING CLIMATES
(7" RISERS, 6'-0" TREADS WITH 12.5% PITCH) — 22.2%

ISOMETRIC

STEPPED RAMP
WITH SINGLE RISERS

Figure 6.59 Stepped ramps.

6.3 Benches

Suggested designs for benches and tables are shown in Figures 6.60 through 6.78.

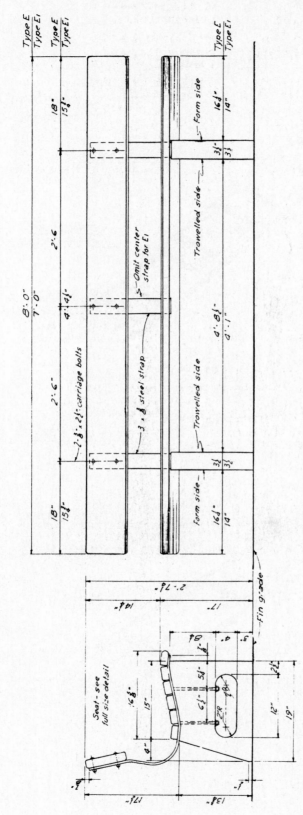

Figure 6.60 Wood bench.

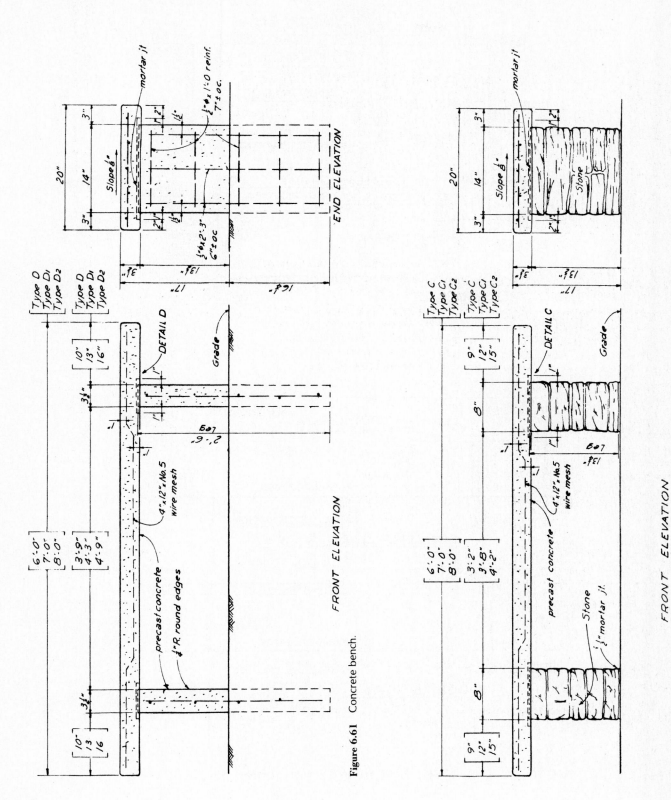

Figure 6.61 Concrete bench.

Figure 6.62 Concrete and stone bench.

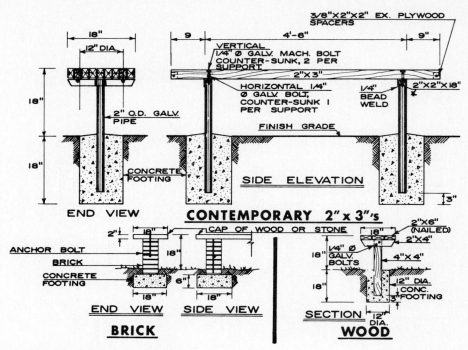

Figure 6.63 Brick and wood benches.

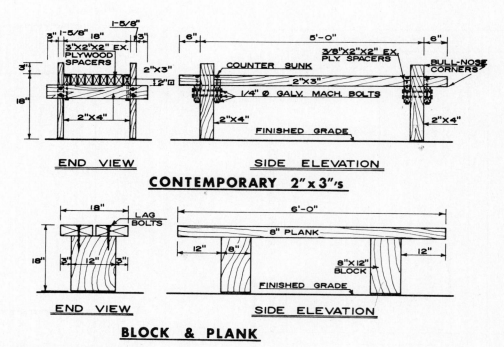

Figure 6.64 Wood benches.

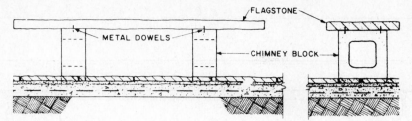

Figure 6.65 Bench. The seat may be flagstone or concrete, 2 to 3 inches thick and 18 to 24 inches wide; height is 14 to 18 inches. Supports are concrete chimney block, 8 by 16 by 16 inches.

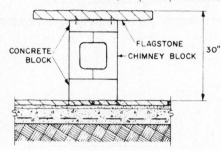

Figure 6.66 Table. The top may be flagstone, slate, or concrete, 1½ to 3 inches thick. Supports are a combination of concrete blocks 4 by 8 by 16 inches and 8 by 8 by 16 inches and chimney block 8 by 16 by 16 inches to obtain a height of 29 to 30 inches.

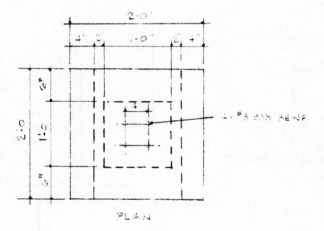

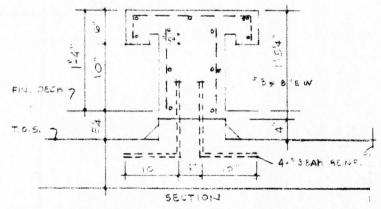

Figure 6.67 Precast concrete bench. All surfaces should be sandblast-finished.

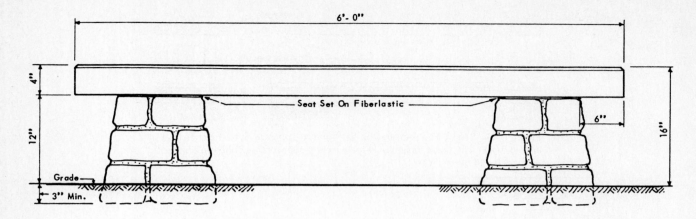

SIDE ELEVATION

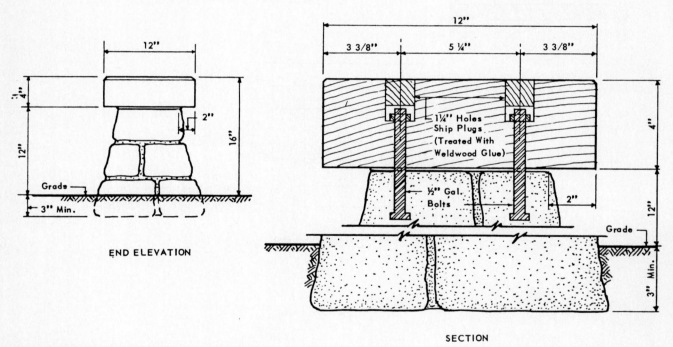

END ELEVATION

SECTION

Figure 6.68 Wood and stone bench.

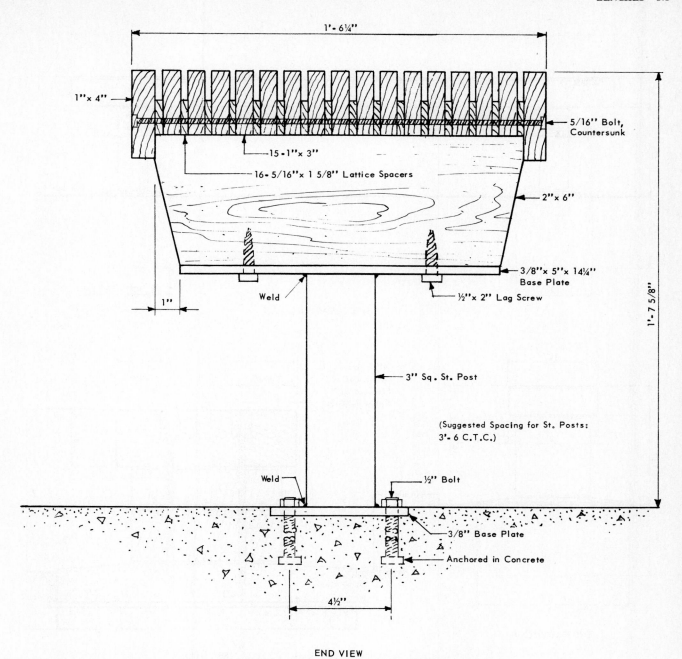

1'- 6¼"

1"x 4"

5/16" Bolt, Countersunk

15 - 1"x 3"

16 - 5/16"x 1 5/8" Lattice Spacers

2"x 6"

3/8"x 5"x 14¼" Base Plate

Weld

½"x 2" Lag Screw

1"

1'- 7 5/8"

3" Sq. St. Post

(Suggested Spacing for St. Posts: 3'- 6 C.T.C.)

Weld

½" Bolt

3/8" Base Plate

Anchored in Concrete

4½"

END VIEW

Figure 6.69 Wood and steel bench. Leave an overhang at each end of bench and make benches as long as desired, using steel posts 3 to 6 feet center to center as needed.

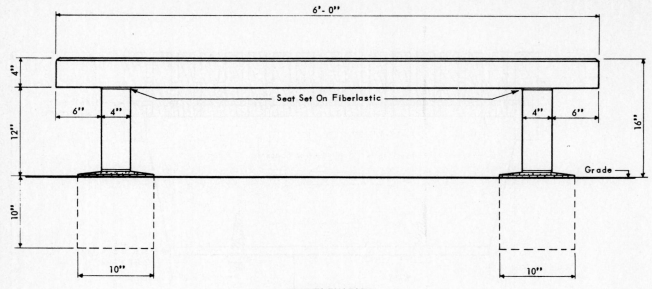

SIDE ELEVATION

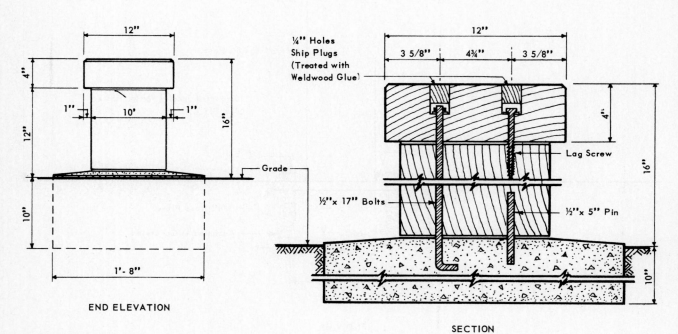

END ELEVATION

SECTION

Figure 6.70 Wood bench.

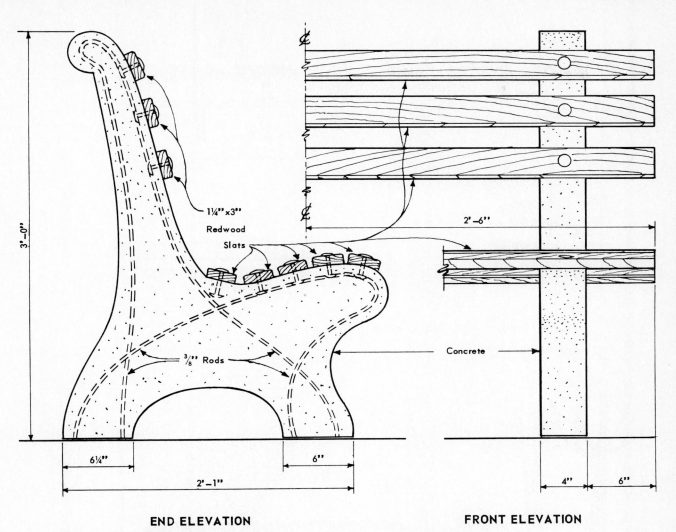

1¼"x3"
Redwood
Slats

3'-0"

⅜" Rods

2'-6"

Concrete

6¼"

6"

4"

6"

2'-1"

END ELEVATION

FRONT ELEVATION

Figure 6.71 Wood and concrete bench.

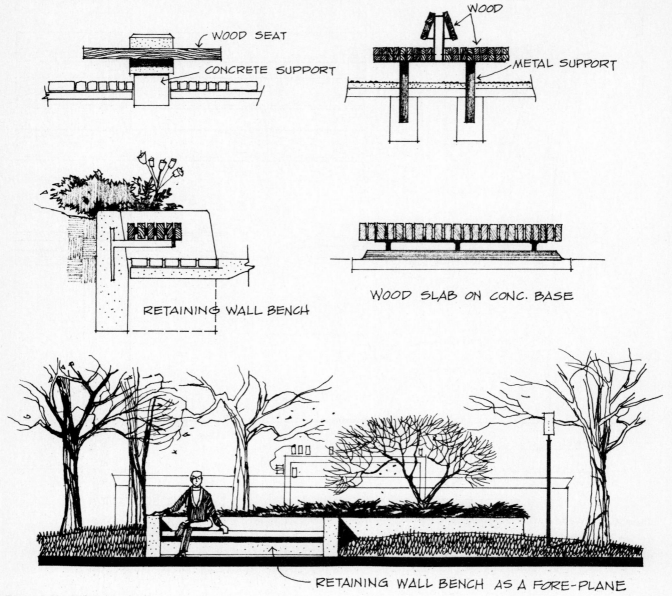

WOOD SEAT

CONCRETE SUPPORT

WOOD

METAL SUPPORT

RETAINING WALL BENCH

WOOD SLAB ON CONC. BASE

RETAINING WALL BENCH AS A FORE-PLANE

Figure 6.72   Built-in benches.

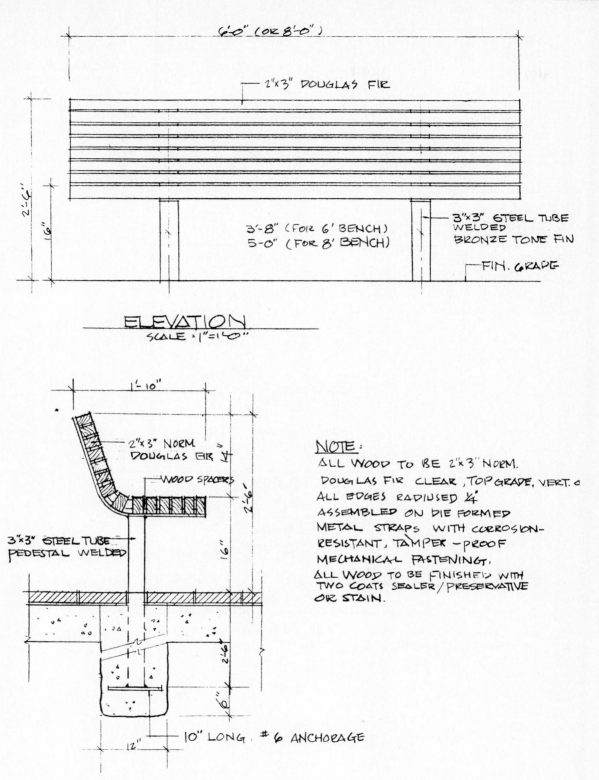

6'-0" (OR 8'-0")

2"x3" DOUGLAS FIR

2'-6"

6"

3'-8" (FOR 6' BENCH)
5'-0" (FOR 8' BENCH)

3"x3" STEEL TUBE
WELDED
BRONZE TONE FIN

FIN. GRADE

ELEVATION
SCALE = 1"=1'-0"

1'-10"

2"x3" NORM.
DOUGLAS FIR

WOOD SPACERS

2'-6"

3"x3" STEEL TUBE
PEDESTAL WELDED

16"

2'-6"

6"

12"

10" LONG. #6 ANCHORAGE

NOTE:
ALL WOOD TO BE 2"x3" NORM.
DOUGLAS FIR CLEAR, TOP GRADE, VERT. c
ALL EDGES RADIUSED ¼"
ASSEMBLED ON DIE FORMED
METAL STRAPS WITH CORROSION-
RESISTANT, TAMPER-PROOF
MECHANICAL FASTENING.
ALL WOOD TO BE FINISHED WITH
TWO COATS SEALER/PRESERVATIVE
OR STAIN.

SECTION

Figure 6.73 Wood and steel bench.

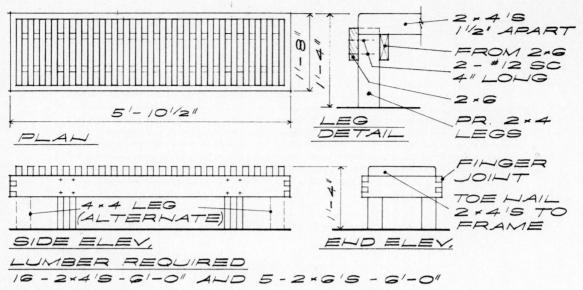

Figure 6.74 Wood bench.

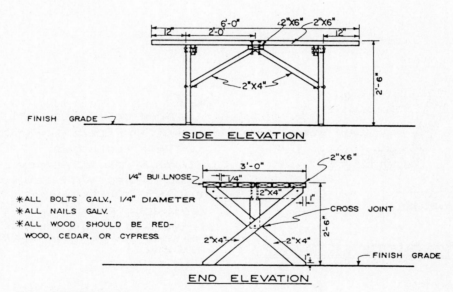

Figure 6.75 Picnic table.

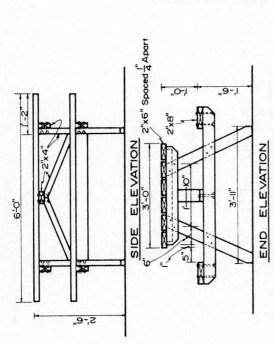

Figure 6.76 Picnic table.

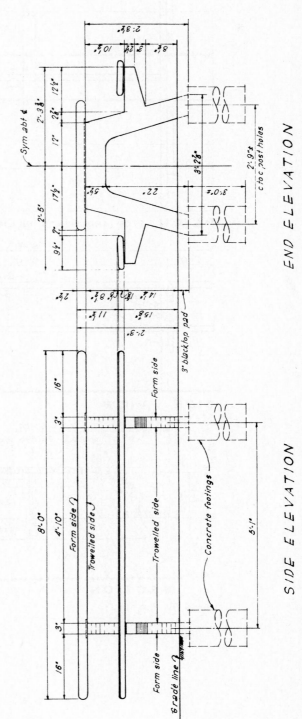

Figure 6.77 Picnic table.

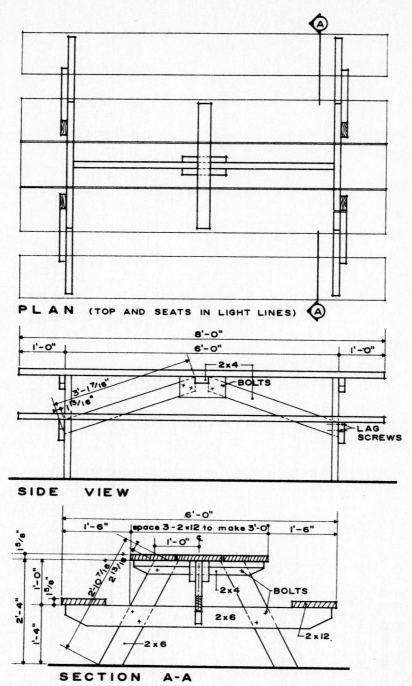

Figure 6.78 Picnic table.

6.4 Fences and Guardrails

Screening Methods Most screening today is accomplished through the use of *fences,* constructed of a variety of materials. The popularity of this screening technique is due to various characteristics of fences when used as screens. In general, fences:

- Require a minimum of land area for the screen itself
- Provide an immediately effective screen when opaque materials are used
- Are commercially available
- Are relatively easy to design and specify for construction
- Are acceptable to and often preferred by the industry
- Are adaptable to various design heights and configurations

Although there are many inherent advantages to the use of fences as screens, they may not provide the most appropriate or aesthetic choice of screen in all instances. Some problems may arise when using fences as screens due to the nature of fences themselves. For example, fences:

- Often are highly visible in the landscape
- May require excessive structural support to allow for wind load when of opaque construction
- May require periodic maintenance
- May be subject to damage from vehicles and equipment
- Can be costly to erect
- Do not adapt easily to undulating terrain

A screen solution offering several distinct advantages is the use of *earth berms.* Because of the nature of their construction, using natural materials such as earth, rock, and gravel, this method:

- Appears natural in the environment
- Requires minimum maintenance
- Is not affected by wind or other adverse site conditions
- May be erected at minimum cost in certain situations
- Offers permanent, immediate, and effective screening

This screening method has not been frequently employed since earth berms:

- Require considerable land area to reach effective heights
- May be costly to erect if fill or ''borrow'' material is not readily or cheaply available
- Do not provide a security barrier
- May affect site drainage patterns

Experience has shown that as a sole method, plantings often provide an unsuccessful and ineffective screen. Both inadequate design and inadequate choice of plant materials have caused some screens to be thin or spotty, allowing a view through or around the plants. This creates an ineffective screen. In many cases the plants specified were very young, requiring a considerable length of time before they reached a height and density sufficient to adequately screen the site.

The use of plantings as screens does offer advantages, however, since plantings:

- Appear most natural in a suburban or rural setting (provided that proper plant types are utilized)
- Require little maintenance once established
- Can be used to create a tall screen (i.e., 20 to 30 feet or taller)
- Become more effective with age (provided proper plant types are utilized)

While plantings may be best for certain suburban or rural settings, there are conditions, particularly in the urban environment, which prevent their use. Some of the disadvantages associated with the use of plantings as screens stem from the characteristics of the material. Plantings:

- May not survive in a hostile or unfavorable environment (i.e., soil, air and water pollution, climatic extremes, etc.)
- May require a period of time to achieve maximum screening ability
- May allow a view through the screen as poorly chosen plants mature and thin out
- May be susceptible to disease and other pests which diminish their effectiveness as screens
- May be destroyed by vehicles, mechanical equipment, and people.

The main advantage of the use of plantings over other types of screens is that wood, metal, and other materials deteriorate with age and may require maintenance or replacement. Plantings, however, become more visually effective with age, and in general require less maintenance during their lifetime. However, to ensure the success of the project, selection of appropriate plant species is critical.

Often the most effective and visually compatible screen is one which employs a combination of two or more screen types. The most common of these is the fence/planting combination, where the major portion of the screening is accomplished by the fence, while the plantings provide additional height and soften the harsh lines of the fence. Quite often it is possible to combine a fence and a berm, particularly when the required screen height is such that it prevents the use of a fence alone. Berms combine well with plantings to produce an effective, natural-looking, and visually attractive screen. The advantages of these combination methods are that they:

Berm/Planting

- Appear most natural in a suburban or rural setting
- May achieve additional height compared with a berm alone
- Require little maintenance once established
- May be inexpensive if borrow material is available nearby

Berm/Fence

- May achieve additional height without the need for additional structural support for fence
- Reduce visual impact of fence in the landscape
- May reduce cost of screen if borrow material is available nearby

In those instances where borrow material is not readily available at the site (or from nearby highway construction perhaps), the cost of building the berm may be prohibitive. In other locations, there may not be sufficient access to the screen site for the necessary earth-moving equipment, making construction of a berm infeasible. The major disadvantage of the combination screen which incorporates an earth berm in conjunction with either plantings or a fence is that it requires sufficient land area to construct the berm itself.

However, the definite advantages of the combination screen in terms of effectiveness and visual quality suggest that its use should be considered wherever possible.

Wood Wood offers a wide range of characteristics, which makes it desirable for use in screening. It is a versatile material which may be obtained in a number of forms and may be erected for costs which vary considerably with the design.

In general, wood fences can be built from one or a combination of four major types of wood:

1. Natural lumber
2. Chemically pretreated lumber
3. Plywood
4. Composition or synthetic wood products

These types vary in their appearance, cost, durability, and use.

Natural Lumber Lumber provides a wide range of design possibilities for screening. Many types of wood are available, including fir, hemlock, spruce, pine, cedar, oak, larch, redwood, etc. Redwood and cedar are the most naturally resistant to decay, although oak is the strongest of this group in terms of structural strength. Locally available types should be specified where possible, but it is absolutely imperative that *kiln-dried* lumber be specified for use in a screen. Otherwise, severe warping is likely to occur, causing gaps between boards and a corresponding reduction in the effectiveness of the screen. Wood which is to be in contact with the ground must be protected by some type of preservative treatment; otherwise, it will deteriorate in a very short time. Most natural woods weather to a light gray color after exposure to the elements; in many cases this helps to blend a wood screen into its surroundings. Bleaching oils, painted on a new wood surface, will create a similar effect in a short time.

Chemically Treated Lumber Various types of chemically treated lumber are available for use in fencing. These treatments are designed to protect the wood from decay, fungi, termites, and moisture absorption. The chemical types vary in their effect on the visual appearance of the wood and in their relative effectiveness. Lumber which is pressure-treated has a longer life than dipped or painted-on treatments. Some treatments incorporate a stain which colors the wood and eliminates the need for additional application of color treatment. Pressure-treated lumber will last up to ten times as long as untreated wood, and is therefore recommended for use in a junkyard screen.

Design Possibilities—Lumber Boards may be attached either vertically, horizontally, or diagonally. The *standard board* fence butts together each board, and is the least costly, but may shrink and allow one to see through the fence, particularly if unseasoned wood is used. It may be made to be more effective if used in combination with plant materials. The stockade fence is a prefabricated fence using split posts which are butted together vertically, and may be constructed with the bark either peeled off or remaining on the lumber. This type of material makes an attractive, inexpensive, and natural-looking fence, particularly useful for suburban and rural applications.

A *board and batten* fence utilizes an extra piece of wood to cover the gap between the boards, and creates an interesting shadow texture at the same time. This eliminates the problem of gaps in the fence caused by shrinkage. Another way to eliminate this problem is through the use of a *shiplap* or *clapboard* design, which incorporates overlapping boards to create a shadow effect. All of these are normally installed in a vertical pattern.

A design which has not been utilized very often is the *louvered* fence, even though it offers distinct advantages to screening. By creating a gap between boards, winds may pass through the fence rather than directing their full force at the fence itself. With proper care in the design phase, a louvered fence may be constructed so as to screen effectively without being structurally opaque. Boards may be attached horizontally in a *basketweave* pattern, a design which is visually attractive but which has several disadvantages. This type of fence is easily climbed, structurally weak as compared with other fences, and may allow a view through the gaps in the boards. It must be combined with plantings to ensure that the screen is visually opaque.

A *board-on-board* fence alternates boards on either side of the framework, and allows air movement through the fence. However, in order to be visually opaque, the boards must overlap a considerable amount. This may increase the cost considerably. It is perhaps better to alternate several

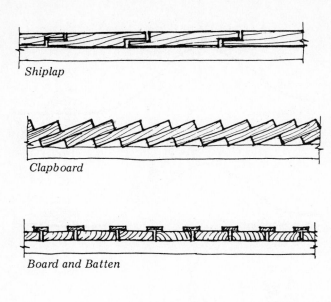

Shiplap

Clapboard

Board and Batten

Standard Board

Louvered

PLAN VIEW

Figure 6.79   Types of wood fences.

boards which are butted together. These may be designed to alternate at each post so as to eliminate the "see-through" aspect of the fence.

Metal   There are two major types of metal fences in use today — *solid panel* and *chain link*.

Metal panels are constructed from either aluminum or sheet steel which is galvanized to resist rust and corrosion. These are available either unpainted or coated with an enamel paint in various colors. A number of manufacturers produce these panels, all of which are corrugated in some configuration to increase the stiffness of the material. Primarily designed for use as roofing and siding, these panels are easily adaptable for fencing, and create an opaque and long-lasting screen. Corrugated roofing is widely available in both aluminum and steel. Aluminum, however, is a weak and easily damaged material unless specified in a heavy gauge. It must also be sturdily fastened to a strong framework in order to withstand the force of the wind. The framework may be either wood or metal, to which the panels are attached with either bolts or sheet-metal screws. To ensure that the panels will be completely opaque, they should be overlapped at each junction.

Colors are usually applied at the factory, using an enamel

paint which is baked on for additional durability. When left unpainted, galvanized steel and aluminum weather to a light gray which in some cases is more desirable than the use of colored panels. Panels which are painted a bright color tend to attract attention, while the dull gray of the unpainted panels tends to be less conspicuous, particularly where the background is also light in color. Some manufacturers produce a type of steel which is designed to oxidize to a uniform natural brown color. This material needs no surface treatment, since the oxidized layer serves as protection for the metal underneath. In addition, the dull brown color blends well in the natural environment.

Several types of aluminum and steel *siding* are available for use in screening. These panels interlock to become completely opaque. These panels may be attached in either a vertical or horizontal configuration, while others are designed specifically for only one method of attachment. They make an effective and durable screen which is easily constructed.

Chain Link   Chain link is perhaps the most widely used type of fencing, and is desirable for a number of reasons. These include long life, high security, flexibility, and relatively low cost. However, the material is not designed to be visually opaque. Chain-link fences which incorporate either wooden or metal slats still provide only marginal screening; when viewed from a moving vehicle, "the motion picture effect" renders the slats almost invisible, while allowing a full view behind the screen. The single slatted chain-link fence design is not recommended because of the "see-through effect," which renders it ineffective as a screen.

The gaps between slats must be closed in order for chain-link fencing to be effective. This has been accomplished through the use of "double slatting," in which slats are inserted at an angle to each other. This, however, is a laborious process which must be accomplished by hand in the field following construction of the fence. This requirement adds considerably to the cost of a chain-link fence, to the point where it is comparable in cost with other screening methods.

Concrete   Concrete is a suitable material for use in the construction of a screen. It is adaptable to a wide range of situations, is readily available in most areas, and can be used in a variety of forms. Perhaps the most appropriate way to utilize this material for screening is through *precast panels*. These panels may be obtained in a variety of sizes and surface textures, ranging from simulated brick or stone textures to a rough sawn wood texture which is difficult to distinguish from the real thing. The panels are generally 4 to 6 inches in thickness and 4 feet in width; lengths of 15 or more feet are possible, since the panels are heavily reinforced. The panels may be supported in a framework of channel steel, or may be directly buried in a trench for support.

While precast panels are relatively expensive to install initially, they are extremely durable and maintenance-free, and may have an application for screening.

Concrete Masonry   Concrete masonry can be used to make an attractive and extremely durable fence. There are a number of decorative concrete blocks available for the

designer to select from, depending on the desired effect and the distance from which the screen is to be viewed. Concrete units such as the "slump block" cast a shadow and are well suited to distant viewing; "split face" blocks and "split rib blocks" are other textures which are particularly well suited for a highway screening situation. Standard concrete blocks can also be used to create an attractive wall by offsetting blocks in a random pattern, and by the addition of vertical pilasters to break up the long expanse of a straight wall. While the cost of a masonry wall is relatively high, there may be situations which are appropriate for their use.

Stucco   Stucco has been used for centuries as a durable and economical building material. It is essentially a plaster mix composed of portland cement, applied over a framework of either wood or metal, to which expanded metal lath is attached. It requires little or no maintenance once it is applied.

The framing material used to support the metal lath may be either lightweight galvanized steel or pressure-treated wood. The primary function of this framing is to support the metal lath until the plaster is applied and hardened. At this point the lath and stucco take over as the primary structural support.

Stucco may also be applied as a finish coat to a concrete block wall. A variety of textural effects may be created by hand troweling the top coat.

Brick   Brick, like masonry, can be used to create an attractive screen, although the cost is relatively high. The natural earth color of brick is well suited to the outdoor environment, and as brick weathers, it actually becomes more attractive. Brick walls combine well with plants and vines to make durable and visually interesting screens. A properly constructed brick wall should be almost maintenance-free for many years.

Brick walls should have a coping which overhangs the wall and which serves a dual purpose; it seals out moisture which can penetrate through the mortar joints; and it creates a shadow effect which gives the wall a finished look. Pilasters should be used to give additional support to the wall. By using reinforcing steel in horizontal brick courses, it is possible to build a wall which consists of only one thickness of brick, and which needs no footing between vertical pilasters. This significantly reduces the amount of brick required to construct the wall.

Combination Screens   A screen which is a combination of several types of screening methods may, in many cases, be the most appropriate and effective solution to the problem. A *combination screen* is usually constructed where site limitations or height requirements make the construction of a single type of screen infeasible or undesirable. There are three types of combination screens which may be utilized for different situations:

Berm/Fence   This combination screen is an effective way to reduce the fence height required to screen the project. This may be desirable for both visual and structural reasons; it serves to reduce the apparent height of the screen, and reduces the structural strength required to support a fence

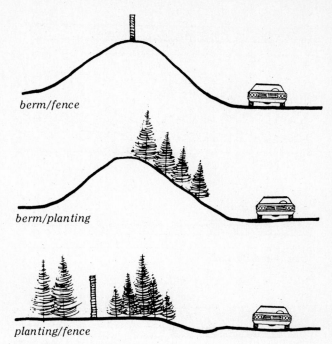

Figure 6.80   Combination screens.

of equal height. A berm/fence combination will be less affected by wind loads than a fence of the same height, and may allow for use of standard sizes and heights of fencing and posts, where special-order posts or fencing would normally be required.

Berm/Planting   This type of combination screen appears the most natural in the suburban or rural environment, yet may be effectively utilized to achieve a high degree of screening. Plantings are used to extend the effective height of a screen, and may be capable of screening a site which cannot be screened in any other manner. Where this type of screen is placed in an environment which is rural and characterized by rolling terrain, it is possible to erect a screen which is as natural as the surrounding countryside.

Planting/Fence   A combination of materials may be used to soften the visual impact of one material used alone, while still providing effective control. Plantings may be used behind, on, or in front of a fence; their placement is dependent upon the desired effect. When in back of a fence, trees serve to soften the outline of the fence against the sky, and may blend it into one more natural form as a silhouette. Trees may also increase the effective height of the fence without the need for additional structural support for a higher fence. A dark fence against a dark background of trees may help hide the presence of the fence.

Vines can be used to cover up a plain and uninteresting fence or to add an interesting texture to a smooth fence. Their color often helps to blend the fence into the background colors of the environment. Where desired, widely

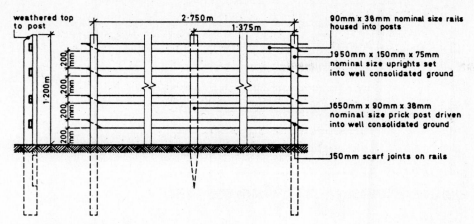

Figure 6.81   Wood fence.

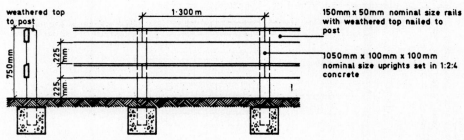

Figure 6.82   Wood fence.

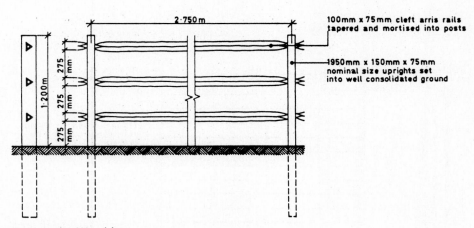

Figure 6.83   Wood fence.

spaced vines can add visual interest as a sculptural element on a plain surface of a fence. A visually attractive screen may be possible without the need for expensive types of fencing, since vines can provide the visual interest on any type of base material—even plain exterior plywood or concrete. However, since vines add weight to a fence, adequate structural support must be incorporated into the design of the fence.

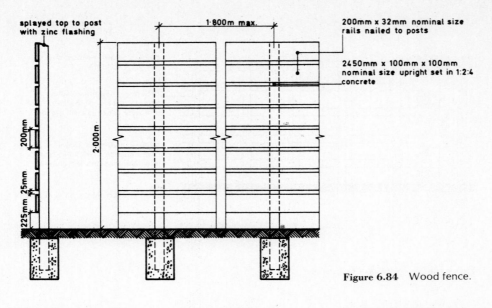

splayed top to post with zinc flashing

1·800m max.

200mm x 32mm nominal size rails nailed to posts

2450mm x 100mm x 100mm nominal size upright set in 1:2:4 concrete

200mm

2·000m

25mm

225mm

Figure 6.84 Wood fence.

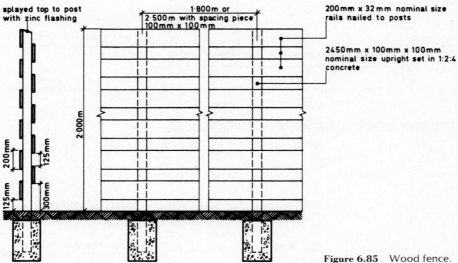

splayed top to post with zinc flashing

1·800m or 2·500m with spacing piece 100mm x 100mm

200mm x 32mm nominal size rails nailed to posts

2450mm x 100mm x 100mm nominal size upright set in 1:2:4 concrete

2·000m

200mm

125mm

125mm

300mm

Figure 6.85 Wood fence.

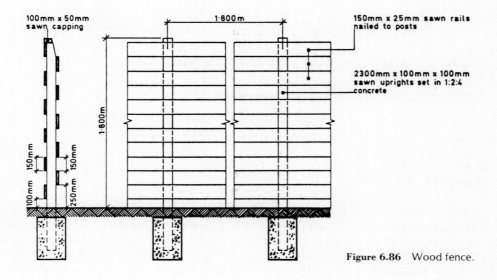

100mm x 50mm sawn capping

1·800m

150mm x 25mm sawn rails nailed to posts

2300mm x 100mm x 100mm sawn uprights set in 1:2:4 concrete

1·800m

150mm

150mm

100mm

250mm

Figure 6.86 Wood fence.

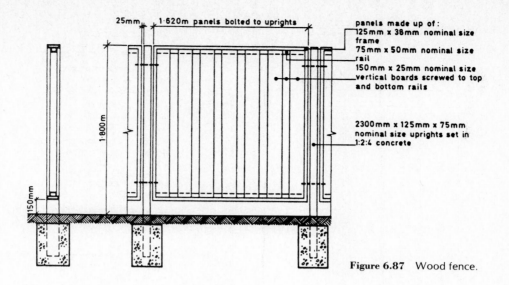

25mm

1·620m panels bolted to uprights

panels made up of:
125mm x 38mm nominal size
frame
75mm x 50mm nominal size
rail
150mm x 25mm nominal size
vertical boards screwed to top
and bottom rails

2300mm x 125mm x 75mm
nominal size uprights set in
1:2:4 concrete

1·800m

150mm

Figure 6.87 Wood fence.

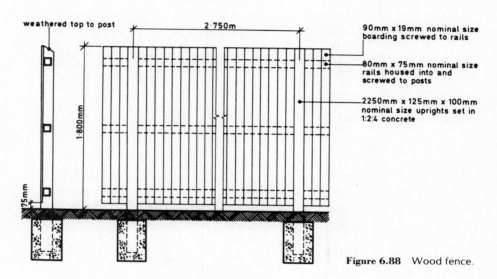

weathered top to post

2·750m

90mm x 19mm nominal size
boarding screwed to rails

80mm x 75mm nominal size
rails housed into and
screwed to posts

2250mm x 125mm x 100mm
nominal size uprights set in
1:2:4 concrete

1·800mm

75mm

Figure 6.88 Wood fence.

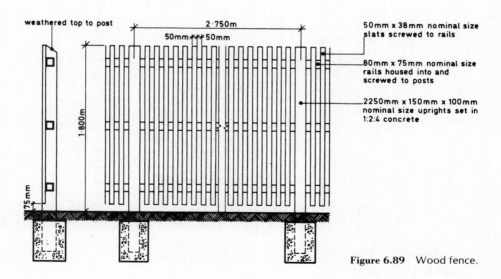

weathered top to post

2·750m

50mm ✕✕✕ 50mm

50mm x 38mm nominal size
slats screwed to rails

80mm x 75mm nominal size
rails housed into and
screwed to posts

2250mm x 150mm x 100mm
nominal size uprights set in
1:2:4 concrete

1·800m

75mm

Figure 6.89 Wood fence.

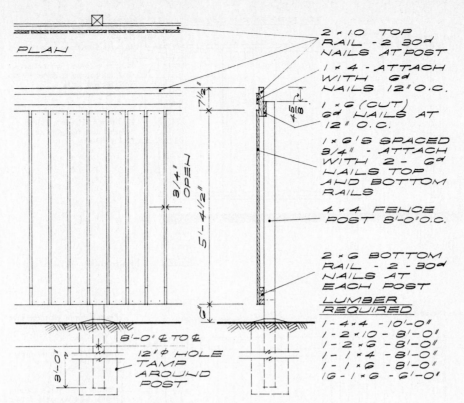

PLAN

2 × 10 TOP
RAIL - 2 30ᵈ
NAILS AT POST

1 × 4 - ATTACH
WITH 6ᵈ
NAILS 12" O.C.

1 × 6 (CUT)
6ᵈ NAILS AT
12" O.C.

1 × 6'S SPACED
3/4" - ATTACH
WITH 2 - 6ᵈ
NAILS TOP
AND BOTTOM
RAILS

4 × 4 FENCE
POST 8'-0" O.C.

2 × 6 BOTTOM
RAIL - 2 - 30ᵈ
NAILS AT
EACH POST

7½"

4-5/8"

3/4"
OPEN

5'-4½"

6"

8'-0" ℄ TO ℄

12" Φ HOLE
TAMP
AROUND
POST

3'-0"

LUMBER
REQUIRED
1 - 4 × 4 - 10'-0"
1 - 2 × 10 - 8'-0"
1 - 2 × 6 - 8'-0"
1 - 1 × 4 - 8'-0"
1 - 1 × 6 - 8'-0"
16 - 1 × 6 - 6'-0"

Figure 6.90 Vertical board fence.

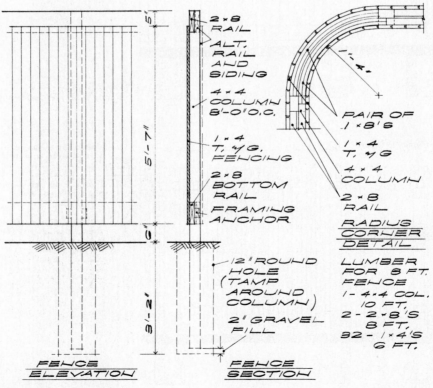

2 × 8
RAIL

ALT.
RAIL
AND
SIDING

4 × 4
COLUMN
8'-0" O.C.

1 × 4
T. & G.
FENCING

2 × 8
BOTTOM
RAIL

FRAMING
ANCHOR

12" ROUND
HOLE
(TAMP
AROUND
COLUMN)

2" GRAVEL
FILL

5'

5'-7"

6"

3'-2"

PAIR OF
1 × 8'S

1 × 4
T. & G.

4 × 4
COLUMN

2 × 8
RAIL

1'-4"

RADIUS
CORNER
DETAIL

LUMBER
FOR 8 FT.
FENCE
1 - 4 × 4 COL.
10 FT.
2 - 2 × 8'S
8 FT.
32 - 1 × 4'S
6 FT.

FENCE
ELEVATION

FENCE
SECTION

Figure 6.91 Curved fence.

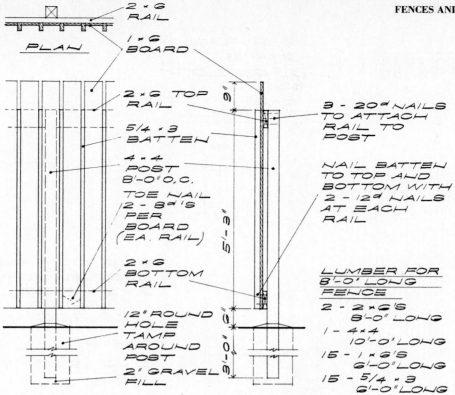

Figure 6.92 Fence with texture.

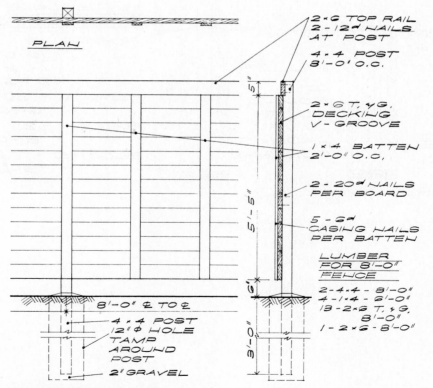

Figure 6.93 Staccato fence texture.

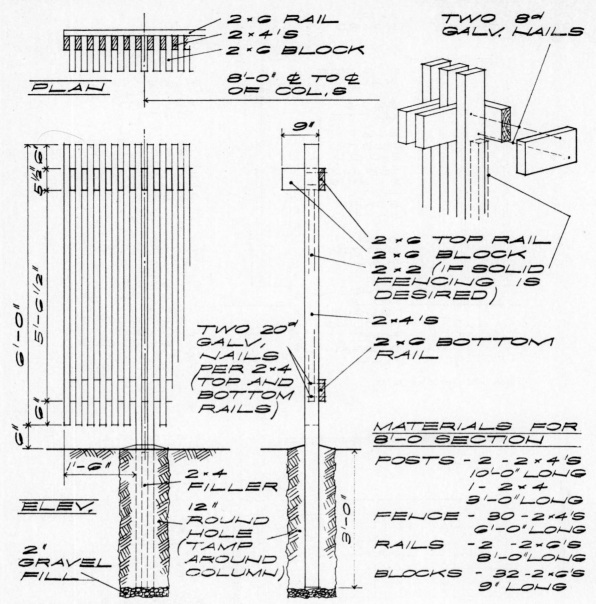

Figure 6.94 Decorative screen fence.

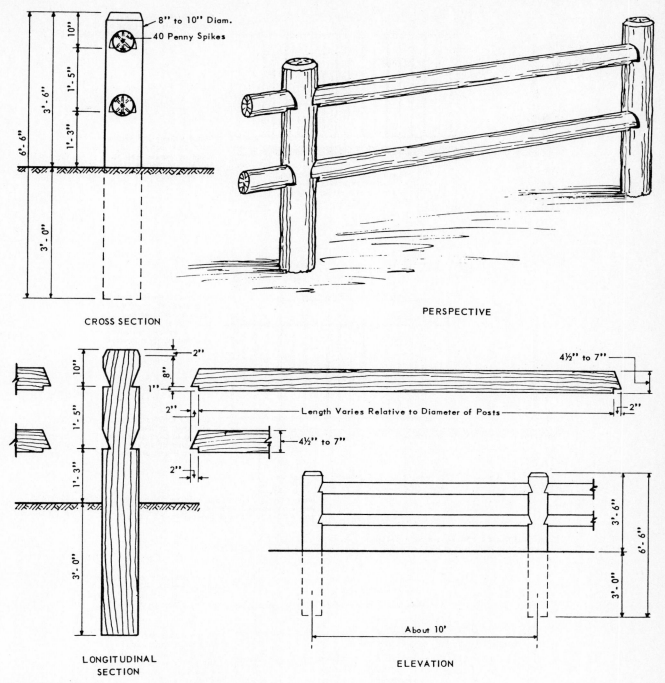

CROSS SECTION

PERSPECTIVE

LONGITUDINAL
SECTION

ELEVATION

Figure 6.95 Rustic rail fence.

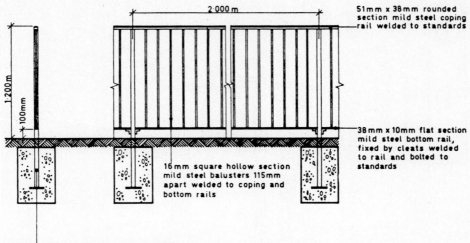

51mm x 38mm rounded section mild steel coping rail welded to standards

38mm x 10mm flat section mild steel bottom rail, fixed by cleats welded to rail and bolted to standards

16mm square hollow section mild steel balusters 115mm apart welded to coping and bottom rails

1700mm x 51mm x 51mm square hollow section mild steel standards set in 1:2:4 concrete

2·000 m

1·200 m

100mm

Figure 6.96 Metal fence.

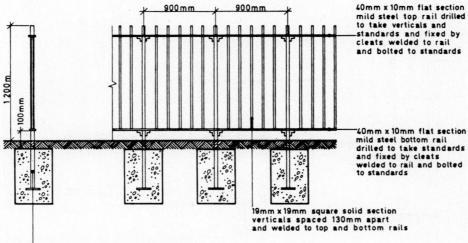

40mm x 10mm flat section mild steel top rail drilled to take verticals and standards and fixed by cleats welded to rail and bolted to standards

40mm x 10mm flat section mild steel bottom rail drilled to take standards and fixed by cleats welded to rail and bolted to standards

19mm x 19mm square solid section verticals spaced 130mm apart and welded to top and bottom rails

1700mm x 25mm x 25mm square hollow section mild steel standards set in 1:2:4 concrete

900mm   900mm

1·200 m

100mm

Figure 6.97 Metal fence.

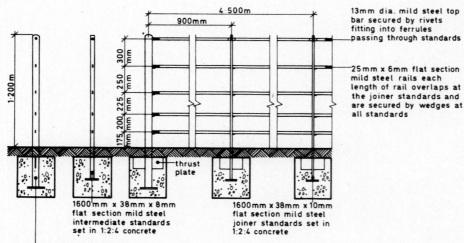

13mm dia. mild steel top bar secured by rivets fitting into ferrules passing through standards

25mm x 6mm flat section mild steel rails each length of rail overlaps at the joiner standards and are secured by wedges at all standards

4·500 m

900mm

300 mm
250 mm
225 mm
200 mm
175 mm

1·200 m

thrust plate

1600mm x 38mm x 8mm flat section mild steel intermediate standards set in 1:2:4 concrete

1600mm x 38mm x 10mm flat section mild steel joiner standards set in 1:2:4 concrete

1700mm x 76mm dia. mild steel main pillar set in 1:2:4 concrete (ends and corners only)   **Figure 6.98** Metal fence.

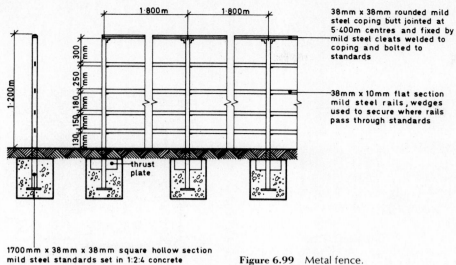

38mm x 38mm rounded mild steel coping butt jointed at 5·400m centres and fixed by mild steel cleats welded to coping and bolted to standards

38mm x 10mm flat section mild steel rails, wedges used to secure where rails pass through standards

thrust plate

1700mm x 38mm x 38mm square hollow section mild steel standards set in 1:2:4 concrete

Figure 6.99 Metal fence.

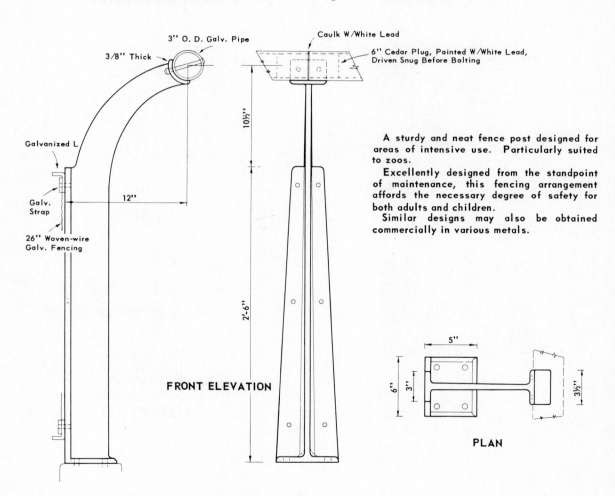

3" O. D. Galv. Pipe

3/8" Thick

Galvanized L

Galv. Strap

26" Woven-wire Galv. Fencing

12"

Caulk W/White Lead

6" Cedar Plug, Painted W/White Lead, Driven Snug Before Bolting

10½"

2'-6"

A sturdy and neat fence post designed for areas of intensive use. Particularly suited to zoos.

Excellently designed from the standpoint of maintenance, this fencing arrangement affords the necessary degree of safety for both adults and children.

Similar designs may also be obtained commercially in various metals.

FRONT ELEVATION

5"

6"

3"

3½"

PLAN

SIDE ELEVATION   **Figure 6.100**   Steel fence post.

FENCES

Various types of fences are illustrated in Figures 6.101 through 6.108.

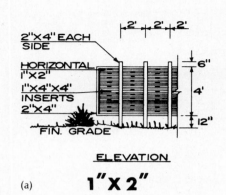

(a) **1"X 2"**

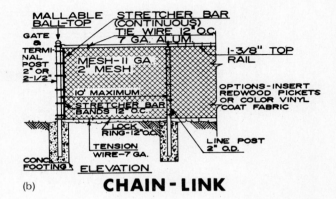

(b) **CHAIN - LINK**

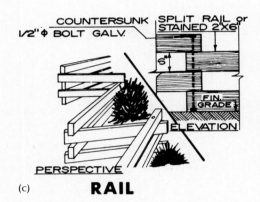

(c) **RAIL**

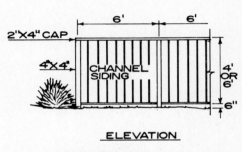

(d) **CHANNEL** (SOLID)

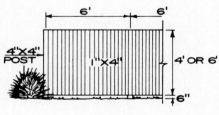

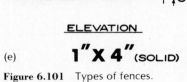

(e) **1"X 4"** (SOLID)

Figure 6.101 Types of fences.

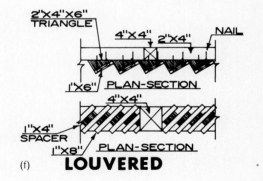

(f) **LOUVERED**

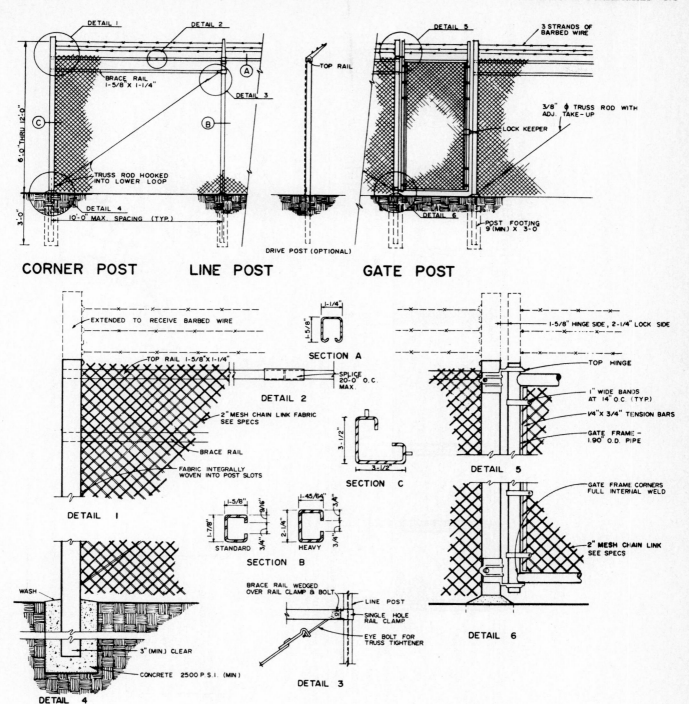

Figure 6.102 Cyclone fence.

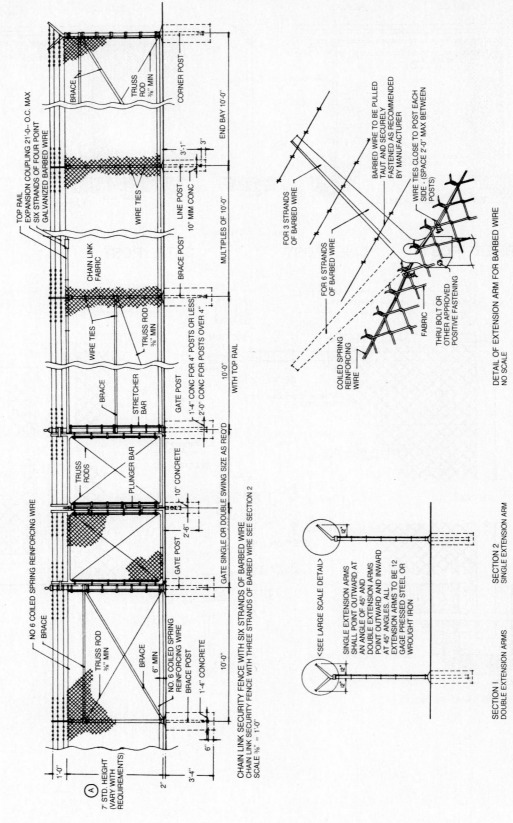

Figure 6.103 Security fence with barbed wire.

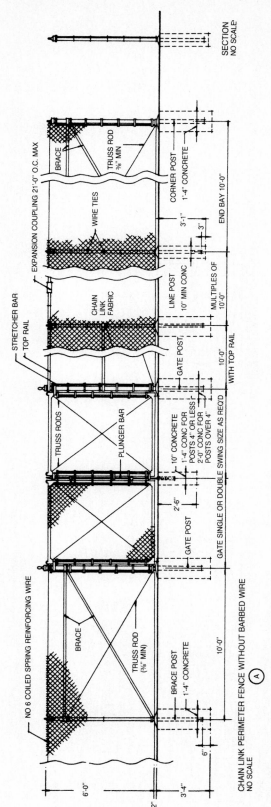

Figure 6.104 Security fence without barbed wire.

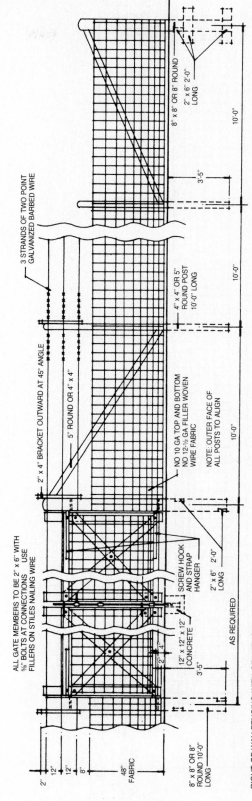

Figure 6.105

–597–

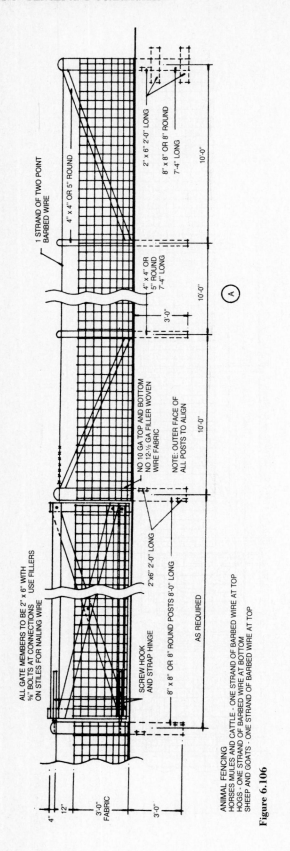

1 STRAND OF TWO POINT BARBED WIRE

4" x 4" OR 5" ROUND

2" x 6" 2'-0" LONG

8" x 8" OR 8" ROUND 7'-4" LONG

10'-0"

4" x 4" OR 5" ROUND 7'-4" LONG

10'-0"

3'-0"

(A)

10'-0"

NO 10 GA TOP AND BOTTOM NO 12½ GA FILLER WOVEN WIRE FABRIC

NOTE: OUTER FACE OF ALL POSTS TO ALIGN

ALL GATE MEMBERS TO BE 2" x 6" WITH ⅝" BOLTS AT CONNECTIONS   USE FILLERS ON STILES FOR NAILING WIRE

2"x6" 2'-0" LONG

8" x 8" OR 8" ROUND POSTS 8'-0" LONG

AS REQUIRED

SCREW HOOK AND STRAP HINGE

4"

12"

3'-0" FABRIC

3'-0"

ANIMAL FENCING
HORSES MULES AND CATTLE - ONE STRAND OF BARBED WIRE AT TOP
HOGS - ONE STRAND OF BARBED WIRE AT BOTTOM
SHEEP AND GOATS - ONE STRAND OF BARBED WIRE AT TOP

Figure 6.106

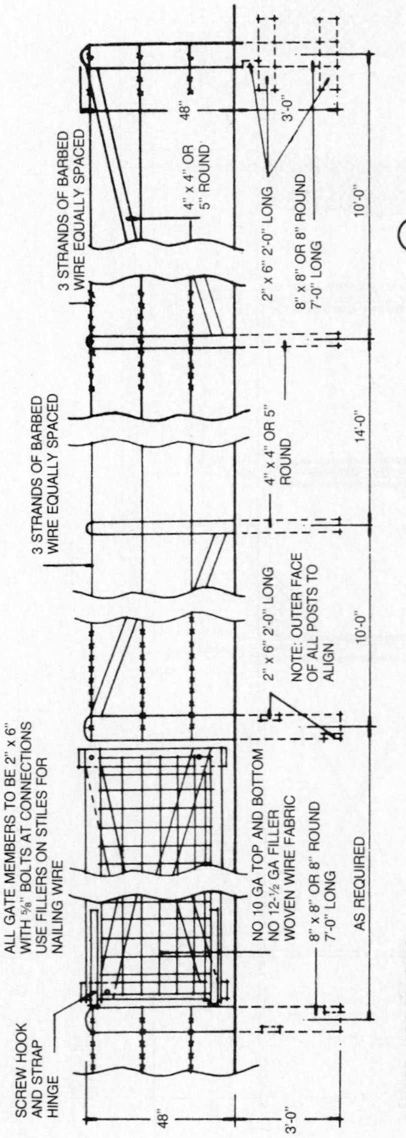

3 STRANDS OF BARBED WIRE EQUALLY SPACED

48"

3'-0"

4" x 4" OR 5" ROUND

2" x 6" 2'-0" LONG

8" x 8" OR 8" ROUND 7'-0" LONG

10'-0"

(A)

3 STRANDS OF BARBED WIRE EQUALLY SPACED

4" x 4" OR 5" ROUND

14'-0"

2" x 6" 2'-0" LONG

NOTE: OUTER FACE OF ALL POSTS TO ALIGN

10'-0"

ALL GATE MEMBERS TO BE 2" x 6" WITH ⅝" BOLTS AT CONNECTIONS USE FILLERS ON STILES FOR NAILING WIRE

NO 10 GA TOP AND BOTTOM NO 12½ GA FILLER WOVEN WIRE FABRIC

8" x 8" OR 8" ROUND 7'-0" LONG

AS REQUIRED

SCREW HOOK AND STRAP HINGE

48"

3'-0"

BARBED WIRE FENCING
48" HIGH - 3 STRANDS

Figure 6.107

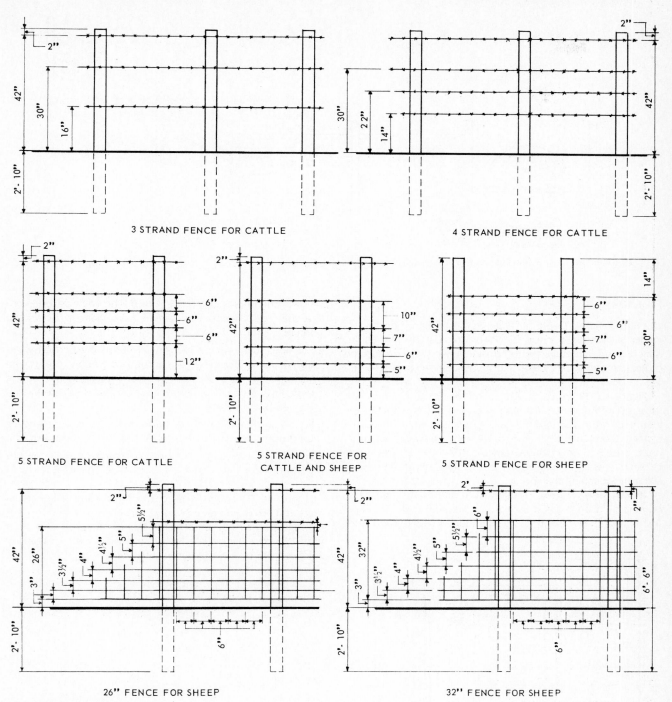

Figure 6.108 Farm fences.

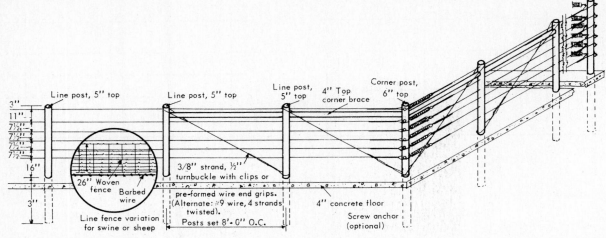

Figure 6.109 Cable fencing — one type of installation.

Farm Fences

Cable Fences Cable fencing consists of heavy galvanized cables attached to metal posts or running through holes drilled through metal or wood posts. When well constructed of good materials, it makes strong, durable fencing. (See Figure 6.108.)

Figure 6.109 shows construction where the cables run through holes drilled through wood posts. Each cable is attached to the anchor post by a spring assembly. The cable is stretched with a block and tackle until the spring begins to open and is then clamped around the next anchor post. When necessary, the tension of the cable is adjusted by tightening or loosening the spring.

Woven Wire Fences Figure 6.110 shows the five most commonly used styles or designs of field or stock fencing — 1155, 1047, 939, 832, and 726 — combined with barbed wire.

- Cattle and horses, use fence *A* or *B*. The single barbed wire at the top prevents the animals from mashing down the fence.

- For hogs, use fence *C*, *D*, or *E* without the barbed wires above the woven wire. The barbed wire below the woven wire discourages the animals from crawling or rooting under the fence. Styles 939 and 832 are available with a barbed bottom wire.

- Style 726 (fence *E* without barbed wire) is convenient for temporarily confining hogs while they hog down corn.

- For sheep, use style 832 or 726 (fence *D* or *E* without barbed wire). Barbs may tear the fleece on sheep.

Figure 6.111 shows a fence for protecting sheep from dogs and coyotes. The extended barbed wire at the top discourages dogs from jumping the fence. An apron of woven wire 18 inches wide along the ground will prevent predatory animals from burrowing beneath the fence.

- Cattle, horses, hogs, and sheep all in the same field, use fence *A*, *B*, *D*, or *E*.

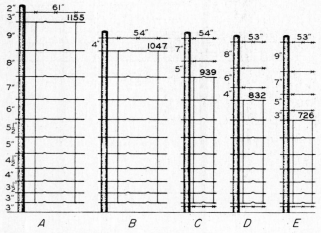

Figure 6.110 Standard styles or designs of woven wire fencing combined with barbed wire. Stay (vertical) wires are spaced 12 inches in fences *A* and *B* and 6 inches in *C*, *D*, and *E*.

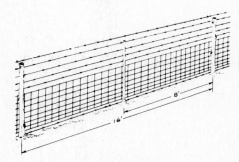

Figure 6.111 Fence for protecting sheep from dogs and coyotes.

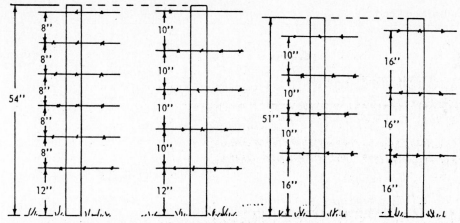

Figure 6.112 Common spacings of wires in barbed-wire fences.

Poultry-garden fencing comes in two standard styles — 2158 and 1948 — and chick fencing comes in three styles — 2672, 2360, and 2048.

Wire netting of 1- or 2-inch mesh is made in nine heights ranging from 12 to 72 inches. The netting is commonly used for fencing small poultry yards, cages, poultry house windows, and tree guards. The 1-inch-mesh wire is recommended for confining baby chicks, turkey poults, and goslings.

Barbed-Wire Fences Figure 6.112 shows the usual wire spacing in three- to six-strand barbed-wire fences. As few as two strands are sometimes used to fence large cattle ranges in the western states.

Barbed-wire suspension fences (Figure 6.113) are often used as cross fencing and boundary fencing of large cattle ranges. They consist of four to six strands of the wire supported by posts spaced 80 to 120 feet apart. Twisted wire stays, spaced about 16 feet apart, hold the wires apart.

Figure 6.114 shows the kinds of standard barbed wire commonly available. The 12½-gauge wire with two-point barbs is the most widely used for cattle ranges. For smaller fields where cattle may subject the fence to considerable pressure, four-point barbs may be more effective. The lighter 14-gauge wire is commonly used for temporary fencing.

You can also buy high-tensile barbed wire, which is stronger and more durable than the comparable sizes of standard wire. The 13½-gauge high-tensile wire, for example, has a breaking strength equal to that of the 12½-gauge standard wire.

Barbed wire, like woven wire, comes with a protective coating of either zinc or aluminum. Thickness of the coating is the same as on comparable sizes of woven wire. Under the same climatic conditions, aluminum-coated wire would be more durable than zinc-coated wire.

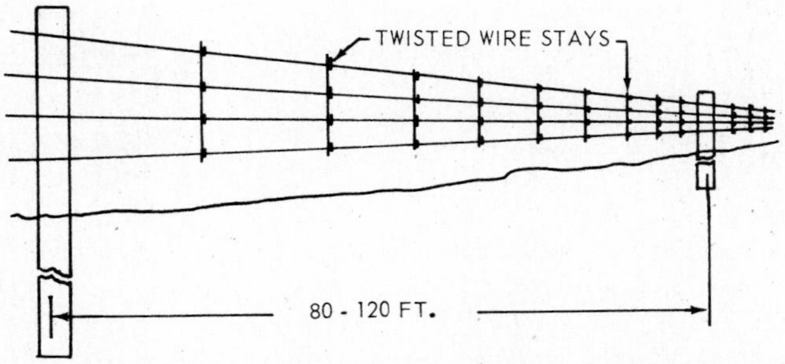

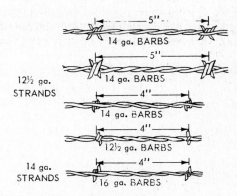

Figure 6.113 Barbed-wire suspension fence, commonly used as cross or boundary fencing on cattle ranges.

Figure 6.114 Common kinds of barbed wire.

Fencing Materials Comparison Chart

Screen type	Relative cost	Durability	Ease of construction	Maximum design height (feet)	Maintenance required	Security offered	Material flexibility
Wood fences:							
Vertical board	Med.	15 years if treated with pressure	Construction on site—relatively simple	10	Paint or stain if desired—5 years	High	Withstand minor damage only
Diagonal board	Med. high	15 years if treated with pressure	Constructed on site or modular—semidifficult	10	Paint or stain if desired—5 yars	High	Withstand minor damage only
Horizontal board	Med.	15 years if treated with pressure	Constructed on site—simple	10	Paint or stain if desired—5 years	High	Slightly flexible
Board and batten	Med.	15 years if treated with pressure	Constructed on site—relatively simple	10	Paint or stain if desired—5 years	High	Withstand minor damage only
Board on board	Med.	15 years if treated with pressure	Constructed on site—simple	10	Paint or stain if desired—5 years	High	Withstand minor damage only
Clapboard	Med. high	15 years if treated with pressure	Constructed on site—semidifficult	10	Paint or stain if desired—5 years	High	Withstand minor damage only
Louvered*	Med.	15 years if treated with pressure	Constructed on site—difficult	8	Paint or stain if desired—5 years	High	Withstand minor damage only
Basketweave*	Med.	15 years if treated with pressure	Constructed on site—modular materials	10	Paint or stain if desired—5 years	Climbable	Withstand minor damage only
Stockade	Med. low	15 years if treated with pressure	Modular, 8-foot sections—relatively simple	10	None—natural finish	High	Withstand minor damage only
Plywood, standard	Med./low	10 years if treated with pressure	Modular—relatively simple	12	Paint or stain—5 years	High	Slightly flexible
Plywood, texture 1-11	Med.	10 years if treated with pressure	Modular—relatively simple	12	Paint or stain—5 years	High	Slightly flexible
Composition (Aspenite)	Med.	10 years if treated with pressure	Modular—relatively simple	12	Paint or stain—5 years	High	Slightly flexible
Metal:							
Aluminum, corrugated	Med./high	5–10 years	2-ft panels—relatively simple	12	May tear in wind, requiring replacement	May be torn off	Semiflexible
Steel, galvanized corrugated	Very high	10–20 years	2-ft panels—relatively simple	12	Painting, 10–15 years	High	Semiflexible
Steel painted corrugated	Very high	10–20 years	2-ft panels—relatively simple	12	Painting, 10 years	High	Semiflexible
Armco panel	Very high	10–20 years	16-in panels—relatively simple	12–15	Painting, 20 years	High	Rigid, could dent easily
Vertical sliding panel	Very high	10–20 years	2-ft panels—relatively simple	12–15	Painting, 20 years	High	Rigid, could dent easily
Steel boards	Very high	15–20 years	Relatively simple	8	Painting 10–15 years	High	Rigid

Design flexibility	Recommended framework 8 feet high	Recommended framework 10 feet high	Post size and spacing	Method of fastening	Structural strength	Visual quality	Comments
Site-adaptable	3 × 4 horiz. at 4-ft center, 4 × 4 post	3 × 4 horiz. at 4-ft center, 6 × 4 post	4 × 4 at 8 ft, 6 × 4 at 8 ft	Nail, screw, bolt	High	Blends well; rustic appearance if unpainted	Good for rural area
Level terrain best	3 × 4 horiz. at 4-ft center, 4 × 4 post	3 × 4 horiz. at 4-ft center, 6 × 4 post	4 × 4 at 8 ft, 6 × 4 at 8 ft	Nail, screw, bolt	Med./high	Attractive; eye catching. Good for close viewing distances	Good for suburban area
Level terrain preferable	3 × 4 horiz. at 4-ft center, 4 × 4 post	3 × 4 horiz. at 4-ft center, 6 × 4 post	4 × 4 at 8 ft, 6 × 4 at 8 ft	Nail, screw, bolt	High	Rustic; blends well	Good for rural areas
Site-adaptable	3 × 4 horiz. at 4-ft center, 4 × 4 post	3 × 4 horiz. at 4-ft center, 6 × 4 post	4 × 4 at 8 ft, 6 × 4 at 8 ft	Nail, screw, bolt	High	Blends well; attractive. Good for close viewing distances	Good for suburban area
Site-adaptable	3 × 4 horiz. at 4-ft center, 4 × 4 post	3 × 4 horiz. at 4-ft center, 6 × 4 post	4 × 4 at 8 ft, 6 × 4 at 8 ft	Nail, screw, bolt	High	Blends well; attractive. Good for close viewing distances	Good for suburban area; allows wind passage
Site-adaptable	3 × 4 horiz. at 4-ft center, 4 × 4 post	3 × 4 horiz. at 4-ft center, 6 × 4 post	4 × 4 at 8 ft, 6 × 4 at 8 ft	Nail, screw, bolt	High	Blends well; attractive. Good for close viewing distances	Good for suburban area
Level terrain	3 × 4 horiz. top and bottom, 4 × 4 post	N/A	4 × 4 at 8 ft o.c.	Nail, screw	Med.	Attractive	Good for suburban area; allows wind passage
Level terrain	3 × 4 horiz. top and bottom, 1 × 6 slats	3 × 4 horiz. top and bottom, 1 × 8 slats	4 × 4	Nail, screw	Med. low	Attractive	Good for suburban area; may allow view through from some angles; allows wind passage
Site-adaptable	3 × 4 horiz. at 4-ft center, 4 × 4 post	3 × 4 horiz. at 4-ft center, 6 × 4 post	4 × 4	Nail, screw, bolt	Med./high	Rustic; blends well	Good for rural area
Site-adaptable	¾-in marine plywood, 3 × 4 horiz. at 4-ft center	¾-in marine plywood, 3 × 4 horiz. at 4-ft center	6 × 4 at 8-ft center	Nail, screw, bolt	High	Good, if distant view; should use in combination with plantings	Good for rural area
Site-adaptable	⅝-in panel, 3 × 4 horiz. at 3-ft center	⅝-in panel, 3 × 4 horiz. at 3-ft center	6 × 4 at 8-ft center	Nail, screw, bolt	High	Good	Good for suburban area
Site-adaptable	⅝-in panel, 3 × 4 horiz. at 3-ft center	⅝-in panel, 3 × 4 horiz. at 3-ft center	6 × 4 at 8-ft center	Nail, screw, bolt	High	Good if distant view	Good for suburban area
Adaptable	Galvanized-steel tube or wood 2 × 6 horizontals	Galvanized-steel tube or wood 2 × 8 horizontals	6 × 4 at 8-ft center	Nail, screw, bolt	Med./low	Fair if distant view; should combine with plantings	Good for rural area; may reflect heat from sun to nearby plants
Adaptable	Galvanized-steel tube or wood 2 × 6 horizontals	Galvanized-steel tube or wood 2 × 8 horizontals	6 × 4 at 8-ft center	Nail, screw, bolt	High	Fair if distant view; should combine with plantings	Good for rural, industrial area; may reflect heat from sun to nearby plants
Adaptable	Galvanized-steel tube or wood 2 × 6 horizontals	Galvanized-steel tube or wood 2 × 8 horizontals	6 × 4 at 8-ft center	Nail, screw, bolt	High	Good if color-compatible; should combine with plantings	Good for rural, industrial area; may reflect heat from sun to nearby plants
Best for level terrain	Interlocking panels, H section post	Interlocking panels, H section post	w 8 × 10 at 12 ft o.c.; w 8 × 13 at 12 ft o.c.	Sheet-metal screws, bolts	Med.	Good if color-compatible	Good for rural, industrial area; may reflect heat from sun to nearby plants
Adaptable	Galvanized-steel tube or manufacturer's standard	Galvanized-steel tube or manufacturer's standard	Galvanized-steel tube 16-gauge at 8 ft o.c.	Sheet-metal screws, bolts	Med.	Good if color-compatible	Good for rural, industrial area; may reflect heat from sun to nearby plants
Level terrain	Manufacturer's standard	N/A	Manufacturer's std.	Boards fit into channels	Med./low	Good—if compatible colors are utilized	Good for rural and suburban areas

Fencing Materials Comparison Chart (*Continued*)

Screen type	Relative cost	Durability	Ease of construction	Maximum design height (feet)	Maintenance required	Security offered	Material flexibility
COR-TEN steel panel	Med./high	25 years	Relatively simple	12	None	High	Semiflexible; dents easily
Chain link/*wood and metal slat	Med./high	25 years	Relatively simple, but high labor to insert slats	10	None/replace broken slats	High	Flexible
Chain link/*double metal slat	High	25 years	Relatively simple, but high labor to insert slats	10	None/replace slats	High	Flexible
Concrete:							
Standard concrete block	Very high	20–30 years	Some preparation required— footings, etc.	12	None	High	Inflexible but strong
Slump block	Very high	20–30 years	Some preparation required— footings, etc.	12	None	High	Inflexible but strong
Split block	Very high	20–30 years	Some preparation required— footings, etc.	12	None	High	Inflexible but strong
Split rib	Very high	20–30 years	Some preparation required— footings, etc.	12	None	High	Inflexible but strong
Precast panel	Med.	20–30 years	No footing required	12	None	High	Inflexible but strong
Fleming Panel-lock	Med./low	20–30 years	No footing required	12	None	May be removed by thieves	Inflexible but strong
Fanwall	Med.	20–30 years	No footing required	12	None	High	Inflexible but strong
Lightweight panel	Med.	20–30 years	No footing required	12	None	High	Inflexible but strong
Stucco over metal lath	High	20–30 years	Requires footings and wood or metal framework	12	None	High	Inflexible but strong
Cast-in-place	Very high	20–30 years	Some preparation required—form work	12	None	High	Inflexible but strong
Brick, common	Very high	20–30 years	Some preparation required, footings, etc.	12	None	High	Inflexible but strong
Brick, fancy	Very high	20–30 years	Some preparation required, footings, etc.	12	None	High	Inflexible but strong

*May not be 100 percent effective if used alone.

Design flexibility	Recommended framework 8 feet high	Recommended framework 10 feet high	Post size and spacing	Method of fastening	Structural strength	Visual quality	Comments
Level terrain	8-ft-wide panels	8-ft-wide panels	16 gauge at 8 ft o.c.	Sheet-metal screws	Med.	Good—oxidizes to natural color	Good for rural, industrial areas
Adaptable to grades	4-in o.d. galvanized posts, tension wire	5 9⁄16-in o.d. galvanized posts, tension wire	Sch. 40 at 8 ft o.c.	Wire/clamps	High	Good; preferably combine with plantings. Weathers to natural color	Good for rural, suburban areas. May allow some view through screen
Adaptable to grades	4-in o.d. galvanized posts, tension wire	5 9⁄16-in o.d. galvanized posts, tension wire	Sch. 40 at 8 ft o.c.	Wire/clamps	High	Fair to poor. Distant view only. Available colors do not blend well	Good for rural, industrial areas. May allow some view through screen
Adaptable	Min. 6-in conc. footing on each side of wall	Min. 6-in conc. footing on each side of wall	N/A	Mortar	High	Fair; good if distant view. Preferably combine with plantings	Good for urban areas
Adaptable	Min. 6-in conc. footing on each side of wall	Min. 6-in conc. footing on each side of wall	N/A	Mortar	High	Good	Good for suburban areas
Adaptable	Min. 6-in conc. footing on each side of wall	Min. 6-in conc. footing on each side of wall	N/A	Mortar	High	Very good	Good for suburban areas
Adaptable	Min. 6-in conc. footing on each side of wall	Min. 6-in conc. footing on each side of wall	N/A	Mortar	High	Very good	Good for suburban areas
Adaptable	Steel-reinforced	Steel-reinforced	Channel iron, varies with panel	Slide into posts	High	Good if textured	Good for urban area
Adaptable	Manufacturer's standard	Manufacturer's standard	Manufacturer's std.	Slide into concrete posts	Med./high	Good, available in colors	Good for urban, suburban areas
Level terrain preferable	Integral	Integral	None	Cables lock together	Med./high	Good distant view; good if textured	Good for urban, some suburban areas
Adaptable	Fiberglass reinforced, integral	Fiberglass reinforced, integral	Channel iron, varies with panel	Slide into posts	Med./high	Good distant view; good if textured	Good for urban area
Adaptable	Galvanized metal frame or treated 2 × 4	A frame design, galvanized metal or 2 × 4	Varies with design	Expanded metal lath	Med./high	Very good if textured; may be colored or painted	Good for urban or suburban locations with similar materials
Adaptable	Steel-reinforced	Steel-reinforced	N/A	N/A	High	Good if textured	Good for urban areas
Adaptable	Min. 6-in conc. footing on each side of wall	Min. 6-in conc. footing on each side of wall	N/A	Mortar	High	Good; natural earth colors blend well	Good for suburban areas
Adaptable	Min. 6-in conc. footing on each side of wall	Min. 6-in conc. footing on each side of wall	N/A	Mortar	High	Very good; natural earth colors blend well	Good for urban or suburban locations with similar materials

PARKING BARRIER

Complete pressure treatment of all posts and rails is recommended for long life. A piece of tarred felt placed in the notch section is valuable in reducing deterioration at that critical point.

Since parking barriers are more subject to vehicle impact than are parallel roadside guardrails (although not often so violently), a log segment laid in the ground a few feet in front of the barrier minimizes this possible damage by serving as a curbing.

All hardware should be galvanized.

See Figures 6.115 and 6.116 for diagrams of guardrails and a parking barrier.

Figure 6.117 shows a parking barrier post with three different horizontal barrier suggestions; the first of 3- by 8-inch timber; the second of the same width and edge thickness, but with a convex outer face; the third of log notched to fit the upright concrete post. All three types of horizontal timbers should be pressure-treated for long life.

Note that steel pipe sleeves are positioned in the concrete posts, flush front and back, to accommodate the ⅝-in bolts holding the barriers. All bolts should be galvanized.

Elevation of the barriers should be equal to automobile bumper height. Spacing of barriers should be not less than 10 feet, center to center.

Figure 6.118 shows a guardrail barrier of limited application. It affords the advantage of using native logs. Pressure treatment is essential to obtain maximum life. Reinforced-concrete posts, with L bolt, may be precast and placed after curing. See also Figures 6.119 through 6.124.

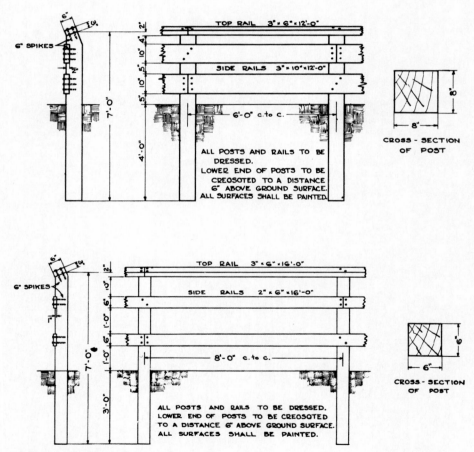

Figure 6.115   Guardrails.

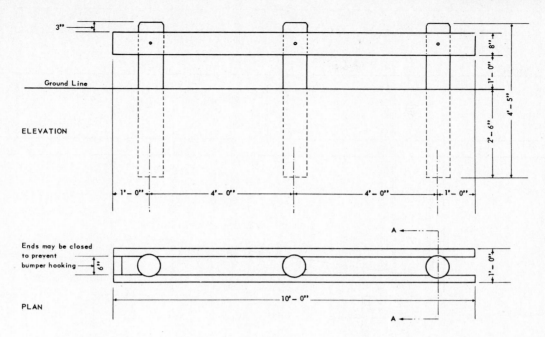

ELEVATION

PLAN

Ends may be closed
to prevent
bumper hooking

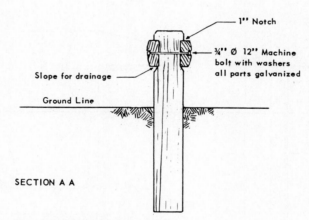

1" Notch

¾" Ø 12" Machine
bolt with washers
all parts galvanized

Slope for drainage

Ground Line

SECTION A A

Figure 6.116 Parking area barrier.

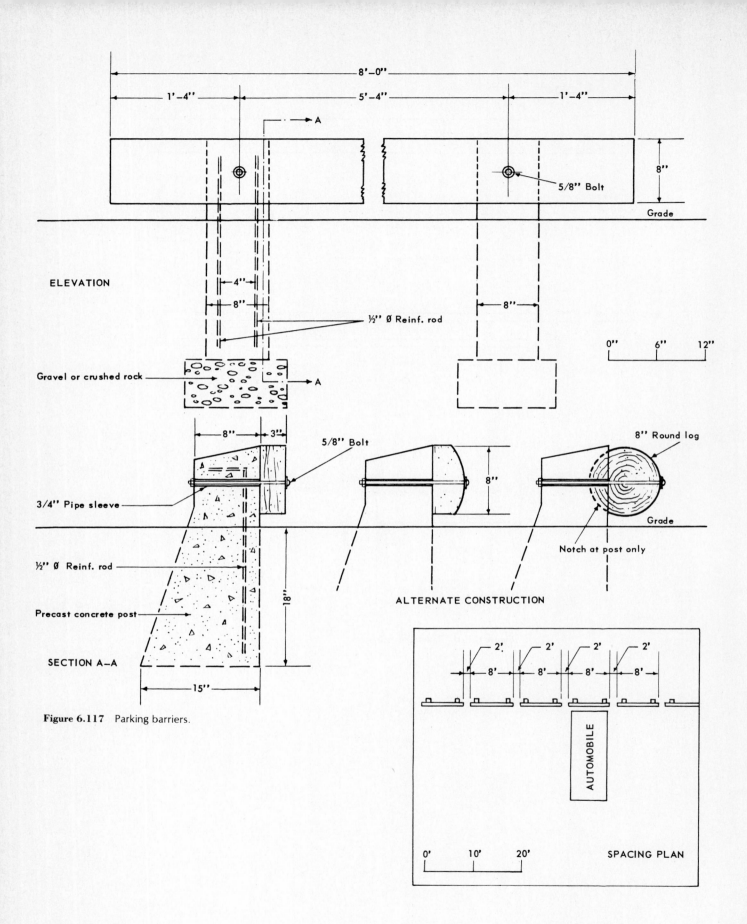

ELEVATION

8'–0''

1'–4'' 5'–4'' 1'–4''

8''

5/8'' Bolt

Grade

4''

8''

½'' Ø Reinf. rod

Gravel or crushed rock

8''

0'' 6'' 12''

SECTION A–A

8'' 3''

5/8'' Bolt

3/4'' Pipe sleeve

½'' Ø Reinf. rod

Precast concrete post

18''

15''

8''

8'' Round log

Grade

Notch at post only

ALTERNATE CONSTRUCTION

Figure 6.117 Parking barriers.

2' 2' 2' 2'

8' 8' 8' 8'

AUTOMOBILE

0' 10' 20'

SPACING PLAN

–608–

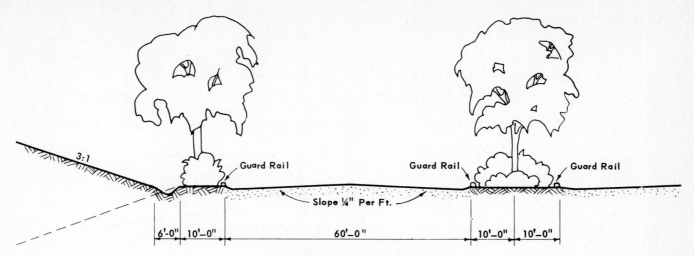

PARKING AREA SECTION

A guard rail barrier of limited application. It affords the advantage of using native logs. Pressure treatment is essential to obtain maximum life. Reinforced concrete posts, with L—bolt, may be precast and placed after curing.

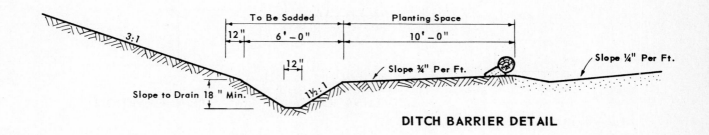

DITCH BARRIER DETAIL

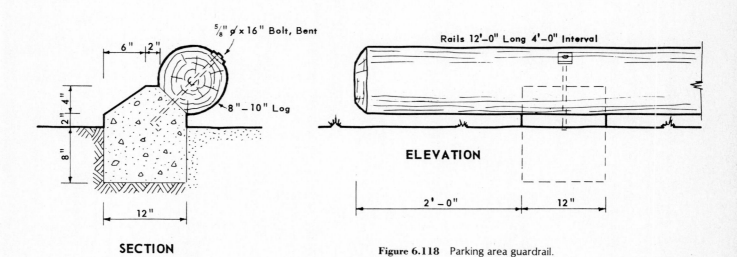

SECTION

ELEVATION

Figure 6.118 Parking area guardrail.

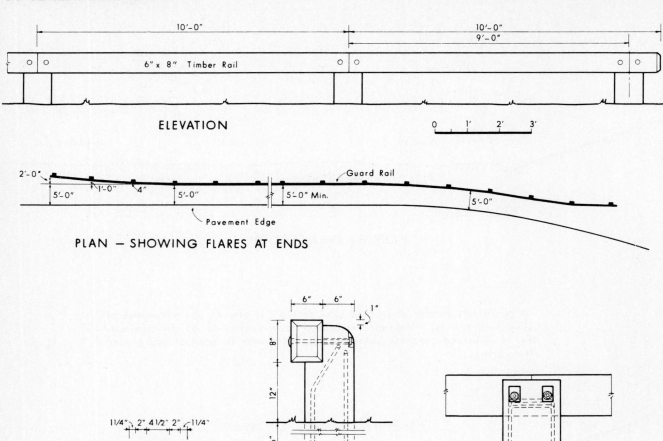

ELEVATION

PLAN — SHOWING FLARES AT ENDS

PLAN

SIDE ELEVATION

REAR ELEVATION

Figure 6.119 Guardrail.

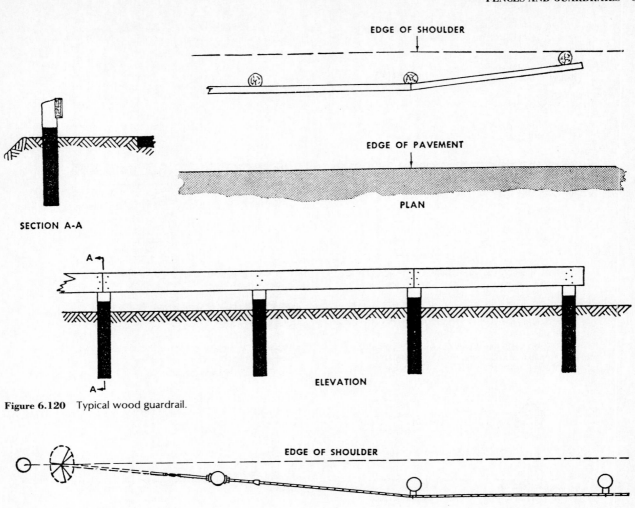

Figure 6.120 Typical wood guardrail.

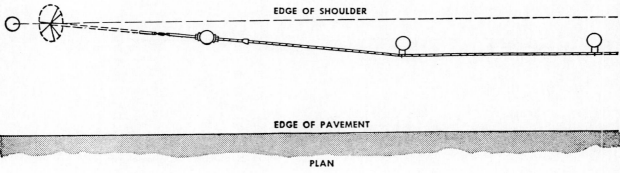

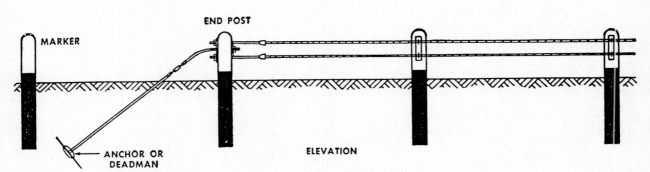

Figure 6.121 Typical two-cable guardrail illustrating method of approaching at end post.

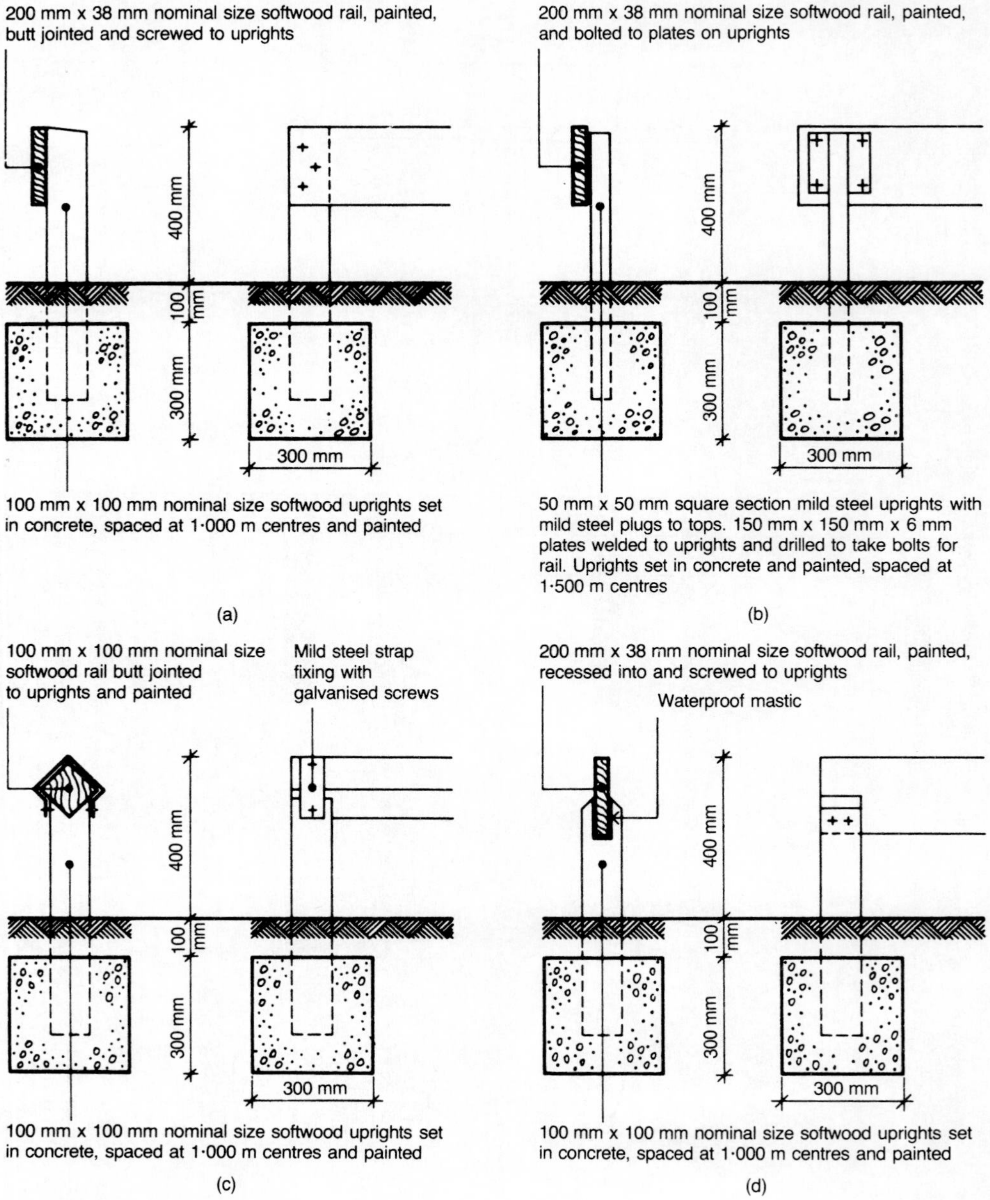
200 mm x 38 mm nominal size softwood rail, painted, butt jointed and screwed to uprights
400 mm
100 mm
300 mm
300 mm
100 mm x 100 mm nominal size softwood uprights set in concrete, spaced at 1·000 m centres and painted
(a)
200 mm x 38 mm nominal size softwood rail, painted, and bolted to plates on uprights
400 mm
100 mm
300 mm
300 mm
50 mm x 50 mm square section mild steel uprights with mild steel plugs to tops. 150 mm x 150 mm x 6 mm plates welded to uprights and drilled to take bolts for rail. Uprights set in concrete and painted, spaced at 1·500 m centres
(b)
100 mm x 100 mm nominal size softwood rail butt jointed to uprights and painted
Mild steel strap fixing with galvanised screws
400 mm
100 mm
300 mm
300 mm
100 mm x 100 mm nominal size softwood uprights set in concrete, spaced at 1·000 m centres and painted
(c)
200 mm x 38 mm nominal size softwood rail, painted, recessed into and screwed to uprights
Waterproof mastic
400 mm
100 mm
300 mm
300 mm
100 mm x 100 mm nominal size softwood uprights set in concrete, spaced at 1·000 m centres and painted
(d)

Figure 6.122 Wood guardrails: low rails – timber, softwood.

150 mm x 50 mm nominal size western red cedar rail
with half checked joints screwed (galvanised or
non-ferrous) to upright with 25 mm rebate and oiled

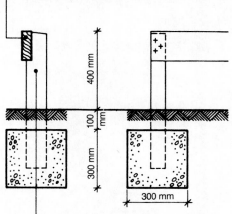

100 mm x 75 mm nominal size western red cedar
uprights, oiled and set in concrete, spaced at 1·000 m
centres

(a)

150 mm x 75 mm nominal size oak rail rebated 15 mm
to receive uprights. Rails to have half checked joints
and to be unpainted

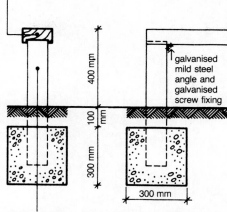

100 mm x 100 mm nominal size oak uprights,
unpainted and set in concrete, spaced at 1·000 m
centres

(b)

125 mm x 38 mm nominal size western red cedar rails,
oiled and screwed to uprights with 13 mm rebates,
joints in rails to be staggered on alternate uprights

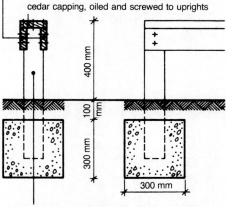

100 mm x 100 mm nominal size western red cedar
uprights, oiled and set in concrete, spaced at 1·000 m
centres

(c)

150 mm x 100 mm nominal size oak rail, unpainted
and bolted to upright with 225 mm x 15 mm coach bolt
in 18 mm hole

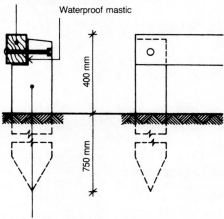

200 mm x 150 mm nominal size oak uprights,
unpainted and driven into firm ground, spaced at
1·000 m centres

(d)

Figure 6.123   Wood guardrails: low rails – timber, hardwood.

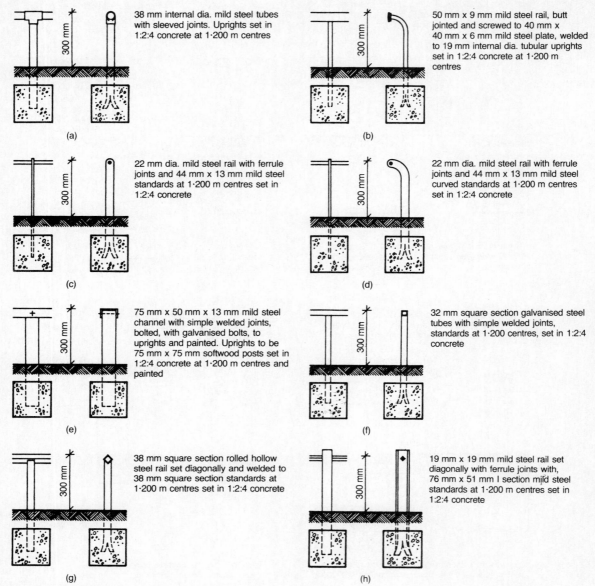

(a) 38 mm internal dia. mild steel tubes with sleeved joints. Uprights set in 1:2:4 concrete at 1·200 m centres

(b) 50 mm x 9 mm mild steel rail, butt jointed and screwed to 40 mm x 40 mm x 6 mm mild steel plate, welded to 19 mm internal dia. tubular uprights set in 1:2:4 concrete at 1·200 m centres

(c) 22 mm dia. mild steel rail with ferrule joints and 44 mm x 13 mm mild steel standards at 1·200 m centres set in 1:2:4 concrete

(d) 22 mm dia. mild steel rail with ferrule joints and 44 mm x 13 mm mild steel curved standards at 1·200 m centres set in 1:2:4 concrete

(e) 75 mm x 50 mm x 13 mm mild steel channel with simple welded joints, bolted, with galvanised bolts, to uprights and painted. Uprights to be 75 mm x 75 mm softwood posts set in 1:2:4 concrete at 1·200 m centres and painted

(f) 32 mm square section galvanised steel tubes with simple welded joints, standards at 1·200 centres, set in 1:2:4 concrete

(g) 38 mm square section rolled hollow steel rail set diagonally and welded to 38 mm square section standards at 1·200 m centres set in 1:2:4 concrete

(h) 19 mm x 19 mm mild steel rail set diagonally with ferrule joints with, 76 mm x 51 mm I section mild steel standards at 1·200 m centres set in 1:2:4 concrete

Figure 6.124   Metal guardrails.

6.5 Outdoor Surfacing

No one surface will satisfactorily meet the needs of all outdoor activities. Each activity has its own surface requirements that dictate which type or types of material can be used.

In the selection of surfacing material for any outdoor area, certain qualities should be sought. These include:

Multiplicity of use

Durability

Resistance of dust and stains

Reasonable initial cost

Ease of maintenance

Low maintenance cost

Pleasing appearance

Nonabrasiveness

Resiliency

All-year usage

Obtaining the proper surface for outdoor recreation areas continues to be a perplexing problem for school administrators, play supervisors, designers, and those responsible for maintenance. Over the years there have been significant developments in surfacing, especially under and around playground apparatus. There has been a gradual change from earth, mud, sand, and turf to bituminous surfacing, which is presently being used throughout the United States. This change has come about because of the consensus that bituminous blacktop surfacing is an improvement over other surfacing types that have proved unsatisfactory.

Types of Surfacing Materials

Group	Type
Earth	Loams, sand, sand and clay, clay and gravel, fuller's earth, stabilized earth, soil and cement
Turf	Bluegrass mixtures, bent, fescue, Bermuda
Aggregates	Gravel, graded stone, graded slag, shell, cinders
Bituminous (asphalt-tar)	Penetration macadam, bituminous or asphaltic concrete (cold and hot-laid), sheet asphalt, natural asphalt, sawdust asphalt, vermiculite asphalt, rubber asphalt, cork asphalt, other patented asphalt mixes
Synthetics	Rubber, synthetic resins, rubber asphalt, chlorinated butyl rubber, mineral fiber, finely ground aggregate and asphalt, plastics, vinyls
Concrete	Monolithic, terrazzo, precast
Masonry	Flagstone (sandstone, limestone, granite, and so on), brick, and so on
Miscellaneous	Tanbark, sawdust, shavings, cottonseed hulls

Turf The advantages of using grass as a surface are its attractiveness, resiliency, and nonabrasiveness and the fact that it is relatively dust-free. Such a surface lends itself very well to activities that require relatively large areas, as most field games do.

Turf is difficult to maintain in areas where usage is intensive. In regions where watering is essential, maintenance costs are high. Turf surfaces are not practical for most activities when the ground is frozen or wet and, in addition, must be given time and care to restore themselves after heavy use.

Since climatic conditions and uses should determine the species of grass selected for a particular locality, careful consideration should be given to the several varieties available.

Soils The use of earth as a surfacing material has been widespread, particularly under apparatus, primarily because it is porous and inexpensive.

Among the difficulties encountered in the use of earth as a surfacing material are dust and the tendency to become rutted. These in turn create drainage problems and relatively high maintenance costs. These difficulties can be partially overcome by mixing the earth with clay or sand. When this is done, the resulting surface is often less resilient and somewhat abrasive.

Natural soils can also be stabilized by the addition of asphalt, resin, or cement, which are the most commonly used stabilizers. The use of stabilized soils is a possibility in many areas where turf is impractical or cannot be grown.

Masonry Natural-stone slabs or blocks and manufactured brick can be used for such installations as walks and terraces where interesting and attractive patterns, colors, and textures are desired.

Concrete Concrete surfaces provide year-round and multiple usage. The costs of maintenance are very low, and the surface is extremely durable.

Bituminous Surfaces The common bituminous surface has many of the advantages sought in any surfacing material. It provides a durable surface that can be used on a year-round schedule. The maintenance of bituminous surface is comparatively easy and inexpensive. Such a surface can also be used for many different activities. When properly installed, the surface is dust-free and drains quickly. Asphalt surfaces can be marked easily and with a relatively high degree of permanence. Asphalt also provides a neat-appearing, no-glare surface that blends well with the landscape.

The disadvantages of bituminous surfaces are their relatively high installation costs and lack of resiliency as compared with some other types of surfaces. However, the high installation cost is offset by low maintenance costs.

Bituminous surfaces vary as to firmness, finish, resiliency, and durability in direct relation to the kinds and proportions of aggregates and other materials used in their mixture. Asphalt can be combined with a variety of other materials to provide a reasonably resilient or extremely hard surface. The use of such materials as cork, sponge, or rubber in combination with asphalt yields a fairly resilient surface. Aggregates such as slag or granite produce an extremely hard surface when they are combined with asphalt.

Synthetics Many different types of surfacing materials have been placed in experimental use, particularly under fixed equipment. Materials such as sponge, sponge rubber, rubber mats, cork, air cell materials, and combinations in conjunction with covers of rubber vinyl, canvas, and asphaltic binder coatings are in experimental use.

Some synthetic materials appear to meet the requirements for durability. Track spikes or cleats do not leave holes in such a material; rather, the material closes around the puncture. Very little maintenance of synthetic materials is required; thus maintenance costs are low. Most synthetics have a very pleasing appearance and are available in different colors. They are nonabrasive and suitable for year-round use.

There are a number of commercially manufactured synthetic compounds such as chlorinated butyl rubber, rubber asphalt, synthetic resins, mineral fiber, finely ground aggregate and asphalt, plastics, and vinyls. They come in a variety of textures, colors, weights, and thicknesses. Most synthetic surfaces have a high degree of resiliency, which makes them desirable for many types of athletic activities. Research is needed, however, to develop a vandal-proof resilient surface.

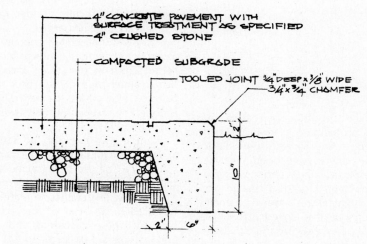

Figure 6.125 Concrete pavement edge.

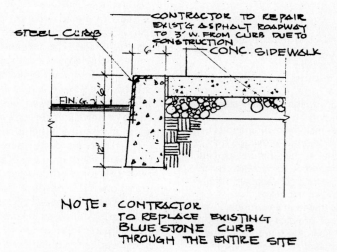

Figure 6.126 Concrete curb.

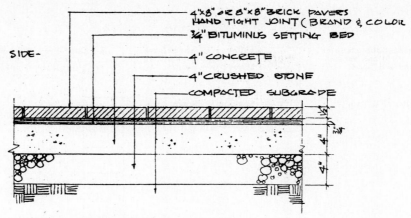

4'×8" OR 8"×8" BRICK PAVERS
HAND TIGHT JOINT (BRAND & COLOR
¾" BITUMINUS SETTING BED
4" CONCRETE
4" CRUSHED STONE
COMPACTED SUBGRADE

SIDE-

Figure 6.127   Brick pavement.

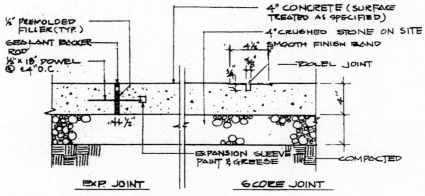

½" PREMOLDED FILLER (TYP.)
SEALANT BACKER ROD
½" × 18" DOWEL @ 24" O.C.

4" CONCRETE (SURFACE TREATED AS SPECIFIED)
4" CRUSHED STONE ON SITE
4½" SMOOTH FINISH BAND
TOOLED JOINT

EXPANSION SLEEVE PAINT & GREASE
COMPACTED

EXP. JOINT   SCORE JOINT

Figure 6.128   Concrete pavement on grade.

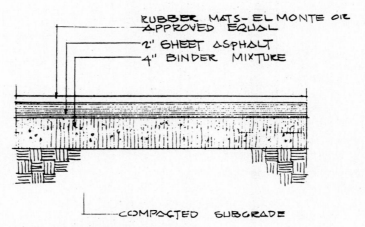

RUBBER MATS - EL MONTE OR APPROVED EQUAL
2' SHEET ASPHALT
4" BINDER MIXTURE

COMPACTED SUBGRADE

Figure 6.129   Safety mats and asphalt.

–617–

Paths: Precast Paving Slabs

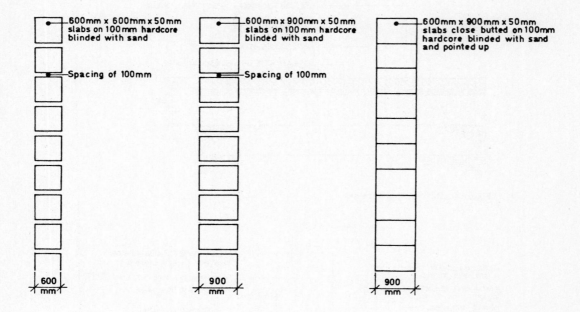

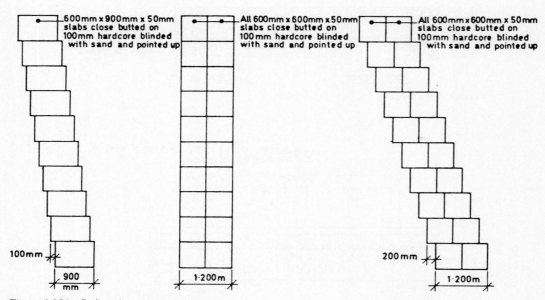

Figure 6.130 Paths using precast paving slabs.

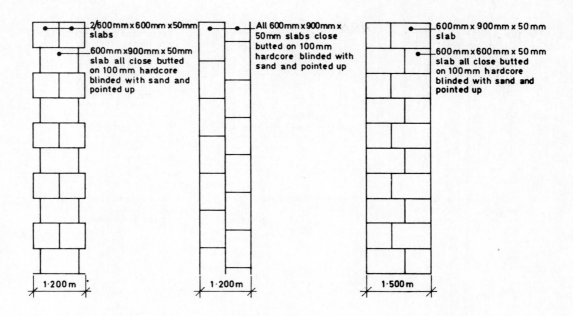

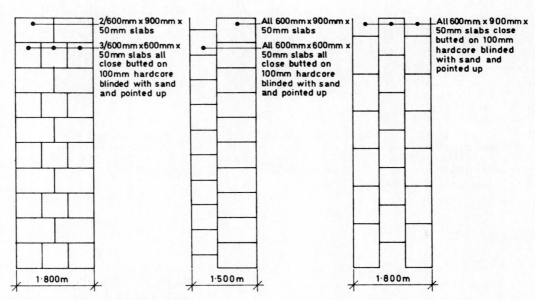

Figure 6.131   Paths using precast paving slabs.

Paths: Brick Paving

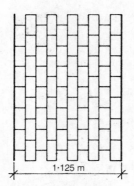

Stretcher bond along path bedded on 1:2:4 concrete blinding on 100 mm hardcore and pointed up

Stretcher bond across path bedded on 1:2:4 concrete blinding on 100 mm hardcore and pointed up

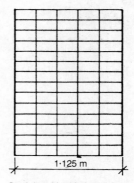

Stack bond bedded on 1:2:4 concrete blinding on 100 mm hardcore and pointed up

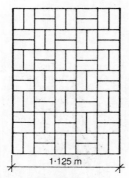

Basket weave pattern bedded on 1:2:4 concrete blinding on 100 mm hardcore and pointed up

Herringbone pattern (flat) bedded on 1:2:4 concrete blinding on 100 mm hardcore and pointed up

Herringbone (brick on edge) bedded on 1:2:4 concrete blinding on 100 mm hardcore and pointed up

Figure 6.132 Paths using brick paving.

Paving: Granite Setts

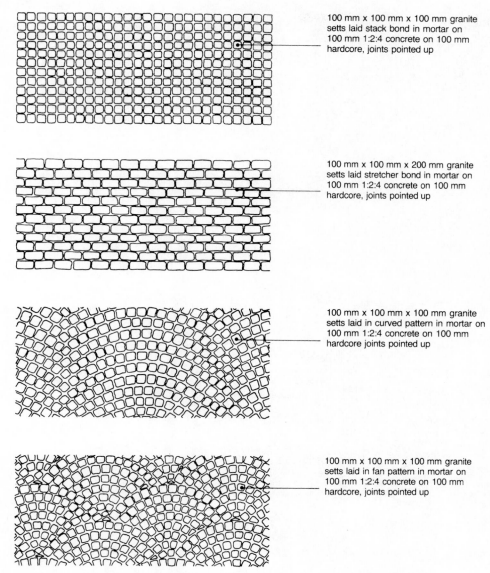

100 mm x 100 mm x 100 mm granite
setts laid stack bond in mortar on
100 mm 1:2:4 concrete on 100 mm
hardcore, joints pointed up

100 mm x 100 mm x 200 mm granite
setts laid stretcher bond in mortar on
100 mm 1:2:4 concrete on 100 mm
hardcore, joints pointed up

100 mm x 100 mm x 100 mm granite
setts laid in curved pattern in mortar on
100 mm 1:2:4 concrete on 100 mm
hardcore joints pointed up

100 mm x 100 mm x 100 mm granite
setts laid in fan pattern in mortar on
100 mm 1:2:4 concrete on 100 mm
hardcore, joints pointed up

Figure 6.133   Paving using granite setts.

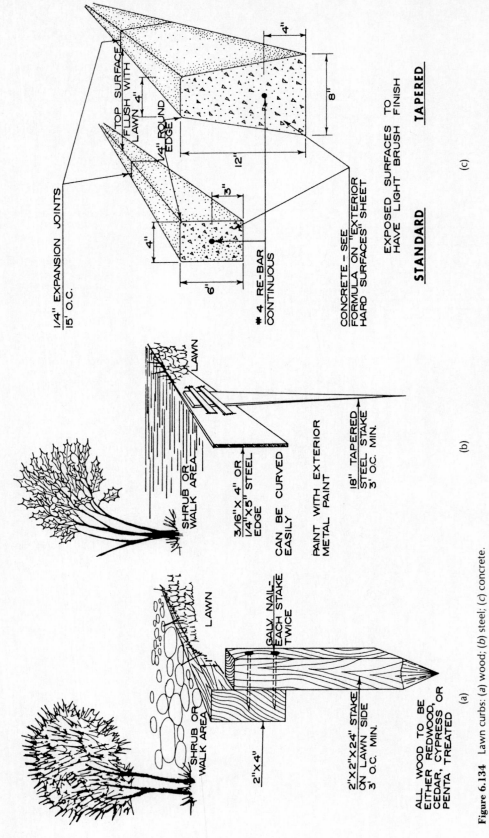

Figure 6.134 Lawn curbs: (a) wood; (b) steel; (c) concrete.

6.6 Bikeways and Trails

BIKEWAYS

In the design of any bikeway system, a number of basic considerations must be taken into account by the designer.

Bicycles and Clearances The dimensions in Figure 6.135 are by no means meant to be finite. The intention is to offer basic dimensions of the common ten-speed racing-touring bicycle and to propose certain "common-sense" design minimums that have emerged from experimentation. Designers in all cases should become aware of local public preferences in bicycle types and adjust their designs accordingly.

Bikeway Surfaces The surface of which a bikeway is constructed is perhaps the single most important feature that the designer must consider. A simple chart showing recommended materials with some basic explanations has been included to help direct the designer in considering surfaces for bikeway construction (see Figure 6.136).

Classification of Bikeways The word "bikeways" has come to be the general term describing any facility reserved for the exclusive or semiexclusive use of bicycles and related vehicles. Current literature on the subject generally accepts that bikeways may assume any of three basic forms (see Figure 6.137).

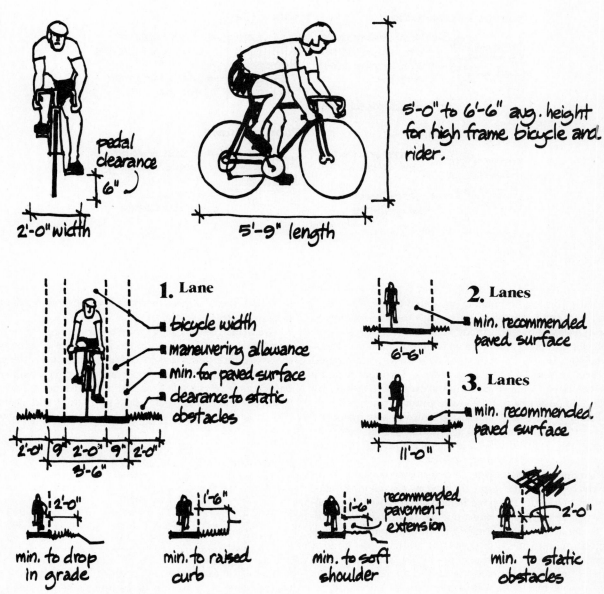

Figure 6.135 General dimensional requirements of bikeways.

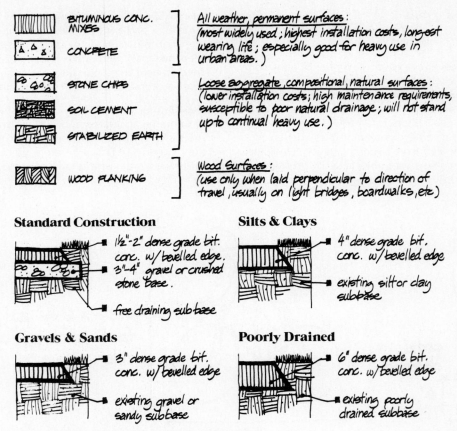

Figure 6.136 Recommended surfaces for bikeways.

Class I This is a completely separated right-of-way designated exclusively for bicyles. Through traffic, whether by motor vehicles or by pedestrians, is not allowed. Cross flows by vehicles and pedestrians are allowed but are minimized.

Class II This is a restricted right-of-way designated exclusively or semiexclusively for bicycles. Through traffic by motor vehicles or pedestrians is not allowed. Cross flows by vehicles and pedestrians are allowed but are minimized.

Class III This is any shared right-of-way designated by signs or stencils. It designates any pathway that shares its through-traffic right-of-way with either moving (but not parked) motor vehicles or pedestrians.

Design Speeds Bicycle speed is determined by several factors, which include the type of bicycle, the gearing ratio, pavement gradients, the pavement surface type and condition, wind velocity and direction, air resistance, and the cyclist's age and physical condition. Although bicyclists have averaged touring speeds in excess of 30 miles per hour, a conservative speed for the average cyclist is around 10 to 11.5 miles per hour, with a range of 3 to 19 miles per hour. In determining minimum widths and radii of curvature on level bikeways, 10 miles per hour is a conservative figure for design speed.

Gradients Since cyclists may be discouraged from using a facility in direct proportion to the amount of energy and the work rate necessary to overcome any given length of grades on the facility, the importance of developing criteria based upon physiological requirements cannot be overemphasized. The following are suggested norms and maximums that should be considered when laying out a bikeway route:

Gradient (percent)	Length (feet)	
	Normal	Maximum
1.5	1600	...
3.0	400	800
4.5	150	300
10.0	30	60

Radius of Curvature At present, accepted radii in the United States vary from 6 to 50 feet. When designing radii

Bikeway Classifications

Class I

Bikeway   Walk   Roadway

Total separation / Dividing strip between right·of·ways on separate surfaces.

Class II

Bikeway/walk   Parking   Roadway
Bikeway

Total or partial separation / Adjacent, but separated right of ways on same surface.

Class III

Bikeway/walk   Bikeway/Roadway

Partial or no separation / Shared right·of·way on same surface

Possible Locations for Bikeways

- Abandoned RR right·of·ways
- Electric and pipeline right·of·ways
- River banks
- Dry washes
- Beach fronts, lake fronts
- Flood control dikes and levees
- Irrigation canal banks and dikes
- Fire breaks

Intersections for Bikeways

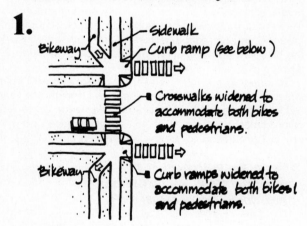

1.

Bikeway — Sidewalk
— Curb ramp (see below)
- Crosswalks widened to accommodate both bikes and pedestrians.
Bikeway — Curb ramps widened to accommodate both bikes and pedestrians.

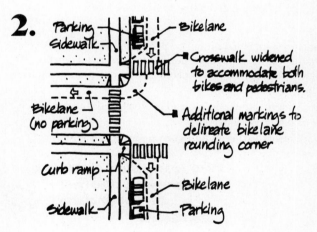

2.

Parking — Bikelane
Sidewalk
- Crosswalk widened to accommodate both bikes and pedestrians.
Bikelane (no parking) — Additional markings to delineate bikelane rounding corner
Curb ramp
Sidewalk — Bikelane
— Parking

Figure 6.137   Bikeway classifications.

for bikeway layouts, the designer should use the following formula. The equivalent radius of curvature as a function of velocity is expressed in the following linear relationship:

$$R = 1.25V + 1.4$$

where R = unbraked radius of curvature (in feet) negotiated by a bicycle on a flat, dry, bituminous concrete surface

V = velocity of bicycle in miles per hour

Example For a Class I bikeway with a use speed of 10 miles per hour, the "comfortable" unbraked radius of curvature is 13.9 feet.

Intersections The most effective way of avoiding conflicts between cyclists and motor vehicles where the bikeway must cross a heavily traveled roadway and heavy bicycle use is anticipated is to employ a total grade separation. This recommendation also holds true when the bikeway must cross heavily traveled intersections where significant bicycle traffic might disrupt the orderly flow of vehicular traffic.

Generally, in densely populated urban areas with little room for underpasses, providing total grade separation at intersections may be completely prohibitive from the standpoint of cost. This being the case, the designer must then route the bikeway across the roadway at grade. Some of the more typical situations are illustrated in Figure 6.137.

Parking When designing or locating bikeway parking areas, the following items should be considered:

1. Secure stanchions should be provided; that is, the bicycle frame rather than a wheel alone should be anchored, since the secured wheel can easily be detached from the frame and the rest of the bike carried off.

2. Stanchions should be located in areas where there is constant visual supervision.

3. Parking areas should be out of pedestrian pathways.

4. Parking areas should be conveniently located near cyclist destinations, adjacent to main entries where possible (preferably within 50 feet or less). If distances become too great cyclists frequently secure bikes to the nearest available permanent object, such as railings, signposts, light posts, flagpoles, and trees.

Various types of parking are illustrated in Figure 6.138.

TRAILS

Exclusive off-street bikeways and hiking and riding trails can be located in a number of fashions and combinations. Perhaps the greatest opportunities lie along the shoreline and along the levees or banks of flood control channels, rivers, creeks, and lakes. Other corridors can be found along railroad and power line right-of-way and fire trails.

Trail Requirements The design of an exclusive trail is determined by the anticipated use, terrain characteristics, and safety requirements. Acceptable standards for exclusive trails are shown in Figures 6.139, 6.140, 6.141, 6.142, and 6.147 through 6.150.

Bikepath (Class I) The safest but most costly system of travel designated exclusively for bicycles and physically separated from motor vehicle traffic, and where cross flows by pedestrians, equestrians, and motorists are limited. This facility is often located in parks or along parkways but may also parallel some major streets and roadways leading from residential areas to major intercommunity destinations including the central business area, shopping centers, schools, playgrounds, and recreation/cultural centers. In some instances the bikepath may also serve nonmotorized recreation, e.g., hiking and jogging. (See Figure 6.143.)

Bikelane (Class II) A restricted segment of a shared street or roadway for bicycles, separated from motor vehicle traffic by a corridor lane designated by a standard, elongated colored marking stripe which gives psychological rather than physical protection to the cyclist referred to as a "second-

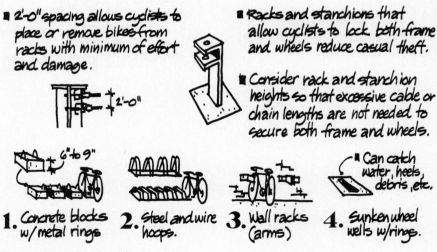

■ 2'-0" spacing allows cyclists to place or remove bikes from racks with minimum of effort and damage.

2'-0"

■ Racks and stanchions that allow cyclists to lock both frame and wheels reduce casual theft.

■ Consider rack and stanchion heights so that excessive cable or chain lengths are not needed to secure both frame and wheels.

6" to 9"

Can catch water, heels, debris, etc.

1. Concrete blocks w/ metal rings **2.** Steel and wire hoops. **3.** Wall racks (arms) **4.** Sunken wheel wells w/ rings.

Figure 6.138 Bicycle parking.

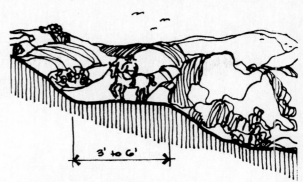

Figure 6.140   Riding trails should have a width of 3 to 6 feet. The grade should be a maximum of 5 percent (10 percent for very short distances). Clearance should be 12 feet. The surface should be natural ground (loose soil).

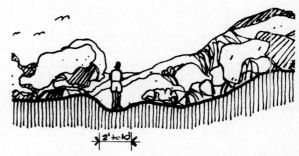

Figure 6.139   Bikeways should have a width of 8 to 10 feet (two-way) or 6 feet (one-way). The grade should be a maximum of 5 percent for lengths up to 1000 feet and 15 percent for very short distances. Clearance should be 10 feet. The surface should be hard, preferably asphalt.

Figure 6.141   Hiking trails vary from 2 to 10 feet in width, depending on their character. The grade should be 5 percent (a maximum of 15 percent for very short distances). The clearance should be 8 feet. The surface should be firm, natural ground.

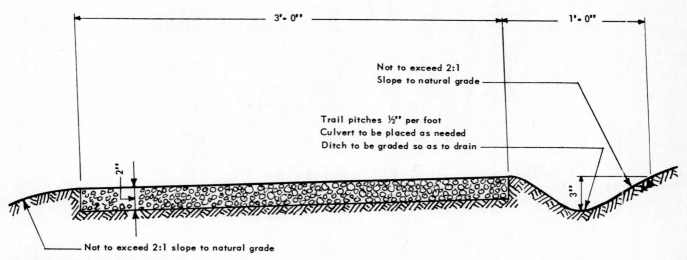

Not to exceed 2:1
Slope to natural grade

Trail pitches ½" per foot
Culvert to be placed as needed
Ditch to be graded so as to drain

Not to exceed 2:1 slope to natural grade

Figure 6.142   Typical trail section.

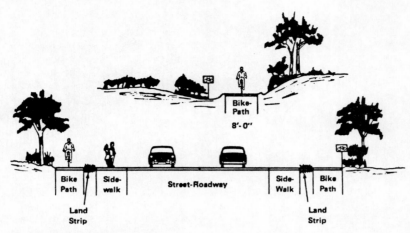

Figure 6.143 Bikepath.

ary'' or ''unprotected'' bikelane; or, a reserved, on-street lane separated from motorized vehicles by a physical barrier such as median buffer landscape area, concrete parking bumper stops, placement of a lane between parked cars and the street curb, or other physical means; referred to as a ''primary'' or ''protected'' bikelane (see Figure 6.144).

Bikeroute (Class III) A shared right-of-way located on lightly traveled streets and roadways; designated solely by the standard ''Bike Route'' signs encouraging cyclists' use and warning motorists to anticipate bicycles on the street. This route is established to provide cyclists with a direct and desir-

able route to specific destinations along low-traffic roadways. The bike route is inexpensive, with costs involving the erection of signs and the redesigning and installation of drainage grates where needed (see Figure 6.145).

Bike-Walk Path A path located within street or roadway rights-of-way for use by both pedestrians and cyclists, where footpath traffic volume is light. This type of path must provide for ramps or curb cuts at street or road crossings or intersections for ease of riding, smooth transition to the street level, and prevention of possible bicycle damage (see Figure 6.146).

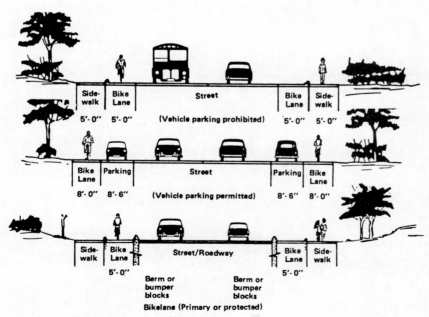

Figure 6.144 Bikelanes.

Figure 6.145   Bikeroute.

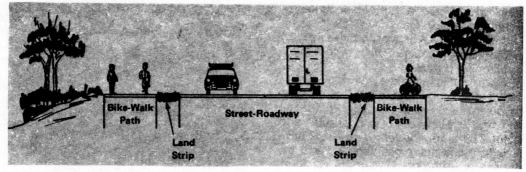

Figure 6.146   Bike-walk path.

Space Requirements for Recreational Road Sections and Trails

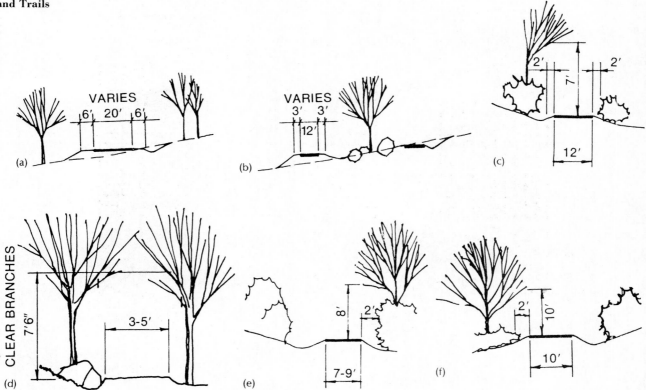

Figure 6.147   Types of roads and trails: (*a*) double lane; (*b*) single lane; (*c*) off-road vehicle (two-way); (*d*) hiking; (*e*) bicycle (two-way); (*f*) horse (two-way).

Trail Design In planning formal teaching trails it would be well to remember that short winding trails reduce fatigue and help sustain interest. The most effective teaching trails are from ⅛ to ¼ mile long and can be walked in 1 to 1½ hours by a naturalist leading a school class. Other reasons for short trails involve school time restrictions and nature center staff limitations.

Trail layout should follow a closed loop design, beginning and ending at approximately the same location — at either an interpretive-orientation structure or a parking lot. Visitor traffic flow on loop trails should be in a one-way direction to avoid retracing steps and to prevent interference between groups using the trails. Long, straight trails should be avoided. Trails with curves and bends at frequent intervals inject an element of surprise and provide an atmosphere of remoteness for the user. Straight stretches of trail should rarely exceed 100 feet.

Interconnecting trails are another important design feature and should be employed where possible. Access from one trail to another, clearly marked, allows the visitor or naturalist a choice of trail use time, expands teaching opportunities, and adds variety and interest to an area. In a formal teaching trail system a figure-eight design has been successfully employed. Another acceptable design involves a basic teaching trail with a number of shorter "special subject" spur loops attached to it at intervals along its length.

The tread (walkway surface) on formal teaching trails should be 4 to 6 feet wide in wooded areas. In open fields it can be 6 to 8 feet wide. Although these dimensions may appear excessive, they are necessary for group use. The naturalist or teacher will need room to gather the group or class around him, or, in general, to keep the group close together. The specifications for hiking trails, because they serve a different purpose, are less rigid, and a trail tread of 3 feet is adequate in most instances.

Construction The interpretive opportunities present, type of program offered, the characteristics of the land, and the needs and anticipated number of visitors will determine

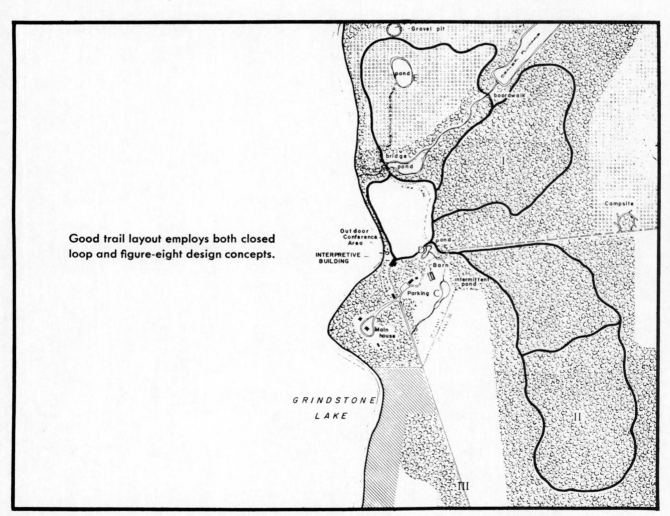

Good trail layout employs both closed loop and figure-eight design concepts.

Figure 6.148

the amount and method of trail construction needed. The following are some basic construction guidelines:

Right-of-Way Where new, heavy-use trails are put through dense growths of bushes, shrubs, or young trees, it is desirable to clear a right-of-way from 8 to 10 feet wide. Such clearing makes trail maintenance easier and encourages ferns, clovers, grasses, and flowering plants. Clearing should be to 10 feet above the trail to allow for drooping branches wet with rain or laden with snow. Large trees should be felled only when it is impractical to build around them. Brush and logs from the clearing process can be stacked into large piles near the trail to serve as wildlife shelters. All stumps and rocks should be cleared from the tread area unless they add to the interpretive program or add an aesthetic touch to the site.

Clearing the trail right-of-way can be done by mechanical means and/or by hand. When mechanical means are employed (i.e., tractors, small bulldozers, bush-hogs, or flails), great care must be taken to avoid unnecessary disruption to trailside areas. The trail planner or a naturalist should be present at all times to supervise the work. Where soil has been exposed, erosion control measures should be immediately employed.

Hand-clearing involves the use of axe, rake, bushhook, or sickle to establish trails. While less disruptive to the trail bed and surrounding area, this method is expensive and time-consuming. Where the danger of erosion is extreme or where topography prohibits using mechanical means, hand-clearing is the only logical alternative. Again, a thorough knowledge of the site — topography and soils — is essential before trail construction begins.

Base In those areas which are perpetually damp or experience shallow flooding (1 to 4 inches) periodically, provision must be made to raise the tread to achieve a dry walking surface. This is done by grading, as illustrated in Figure 6.149, or by constructing a boardwalk.

Another possibility for a trail through a wet area is one in which railroad ties (small logs can be substituted), treated to resist decay, are used to retain a shale or gravel fill. Drain tiles are buried across the trail at suitable intervals to prevent flooding of the land on one side of the trail.

Trails cut into hillsides present innumerable construction problems, and expert engineering advice will be required in each instance. Basic considerations which generally apply to hillside trails include: drain tiles or culverts across the trail at suitable intervals, drainage ditch on the uphill side of the trail, erosion control plantings on exposed soil above and below the trail; the trail should be sloped toward the downhill edge with this edge 3 inches lower than the hillside edge.

Grade When possible, trails should be built to follow the contour of the land. When inclines present themselves and must be negotiated, care should be exercised so that the slope does not exceed 15 percent. Such grades are difficult to climb and are rarely acceptable on heavy-use teaching trails. An average grade of 10 percent is generally considered the maximum for comfortable walking. The specifications for hiking trails are not as rigid, but when possible, steep trail grades should be avoided.

Steps are a more acceptable construction technique on steeper slopes. Slabs of native stone, old railroad ties, or hewn logs should be considered for step construction. If wood is used, treatment with a water repellent and preservative is essential. Untreated wood will not last long when in contact with the ground because of rot and insect attack.

Drainage Good drainage is one of the most important considerations in trail construction, especially if trails are heavily used or built on a steep grade. Small logs or rows of rocks placed diagonally across a trail will intercept and divert water off the trail surface into plant-covered areas. Drain tiles, culverts, and sloping the trail tread, as previously discussed for hillside trails, apply here.

Surface Unless trails are heavily used, no special surfacing is required. Formal teaching trails and certain special-use trails will require surfacing, while hiking trails usually will not. Surfacing, where needed, includes a variety of materials, as is indicated in the following list:

1. Wood chips — often available from pruning, trimming, or thinning operations or from telephone and power companies. Wood chips make a durable and natural-appear-

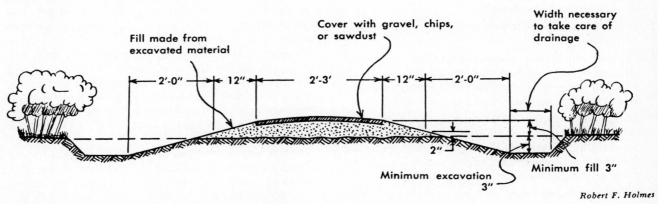

Cover with gravel, chips, or sawdust

Fill made from excavated material

Width necessary to take care of drainage

2'-0" 12" 2'-3' 12" 2'-0"

2"

Minimum excavation 3"

Minimum fill 3"

Robert F. Holmes

Figure 6.149 Design of a built-up trail through a wet area.

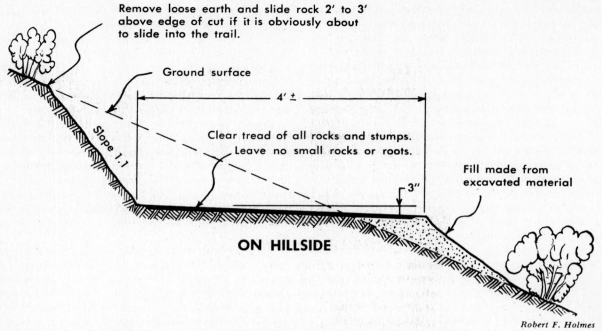

Remove loose earth and slide rock 2' to 3' above edge of cut if it is obviously about to slide into the trail.

Ground surface

Slope 1:1

4' ±

Clear tread of all rocks and stumps. Leave no small rocks or roots.

Fill made from excavated material

3"

ON HILLSIDE

Robert F. Holmes

Figure 6.150 Design of a trail along a slope. Exposed surfaces on sides should be planted to prevent erosion.

ing surface. Chips made from hardwood limbs when leaves are off the trees are best.

2. Shredded bark — similar to wood chips as a surfacing material, but not as readily available. Sawmills and pulpmills are main sources of supply.

3. Grass — contrary to the belief of many, grass has proved to be a good trail surface material in open areas. For best results the trail should be wide to better distribute traffic, be fertilized in the spring and fall, and be closed to visitors periodically to allow the grass to recover from heavy use. The latter requires a fairly extensive trail system in which alternate trails are available to replace those closed.

4. Fine shale — good for covering on rocky or wet areas and for very heavy visitor traffic.

5. Run-of-bank sand and small gravel — sometimes available on the site.

6. Crushed native rock — sometimes available from local sources; tends to wash off trails readily and is generally not liked by those who use paths. Such material is difficult to walk on, scars shoes, and gets in footwear.

7. Blacktop and concrete — almost a necessity when visitors are counted in the thousands daily or on weekends. In a general way, the building of a hardtop trail is not unlike building an improved road. Only the nature of the traffic to be carried and the width are different.

8. Soil cement — very similar to concrete; only sandy soil, often from the site, is used in place of sand and gravel.

The mix is sometimes poured into a dug-out tread area without the use of forms. The walks are usually built 4 to 5 inches thick with a slight crown for drainage.

9. Concrete blocks — for heavy use areas on sites where the ground does not freeze. Regular 4-8-16 blocks can be laid in a prepared bed of sand 6 to 8 feet wide.

10. Special sand dune trail surfaces — for use on sloping sandy areas and sand dunes.

6.7 Driveways

Purpose Use of the accompanying diagrams, dimensions, and formulas will enable the designer to lay out straight or curved driveways to suit any condition between extremes of (1) minimum practicability and safety, and (2) maximum ease of driving.

General Unless an automobile is driven in a straight line, rear wheels do not follow exactly in the tracks of front wheels, because front wheels only are controlled by the steering gear. Hence, on curved driveways, the inner rear wheel may track off a roadway if the inner radius of the drive is too great. The outer front wheel may track off if the outer radius is too small.

Determination of the minimum width of driveway for various radii (and vice versa) depends on three properties of an automobile: tread, wheelbase, and turning radius. To these properties are added inside and outside clearances to provide a margin of safety, so that both front and rear bumpers,

fenders, trunks, etc., will safely clear shrubbery or walls bordering the drive.

The tread T of a car is the distance center-to-center of the front or rear wheels. The tread varies both between the front and the rear wheels and with the make or year of the car. The tread of the rear wheels, being a constant on curves and normally greater than that of the front wheels, is used in driveway calculations.

The wheelbase B is the distance center-to-center between front and rear axles. It also varies.

The turning radius X is the radius of the circular track of the outer front wheel. It is variable not only with the car but also with other factors discussed below.

Inside and outside clearances, as used here, are fixed dimensions which have been calculated to meet necessities of all types of cars. Lesser clearances are not advisable, as their use requires more caution than the average driver habitually employs.

Calculations Use of the values of T, B, and X given in the accompanying table will result in driveways adequate for any passenger car. Basic principles of driveway design may be applied geometrically by following Figure 6.151 or using the formulas in the table. In either case, turning radius X may be any dimension not less than the minimum.

Minimum Dimensions and Formulas

Dimensions	Formulas
T = tread = 5 feet 2 inch	R = outside radius of drive = $X + F$
B = wheelbase = 12 feet 0 inch	D = divergence between front and rear wheels = $X - \sqrt{X^2 - B^2}$
X = turning radius = 27 feet 0 inch	
E = inside radius = 1 foot 3 inch	W = width of drive for a given radius = $T + D + E + F$
F = outside clearance = 1 foot 9 inch	r = inside radius of drive = $R - W$
Minimum landing = 22 feet 0 inch	

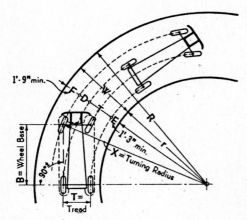

Figure 6.151 Basic principles.

Straight Driveways The minimum width of straight driveways can be calculated from the formula, $W = T + 2E$, or 7 feet 8 inches, but should not be less than 8 feet 0 inch. A width of 9 feet 0 inch is recommended for comfortable driving.

Landings Straight portions of driveway interposed between two curved quadrants will permit a car to be brought nearly alongside a curb or step at an entrance. Theoretically a car cannot be brought exactly parallel to the curb without backing and moving forward again at least once. Actually, a landing 22 feet 0 inch in length will permit driving close enough to the curb to prevent discomfort in alighting, without "jockeying."

Curved Driveways By substituting the values in the first column of the table in the formulas in the second column, the least practical outside radius for a curved drive is found to be 27 feet 0 inch plus 1 foot 9 inches or 28 feet 9 inches. Similarly, the least practical width of a drive of 29 feet 0 inch outer radius is 11 feet 0 inch; the minimum inner radius, 18 feet 0 inch.

The larger the turning radius X, the more may riding comfort and speed be increased. Therefore, it is advisable to use

Minimum Standard Dimensions for Curved Driveways

R outer radius	W minimum width	r inner radius
29'0" to 30'0"	11'0"	18'0" to 19'0"
30'0" to 31'0"	10'11"	19'1" to 20'1"
31'0" to 32'0"	10'10"	20'2" to 21'2"
32'0" to 33'0"	10'9"	21'3" to 22'3"
33'0" to 34'0"	10'8"	22'4" to 23'4"
34'0" to 35'0"	10'7"	23'5" to 24'5"
35'0" to 36'0"	10'6"	24'6" to 25'6"
36'0" to 37'0"	10'5"	25'7" to 26'7"
37'0" to 38'0"	10'4"	26'8" to 27'8"
38'0" to 39'0"	10'3"	27'9" to 28'9"
39'0" to 41'0"	10'2"	28'10" to 30'10"
41'0" to 43'0"	10'1"	30'11" to 32'11"
43'0" to 45'0"	10'0"	33'0" to 35'0"
45'0" to 47'0"	9'11"	35'1" to 37'1"
47'0" to 49'0"	9'10"	37'2" to 39'2"
49'0" to 51'0"	9'9"	39'3" to 41'3"
51'0" to 54'0"	9'8"	41'4" to 44'4"
54'0" to 57'0"	9'7"	44'5" to 47'5"
57'0" to 61'0"	9'6"	47'6" to 51'6"
61'0" to 65'0"	9'5"	51'7" to 55'7"
65'0" to 70'0"	9'4"	55'8" to 60'8"
70'0" to 75'0"	9'3"	60'9" to 65'9"
75'0" to 82'0"	9'2"	65'10" to 72'10"
82'0" to 89'0"	9'1"	72'11" to 79'11"
89'0" to 99'0"	9'0"	80'0" to 90'0"
99'0" to 111'0"	8'11"	90'1" to 102'1"
111'0" to 126'0"	8'10"	102'2" to 117'2"
126'0" to 147'0"	8'9"	117'3" to 138'3"
147'0" to 176'0"	8'8"	138'4" to 167'4"
176'0" to 219'0"	8'7"	167'5" to 210'5"
219'0" to 300'0"	8'6"	210'6" to 291'6"

curves having radii as large as practical considerations of site and economy permit. The accompanying table is a tabulation of the results of substituting varying values of X in the basic formulas. Any one of the three factors, R, W, or r may govern. For instance: R (outer radius) may be determined by lot lines; W (width of drive) may be prescribed by the distance between two obstructions; r (inner radius) may be the radius of a circular flower bed.

On a minimum circular curve (Figure 6.152) automobiles stop in a parking position. Radius of the flare from the property line to the curb should be the same as the inner radius of curve. *Minimum values:* R = 29 feet 0 inch, W = 11 feet 0 inch, r = 18 feet 0 inch.

A straight portion, or "landing," at the entrance step (Figure 6.153) will lessen the rake of the car. As the tangent in advance of the entrance is lengthened, the angle of rake is lessened. *Minimum values:* R = 29 feet 0 inch, W = 11 feet 0 inch, r = 18 feet 0 inch.

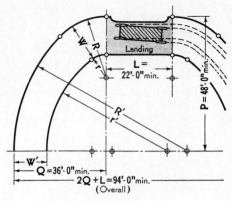

Figure 6.154 Compound curves.

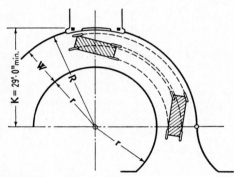

Figure 6.152 Circular curves.

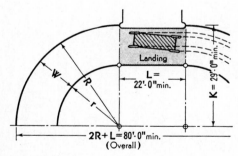

Figure 6.153 Landings.

Compound Curves Since ease of driving is dependent primarily on uniformity of curvature, and speed on radius, a drive formed of circular curves of large radius is theoretically most nearly perfect. Practical considerations of site and expense, however, normally limit the radius. Hence, compound curves approach the maximum of riding ease compatible with practicality. Portions of the drive may be laid out at great radius and other portions at small radius. Relation-

ships of these portions should be carefully studied. Shorter radii may be placed near the landings, where speed is reduced, or greater radii may be so placed when it is desirable to minimize the raking position of the car when stopped. Circumstances of each problem will determine its solution. Figure 6.154 shows the development and minimum dimensions of compound curves. Compound curves of this type, with the short radius near the landing, permit greater speed upon entering the drive but require more manipulation of the steering wheel, and consequently a slower speed near the landing. *Minimum values:* P = 48 feet 0 inch, Q = 36 feet 0 inch, R = 30 feet 0 inch, R' = 60 feet 0 inch, W' = 9 feet 6 inch, W = 11 feet 0 inch, r = 19 feet 0 inch, r' = 44 feet 4 inches. Basic formulas are the same as for circular curves. Radii and widths of drive for quadrants of varying sizes are given in the accompanying table.

Double Driveways Minimum safe clearance between two moving cars is 2 feet 0 inches. To determine the total width of a double drive: (1) Establish the inner or outer radius of either lane; (2) determine the minimum width of that lane from the table; (3) add the necessary clearance and obtain the inner or outer radius of the other lane; (4) from the table determine the width of the second lane; and (5) add this to the width-plus-clearance already obtained. The result is the total width.

The minimum circular turnaround (Figure 6.155) requires great manipulation of the steering wheel where curves are reversed. Nevertheless, uniform width is permissible even at this point. Cars stop in a raking position. *Minimum values:* R = 29 feet 0 inch, W = 11 feet 0 inch, r = 18 feet 0 inch.

Formulas:

$$H = \sqrt{G(2R + 2r - G)}$$

where $G = R - \tfrac{1}{2}W$

Easings, or tangents at points of reversal of curvature, make driving easier (see Figure 6.156). Note that a great increase in length of the tangent requires only a small increase in overall distance from entrance to curb. Use of a landing lessens the

Dimensions of Compound Driveway Quadrants

Rectangular dimensions		Outer radii		Widths of driveway (from preceding table)		Inner radii	
P	Q	R	R'	W	W'	r	r'
48'0"	36'0"	30'0"	60'6"	11'0"	9'6"	19'0"	44'4"
50'0"	37'6"	31'3"	62'6"	10'10"	9'5"	20'5"	47'2"
52'0"	39'0"	32'6"	65'0"	10'9"	9'4"	21'9"	49'9"
54'0"	40'6"	33'9"	67'6"	10'8"	9'4"	23'1"	52'6"
56'0"	42'0"	35'0"	70'0"	10'6"	9'3"	24'0"	55'4"
58'0"	43'6"	36'3"	72'6"	10'5"	9'3"	25'10"	58'2"
60'0"	45'0"	37'6"	75'0"	10'4"	9'2"	27'2"	60'9"
62'0"	46'6"	38'9"	77'6"	10'3"	9'2"	28'6"	63'6"
64'0"	48'0"	40'0"	80'0"	10'2"	9'2"	29'10"	66'4"
66'0"	49'6"	41'3"	82'6"	10'1"	9'1"	31'2"	68'11"
68'0"	51'0"	42'6"	85'0"	10'1"	9'1"	32'5"	71'5"
70'0"	52'6"	43'9"	87'6"	10'0"	9'1"	33'9"	74'3"
72'0"	54'0"	45'0"	90'0"	9'11"	9'0"	35'1"	76'10"
74'0"	55'6"	46'3"	92'6"	9'11"	9'0"	36'4"	79'3"
76'0"	57'0"	47'6"	95'0"	9'10"	9'0"	37'8"	82'2"
78'0"	58'6"	48'9"	97'6"	9'10"	9'0"	38'11"	84'8"
80'0"	60'0"	50'0"	100'0"	9'9"	8'11"	40'3"	87'3"

See Figs. 6.154 and 6.157 for applications and reference letters. Dimensions are taken to the nearest inch. The values of r' in this table were computed by means of the following general formula:

$$r' = \frac{(P - W)^2 - (Q - W')^2 - 2r[(P - W) - (Q - W')]}{2[(Q - W') - r]} + (Q - W')$$

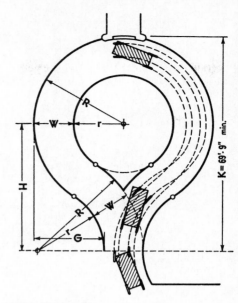

Figure 6.155 Circular driveways.

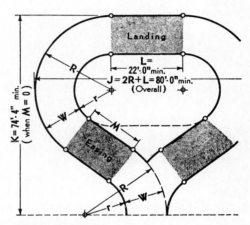

Figure 6.156 Landings and easings.

K	M
74'4"	0
75'0"	8'0"
76'0"	12'1"
77'0"	15'10"
78'0"	18'8"
79'0"	21'2"
80'0"	23'5"

rake of the car upon stopping. *Minimum values*: R = 29 feet 0 inch, r = 18 feet 0 inch, W = 11 feet 0 inch.

Ellipses formed of compound curves permit the maximum of speed compatible with both riding and driving ease. In the type shown in Figure 6.157 the shorter radii occur away from the landing, which still further reduces the rake of the car upon stopping.

Formulas:

$$H' = \sqrt{G(2R' + 2r' - G)} - (W' + r' - Q)$$

where $G = r' + \frac{1}{2}(L + W')$

$$r'' = \frac{g^2 + h^2}{2g} - R'$$

where $g = R' + \frac{1}{2}(L - W')$, $h = H' + R' - Q$

Minimum values: P = 48 feet 0 inch, Q = 36 feet 0 inch, R = 30 feet 0 inch, R' = 60 feet 0 inch, W' = 9 feet 6 inch, W = 11 feet 0 inch, r = 19 feet 0 inch, r' = 44 feet 4 inch, r'' = 49 feet 7 inches.

Construction Widths range from 9 to 12 feet for single-lane driveways and from 15 to 18 feet for double-lane driveways. If the 9-foot width is used, it should be increased to at least 10 feet on curves.

Drainage is essential. The minimum longitudinal pitch for proper drainage is 1 percent. If gutters are used, runoffs or catch basins should be provided at suitable intervals. Under-ground drains should be employed if they are required by soil conditions (see Figures 6.160, 6.161, and 6.162).

Bases must be well compacted and well drained. Soft clay soil or heavy truck traffic necessitates a heavier base. Typical bases for driveways are as follows:

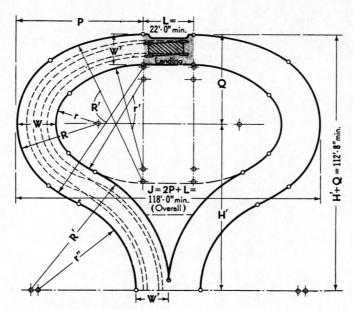

Figure 6.157 Elliptical driveways.

1. For dry construction: bank-run gravel or sandy gravel; no air pockets

2. For concrete surfaces: sandy gravel, cinders, crushed stone, slag, or other inorganic porous material

3. Crushed stone, 1½ inches in size, rolled with a 6- to 10-ton roller

Bank-run gravel consists of stones, sand, and some clay for binder; fine bank-run gravel, of stones up to 1 inch in size; coarse bank-run gravel, of stones up to 4 inches in size.

RESIDENTIAL DRIVEWAYS

The grade, width, and radius of curves in a driveway are important factors in establishing a safe entry to the garage. Driveways for attached garages that are located near the street on relatively level property need only be sufficiently wide to be adequate. Driveways that have a grade of more than 7 percent (7-foot rise in 100 feet) should have some type of pavement to prevent wash. Driveways that are long and require an area for a turnaround should be designed carefully. Figure 6.163 shows a driveway and turnaround that allow the driver to back out of a single or double garage into the turn and proceed to the street or highway in a forward direction. In areas of heavy traffic this is much safer than having to back into the street or roadway. A double garage should be serviced by a wider entry and turnaround.

For safety, driveways that are of necessity quite steep should have a near-level area from 12 to 16 feet long in front of the garage.

Two types of paved driveways may be used, the more common slab or full-width type and the ribbon type (see Figures 6.164 through 6.166). When driveways are fairly long or steep, the full-width type is the most practical. The ribbon driveway is cheaper and perhaps less conspicuous because of the grass strip between the concrete runners. However, it is not practical for all locations.

The width of the single-slab type of driveway should be 9 feet for the modern car, although 8 feet is often considered the minimum. When the driveway is also used as a walk, it should be at least 10 feet wide to allow for a parked car as well as a walkway. The width should be increased by at least 1 foot at curves. The radius of the drive at the curb should be at least 5 feet. Relatively short double driveways should be at least 18 feet wide and be 2 feet wider when they are also to be used as a walk from the street.

The concrete strips in a ribbon driveway should be at least 2 feet in width and located so that they are 5 feet on center. When the ribbon is also used as a walk, the width of the strips should be increased to at least 3 feet. This type of driveway is not practical if a curb or a turn is involved or the driveway is long.

Where driveways are not hard-surfaced, the maximum gradient should be limited to 7 percent.

Ribbon drives are not recommended because of increased maintenance needs.

Alignment should be convenient for backing a car out of a drive, or adequate turning space should be provided.

Gutters and Curbs Designs for various kinds of gutters and curbs are shown in Figures 6.168 through 6.175.

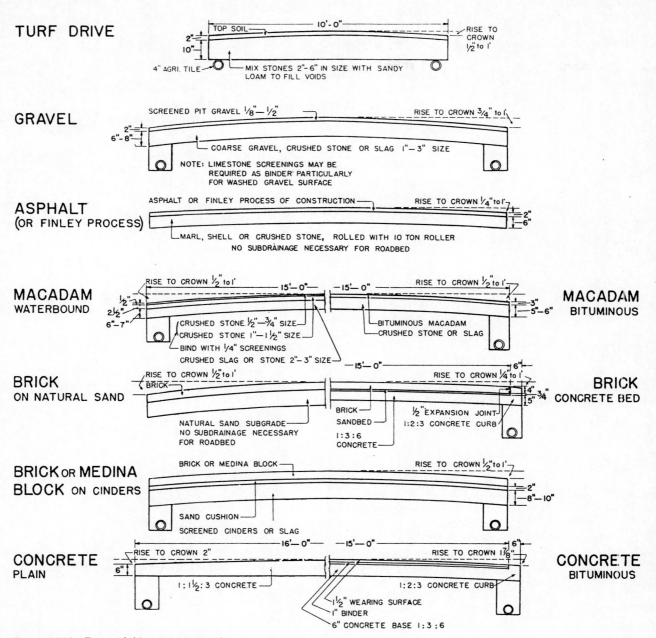

Figure 6.158 Types of driveway construction.

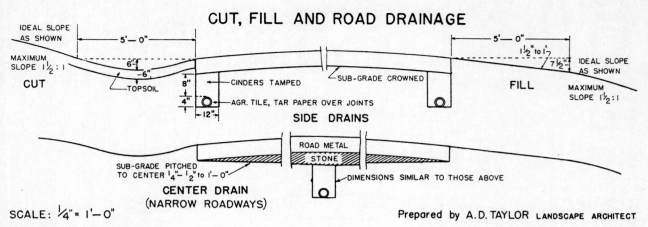

Figure 6.159 Cut, fill, and road drainage.

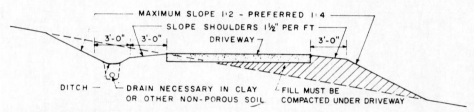

Figure 6.160 Cut-and-fill grading for driveway construction.

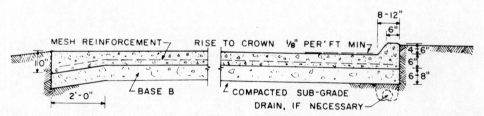

Figure 6.161 Concrete driveway. An asphalt expansion joint should be provided every 30 feet. The concrete may be covered with blacktop 1 to 2 inches thick.

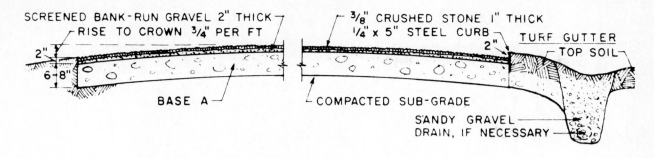

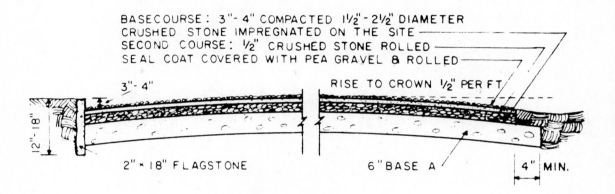

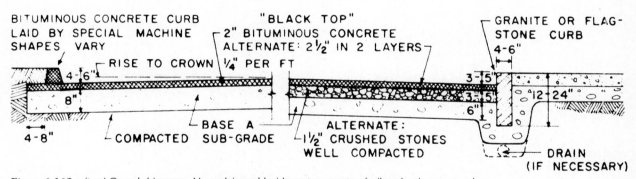

Figure 6.162 *(top)* Gravel driveway. Use calcium chloride or two coats of oil on bank-run gravel to keep surface from dusting. For heavy traffic, use two 6-inch courses for the base. *(center)* Penetration asphalt driveway. *(bottom)* Bituminous-concrete (blacktop) driveway.

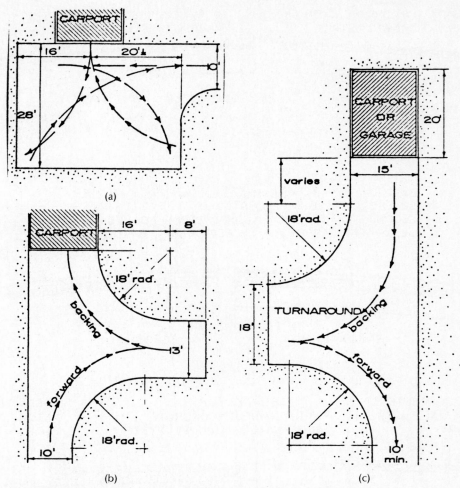

Figure 6.163 Garage drives: (*a*) double Y, limited space; (*b*) Y turn, back in; (*c*) Y turn, back out. All turns require 2 feet 0 inch clearance beyond edge of surfacing.

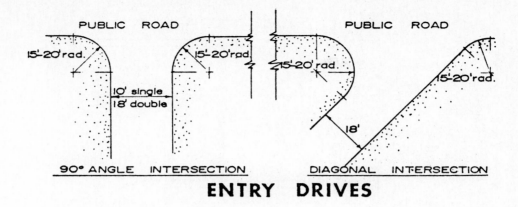

ENTRY DRIVES

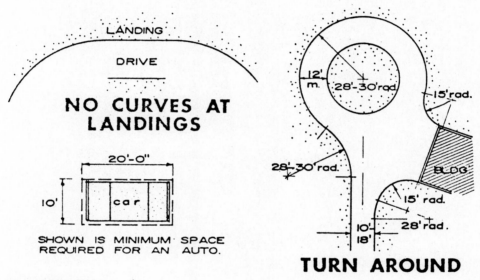

NO CURVES AT LANDINGS

SHOWN IS MINIMUM SPACE
REQUIRED FOR AN AUTO.

TURN AROUND

Figure 6.164 Private roads.

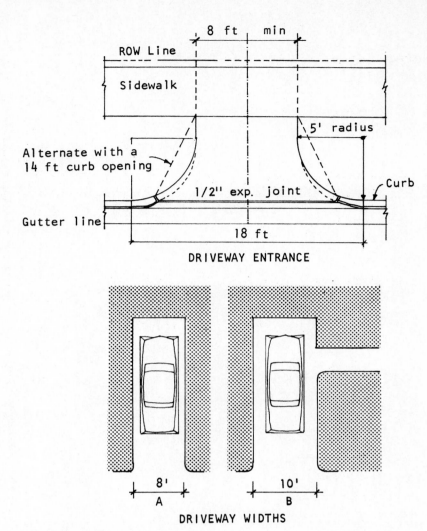

Figure 6.165   Residential driveways.

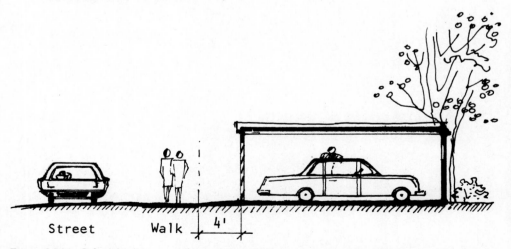

Figure 6.166   Safe sight distance. 4 feet is the desirable minimum between pedestrian ways or traffic lanes and rear of parking spaces where vision is clear. 8 feet is needed between pedestrian ways or traffic lanes and the rear of parking spaces where solid vision barriers such as garage walls exist.

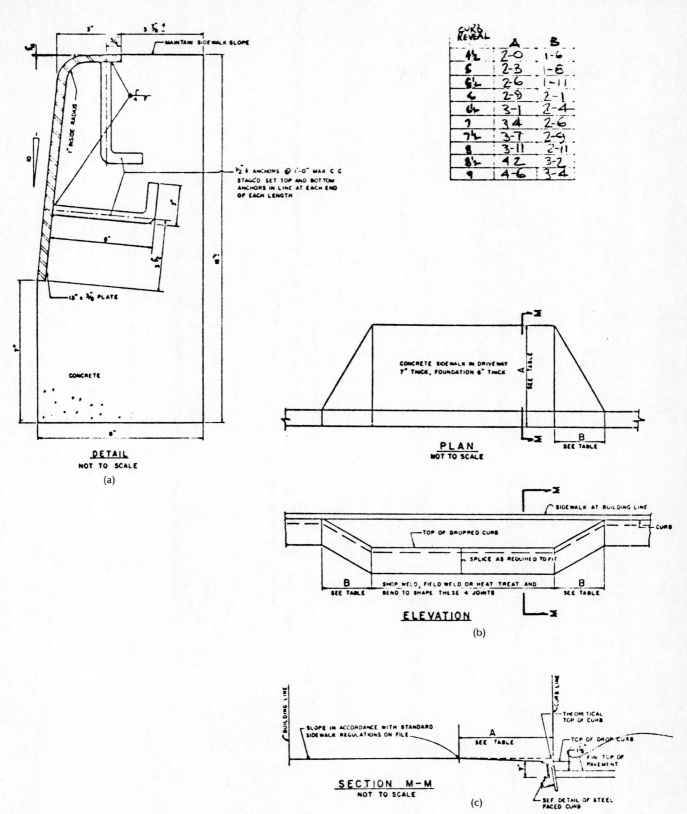

CURB REVEAL	A	B
4½	2-0	1-6
5	2-3	1-8
5½	2-6	1-11
6	2-9	2-1
6½	3-1	2-4
7	3-4	2-6
7½	3-7	2-9
8	3-11	2-11
8½	4-2	3-2
9	4-6	3-4

Figure 6.167 Curb cuts.

Curb Types

Comments

1. Vertical Face Curb

a. HEIGHTS GREATER THAN 6" ARE AWKWARD FOR HANDICAPPED AND OTHERS.

b. REQUIRES CURB CUT RAMP FOR WHEELCHAIR.

c. CONTRASTING COLOR TO ADJACENT PAVEMENT INCREASES VISABILITY.

6" MAX. HT.

AVOID LIP

2. Sloped Face Curb

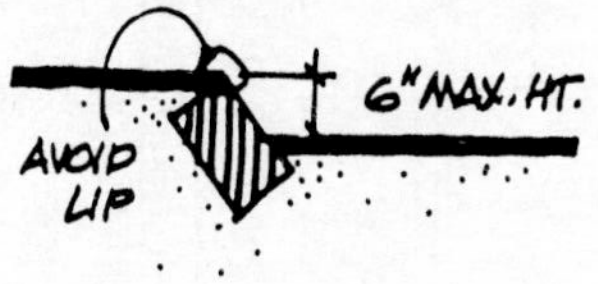

a. HEIGHTS GREATER THAN 6" ARE AWKWARD.

b. PROVIDE CURB RAMP FOR WHEELCHAIRS.

c. SLOPING FACE MAY PRESENT HAZARDOUS SURFACE IF STEPPED UPON.

d. CONTRAST COLOR WITH ADJACENT PAVEMENT.

6" MAX. HT.

AVOID LIP

3. Pre-Made Wheel Stops

a. STANDARD LENGTH 8'-0".

b. ANCHOR SECURELY TO PAVEMENT TO AVOID MIS-ALIGNMENT.

c. PROVIDE MIN. 32" CLEAR SPACE BETWEEN UNITS FOR WHEELCHAIRS AND OTHERS.

d. CONTRAST COLOR WITH ADJACENT PAVEMENT.

6" MAX. HT.

ANCHOR TO PAVEMENT

4. Posts and Bollards

a. 2'-0" HT. MIN. FOR VISABILITY FROM APPROACHING VEHICLES.

b. ANCHOR SECURELY TO PAVEMENT TO AVOID MIS-ALIGNMENT.

c. PROVIDE MIN. 32" CLEAR SPACE BETWEEN UNITS FOR WHEELCHAIRS AND OTHERS.

d. CONTRAST COLOR WITH ADJACENT PAVEMENT.

ALLOWS SITTING

24" MIN.

12"

5. Guard Rails

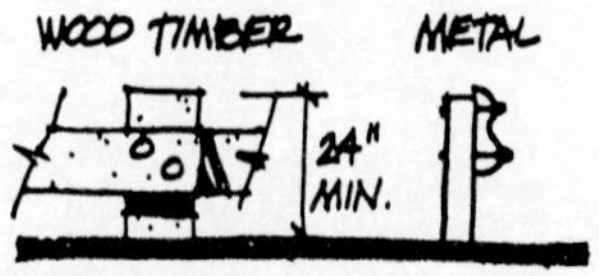

a. PROVIDES MAX. CONTROL OF VEHICLES.

b. 2'-0" HT. MIN. FOR VISABILITY FROM VEHICLES.

c. USEFUL ALONG PERIMETERS OF ROADWAY OR PARKING AREAS.

d. PROVIDE OPENINGS FOR WHEELCHAIRS AND OTHERS WHERE NECESSARY, 32" MIN.

WOOD TIMBER   METAL

24" MIN.

6. Posts and Chains

a. HAZARDOUS IF CHAIN SAGS BELOW 32" OR HIGHER THAN 42".

b. PROVIDE MIN. 32" CLEAR SPACE BETWEEN UNITS WHERE NECESSARY.

c. CONTRAST COLOR WITH ADJACENT PAVEMENT.

32" MIN.

Figure 6.168 Curb types.

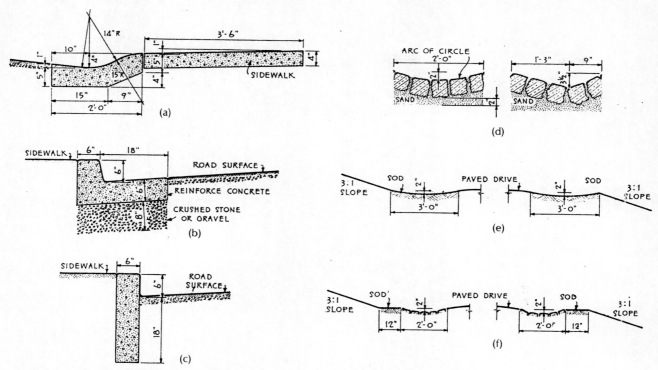

Figure 6.169 Curbs and gutters: (*a*) rolled curb and walk; (*b*) typical curb of reinforced concrete — gutter becomes part of road area; (*c*) typical traditional vertical curb of reinforced concrete; (*d*) cobble gutters, left section used for straight road alignment, right section used on curved road or bend in road; (*e*) sections of sod gutter at top and bottom of bank for road on sloping terrain; (*f*) paved gutter showing shoulder and desirable maximum ratio of slope.

Figure 6.170 Turf gutter or drainage ditch. Over a base of sandy gravel install topsoil and grade to smooth the contour. Apply seed and cover with glass-fiber blanket held in place by steel T pins; grass will come up through the blanket. Blankets are available in a 1-inch thickness in rolls measuring 6 by 150 feet and weighing 56 pounds.

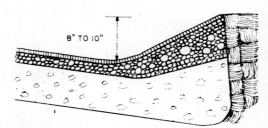

Figure 6.171 Oiled gutter for an oiled-gravel driveway.

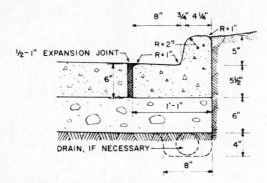

Figure 6.172 Precast concrete gutter and curb.

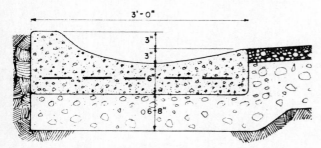

Figure 6.173 Precast concrete gutter.

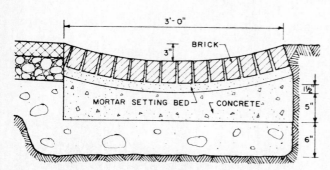

Figure 6.174 Brick gutter. Cobblestones or Belgian block may be used with similar details.

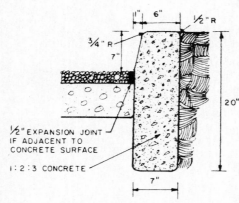

Figure 6.175 Precast concrete curb.

DRAINAGE DETAILS

Details for the construction of drains are shown in Figures 6.176 through 6.180.

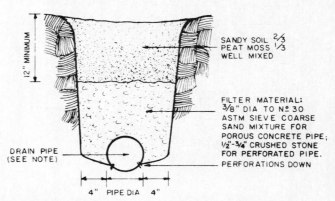

Figure 6.176 Details of a typical drain. The drainpipe may be perforated asphalt, perforated or porous concrete, or perforated galvanized-steel pipe. Under driveways use pipe strong enough to support heavy trucks. The minimum pitch for the drainpipe is 0.5 percent.

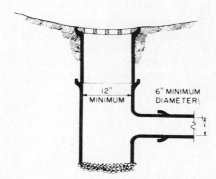

Figure 6.177 Lawn drain of concrete or vitrified-clay pipe with a standard round cast-iron grate.

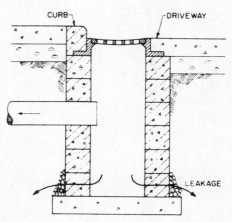

Figure 6.178 Curb inlet. The standard cast-iron grate measures 22¾ by 72½ inches.

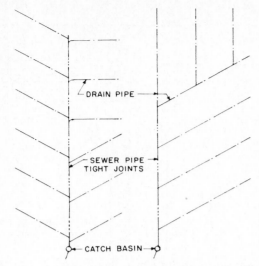

Figure 6.179 Typical drainage field layout (not drawn to scale).

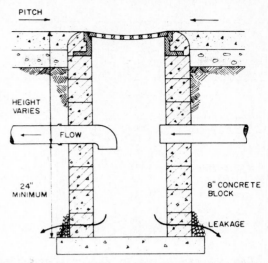

Figure 6.180 Catch basin, square or rectangular, with inside dimensions of 12, 18, 24, or 32½ inches. Use a standard cast-iron grate, light- or heavy-duty, or a sidewalk grate.

Typical Access Designs

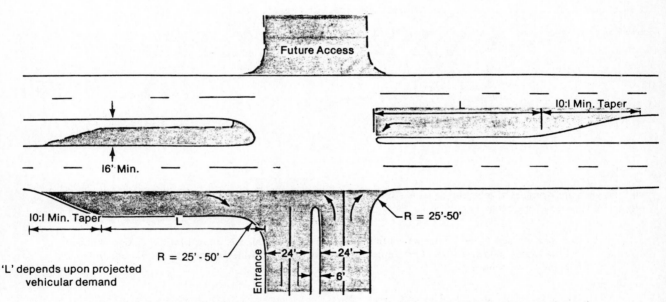

'L' depends upon projected vehicular demand

This design is the most commonly used treatment for a major entrance/exit drive. As noted in the figure, it is desirable to locate access facilities for undeveloped properties opposite the existing center's access. This allows for efficient signalization and overall safe traffic operation.

The divided cross section of the entrance provides the desired separation. As shown, the turning radii of 25 to 50 feet will permit higher turning speeds; however, if pedestrian movements are a consideration, smaller radii may be desirable. Protected left-turn storage lanes are provided to separate left-turning traffic from through traffic. The length of the storage lane is based on a capacity analysis. The transition taper should not be less than 10:1, and the storage lanes should be at least 12 feet wide.

Figure 6.181

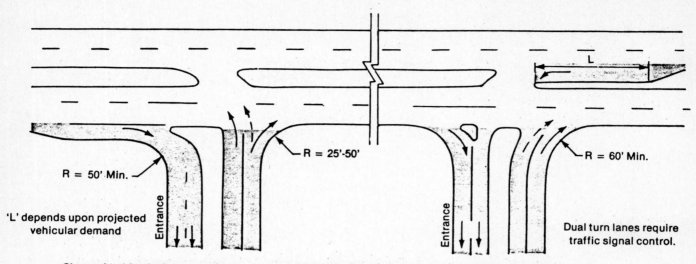

R = 50' Min.

R = 25'-50'

R = 60' Min.

'L' depends upon projected vehicular demand

Entrance

Entrance

Dual turn lanes require traffic signal control.

Shown in this design are directional entrance/exits that can be used for major inbound/outbound movements. The design would best serve where drivers have limited access to other entrances and thus, the use of the entrance serving each direction is maximized. Skewing of entrances can enhance operation and increase capacity.

Figure 6.182

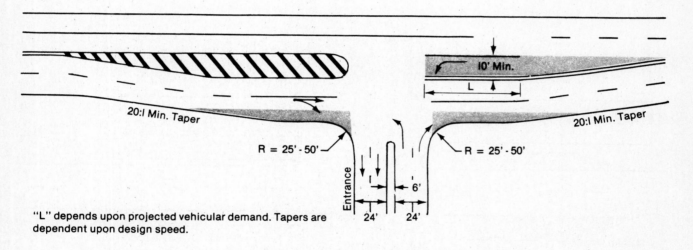

10' Min.

L

20:1 Min. Taper

20:1 Min. Taper

R = 25' - 50'

R = 25' - 50'

Entrance

24'   6'   24'

"L" depends upon projected vehicular demand. Tapers are dependent upon design speed.

Left-turn storage on an undivided route created by road widening is shown in this design. Although the roadway is shown widened on the entrance drive side, the widening may be on the opposite side or equal on both sides.

Figure 6.183

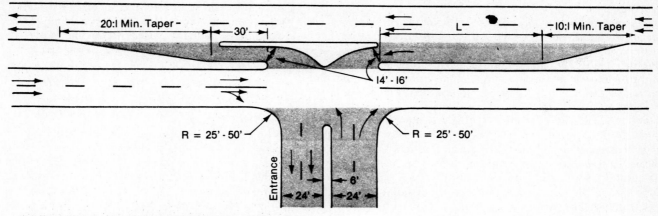

"L" depends upon projected vehicular demand. Tapers are
dependent upon design speed.

**This design illustrates a left-turn treatment for use on a major road with a median wider than 20 feet.
It may also be used successfully on two-lane roadways where adequate width is available to flare the
intersection.**

Figure 6.184

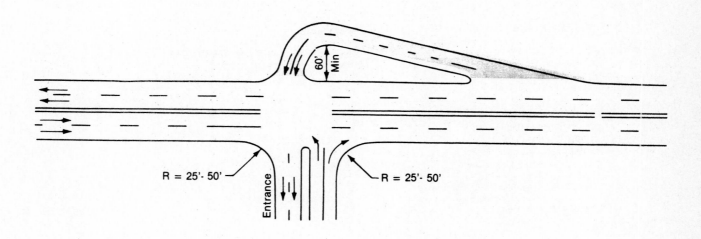

**This jug handle design provides high left-turn capacity and usually re-
quires less right-of-way than a dual left turn.**

Figure 6.185

Truck Loading Areas

Operational Design In general, off-street truck loading areas can be classified as either "off-street maneuver" facilities or "on-street maneuver" facilities. In the former design, the truck maneuver into and out of a loading space is conducted entirely in an off-street area. In the latter design, the outside lane of the street is used as part of the truck maneuvering area.

Off-Street Maneuver These facilities are physically separated from the street and are frequently built below street level in conjunction with underground auto parking areas. In regard to traffic operations and aesthetics, this alternative design approach is clearly superior to the on-street maneuver approach. Conflicts with other vehicles and pedestrians are significantly reduced.

It is desirable to provide sufficient maneuvering area to permit a vehicle to back into the space in a single maneuver. This approach, for a facility designed to accommodate a single-unit truck, requires devoting a distance into the building from the curb of at least 100 feet (30 meters) if right-angle truck loading space is provided.

Figure 6.186 presents four alternative design approaches for providing off-street maneuver loading spaces to accommodate trucks. Design dimensions for single-unit trucks are presented. Design criteria appear in the accompanying table.

- If a choice in vehicle approach direction exists, it is desirable that the vehicle approach the dock in a counterclockwise direction to permit the driver to be able to see the rear of the truck when backing into the loading space.
- Column location should be evaluated to ensure that it does not unnecessarily hinder or prohibit critical maneuvers.
- Access from a one-way street may be desirable.

On-Street Maneuver This type of design, from both a traffic operations and aesthetics viewpoint, is inferior to the off-street maneuver design. However, in certain facilities that must accommodate tractor-trailers, it may be necessary to utilize this design approach. Permitting the truck to stop in a moving traffic lane and back into a loading facility is not desirable.

When on-street maneuver facilities have been provided, consideration must be given to the maneuvering capabilities of the truck. As a result, a truck attempting to enter these facilities often is forced to block several lanes of traffic or, in some cases, an entire street. Desirably, a truck entering an on-street maneuver facility should not block any lane of traffic except that lane from which the backing maneuver is initiated; the vehicle should be able to back into the loading space in a single maneuver. A minimum of approximately 62

Minimal Design Criteria for Off-Street Loading Spaces

| | Type of vehicle to be accommodated | | |
Design criteria	Auto pickup panel	Single-unit truck	Tractor-trailer truck
Vertical clearance (feet)		13	14
Depth of space (feet)	25	35	60*
Width of space (feet)	11	12	12
Depth of loading dock (feet)	15	15	15
Height of loading dock (inches)	24–30	35–50	48–52

1 foot = 0.3 meter; 1 inch = 2.5 centimeters
* This depth required if tractor is not separated from trailer after the truck is parked. If tractor is separated, a 50-foot (15-meter) depth would be adequate.

Other Operational Considerations The factors listed below should be given consideration in designing an off-street truck loading facility.

- If possible, access to the facility should be from a minor street.
- Access points to off-street facilities should generally be at about mid-block in order to minimize conflicts with intersections.
- A vehicle parked in an off-street loading space should be entirely within the property line.
- Sight distance must be evaluated at access points. If sight distance is not adequate, traffic control devices may be needed.

feet (18.6 meters) of maneuvering area is required to accommodate the tractor-trailer. Figure 6.187 depicts a possible design for such a facility.

Yard and Dock Standards

Traffic Flow The traffic plan for off-street loading spaces should incorporate ease of ingress and egress, safety, and efficiency. For instance, counterclockwise vehicle circulation patterns facilitate easier turning maneuvers.

Service roads should be 12 feet wide for one-way usage and 22 to 23 feet for two-way operations. Gates should be 20 feet wide for the former and 30 feet for the latter, while intersection radii should not be less than 50 feet in both cases. Pedestrian circulation should be carefully planned with safety as the primary objective.

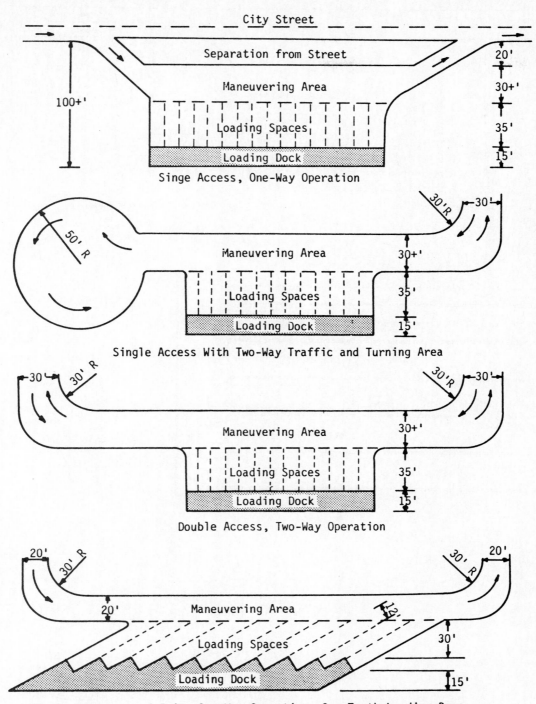

Single Access, One-Way Operation

Single Access With Two-Way Traffic and Turning Area

Double Access, Two-Way Operation

Single Entrance and Exit, One-Way Operation, Saw-Tooth Loading Bays

Note: 1 foot = 0.3 m

Source: References 1 and 11

Figure 6.186 Alternative designs for off-street maneuver facilities, single-unit truck design vehicle.

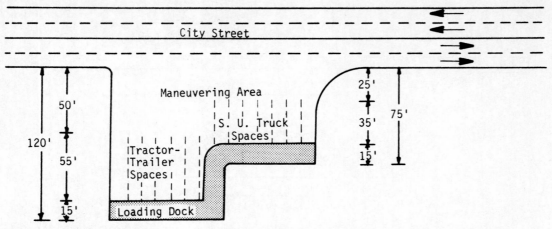

Figure 6.187 Alternative design for an on-street maneuver facility, combined tractor-trailer and single-unit truck design.

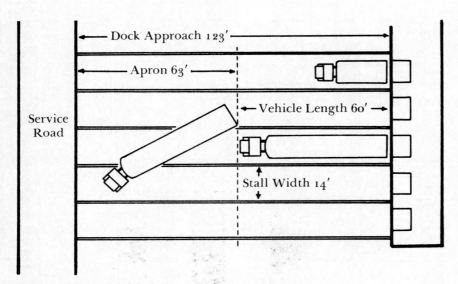

Figure 6.188 Yard and dock dimensions.

Overall length of tractor-trailer (feet)	Berth width (feet)	Recommended apron length (feet)	Dock approach (feet)
40	10	46	86
	12	43	83
	14	39	79
45	10	52	97
	12	49	94
	14	46	91
50	10	60	110
	12	57	107
	14	54	104
55	10	65	120
	12	62	117
	14	58	113
60	10	72	132
	12	63	123
	14	60	120

Dock Design Docks must be designed to facilitate efficient loading and unloading. Platform heights should be 44 inches for light pickup and delivery trucks and 48 to 52 inches for heavy trucks and trailers. The dock area should be at least twice the total body floor area of the largest number of trucks that can be docked at one time. Minimum dock overhead clearances (including pipes, lights, and so on) should be 12 feet.

Other Design Features Proper drainage, adequate paving, and protection of dock facilities from inclement weather lead to greater efficiency in off-street loading activities. Pits, ramps, and steep grades should be avoided, and wide aisles and doorways encouraged. Noise, fumes, and visual intrusion should be kept to a minimum.

Size of Required Loading Berths The dimensions of off-street berths should not include driveways, or entrances to or exits from such off-street berths (see accompanying table).

Minimum Dimensions for Accessory Off-Street Loading Berths (in Feet)

	Length	Width	Vertical clearance
Hospitals and related facilities or prisons	33	12	12
Funeral establishments	25	10	8
Hotels, offices, or court houses	33	12	12
Commercial uses	33	12	14
Wholesale, manufacturing, or storage uses:			
With less than 10,000 square feet of floor area	33	12	14
With 10,000 square feet of floor area or more	50	12	14

Recommended Minimum Standards for Off-Street Loading

Land use	Gross floor area	
	At which first berth is required	At which second berth is required
Industrial:		
Manufacturing	5,000	40,000
Warehouse	5,000	40,000
Storage	10,000	25,000
Commercial:		
Wholesale	10,000	40,000
Retail	10,000	20,000
Service establishments	10,000	40,000
Commercial recreational (incl. bowling alleys)	10,000	100,000
Restaurants	10,000	25,000
Laundry	10,000	25,000
Office building	10,000	100,000
Hotel	10,000	100,000
Residential:		
Apartment buildings	25,000	100,000
Apartment hotels	25,000	100,000
Institutional:		
Schools	10,000	100,000
Hospitals	10,000	100,000
Sanitariums (homes)	10,000	100,000
Public buildings:		
Terminals	5,000	40,000
Auditoriums	10,000	100,000
Arenas	10,000	100,000
Funeral homes	10,000	100,000

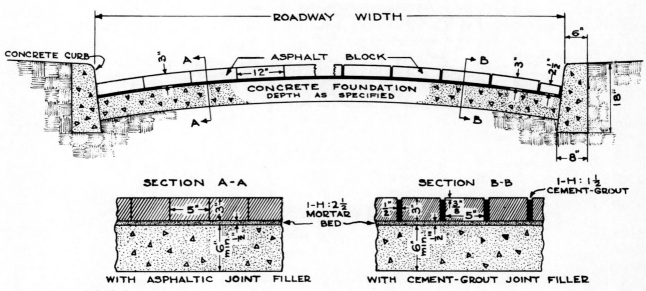

Figure 6.189 Asphalt block.

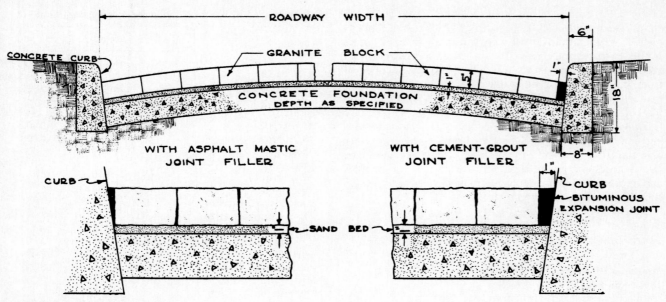

Figure 6.190 Granite block.

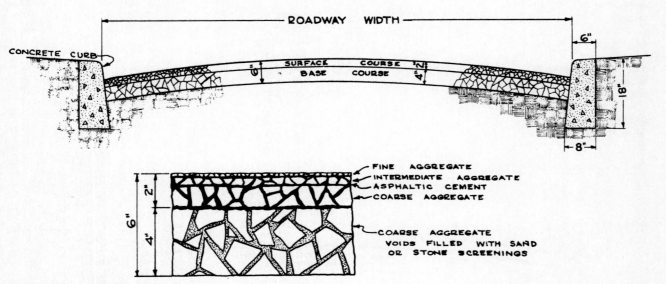

Figure 6.191 Asphalt macadam.

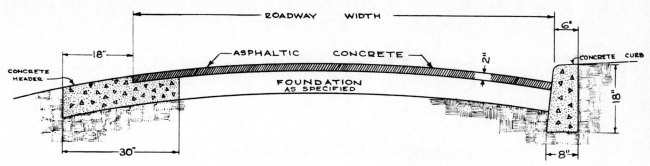

Figure 6.192 Asphaltic concrete.

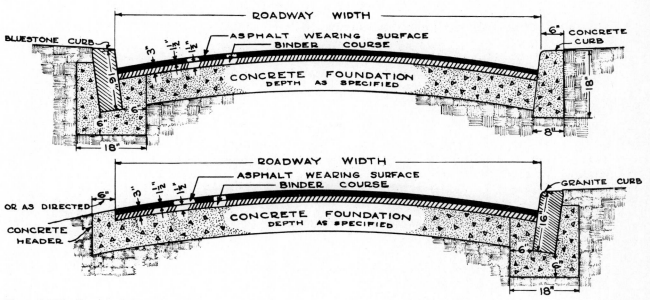

Figure 6.193 Sheet asphalt.

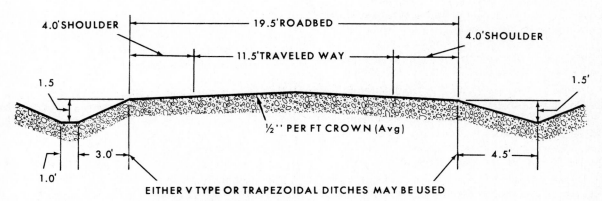

EITHER V TYPE OR TRAPEZOIDAL DITCHES MAY BE USED

Figure 6.194 Cross-section design of one-way earth road.

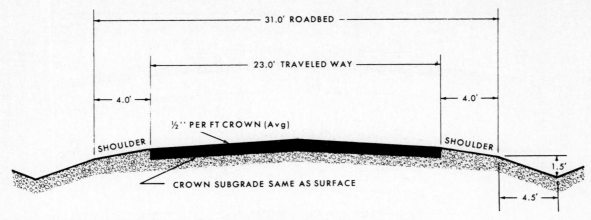

Figure 6.195 Cross-section design for two-way road using single-course construction.

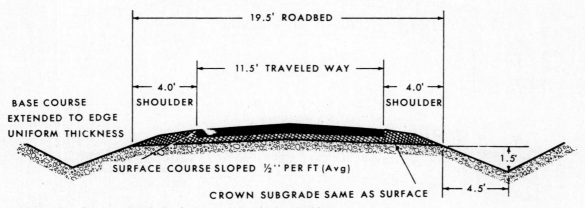

Figure 6.196 Cross-section design for one-way road using double-course construction.

6.8 Parking

A good parking lot should meet certain standards of design, location, and serviceability.

1. It should not detract from the attractiveness of the surrounding area. It should harmonize with adjacent land developments.

2. It should have utility or "workability" and attract customers. If the lot is operated by attendants, the motorist should expect courteous and efficient service. If the lot offers self-parking, parking stalls should be comfortably large, plainly marked, and easily accessible.

3. It should be designed to coordinate its use with nearby traffic flow. Entrances and exits should be on minor streets or in alleys.

PARKING LOT DESIGN

Some parking lots permit self-parking; others have attendant parking. Attendant parking is used at most commercially operated lots because it permits more efficient use of space.

A commonly used plan of attendant parking is shown in Figure 6.197.

A self-parking lot offers a minimum of inconvenience and a maximum of ease. Although the following principles of parking lot design and operation pertain largely to self- or customer parking, they apply also to the development of attendant-parking lots.

Size of Lots Parking lot sizes, measured in terms of cars accommodated, generally range from 25 to 500 or more. Lots that accommodate from 100 to 200 cars are efficient and practical. A few small lots, strategically sited, will usually serve better than a single large one.

Parking lots of large capacity can cause congestion on bordering streets, especially during peak-traffic hours. If capacity is small and the number of lots large, the potential traffic congestion tends to spread over several areas and thus minimize its effect.

If attendant-operated, moderately sized lots develop better operational efficiency than large ones. Parking and unparking of cars are accomplished much more rapidly in lots of low capacity. The distance traveled by attendants is shorter; service is faster.

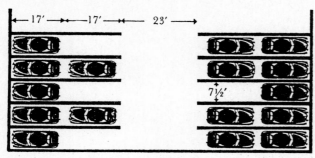

Figure 6.197 Ninety-degree double-stall attendant parking uses about 170 square feet per car. It is often the only feasible parking plan where land cost is high.

For efficient land use, the self-parking lot should provide 300 square feet for each car space. Depending on design features, 300 square feet is an acceptable standard for quickly estimating capacity for possible development of a convenient parking facility. A smaller per-car space area will necessitate narrower stalls and aisles. A higher figure will allow wider stalls and aisles, more maneuvering space, greater safety and convenience, and simple, faster operation.

The size and capacity of a parking lot should be tailored to the area it is to serve.

Entrances and Exits To minimize conflicts with street traffic, parking lot exits and entrances should be well defined and as few in number as practicable to provide for peak-hour operation. They should be at least 50 feet from intersections. When the lot fronts on a heavy-traffic street, separate entrances and exits are best.

Where possible, parking lot openings should be oriented to favor right-hand turns for entering and exiting traffic. Where such design is not possible and there is considerable street traffic, it may be necessary to prohibit left turns into and out of the parking lot.

Reservoir space at entrances and exits is of particular importance in lots on busy traffic streets. Space to accommodate accumulation of incoming cars prevents back-ups into traffic lanes if claim-ticket issuance delay, a confused driver, or other conditions temporarily block entering lanes. Area within the parking lot to accommodate some cars at the exit permits groups of leaving cars to take advantage of gaps in traffic.

Single-lane entrances to parking lots should be at least 14 feet wide. Exits should be a minimum of 10 feet. A combined entrance-exit should be not less than 26 feet wide. All should have suitable curb returns.

Grading and Surfacing If a parking lot is to offer year-round service, it should be pleasing in appearance and completely finished. Grading should provide good drainage; surfacing should abate dust. Inadequate drainage and inferior surfacing will discourage parking lot use and increase the difficulty of aisle and stall marking.

Gravel or crushed stone paving gives satisfactory service if adequate drainage is provided. Concrete is expensive for parking lot surfacing. Blacktop pavement has been used extensively as a satisfactory all-weather surface. It should be applied with the same care used in city street work or highway construction.

Lot Enclosure For safety and aesthetic value, lots should be fully enclosed by a fence, guardrail, wall, or hedge. This precaution adds to the lot appearance as well as ensuring that vehicle movements are kept within the lot.

In designing lot enclosures, allowance should be made for the overhang of car bumpers over the parking curb. With head-in parking, the overhang will average about 2½ feet. With back-in parking, the average overhang will be about 4 feet (Figures 6.198 and 6.199).

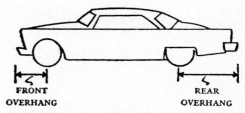

Figure 6.198 In designing parking lots, allowance should be made for the overhang of car bumpers over the parking curb.

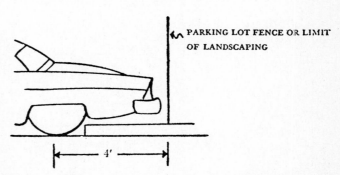

Figure 6.199 With back-in parking, the average overhang will be about 4 feet.

Signs Signs identifying the lot as a parking facility should be posted at all entrances. If adequate in size, and lighted at night, they speed up the entry of cars and thus minimize interference with traffic on adjacent streets. Signs should be used to indicate one-way aisles, location of exits, pedestrian walkways, and other helpful information.

Parking Stall and Aisle Dimensions The minimum area required to park a car is 144 square feet — a rectangle 18 feet long and 8 feet wide. These dimensions are minimum size for self-parking lots. Stalls 8½ feet wide are preferable. Most shopping centers use 9-foot widths.

Comparison of per-car area required in lots using self-parking, and those using attendant parking shows that driver parking requires almost 25 percent more space.

Varying the angle of parking changes the length of curb and width of aisle required for each car. The parking angle determines the efficiency of lot design because it governs the car space area.

Off-Street Parking Requirements for Residential Land Uses

| | Building type | | | | | | | | | | | | | |
| | Single-family dwelling | Duplexes | Multiple-family dwellings | Dormitories, sororities, and fraternities | | Hotels | | Motels | | Clubs and lodges | | Trailer courts | Apartment hotels and rooming houses | |
Requirement	Spaces per dwelling unit	Spaces per dwelling unit	Spaces per dwelling unit	Spaces per room	Spaces per occupant	Spaces per bedroom	Spaces per unit	Spaces per bedroom	Spaces per unit	Spaces per 100 sq ft GFA	Spaces per member	Spaces per unit	Spaces per dwelling unit	Spaces per room
Minimum	0.50	0.50	0.50	0.16	0.07	0.16	0.25	0.25	0.25	0.10	0.07	1.00	0.25	0.20
Maximum	3.00	2.00	2.00	1.00	1.00	2.00	1.33	1.25	1.33	3.00	1.00	2.00	1.50	1.00
Modal	1.00	1.00	1.00	0.50	0.33	1.00	1.00	1.00	1.00	0.50	0.20	1.00	1.00	0.50
Mean	1.28	1.26	1.20	0.56	0.44	0.64	0.97	0.90	1.00	0.71	0.23	1.23	0.94	0.62

Off-Street Parking Requirements for Commercial Land Uses

| | Building type | | | | | | | | |
| | Office buildings and banks | Business and professional services | Commercial recreational facilities | Bowling alleys | Shopping goods, retail | Convenience goods, retail | Restaurants | | Personal services and repair |
Requirement	Spaces per 100 sq ft floor area	Spaces per 100 sq ft floor area	Spaces per 100 sq ft floor area	Spaces per alley	Spaces per 100 sq ft floor area	Spaces per 100 sq ft floor area	Spaces per 100 sq ft floor area	Spaces per seat	Spaces per 100 sq ft floor area
Minimum	0.08	0.08	0.16	0.33	0.06	0.10	0.06	0.08	0.08
Maximum	1.33	1.33	2.00	10.00	3.00	1.33	2.00	0.50	1.00
Modal	0.25	0.33	1.00	5.00	0.50	0.50	1.00	0.25	0.50
Mean	0.33	0.37	0.79	4.50	0.44	0.44	0.75	0.28	0.40

Off-Street Parking Requirements for Educational Buildings

| | Building type | | | | | | | | | | |
| | Elementary and junior high schools | | | High schools | | | | Colleges and universities | | | |
Requirement	Spaces per auditorium seat	Spaces per classroom	Spaces per employee	Spaces per classroom seat	Spaces per auditorium seat	Spaces per employee	Spaces per classroom	Spaces per auditorium seat	Spaces per student	Spaces per 100 sq ft floor area	Spaces per employee
Minimum	0.05	0.50	0.33	0.05	0.05	0.33	0.50	0.06	0.10	0.12	0.33
Maximum	0.25	3.00	1.20	0.15	0.25	3.00	8.00	0.25	0.75	1.00	1.00
Modal	0.10	1.50	1.00	0.10	0.10	1.65	5.50	0.10	0.10	1.00	1.00
Mean	0.16	1.60	0.78	0.09	0.14	0.95	4.55	0.15	0.26	0.54	0.59

Off-Street Parking Requirements for Public Buildings

| | Building type | | | | | | | | |
| Requirement | Auditoriums and theaters | Museums and libraries | | Public utilities | | Stadiums and arenas | Welfare institutions | | |
	Spaces per seat	Spaces per 100 sq ft floor area	Spaces per seat	Spaces per 100 sq ft floor area	Spaces per employee	Spaces per seat	Spaces per bed	Spaces per 100 sq ft floor area	
Minimum	0.06	0.10	0.06	0.10	0.25	0.05	0.06	0.10	
Maximum	0.33	3.33	0.25	1.00	0.50	0.33	1.00	0.67	
Modal	0.25	0.33	0.10	0.33	0.40	0.25	0.16	0.25	
Mean	0.20	0.42	0.13	0.29	0.40	0.20	0.30	0.30	

Parking Spaces for Manufacturing Uses

Type of use	Parking spaces required, in relation to specified unit of measurement	Minimum
Manufacturing or semi-industrial	Spaces per square foot of floor area or spaces per employee	1 per 1000 square feet of floor area, or 1 per 3 employees, whichever will require a larger number of spaces
Light manufacturing	Spaces per square foot of floor area or spaces per employee	1 per 1000 square feet of floor area, or 1 per 3 employees, whichever will require a larger number of spaces
Storage or miscellaneous	Spaces per square foot of floor area or spaces per employee	1 per 2000 square feet of floor area, or 1 per 3 employees, whichever will require a lesser number of spaces

Off-Street Parking Requirements for Industrial Land Uses

| | Building type | | | | | | | |
| Requirement | Manufacturing plants | | | Warehouses | | | Wholesale | |
	Spaces per employee	Spaces per maximum shift employee	Spaces per 100 sq ft floor area	Spaces per employee	Spaces per maximum shift employee	Spaces per 100 sq ft floor area	Spaces per employee	Spaces per 100 sq ft floor area
Minimum	0.10	0.20	0.08	0.10	0.20	0.02	0.10	0.03
Maximum	1.00	1.50	1.00	2.00	1.50	0.67	1.00	1.33
Modal	0.50	0.50	0.20	0.33	0.50	0.10	0.50	0.15
Mean	0.44	0.50	0.25	0.48	0.52	0.16	0.45	0.30

Off-Street Parking Requirements for Manufacturing Plants

Measurement basis: spaces per	Average requirement in spaces by city population		
	25–50M	50–100M	Over 100M
Employee	0.57	0.40	0.38
Employee in maximum shift	0.67	0.46	0.33
100 square feet GFA	0.42	0.22	0.16

Off-Street Parking Spaces for Community Facility Uses

Type of use	Parking spaces required in relation to specified unit of measurement	Minimum	Maximum
Hospitals and related facilities	Spaces per bed	1 per 10 beds	1 per 5 beds
Medical offices or group medical centers	Spaces per sq ft floor area	1 per 800	1 per 400
Churches	Spaces per fixed seat	1 per 20 fixed seats	1 per 10 fixed seats
Libraries, museums, or noncommercial art galleries	Spaces per sq ft floor area	1 per 2000	1 per 1000
Colleges, universities, or seminaries:			
Classrooms, laboratories, student centers, or offices	Spaces per sq ft floor area	1 per 2000	1 per 1000
Theaters, auditoriums, gymnasiums, or stadiums	Spaces per fixed seat	1 per 12	1 per 8
Agricultural uses, including greenhouses, nurseries, or truck gardens	Spaces per sq ft floor area	1 per 2500	1 per 1000
Outdoor skating rinks	Spaces per feet of lot area	1 per 2000	1 per 800
Outdoor tennis courts	Spaces per court	1 per 5 courts	1 per 2 courts
Philanthropic or nonprofit institutions with sleeping accommodations; all types of nursing homes, health-related facilities, domiciliary care facilities or sanitariums	Spaces per bed	1 per 20 beds	1 per 10 beds
Schools	Spaces per sq ft area	1 per 3000	1 per 1500
College dormitories, fraternity or sorority houses	Spaces per bed	1 per 12 beds	1 per 6 beds
Prisons	Spaces per bed	1 per 20 beds	1 per 10 beds
Court houses	Spaces per sq ft of floor area	1 per 1000	1 per 500
Clubs, community centers, or settlement houses; philanthropic or nonprofit institutions	Spaces per person	1 per 20 persons	1 per 10 persons
Places of assembly	Spaces per person	1 per 12 persons	1 per 4 persons
Open commercial amusements	Spaces per sq ft of lot area	1 per 2000	1 per 500

Off-Street Parking Spaces for Commercial Uses

Type of use	Parking spaces required in relation to specified unit of measurement	Minimum	Maximum
General retail or service uses— food stores with less than 2000 sq ft of floor area	Spaces per sq ft of floor area	None required	1 per 300
Food stores with 2000 or more square feet of floor area per establishment	Spaces per sq ft of floor area	1 per 600	1 per 400
Hotels	Spaces per guest room or suite	1 per 8	1 per 4
Motels or tourist cabins	Space per guest room	1 per guest room	1 per guest room
Refreshment stands, drive-in	Space per sq ft of floor area	1 per 100	1 per 50
Funeral establishments	Space per sq ft of floor area	1 per 600	1 per 200
Boat docks or boat rental establishments		1 per 2 boat berths	

Screening Screening should be provided by a strip at least 4 feet wide, densely planted with shrubs or trees at least 4 feet high at the time of planting, and which are of a type which may be expected to form a year-round dense screen at least 6 feet high within 3 years, or a wall or barrier or uniformly painted fence of fire-resistant material at least 6 feet high, but not more than 8 feet above finished grade (or above the roof level, if on a roof). Such wall, barrier, or fence may be opaque or perforated provided that not more than 50 percent of the face is open.

Size of Parking Spaces Each 300 square feet of unobstructed standing or maneuvering area should be considered one parking space. In no event should the dimensions of any parking stall be less than 18 feet long and 8 feet 6 inches wide.

Location of Access to the Street The entrances and exits of all parking lots or public parking garages with 10 or more spaces should be located not less than 50 feet from the intersection of any two street lines.

Location and Type of Parking Four types of parking facilities are shown in Figures 6.200 through 6.203.

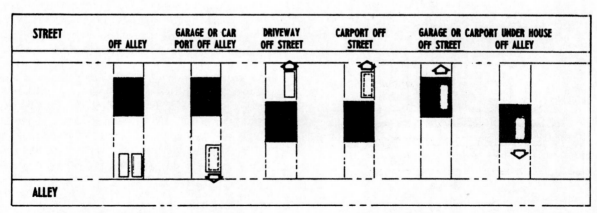

Figure 6.200 Parking on private lot.

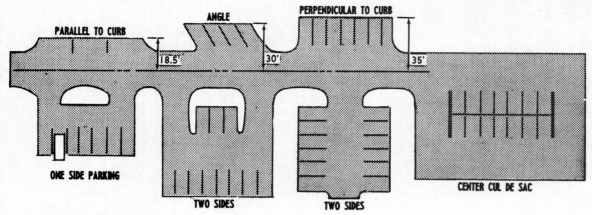

PARALLEL TO CURB · ANGLE · PERPENDICULAR TO CURB

18.5' · 30' · 35'

ONE SIDE PARKING · TWO SIDES · TWO SIDES · CENTER CUL DE SAC

Figure 6.201 Parking off private lot.

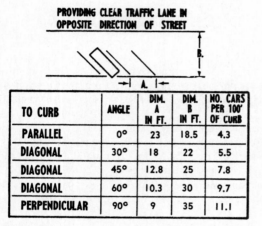

PROVIDING CLEAR TRAFFIC LANE IN OPPOSITE DIRECTION OF STREET

TO CURB	ANGLE	DIM. A IN FT.	DIM. B IN FT.	NO. CARS PER 100' OF CURB
PARALLEL	0°	23	18.5	4.3
DIAGONAL	30°	18	22	5.5
DIAGONAL	45°	12.8	25	7.8
DIAGONAL	60°	10.3	30	9.7
PERPENDICULAR	90°	9	35	11.1

Figure 6.202 Providing clear traffic lane in opposite direction of street.

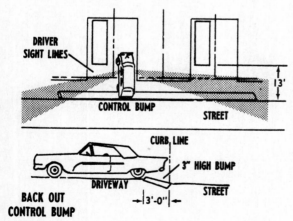

DRIVER SIGHT LINES · CONTROL BUMP · STREET · 13'

CURB LINE · 3" HIGH BUMP · DRIVEWAY · STREET · 3'-0"

BACK OUT CONTROL BUMP

Figure 6.203 Back-out control bump.

PARKING LOT LAYOUTS

Various types of layouts for parking lots are shown in Figures 6.204 and 6.205.

The herringbone pattern, shown in two variations in Figure 6.204, permits economies in some cases where space limitations prevent 90-degree parking. It will be noted that both patterns call for one-way aisles (unless cars are backed into stalls) but that type A requires the same direction of travel in all aisles while type B requires opposite directions in alternate aisles. The selection will depend upon the plan for circulation and the positions of entrances and exits. It will also be noted that either pattern is more economical of space than plain angle parking, which is used when only one aisle or row is possible.

Figure 6.205 shows the stall and aisle dimensions required for several different angles of parking, in each case giving an aisle wide enough to permit direct entrance without maneuvering for more than 75 percent of the cars. For any parking pattern, particularly one in which any angle other than 90 degrees is used, stall marking lines or curbs are necessary to achieve safety and to maintain the planned capacity.

The angle of parking affects the dimensions of a parking layout that contemplates stalls 8½ feet wide and 18 feet long as shown in Figure 6.206, in which l is the curb length required per car for various parking angles. Ninety-degree parking requires the least — 8½ feet; parallel parking requires the most. Although 18-foot stalls are long enough for angle parking, 22-foot stalls are needed for parallel parking to permit reasonably easy entrance and exit.

Depth of the parking stall d is measured perpendicular to the aisle. Parallel parking requires the least depth; 70-degree parking requires the most — almost 2 feet more than 90-degree stalls.

Under w is width of aisle needed to maneuver into and out of stalls with reasonable ease and convenience. Aisle width shown for the parking angles from 0 to 50 degrees is

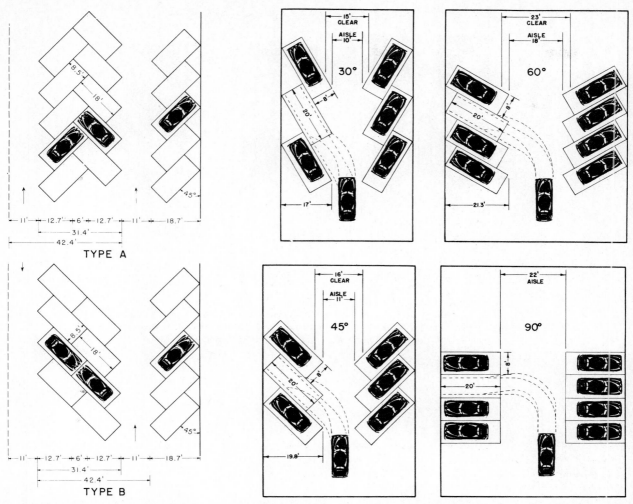

Figure 6.204 Herringbone-pattern parking layouts.

Figure 6.205 Space and aisle requirements for lot or garage parking at various angles.

greater than that required for the parking and unparking maneuver. With shallow parking angles, minimum aisle width is based on the needs of cars moving in the aisles. Two-foot clearance between parked and moving cars should be provided along parking-lot aisles. Minimum aisle widths thus become 10 feet for one-way and 20 feet for two-way aisles. The minimum desirable aisle widths are 12 and 24 feet, respectively.

A is the average gross area required per car. It varies for different angles of parking because of the parking angle–aisle width interrelationship. Angled parking permits narrower aisles than perpendicular parking; flatter angles allow narrower aisles than do wider angles. Aisle widths must be considered to get an accurate comparison of area requirements for parking at different angles. Thus, the average area per car space listed in the fifth column of the accompanying table is

the area of a parking stall plus one-half the area of aisle in front of it. Waste space at the end of parking lines and the area of circumferential access roadway space are not included. However, the figures listed provide a reasonably accurate basis for comparison.

Unit parking depth *upd* is a convenient unit of measure for quickly determining the best parking layout for a lot. It is the width of a parking aisle plus the perpendicular depth of a stall on each side. It is a definite value for each angle of parking.

Determining the best parking angle depends on the size and outline of the parking lot. It may be necessary to use two or more parking angles in the same lot to make best use of the space. Another space-saving method uses the herringbone and interlocking patterns of angle parking. illustrated in Figure 6.206 and tabulated in the last four columns of the

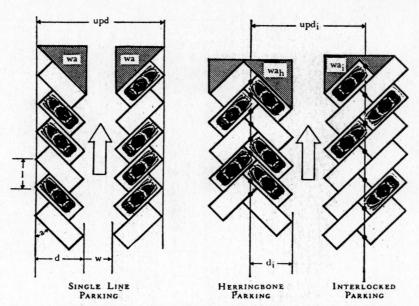

Figure 6.206 Changing the angle of parking affects the dimensions of a parking layout. Curb length of car stall *l* = measured parallel to parking line; depth of car stall (single parking line) *d* = measured perpendicular to the parking line; average depth of car stall (interlocked pair of parking lines) d_i = measured perpendicular to the parking line.

Typical Parking Lot Capacity Figures

	Single parking lines						Intermeshed multiple parking lines			
α, parking angle (degrees)	L, curb length per car (feet)	D, depth of stall (feet)	W, width of aisle (feet)	A, gross area per car (sq ft)	upd, unit parking depth (feet)	N, approximate number of cars per acre	D_h depth of stall (feet)	A_i, gross area per car (sq ft)	upd, unit parking depth (feet)	N_h approximate number of cars per acre
0	22	8	12	308	28	141	8	308	28	141
20	24.9	14.2	12	502.9	40.4	87	10.1	400.9	32.2	109
25	20.1	15.4	12	430.1	42.8	101	11.4	349.7	34.8	125
30	17.0	16.4	12	380.8	44.8	114	12.7	309.8	37.4	141
35	14.8	17.3	12	344.8	47.6	126	13.7	291.6	39.4	149
40	13.2	18.1	12	318.1	48.2	137	14.8	274.6	41.6	159
45	12.0	18.7	12	296.4	49.4	147	15.8	261.6	43.6	167
50	11.1	19.2	12	279.7	50.4	156	16.6	250.9	45.2	174
55	10.4	19.6	12	266.2	51.2	164	17.2	241.3	46.4	181
60	9.8	19.8	14.5	265.1	54.1	164	17.8	245.5	49.6	177
65	9.4	19.9	17	267.0	56.8	163	18.2	250.9	53.4	174
70	9.0	19.8	20	268.2	59.6	162	18.4	255.6	56.8	170
75	8.8	19.6	23	273.7	62.2	159	18.6	264.9	60.2	164
80	8.6	19.2	24	268.3	62.4	162	18.4	261.4	60.8	167
85	8.5	18.7	24	260.9	61.4	167	18.3	257.6	60.6	169
90	8.5	18.0	24	255.0	60.0	171	18.0	255.0	60.0	171

accompanying table. This intermeshing, or fitting together, at adjacent parking lines is also illustrated in Figure 6.207.

The herringbone pattern is most efficient with 45-degree parking. Made up of complementary 30- and 60-degree stalls, it is a means of gaining more car spaces. It is not an efficient or particularly space-saving method of angle parking. The interlocking parking pattern is equally effective at any angle of parking.

Normally the most efficient use of space is obtained when parking stalls are laid out perpendicular to the aisles. This plan provides more stalls per unit area and permits parking and unparking in either direction. Angled parking yields fewer

stalls for a given length of parking curb (see the fourth column in the accompanying table) and permits parking and unparking in one direction only. However, angle parking is more convenient. Drivers find it easier to maneuver in and out of angled stalls; they also make it easier to spot empty spaces.

Back-in parking requires narrower aisles than drive-in, but backing in requires more maneuvering ability than drive-in angle parking and takes more time. Most designers are reluctant to use back-in parking in customer-parking facilities because its difficulty may neutralize the driver convenience that so often motivates parking lot development.

Dimensions for Parking Angles with Varying Stall Sizes

α, parking angle	W and L, width and length of stall	N, number of car stalls per unit length of parking line (PL)	N_{100}, number of car stalls per 100 linear feet of parking length	Wa, wasted area (in square feet) at each end of single parking line	Wa_i, wasted area (in square feet) at each end of a pair of interlocked parking lines	Wa_h, wasted area (in square feet) at each end of a pair of herringbone parking lines
30	7'6" × 18'	$n = \dfrac{PL - 4'4''}{15'0''}$	6.4	207.8	336.0	
	8' × 18'	$n = \dfrac{PL - 3'7''}{16'0''}$	6.0	219.1	341.8	
	8'6" × 18'	$n = \dfrac{PL - 2'10''}{17'0''}$	5.7	230.3	343.1	
45°	9' × 18'	$n = \dfrac{PL - 2'1''}{8'0''}$	5.4	244.2	344.3	
	7'6" × 18'	$n = \dfrac{PL - 7'6''}{10'7''}$	8.7	162.0	311.2	311.2
	8' × 18'	$n = \dfrac{PL - 7'1''}{11'4''}$	8.2	169.5	310.7	322.1
	8'6" × 18'	$n = \dfrac{PL - 6'9''}{12'0''}$	7.8	176.0	306.9	333.1
	9' × 18'	$n = \dfrac{PL - 6'4''}{12'9''}$	7.3	182.5	303.1	344.3
60°	7'6" × 18'	$n = \dfrac{PL - 6'10''}{8'8''}$	10.7	107.4	202.4	
	8' × 18'	$n = \dfrac{PL - 6'8''}{9'3''}$	10.1	110.8	195.2	
	8'6" × 18'	$n = \dfrac{PL - 6'6''}{9'10''}$	9.5	113.1	189.9	
	9' × 18'	$n = \dfrac{PL - 6'4''}{10'5''}$	9.0	116.6	180.4	
90°	7'6" × 18'	$n = \dfrac{PL}{7'6''}$	13.3	0		
	8' × 18'	$n = \dfrac{PL}{8'0''}$	12.5	0		
	8'6" × 18'	$n = \dfrac{PL}{8'6''}$	11.7	0		
	9' × 18'	$n = \dfrac{PL}{9'0''}$	11.1	0		

Parking Lot Capacity (8½-by 18-foot stalls)

Width of area (feet)	Parking plan	Number and width of aisles	Car capacity per 100 feet of lot length
40	1 row of 90° stalls	1—22 feet	12
50	2 rows of 45° stalls	1—12 feet	14
60	2 rows of 90° stalls	1—24 feet	24
70	1 row of 90° stalls	1—24 feet	23
	2 rows of 30° stalls	1—12 feet	
80	1 row of 60° stalls	2—12 feet	24
	2 rows of 45° stalls	2—12 feet	
90	1 row of 45° stalls		
	2 rows of 45° stalls, interlocked	2—12 feet	28
	1 row of 45° stalls		
100	2 rows of 90° stalls	1—24 feet	38
	2 rows of 45° stalls, interlocked	1—12 feet	
	1 row of 60° stalls		36
110	2 rows of 60° stalls, interlocked	2—14.5 feet	36
	1 row of 60° stalls		
120	4 rows of 90° stalls	2—24 feet	48

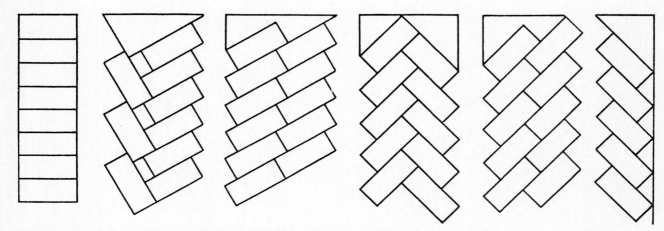

Figure 6.207 Selection of the best parking angle depends primarily on the size and outline of the parking lot.

Stall Markings Parking stalls should be marked by surface paint lines. Where stalls head into a wall or fence, it helps the driver if stalls markings are extended up the wall or fence.

Stall markings are particularly important on self-parking lots. They minimize inefficient space use caused by straddling of stalls, and encourage orderly parking where attendants are used.

Circulation Lots with angle stalls require continuous aisles because unparking cars are always headed in their original direction. The best aisle plan for such lots is a series of continuous one-way aisles that alternate in direction. This requires that the angled stalls be laid out in an interlocked rather than herringbone pattern. See Figure 6.208. One-way aisles are desirable because they are most economical of space and eliminate head-on and side-swipe accidents.

When 90-degree parking is used, cars can unpark to the right or left and may use the aisle in either direction. Two-way aisles reduce travel distance—parking and unparking cars can take the most direct route to their destinations. Some lots may necessitate a few dead-end aisles to use all available area. In those cases, 90-degree parking must be used.

Circulation aisles within the parking lot should be laid out to reduce travel distance and the number of turns. A poorly designed system of aisles, requiring excessive travel and turn-

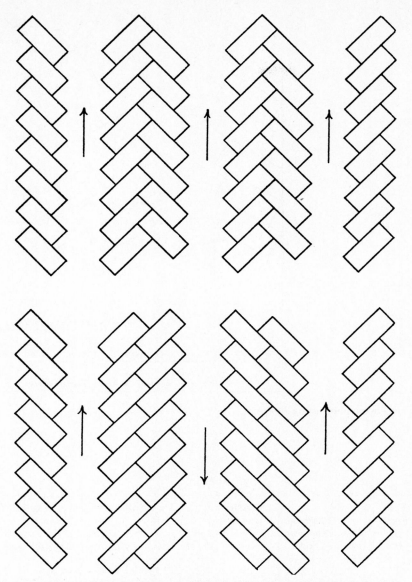

Figure 6.208 The 45-degree interlock layout allows driver to return via the adjacent aisle if all stalls are occupied in an aisle. Stall-bumper layout is fairly simple; even if the bumpers are not especially effective, there is little chance for collision damages.

ing for drivers to find an empty stall, develops confusion and hazard. Directional signs, prominently displayed, can help create orderly and safe interlot circulation.

Parking lot aisles should be as wide as practical. Wide aisles permit the entering driver to spot empty stalls quickly and encourage quick, easy parking. The faster an entering car is parked, the less it contributes to congestion.

Illumination Lighting should be provided where condi-

tions indicate a reasonable amount of nighttime parking. Illumination will discourage thievery and minimize pedestrian and property-damage accidents.

If the parking lot borders a residential neighborhood, landscaping, underground wiring, and attractive lighting standards may be justified on aesthetic considerations. For a lot surrounded by business property, these may be considered extravagant but nevertheless desirable.

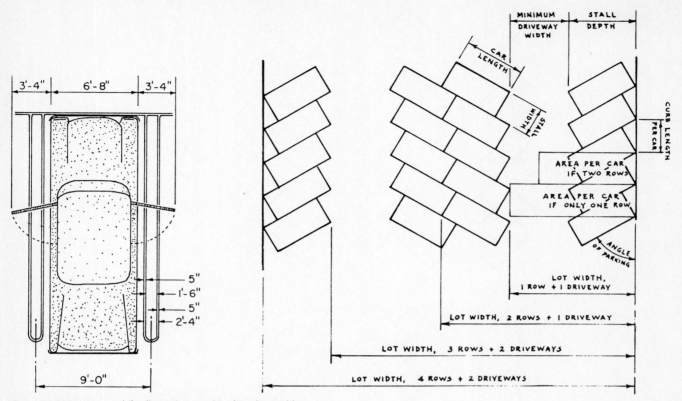

Figure 6.209 Automobile dimensions and parking lot widths.

Parking Lot Dimensions

Angle of parking	Stall width	Curb length per car	Stall depth	Minimum driveway width	Lot width 1 row + 1 driveway	Sq ft per car	Lot width 2 rows + 1 driveway	Sq ft per car	Lot width 3 rows + 2 driveways	Sq ft per car	Lot width 4 rows + 2 driveways	Sq ft per car
Along curb = 0°	9′	23′	9′	12′	21′	483	30′	345	51′	391	60′	345
	10′	23′	10′	12′	22′	506	32′	368	54′	414	64′	368
30°	9′	18′	17′4″	11′	28′4″	510	45′8″	411	66′2″	397	83′6″	376
	10′	20′	18′3″	11′	29′3″	585	47′6″	475	68′0″	453	86′2″	431
45″	9′	12′9″	19′10″	13′	32′10″	420	52′8″	336	79′0″	376	98′10″	315
	10′	14′2″	20′6″	13′	33′6″	490	54′0″	383	80′4″	379	100′10″	358
60°	9′	10′5″	21′0″	18′	39′0″	407	60′	313	95′0″	330	116′0″	305
	10′	11′6″	21′6″	18′	39′6″	455	61′	351	95′6″	366	116′6″	335
90°	9′	9′	19′	24′	43′	387	62′	279	105′	315	124′	279
	10′	10′	19′	24′	43′	430	62′	310	105′	350	124′	310

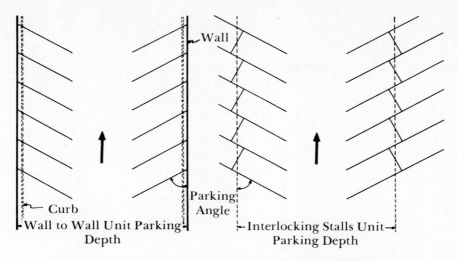

Figure 6.210 Recommended parking layout dimensions.

Parking Angle	Stall Width	Unit Parking Depth	
		Wall-to-Wall	*Interlocking Stalls*
45°	8'6"-10'0"	48'0"-52'0"	42'0"-45'0"
60°	8'6"- 9'0"	57'0"-60'0"	53'0"-56'0"
	9'6"-10'0"	55'0"-58'0"	51'0"-54'0"
75°	8'6"- 9'0"	60'0"-62'0"	56'0"-58'0"
	9'6"-10'0"	58'0"-60'0"	54'0"-56'0"
90°	8'6"- 9'0"	62'0"-66'0"	62'0"-66'0"
	9'6"-10'0"	62'0"-64'0"	62'0"-64'0"

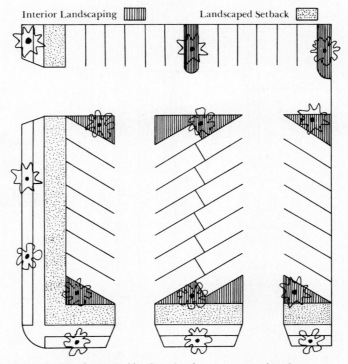

Figure 6.211 Suggested landscaping for a corner parking lot.

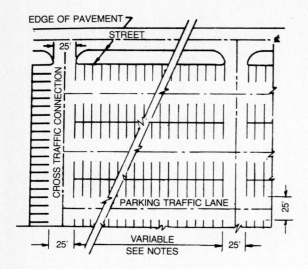

PERPENDICULAR OFF-STREET
PARKING — PLAN "A"
SEE DETAIL "A"

1. MULTIPLE LANE, OFF-STREET PARKING AREAS SHALL BE DESIGNED FOR 90 DEGREE PARKING, WITH TWO-WAY TRAFFIC LANES, IN ACCORDANCE WITH PLANS "A" AND "B".

2. NORMALLY, THE TOTAL SIZE OF THE PARKING AREA SHALL BE BASED UPON MAXIMUM ALLOWANCE OF 35 SQUARE YARDS PER VEHICLE, AND WILL INCLUDE SPACE FOR PARKING AND PARKING AREA ENTRANCES. SPACE NEEDED FOR LONG ACCESS ROADS SHALL NOT BE INCLUDED IN THE ALLOWANCE. THE AREA ALLOWANCE FOR INDIVIDUAL PARKING SPACES, WITHIN THE PARKING LANES, SHALL USUALLY BE 31.5 SQUARE YARDS.

PLAN "A"

1. NORMALLY, THE USE OF PLAN "A" IS RECOMMENDED. THE WIDE SPACING OF ENTRANCES REDUCES TRAFFIC INTERFERENCE ON THE ACCESS ROAD.

2. WHEN PARKING AREAS ARE OF EXCEPTIONAL LENGTH, CROSS TRAFFIC CONNECTIONS SHOULD BE PROVIDED ON THE BASIS OF ONE FOR ABOUT 40 VEHICLE SPACES OR 360 FEET. HOWEVER, FOR THE SMALLER ODD SHAPED LOTS THIS LIMIT MAY VARY SOMEWHAT TO PROVIDE FOR THE MOST EFFICIENT ARRANGEMENT.

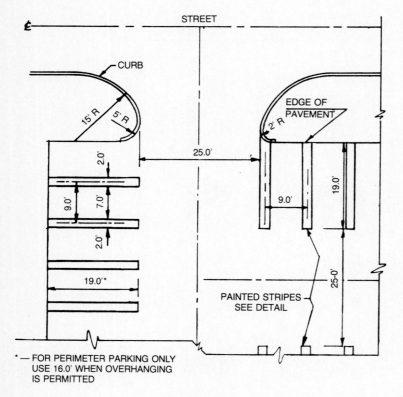

* — FOR PERIMETER PARKING ONLY
USE 16.0' WHEN OVERHANGING
IS PERMITTED

DETAIL "A"
SEE PLAN "A"
SCALE: 1" = 10'

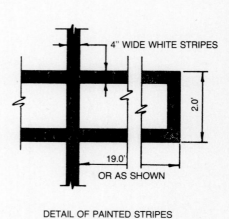

DETAIL OF PAINTED STRIPES

Figure 6.212 Off-street parking.

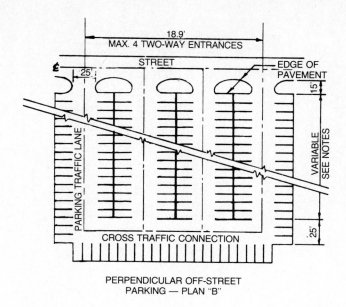

PERPENDICULAR OFF-STREET
PARKING — PLAN "B"

PLAN "B"

1. THERE SHALL BE A MAXIMUM OF FOUR TWO-WAY TRAFFIC LANES.

2. IF ACCESS IS FROM ONE ROAD ONLY, THE MAXIMUM DEPTH OF THE PARKING LOT SHALL BE APPROX. 400 FEET OR 40 VEHICLE SPACES WITH A CROSS TRAFFIC CONNECTION AT THE END OF THE TRAFFIC LANES.

3. WHEN ACCESS TO THE LOT IS FROM MORE THAN ONE ROAD, SEVERAL CROSS TRAFFIC CONNECTIONS WITH APPROX. 40 SPACES BETWEEN THEM MAY BE USED.

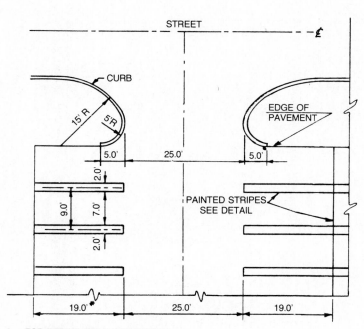

* — FOR PERIMETER PARKING ONLY
USE 16.0' WHEN OVERHANG
IS PERMITTED

DETAIL "B"

Figure 6.213 Off-street parking.

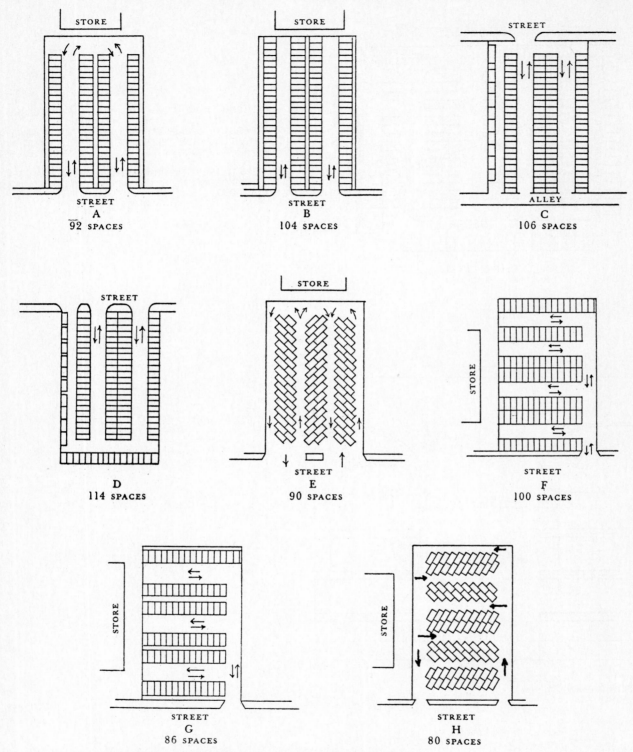

Figure 6.214 Typical parking lot layouts for a commercial building.

Pedestrian Safety Walkways within the parking lot are not essential, but their advantages often outweigh the additional space they require. Pedestrian walkways — never less than 4 feet wide — should be protected by a fence or barrier, and properly identified by guide signs.

Proper orientation of parking lines can promote pedestrian safety. If they are set at right angles to the major destination of most parkers, pedestrians are provided with direct access and are not required to cross aisles.

Where it is important to align parking stalls perpendicular to the most important pedestrian objective, raised pedestrian walkways between adjacent parking lines may be used. A raised pedestrian walk should have a clear width of 4 feet after allowing for the overhang of cars parked against it.

Stall Bumpers Stall bumpers are desirable aids for quick, safe parking. They encourage drivers to pull all the way into a parking stall and prevent them from overrunning the stall. Because of the difference in bumper overhang in various makes and models of cars, stall bumpers high enough to meet the car bumper are more efficient than low curbs that stop the car wheels. They also prevent encroachment on pedestrian walkways, and areas where parking meters may be installed.

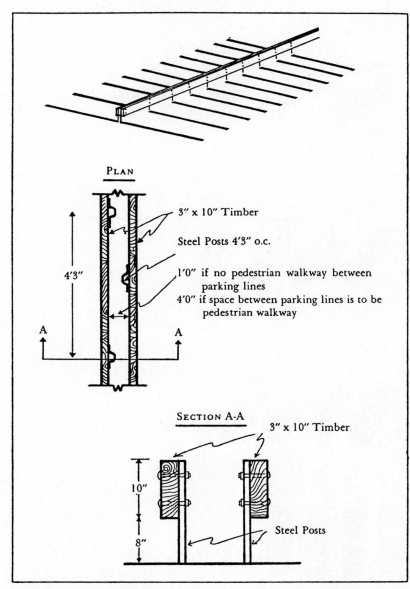

Figure 6.215 Stall bumpers high enough to meet the car bumpers are more efficient than low curbs that stop the car wheels.

CURB PARKING

Parking spaces, whether at the curb, in a lot, or in a garage, may be parallel, at an acute angle, or at right angles to the curb, wall, or aisle. The choice depends upon the shape and dimensions of the available area. At the curb, parallel parking is the rule unless traffic is extremely light or the street is extremely wide. Figure 6.216 shows the curb and street space required for various angles of parking. Spaces for parallel parking should be at least 20 feet long (preferably 22 feet); if the allowed parking period is short (say, 15 minutes), such spaces should be 22 feet long or longer so that the parking operation of entering or leaving the stall may be completed in one maneuver.

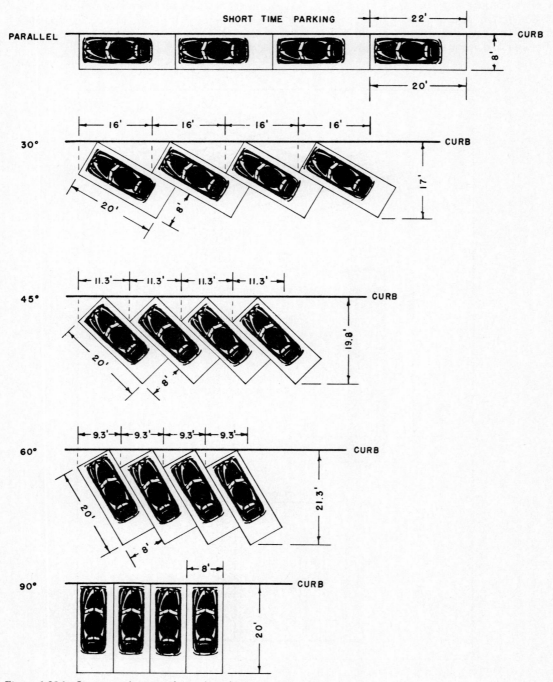

Figure 6.216 Space requirements for curb parking at various angles.

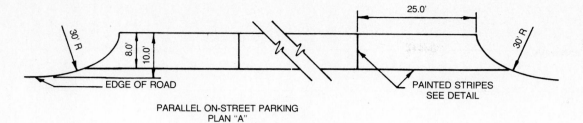

PARALLEL ON-STREET PARKING
PLAN "A"

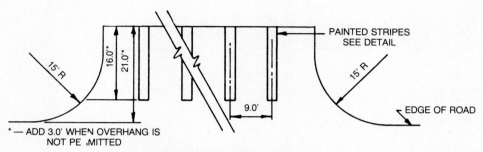

* — ADD 3.0' WHEN OVERHANG IS
NOT PERMITTED

90° PERPENDICULAR ON-STREET PARKING
PLAN "B"

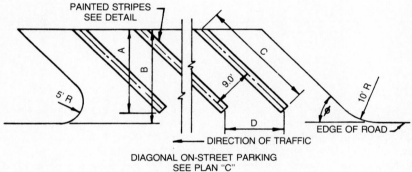

DIAGONAL ON-STREET PARKING
SEE PLAN "C"

1. ON-STREET PARKING SHALL BE DESIGNED FOR EITHER PARALLEL, PERPENDICULAR, OR DIAGONAL PARKING, IN ACCORDANCE WITH RESPECTIVE PLANS "A" OR "B", DEPENDENT UPON AVAILABLE SPACE AND SAFETY OF OPERATIONS.

2. THE SIZE OF THE PARKING AREA SHALL BE BASED UPON THE FOLLOWING ALLOWANCE:
   PLAN "A"   27.6 S.Y. PER CAR
   PLAN "B"   21.0 S.Y. PER CAR
   PLAN "C"   SEE TABLE

Figure 6.217 On-street parking.

45 and 60 Degree Diagonal On-Street Parking

Situation	0 (degrees)	A (feet)	B (feet)	C (feet)	D (feet)	Sq yd per/space
Overhang not	45	19.8	21.8	28	12.73	30.8
permitted	60	21	23	24.2	10.4	32.5
Overhang	45	17.7	19.7	25	12.75	27.66
permitted	60	18.4	20.4	21.2	10.4	28.8

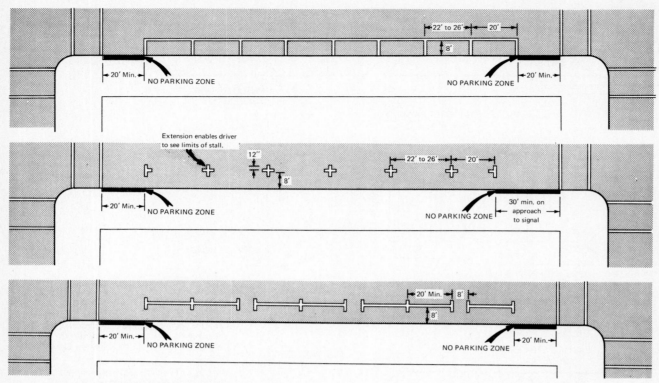

Figure 6.218 Typical parking-space-limit markings.

The normal and usually the most efficient layout in larger lots is to place the stalls at right angles to the aisles for as much of the area as possible. This permits entrance or exit in either direction and is the most economical use of space. With stalls 8 feet 6 inches wide and aisles 25 feet wide, the stalls may easily be entered by a driver-parker without maneuvering. Acute-angle parking gives fewer stalls for a given length of curb or aisle than right-angle parking does, and it requires one-way aisles; but entrance is easier for drivers, and the aisle may be narrower, thus permitting the use of a lot too narrow for right-angle parking.

Parking for the Physically Handicapped Parking spaces of greater-than-normal width are necessary for people who are disabled and use mechanical aids such as wheelchairs, crutches, and walkers. For example, a person who is chairbound must have a wider aisle in which to set up a wheelchair.

A minimum of two spaces per parking lot should be designed for use by physically restricted people, or at least one space per 20 cars, whichever is greater. These spaces should be placed as close as possible to a major entrance of a building or function, preferably no more than 100 feet away.

Parallel Parking Parallel parking spaces should be placed adjacent to a walk system so that access from the car to the destination is over a hard surface. Such spaces should

be made 12 feet wide and 24 feet long and should either have a 1:6 ramp up to the walk or be separated from it by bollards or some other device if the road level is at the same elevation as the walk. These areas should be designated as special parking, since otherwise they may appear to be a drop-off zone.

Ninety-Degree and Angled Parking Spaces designed for use by disabled people functioning with large mechanical aids should be 9 feet wide as a minimum. In addition, an aisle from 3 feet 6 inches to 4 feet wide should be provided between cars for access alongside the vehicle. It is important that there be plenty of room to open the car door entirely and, in the case of a dependent chair-bound person, that there be room for friends or attendants to assist him or her out of the car, into the chair, and away from the car.

The 9-foot standard width for a parking stall, with no aisle between spaces, does not drastically hinder semiambulant people with minor impairments, but an 8-foot width, unless used exclusively for attendant parking, is too narrow and should be avoided.

A 4-foot minimum clear aisle width should be provided between rows of cars parked end to end. The overhang of the automobile should be taken into account so that the island strip is wide enough to leave a 4-foot clear aisle when the stalls are filled. A strip 8 feet wide is a recommended minimum for an on-grade aisle, and 10 feet is a recommended

minimum if the aisle is raised 6 inches above the parking level.

If the aisle between rows of cars is not at the same grade level as the cars, ramps must be provided to mount the curbs. A 1:6 (17 percent) ramp is suitable for such a short distance.

Economically, the installation of an on-grade pathway 4 feet wide is less expensive than a raised walk. Precast car stops to delineate the passage can be used provided a 4-foot-wide space between the ends of the stops is maintained to allow access to the main passageway.

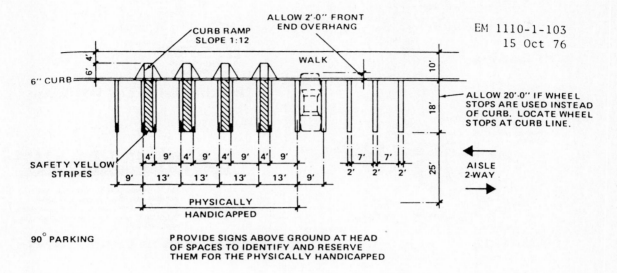

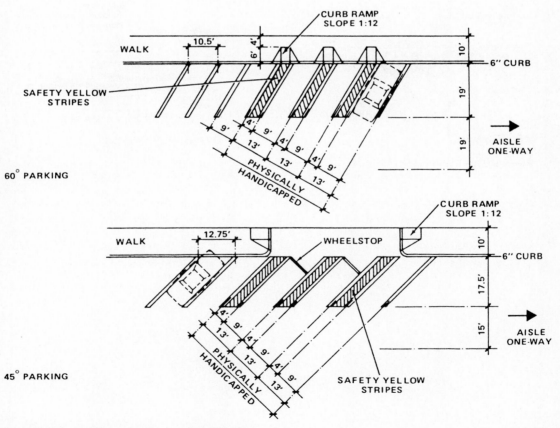

Figure 6.219 Parking spaces for use by the handicapped.

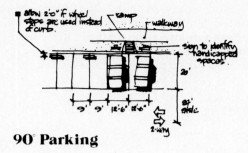

90° Parking

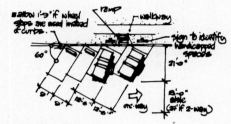

60° Parking

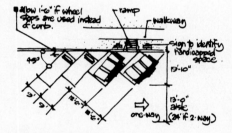

45° Parking

Figure 6.220 Types of parking.

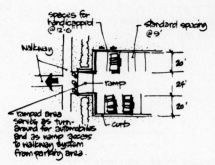

Parking Using End-Lot Access

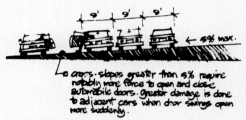

Cross-Slope in Parking Areas

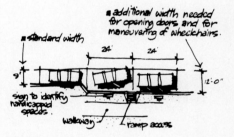

Parallel Parking

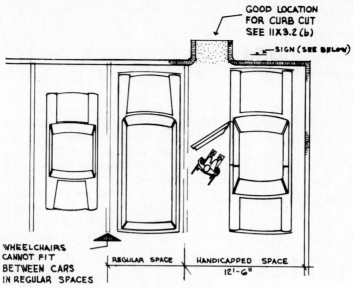

GOOD LOCATION
FOR CURB CUT
SEE IIX3.2 (b)

SIGN (SEE BELOW)

WHEELCHAIRS
CANNOT FIT
BETWEEN CARS
IN REGULAR SPACES

REGULAR SPACE

HANDICAPPED SPACE
12'-6"

Figure 6.221   Typical parking stall for the handicapped.

SUGGESTED SIGNS DISPLAYING
THE INTERNATIONAL SYMBOL
FOR ACCESSIBILITY

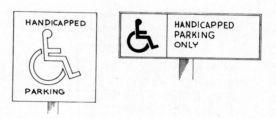

HANDICAPPED

PARKING

HANDICAPPED
PARKING
ONLY

Figure 6.222   Sign for handicapped parking.

Accessible Parking Spaces   Provide accessible parking spaces that:

1. Are at least 8 feet 0 inch (2440 millimeters) wide

2. Have an adjacent access aisle at least 5 feet 0 inch (1525 millimeters) wide

Exception: If accessible parking spaces for vans designed for handicapped persons are provided, each should have an adjacent access aisle at least 8 feet 0 inch (2440 millimeters) wide.

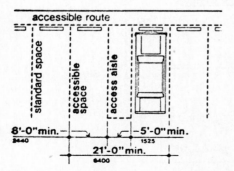

accessible route

standard space

accessible space

access aisle

8'-0"min.
2440

5'-0"min.
1525

21'-0"min.
6400

Figure 6.223   Accessible parking: two spaces sharing one aisle.

Total parking in lot	Required minimum number of accessible spaces
1–25	1
26–50	2
51–75	3
76–100	4
101–150	5
151–200	6
201–300	7
301–400	8
401–500	9
501–1000	2% of total
Over 1000	20 plus 1 for each 100 over 1000

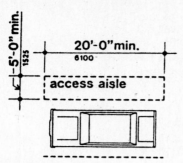

Figure 6.224 Unloading zone.

Passenger Loading Zones Provide accessible passenger loading zones that:

1. Have an access aisle at least 5 feet 0 inch (1525 millimeters) wide by 20 feet 0 inch (6 meters) long adjacent, parallel, and level with the vehicle standing space

2. Have vehicle standing spaces and access aisles with surface slopes not exceeding 1:48 (¼ inch per foot) in all directions

Provide minimum vertical clearances of 9 feet 6 inches at accessible passenger loading zones and along vehicle access routes to such areas from site entrances. Minimum vertical clearances of 9 feet 6 inches (3.45 meters) at accessible van parking spaces should be provided.

Garage or Carport Parking spaces should have a minimum width of 13 feet 6 inches.

Garages or carports must allow a clear area with a minimum width of 5 feet 0 inch on at least one side of the car. To accommodate this area, a single-car garage should have a minimum width of 14 feet 6 inches. A passageway 4 feet 0 inch wide should be provided in front of or behind the automobile. A garage or carport should therefore have a minimum length of 24 feet 0 inch.

A covered passageway with adequate overhang for protection against inclement weather should be provided when the garage is detached from the residence.

Automatically operated garage doors are a great convenience and should be incorporated whenever possible.

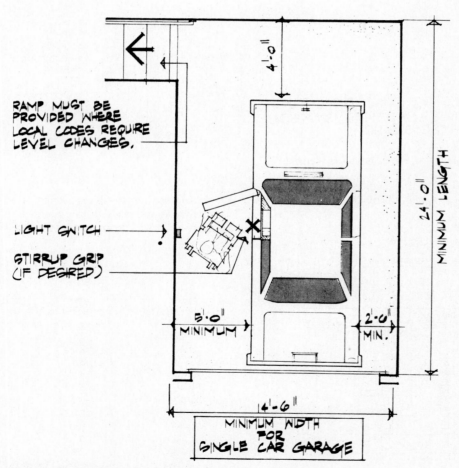

Figure 6.225 Dimensions of garage for use by the handicapped.

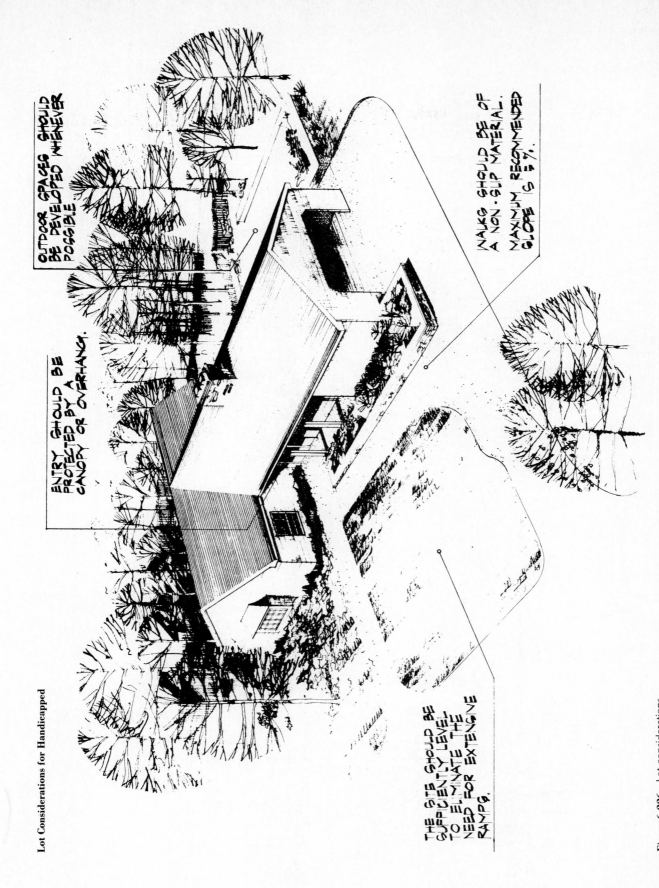

OUTDOOR SPACES SHOULD BE DEVELOPED WHENEVER POSSIBLE.

WALKS SHOULD BE OF A NON-SLIP MATERIAL. MAXIMUM RECOMMENDED SLOPE IS 5%.

ENTRY SHOULD BE PROTECTED BY A CANOPY OR OVERHANG.

THE SITE SHOULD BE SUFFICIENTLY LEVEL TO ELIMINATE THE NEED FOR EXTENSIVE RAMPS.

Figure 6.226 Lot considerations.

6.9 Drinking Fountains

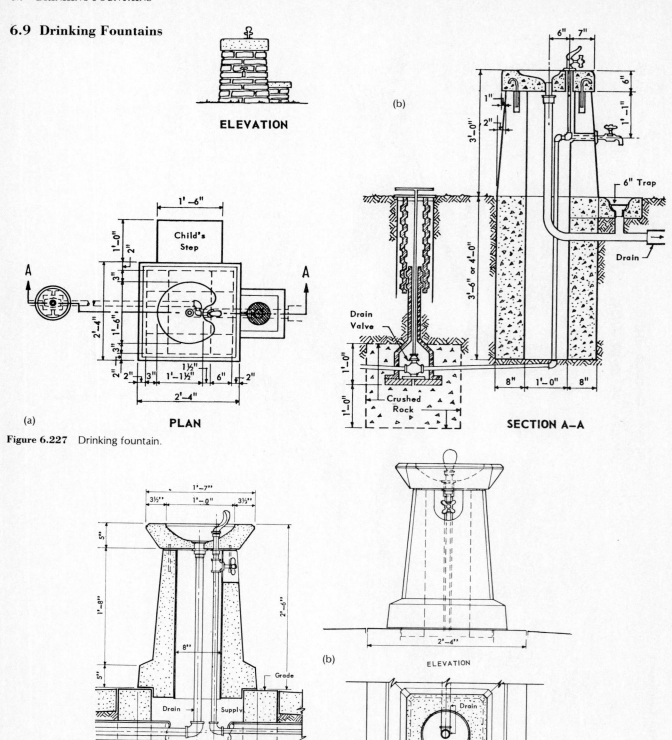

Figure 6.227 Drinking fountain.

Figure 6.228 Drinking fountain.

6.10 Outdoor Fireplace

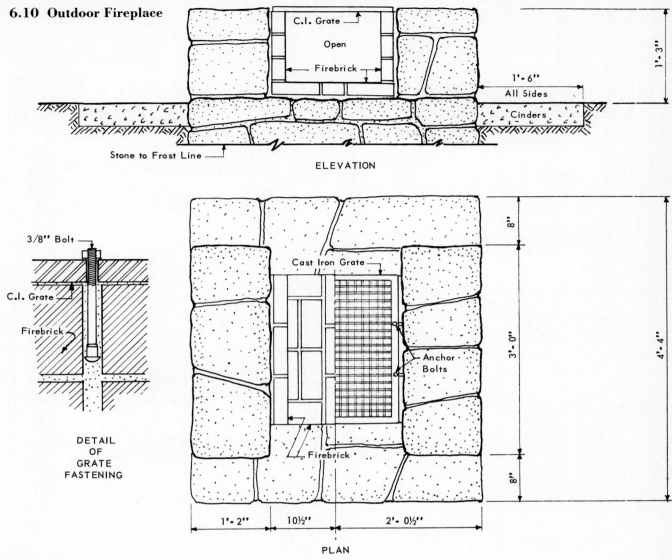

Figure 6.229 Outdoor fireplace.

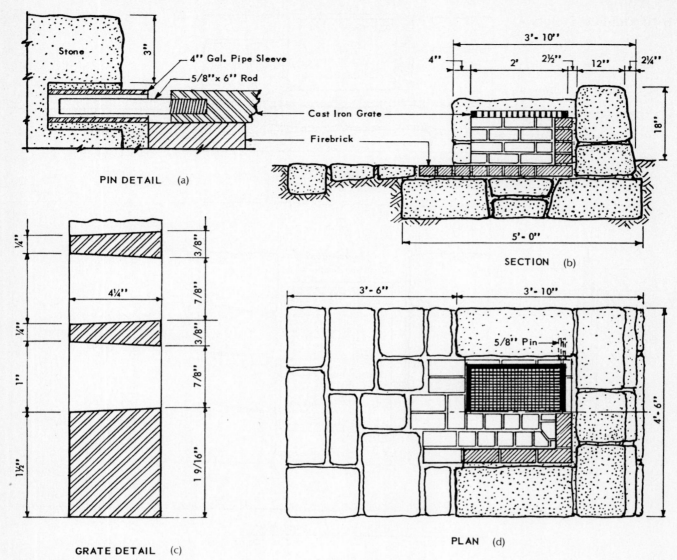

PIN DETAIL (a)

SECTION (b)

GRATE DETAIL (c)

PLAN (d)

Figure 6.230 Outdoor fireplace.

6.11 Retaining Walls

General Retaining walls should be designed in accordance with sound engineering principles, to resist the pressures of the retained materials, including both dead and live load surcharge to which they may be subjected.

These forces include both dead and live loads resting either directly on the structures or the soils behind it, lateral soil and water pressures, and the forces which the structure transmits to the soil. Retaining structures with satisfactory performance characteristics may be designed and constructed of most construction materials, such as treated wood, concrete, stone, brick, steel, or combinations.

The wall should be designed for stability against overturning and sliding, as well as for general structural integrity.

Seismic Design of Retaining Walls Retaining walls should be designed to resist seismic induced forces. Earth pressures assumed acting on the wall should be increased at least 20 percent in seismic zones 2 and 3 to provide for the seismic effect. No theoretical tension should exist between the wall footing or wall and the supporting soil.

ACCEPTABLE TYPES OF RETAINING WALLS

Gravity Walls Gravity-type retaining walls are normally constructed with a cross section similar to the illustration and

may be of stone, brick, masonry, or unreinforced concrete. Wall dimensions should be sufficiently massive to ensure stability and that no tensile stresses will develop within it as a result of forces which may act upon it. See Figures 6.231 (type 1) and 6.233.

Cantilever Walls   Cantilever-type retaining walls are normally constructed with relatively thin cross sections similar to the illustrations and are of reinforced masonry or concrete. Weight needed for stability is provided primarily by the backfill behind the vertical stem of the wall. The magnitude and direction of forces to be resolved depend upon many variables which must be accounted for in the design so that sufficient reinforcing steel may be properly positioned during construction to assure satisfactory performance. Concrete walls are illustrated.

Cantilever walls reinforced types, Figure 6.231 (2*a*, 2*b*, and 2*c*) and Figure 6.234, are of relatively thin sections and are generally more economical than walls of type 1. Conditions influencing a choice of the several shapes of this type of wall are:

Type 2a walls are the most economical and should be used except where the footing would extend beyond the property line.

Type 2b should be used at property lines when the grade of the adjoining property is below the grade of the site.

Type 2c walls should be used at property lines when the grade of the adjoining property is higher than the grade of the site. The height of this type of wall should preferably not exceed 16 feet, since beyond this height it becomes uneconomical to provide adequate resistance to overturning and sliding.

Terrace Walls   The use of low terrace walls (type 3, Figure 6.232) should be limited to cases in which the difference in grade on the two faces of the wall does not exceed 3 feet and where there is no probability that the wall will have to support a surcharge in excess of 50 pounds per square foot.

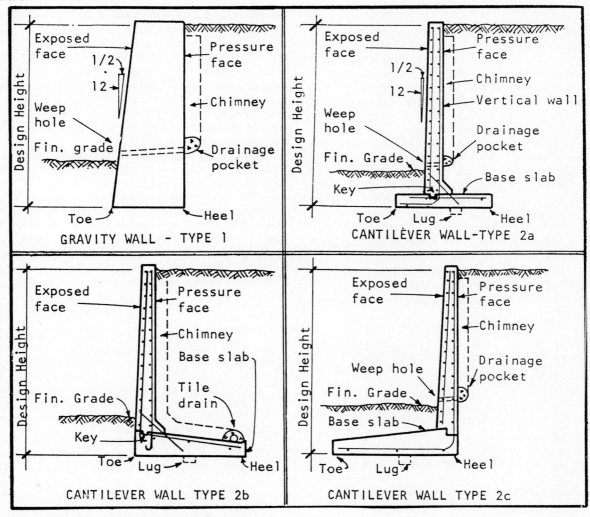

Figure 6.231   Two types of retaining walls.

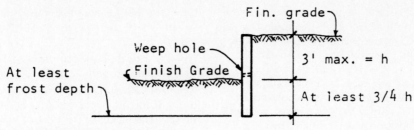

Figure 6.232   Terrace wall — type 3.

This type of wall may be of brick, stone, concrete, or combinations of these materials. The wall should completely fill the trench into which it is built by providing concrete backfill as required, because its stability is dependent upon full bearing on both trench faces. The earth below the lower grade should be firm, reasonably dry, and not readily compressible under horizontal loading.

DRAINAGE BEHIND RETAINING WALLS

General   Drainage of the earth behind a retaining wall is highly important, since lack of drainage may cause either complete or partial failure due to water filling the pores of the soil and in cold climates freezing and thereby through frost action or hydrostatic pressure generating lateral forces for which the wall was not designed. Trapped water may also result in uplift under the base or changes in soil "composition," either of which may induce failure.

General Requirements for Free-Standing Retaining Walls   The following information applies to all free-standing retaining walls except: walls more than 10 feet in height; walls with clay backfill; wall bases bearing on clay, silt, or quicksand; walls with backfill weighing more than 100 pounds per cubic foot; walls on piles; walls with surcharge such as from a roadway or building nearer to the wall than the height of wall. These exceptions should be designed in accordance with sound engineering principles.

Proper drainage of retaining walls should be provided. A layer of coarse stone 12 inches thick should be placed against the back of wall and weep holes through the stem installed at frequent intervals to empty on the ground at the front of the wall; 4-inch-diameter tile drains spaced 10 feet apart are usually sufficient.

Resistance to sliding is obtained by frictional resistance between the base and the soil. Often a lug or offset is cast integrally with the base slab and under it to assist in resisting the tendency to slide. For the height of walls covered here the same resistance is achieved by requiring that the base slab be well below the ground surface in addition to being below the frost line. Footings of retaining walls are to be placed on firm undisturbed soil.

Vertical tongue-and-groove contraction joints should be placed in the wall at 20- to 30-foot intervals to maintain alignment of adjacent sections and prevent the occurrence of irregular cracks due to temperature change and shrinkage. It is advisable to cover contraction joints with a strip of membrane waterproofing on the back of the wall to prevent seepage through the joint. Expansion joints should be placed not more than 75 feet apart.

Backfill should be placed in such manner as not to produce impact, as from large stones rolling down a slope or dropping on the wall. It is good practice to bring up the fill material along the wall at a rate as uniform as possible. The allowable soil bearing value should not be exceeded; the retaining walls shown produce soil pressures of 1.1 and 1.5 tons per square foot for level fill and sloping fill, respectively, at the toe.

For gravity walls, the upper inclined form should be weighted to prevent flotation. For cantilever walls, the reinforcement should be protected by concrete cover as shown in the illustrations.

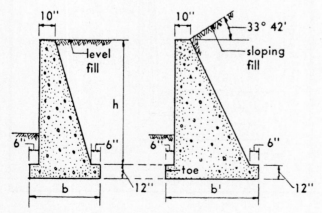

Figure 6.233   Gravity-type retaining walls.

h feet	b	b'
3	2'-3"	2'-6"
4	2 -7	3 -0
5	3 -0	3 -6
6	3 -4	4 -2
7	3 -9	4 -10
8	4 -3	5 -6
9	4 -9	6 -3
10	5 -3	7 -0

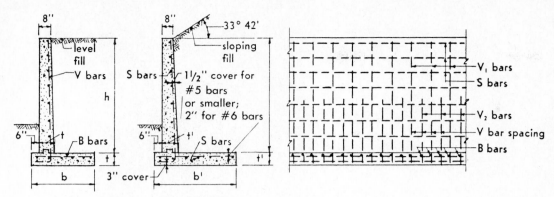

Figure 6.234 Cantilever-type retaining wall.

			V bars—lengths				B bars			S bars*	
h	b	t	Size	Spacing	V₁'s	V₂'s	Size	Spacing	Length	Size	Number
5′	2′9″	10″	#2	12″	6′6″		#2	12″	2′4″	#3	8
6′	3′4″	10″	#2	7″	7′6″		#2	7″	3′0″	#3	10
7′	3′10″	10″	#3	9″	8′4″		#3	9″	3′6″	#3	12
8′	4′6″	12″	#4	12″	9′8″	5′0″	#4	12″	4′2″	#3	13
9′	5′10″	12″	#4	9″	10′8″	5′4″	#4	9″	4′8″	#3	15
10′	5′6″	12″	#5	11″	11′8″	6′2″	#5	11″	5′2″	#3	16

			V bars—lengths				B bars			S bars*	
h′	b′	t′	Size	Spacing	V₁'s	V₂'s	Size	Spacing	Length	Size	Number
5′	3′6″	10″	#3	10″	6′4″		#3	10″	3′2″	#3	9
6′	4′3″	10″	#4	10″	7′4″		#4	10″	3′10″	#3	11
7′	5′0″	10″	#4	7″	8′4″		#4	7″	4′8″	#3	13
8′	6′0″	12″	#5	9″	9′8″	6′0″	#5	9″	5′8″	#3	15
9′	7′0″	13½″	#5	7″	11′0″	6′8″	#5	7″	6′8″	#3	17
10′	7′9″	15″	#6	9″	12′2″	6′10″	#6	9″	7′4″	#3	19

*Space S bars evenly.

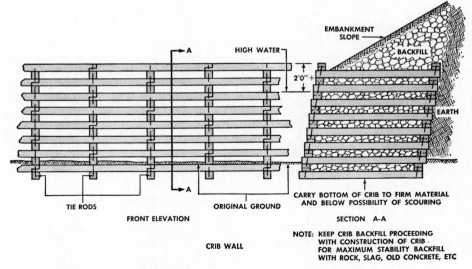

Figure 6.235 Crib-type retaining wall.

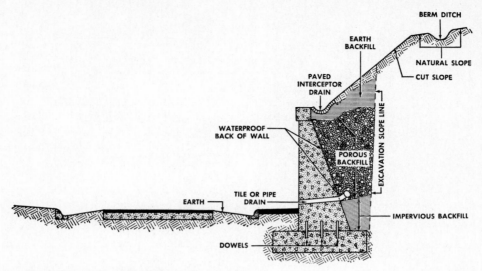

Figure 6.236 Concrete retaining wall.

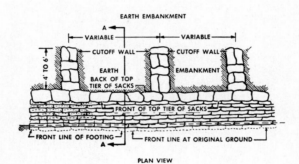

Figure 6.237 Sacked-concrete retaining wall.

Diagrams of free-standing and retaining walls are shown in Figures 6.238 through 6.250.

Dry retaining walls are illustrated in Figure 6.241, and various types of free-standing masonry walls in Figures 6.242 through 6.245. For the latter, foundations must extend below the frost line.

The main criteria for construction of retaining walls are a foundation to prevent damage through frost action and the provision for drainage behind the wall by means of gravel and seepage holes.

Limit dry stone retaining walls to a height of 4 feet. The economical use of this type of wall depends greatly on the availability of suitable rock. A flat facing for appearance and flat sides for close fitting are important. Mortared joints will increase strength, not through adherence but by ensuring a close fit (Figure 6.248).

An 8- to 10-inch gravel base and backfill with permeable material will give constant drainage and will also reduce frost action.

The wall is kept in place by sheer weight, and a 1 inch in 1 foot batter will give added strength.

Carry surface water parallel to dry stone walls. Any drainage allowed over the top will damage the structure.

Low retaining walls or walls for planters are sometimes constructed with brick or concrete with brick facing.

It is illogical to construct a 4-foot-deep foundation to prevent frost damage to a 2- to 3-foot wall.

Low walls can be built on a wide trench filled with compacted pervious material to a depth well below frost level. This trench must be well drained, either naturally or through the installation of drain tile (Figure 6.249).

The foundation of low walls can also be provided by using concrete piles, supporting a reinforced-concrete beam on which the wall is built (Figure 6.250).

Sufficient space is left between the level of the soil and the bottom of the beam to allow for frost action. This space could be filled with polystyrene, so as to provide insulation. Proper drainage of this area is essential.

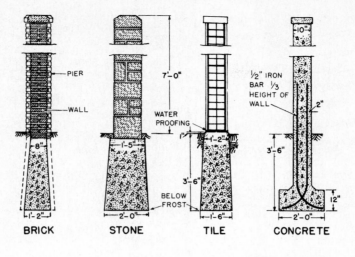

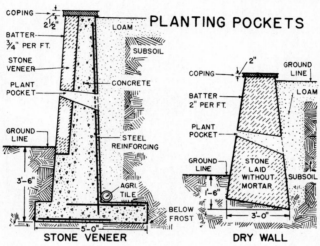

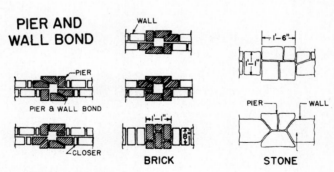

Figure 6.238 Free-standing walls.

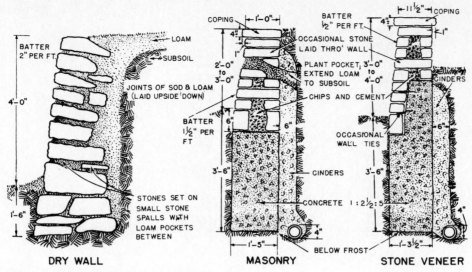

Figure 6.239 Unreinforced retaining walls.

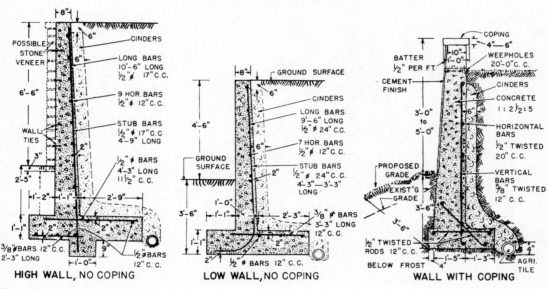

Figure 6.240 Reinforced retaining walls.

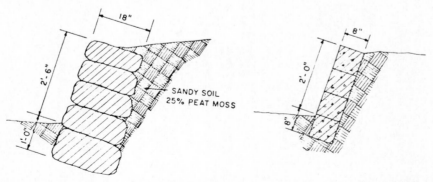

Figure 6.241 Dry retaining walls *(left)* Stone wall. Batter 3 inches per foot in cold regions and 1½ inches per foot in frost-free regions. Spaces between stones may be filled with topsoil and plants, if desired. *(right)* Concrete-block walls; solid block 8 by 8 by 16 inches. Batter 2½ inches per foot. The wall will be slowly displaced by frost.

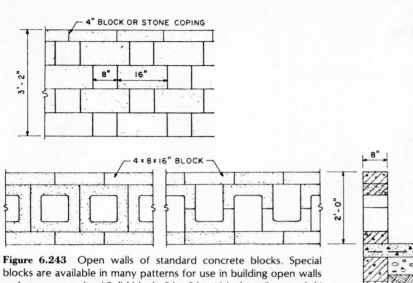

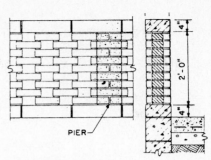

Figure 6.242 Open wall of brick. The minimum overlap of bricks is 1½ inches. Provide 8- by 8-inch piers 6 to 8 feet on center. Coping and base may be of precast concrete or cut stone.

Figure 6.243 Open walls of standard concrete blocks. Special blocks are available in many patterns for use in building open walls and sunscreens. *(top)* Solid block, 8 by 8 by 16 inches. *(bottom left)* Chimney block 8 by 16 by 16 inches. *(bottom right)* Chimney block 8 by 8 by 16 inches.

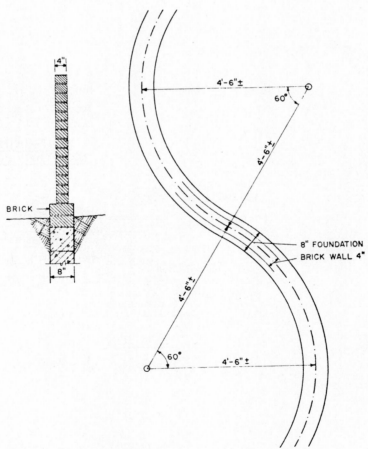

Figure 6.244 Serpentine brick wall.

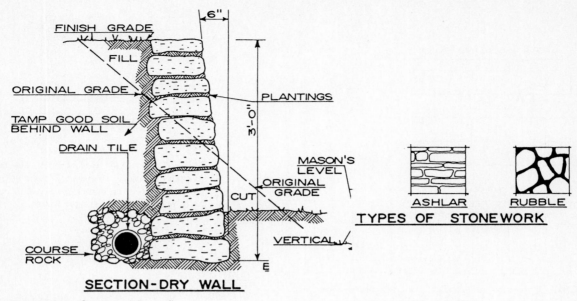

Figure 6.245 Stone retaining wall.

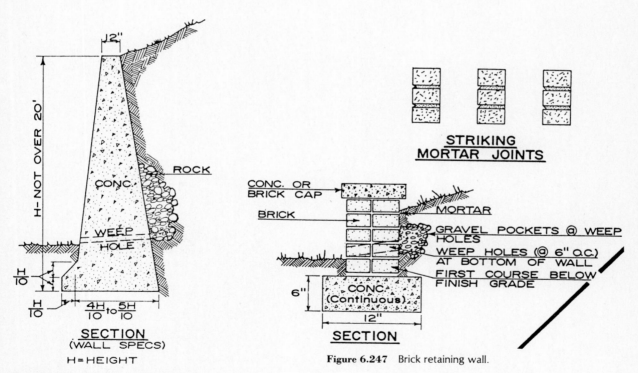

Figure 6.246 Concrete retaining wall.

Figure 6.247 Brick retaining wall.

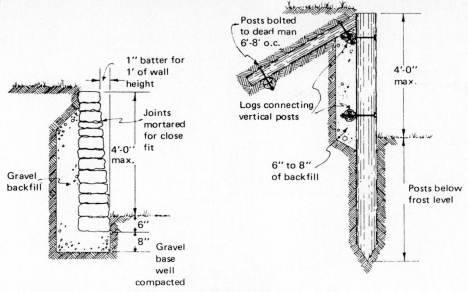

Figure 6.248 Dry stone and cedar post retaining walls.

1" batter for 1' of wall height

Joints mortared for close fit

4'-0" max.

Gravel backfill

6"

8"

Gravel base well compacted

Posts bolted to dead man 6'-8' o.c.

Logs connecting vertical posts

4'-0" max.

6" to 8" of backfill

Posts below frost level

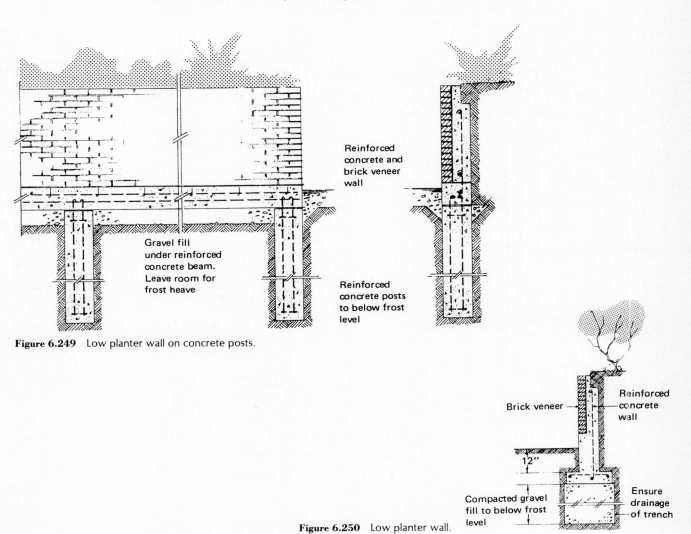

Figure 6.249 Low planter wall on concrete posts.

Reinforced concrete and brick veneer wall

Gravel fill under reinforced concrete beam. Leave room for frost heave

Reinforced concrete posts to below frost level

Figure 6.250 Low planter wall.

Brick veneer

Reinforced concrete wall

12"

Compacted gravel fill to below frost level

Ensure drainage of trench

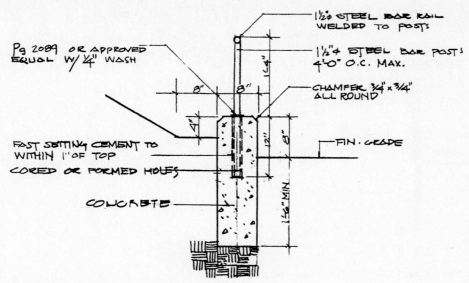

Figure 6.251   Barrier fence and concrete curb wall.

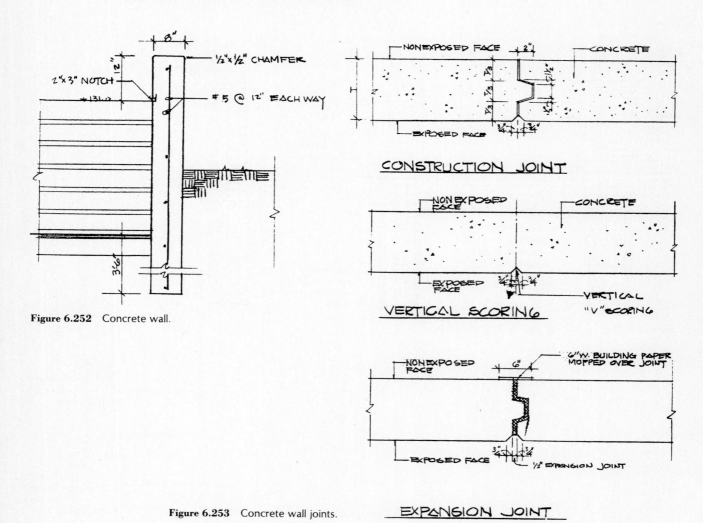

Figure 6.252   Concrete wall.

Figure 6.253   Concrete wall joints.

6.12 Berms and Banks

Figures 6.254 and 6.255 illustrate types of berms and banks.

One of the advantages of earth berms is that they may be constructed of a variety of naturally occurring materials, such as rock, earth, and gravel. Where the berm is to utilize plantings, the top layers of earth must be capable of supporting plant life; however, the bulk of the berm may be constructed of any available material. This material need only be capable of retaining its stability without deteriorating or eroding. Consequently, it is possible to construct a berm using inexpensive and readily available scrap material, such as broken concrete or asphalt from roadway repairs, mining "spoil," rubble from building demolition, or any other solid fill that will compact well. In some cases it may be possible to construct an earth berm in conjunction with highway construction activities, utilizing excess borrow material for the berm.

An earth berm, as a natural-appearing element in the landscape, should be visually compatible and related to its surroundings. This may require a gradual transition in elevation, beginning with a slope ratio of 4:1 or more, and approaching 2:1 at its steepest. Slopes which are steeper than 2:1 are difficult to plant and maintain, and are likely to erode; a 3:1 slope is preferred if it is to be mowed.

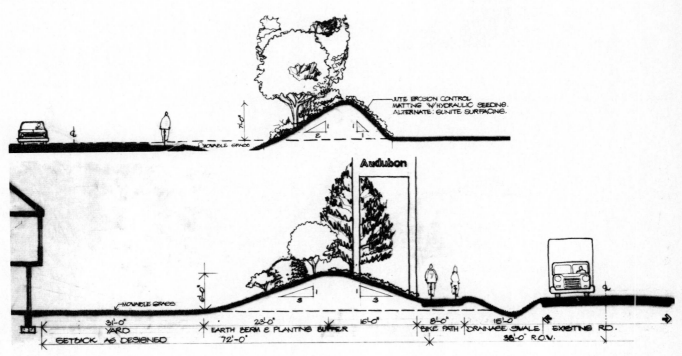

Figure 6.254 Berms.

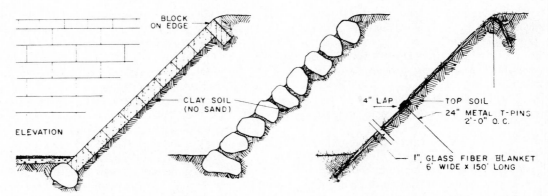

Figure 6.255 Banks. *(left)* Concrete-block riprap; solid block 4 by 8 by 16 inches, laid dry. *(center)* Stone riprap *(right)* Grass. Sow grass seed and cover with a glass-fiber blanket, held in place with T pins.

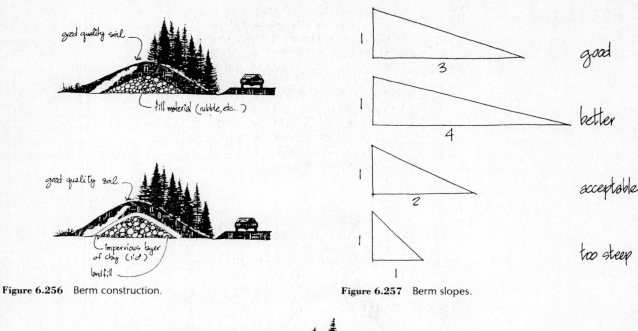

Figure 6.256   Berm construction.

Figure 6.257   Berm slopes.

Figure 6.258   Providing positive drainage for swale.

Construction of an earth berm may affect surface water drainage patterns, by acting as a dam or by redirecting runoff to other areas. This can lead to unwanted flooding or ponding of surface water during or after storms. In addition, the natural flow of water across a site may be affected, leading to a change in groundwater conditions elsewhere. When designing an earth berm, it is important for the designer to thoroughly examine the existing topography of the site, and to alter the existing drainage as little as possible. In areas where there is likely to be a complete damming of surface runoff, installation of a culvert through the berm may be required.

Riprap   Soil stabilization on slopes can be obtained through the placement of large flat rocks or special precast concrete slabs.

Large rocks are hand-placed, presenting a smooth surface. Rocks can be placed on cement with mortared joints, or the openings can be filled with sod or groundcover.

Plants, used among riprap, must be drought-resistant and be able to withstand heat reflection from the stones.

Perforated concrete slabs make it possible to combine the benefits of vegetation with a durable concrete cover. The blocks are poured 4 inches thick, in sizes that are easy to handle. They have 2- by 2-inch holes (a 16- by 24-inch block has 24 holes) in which grass or ground cover can be grown. Blocks can be locked together by means of special keys, while steel pins can be used to hold them more firmly in place (Figure 6.259).

Cribbing   Pressure-treated pine beams or logs can be used in a criblike fashion. Native cedar can be used without pressure treatment. Beams are stacked on top of one another with a 1-inch in 1-foot batter. Galvanized spikes hold the beams in place. Ties with a "dead man" are used to hold the stretchers in place.

Size and length of stretchers will depend on the height of the soil to be retained. They are generally 8 to 10 feet long with the ties fastened at the joints. Stagger the joints to avoid weak spots and to distribute ties evenly.

Cedar logs are suitable when a rustic effect is desired.

Concrete precast cribbing with horizontal beams and ties for anchoring can be made available in specific sizes as required.

Because wooden and concrete beams are loosely placed so as to allow for individual movement, no foundations are

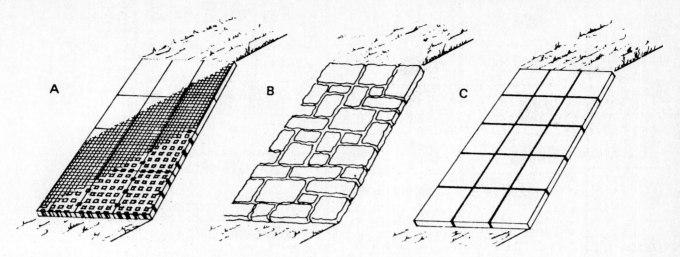

"A" — Precast Concrete Waffles, permits vegetation growth through openings in slabs.
"B" — Stone placement.
"C" — Precast Concrete Slabs.
   "B" and "C" on concrete with grout filled joints, or over soil with open joints,
   if vegetation growth is desirable.

Figure 6.259   Riprap treatment for erosion control on slopes.

required. Unless concrete beams can be poured on site, the cost of transportation may be prohibitive.

The choice of material depends to a great extent on the effects required.

6.13 Trees and Shrubs

A number of trees and shrubs are shown with their dimensions in Figure 6.260 on page 698.

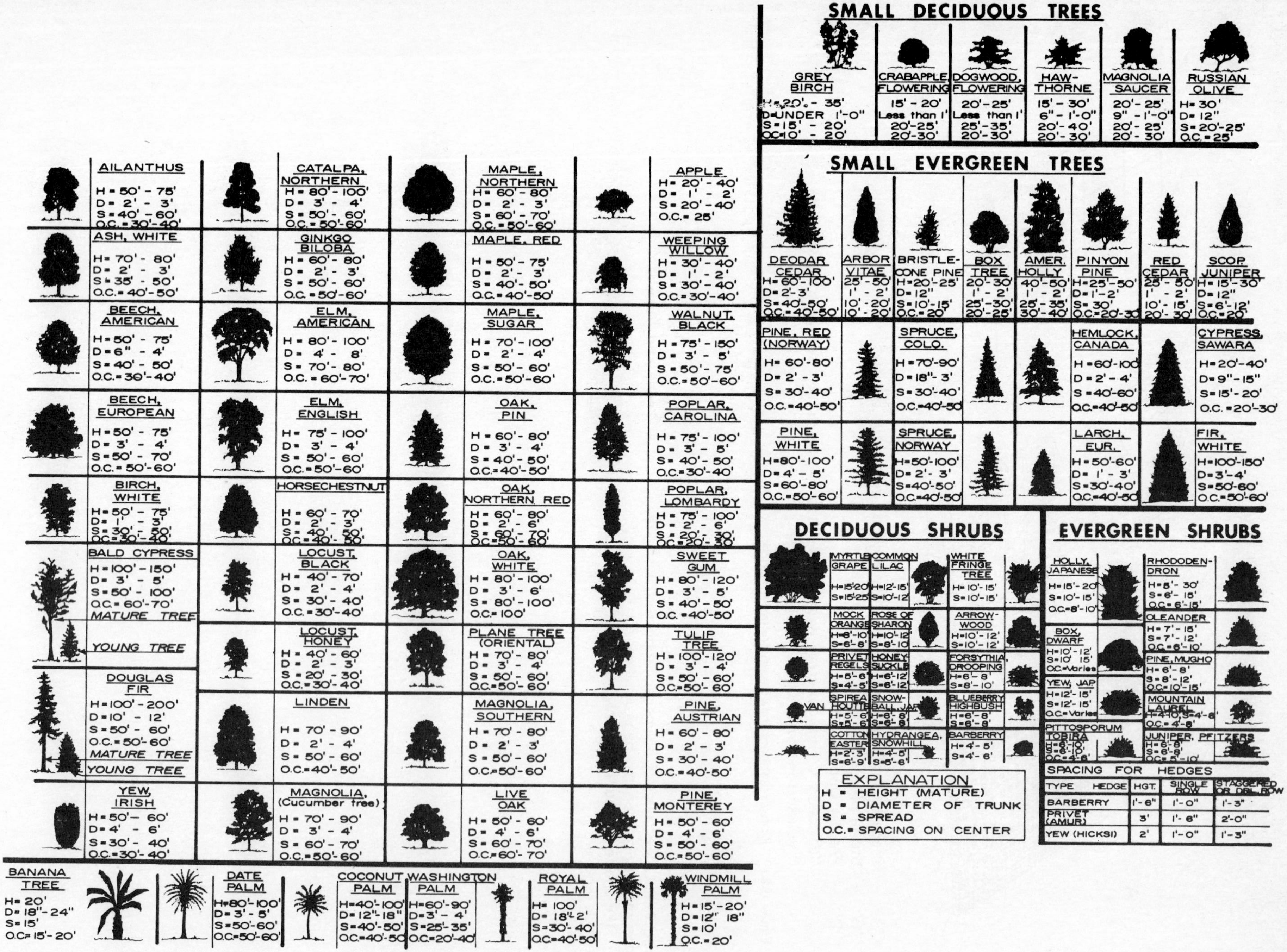

Figure 6.260 Typical sizes and shapes of various trees.

SMALL DECIDUOUS TREES

GREY BIRCH	CRABAPPLE FLOWERING	DOGWOOD FLOWERING	HAW-THORNE	MAGNOLIA SAUCER	RUSSIAN OLIVE
H=20'-35'	15'-20'	20'-25'	15'-30'	20'-25'	H=30'
D=UNDER 1'-0"	Less than 1'	Less than 1'	6'-1'-0"	9'-1'-0"	D=12"
S=15'-20'	20'-30'	25'-30'	20'-25'	20'-30'	S=20'-25'
OC=10'-20'	20'-30'	20'-30'	20'-25'	20'-30'	OC=25'

SMALL EVERGREEN TREES

DEODAR CEDAR	ARBOR VITAE	BRISTLE-CONE PINE	BOX TREE PINE	AMER. HOLLY	PINYON PINE	RED CEDAR	SCOP. JUNIPER
H=60'-100'	25'-50'	H=20'-25'	H=20'-25'	40'-50'	H=25'-50'	25'-50'	H=15'-30'
D=2'-3'	1'-2'	D=12"	D=12"	1'-2'	D=1'-2'	10'-15'	D=12"
S=40'-50'	10'-20'	S=10'-15'	S=20'	25'-30'	S=10'-15'	10'-15'	S=6'-12'
OC=40'-50'	OC=10'-20'	OC=20'-25'	OC=20'-25'	25'-30'	OC=20'	OC=20'-30'	OC=25'

PINE, RED (NORWAY)	SPRUCE, COLO.	HEMLOCK CANADA	CYPRESS, SAWARA
H=60'-80'	H=70'-90'	H=60'-100'	H=20'-40'
D=2'-3'	D=18"-3'	D=2'-4'	D=9"-15"
S=30'-40'	S=30'-40'	S=40'-60'	S=15'-20'
OC=40'-50'	OC=40'-50'	OC=30'	OC=20'-30'

PINE, WHITE	SPRUCE, NORWAY	LARCH EUR.	FIR, WHITE
H=80'-100'	H=50'-100'	H=50'-60'	H=100'-150'
D=4'-5'	D=2'-3'	D=1'-3'	D=3'-4'
S=50'-60'	S=40'-50'	S=40'-50'	S=50'-60'
OC=50'-60'	OC=40'-50'	OC=40'-50'	OC=40'-50'

EVERGREEN SHRUBS

RHODODEN-DRON	HOLLY JAPANESE	OLEANDER	PINE, MUGHO	MOUNTAIN LAUREL
H=15'-20'	H=8'-30'	H=7'-15'	H=6'-8'	H=20'-40'
S=6'-15'	S=6'-15'	S=7'-12'	S=8'-10'	S=12'-15'
OC=6'-15'	OC=6'-10'	OC=8'-10'	OC=Varies	OC=Varies

BOX DWARF	YEW JAP	PITTOSPORUM	JUNIPER PFITZER
H=10'-12'	H=12'-15'	H=10'	H=4'-5'
S=6'-8'	S=8-10	S=6'-10'	S=6'-9'
OC=6'-8'	OC=Varies	OC=4'-5'	OC=4'-8

SPACING FOR HEDGE

TYPE	HEDGE HGT.	SINGLE ROW	STAGGERED OR DBL. ROW
BARBERRY	1'-6"	1'-3"	
PRIVET (AMUR)	3'	2'-0"	1'-6"
YEW (HICKSII)	2'	1'-0"	1'-3"

DECIDUOUS SHRUBS

MYRTLE COMMON GRAPE LILAC	WHITE FRINGE TREE	MOCK ORANGE	ROSE OF SHARON	ARROW-WOOD	PRIVET REGELS	FORSYTHIA DROOPING	HONEY-SUCKLE
H=15'-20'	H=10'-15'	H=8'-10'	H=10'-12'	H=10'-12'	H=5'-6'	H=6'-8'	H=6'-12'
S=15'-25'	S=10'-12'	S=6'-8'	S=6'-8'	S=8'-10	S=4'-6'	S=6'-8'	S=6'-12

SPIREA VAN	SNOW-BALL	BIG-LEAF HYDRANGEA	BLUEBERRY HIGHBUSH	BARBERRY	COTTON HYDRANGEA, EASTER SNOWHILL
H=5'-6'	H=6'-8'	H=8-10	H=6'-8'	H=4'-6'	H=2'-3'
S=5'-6'	S=6'-8'	S=8'-10	S=8'	S=4'-6'	S=2'-3'

EXPLANATION

- H = HEIGHT (MATURE)
- D = DIAMETER OF TRUNK
- S = SPREAD
- O.C. = SPACING ON CENTER

AILANTHUS	CATALPA, NORTHERN	APPLE
H=50'-75'	H=60'-100'	H=20'-40'
D=2'-3'	D=3'-4'	D=2'-3'
S=40'-60'	S=50'-60'	S=20'-40'
OC=30'-40'	OC=50'-60'	OC=25'

ASH, WHITE	GINKGO BILOBA	MAPLE, NORTHERN	WEEPING WILLOW
H=70'-80'	H=50'-80'	H=60'-80'	H=30'-40'
D=2'-3'	D=2'-3'	D=2'-3'	D=1'-2'
S=35'-50'	S=50'-60'	S=60'-70'	S=30'-40'
OC=40'-50'	OC=50'-60'	OC=50'-70'	OC=30'-40'

BEECH AMERICAN	ELM, AMERICAN	MAPLE, RED	WALNUT BLACK
H=50'-75'	H=80'-100'	H=50'-75'	H=75'-150'
D=2'-4'	D=4'-8	D=2'-3'	D=3'-5'
S=40'-50'	S=70'-80'	S=40'-50'	S=50'-60'
OC=30'-40'	OC=60'-70'	OC=40'-50'	OC=50'-60'

BEECH EUROPEAN	ELM, ENGLISH	MAPLE, SUGAR	POPLAR, CAROLINA
H=50'-75'	H=75'-100'	H=70'-100'	H=75'-100'
D=2'-3'	D=3'-5'	D=2'-4'	D=3'-5'
S=50'-60'	S=50'-60'	S=50'-60'	S=40'-50'
OC=50'-60'	OC=60'-70'	OC=50'-60'	OC=50'-60'

BIRCH WHITE	HORSECHESTNUT	OAK, PIN	POPLAR, LOMBARDY
H=50'-75'	H=60'-70'	H=60'-80'	H=75'-100'
D=2'-3'	D=3'-4'	D=3'-5'	D=2'-6'
S=30'-40'	S=50'-60'	S=40'-50'	S=20'-30'
OC=30'-40'	OC=40'-50'	OC=50'-60'	OC=20'-30'

BALD CYPRESS	LOCUST BLACK	OAK, NORTHERN RED	SWEET GUM
H=100'-150'	H=40'-70'	H=60'-80'	H=80'-120'
D=3'-5	D=2'-4'	D=2'-4'	D=3'-5'
S=50'-100'	S=30'-40'	S=50'-60'	S=40'-50'
OC=50'-100'	OC=30'-40'	OC=50'-60'	OC=40'-50'

DOUGLAS FIR	LOCUST HONEY	OAK, WHITE	TULIP TREE
H=100'-200'	H=40'-60'	H=80'-100'	H=100'-120'
D=10'-12'	D=2'-3'	D=3'-5'	D=3'-4'
S=50'-60'	S=20'-30'	S=80'-100'	S=50'-60'
OC=50'-60'	OC=30'-40'	OC=80'-100'	OC=50'-60'

YEW IRISH	LINDEN	PLANE TREE (ORIENTAL)	PINE, AUSTRIAN
H=50'-75'	H=70'-90'	H=70'-80'	H=60'-80'
D=4'-6'	D=3'-4'	D=3'-4'	D=2'-3'
S=30'-40'	S=50'-60'	S=50'-60'	S=50'-60'
OC=30'-40'	OC=40'-50'	OC=50'-60'	OC=50'-60'

MAGNOLIA, (Cucumber Tree)	MAGNOLIA, SOUTHERN	PINE, MONTEREY
H=70'-90'	H=70'-80'	H=50'-60'
D=3'-4'	D=2'-3'	D=2'-3'
S=60'-70'	S=50'-60'	S=50'-60'
OC=40'-50'	OC=50'-60'	OC=40'-50'

LIVE OAK
H=50'-60'
D=4'-6'
S=60'-70'
OC=60'-70'

BANANA TREE	DATE PALM	COCONUT PALM	WASHINGTON PALM	ROYAL PALM	WINDMILL PALM
H=20'	H=80'-100'	H=80'-100'	H=60'-90'	H=100'	H=15'-20'
S=18"-24"	D=3'-5	D=12"-18"	D=3'-4	D=18"-2'	D=12"-18"
S=15'	S=50'-60'	S=40'-50'	S=25'-35'	S=30'-40'	S=10'
OC=15'-20'	OC=50'-60'	OC=40'-50'	OC=20'-40	OC=40'-50'	OC=20'

MATURE TREE / *YOUNG TREE*

TREE-PLANTING DETAILS

Detailed drainage of tree-planting methods is shown in Figures 6.261 through 6.272.

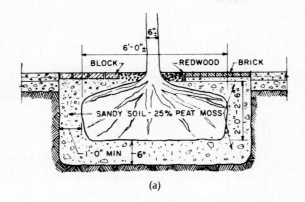

(a)

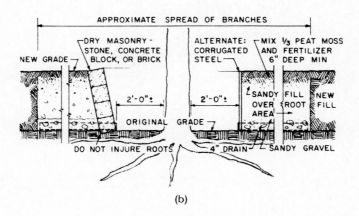

(b)

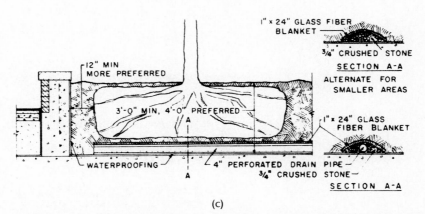

(c)

Figure 6.261 (*a*) Tree roots need air. Sandy soil is porous to air; it should be mixed with peat moss to retain moisture. Do not compress the soil around trees with bulldozers or other heavy equipment. Disturb the roots as little as possible. *(b)* Tree in pavement. Over porous backfill of sandy soil with 25 to 30 percent peat moss, lay dry paving units with tight joints. Belgian block, granite block, asphalt block, concrete block, brick, or crushed stone may be used. If brick is used, install an inner edging of 2- by 4-inch redwood. *(c)* Tree in filled ground. Install eight to ten drain lines, placed radially around the tree. Drains may be porous or perforated pipe or agricultural tile with open joints covered by a 1-inch fiber blanket.

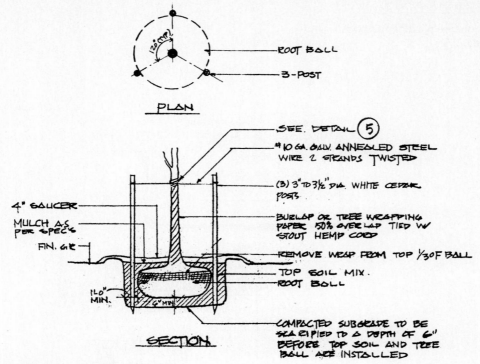

PLAN

ROOT BALL

3-POST

SEE DETAIL (5)

#10 GA. GALV. ANNEALED STEEL WIRE 2 STRANDS TWISTED

(3) 3" TO 3½" DIA. WHITE CEDAR POSTS

BURLAP OR TREE WRAPPING PAPER 50% OVERLAP TIED W/ STOUT HEMP CORD

REMOVE WRAP FROM TOP ⅓ OF BALL

TOP SOIL MIX.

ROOT BALL

COMPACTED SUBGRADE TO BE SCARIFIED TO A DEPTH OF 6" BEFORE TOP SOIL AND TREE BALL ARE INSTALLED

4" SAUCER

MULCH AS PER SPEC'S

FIN. GR.

1'-0" MIN.

6" MIN

SECTION

Figure 6.262   Tree planting.

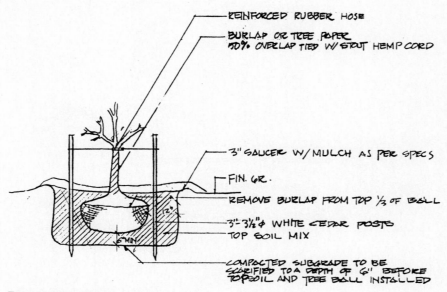

REINFORCED RUBBER HOSE

BURLAP OR TREE PAPER 50% OVERLAP TIED W/ STOUT HEMP CORD

3" SAUCER W/ MULCH AS PER SPECS

FIN. GR.

REMOVE BURLAP FROM TOP ⅓ OF BALL

3"-3½" Ø WHITE CEDAR POSTS

TOP SOIL MIX

COMPACTED SUBGRADE TO BE SCARIFIED TO A DEPTH OF 6" BEFORE TOPSOIL AND TREE BALL INSTALLED

2"

6" MIN

Figure 6.263   Minor tree planting.

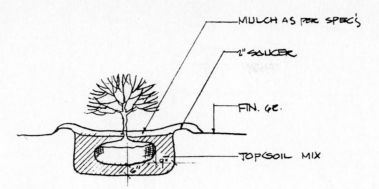

Figure 6.264 Shrub planting.

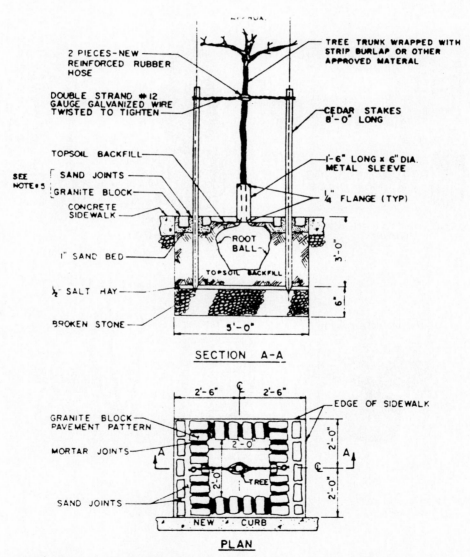

Figure 6.265 Tree planting, staking, and tree pit. Pavement details for sidewalk area.

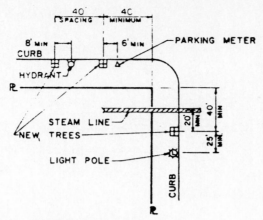

Figure 6.266 Street tree spacing.

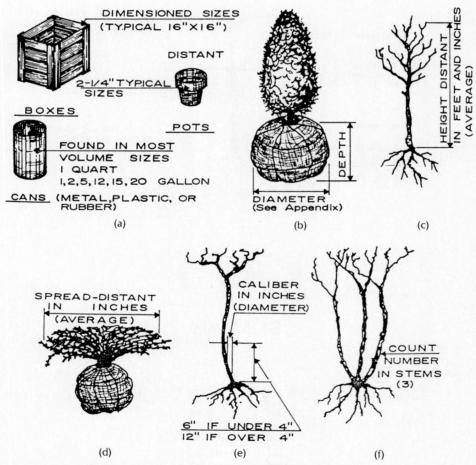

Figure 6.267 Tree planting data: (*a*) containers; (*b*) balls; (*c*) height; (*d*) spread; (*e*) caliber; (*f*) stalk number.

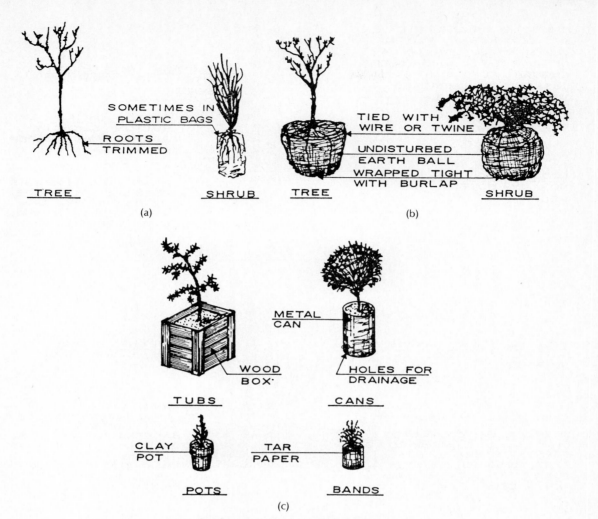

Figure 6.268 (*a*) Bare root; (*b*) balled and burlapped; (*c*) containered.

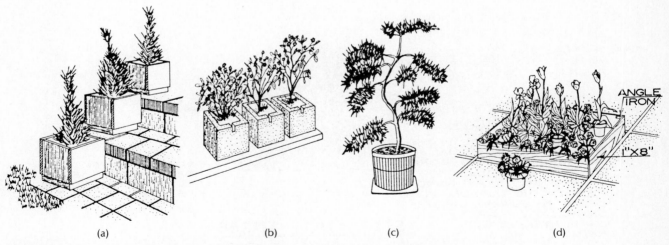

Figure 6.269 Types of containers: (*a*) wood boxes; (*b*) concrete block; (*c*) barrel; (*d*) clay pots.

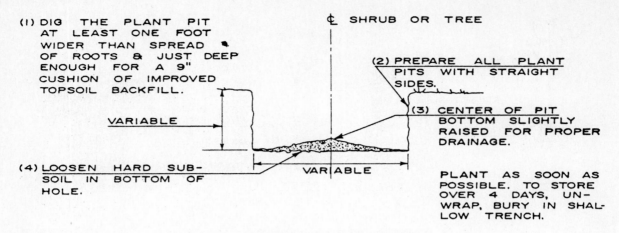

(1) DIG THE PLANT PIT AT LEAST ONE FOOT WIDER THAN SPREAD OF ROOTS & JUST DEEP ENOUGH FOR A 9" CUSHION OF IMPROVED TOPSOIL BACKFILL.

VARIABLE

(4) LOOSEN HARD SUB-SOIL IN BOTTOM OF HOLE.

₵ SHRUB OR TREE

(2) PREPARE ALL PLANT PITS WITH STRAIGHT SIDES.

(3) CENTER OF PIT BOTTOM SLIGHTLY RAISED FOR PROPER DRAINAGE.

PLANT AS SOON AS POSSIBLE. TO STORE OVER 4 DAYS, UN-WRAP, BURY IN SHAL-LOW TRENCH.

VARIABLE

STEP NO. ONE

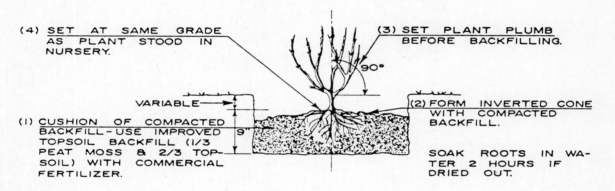

(4) SET AT SAME GRADE AS PLANT STOOD IN NURSERY.

VARIABLE

(1) CUSHION OF COMPACTED BACKFILL-USE IMPROVED TOPSOIL BACKFILL (1/3 PEAT MOSS & 2/3 TOP-SOIL) WITH COMMERCIAL FERTILIZER.

(3) SET PLANT PLUMB BEFORE BACKFILLING.

90°

(2) FORM INVERTED CONE WITH COMPACTED BACKFILL.

SOAK ROOTS IN WA-TER 2 HOURS IF DRIED OUT.

9"

STEP NO. TWO

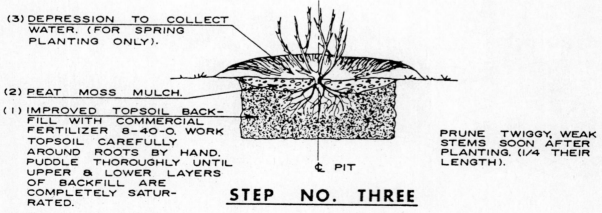

(3) DEPRESSION TO COLLECT WATER. (FOR SPRING PLANTING ONLY).

(2) PEAT MOSS MULCH.

(1) IMPROVED TOPSOIL BACK-FILL WITH COMMERCIAL FERTILIZER 8-40-0. WORK TOPSOIL CAREFULLY AROUND ROOTS BY HAND. PUDDLE THOROUGHLY UNTIL UPPER & LOWER LAYERS OF BACKFILL ARE COMPLETELY SATUR-RATED.

PRUNE TWIGGY, WEAK STEMS SOON AFTER PLANTING. (1/4 THEIR LENGTH).

₵ PIT

STEP NO. THREE

Figure 6.270 Bare root planting.

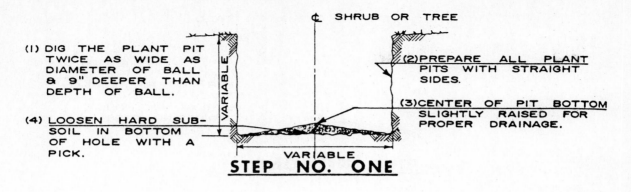

(1) DIG THE PLANT PIT TWICE AS WIDE AS DIAMETER OF BALL & 9" DEEPER THAN DEPTH OF BALL.

(4) LOOSEN HARD SUB-SOIL IN BOTTOM OF HOLE WITH A PICK.

Ç SHRUB OR TREE

(2) PREPARE ALL PLANT PITS WITH STRAIGHT SIDES.

(3) CENTER OF PIT BOTTOM SLIGHTLY RAISED FOR PROPER DRAINAGE.

VARIABLE

STEP NO. ONE

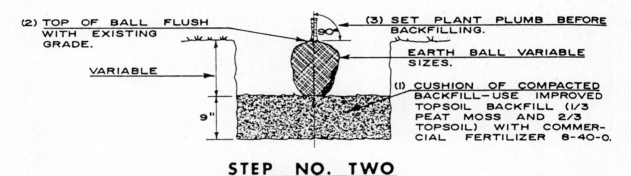

(2) TOP OF BALL FLUSH WITH EXISTING GRADE.

VARIABLE

9"

(3) SET PLANT PLUMB BEFORE BACKFILLING.

EARTH BALL VARIABLE SIZES.

(1) CUSHION OF COMPACTED BACKFILL—USE IMPROVED TOPSOIL BACKFILL (1/3 PEAT MOSS AND 2/3 TOPSOIL) WITH COMMERCIAL FERTILIZER 8-40-0.

STEP NO. TWO

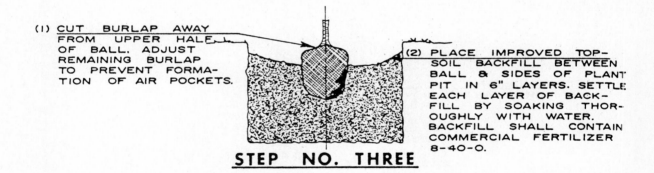

(1) CUT BURLAP AWAY FROM UPPER HALF OF BALL. ADJUST REMAINING BURLAP TO PREVENT FORMATION OF AIR POCKETS.

(2) PLACE IMPROVED TOP-SOIL BACKFILL BETWEEN BALL & SIDES OF PLANT PIT IN 6" LAYERS. SETTLE EACH LAYER OF BACK-FILL BY SOAKING THOR-OUGHLY WITH WATER. BACKFILL SHALL CONTAIN COMMERCIAL FERTILIZER 8-40-0.

STEP NO. THREE

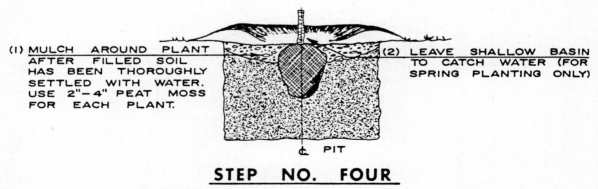

(1) MULCH AROUND PLANT AFTER FILLED SOIL HAS BEEN THOROUGHLY SETTLED WITH WATER. USE 2"-4" PEAT MOSS FOR EACH PLANT.

(2) LEAVE SHALLOW BASIN TO CATCH WATER (FOR SPRING PLANTING ONLY)

Ç PIT

STEP NO. FOUR

Figure 6.271   Balled and burlapped planting.

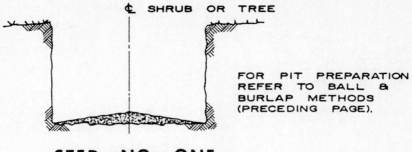

FOR PIT PREPARATION
REFER TO BALL &
BURLAP METHODS
(PRECEDING PAGE).

STEP NO. ONE

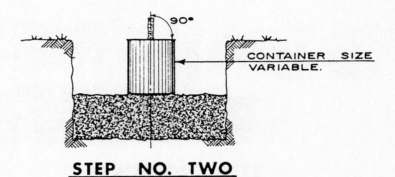

90°

CONTAINER SIZE
VARIABLE.

STEP NO. TWO

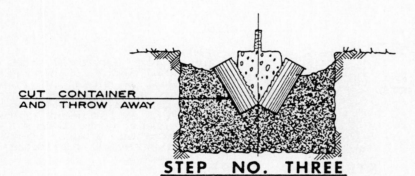

CUT CONTAINER
AND THROW AWAY

STEP NO. THREE

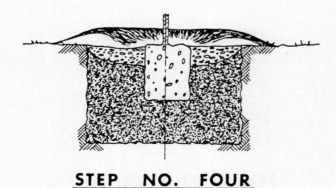

STEP NO. FOUR

Figure 6.272 Containered planting.

Tree Preservation and Protection   Indiscriminate preservation of trees can be just as detrimental to a satisfactory final solution as the complete destruction of all plant material and leveling of all contours.

Large expenses are often incurred to preserve trees whose life span is coming to an end.

Roadways, grades, and sometimes buildings are adapted to the intention to retain some trees.

During the construction, root systems are damaged and trees die. It can take several years for damage to show, thus creating a false impression that the trees have survived.

Protection applies not only to the trunk and branches but to the root system as well.

Aside from damage during excavations, root systems are injured through continuous passage of heavy equipment, storage of gravel, soil, or building material in the immediate vicinity of the trunk, or spillage of gasoline, oil, solvent, and other chemicals.

Encircle the tree trunk with heavy planks up to a height of 8 feet or up to the first branch. Fasten planks with wire without causing abrasion (Figure 6.273).

The area within the spread of the tree crown generally contains the root systems. Protect this area with temporary fencing.

If circumstances dictate a change in elevation, cutting of roots, or other actions detrimental to the tree, it is imperative that precise instructions be laid down in the grading specifications.

Lowering of the grade may require the shaping of tree mounds or the construction of retaining walls protecting the root system.

The water table may drop in relation to the roots, and leave the tree with a reduced water supply.

Provide water during the first growing seasons so as to assist trees while they are adapting to the new growing conditions (Figure 6.274).

Raising of grades in excess of 6 inches will smother the roots, depriving the system of air.

Place 4-inch agricultural tiles on the existing grade, radiating like spokes from a wheel from the trunk, and fill the area with a granular material. Vertical tiles connected with the horizontal tiles with provide a supply of air.

Tarpaper or a drystone retaining wall will prevent moisture from causing rot to the base of the tree.

Where circumstances permit, a retaining wall at the periphery of the branch spread will suffice, but only if proper drainage can be provided (Figure 6.275).

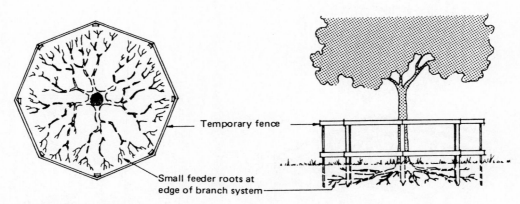

Figure 6.273   Temporary tree protection.

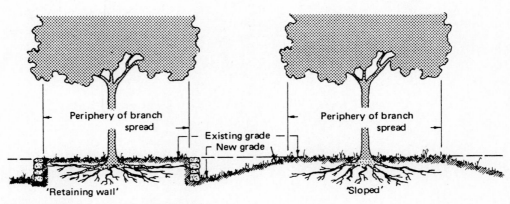

Figure 6.274   Lowering of grade around trees.

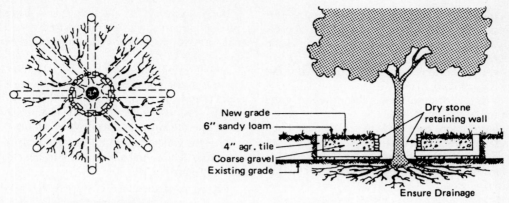

Figure 6.275 Raising of grade around trees.

Temporary Protective Fence It is necessary when preserving existing trees in areas of construction activity to avoid disturbing the ground area beneath the spread of the trees' branches where most of the roots lie.

A 10- by 10-foot fence 3 feet high will protect most individual trees adequately. See Figures 6.276 through 6.284.

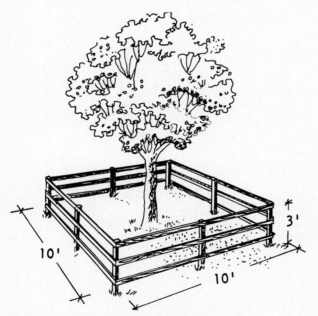

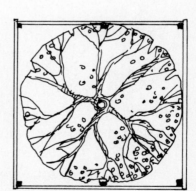

Figure 6.276 Temporary protective fence plan.

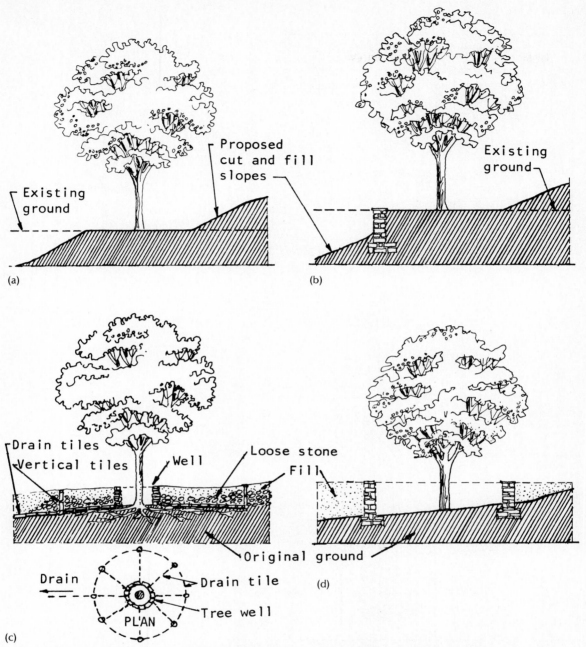

Figure 6.277   Tree protection with grade changes: (*a*) permanent protective slope; (*b*) permanent protective wall; (*c*) tree well with raised grade; (*d*) open tree well—where space allows, this is a preferred method of tree well design.

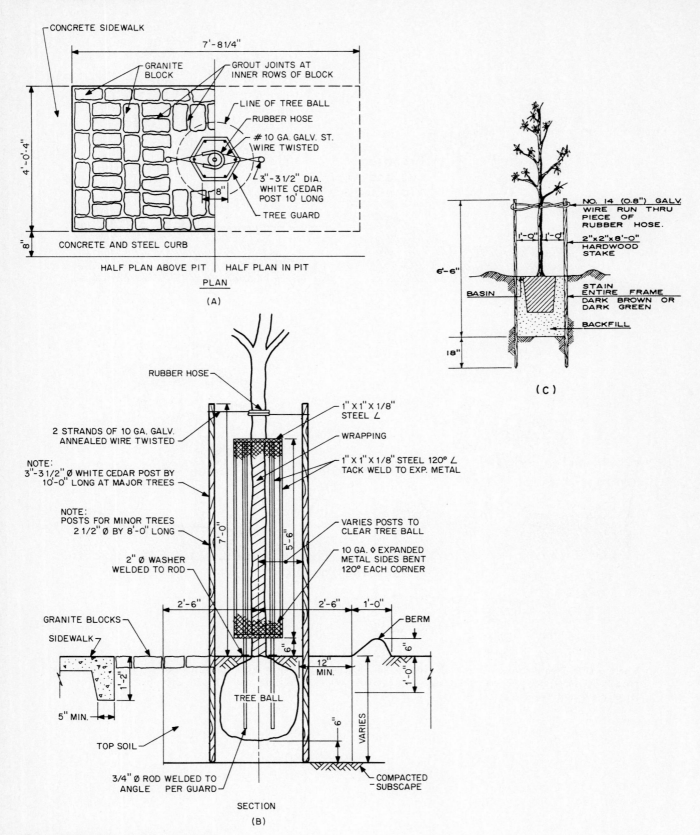

Figure 6.278 Tree pit and staking detail.

(a)

(b)

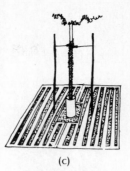
(c)

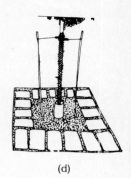
(d)

Figure 6.279 Types of tree protection: (*a*) walls protect trees from salt and urine; (*b*) mulch protects trees from cold and weeds; (*c* and *d*) grating blocks or bricks protect roots from being trampled.

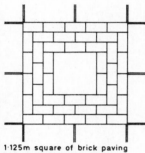

1·125m square of brick paving bedded on 50mm sand, with 450mm square hole left for tree

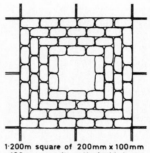

1·200m square of 200mm x 100mm x100mm granite setts bedded on 50mm sand, with 400mm square hole left for tree

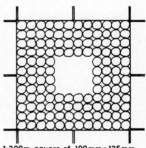

1·200m square of 100mm–125mm kidney flint cobbles bedded on 50mm sand, with 400mm square hole left for tree

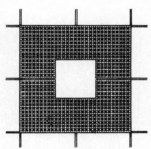

1·200m square cast iron tree grid made in two sections, bedded on 50mm sand, with 400mm square hole left for tree

Figure 6.280 Tree grates.

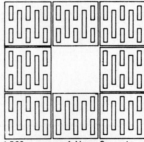

1·500m square of Mono Concrete Metric 4 square tree grilles bedded on 50mm sand, with 500mm square hole left for tree

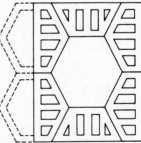

1·520m x 1·140m rectangle of Mono Concrete Monohex tree grilles bedded on 50mm sand

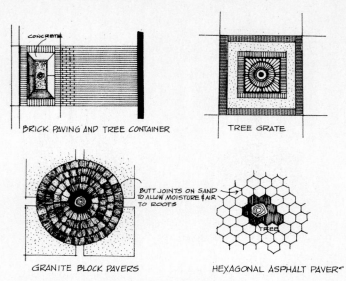

Figure 6.281 Pavement treatment at tree base.

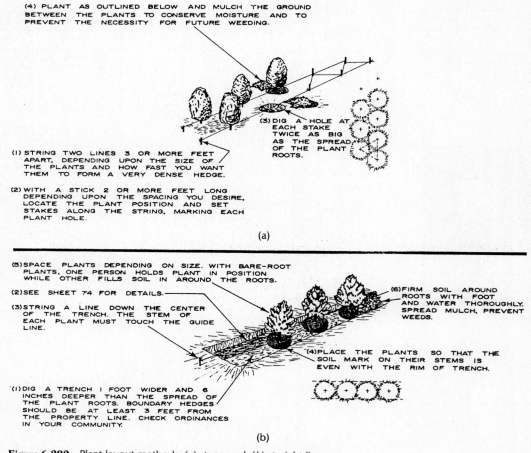

(4) PLANT AS OUTLINED BELOW AND MULCH THE GROUND BETWEEN THE PLANTS TO CONSERVE MOISTURE AND TO PREVENT THE NECESSITY FOR FUTURE WEEDING.

(3) DIG A HOLE AT EACH STAKE TWICE AS BIG AS THE SPREAD OF THE PLANT ROOTS.

(1) STRING TWO LINES 3 OR MORE FEET APART, DEPENDING UPON THE SIZE OF THE PLANTS AND HOW FAST YOU WANT THEM TO FORM A VERY DENSE HEDGE.

(2) WITH A STICK 2 OR MORE FEET LONG DEPENDING UPON THE SPACING YOU DESIRE, LOCATE THE PLANT POSITION AND SET STAKES ALONG THE STRING, MARKING EACH PLANT HOLE.

(a)

(5) SPACE PLANTS DEPENDING ON SIZE. WITH BARE-ROOT PLANTS, ONE PERSON HOLDS PLANT IN POSITION WHILE OTHER FILLS SOIL IN AROUND THE ROOTS.

(2) SEE SHEET 74 FOR DETAILS.

(3) STRING A LINE DOWN THE CENTER OF THE TRENCH. THE STEM OF EACH PLANT MUST TOUCH THE GUIDE LINE.

(6) FIRM SOIL AROUND ROOTS WITH FOOT AND WATER THOROUGHLY. SPREAD MULCH, PREVENT WEEDS.

(4) PLACE THE PLANTS SO THAT THE SOIL MARK ON THEIR STEMS IS EVEN WITH THE RIM OF TRENCH.

(1) DIG A TRENCH 1 FOOT WIDER AND 6 INCHES DEEPER THAN THE SPREAD OF THE PLANT ROOTS. BOUNDARY HEDGES SHOULD BE AT LEAST 3 FEET FROM THE PROPERTY LINE. CHECK ORDINANCES IN YOUR COMMUNITY.

(b)

Figure 6.282 Plant layout methods: (*a*) staggered; (*b*) straight-line.

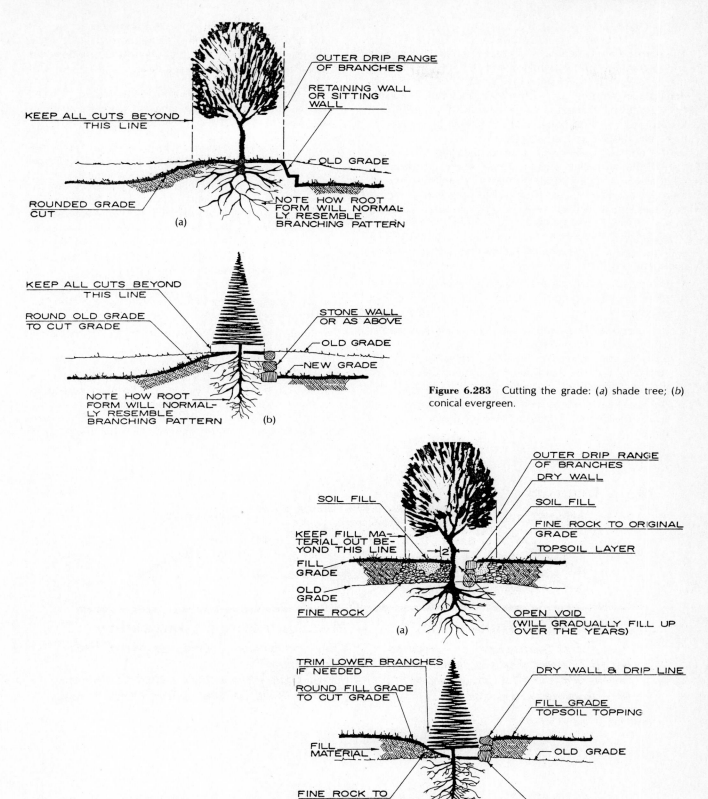

KEEP ALL CUTS BEYOND THIS LINE

OUTER DRIP RANGE OF BRANCHES

RETAINING WALL OR SITTING WALL

OLD GRADE

ROUNDED GRADE CUT

NOTE HOW ROOT FORM WILL NORMAL-LY RESEMBLE BRANCHING PATTERN

(a)

KEEP ALL CUTS BEYOND THIS LINE

ROUND OLD GRADE TO CUT GRADE

STONE WALL OR AS ABOVE

OLD GRADE

NEW GRADE

NOTE HOW ROOT FORM WILL NORMAL-LY RESEMBLE BRANCHING PATTERN

(b)

Figure 6.283 Cutting the grade: (*a*) shade tree; (*b*) conical evergreen.

OUTER DRIP RANGE OF BRANCHES

DRY WALL

SOIL FILL

SOIL FILL

KEEP FILL MA-TERIAL OUT BE-YOND THIS LINE

FINE ROCK TO ORIGINAL GRADE

TOPSOIL LAYER

2'

FILL GRADE

OLD GRADE

FINE ROCK

OPEN VOID (WILL GRADUALLY FILL UP OVER THE YEARS)

(a)

TRIM LOWER BRANCHES IF NEEDED

DRY WALL & DRIP LINE

ROUND FILL GRADE TO CUT GRADE

FILL GRADE TOPSOIL TOPPING

FILL MATERIAL

OLD GRADE

FINE ROCK TO OUTER DRIP LINE

DRIP LINE

(b)

Figure 6.284 Raising the grade: (*a*) shade tree; (*b*) conical evergreen.

6.14 Outdoor Lighting

The purpose of site lighting is basically twofold: to illuminate and to provide security. Lighting should be provided in areas that receive heavy pedestrian or vehicular use and in areas that are dangerous if unlit, such as stairs and ramps, intersections, or abrupt changes in grade. Likewise, areas that have high crime rates should be well lit so that people traveling at night may feel secure from attack.

The phrase "well lit" has a wider meaning than simple higher light levels. Unless light is placed where it is most useful, the expense of increasing footcandle levels is wasted. An area may need only the addition of a few lights to correct its problems, not an increase in light levels from fixtures that are too few or are poorly located.

When considering the installation or renovation of lighting systems, the designer should be aware of the following considerations:

1. Overhead lamps have the advantage over low-level fixtures of providing better economy and more even light distribution.

2. Fixtures should be placed so that light patterns overlap at a height of 7 feet, which is sufficiently high to illuminate a person's body vertically. This is a particularly important consideration now that lighting fixture manufacuers are designing luminaires with highly controlled light patterns.

3. At hazardous locations such as changes of grade, lower-level supplemental lighting or additional overhead units should be used.

4. Where low-level lighting (below 5 feet) is used, fixtures should be placed so that they do not produce glare. Most eye levels occur between 3 feet 8 inches (for wheelchair users) and 6 feet for standing adults.

5. Posts and standards along thoroughfares should be placed so that they do not present hazards to pedestrians or vehicles.

6. A minor consideration is the use of shatterproof coverings on low-level lighting where there is a chance of breakage.

7. When walkway lighting is provided primarily by low fixtures, there should be sufficient peripheral lighting to illuminate the immediate surroundings. Peripheral lighting provides for a better feeling of security for individuals because they can see into their surroundings to determine whether passage through an area is safe. Such lighting should be approached from one of two ways:

   a. By lighting the area so that an object or a person may be seen directly

   b. By lighting the area to place an object or a person in silhouette.

Figures 6.285 and 6.286 show recommended light intensity and distribution, and Figure 6.287 shows how trees should be pruned to avoid interference with light.

Figures 6.288 and 6.290 show the characteristics of various types of lamps and recommended installations.

Light Intensity

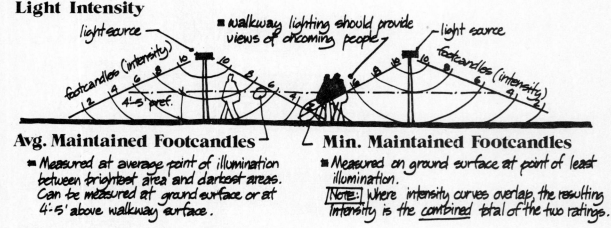

Figure 6.285 Lighting intensity and distribution.

Lateral Light Distribution

- *Light patterns can be varied according to the needs of a particular situation. Choose the proper pattern and fixture for your specific requirements.*

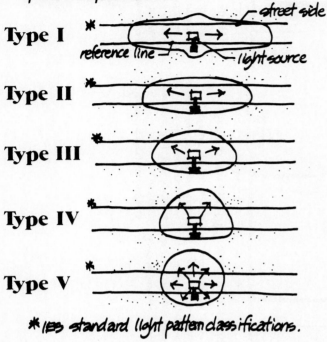

Type I

Type II

Type III

Type IV

Type V

* IES standard light pattern classifications.

Figure 6.286 Lamp types and heights of lighting fixtures.

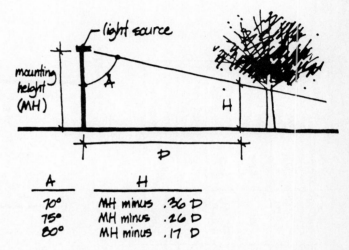

A	H
70°	MH minus .36 D
75°	MH minus .26 D
80°	MH minus .17 D

Figure 6.287 Illuminating Engineering Society tree-pruning recommendations.

Lamp Types & Characteristics

LAMP TYPE	WATTAGE RANGE	EFFICIENCY (lumens/watt)	LIFE (hours)	COLORS STRENGTHENED	COLORS DIMINISHED	REMARKS
Incandescent	15-1000	low	750-2000	yellow, red, orange	blue	good color rendition
Deluxe Cool-White Fluorescent	15-215	medium	7,500-15,000	all	none	best overall color rendition
Deluxe White Mercury	90-1000	medium	10,000-24,000	blue, red, yellow	green	good color rendition
Metal Halide	175-1000	high	7500-10,500	yellow, blue green	red	good color rendition
High-Pressure Sodium	250-1000	high	10,000-15,000	yellow, green orange	red, blue	poor color rendition

High-Intensity Discharge

NOTE: All exterior installations must be provided with ground-fault interruption circuit.

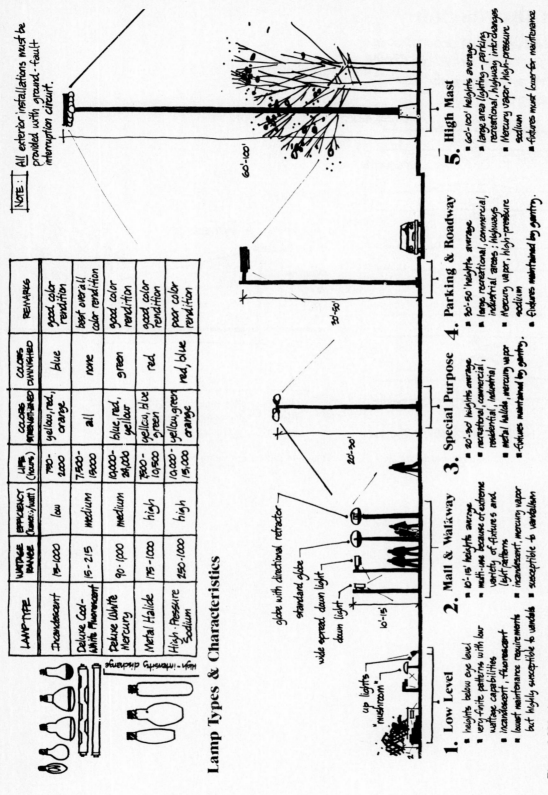

globe with directional refractor
standard globe
wide spread down light
down light
up lights
"mushroom"

1. Low Level
- heights below eye level
- very finite patterns with low wattage capabilities
- incandescent, fluorescent
- lowest maintenance requirements but highly susceptible to vandals

2. Mall & Walkway
- 10'-15' heights average
- multi-use because of extreme variety of fixtures and light patterns
- incandescent, mercury vapor
- susceptible to vandalism

3. Special Purpose
- 20'-30' heights average
- recreational, commercial, residential, industrial
- metal halide, mercury vapor
- fixtures maintained by gantry.

4. Parking & Roadway
- 20'-50' heights average
- large recreational, commercial, industrial areas; highways
- mercury vapor, high-pressure sodium
- fixtures maintained by gantry.

5. High Mast
- 60'-100' heights average
- large area lighting - parking, recreational, highway interchanges
- mercury vapor, high-pressure sodium
- fixtures must lower for maintenance

Figure 6.288 Lamp types and characteristics.

Lighting levels recommended by the Illuminating Engineering Society are shown in the accompanying table, and details of lighting installations in Figure 6.289.

Recommended Lighting Levels*

	Commercial	Industrial	Residential
Pedestrian areas:			
Sidewalks	0.9	0.6	0.2
Pedestrian ways	2.0	1.0	0.5
Roadways:			
Freeways	0.6	0.6	0.6
Major roads and expressways	2.0	1.4	1.0
Collectors	1.2	0.9	0.6
Local streets	0.9	0.6	0.4
Alleys	0.6	0.4	0.2
Parking areas:			
Self-parking	1.0	. . .	. . .
Attendant parking	2.0	. . .	. . .
Buildings:			
Entrance and doorway areas	5.0	. . .	. . .
General grounds	1.0	. . .	. . .

*Values are given in minimum average maintained horizontal footcandles.
SOURCE: *IES Lighting Handbook*, 4th ed., Illuminating Engineering Society, New York.

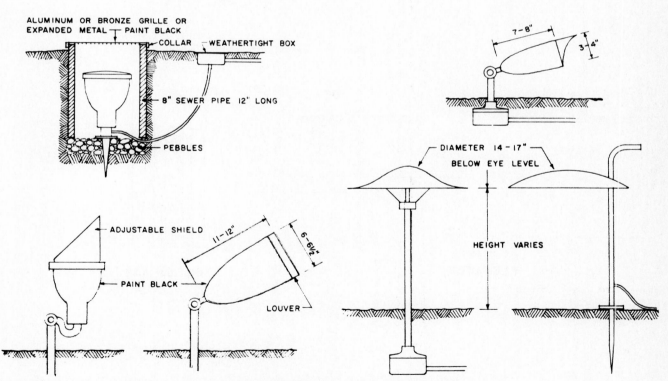

Figure 6.289 Lighting details. *(top left)* Underground floodlight to light trees; 75 to 150 watts. *(top right)* Small spotlight or floodlight to light statuary; 30 watts. *(bottom left)* Aboveground floodlights, hidden by stones or bushes, to light trees or house; 75 to 150 watts. *(bottom right)* Low-level lighting for walks or flowers; 75 watts. It may be permanent or portable.

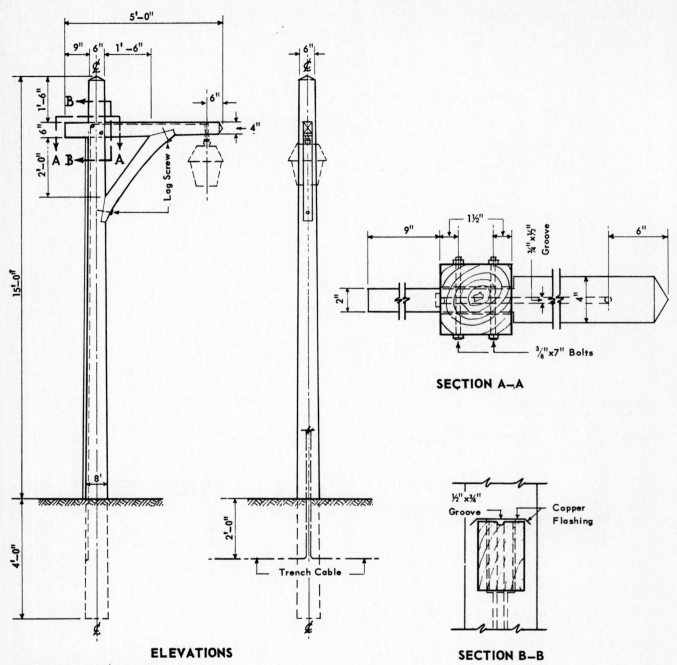

ELEVATIONS

SECTION A–A

SECTION B–B

Figure 6.290 Wood lamp post.

6.15 Pools

WADING POOLS

The slope of the bottom of any wading pool should be not greater than 1 foot in 15 feet. All side and end walls should be vertical. The maximum depth of the water should not exceed 24 inches.

The complete water turnover time for wading pools should not exceed 2 hours. This turnover may be accomplished by recirculation through a filter system or by the use of an automatic valve that regulates the freshwater addition. If a recirculation and filtration system is used for a wading pool, it should be completely separate from the system serving a swimming pool.

Skimming action should be provided by means of scum gutter, deck-level, or suction-type floating weir overflows, all of which should discharge to waste. The water level should be maintained near the overflow level by means of an automatic, level regulator valve.

See Figures 6.291 and 6.292 for the design of rectangular and circular wading pools.

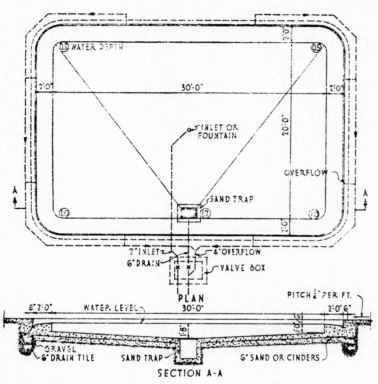

Figure 6.291   Rectangular wading pool.

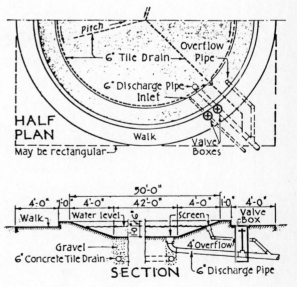

Figure 6.292   Circular wading pool.

POOLS AND FOUNTAINS

Figures 6.293, 6.294, 6.295 and 6.296 show the design of concrete and asphalt-lined pools. All concrete pools must be poured or sprayed in one continuous operation without joints. Inside forms, where required, must be supported from the outside and suspended. Painting the inside of the pool very dark with cement- or rubber-based paint will make the pool look deeper, improve surface reflections, and show less dirt. The water supply outlet must be a minimum of 6 inches above the surface of the water.

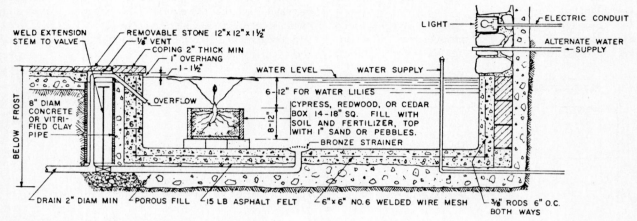

Figure 6.293   Concrete pool with vertical sides.

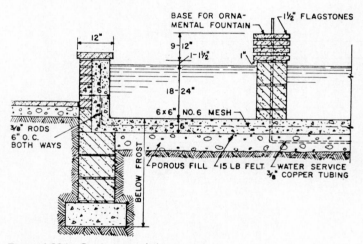

Figure 6.294   Concrete pool above grade.

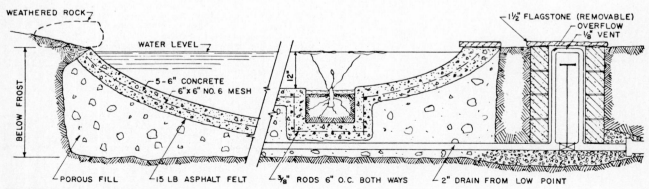

Figure 6.295   Concrete pool with sloping sides.

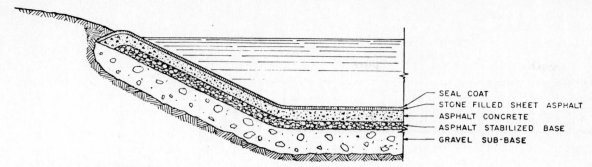

Figure 6.296 Asphalt-lined pool.

SEAL COAT
STONE FILLED SHEET ASPHALT
ASPHALT CONCRETE
ASPHALT STABILIZED BASE
GRAVEL SUB-BASE

6.16 Footbridges

Designs for two footbridges are shown in Figures 6.297 and 6.298.

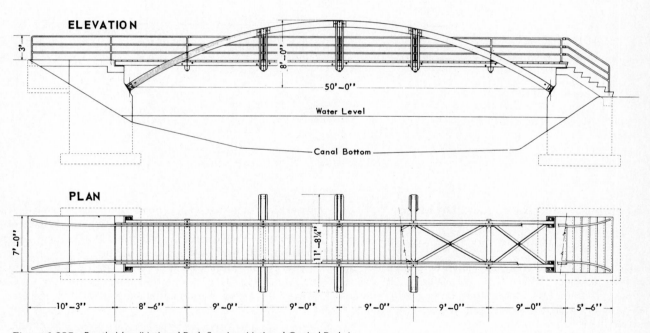

Figure 6.297 Footbridge (National Park Service, National Capital Parks).

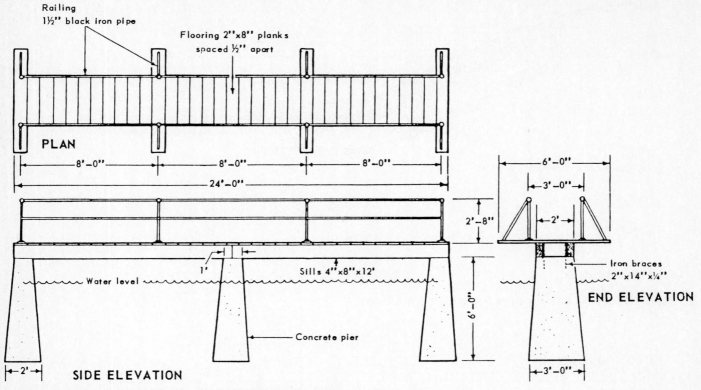

Figure 6.298 Wood footbridge (Frank Bunce, Bernheim Forest, Ky.).

6.17 Inlets, Catch Basins, and Drainage

STORM INLETS AND CATCH BASINS

All storm inlets and catch basins should be adequate in size and design to accept and carry the calculated potential runoff without overflow. They should be constructed of durable materials that will not erode and will accept potentially imposed loads without failure.

Where inlets are accessible to small children, openings should have one dimension limited to a maximum of 6 inches. Access for cleaning should be provided to all inlet boxes and catch basins. Where inlets are located in areas of potential pedestrian use, the design of openings and exposed surfaces should be arranged to minimize the dangers of tripping or slipping. Horizontal inlet openings in paved areas should be designed to avoid entrapment or impedance of bicycles, baby carriages, and so on. See also Figure 6.299.

Manholes and Junction Boxes Manholes and junction boxes should be spaced at optimum intervals as indicated by analysis and be provided in places where there is a particular hazard of blockage, at abrupt changes in alignment, and at junctions with mains and principal laterals.

Headwalls Headwalls and other appropriate construction should be placed at the open ends of storm sewers to prevent excessive erosion and undermining of conduits.

SURFACE INLETS AND JUNCTION BOXES

Surface inlets should be used in low areas where surface drainage cannot otherwise be provided (Figure 6.300). They must be properly constructed to prevent washouts and silting. However, surface inlets should be avoided wherever possible. If silt is a hazard, place a silt trap (Figure 6.301) at a convenient location immediately downstream from inlet.

Junction boxes should be used when two or more main or submain drains join or when several laterals join at different elevations. (See also Figures 6.302 and 6.303.)

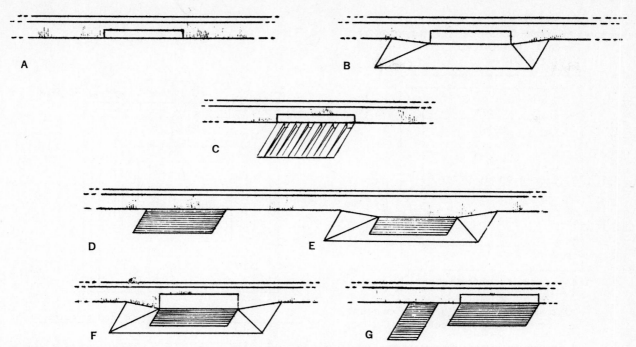

Figure 6.299   Curb inlets: (*a*) undepressed; (*b*) depressed; (*c*) deflector inlet. Gutter inlets: (*d*) undepressed; (*e*) depressed. Combination inlet: (*f*) grate placed directly in front of open depressed curb. Multiple inlet: (*g*) undepressed.

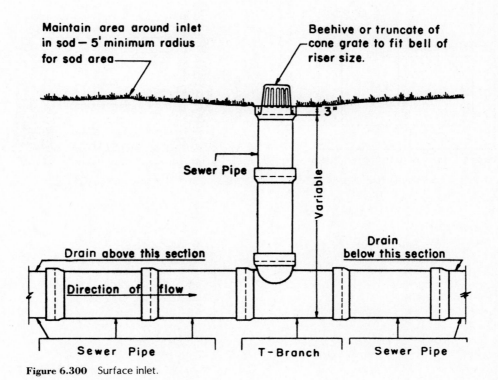

Figure 6.300   Surface inlet.

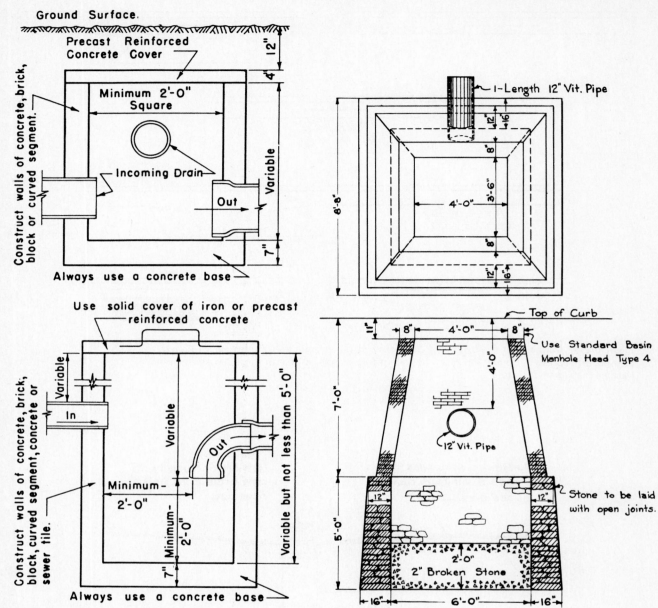

Figure 6.301   Junction box, catch basin, and silt trap.

Figure 6.302   Rectangular sewer.

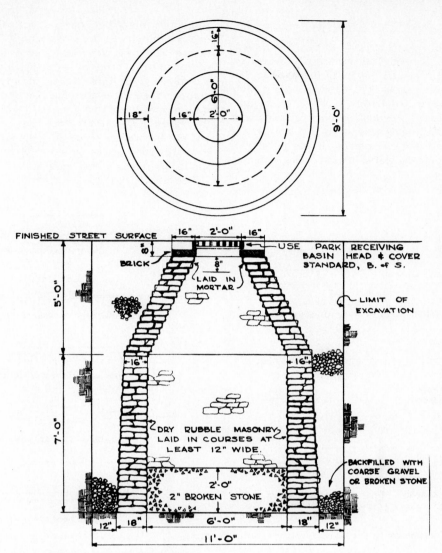

Figure 6.303   Circular masonry sewer.

DRAIN CROSSINGS AND OUTLETS

Culverts should be located to coincide as closely as practicable with stream alignment. The inlet should be placed where cross drainage intersects the toe of the slope so that water is not forced to travel out of its natural course to enter the opening (see Figure 6.304). Relief culverts should be provided at frequent intervals on sidehill sections to drain water to the low side, thus minimizing erosion. Install culverts at not more than 300-foot intervals on 8 percent grades and at 500-foot intervals on 5 percent grades unless paving or other ditch lining is provided.

Inlet and outlet elevations normally coincide with the cor-responding streambed, ditch, or natural ground elevation at ends of the culvert. Minor variations, involving ditching to secure proper fall, are required to obtain a minimum cover of from 6 to 12 inches over the top of culvert or to increase the culvert grade to the desired minimum. When the culvert grade must be flatter than the inlet-ditch grade, a sump or a trap may be required at the inlet end to avoid silting inside the culvert. If the culvert is too high, drainage is obstructed, and undermining or scouring of the ends may occur; if it is too low, silting occurs, and the effective opening is correspondingly reduced. See Figure 6.305.

See also Figure 6.306 for illustrations of drain crossings.

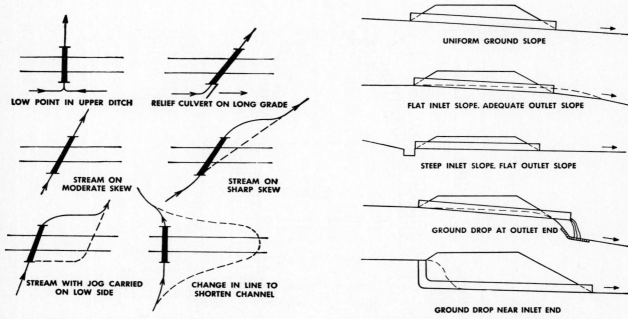

Figure 6.304 Proper location and alignment of culverts.

Figure 6.305 Proper elevation and grade of culverts.

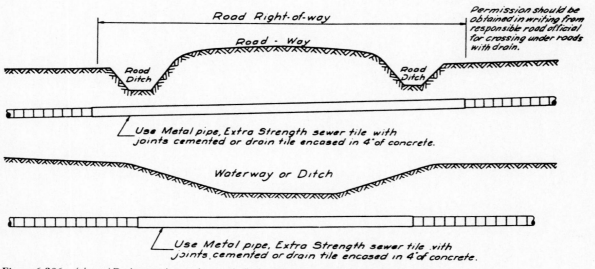

Figure 6.306 *(above)* Drain crossing under road. *(below)* Drain crossing under waterway or ditch.

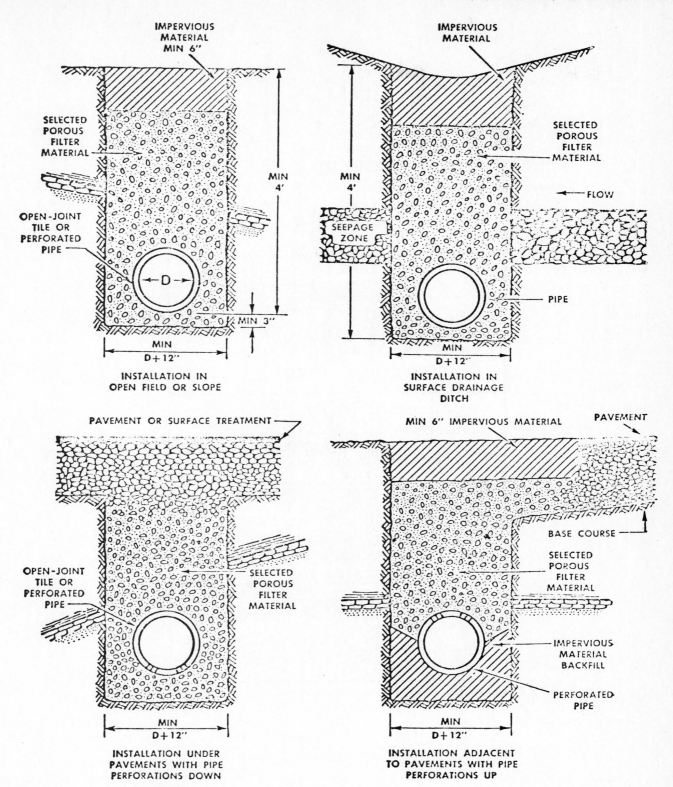

Figure 6.307 Typical subdrain installation.

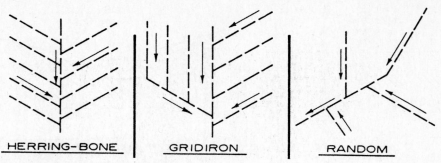

Figure 6.308 Patterns for drainage field.

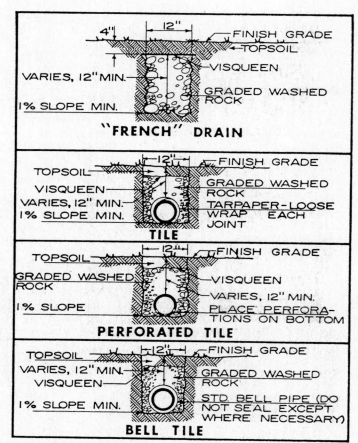

Figure 6.309 Types of drains.

Swales and Flumes Swales are wide shallow ditches transferring surface water to a catch basin or other outlet.

Longitudinal slopes of a swale must be not less than 2 percent.

Swales may be required to speed up the flow of water. Their location, however, may create an interruption in the smooth surface of lawn areas.

The location and ultimate shape of swales therefore are subject not only to technical but also to design considerations. Introduce catch basins where the depth of swales tends to become excessive.

Indicate the runoff through swales and flumes with spot elevations at approximately 25-foot intervals (Figure 6.310).

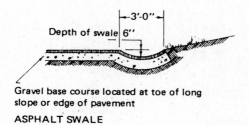

Gravel base course located at toe of long slope or edge of pavement

ASPHALT SWALE

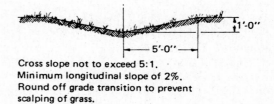

Cross slope not to exceed 5:1.
Minimum longitudinal slope of 2%.
Round off grade transition to prevent scalping of grass.

SWALE IN LAWN

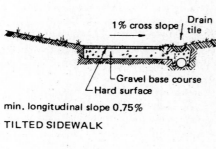

min. longitudinal slope 0.75%

TILTED SIDEWALK

Concrete base or
soil with grass joints

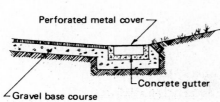

CONCRETE GUTTER

Quarter of half pipe
section on sub soil

FLUMES OF FIELD STONES OR PIPE

Figure 6.310 Swales and flumes.

Drain Tile In low-lying areas or on heavy soils, agricultural tile may be required to obtain satisfactory drainage. Agriculture tiles to ASTM C-4 are the most common. Where available, concrete tile to ASTM C-412 can also be used. Plastic drain tubing may be utilized if prices are competitive.

Tile runs start at a rock filter at the highest point, a minimum of 18 inches below finished grade. Tile can be laid on the subsoil or an even bed of coarse sand, with a gradient of not less than 1 inch in 30 feet.

Where settling is likely to occur, tiles are to be supported with 1- by 8-inch cedar boards. Saddles of tarpaper over the upper half of each joint will prevent dirt from washing in. Open joints at the bottom of the tiles must be not less than ⅜ inch and not more than ½ inch.

Backfill of clean 1- to 1½-inch rock, 6 to 8 inches deep will keep tiles in place and filter the water. Only light sand-loam should be used for the final backfilling; heavy clay, for example, would make the system of drainage useless.

Except for extreme seasonal fluctuations, keep the water table at 24 inches below the soil surface.

Size of tiles and distances between runs will depend on the quantity of water to be drained off.

French drains are trenches 18 to 24 inches deep and approximately 18 inches wide, filled with coarse gravel covered with 6 inches of sandy loam topsoil.

A system of drain tiles or French drains with the last 20 feet converted to drain tile can be connected with the regular storm drains on the property. See Figure 6.311.

When an extensive system is being installed, it becomes imperative that silt basins be constructed to permit occasional inspection and possible cleaning.

Backfill Use of granular material for backfill of service trenches and near foundation walls will reduce the chance of soil settlement requiring costly repairs.

Place backfill in layers 8 inches thick, and compact each layer with power equipment to a density equal to that existing in adjacent material. Sprinkling of fill with water will bring it to its optimum moisture content suitable for proper compaction.

Drawings showing the as built location of underground services are to be made available before any grading operation starts.

Fill under sodded or seeded areas can be compacted by the passage of heavy earth-moving equipment over each layer of fill.

Avoid heavy compaction of the final 6 inches of topsoil.

6.18 CISTERNS

Catchment Area The required catchment area depends on the amount of rainfall in your area and the amount of water needed. Proceed as follows:

1. Determine your yearly water needs by using the information given below.

2. Determine the normal annual rainfall in your area from local weather people.

3. Enter the graph in Figure 6.312 with "total annual water needs" and "normal annual precipitation."

4. Read the bottom line of the graph for "area of catchment needed."

Example A family of four estimates its water needs to be 200 gallons per day, or 73,000 gallons per year. The normal annual rainfall is 40 inches. From the graph the needed catchment area is 4400 square feet. If the cistern must supply only half of the family's water needs, the needed catchment area would be 2200 square feet.

Catchment Yields Each square foot of catchment area will provide the quantities of water listed below.

See also the table "Capacity of Cisterns" on page 732 and the typical cistern illustrated in Figure 6.313.

Annual rainfall (in inches)	Water available (in gallons)
10	4.2
15	6.2
20	8.3
25	10.4
30	12.5
35	14.5
40	16.6
45	18.8
50	20.8

Cistern leakage, roof washer, and evaporation will use up about one-third of the annual rainfall. These losses have already been subtracted from the numbers listed in the table.

TILE DRAIN   FRENCH DRAIN

Figure 6.311 Types of drains.

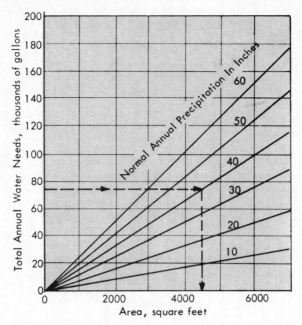

Figure 6.312 Required area of catchment.

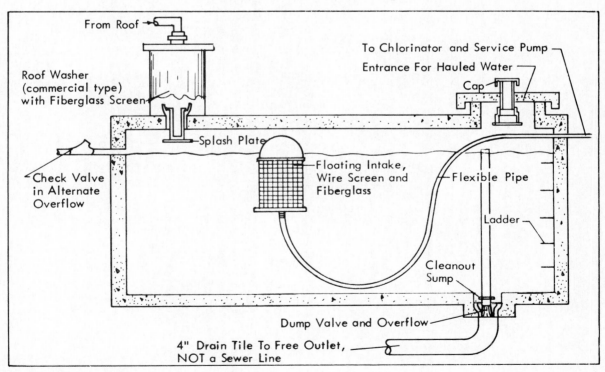

Figure 6.313 A typical cistern.

Capacity of Cisterns (in Gallons)

Depth in feet	Diameter of round type and length of side of square type (in feet)													
	5	6	7	8	9	10	11	12	13	14	15	16	17	18
Round Type														
5	735	1,055	1,440	1,880	2,380	2,935	3,555	4,230	4,965	5,755	6,610	7,515	8,435	9,510
6	882	1,266	1,728	2,256	2,856	3,522	4,266	5,076	5,958	6,906	7,932	9,018	10,182	11,412
7	1,029	1,477	2,016	2,632	3,332	4,109	4,977	5,922	6,951	8,057	9,254	10,521	11,879	13,314
8	1,176	1,688	2,304	3,008	3,808	4,696	5,688	6,768	7,944	9,208	10,576	12,024	13,576	15,216
9	1,323	1,899	2,592	3,384	4,284	5,283	6,399	7,614	8,937	10,359	11,898	13,527	15,273	17,118
10	1,470	2,110	2,880	3,760	4,760	5,870	7,110	8,460	9,930	11,510	13,220	15,030	16,970	19,020
12	1,764	2,532	3,456	4,512	5,712	7,044	8,532	10,152	11,916	13,812	15,864	18,036	20,364	22,824
14	2,058	2,954	4,032	5,264	6,664	8,218	9,954	11,844	13,902	16,114	18,508	21,042	23,758	26,628
16	2,342	3,376	4,608	6,016	7,616	9,392	11,376	13,536	15,888	18,416	21,152	24,048	27,152	30,432
18	2,646	3,798	5,184	6,768	8,568	10,566	12,798	15,228	17,874	20,718	23,796	27,054	30,546	34,236
20	2,940	4,220	5,760	7,530	9,520	11,740	14,220	16,920	19,860	23,020	26,440	30,060	33,940	38,040
Square Type														
5	935	1,345	1,835	2,395	3,030	3,740	4,525	5,385	6,320	7,330	8,415	9,575	10,810	12,112
6	1,122	1,614	2,202	2,874	3,636	4,488	5,430	6,462	7,584	8,796	10,098	11,490	12,974	14,534
7	1,309	1,883	2,569	3,353	4,242	5,236	6,335	7,539	8,848	10,262	11,781	13,405	15,134	16,956
8	1,496	2,152	2,936	3,832	4,848	5,984	7,240	8,616	10,112	11,728	13,464	15,320	17,296	19,378
9	1,683	2,421	3,303	4,311	5,454	6,732	8,145	9,693	11,376	13,194	15,147	17,235	19,458	21,800
10	1,870	2,690	3,670	4,790	6,060	7,480	9,050	10,770	12,640	14,660	16,830	19,150	21,620	24,222
12	2,244	3,228	4,404	5,748	7,272	8,976	10,860	12,924	15,168	17,592	20,196	22,980	25,944	29,068
14	2,618	3,766	5,138	6,706	8,484	10,472	12,670	15,078	17,696	20,524	23,562	26,810	30,268	33,912
16	2,992	4,204	5,872	7,664	9,696	11,968	14,480	17,232	20,224	23,456	26,928	30,640	34,592	38,756
18	3,366	4,842	6,606	8,622	10,908	13,464	16,290	19,386	22,752	26,388	30,294	34,470	38,916	42,600
20	3,740	5,380	7,340	9,580	12,120	14,960	18,100	21,540	25,280	29,320	33,660	38,300	43,240	48,444

6.19 Shoreline Protection

SEAWALLS

The distinction between seawalls, bulkheads, and revetments is mainly a matter of purpose. Design features are determined at the functional planning stage, and the structure is named to suit its intended purpose. In general, seawalls are the most massive of the three types because they resist the full force of the waves. Bulkheads are next in size; their function is to retain fill, and they are generally not exposed to severe wave action. Revetments are the lightest because they are designed to protect shorelines against erosion by currents or light wave action.

A curved-face seawall is illustrated in Figure 6.314. This massive structure is built to resist high wave action and reduce scour. The stepped seawall (Figure 6.315) is designed for stability against moderate waves. The rubble-mound seawall (Figure 6.316) is built to withstand severe wave action. Although scour of the fronting beach may occur, the rock comprising the seawall can readjust and settle without causing structural failure.

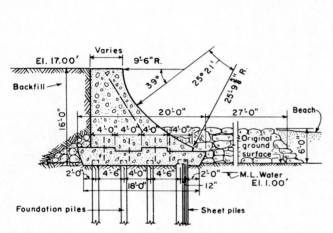

Figure 6.314 Concrete curved-face seawall.

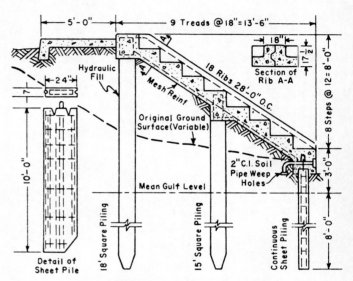

Figure 6.315 Concrete stepped-face seawall.

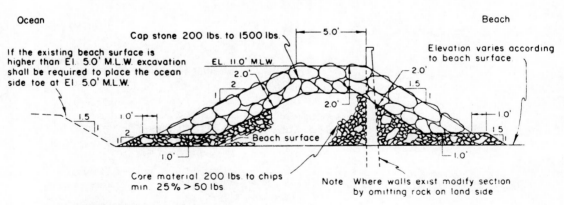

Figure 6.316 Rubble-mound seawall.

REVETMENTS

Structural types of revetments used for coastal protection in exposed and sheltered areas are illustrated in Figures 6.317 and 6.318. There are two types of revetments: the rigid, cast-in-place concrete type illustrated in Figure 6.317 and the flexible or articulated armor unit type illustrated in Figure 6.318. A rigid concrete revetment provides excellent bank protection, but the site must be dewatered during construction to pour the concrete. A flexible structure also provides excellent bank protection, and it can tolerate minor consolidation or settlement without structural failure. This is true for the riprap revetment and to a lesser extent for the interlocking concrete-block revetment. Both the articulated block structure and the riprap structure permit relief of hydrostatic uplift pressure generated by wave action. The underlying plastic filter cloth and gravel or a crushed-stone filter and bedding

layer provide relief of pressure over the entire foundation area rather than through specially constructed weep holes.

Interlocking concrete blocks have been used extensively for shore protection in the Netherlands and England and have recently become popular in the United States. Typical blocks are generally square slabs with the shiplap type of interlocking joints. Joints of the shiplap type provide a mechanical interlock with adjacent blocks.

The stability of an interlocking concrete block depends largely on the type of mechanical interlock. It is impossible to analyze block stability under specified wave action on the basis of the weight alone. However, prototype tests at the Coastal Engineering Research Center on blocks having shiplap joints and tongue-and-groove joints indicate that the stability of tongue-and-groove blocks is much greater than that of shiplap blocks.

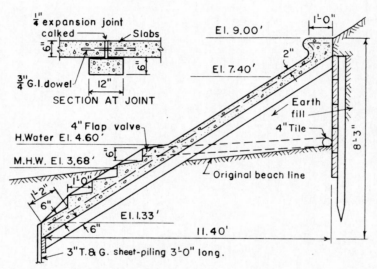

Figure 6.317 Concrete revetment.

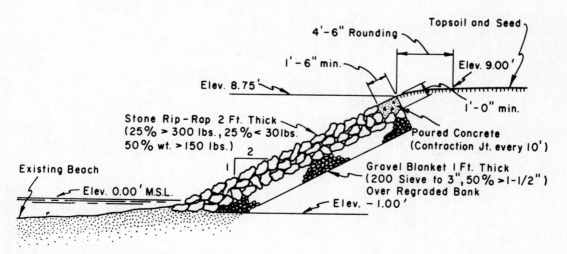

Figure 6.318 Riprap revetment.

GROINS

Groins are classified principally as to permeability, height, and length. Groins built of common construction materials can be made permeable or impermeable and high or low in profile. The materials used are stone, concrete, timber, and steel. Asphalt and sand-filled nylon bags have also been used to a limited extent. Structural types of groins built with different construction materials are illustrated in Figures 6.319 and 6.320.

Timber Groins A common type of timber groin is an impermeable structure composed of sheet piles supported by wales and round piles. Some permeable timber groins have been built by leaving spaces between the sheeting. A typical timber groin is shown in Figure 6.319. The round tim-

ber piles forming the primary structural support should be at least 12 inches in diameter at the butt. Stringers or wales, bolted to the piling, should be at least 8 by 10 inches, preferably cut and drilled before being pressure-treated with creosote. The sheet piles are usually of the Wakefield, tongue-and-groove, or shiplap type, supported in a vertical position between the wales and secured to the wales with nails. All timbers and piles used for marine construction should be given the maximum recommended pressure treatment of creosote or creosote and a coal-tar solution.

Steel Groins A typical design for a timber-steel sheet-pile groin is shown in Figure 6.320. Steel sheet-pile groins have been constructed with straight web, arch web, or Z piles. Some have been made permeable by cutting openings in the

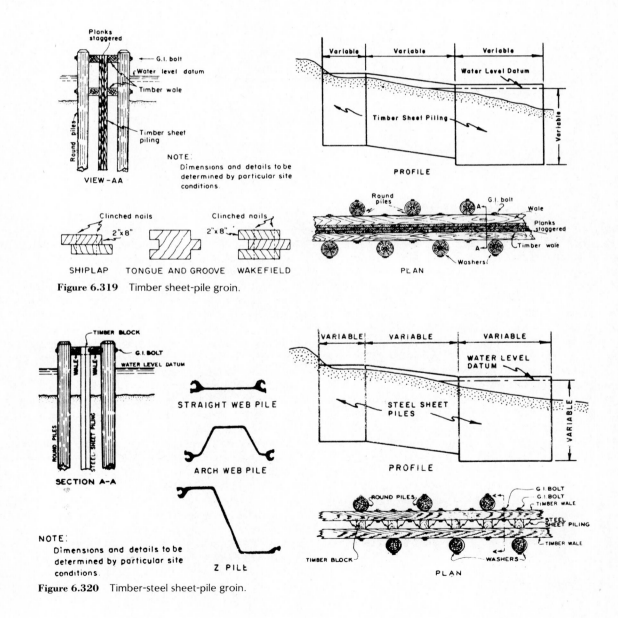

Figure 6.319 Timber sheet-pile groin.

Figure 6.320 Timber-steel sheet-pile groin.

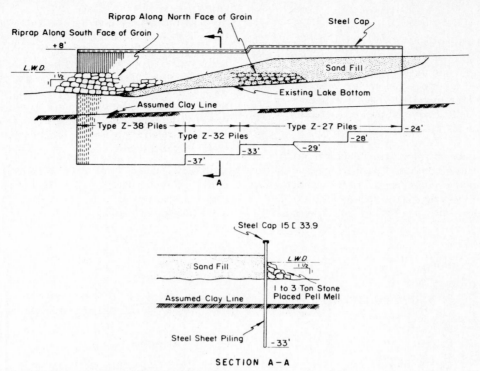

Figure 6.321 Cantilever steel sheet-pile groin.

piles. The interlock type of joint of steel sheet piles provides a sandtight connection. The selection of the type of sheet piles depends on the earth forces to be resisted. Where the forces are small, straight web piles can be used. Where the forces are great, deep-web Z piles should be used. The timber-steel sheet-pile groins are constructed with horizontal timber or steel wales along the top of the steel sheet piles, and vertical round timber piles or brace piles are bolted to the outside of the wales for added structural support. The round piles may not always be required with the Z pile but ordinarily are used with the flat or arch web sections. The round piles and timbers should be creosoted to maximum treatment for use in waters with marine borers.

Figure 6.321 illustrates the use of a cantilever steel sheet-pile groin. A groin of this type may be employed where the wave attack and earth loads are moderate. In this structure the sheet piles are the basic structural members; they are restrained at the top by a structural-steel channel welded to the piles.

A typical cellular type of groin is shown in Figure 6.322. This groin is composed of cells of varying sizes, each consisting of semicircular walls connected by cross diaphragms. Each cell is filled with sand or stone to provide structural stability. Concrete, asphalt, or stone caps are used to retain the fill material.

Rubble-Mound Groins   The rubble-mound groins are constructed with a core of quarry-run material, including fine material to make them sandtight, and covered with a layer of armor stone. The armor stone should weigh enough to be stable against the design wave. A typical rubble-mound groin is illustrated in Figure 6.323.

If permeability of a rubble-mound groin is a problem, the voids between stones can be filled with concrete or asphalt grout. This sealing also increases the stability of the entire structure against wave action.

Concrete Groins   Previously, the use of concrete in groins was generally limited to permeable-type structures that permitted the passage of sand through the structure. A more recent development in the use of concrete for groin construction is illustrated in Figure 6.324. This groin is an impermeable, prestressed-concrete pile structure with a cast-in-place concrete cap.

Selection of Groin Type   After planning has indicated that the use of groins is practicable, the selection of groin type is based on varying interrelated factors. No universal type of groin can be prescribed because of the wide variation in conditions at each location. A thorough investigation of foundation materials is essential to selection. Borings or probings should be taken to determine the subsurface conditions for penetration of piles. Where foundations are poor or where little penetration is possible, a gravity-type structure such as a rubble or a cellular steel sheet-pile groin should be considered. Where penetration is good, a cantilever type of struc-

ture of concrete, timber, or steel sheet piles should be considered.

The availability of materials affects the selection of the type of groin because of costs. The economic life of the material and the annual cost of maintenance to attain that economic life are also selection factors. The first costs of timber and steel sheet-pile groins, in that order, are often less than those of other types of construction. Concrete sheet-pile groins are generally more expensive than either timber or steel groins but may cost less than a rubble-mound groin does. However, concrete and rubble-mound groins require less maintenance and have a much longer life than do the timber of steel sheet-pile groins. The funds available for initial construction, the annual charges, and the period during which protection will be required must all be studied before deciding on a particular type.

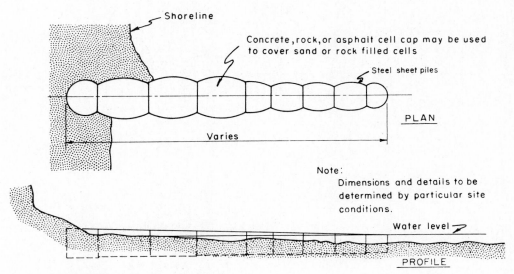

Figure 6.322   Cellular steel sheet-pile groin.

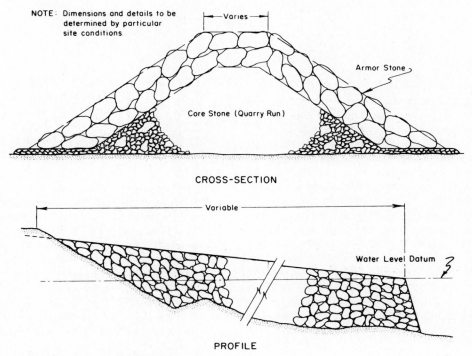

Figure 6.323   Rubble-mound groin.

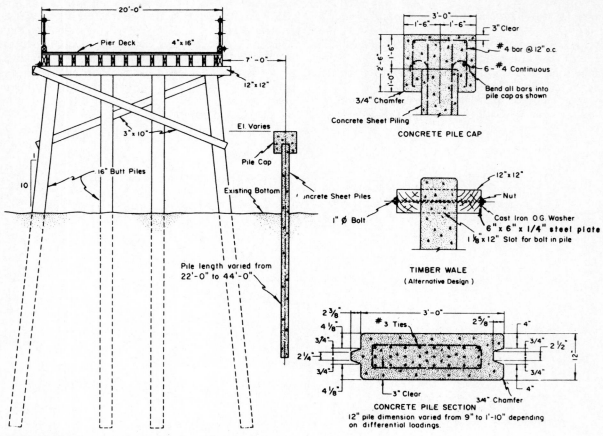

Figure 6.324 Prestressed-concrete sheet-pile groin.

JETTIES

The principal construction materials of jetties are stone, concrete, steel, and timber. Asphalt has occasionally been used as a binder. Some structural types of jetties are illustrated in Figures 6.325, 6.326, and 6.327.

Rubble-Mound Jetties The rubble-mound structure is a mound of stones of different sizes and shapes either dumped at random or placed in courses. Side slopes and stone sizes are designed so that the structure will resist the expected wave action. Rubble-mound jetties illustrated in Figures 6.325 and 6.326 are adaptable to any depth of water and to most foundation conditions. Rubble-mound structures are used extensively. Their chief advantages are that settlement of the structure results in readjustment of component stones and increased stability rather than in failure of the structure, damage is easily repaired, and rubble absorbs rather than reflects wave action. Their chief disadvantages are the large quantity of material required, the high first cost if satisfactory material is not locally available, and the wave energy propagated through the structure if the core is not high and impermeable.

Where rock armor units in adequate quantities or size are not economically available, concrete armor units are used. Figure 6.326 illustrates the use of quadripod armor units on a rubble-mound jetty. Figure 6.325 illustrates the use of the more recently developed dolos armor unit in which 42- and 43-ton dolos were used.

Sheet-Pile Jetties Timber, steel, and concrete sheet piles have been used for jetty construction where waves are not severe. Steel sheet piles are used for jetties in various ways. These include a single row of piling with or without pile buttresses; a single row of sheet piles arranged so that the row of piles acts as a buttressed wall; double walls of sheet piles held together with tie rods with the space between the wall filled with stone or sand, usually separated into compartments by cross walls if sand is used; and cellular steel sheet-pile structures that are modifications of the double-wall type. An example of a cellular steel sheet-pile jetty is shown in Figure 6.327.

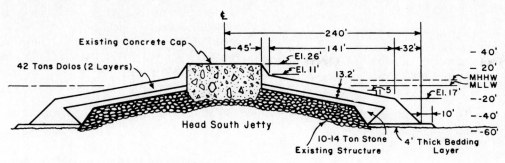

Figure 6.325 Dolos rubble-mound jetty.

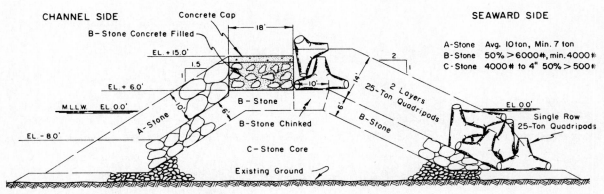

Figure 6.326 Quadripod rubble-mound jetty.

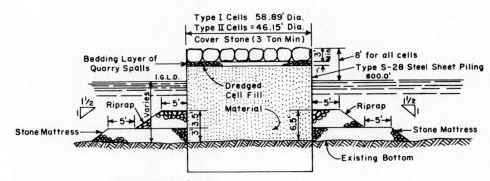

Figure 6.327 Cellular steel sheet-pile jetty.

BREAKWATERS

In exposed locations breakwaters are generally some variation of a rubble-mound structure. In less severe exposures both cellular steel and concrete caissons have been used. Figures 6.328 and 6.329 illustrate structural types of shore-connected breakwaters used for harbor protection. The rubble-mound breakwaters depicted in these figures are adaptable to almost any depth and can be designed to withstand severe waves. Figure 6.329 illustrates the use of tribar armor units on a rubble-mound structure.

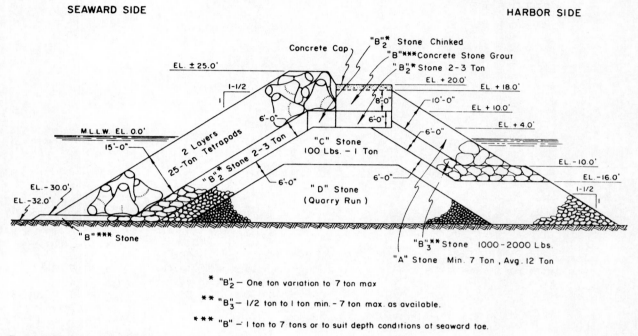

Figure 6.328   Tetrapod rubble-mound breakwater.

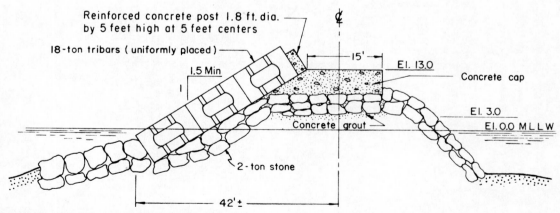

Figure 6.329   Tribar rubble-mound breakwater.

BULKHEADS

Three structural types of bulkheads (concrete, steel, and timber) are shown in Figures 6.330, 6.331, and 6.332. Cellular steel sheet-pile bulkheads are used where rock is near the surface and adequate penetration is impossible for the anchored sheet-pile bulkhead illustrated in Figure 6.331. When vertical or nearly vertical bulkheads are constructed and the water depth at the wall is less than twice the anticipated maximum wave height, the design should provide for riprap armoring at the base to prevent scouring. Excessive scouring may endanger the stability of the wall.

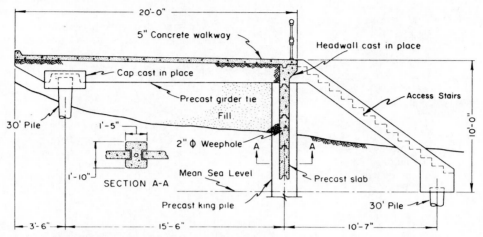

Figure 6.330   Concrete slab and king-pile bulkhead.

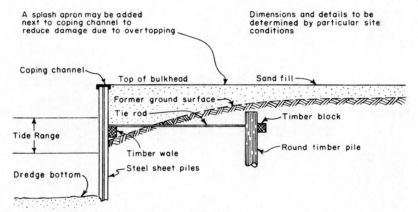

Figure 6.331   Steel sheet-pile bulkhead.

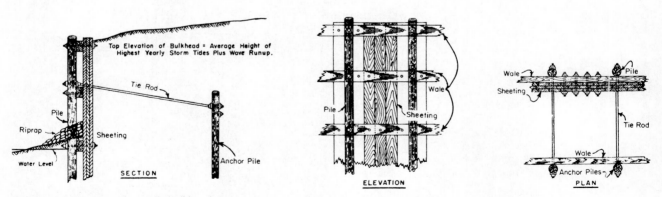

Figure 6.332   Timber sheet-pile bulkhead.

BULKHEAD WALLS

Walls of types D, E, and F are sheet-pile walls (see Figure 6.335). They generally show little settlement but at times may not be considered as permanent as walls on riprap, filled cribs, or relieving platforms. Type D, a wood sheet-pile wall, is probably the most widely used at marinas if there is a firm foundation material into which the sheet piles may be driven to the required depth. This type of wall is often the most economical. Type E, a steel sheet-pile wall, can be modified to fit many varied conditions. In some instances it may compare favorably in cost with a wood sheet-pile wall. Type F is a steel sheet-pile wall with a concrete cap or coping. In addition to enhancing the appearance of the wall, the concrete may increase the life of the wall if it extends from above the high-water line to below the low-water line, thus encasing the area where disintegration is greatest.

Walls of types A and C have economical original construction costs but are subject to large settlements and sliding. They are often the source of future trouble and costly maintenance in recapping the walls and repairing adjacent structures. When settling and sliding are of small consequence and the encroachment of riprap or crib base does not unduly limit the use of the marina, they are recommended because of their economy. The bottom should be reasonably solid for this type of construction; otherwise the entire wall may be lost. It is suggested that the top of the riprap or crib base be built up approximately 1 foot higher than the desired grade to allow for settlement. It is also recommended that a year or two elapse before constructing the wall on the base. The walls on these types of bases can be varied as required by architectural appearance or other local conditions. The walls might be cast-in-place concrete, precast-concrete blocks, rubble masonry, cut-stone masonry, or concrete walls with stone facings and copings.

Type B, a platform type of bulkhead, consists of a concrete wall resting on a timber relieving platform supported by bearing piles. This type of wall is suitable for greater depths of water and softer underlying strata than are the sheet-pile walls.

See also Figures 6.333 and 6.334.

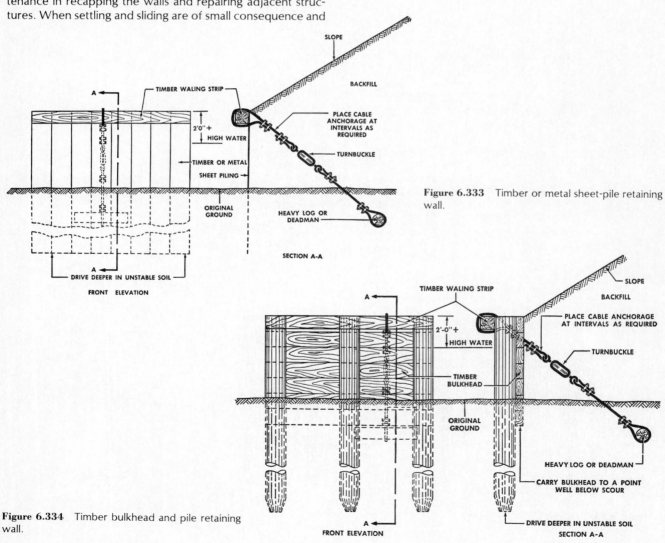

Figure 6.333 Timber or metal sheet-pile retaining wall.

Figure 6.334 Timber bulkhead and pile retaining wall.

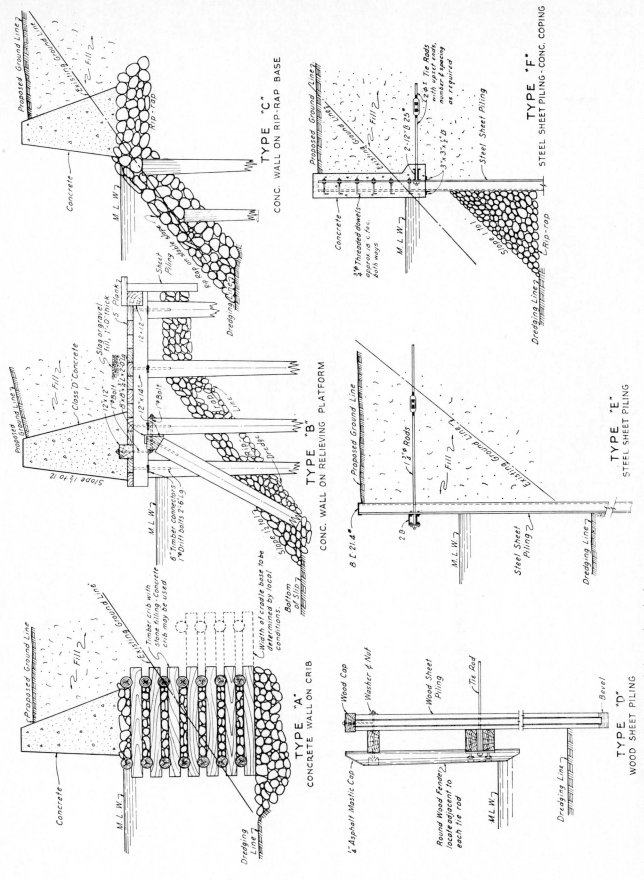

Figure 6.335 Bulkhead walls for typical pleasure boat basin.

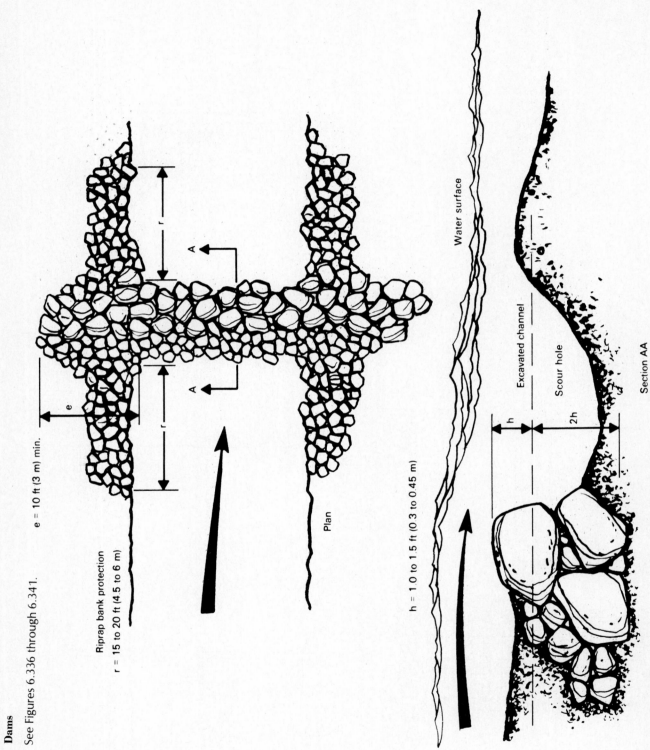

Dams

See Figures 6.336 through 6.341.

e = 10 ft (3 m) min.

Riprap bank protection
r = 15 to 20 ft (4.5 to 6 m)

Plan

h = 1.0 to 1.5 ft (0.3 to 0.45 m)

Water surface

Excavated channel

Scour hole

Section AA

Figure 6.336 Rock check dam.

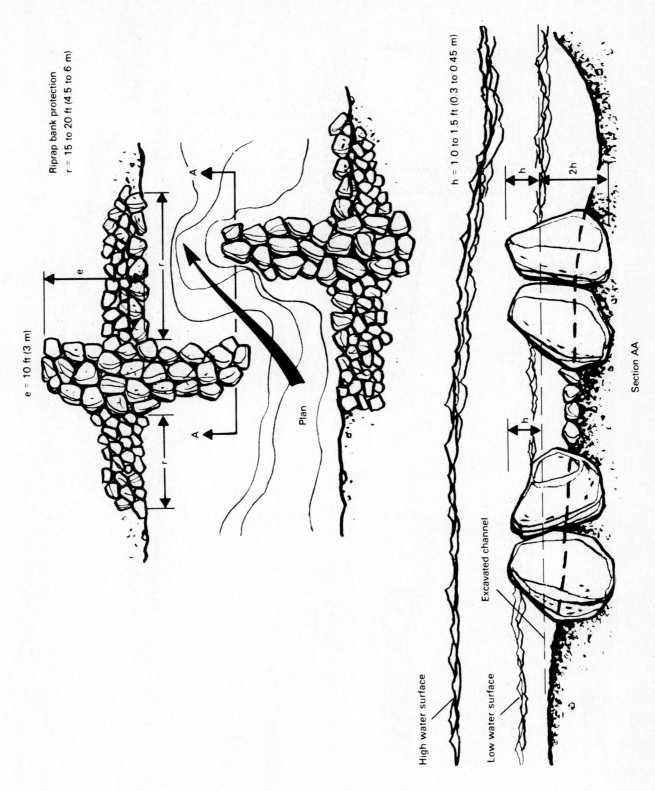

Riprap bank protection
r = 15 to 20 ft (4.5 to 6 m)

e = 10 ft (3 m)

Plan

h = 1.0 to 1.5 ft (0.3 to 0.45 m)

High water surface

Low water surface

Excavated channel

Section AA

Figure 6.337   Split rock check dam.

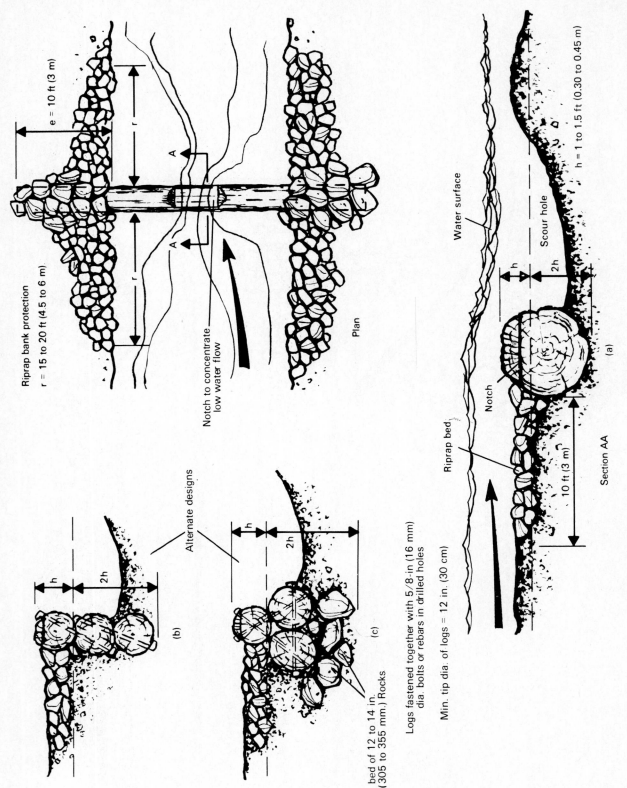

Riprap bank protection
r = 15 to 20 ft (4.5 to 6 m)

e = 10 ft (3 m)

Notch to concentrate
low water flow

Plan

Alternate designs

bed of 12 to 14 in.
(305 to 355 mm.) Rocks

(b)

(c)

Logs fastened together with 5/8-in (16 mm)
dia. bolts or rebars in drilled holes

Min. tip dia. of logs = 12 in. (30 cm)

Water surface

Scour hole

h = 1 to 1.5 ft (0.30 to 0.45 m)

Notch

Riprap bed

10 ft (3 m)

Section AA

(a)

Figure 6.338 Log check dam.

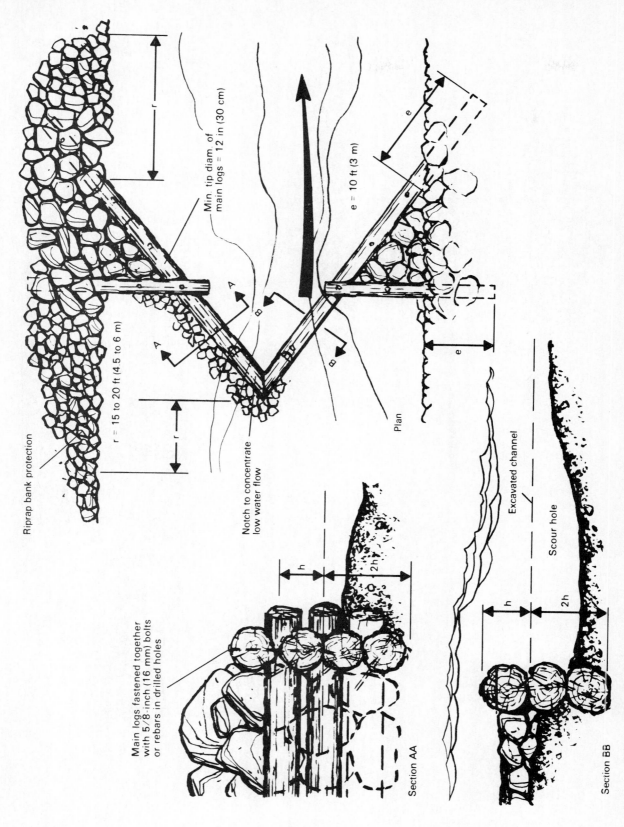

Riprap bank protection

r = 15 to 20 ft (4.5 to 6 m)

Notch to concentrate
low water flow

Min. tip diam. of
main logs = 12 in (30 cm)

e = 10 ft (3 m)

Plan

Main logs fastened together
with 5/8-inch (16 mm) bolts
or rebars in drilled holes

Section AA

Excavated channel

Scour hole

Section BB

Figure 6.339   Log V dam.

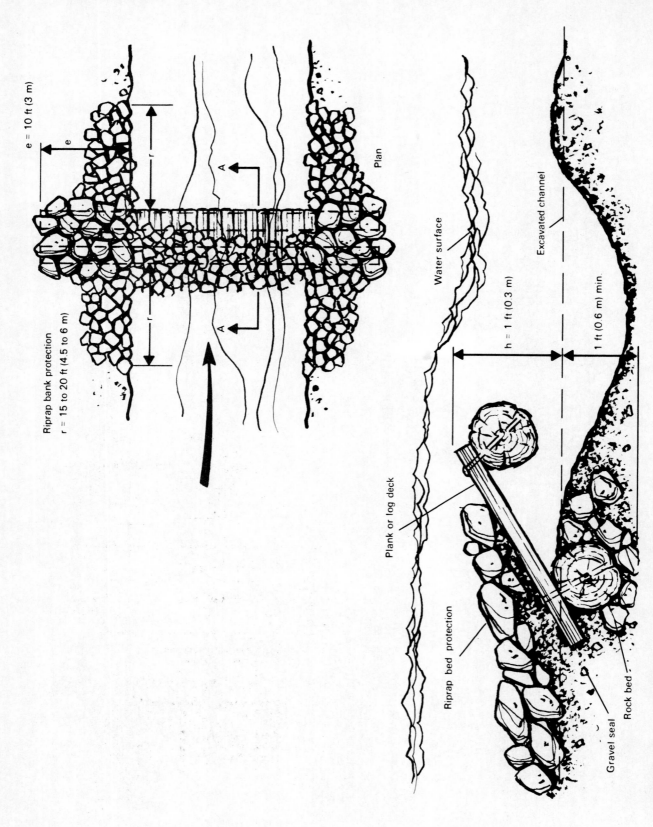

Figure 6.340 Log and plank dam.

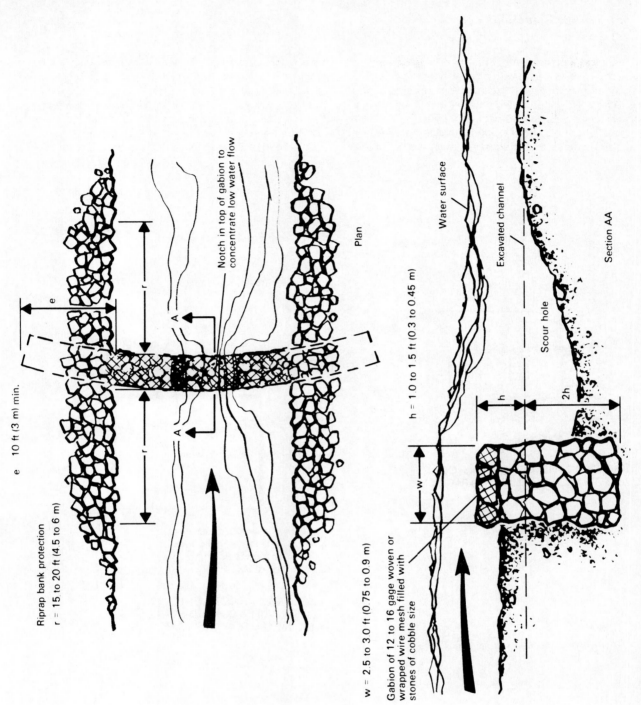

Plan

Notch in top of gabion to concentrate low water flow

e ≤ 10 ft (3 m) min.

Riprap bank protection
r = 15 to 20 ft (4.5 to 6 m)

w = 2.5 to 3.0 ft (0.75 to 0.9 m)

Gabion of 12 to 16 gage woven or wrapped wire mesh filled with stones of cobble size

Water surface

Excavated channel

Scour hole

Section AA

h = 1.0 to 1.5 ft (0.3 to 0.45 m)

Figure 6.341 Gablon check dam.

6.20 Gas Storage Containers and Overhead Electrical Service

An aboveground liquefied petroleum gas storage container (see Figures 6.342 and 6.344) with a capacity of less than 125 gallons may be placed directly against the exterior wall of a building. When two gas storage container units are installed, no clearance need be provided between them.

The space where the aboveground containers are located should be arranged so that the lowest level of such space, whether enclosed or open, is ventilated horizontally to the outside air and is at least 5 feet in horizontal distance from any openable window, door, or other ventilating opening that is wholly or in part at a lower level. Such an arrangement should also be maintained for the discharge of cylinder and regular relief valves.

An aboveground installation should be at least 5 feet from any driveway. Underground containers should be buried at least 2 feet below grade.

Overhead electrical service is illustrated in Figure 6.343.

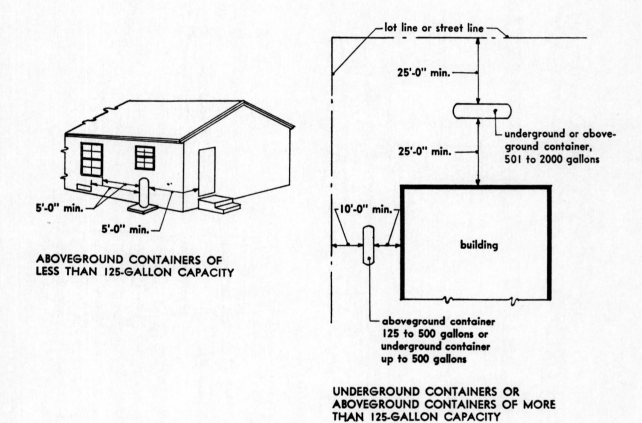

ABOVEGROUND CONTAINERS OF LESS THAN 125-GALLON CAPACITY

UNDERGROUND CONTAINERS OR ABOVEGROUND CONTAINERS OF MORE THAN 125-GALLON CAPACITY

Figure 6.342 Liquefied petroleum gas storage containers.

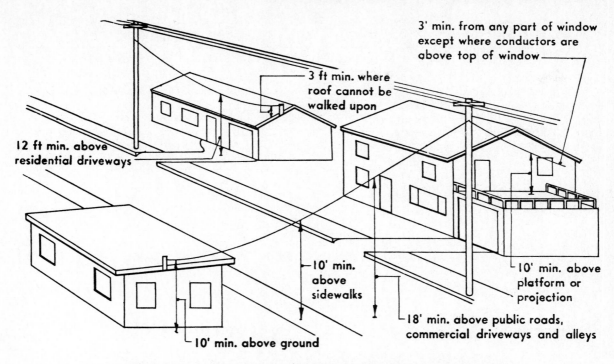

3' min. from any part of window except where conductors are above top of window

3 ft min. where roof cannot be walked upon

12 ft min. above residential driveways

10' min. above sidewalks

10' min. above platform or projection

18' min. above public roads, commercial driveways and alleys

10' min. above ground

MINIMUM CLEARANCES OF OVERHEAD SERVICE CONDUCTORS

Figure 6.343   Overhead electrical service.

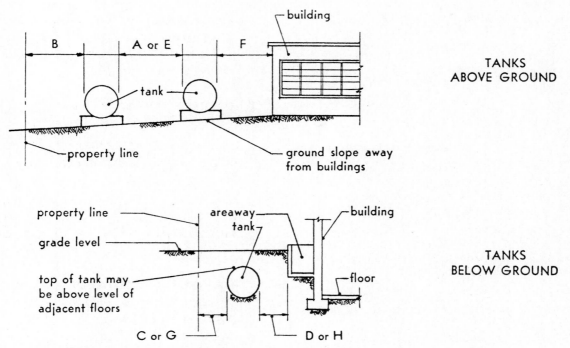

building

B   A or E   F

tank

property line

ground slope away from buildings

TANKS ABOVE GROUND

property line

areaway tank

building

grade level

top of tank may be above level of adjacent floors

floor

C or G   D or H

TANKS BELOW GROUND

Figure 6.344   Location of outside storage tanks for flammable liquids.

	Above-ground and below-ground tanks, distances in feet*									
	Liquid flash point of 70°F or less					Liquid flash point over 70°F				
	Above ground			Below ground		Above ground			Below ground	
Individual tank capacity, in gallons	Between tanks A	From property line B	From building on same premises F	From property line C	From building on same premises D	Between tanks E	From property line B	From building on same premises F	From property line G	From building on same premises H
0–275	3	5	5	3	1	3	5	5	1	1
276–750	3	10	5	3	1	3	10	5	1	1
751–12,000	3	15	5	3	1	3	15	5	1	1
12,001–30,000	3	20	5	3	1	3	20	5	1	1
30,001–50,000	3	30	10	3	1	3	30	10	1	1

*Applies to tanks vented to prevent development of pressure exceeding 2.5 pounds per square inch gauge.

6.21 Outdoor Signs

Essentially, signs should perform three functions. They should (1) identify a place and indicate whether it is accessible to everyone, (2) indicate warnings where necessary, and (3) give routing information (see Figure 6.345). The information given on signs should always be clear and precise, and sign locations should never present unnecessary hazards for pedestrian or vehicular traffic.

Identification and Accessibility

1. Key site-related areas that should be identified by sign posting are as follows:
   a. Traffic signs announcing public rest stops with accessible facilities
   b. Public lavatories accessible to pedestrians
   c. Special car parking
   d. Directional signs for vehicles and pedestrians such as one-way street signs
   e. Signs identifying accessible entrances to buildings or facilities
   f. Informative signs on buildings
2. So that signs can be made more useful to everyone, they should be designed to be readable by all people, including the visually handicapped. This can be accomplished in a number of different ways:
   a. Braille strips can be placed along sign edges.
   b. Raised or routed letters are readable by the blind or the partially sighted.
   c. Graphic symbols are useful in transmitting messages quickly but should be avoided as the sole means of

1. Directional

Usually included with an arrow; are used for indication of a change in route, or confirmation of a correct direction.

2. Informational

Used for overall information for general organization of a series of elements; i.e. campus plan, bus routes, building layout, shopping mall plan, etc.

3. Identification

Gives specific location information, identifies specific items; i.e. parking lot "b", building #5, First Aid, etc.

4. Regulatory

Gives operational requirements, restrictions, or gives warnings. Usually used for traffic delineation or control; i.e. "stop" signs, "no parking," "one way," etc.

Figure 6.345 Types of signs.

imparting information because they can be confusing to the blind.

    d. Signs that will be used by the visually handicapped must be located in a manner that first allows the signs to be recognized and, second, allows sign surfaces to be touched by the reader's hand.

    e. Signs along walkways or corridors should be set back a minimum of 18 inches and placed at a height of from 4 feet to 5 feet 6 inches.

3. The international symbol for access, the abstract man in a wheelchair, is already in extensive use in the United States. It is used to show where special provisions have been made to allow access for restricted persons.

Warnings Textural paving may be used to warn of imminent hazards such as abrupt changes of grade, stairs, ramps, and walk intersections and the locations of special information. However, the use of textural paving as a warning device for the blind is extremely impractical because of the widely varying nature of walkways in the country. The only effective use for such a system would be in a closed environment such as a school for the blind. Unfortunately, once away from their protected surroundings, blind persons would be vulnerable to a world full of unforewarned hazards.

Routing Information Where it is critical that people be able to travel quickly and unhindered to their destinations, routing information should be given.

1. Hospitals, college campuses, institutions, and so on should have posted signs, lines, or arrows painted on walk systems that are accessible to wheeled vehicles, particularly if such path systems are limited in number.

2. Access to buildings with only one or two entrances that are accessible to wheeled vehicles should be clearly indicated by routing signs.

Readability The readability of any sign is a function of many factors. In designing or choosing the format of a sign, the following points should be considered.

1. Information should be as concise and direct as possible.

2. Lettering styles and graphic symbols should be as bold and simple as possible. Fancy styles become cluttered, are time-consuming, and are confusing to read.

3. Schemes of contrasting colors with light images on dark backgrounds make signs both easier to read and more readable from longer distances.

Placement Placement is important because wrongly located signs may present an obstacle or a hazard. Unless intended to be read by the blind or the partially sighted, they should be set far enough off a traveled way or high enough off the ground, or both, so as not to be inadvertently walked into.

See Figure 6.346 for information on the design and location of signs.

Standardization of Signs The basic requirements of a highway sign are that it be legible to those using it and that it be understood in time to permit a proper response. This means high visiblity, lettering or symbols of adequate size, and a short legend for quick comprehensions by a driver approaching a sign at high speed. Standardized colors and shapes are specified so that the several classes of traffic signs can be promptly recognized. Simplicity and uniformity in design, position, and application are important.

Design Uniformity in design includes shape, color, dimensions, legends, and illumination or reflectorization. Detailed drawings of the standard signs are available to state and local highway and traffic authorities, sign manufacturers, and similarly interested agencies.

Shapes Standard sign shapes are (see Figure 6.347):

The octagon is reserved exclusively for the stop sign.

The equilateral triangle, with one point downward is reserved exclusively for the yield sign.

The round shape is used for the advance warning of a railroad crossing and for the civil defense evacuation route marker.

The pennant shape, an isosceles triangle, with its longest axis horizontal, is used to warn of no passing zones.

The diamond shape is used only to warn of existing or possible hazards either on the roadway or adjacent thereto.

The rectangle, ordinarily with the longer dimension vertical, is used for regulatory signs, with the exception of stop signs and yield signs.

The rectangle, ordinarily with the longer dimension horizontal, is used for guide signs, with the exception of certain route markers and recreational area guide signs.

The trapezoid shape may be used for recreational area guide signs.

The pentagon, point up, is used for school advance and school crossing sign.

Other shapes are reserved for special purposes; for example, the shield or other characteristic design for route markers and crossbuck for railroad crossings.

Sign Colors The colors to be used on standard signs shall be as follows:

Red is used only as a background color for stop signs, multiway supplemental plates, do-not-enter messages, wrong way signs, and on interstate route markers; as a legend color for yield signs, parking prohibition signs, the circular outline and diagonal bar prohibitory symbol.

Black is used as a background on one-way signs, certain weigh station signs, and night speed limit signs. Black is used as a message on white, yellow, and orange signs.

Design and Location

- When possible, gather signs together into unified systems. Avoid sign clutter in the landscape.

- Combine signs with lighting fixtures to reduce unnecessary posts and to illuminate signs – signage can't be effective in dark areas.

- Low-level informational signs can also illuminate paving below.

- Information signs should be placed at natural gathering spots and included into the design of site furniture.

- Avoid placement of signs where they may conflict with pedestrian traffic.

- Sign location should avoid conflict with door opening or vehicular operation.

- Signs should be placed to allow safe pedestrian clearance, vertically and laterally.

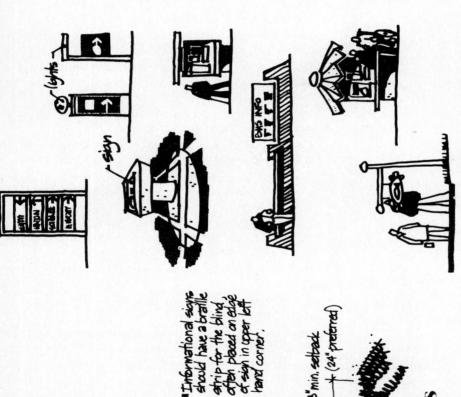

Braille on Signs

- Raised or routed letters are also helpful for the blind in reading signs.

- Informational signs should have a braille strip for the blind, often placed on edge of sign in upper left hand corner.

- Signs should never interfere with adjacent pedestrian traffic.

Figure 6.346 Outdoor signs.

–754–

White is used as the background for route markers, guide signs, the fallout shelter directional sign, and regulatory signs, except stop signs, and for the legend on brown, green, blue, black, and red signs.

Orange is used as a background color for construction and maintenance signs and should not be used for any other purpose.

Yellow is used as a background color for warning signs, except where orange is specified, and for school signs.

Brown is used as a background color for guide and information signs related to points of recreational or cultural interest.

Green is used as a background color for guide signs (other than those using brown or white), mileposts, and as a legend color with a white background for permissive parking regulations.

Blue is used as a background color for information signs related to motorist services (including police services and rest areas) and the Evacuation Route Marker.

Four other colors — purple, light blue, coral, and strong yellow-green — have been identified as suitable for highway use and are being reserved for future needs.

Wherever white is specified, it is understood to include silver colored reflecting coatings or elements that reflect white light.

Color Code

Red: stop or prohibition

Green: indicated movements permitted, direction guidance

Blue: motorist services guidance

Yellow: general warning

Black: regulation

White: regulation

Orange: construction and maintenance warning

Purple: unassigned

Brown: public recreation and scenic guidance

Strong yellow-green: unassigned

Light blue: unassigned

Coral: unassigned

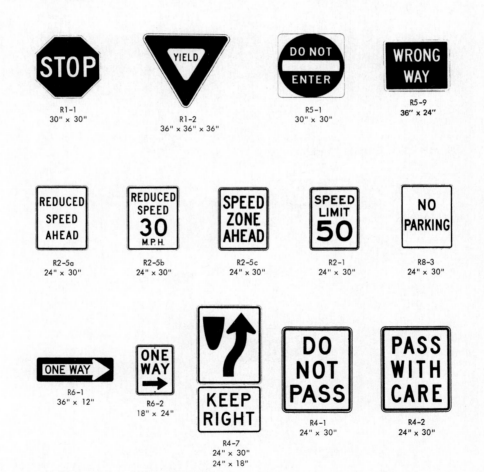

Figure 6.347 Commonly used regulatory signs.

Standardization of Location Standardization of position cannot always be attained in practice; however, the general rule is to locate signs on the right-hand side of the roadway, where the driver is looking for them. On wide expressways, or where some degree of lane-use control is desirable, or where space is not available at the roadside, overhead signs are often necessary. Signs in any other locations ordinarily should be considered only as supplementary to signs in the normal locations. Under some circumstances signs may be placed on channelizing islands or (as on sharp curves to the right) on the left-hand shoulder of the road, directly in front of approaching vehicles. A supplementary sign located on

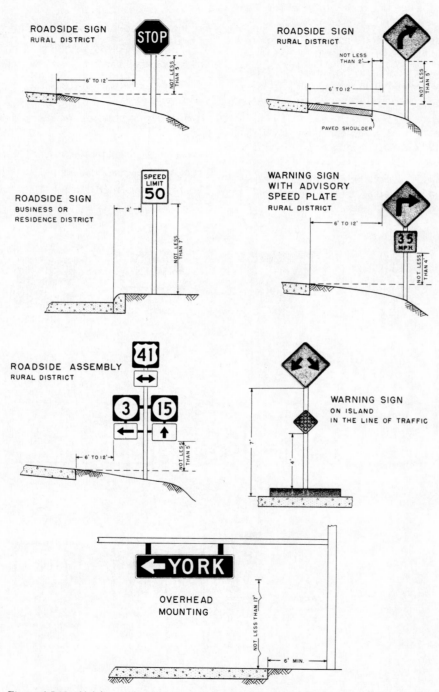

Figure 6.348 Height and lateral location of signs, typical installations.

the left of the roadway is often helpful on a multilane road where traffic on the right-hand lane may obstruct the view to the right.

Special care should be taken in sign location to ensure an unobstructed view of each sign.

Signs should be set to provide as much lateral clearance as existing conditions permit but should not exceed maximum clearances as shown in the lateral placement table. The lateral placement for all signs may be reduced to a minimum of 2 feet outside of the shoulder edge or face of curb where necessary to fit field conditions or improve sight distance. See Figures 6.348 through 6.357.

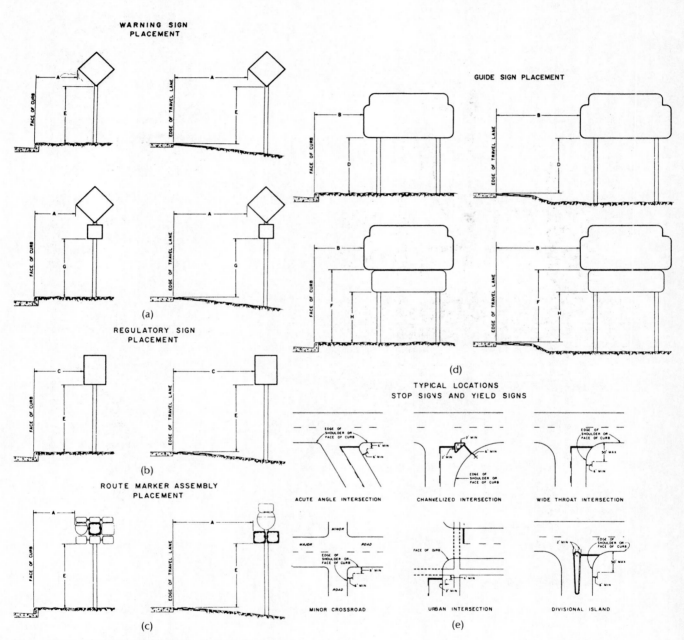

Figure 6.349   Placement and location of street signs.

Placement Tables

	Lateral	
	All systems	
Key	Min	Max
A	*	22'0"
B	*	30'0"
C	*	16'0"

*In no case shall any part of a sign be less than 2 ft. beyond any surface prepared for normal or emergency travel of vehicles.

		Vertical			
		Expressway		Nonexpressway	
Key	Interstate	Urban	Rural	Urban	Rural
D	5'0"†	7'0"	6'0"	7'0"	5'0"
E	6'0"	7'0"	5'0"	7'0"	5'0"
F	8'0"				
G	6'0"	6'0"	4'0"	6'0"	4'0"
H	5'0"	4'0"	4'0"	4'0"	4'0"

†When lateral placement is minimum, D should be 7'0"

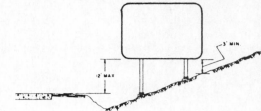

Figure 6.350 Ground clearance.

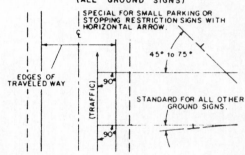

Figure 6.351 Angular placement.

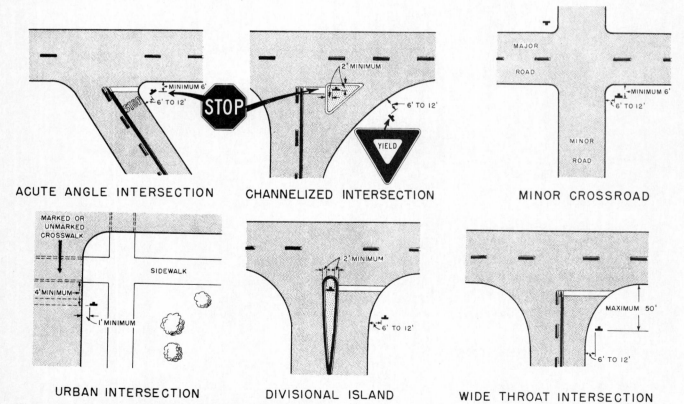

ACUTE ANGLE INTERSECTION CHANNELIZED INTERSECTION MINOR CROSSROAD

URBAN INTERSECTION DIVISIONAL ISLAND WIDE THROAT INTERSECTION

Figure 6.352 Typical locations for stop signs and yield signs.

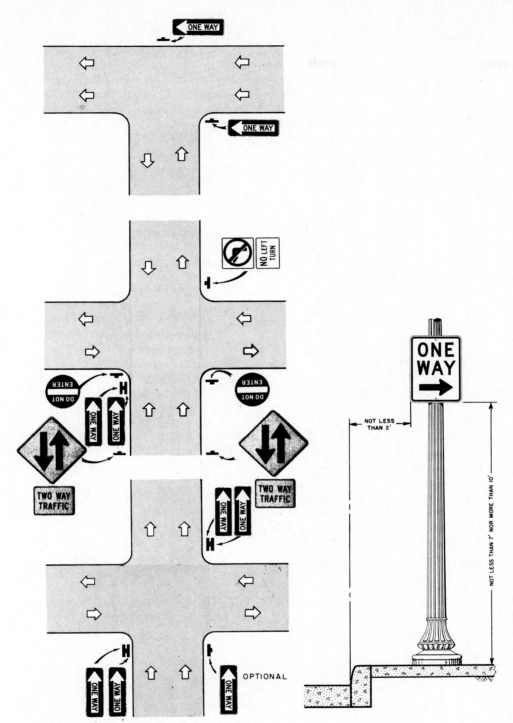

Figure 6.353 Location of one-way signs.

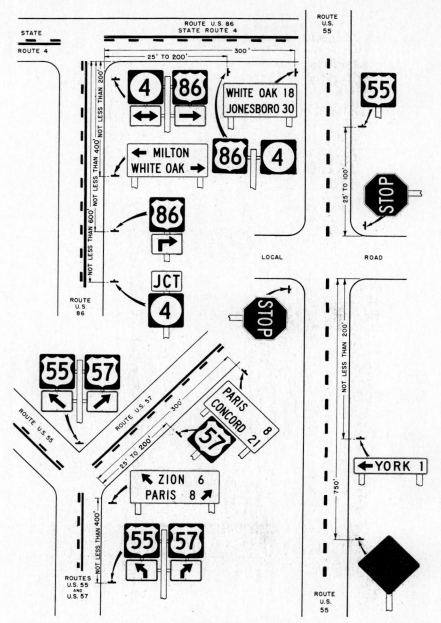

Figure 6.354 Typical route markings of rural intersections (for one direction of travel only).

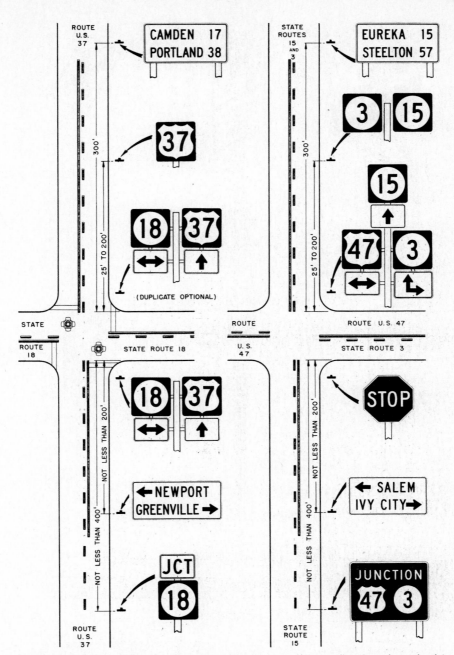

Figure 6.355 Typical route markings at rural intersections (for one direction of travel only).

Figure 6.356   Typical ground signs.

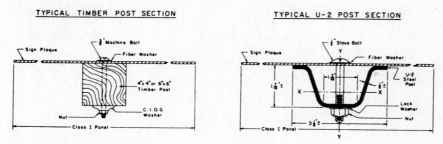

Figure 6.357   Sign details.

6.22 Street Crosswalk Markings

Crosswalks and Crosswalk Lines Crosswalk markings at signalized intersections and across intersectional approaches on which traffic stops serve primarily to guide pedestrians in the proper paths. Crosswalk markings across roadways on which traffic is not controlled by traffic signals or stop signs must also serve to warn the motorist of a pedestrian crossing point. At nonintersectional locations, these markings legally establish the crosswalk.

Crosswalk lines should be solid white lines, marking both edges of the crosswalk. They should be not less than 6 inches in width and should not be spaced less than 6 feet apart. Under special circumstances where no advance stop line is provided or where vehicular speeds exceed 35 miles per hour or where crosswalks are unexpected, it may be desirable to increase the width of the crosswalk line up to 24 inches in width. Crosswalk lines on both sides of the crosswalk should extend across the full width of pavement to discourage diagonal walking between crosswalks (Figure. 6.358a).

For added visibility, the area of the crosswalk may be marked with white diagonal lines at a 45 degree angle or with white longitudinal lines at a 90 degree angle to the line of the crosswalk (Figure 6.358b and c). These lines should be approximately 12 to 24 inches wide and spaced 12 to 24 inches apart. When diagonal or longitudinal lines are used to mark a crosswalk, the transverse crosswalk lines may be omitted. This type of marking is intended for use at locations where substantial numbers of pedestrians cross without any other traffic control device, at locations where physical conditions are such that added visibility of the crosswalk is desired, or at places where a pedestrian crosswalk might not be expected. Care should be taken to ensure that crosswalks with diagonal or longitudinal lines used at some locations do not weaken or detract from other crosswalks (where special emphasis markings are not used) (Figure 6.358a).

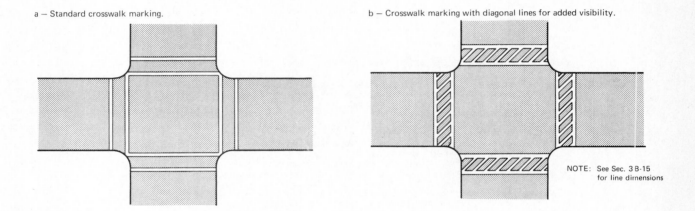

a — Standard crosswalk marking.

b — Crosswalk marking with diagonal lines for added visibility.

NOTE: See Sec. 3 B-15 for line dimensions

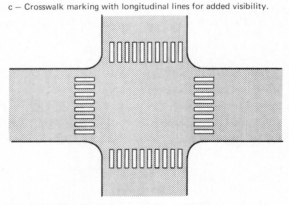

c — Crosswalk marking with longitudinal lines for added visibility.

Figure 6.358 Typical crosswalk markings.

6.23 Flagpoles

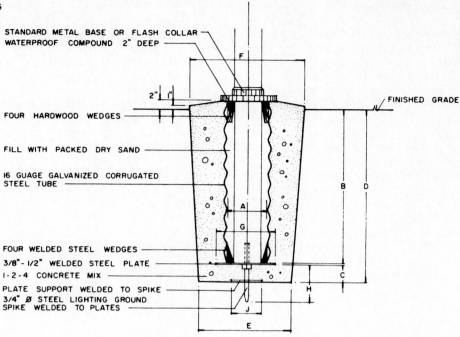

STANDARD METAL BASE OR FLASH COLLAR
WATERPROOF COMPOUND 2" DEEP

FINISHED GRADE

FOUR HARDWOOD WEDGES

FILL WITH PACKED DRY SAND

16 GUAGE GALVANIZED CORRUGATED STEEL TUBE

FOUR WELDED STEEL WEDGES
3/8"- 1/2" WELDED STEEL PLATE
1-2-4 CONCRETE MIX

PLATE SUPPORT WELDED TO SPIKE
3/4" Ø STEEL LIGHTING GROUND SPIKE WELDED TO PLATES

KEY TO STANDARD FOUNDATION DIMENSIONS

A INSIDE DIAMETER OF FOUNDATION TUBE AND SHOULD BE ABOUT 3" LARGER THAN OUTSIDE BUTT DIAMETER OF FLAGPOLE

B DISTANCE POLE SETS IN FOUNDATION AND SHOULD BE 10% OF POLE HEIGHT ABOVE GROUND

C THICKNESS OF FOOTING UNDER BASE PLATE AND SHOULD BE AT LEAST 15% IN INCHES OF POLE HEIGHT IN FEET WITH 4" MINIMUM THICKNESS

D DEPTH OF EXCAVATION AND SHOULD BE DISTANCE POLE SETS IN FOUNDATION PLUS THICKNESS OF FOOTING

E DIAMETER OF FOUNDATION AT BOTTOM AND SHOULD BE AT LEAST FOUR TIMES OUTSIDE BUTT DIAMETER OF POLE AND IN NO CASE LESS THAN 24" DIAMETER

F DIAMETER OF FOUNDATION AT TOP AND SHOULD BE AT LEAST FIVE TIMES OUTSIDE BUTT DIAMETER OF POLE AND IN NO CASE LESS THAN 30" DIAMETER

G SIZE OF BASE PLATE OF FOUNDATION TUBE AND SHOULD BE AT LEAST 6" LARGER SQUARE THAN INSIDE DIAMETER OF FOUNDATION TUBE

H LENGTH OF LIGHTING GROUND SPIKE ATTACHED TO BASE PLATE OF FOUNDATION TUBE AND SHOULD EXTEND BELOW FOOTING FOR A DISTANCE AT LEAST EQUIVALENT TO THE FOOTING THICKNESS

J SIZE OF SUPPORT PLATE ATTACHED TO GROUND SPIKE AND SHOULD BE ABOUT 25% OF AREA OF BASE PLATE

Figure 6.359 Detail of foundation for flagpole with metal base.

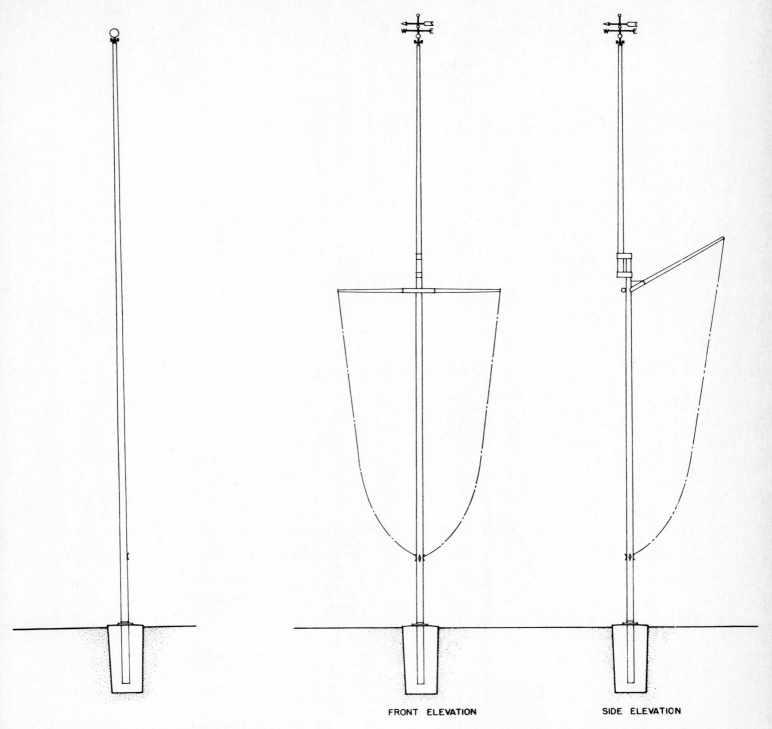

CONE TAPPERED METAL FLAGPOLE
NO SCALE

FRONT ELEVATION   SIDE ELEVATION

DOUBLE MAST NAUTICAL TYPE METAL FLAGPOLE
NO SCALE

Figure 6.360   Types of flagpoles.

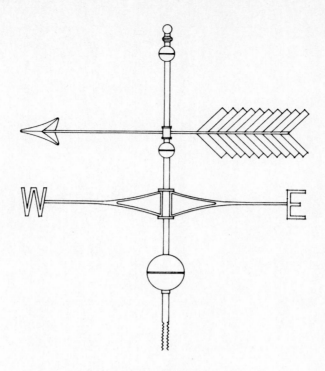

DETAIL OF WEATHER VANE
SCALE 3"= 1'-0"

Figure 6.361 Detail of weather vane, scale 3 inches to 1 foot 0 inch.

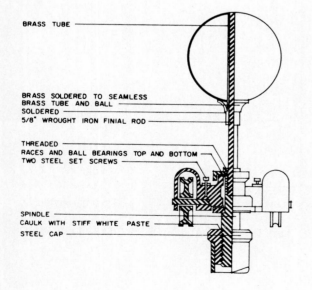

BRASS TUBE

BRASS SOLDERED TO SEAMLESS
BRASS TUBE AND BALL
SOLDERED
5/8" WROUGHT IRON FINIAL ROD

THREADED
RACES AND BALL BEARINGS TOP AND BOTTOM
TWO STEEL SET SCREWS

SPINDLE
CAULK WITH STIFF WHITE PASTE
STEEL CAP

BALL
8" DIAMETER 20 OZ SPUN COPPER FOR
50'-0" AND 60'-0" POLE
6" DIAMETER 20 OZ SPUN COPPER FOR
30'-0" AND 40'-0" POLE
BALL FINIAL SHALL BE SOLDERED TO
3/4" SEAMLESS BRASS TUBE
5/8" WROUGHT IRON FINIAL ROD SCREWED
INTO SPINDLE OF TRUCK
GILD BALL WITH GOLD LEAF AS
SPECIFIED

TRUCK
BRONZE SHEAVES HOUSING AND PETTICOAT
MANGANESE BRONZE SPINDLE AND BALL
BEARINGS
EITHER STAINLESS STEEL OR MONEL
METAL SHEAVE AXLES

HALYARDS
TWO 3/8" DIAMETER EXTRA LONG FIBER
MANILA BOLT ROPE

Figure 6.362 Detail of revolving truck and ball.

6.24 Standard Symbols for Site Drawings

- P.P.L. — Project Property Line
- EX.P.L. — Existing Property Line
- C.L.L.#12 — Contract Limit Line (Indicate Contract # where differences occur.)
- Limit of Rough Grading, Contract 5
- Future Curb Line
- Established Legal Grade
- 0.00
- Existing Curb Line
- Existing Curb Grade
- Existing Contour / Proposed Contour
- 100 / 101 / 100
- +102.3 — Spot Elevation
- TW 108.5 / BW 104.75 — Top & Bottom of Wall Elevations
- TC 108.6 / BC 108. — Top and Bottom of Curb Elevations
- FL. EL. 113.2 — Floor Elevation
- TF110.5 DB(1) — Top of Frame Elevation / New Drain Basin
- (TO REMAIN) DB(1) — Existing Drain Basin (to remain)
- TF112.3 DI — New Drain Inlet (for project playground only)
- FLAGPOLE — Flagpole
- 12 U — Bench (Arrow indicates direction bench faces) (12 U indicates No. of 3'-3" Units)

- ASPH. — Asphaltic Concrete
- H.D. — Heavy Duty
- CONC. — Concrete Yard Walks
- REINF. CONC. — Reinforced Concrete
- Construction Joint (Panel Marking) / Dummy Joint
- EX.JT. — Expansion Joint
- Granite Block
- BR — Brick Pavement
- Ft. — Flagstone
- ASPH. BL. — Asphaltic Block
- PL. — Planted Areas
- (3) (F) — Raised Curb (No. indicates height) / Flush Curb
- TC 87.2 / BC 86.5 (V) — Variable Height Curb (Use T.C. & B.C. at all changes of height or grade)
- (R1) (R2) + BW 108.5 — Retaining Wall (No. indicates changes in cross-section details)
- TW 113.0 (W) — Free Standing Wall
- Stepped Ramp (with pipe rails)

- Guard Fence in piers
- Guard Fence on Curb
- 3½ — Chain Link Fence on piers (No. designates height)
- 3½ — Chain Link Fence on Curb (No. designates height)
- WIF — Wrought Iron Fence
- Timber Rail Fence
- Fixed Stanchions w/chain
- Fixed Stanchion
- o R. — Removable Stanchion
- 18"MAPLE #21 / 79.6 — Existing Tree (to remain) Designate Caliper, Name Number & Grade Elevation
- 24" #22 — Existing Tree (to be cut) Designate Caliper and Number
- 8"N.MAPLE #23 — Existing Tree (to be moved) Designate Caliper, Name and Number
- *23 / 89.7 — Existing Tree (new location) Designate Number and Grade Elevation
- Tree Pit in lawn area
- Tree Pit in paved area
- Tree in Tree Well
- Tree in Half Well
- Tree with high curb

- Proposed Utility (Solid Line)
- Existing Utility (Broken Line)
- ST — Steam Main
- 24"SS — Storm Sewer
- 8"SAN — Sanitary Sewer
- 6"YD — Yard Drain Line
- 4"W — Water Main
- 3"G — Gas Main
- E — Electrical Conduit
- T — Telephone (Underground)
- P — Police Alarm
- MH — Man Hole
- TV — Transformer Vault
- (S) (Y) — Street Light – Yard Light (Free-Standing)
- (Y) — Bracket Yard Light
- PB — Electrical Pull Box
- New Hydrant
- Existing Hydrant
- SC — Hydrant – Siamese / Sill Cock On Building
- SW — Street Washer

(1) Use CB in project playground

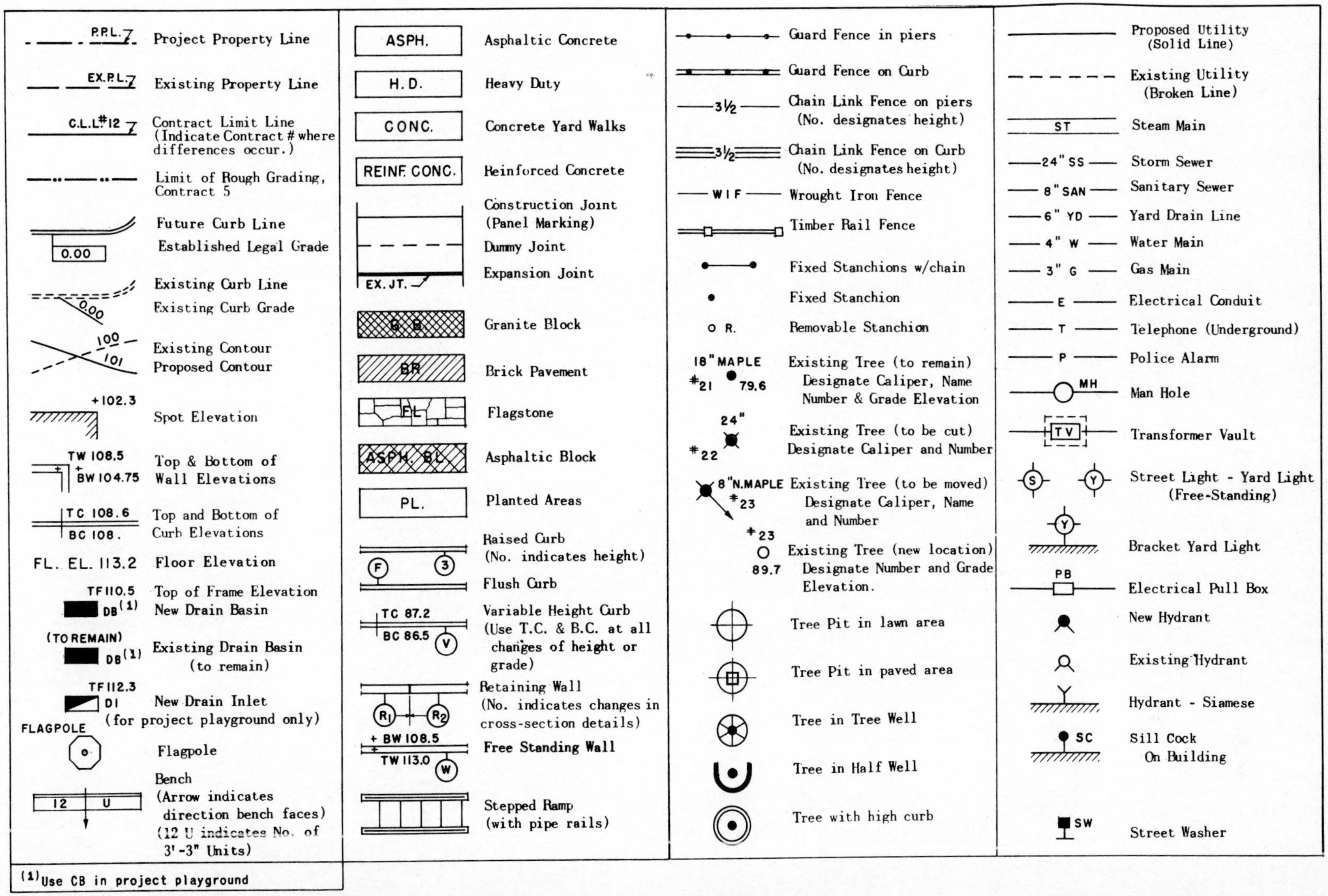

Figure 6.363 Standard symbols for site drawings.

SECTION
SEVEN

Illustrative Site Plans

The site plans presented in this section provide a small sampling of different approaches to the site planning process. In them can be seen an amalgamation of the many and varied aspects of site planning discussed in previous sections.

The town house development illustrates the various steps required to evolve properly a final design solution. It involves a typical low-density development in a suburban area.

The planned unit development shows a similar type of analysis and solution in a more heavily built-up area. Also, it illustrates various alternative approaches prior to its final scheme and even delves into typical housing types as part of the site planning process.

In addition to residential development, a school complex and a beach park development are included. Both again demonstrate the different approaches and solutions to the specific problems involved.

As previously stated, these site plans are presented not as definitive site designs but as guides to the site designer for greater awareness and understanding of the problems and their subsequent solution.

7.1 Town House Development Process

Site Analysis and Development Program   The phase of site analysis should identify the character, structure, and potential of the site. In discovering these characteristics and relying upon them to inspire proper land use, the analyst should consider and record the items listed below. The analysis should be done on a site topographic map at a scale no smaller than 1 inch = 100 feet. The map should cover not only the site but also surrounding areas that may influence site uses and design. See Figure 7.1.

1. Contiguous land use
   Indicate type and impact of adjoining land use.
   Indicate direction and distance to community services, hospitals, shopping, and so on.
   Show public transportation route and stops.

2. Topography
   Basic topography.
   Special or unique ground forms.
   Percentage of slope.

3. Drainage
   Natural watershed (direction).
   Drainage swales.
   Bog and swamp areas.

4. Soils
   Show depth and analysis of topsoil.
   Locate soil borings and give data (this can be presented in a separate report).

5. Vegetation
   Locate and identify existing tree masses.
   Locate and identify specimen plant material.
   Indicate type of ground cover.

6. Climatology
   Prevailing wind direction.
   Sun angles.

7. Existing conditions
   Structures.
   Utilities.
   Circulation.

8. Special features
   Lakes and ponds.
   Special land features, rock outcroppings, and so on.
   Dramatic views.

In addition to the site analysis and conceptual plan, the first phase must include a written program statement for the project. This statement provides the guidelines for development of the project and, in conjunction with the site analysis, the basis for the conceptual plan. The program should include the following:

1. Budget
   Project-design budget allocated for on- and off-site improvements, residential and accessory buildings, and so on. This budget will guide and control the development of plans.

2. Time schedule
   Target time periods for completion of subsequent phases.
   Anticipated construction starting date.

3. Dwelling units
   Type of ownership or tenancy (rental, cooperative, condominium).
   Total number of units anticipated.
   Allowable site density.
   Types of dwelling units (differentiated by number of bedrooms, floor areas, and configurations) and distribution of total number of units among the various types.
   Special requirements, conditions, or features.

4. Community facilities
   Methods and requirements of project marketing, management, and maintenance.
   Management and maintenance spaces and facilities anticipated.
   Indoor and outdoor community recreation and social spaces and facilities anticipated.

5. Nonresidential facilities
   Nonresidential facilities anticipated.

Conceptual Plan   The phase of the conceptual plan should relate in a logical manner the needs in the development program to the physical site structure as clarified by the site analysis. The conceptual plan should indicate the general building masses, circulation (vehicular and pedestrian), parking, open space, and special facilities. In general, it should convey the general intent, spatial form, and system of development. It should be prepared on a site topography base map at a scale no smaller than 1 inch = 100 feet. See Figure 7.2.

1. Structures
   Indicate location, arrangement, and general massing groupings.
   Locate any recreation or service structures.

2. Circulation

   a. Vehicular
      Show system of roads, parking, and service.

   b. Pedestrian
      Indicate general walkway system and connection to common facilities.

3. Utilities
   Indicate major trunk lines and connection points to existing utilities.

4. Recreation
   Show open space and facilities for recreational use.
   Indicate parking and service for common facilities.

5. Parking
   Location.

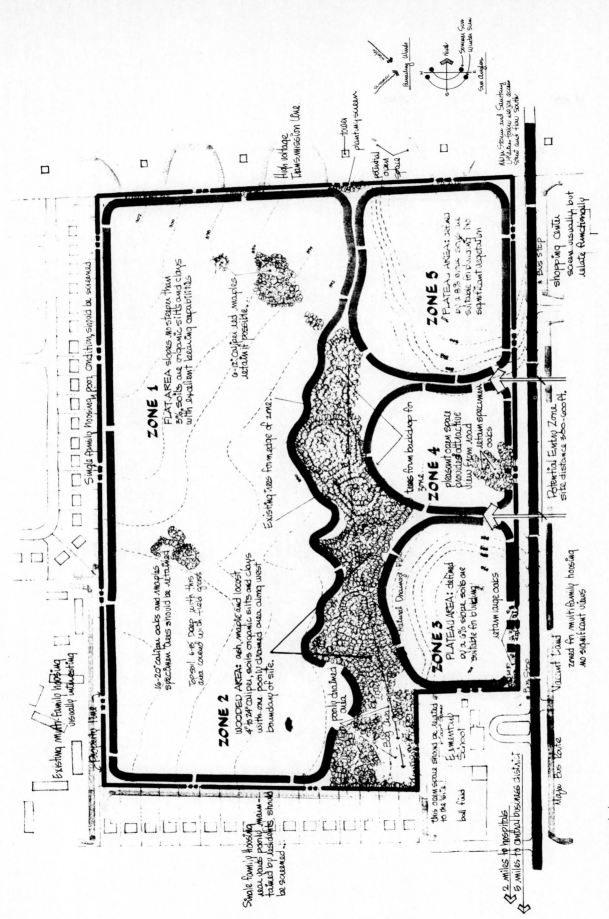

Figure 7.1 Site analysis.

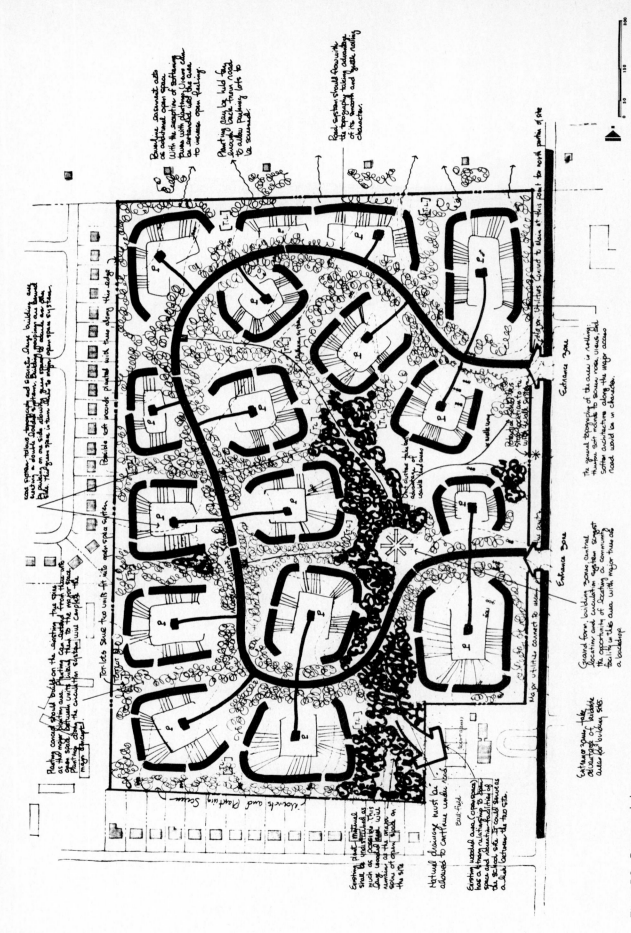

Figure 7.2 Conceptual plan.

6. Grading
General character.
Mounds and berms.
Any special grading problems.

7. Planting
Consider existing vegetation in the development of the
concept.
Indicate the general planting concept.

Schematic Site Plan Schematics should be developed
from careful study and observation of the conceptual plan.
The plan should refine to a more exact scale the arrangement
and functional groupings of units to create a meaningful
sequence of usable spaces. The specific relationship of unit
arrangement, the relationship of structure to site, site grading, circulation, lighting, paving, screening, setbacks, parking,
play areas, and recreation areas should be communicated in
this phase. This plan should be produced at a scale of 1 inch
= 50 feet where practicable. In addition to the required plan,
additional sections, sketches, and study models to convey
the intent should be made. See Figures 7.3 and 7.4.

1. Structures
Location, shape, size, arrangements, and groupings.

2. Circulation
Indicate location and materials of vehicular and pedestrian
routes.
Indicate parking-dwelling unit relationships.

3. Utilities
Indicate general major utility layout and connections. This
can be done as an overlay.

4. Recreation
Location.
Type of facilities.

5. Parking
Location, material, number of spaces, and parking ratio.

6. Grading
Resolve special and typical grade relationships.
Indicate general grading character, proposed contours,
sections, and so on.
Show berms and mounds.

7. Planting
Indicate character.
Indicate screening concepts, planting relationship to units,
open space, and so on with sections or sketches.

8. Lighting
Location.
Character (this can be shown with a catalog illustration).

Preliminary Site Plan This plan should be a detailed document of the schematic plan, refined to the point at which
accurate bids can be received from bidders. It should include
all elements of the site with necessary drawings and details
to convey the intent. The preliminary plan should be prepared at a scale of not less than 1 inch = 50 feet. See Figure
7.5.

1. Structures
Final location to exact scale.

2. Circulation
Location, key dimensions, and materials of roads and
walks.

3. Utility (separate drawing)
Show all utility lines and other construction with grades.
Locate all storm inlets, catch basins, fire hydrants, and so
on, and give grades.

4. Recreation
Locate open space or special use areas.
Indicate play areas and equipment.
Show equipment and facilities.

5. Grading
Indicate spot grades, proposed and existing contours,
berms, and mounds.
Show floor grades of all structures.

6. Planting (separate drawing)
Locate plant material.
Indicate areas to receive seed or sod.
Indicate plant list, showing quantity, size, and root
specification.

7. Legend
Units, density, parking, acreage and similar information.
Symbol key relating to drawing.

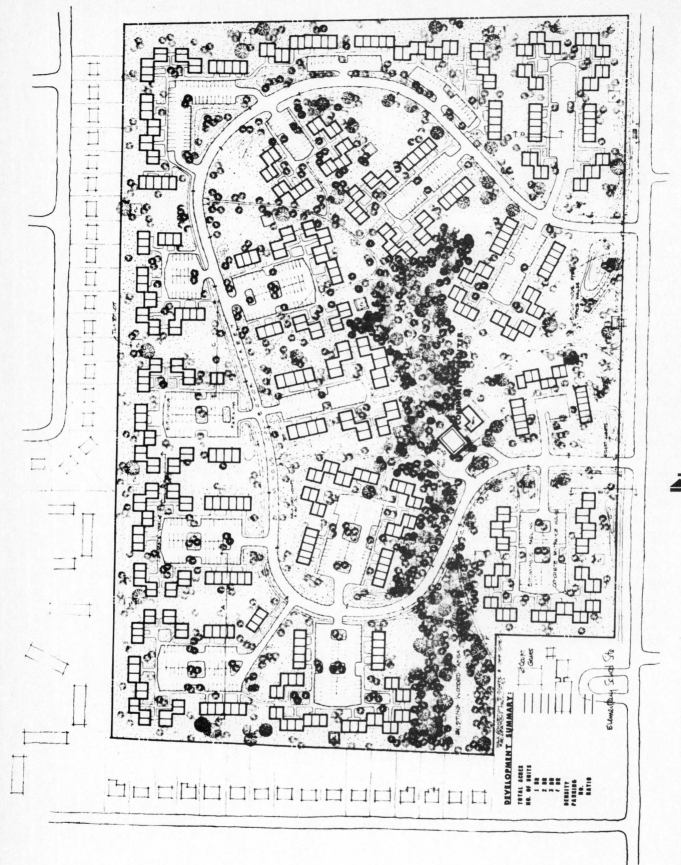

Figure 7.3 Schematic site plan.

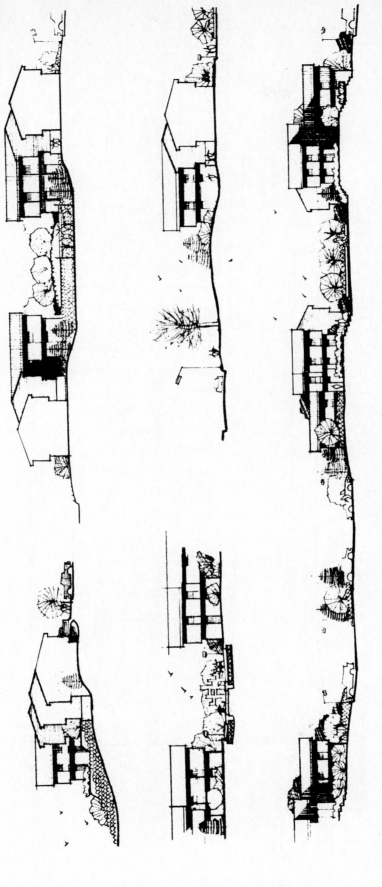

SCHEMATIC SECTIONS

Figure 7.4 Schematic sections.

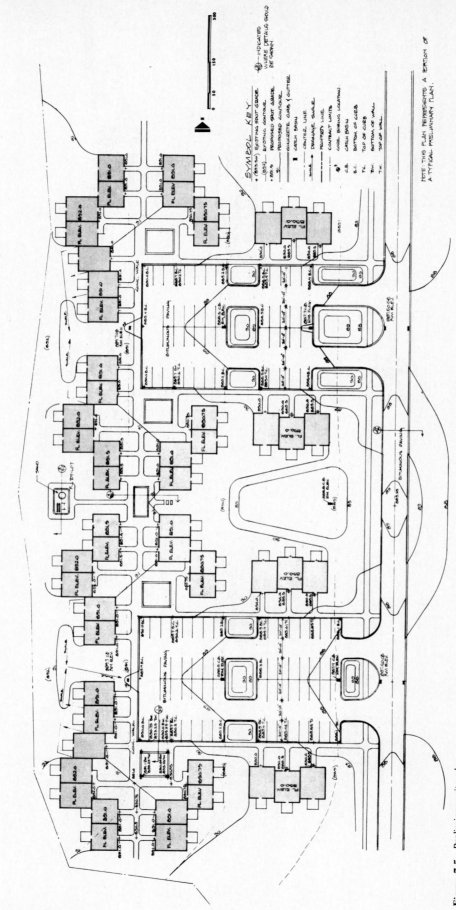

Figure 7.5 Preliminary site plan.

7.2 Planned Unit Development

Comparative Advantages An actual site has been selected to illustrate the full range of site planning opportunities possible under the planned unit development amendment. The following pages show a drawing accentuating the site topography (Figure 7.7), a conventional proposal for development of the site (Figure 7.19), and four alternatives for development under planned unit development (Figures 7.12, 7.16, 7.17, and 7.18). The first two alternatives show developments of about the same density as the conventional proposal. The third alternative shows a scheme of somewhat higher density that could accommodate an increased demand for apartment units. The fourth alternative shows how a maximum number of units could be developed by using apartment houses with an urban scale but within the permissible density. A comparison of the advantages of planned unit development schemes is shown in Figure 7.6.

Contour Map The contour map shown in Figure 7.7 accentuates the topography and water resources of the site. The high ground to the south pitches sharply and irregularly down to rather flat ground in the north.

Typical Site Development Figure 7.8 shows a site plan of existing features, contours, including trees, ponds, and creek, and bounding roads. Figure 7.9 shows site sections, taken where indicated on the plan, with the vertical scale exaggerated by 2:1.

Figure 7.10 is a preliminary street and drainage plan showing the alignment of streets and the direction of flow of storm and sanitary sewer lines. Figure 7.11 is an architectural site plan showing streets, parking, curb cuts, pedestrian ways, the placement of buildings, shopping center, and school, and open space treatment.

Town House Clusters on Collector Streets The scheme of town house clusters in Figure 7.12 shows 1431 homes on the same site. It is characterized by a horseshoe-loop street system that eliminates through traffic, uses a minimum amount of land for streets, furnishes an enormous common open area in addition to the open spaces defined by individual clusters, and provides shopping and a school, accessible from the cul-de-sac street that enters from the eastern boundary. The strongest natural features of the site are preserved as a recreational and scenic amenity.

Figure 7.13 is an axonometric projection of one loop that indicates the three-dimensional character of the scheme of town house clusters.

Detached Houses and Town Houses Figure 7.14 shows plans of the types of the detached houses and town houses to be built in the development, indicating the relationships of yards to house interiors.

Garden Apartments and Detached-Attached Houses Figure 7.15 shows plans of the types of garden apartments and detached-attached houses to be built in the development, indicating the relationships of yards to house interiors.

Varied House Types on Loop Streets The scheme shown in Figure 7.16 includes a full range of house types; detached houses, semidetached houses, town houses, and town house apartments, totaling 1445 dwelling units. This mix of houses is organized around a P-loop street system in which each loop serves only the houses on it. Within each loop a sizable common open space with recreational facilities is formed, and all the loops together enclose a large parkland running along the ridge and into the center of the site. Again, school and shopping facilities are provided, and the natural beauty of the site is respected.

House Clusters and Apartment Houses The scheme shown in Figure 7.17 anticipates changing market conditions during the course of a development time span. It assumes an initial development similar to the scheme in Figure 7.16 in the northern, lower portion of the site and then a transformation of the market into apartment units. Three- and four-story maisonette apartments perch on the crest of the ridge, backed up by six-story terraced apartment buildings. Rising out of the expanded shopping center at the south center of the site is a square apartment tower of twenty stories. A total of 2800 dwelling units is thus accommodated on the site. In addition to the surface parking shown, a sizable amount of parking is contained underneath the terraced apartments and shopping center. More open space is provided than in Figures 7.12 and 7.16, in recognition of the increased density.

Apartments for Maximum Open Space The scheme in Figure 7.18 represents an entirely urban, apartment-type residential environment in which a maximum number of dwelling units is realized. The apartment types consist of three- and four-story town house and maisonette apartments, five-, six-, and seven-story terraced apartments, and major apartment structures stepping up from four stories to as many as twenty-eight or thirty stories, for an overall total of about 4500 dwelling units. The center of the site is the focus of the development, with extensive shopping facilities at the first-floor level of the apartment structures. The terraced structure at the south end of the center houses an elementary school. Around the periphery of the site are extensive recreational facilities as well as a considerable amount of land left in its natural state.

Conventional Street Grid The conventional scheme for the development of the site, illustrated in Figure 7.19, shows 1427 single-family houses on about 205 acres. Houses are placed according to zoning lot requirements on a street system previously adopted by the city. The result is a disastrously barren environment, and all the natural character of the site has been destroyed.

Table shows land use and site utilization advantages of Planned Unit Development schemes (numbered 2, 3, and 4) over conventional subdivision scheme, 1.

	GROSS SITE AREA	STREET AREA	STREET AREA % OF GROSS SITE AREA	NET SITE AREA	COMMON OPEN SPACE	NUMBER OF DWELLING UNITS	ALLOWABLE FLOOR AREA PER DWELLING UNIT	ALLOWABLE COVERAGE PER DWELLING UNIT	ALLOWABLE NUMBER OF ROOMS PER DWELLING UNIT	
1 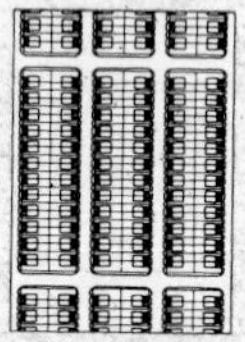	20 ACRES	6.3 ACRES	31.4%	13.7 ACRES	NONE	semi-det: 198	1400 sq. ft.	700 sq. ft.	7.5	Figures based on: typical zoning lot of 2800 sq. ft. F.A.R. (Floor Area Ratio) of .5 O.S.R. (Open Space Ratio) of 150 lot area per room of 375 sq. ft.
2 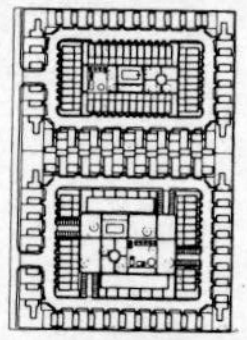	20 ACRES	5.6 ACRES	28%	14.4 ACRES	2.3 ACRES	detached: 59 semi-det: 23 townhouses: 62 garden apts: 56 total: 200	1840 sq. ft.	940 sq. ft.	9.5	Figures based on: net site area divided by number of dwelling units, and application of full bonuses resulting in: F.A.R. (Floor Area Ratio) of .575 O.S.R. (Open Space Ratio) of 120 lot area per room of 337 sq. ft.
3 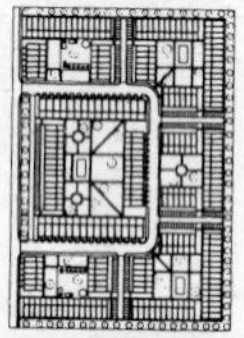	20 ACRES	4.1 ACRES	20.5%	15.9 ACRES	8.6 ACRES	townhouses: 213	1900 sq. ft.	980 sq. ft.	9.8	
4 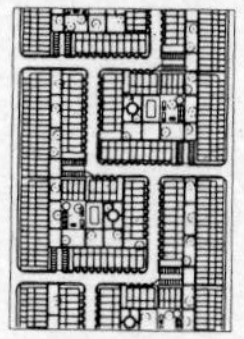	20 ACRES	5 ACRES	25%	15.0 ACRES	4.0 ACRES	townhouses: 210	1820 sq. ft.	975 sq. ft.	9.35	

Figure 7.6 Table of comparative advantages.

Figure 7.7   Contour map.

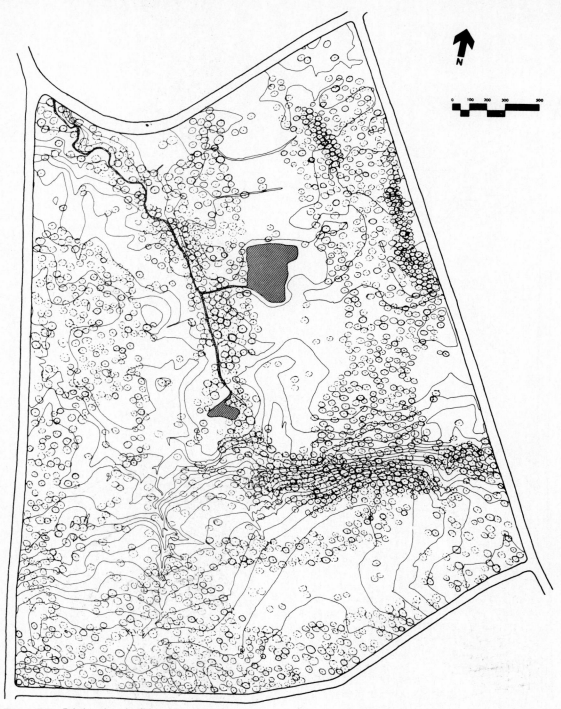

Figure 7.8 Existing site conditions.

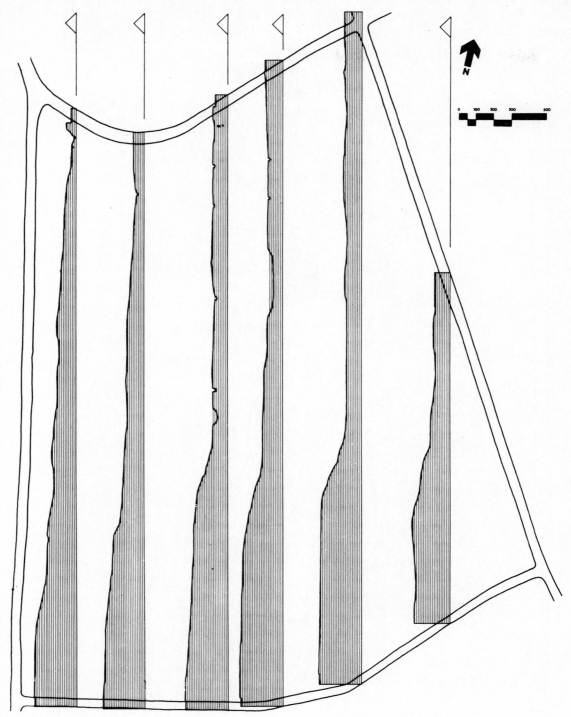

Figure 7.9 Site sections.

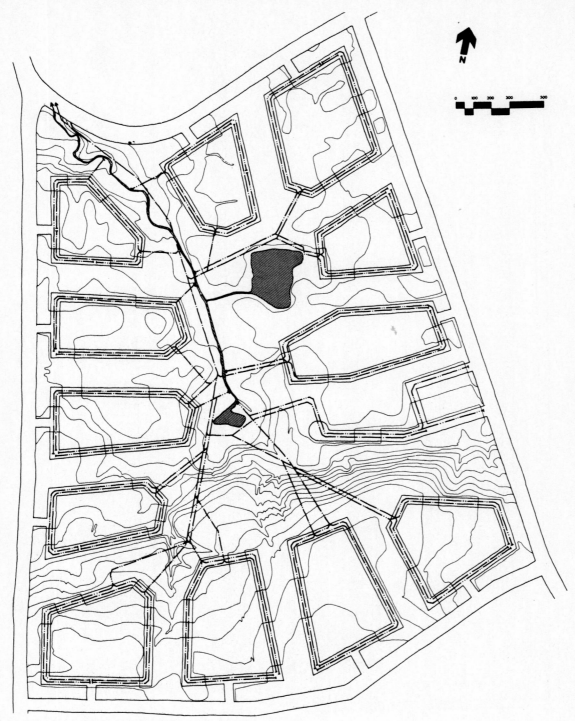

Figure 7.10   Preliminary street and drainage plan.

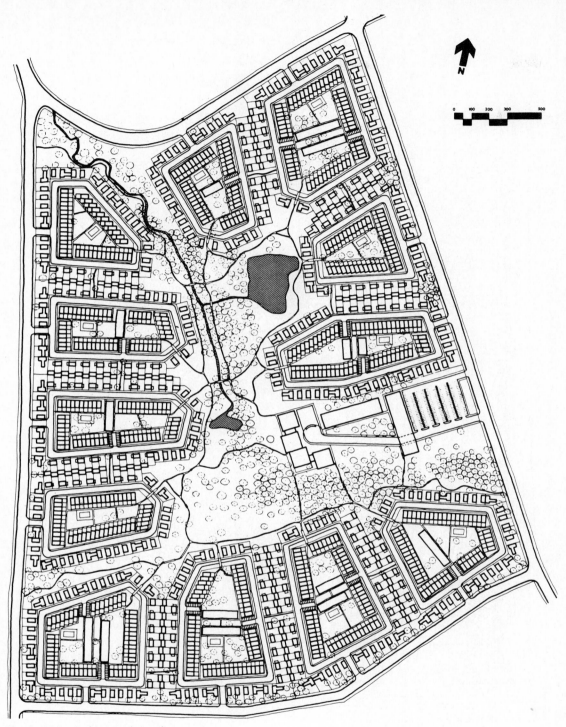

Figure 7.11 Architectural site plan.

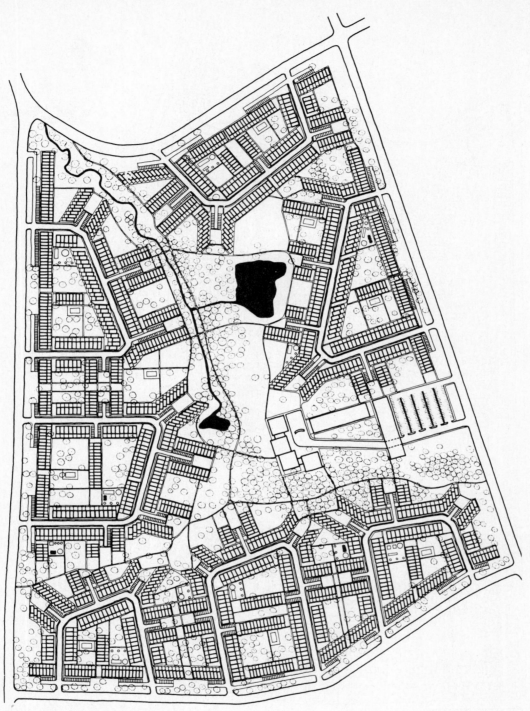

Figure 7.12 Town house clusters.

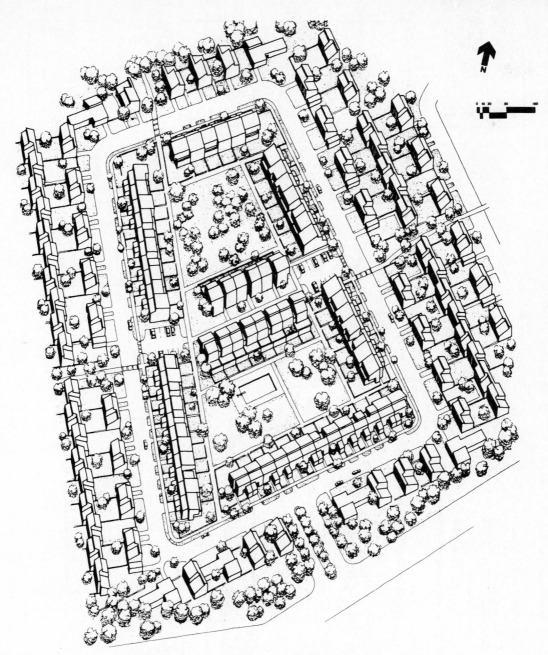

Figure 7.13 Axonometric projection of a loop of the town house scheme.

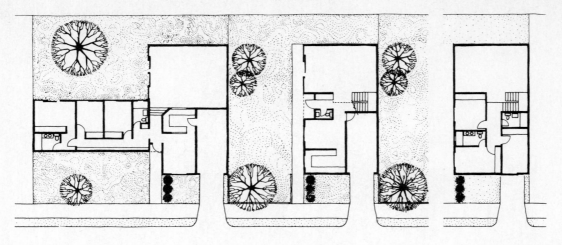

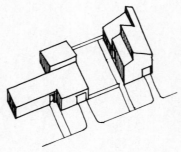

Detached houses

Two detached house plans, one showing a one story, three bedroom house, the other showing a two story, three bedroom house with a "cathedral" ceiling over the living room.

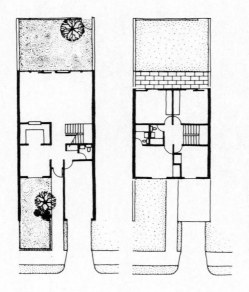

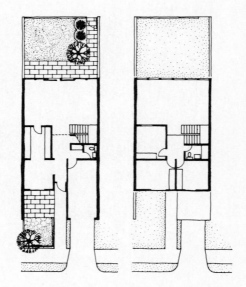

Townhouses

Two alternative townhouse plans, one showing a four-bedroom scheme in which two of the bedrooms open out onto a second story deck, the other showing a three bedroom scheme which permits a "cathedral" ceiling over the living room.

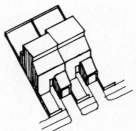

Figure 7.14

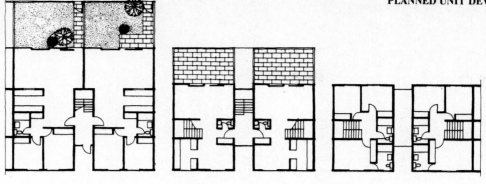

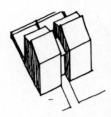

Garden apartments

Apartment plans, showing one story, two bedroom units at the ground floor and three bedroom duplexes at the second and third stories.

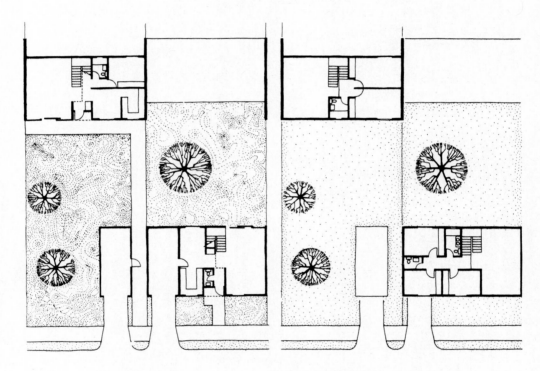

Detached-attached houses

Front house is a three bedroom unit with a "cathedral" ceiling over the living room and a large rear yard containing nearly all the open space on the lot. Back house, also a three-bedroom unit with a "cathedral" ceiling over the living room, has its yard in front.

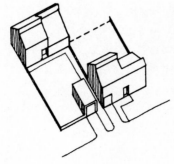

Figure 7.15

–787

Figure 7.16 Varied house types on loop streets.

Figure 7.17 House clusters and apartment houses.

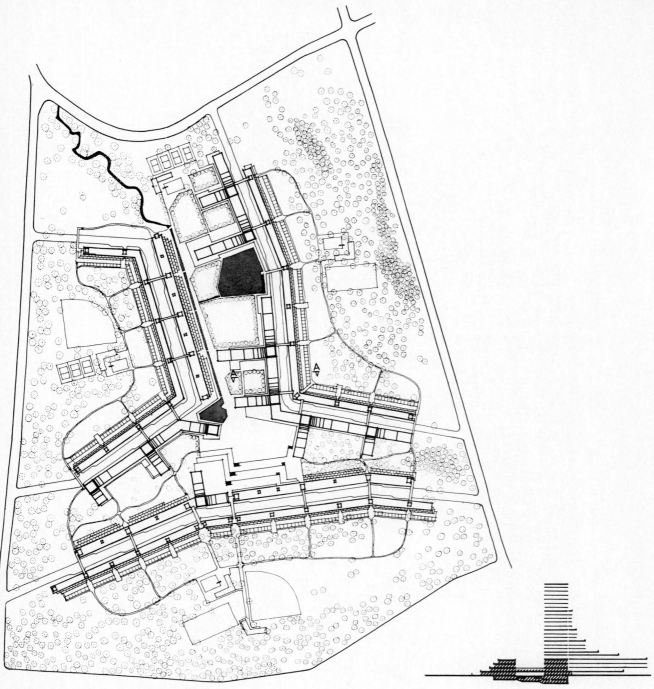

Figure 7.18 Apartments for maximum open space.

Figure 7.19 A conventional street grid.

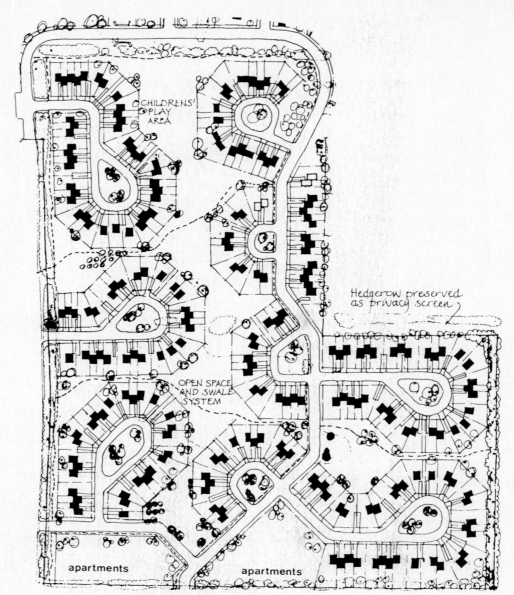

Figure 7.20 Maryfield, Charlottetown, Prince Edward Island.

7.3 School-Park Complex

Site Analysis   Figure 7.21 presents an analysis of a site to
be used for a school-park complex.

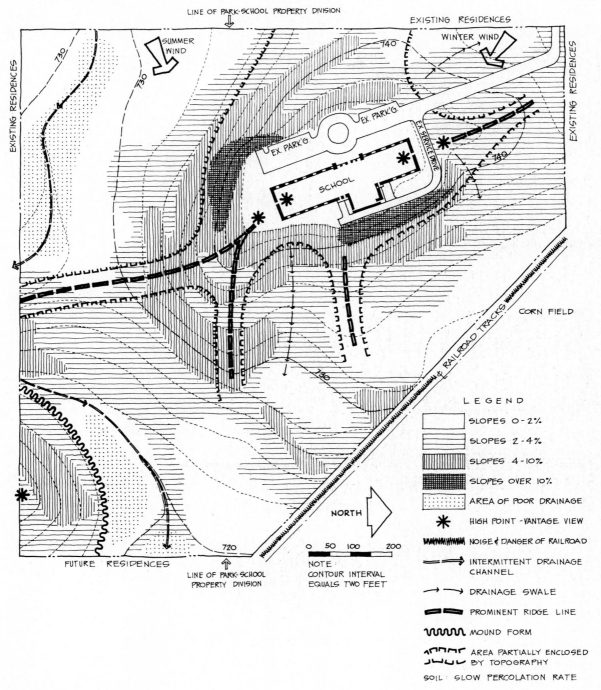

LINE OF PARK-SCHOOL PROPERTY DIVISION

EXISTING RESIDENCES

SUMMER WIND

WINTER WIND

EXISTING RESIDENCES

EXISTING RESIDENCES

EX. PARK'G

EX. PARK'G

EX. SERVICE DRIVE

SCHOOL

CORN FIELD

RAILROAD TRACKS

NORTH

0   50   100   200

FUTURE RESIDENCES

LINE OF PARK-SCHOOL
PROPERTY DIVISION

NOTE:
CONTOUR INTERVAL
EQUALS TWO FEET

LEGEND

SLOPES 0-2%
SLOPES 2-4%
SLOPES 4-10%
SLOPES OVER 10%
AREA OF POOR DRAINAGE
HIGH POINT - VANTAGE VIEW
NOISE & DANGER OF RAILROAD
INTERMITTENT DRAINAGE CHANNEL
DRAINAGE SWALE
PROMINENT RIDGE LINE
MOUND FORM
AREA PARTIALLY ENCLOSED BY TOPOGRAPHY
SOIL: SLOW PERCOLATION RATE

Figure 7.21   Site analysis.

Design Concept—Scheme 1 A design concept for a scheme for development of the school-park complex is shown in Figure 7.22.

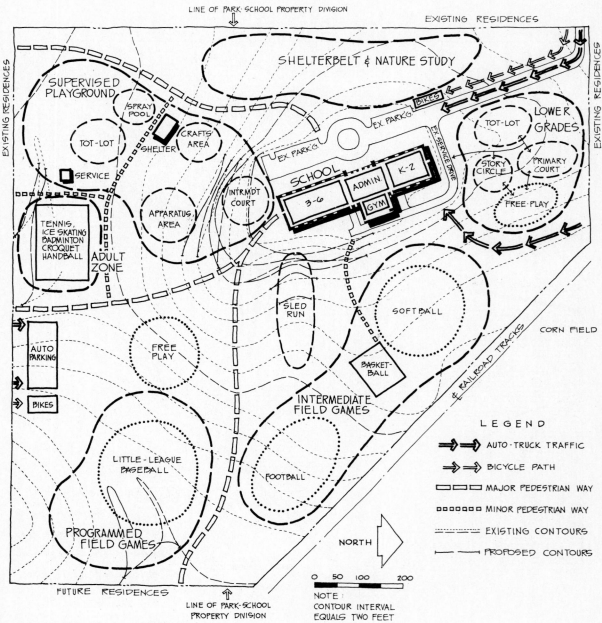

Figure 7.22 Design concept – scheme 1.

Site Plan—Scheme 1 Figure 7.23 shows a plan for developing the site under the scheme in the design concept of Figure 7.22.

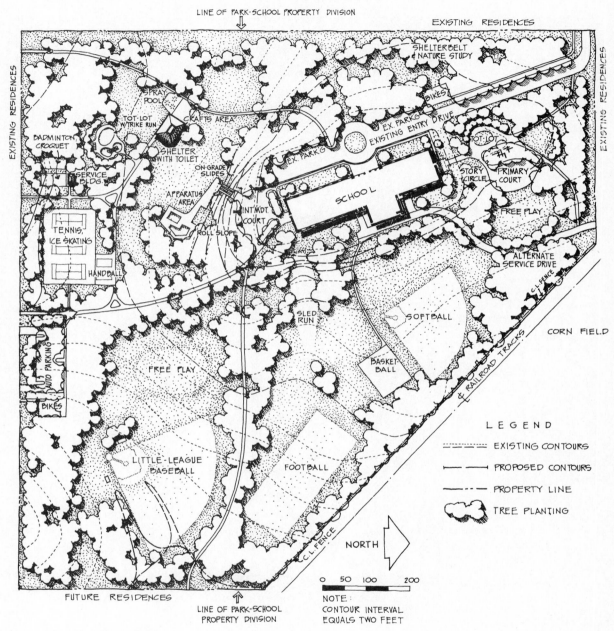

Figure 7.23 Site plan — scheme 1.

Site Plan — Scheme 2   The site plan of a second scheme is shown in Figure 7.24.

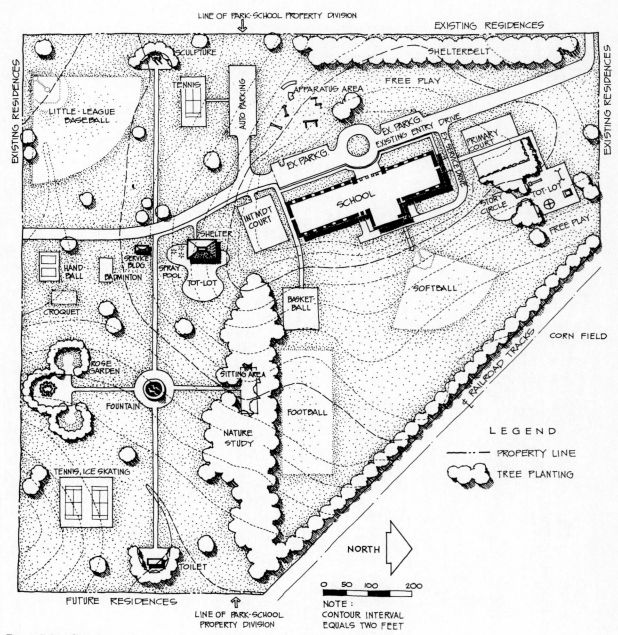

Figure 7.24   Site plan — scheme 2.

7.4 Beach-Park Development

A comprehensive plan for a beach-park development is shown in Figure 7.25, and details of the plan in Figures 7.26 and 7.27.

Figure 7.28 shows further plans and details of the beach-park development. Figure 7.29 shows the plan and details for the bathhouse, and Figure 7.30 the plan and details for the beach pavilion.

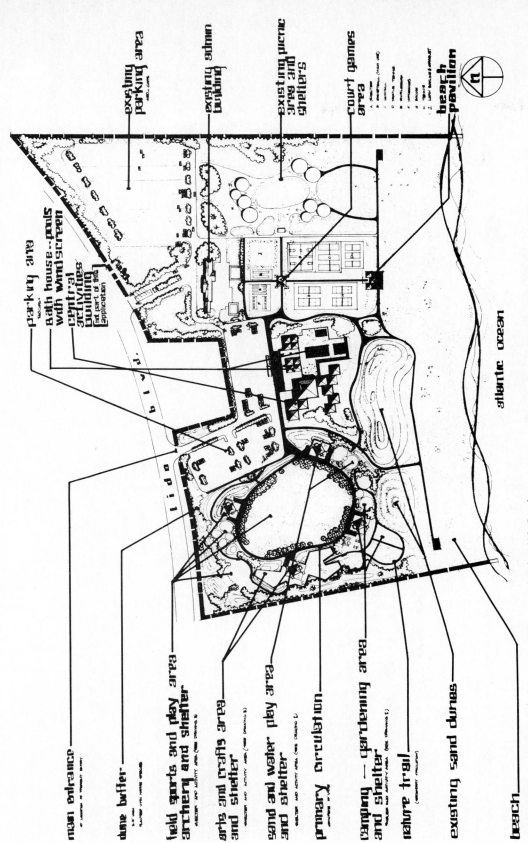

existing parking area

existing admin building

existing picnic area and shelters

court games area

beach pavilion

parking area

bath house—pools with wind screen

central activities building [not part of this application]

main entrance

dune buffer

field sports and play area archery and shelter

arts and crafts area and shelter

sand and water play area and shelter

primary circulation

camping—gardening area and shelter

nature trail

existing sand dunes

beach

atlantic ocean

Figure 7.25   Comprehensive plan, town of Hempstead Park at Lido Beach.

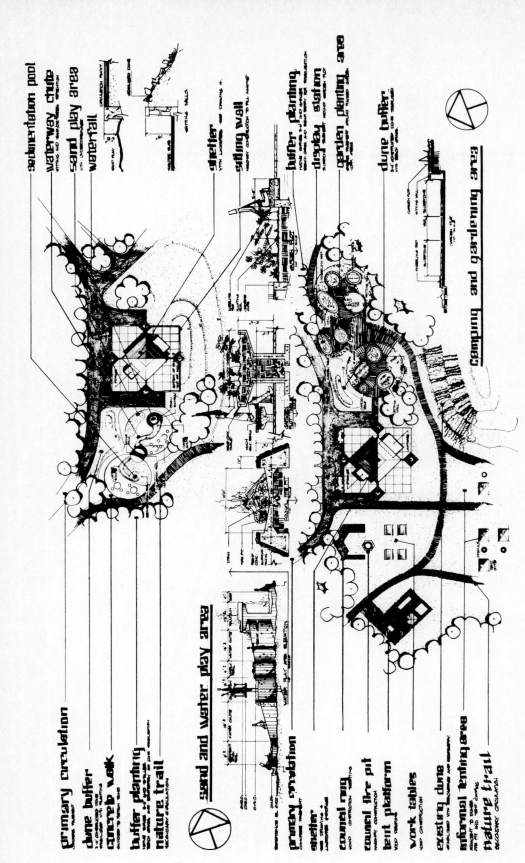

sedimentation pool
waterway chute
sand play area
waterfall

shelter
sitting wall

buffer planting
display station
garden planting area

dune buffer

camping and gardening area

primary circulation
dune buffer
concrete walk
buffer planting
nature trail

sand and water play area

primary circulation
shelter
council ring
council fire pit
tent platform
work tables
existing dune
informal tenting area
nature trail

Figure 7.26 Details of the beach-park plan.

-798-

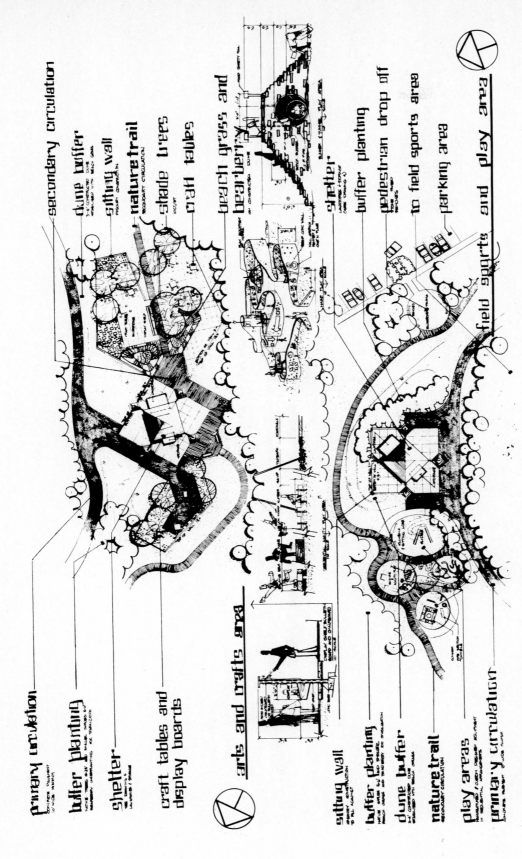

primary circulation

buffer planting

shelter

craft tables and
display boards

arts and crafts area

sitting wall

buffer planting

dune buffer

nature trail

play areas

primary circulation

secondary circulation

dune buffer

sitting wall

nature trail

shade trees

craft tables

beach grass and
bearberry

shelter

buffer planting

pedestrian drop off

to field sports area

parking area

field sports and play area

Figure 7.27 Details of the beach-park plan.

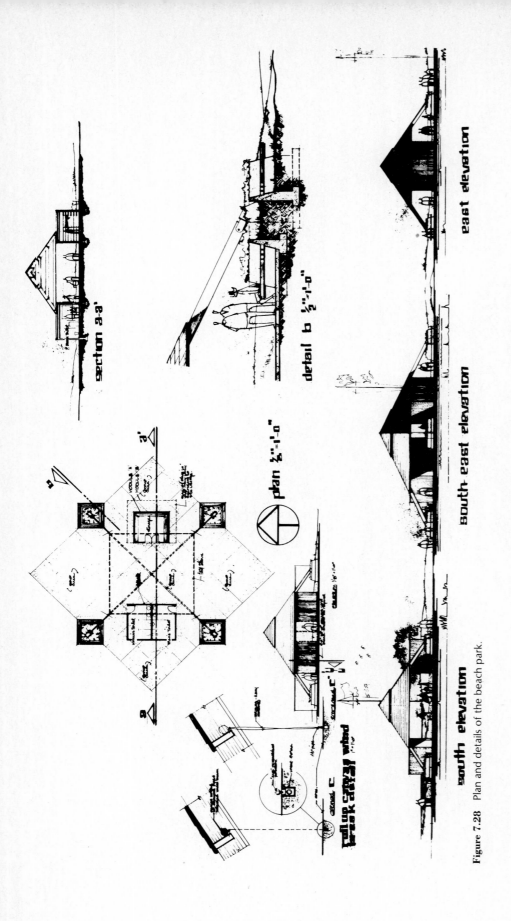

Figure 7.28 Plan and details of the beach park.

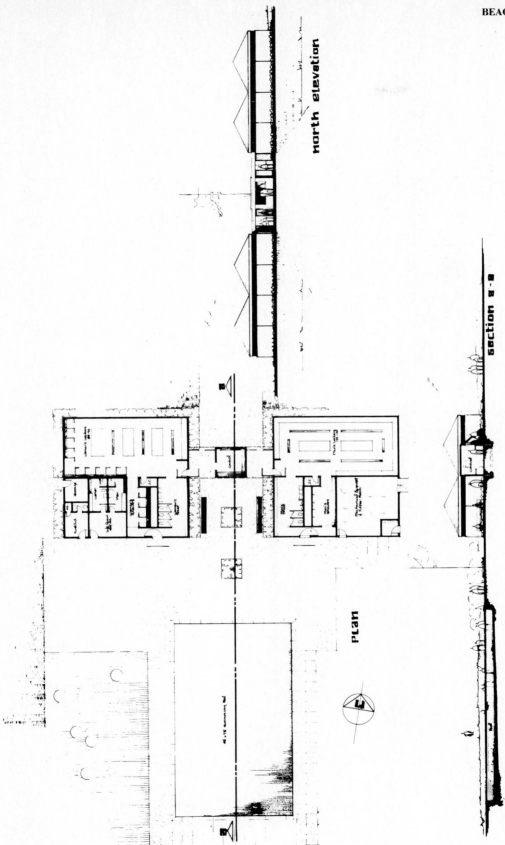

Figure 7.29 Plan and details of the bathhouse.

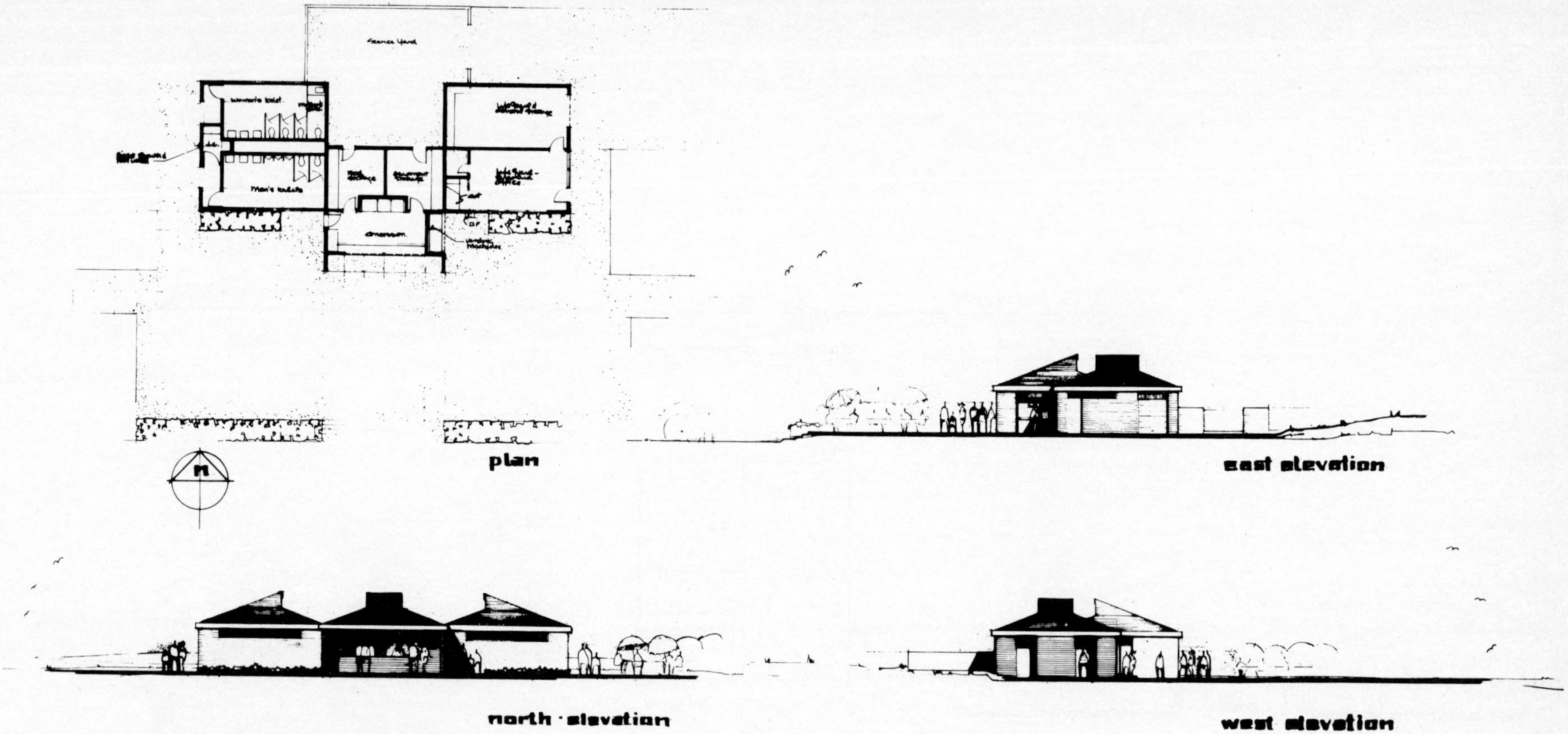

Figure 7.30 Plan and details of the beach pavilion.

7.5 Mobile Home Park

Figure 7.31 shows a plan for a typical mobile home park.

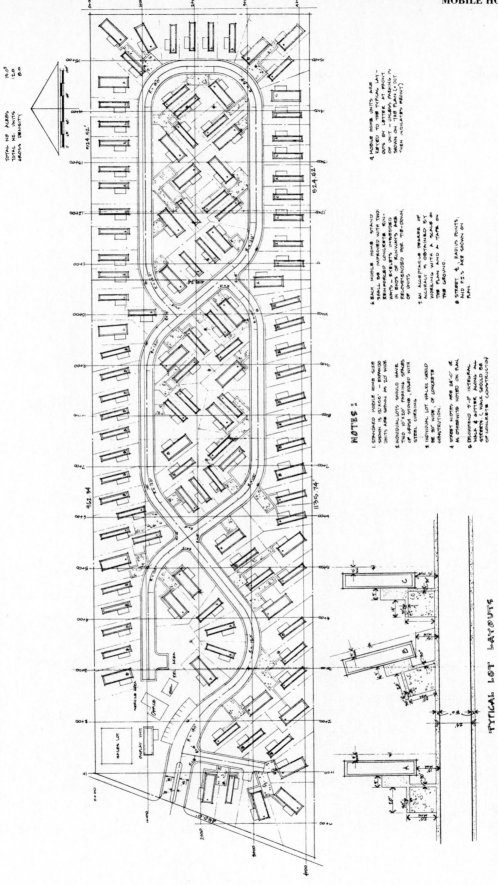

Figure 7.31 A mobile home park.

7.6   Schematic Plan for Recreational Facilities

Planning for recreational facilities is illustrated in Figure 7.32.

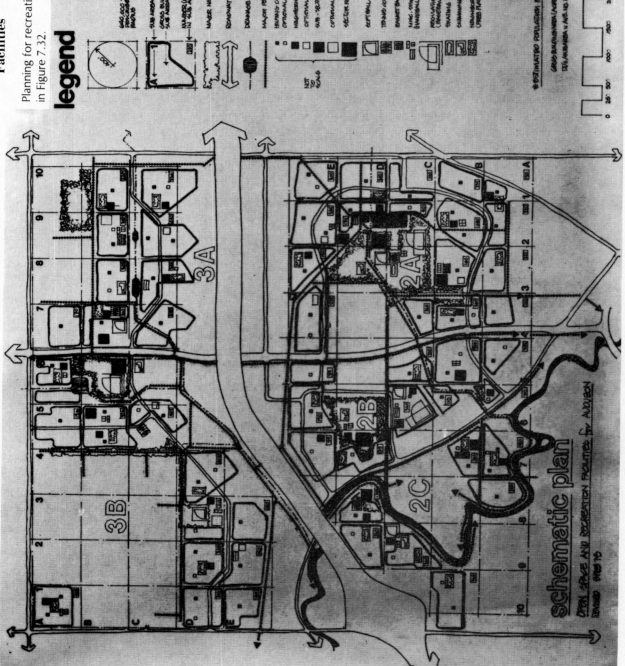

Figure 7.32   A schematic plan for recreational facilities.

7.7 Site Planning for Solar Energy Utilization

The building site is an extremely important solar design consideration. Together with the solar design factors, the conditions and charactertistics of the building site will influence both solar dwelling and system design. Existing vegetation geology, topography, and climate are the primary site characteristics considered during site planning and design. These factors will influence not only the design of dwellings incorporating solar heating and cooling systems but also the layout and organization of groups of neighborhoods of solar dwellings.

Every building site will have a unique combination of site conditions. As a result, the same solar dwelling placed on various sites will generally require completely different site planning and design decisions. Therefore, the site for a solar heated or cooled dwelling should be selected with care and modified as necessary to maximize the collection of solar energy and to minimize the dwelling's need for energy.

The Site Planning Process   Site planning is concerned with applying an objective analysis and design process to specific site-related problems at increasingly smaller scales. While the building site and the dwelling design may vary significantly from one project to another, the process of site planning is replicable and easily adapted to the requirements of most projects. In the case of solar dwelling design, the process is altered to include design criteria related specifically to the use of solar heating and cooling systems.

Site planning for the utilization of solar energy is concerned with two major issues: (1) access to the sun and (2) location of the building on the site to reduce its energy requirement. The placement and integration of the solar dwelling on the site in response to these concerns entails numerous decisions made at a variety of scales. The process may commence at a regional climatic and geographic scale and terminate at a specific location on the building site. At every scale, decisions regarding site selection, building orientation, and placement, and site planning and design are made.

Site Selection   At times, a builder, developer, or designer may have the option of selecting a site or of determining the precise location on a larger site for the placement of the solar dwelling or dwellings. In such instances, the best site for effective solar energy utilization should be chosen by analyzing and evaluating carefully all of the following factors.

Geography of the area surrounding the site:

- The daily and seasonal path of the sun across the site
- The daily and seasonal windflow patterns around or through the site
- The presence of earthforms which may block the sun or wind
- The presence of low areas where cold air could settle

Topography of site:

- Steepness of the slope — can it be built upon economically?
- The presence of slopes beneficial or detrimental to energy conservation and solar energy utilization

Orientation of slopes on the site:

- South-facing slopes for maximum solar exposure
- West-facing slopes for maximum afternoon solar exposure
- East-facing slopes for maximum morning solar exposure
- North-facing slopes for minimum solar exposure

Geology underlying the site:

- Depth and type of rock on the site
- Unbuildable areas on the site

Existing soil potential and constraints:

- Soils with engineering limitations unable to support structures
- Soils with agricultural limitations, unable to support vegetation

Existing vegetation:

- Size, variety, and location of vegetation which would impair solar collection
- Building sites which would disturb existing vegetation to a minimum
- Size, variety, and location of vegetation which would assist in energy conservation

Climatically protected areas on the site:

- Areas protected at certain times of the day or year
- Areas protected by topography
- Areas protected by vegetation

Climatically exposed locations on the site:

- Areas exposed to sun or wind
- Areas exposed primarily in winter
- Areas exposed primarily in summer
- Areas exposed all seasons of the year

Natural access routes to and through the sites:

- Adjacent streets for vehicular access to the site
- Adjacent walkways for pedestrian access to the site

Solar radiation patterns on the site:

- Daily and monthly
- Seasonal
- Impediments (e.g., vegetation that may cover the site or shadow buildable areas on the site)

Wind patterns on the site:

- Daily and monthly
- Seasonal
- Impediments (e.g., thick vegetation or underbrush that may block air movement on or through the site)

Precipitation patterns on the site:

- Fog movement, collection or propensity patterns
- Snow drift and collection patterns
- Frost "pockets"

Temperature patterns on the site:

- Daily and monthly
- Seasonal
- Warm areas
- Cold areas

Water or air drainage patterns on or across the site:

- Seasonal air or water flow patterns
- Daily air or water flow patterns
- Existing or natural impediments to air or water flow patterns

Tools for site analysis include air photos, topographic maps, climatic charts, or direct observations on the site. Site selection at whatever scale must take into account the distinctive characteristics of the major climatic regions of the United States mentioned earlier. Once the data are collected and organized, they can be used to evaluate, rate, and eventually select a specific location or site for the placement of the dwelling, solar system, and other site-related activities.

A simplified example of the site analysis process for determining preferred locations for solar dwellings in western temperate climates is shown in the following illustrations.

Altitude and Scope The topography is analyzed in both plan and cross section to locate buildable areas on upper and middle slopes (Figure 7.33).

Orientation and Winds The site is next assessed for areas oriented in a southerly direction for maximum solar exposure. Also, the prevailing and storm winds which move regularly or occasionally across the site are plotted (Figure 7.34).

Vegetation and Moisture Existing vegetation and moisture patterns on the site are related to their potential for assistance in the creation of sun pockets and for providing wind protection. The density and type of vegetation are ana-

lyzed and graphically depicted in order to gain an understanding of the patterns of shade or protection and air or moisture flow (Figure 7.35).

Composite Showing Preferred Sites A composite is prepared from the preceding factors showing a ranking or a rating of the preferred sites for placement of a solar dwelling (Figure 7.36, a being best, b next best, and so on).

Siting and Orientation Optimum solar energy utilization is achieved by the proper placement and integration of the dwelling, solar collectors, and other site-related activities and elements on the building site.

In addition to the dwelling, the most common activity areas found on residential sites include:

- Means of access (entrances to the site and the dwelling)
- Means of service (service and storage areas)
- Areas for outdoor living (patios, terraces, etc.)
- Areas for outdoor recreation (play areas, pools, courts, etc.)

On sites where the dwelling(s) will be heated or cooled by solar energy, additional site planning factors must be considered for accommodating solar collection — by either dwelling or on-site collectors.

Each of the four major climatic regions in the United States has different siting and orientation considerations. The following is an overview of the major determinants for each region.

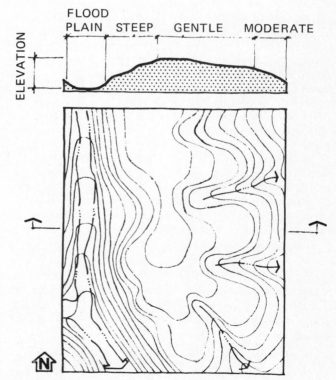

Figure 7.33

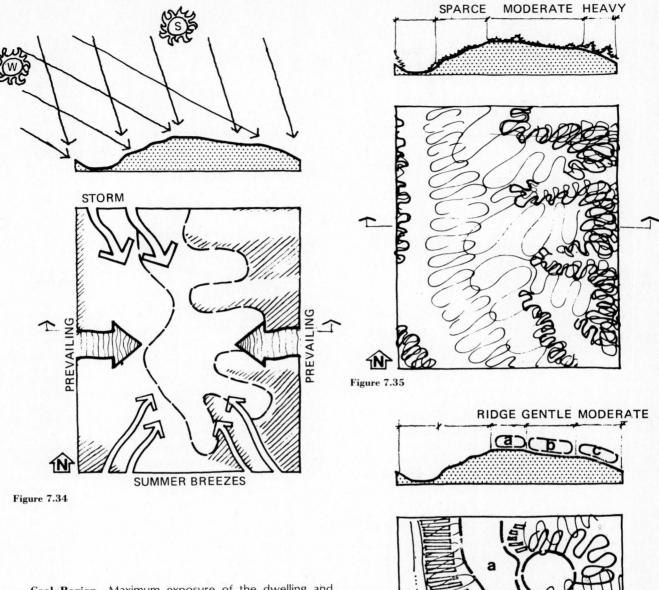

SPARCE   MODERATE HEAVY

Figure 7.35

STORM

PREVAILING

PREVAILING

SUMMER BREEZES

Figure 7.34

RIDGE GENTLE MODERATE

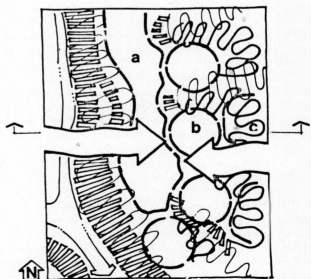

Figure 7.36

Cool Region   Maximum exposure of the dwelling and solar collector to the sun is the primary objective of site planning in cool regions. Sites with south-facing slopes are advantageous because they provide maximum exposure to solar radiation. Outdoor living areas should be located on the south sides of buildings to take advantage of the sun's heat. Exterior walls and fences can be used to create sun pockets and to provide protection from chilling winter winds.

Locating the dwelling on the leeward side of a hill or in an area protected from prevailing cold northwest winter winds—known as a wind shadow—will conserve energy. Evergreen vegetation, earth mounds (berms), and windowless insulated walls can also be used to protect the north and northwest exterior walls of buildings from cold winter winds.

Structures can be built into hillsides or partially covered with earth and planting for natural insulation.

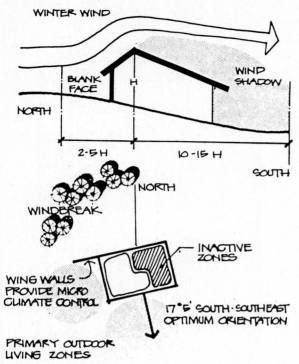

Figure 7.37

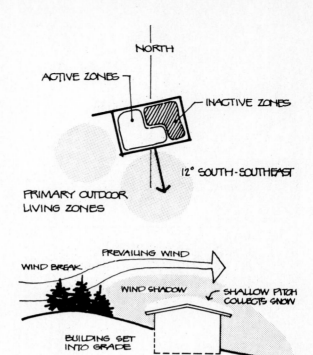

Figure 7.38

Temperate Region In the temperate region it is vital to assure maximum exposure of the solar collectors during the spring, fall, and winter months. To do so, the collector should be located on the middle to upper portion of any slope and should be oriented within an arc 10° either side of south. The primary outdoor living areas should be on the southwest side of the dwelling for protection from north or northwest winds. Only deciduous vegetation should be used on the south side of the dwelling since this provides summer shade and allows for the penetration of winter sun. (See Figure 7.40.)

The cooling impact of winter winds can be reduced by using existing or added landforms or vegetation on the north or northwest sides of the dwelling. The structure itself can be designed with steeply pitched roofs on the windward side, thus deflecting the wind and reducing the roof area affected by the winds. Blank walls, garages, or storage areas can be placed on the north sides of the dwelling. To keep cold winter winds out of the dwelling, north entrances should be protected with earth mounds, evergreen vegetation, walls, or fences. Outdoor areas used during warm weather should be designed and oriented to take advantage of the prevailing southwest summer breezes.

Hot, Humid Region In hot, humid regions where the heating requirement is small, solar collectors for heating only systems require maximum exposure to solar radiation primarily during the winter months. During the remainder of the year air movement in and through the site and shading are the most important site design considerations. However, for solar cooling or domestic water heating, year-round solar

collector exposure will be required. Collector orientation within an arc 10° either side of south is sufficient for efficient solar collection. Figures 7.41 and 7.42 illustrate a number of site planning and design considerations for solar energy utilization and energy conservation.

Hot, Arid Region The objectives of siting, orientation, and site planning in hot, arid regions are to maximize duration of solar radiation exposure on the collector and to provide shade for outdoor areas used in late morning or afternoon. To accomplish these objectives, the collector should be oriented south-southwest and the outdoor living areas should be located to the southeast of the dwelling in order to utilize early morning sun and take advantage of shade provided by the structure in the afternoon. (See Figures 7.43 through 7.44.)

Indoor and outdoor activity areas should take maximum advantage of cooling breezes by increasing the local humidity level and lowering the temperature. This may be done by locating the dwelling on the leeward side of a lake, stream, or other bodies of water. Also, lower hillside sites will benefit from cooler natural air movement during early evening and warm air movement during early morning.

Excessive glare and radiation in the outdoor environment can be reduced by providing:

- Small shaded parking areas or carports
- Turf adjacent to the dwelling unit
- Tree-shaded roadways and parking areas
- Parking areas removed from the dwelling units
- East-west orientation of narrow roadways

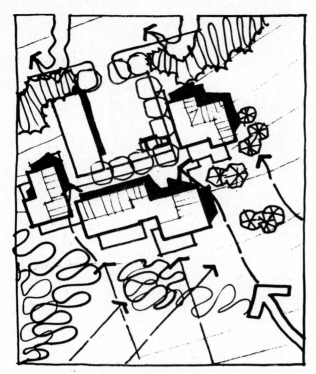

FOR CLUSTERED MULTI-FAMILY DWELLINGS,
TERRACES AND OUTDOOR LIVING AREAS
SHOULD BE INTEGRATED WITHIN THE BUILDING
CLUSTERS. THIS WILL REDUCE COLD AIR
MOVEMENT IN WINTER AND WILL CHANNEL
AND DIRECT BREEZES IN SUMMER.

STREETS AND PARKING AREAS SHADED WITH
DECIDUOUS VEGETATION WILL ALSO CHANNEL
SUMMER BREEZES AND REDUCE RADIATION
REFLECTION WHILE ALLOWING THE SUN TO
PENETRATE DURING THE WINTER.

Figure 7.39

Exterior wall openings should face south but should be
shaded either by roof overhangs or by deciduous trees in
order to limit excessive solar radiation into the dwelling. The
size of the windows on the east and west sides of the dwelling should be minimized in order to reduce radiation heat
gain into the house in early morning and late afternoons. Multiple buildings are best arranged in clusters for heat absorption, shading opportunities, and protection from east and
west exposures.

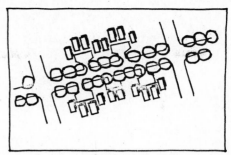

ROADWAYS CAN SERVE TO CHANNEL AND
DIRECT DESIRABLE BREEZES OR BLOCK
UNWANTED COLD WINDS. FOR TEMPERATE
REGIONS, AN EAST-WEST STREET ORIENTATION
CAN BEST SERVE THESE PURPOSES.

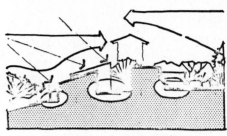

"SUN POCKETS" OR "SOLAR NOOKS" LOCATED
ON THE SOUTH SIDES OF BUILDINGS MAY HELP
EXTEND PERIODS OF SEDENTARY OUTDOOR
LIVING DURING COOLER MONTHS.

Figure 7.40

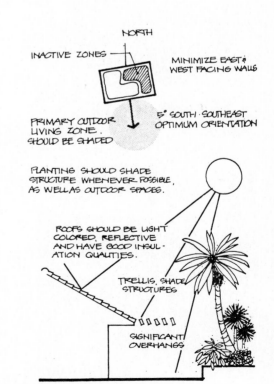

Figure 7.41

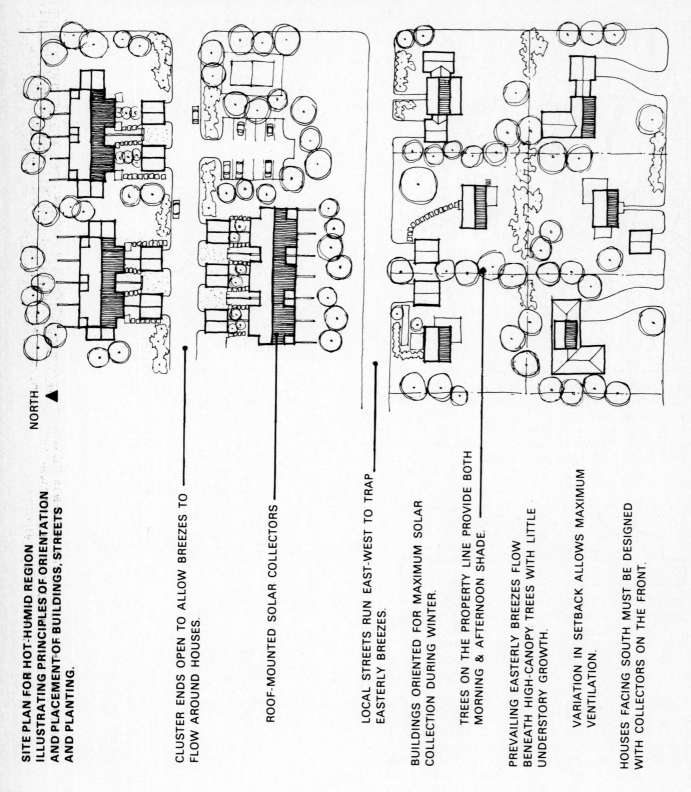

NORTH ▲

SITE PLAN FOR HOT-HUMID REGION ILLUSTRATING PRINCIPLES OF ORIENTATION AND PLACEMENT OF BUILDINGS, STREETS AND PLANTING.

CLUSTER ENDS OPEN TO ALLOW BREEZES TO FLOW AROUND HOUSES.

ROOF-MOUNTED SOLAR COLLECTORS

LOCAL STREETS RUN EAST-WEST TO TRAP EASTERLY BREEZES.

BUILDINGS ORIENTED FOR MAXIMUM SOLAR COLLECTION DURING WINTER.

TREES ON THE PROPERTY LINE PROVIDE BOTH MORNING & AFTERNOON SHADE.

PREVAILING EASTERLY BREEZES FLOW BENEATH HIGH-CANOPY TREES WITH LITTLE UNDERSTORY GROWTH.

VARIATION IN SETBACK ALLOWS MAXIMUM VENTILATION.

HOUSES FACING SOUTH MUST BE DESIGNED WITH COLLECTORS ON THE FRONT.

Figure 7.42

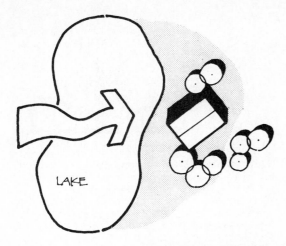

BODIES OF WATER OFFER THE
OPPORTUNITY TO PLAN FOR
THE COOLING EFFECTS OF
EVAPORATION.

Figure 7.43

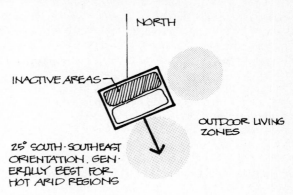

25° SOUTH·SOUTHEAST
ORIENTATION. GEN·
ERALLY BEST FOR
HOT ARID REGIONS

Figure 7.44

Each climatic region has its own distinctive characteristics and conditions that influence site planning and dwelling design for solar energy utilization and for energy conservation. The following chart suggests the general objectives of site planning and dwelling design for each climatic region as well as some methods for achieving these objectives. The chart reflects the seasonal trade-offs made between climatic optimums. In all cases, a detailed analysis should be undertaken to identify the site trade-offs between optimums for solar energy collection and optimums for energy conservation.

Site Orientation Chart

Objectives	Cool	Temperate	Hot, humid	Hot, arid
Adaptations	Maximize warming effects of solar radiation. Reduce impact of winter wind. Avoid local climatic cold pockets	Maximize warming effects of sun in winter. Maximize shade in summer. Reduce impact of winter wind but allow air circulation in summer	Maximize shade. Maximize wind	Maximize shade late morning and all afternoon. Maximize humidity. Maximize air movement in summer
Position on slope	Low for wind shelter	Middle-upper for solar radiation exposure	High for wind	Low for cool air flow
Orientation on slope	South to southeast	South to southeast	South	East-southeast for P.M. shade
Relation to water	Near large body of water	Close to water, but avoid coastal fog	Near any water	On lee side of water
Preferred winds	Sheltered from north and west	Avoid continental cold winds	Sheltered from north	Exposed to prevailing winds
Clustering	Around sun pockets	Around a common, sunny terrace	Open to wind	Along E-W axis, for shade and wind
Building orientation*	Southeast	South to southeast	South, toward prevailing wind	South
Tree forms	Deciduous trees near building, evergreen for windbreaks	Deciduous trees nearby on west. No evergreens near on south	High canopy trees. Use deciduous trees near building	Trees overhanging roof if possible
Road orientation	Crosswise to winter wind	Crosswise to winter wind	Broad channel, E-W axis	Narrow, E-W axis
Materials coloration	Medium to dark	Medium	Light, especially for roof	Light on exposed surfaces, dark to avoid reflection

*Must be evaluated in terms of impact on solar collector, size, efficiency, tilt.

Integration of the Building and Site   Ideally, a building is designed for the specific site on which it is to be placed. Commonly, however, a building design may be replicated with only minor changes on different sites and in different climates.

Site planning solutions are not as easy to replicate because each site has a unique geography, geology, and ecology. The most appropriate way to integrate any building and its site is first to analyze the site very carefully and then to place the building on the site with a minimum of disruption and the greatest recognition of the site's distinctive features.

It is possible, however, to provide general techniques for integrating buildings with their sites. Historically, a number of such techniques have evolved, among which are indigenous architectural characteristics adapted to local site conditions, architectural extensions to the building such as walls and covered walks, the use of native materials found on the site, and techniques for preserving or enhancing the native ecology.

In each climatic region, guidelines can be determined to help apply the many techniques available for integrating a building and its site in ways appropriate to the particular region. These guidelines can be particularly helpful in maximizing energy conservation and increasing the opportunity for successful use of solar heating and cooling. The bibliography contains several documents pertaining to the integration of the building and site.

Detailed Site Design   The detailed design of a site for optimum solar energy utilization and energy conservation entails the use of a variety of types of vegetation, paving, fences, walls, overhead canopies, and other natural and man-made elements. These elements are used to control the solar exposure, comfort, and energy efficiency of the site and the dwelling.

The materials used in site design have the ability to absorb, store, radiate, and deflect solar radiation as well as to channel warm or cool air flow. For instance, trees of all sizes and types block incoming and outgoing solar radiation, deflect and direct the wind, and moderate precipitation, humidity, and temperature in and around the site and dwelling. Shrubs deflect wind and influence site temperature and glare. Groundcovers regulate absorption and radiation. Turf influences diurnal temperatures and is less reflective than most paving materials. Certain paving surfaces, fences, walls, canopies, trellises, and other site elements may be located on the site to absorb or reflect solar radiation, channel or block winds, and expose or cover the dwelling or solar collector. (See Figures 7.45–7.51.)

Site planning for solar energy utilization and energy conservation based on the vegetation, topographic, and climatic analysis of the site suggests southern exposure, northern protection, and unimpaired air movement for multifamily housing projects located in a cool climate (Figures 7.52 and 7.53).

The site plan and dwelling design concept in Figure 7.54 are indicative of the solar energy utilization and energy conservation considerations for the cool and temperate regions. The techniques employed include:

DECIDUOUS TREES CAN BE USED FOR SUMMER SUN SHADING OF THE DWELLING AND YET ALLOW WINTER SUN PENETRATION THROUGH THEIR BARE BRANCHES FOR SOLAR COLLECTION. BARE BRANCHED DECIDUOUS TREES DO, HOWEVER, CAST A SUBSTANTIAL SHADOW AND WILL REDUCE COLLECTION EFFICIENCY. EVERGREENS SHADE COLLECTORS HEAVILY ALL YEAR.

Figure 7.45

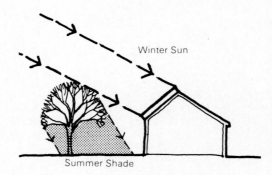

BECAUSE TREE IS AWAY FROM HOUSE TO AVOID SHADING ROOF COLLECTOR IN WINTER. IT CANNOT SHADE THE SOUTH WALL OF THE HOUSE IN SUMMER.

Figure 7.46

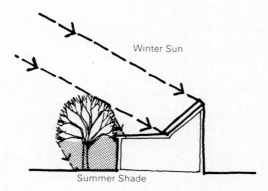

BY MOVING COLLECTOR BACK, TREES CAN BE GROWN NEAR THE HOUSE TO PROVIDE SUMMER SHADE. WITHOUT ALSO SHADING COLLECTOR IN WINTER.

Figure 7.47

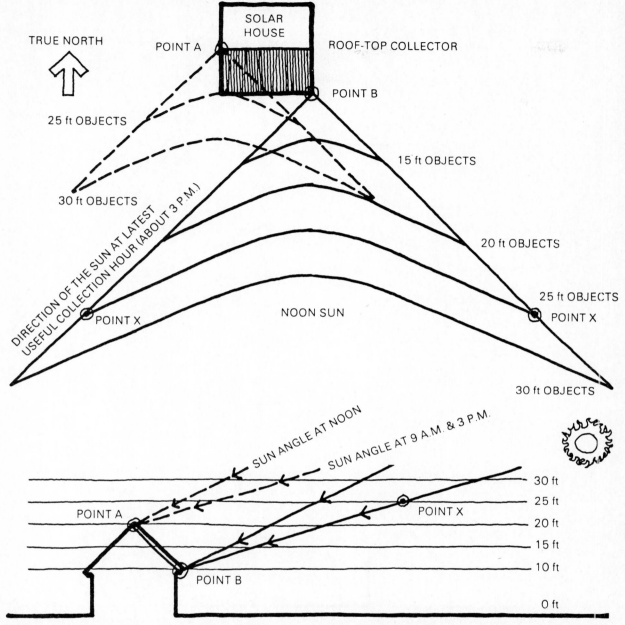

Figure 7.48   Solar interference boundaries of individual points: latitude, 40; December 21st, winter solstice. Every point on the collector for a given latitude and day of the year has a set of solar interference boundaries. These define the areas within which objects of a given height above a flat site will cast a shadow on the collector areas before and after useful collection. Solar interference boundaries (*bottom*) are drawn by plotting in plan the points of an intersection between the sun angles and the various elevations above the zero grade (such as point X).

- The use of windbreak planting
- The orientation of road alignment with planting on either side to channel summer breezes
- The location of units in a configuration suggested by the topography

- The use of the garage to buffer the dwelling from northwest winter winds
- The use of berms to shelter outdoor living terraces
- The use and location of deciduous trees to block or filter afternoon summer sun

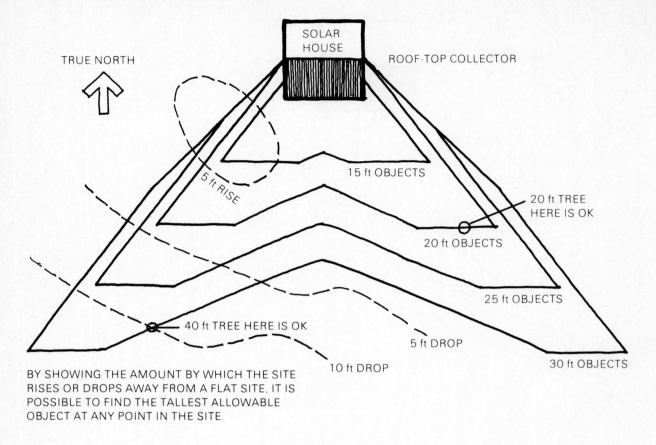

TRUE NORTH

SOLAR HOUSE

ROOF-TOP COLLECTOR

5 ft RISE

15 ft OBJECTS

20 ft TREE HERE IS OK

20 ft OBJECTS

25 ft OBJECTS

40 ft TREE HERE IS OK

5 ft DROP

10 ft DROP

30 ft OBJECTS

BY SHOWING THE AMOUNT BY WHICH THE SITE RISES OR DROPS AWAY FROM A FLAT SITE, IT IS POSSIBLE TO FIND THE TALLEST ALLOWABLE OBJECT AT ANY POINT IN THE SITE.

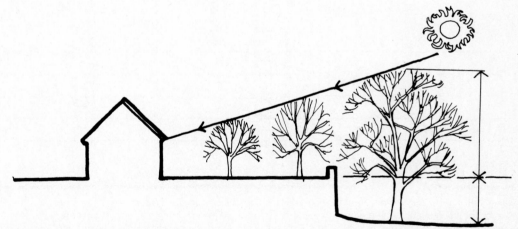

Figure 7.49 Composite solar interference boundaries for entire collector: latitude, 40; December 21st, winter solstice. A composite plan of the solar interference boundaries for every point on the collector can be made relatively simple. If the site falls away to the south (*bottom*), large trees can be planted without shading the collector. The extra height allowable can be shown in the plan.

MULTI-LAYERED VEGETATION INCLUDING CANOPY TREES AND UNDERSTORY TREES OR SHRUBS PROVIDES A MULTIPLE BRAKING EFFECT. SUBSTANTIALLY DECREASING THE WIND VELOCITY MOVING OVER A SITE.

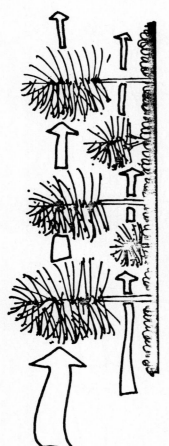

MULTIPLE BRAKING EFFECT

A MASS PLANTING OF TREES PROVIDES A DEAD AIR SPACE UNDER AND AROUND ITSELF. IT ALSO DECREASES THE AIR VELOCITY 5 TIMES ITS HEIGHT TO WINDWARD AND 25 TIMES ITS HEIGHT TO LEEWARD OF THE PLANTING.

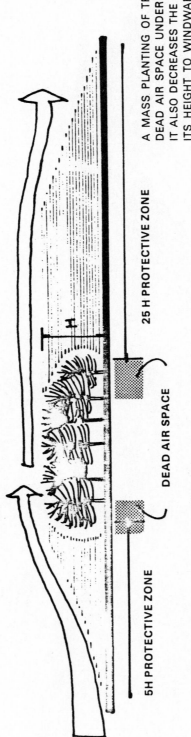

25 H PROTECTIVE ZONE

DEAD AIR SPACE

5H PROTECTIVE ZONE

PLANTING ON THE LEEWARD SIDE OF A HILL SUBSTANTIALLY INCREASES THE DOWNWIND ZONE OF REDUCED AIR VELOCITY, WHILE PLANTING ON THE WINDWARD SIDE CORRESPONDING DECREASES THE ZONE.

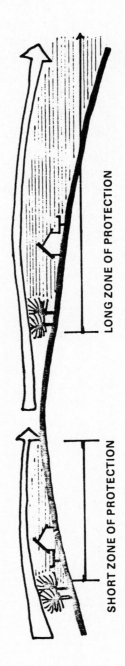

LONG ZONE OF PROTECTION

SHORT ZONE OF PROTECTION

Figure 7.50

-815-

VEGETATION MAY BE PRESERVED OR PLACED IN SUCH A WAY AS TO CHANNEL OR BLOCK DAILY OR SEASONAL AIR FLOW PATTERNS.

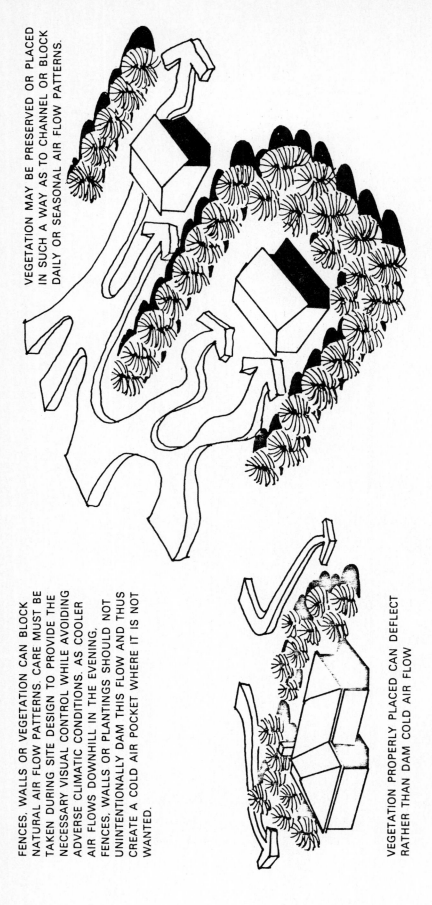

FENCES, WALLS OR VEGETATION CAN BLOCK NATURAL AIR FLOW PATTERNS. CARE MUST BE TAKEN DURING SITE DESIGN TO PROVIDE THE NECESSARY VISUAL CONTROL WHILE AVOIDING ADVERSE CLIMATIC CONDITIONS. AS COOLER AIR FLOWS DOWNHILL IN THE EVENING, FENCES, WALLS OR PLANTINGS SHOULD NOT UNINTENTIONALLY DAM THIS FLOW AND THUS CREATE A COLD AIR POCKET WHERE IT IS NOT WANTED.

VEGETATION PROPERLY PLACED CAN DEFLECT RATHER THAN DAM COLD AIR FLOW

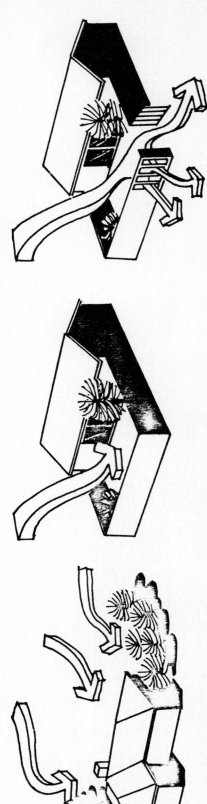

FENCE DESIGN CAN PROVIDE FOR COLD AIR DRAINAGE

COLD AIR TRAPPED BY FENCE

DWELLING UNPROTECTED FROM COLD AIR FLOW

Figure 7.51

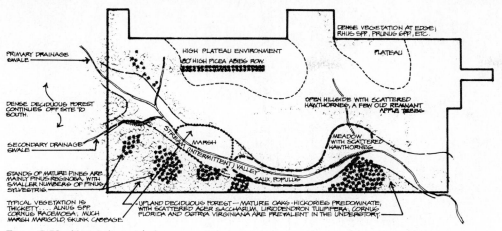

Figure 7.52   Vegetation analysis

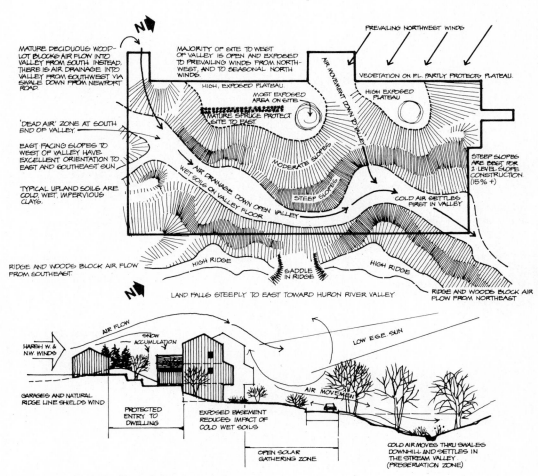

Figure 7.53   Topographic and climatic analysis.

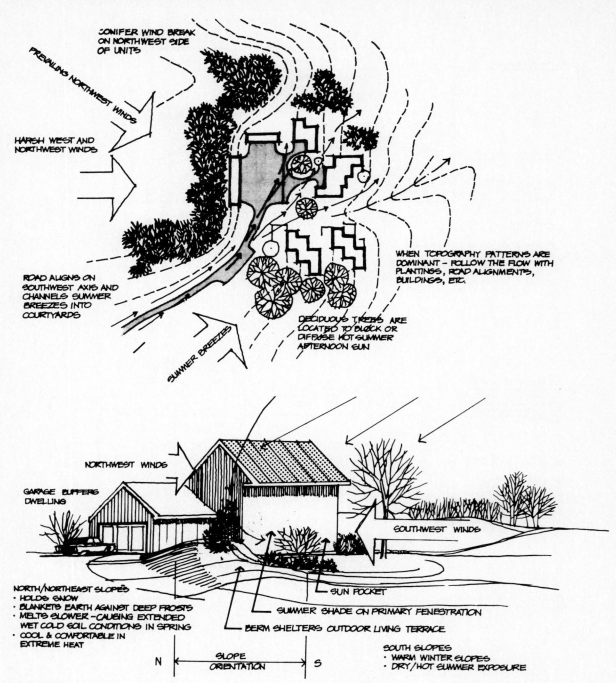

CONIFER WIND BREAK ON NORTHWEST SIDE OF UNITS

PREVAILING NORTHWEST WINDS

HARSH WEST AND NORTHWEST WINDS

WHEN TOPOGRAPHY PATTERNS ARE DOMINANT — FOLLOW THE FLOW WITH PLANTINGS, ROAD ALIGNMENTS, BUILDINGS, ETC.

ROAD ALIGNS ON SOUTHWEST AXIS AND CHANNELS SUMMER BREEZES INTO COURTYARDS

SUMMER BREEZES

DECIDUOUS TREES ARE LOCATED TO BLOCK OR DIFFUSE HOT SUMMER AFTERNOON SUN

NORTHWEST WINDS

GARAGE BUFFERS DWELLING

SOUTHWEST WINDS

SUN POCKET

NORTH/NORTHEAST SLOPES
• HOLDS SNOW
• BLANKETS EARTH AGAINST DEEP FROSTS
• MELTS SLOWER — CAUSING EXTENDED WET COLD SOIL CONDITIONS IN SPRING
• COOL & COMFORTABLE IN EXTREME HEAT

SUMMER SHADE ON PRIMARY FENESTRATION

BERM SHELTERS OUTDOOR LIVING TERRACE

N   SLOPE ORIENTATION   S

SOUTH SLOPES
• WARM WINTER SLOPES
• DRY/HOT SUMMER EXPOSURE

Figure 7.54

On-Site Collectors   In some situations, it may be desirable or necessary to place the solar collector at some distance from the dwelling. When this happens, there are three alternative approaches available to the site designer. The first approach is to screen the solar collector from view. The second approach is to integrate the collector with the site by the use of earthforms, vegetation, or architectural elements. A third approach is to emphasize the collector as a design feature.

Regardless of the approach taken, on-site solar collectors together with any structural supports or additional equipment may be unsightly, hazardous, and subject to vandalism. Earthforms, planting, and other site elements can be used effectively to hide the collector and associated apparatus and to prevent easy access. In some cases it is possible to use the space under the collector if this can be done without interfering with its performance.

Figures 7.55 and 7.56 indicate several methods by which on-site solar collectors can be screened, integrated, or emphasized, and the space under the collector used, through the manipulation of earthforms, planting, and architectural elements.

Summary   Site selection, planning, and design can significantly influence the effective use of solar energy for residential heating and cooling. The topography, geology, soils, vegetation, and local climate of a building site should be considered prior to site selection or building placement. Each climatic region has its own distinctive characteristics and conditions that influence site planning and design for solar energy utilization and for energy conservation. These should be recognized during the development of design objectives for each proposed site in terms of building-collector placement and orientation; relation of building-collector to wind, water, and existing vegetation; and the merits of clustering, new vegetation, and material selection.

Besides building design and orientation, the careful selection and location of all forms of planting, paving, fences, canopies, and earthforms can contribute to the effective and efficient use of solar energy. Dwelling design and site design for the utilization of solar energy are complementary and should be considered simultaneously throughout the design process.

7.8 Solar Site Planning

This section demonstrates how a development plan with solar access protection as a primary objective might be developed for a conventional subdivision and for a planned unit development. A PUD offers a number of advantages in designing a project for good solar access, but a more conventional development also can be laid out to reach this end. Both the examples here use the same 20-acre site in the southern end of California's Central Valley, near Fresno, at 37 degrees north latitude. The conventional plan shows all single-family residences, each having adequate solar access. The PUD plan will have a somewhat higher density and a mix of housing types.

The decision-making process for planning the development follows this format:

Determination of planning criteria

Site analysis and preparation of a preliminary site plan

Development of a more detailed site plan, showing individual lots and dwelling locations

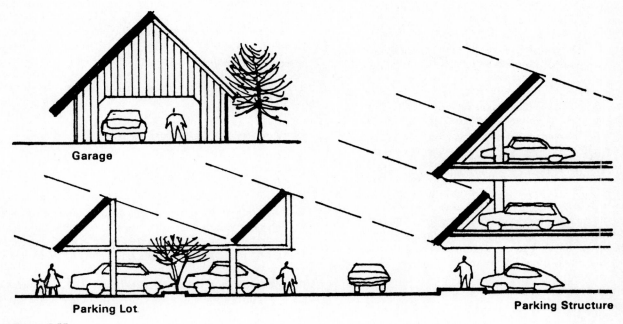

Garage

Parking Lot

Parking Structure

Figure 7.55

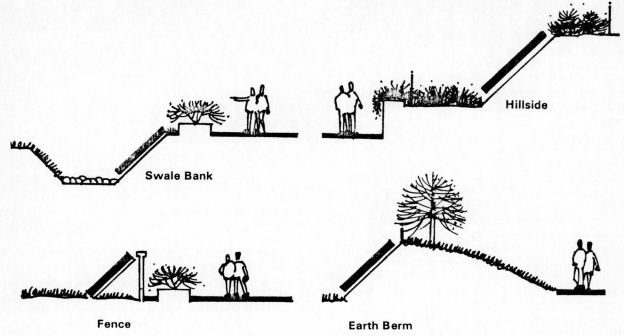

Swale Bank

Hillside

Fence

Earth Berm

Figure 7.56

Determining Planning Criteria The site is located in a county with fairly conventional land use controls. The parcel is zoned AR-1, agricultural/residential with single-family detached housing and accessory structures allowed by right. The zoning establishes large lot-size requirements — 1 dwelling unit per 7000 square feet of lot area, resulting in an overall density of 5.8 dwelling units per acre, or a total development potential of about 120 dwelling units on the site. Subdivision regulations have typical road and infrastructure standards governing utilities, sewage, water supply, and roads. In addition, there are requirements for open-space dedication and for the preservation of existing trees and other major vegetation.

The community also has a PUD provision in its zoning, which can be invoked by application to the city council for a rezoning. The PUD provision is also typical, allowing both a variety of housing types (single-family detached, attached low-rise, and mid- and high-rise buildings) on the same parcel, provided that the project complies with specified standards. As an incentive, the PUD provision allows a maximum density of 7 dwelling units per acre, creating a development potential of up to 140 dwelling units for the 20-acre parcel. As with the conventional zoning provisions, there are requirements for environmental standards, tree preservation, and open space.

Based on these regulatory standards, the developer has selected the following development objectives for the site:

South-wall access for all dwellings

Maximum energy efficiency, with solar collectors providing most of the seasonal heating requirements and natural cooling

The preservation of views to the Sierra Nevada mountains

A central recreation facility and common open space, meeting the minimum standards required for project approval

Maximum allowable density under both ordinance provisions (conventional and PUD) consistent with market demand in the region. For the PUD development, this means a housing mix of single-family detached, low-rise attached, and mid-rise apartments.

Site Selection The site shown in Figure 7.57 was chosen because it is essentially flat, with a long east/west axis that maximizes its southern exposure and presents no solar access obstructions. Close to downtown, schools, and a commercial center, it is served by utilities and has a superb view of the mountains.

Site Analysis and Preliminary Site Plan

Climate California's Central Valley has mild but cloudy and foggy winters. In the hot season, June through September, the temperature averages 100°F. It is dry, and relief from heat can be provided by shading, ventilation, and roof pond coolers. The winter winds and the north winds in the spring, when air temperatures are moderate, can be a significant climatic factor. Summer breezes are light and variable. The growing season lasts all year.

Vegetation and Site Characteristics The site surface is composed of grass and dark earth; the only trees are along the north border, where shadow conflicts are minimal. There are neither trees nor tall structures on the adjacent lots. This site enjoys beautiful views of the Sierra Nevada to the north

and east. The major roadway, on the east edge of the property, connects the site with the local school and a downtown area less than a mile away. The combination of flat terrain, good weather, and short distances makes bike riding an attractive mode of transportation. Figure 7.58 shows the details of the site analysis.

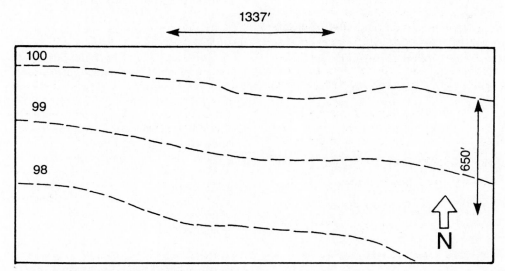

Figure 7.57   Site topography; total area 20 acres.

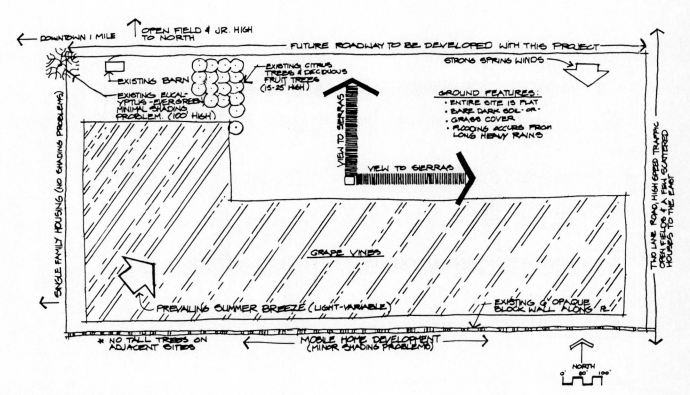

Figure 7.58   Site analysis.

Conventional Development

Preliminary Site Plan The overall concept shown in Figure 7.59 stresses the key points.

The entire site is suitable for housing; there are no solar access conflicts.

All local streets run east/west to allow houses to be oriented south. Straight street layout is used for simplification, but curvilinear streets would also be appropriate.

The roadways form view corridors to the north and east.

Detailed Site Plan

Streets, Lots, and Building Siting The major considerations are:

The street and lot layout allow south orientation of houses.

The houses are located to the north end of each lot to minimize solar access conflicts with other buildings.

The layout of individual buildings and trees is based on shadow patterns.

The general types of single-family housing shown in Figure 7.60 are planned for this project, all using (or having the potential to use) passive south-facing collector area for space heating, rooftop active systems for water heating, and roof ponds for natural cooling.

Figure 7.61 shows one possible layout of housing on the site. Shadow patterns indicate that all dwellings have south-wall solar access. Allowing solar access to the south wall ensures adequate sun for rooftop water heating systems. South yards are made as large as possible on most lots.

Landscaping For planning purposes, a few types of trees and shrubs were chosen; ranging in height from 15 to 40 feet, each group has both deciduous and evergreen species. Figure 7.62 shows the December tree shadows for these types.

For planning purposes, a deciduous tree should be evaluated as if it were evergreen, because its bare branches block a significant amount of sunlight. This is a very conservative approach, since many passive systems still work even when shaded by bare branches. At higher latitudes, this approach may be too restrictive.

Figure 7.63 shows a layout for trees superimposed on the housing layout. A significant number of trees have been used without shading any south walls of the dwellings.

The site plan is completed by combining the building and tree shadow plans in a manner consistent with the development objectives. The resulting plan is shown in Figure 7.64.

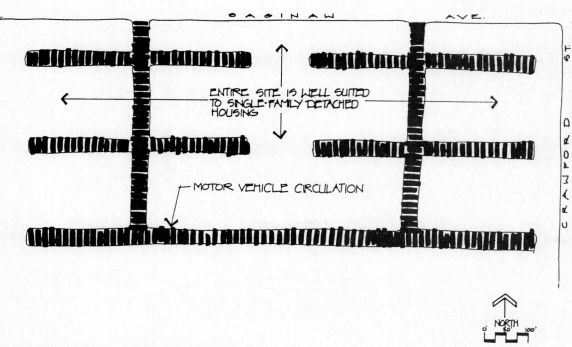

Figure 7.59 Preliminary site plan: conventional development.

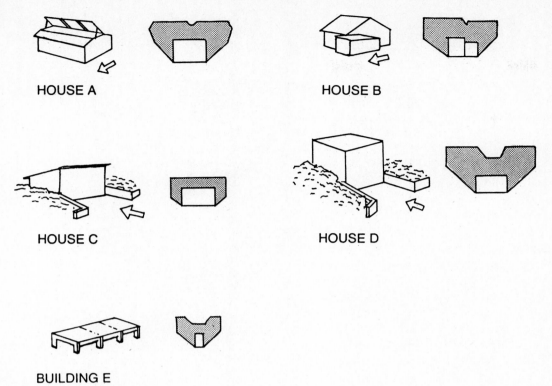

HOUSE A

HOUSE B

HOUSE C

HOUSE D

BUILDING E

Figure 7.60 Housing types.

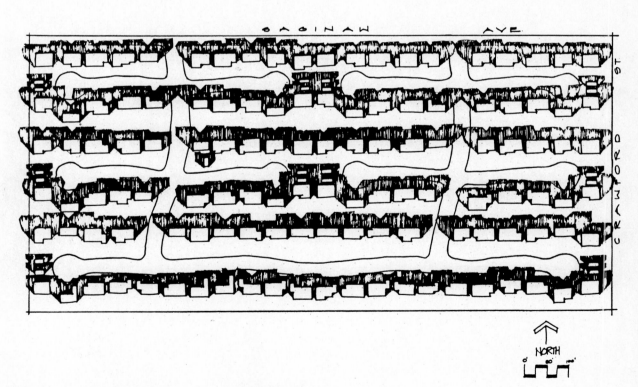

Figure 7.61 Building shadow plan: conventional development.

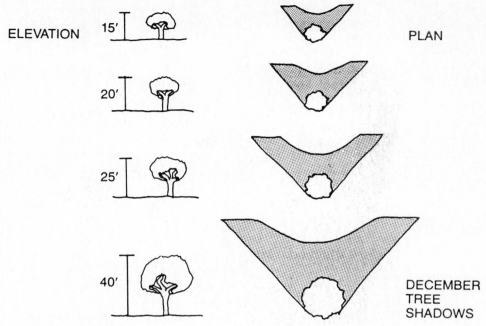

ELEVATION   15'   PLAN

20'

25'

40'

DECEMBER
TREE
SHADOWS

Figure 7.62   Tree types.

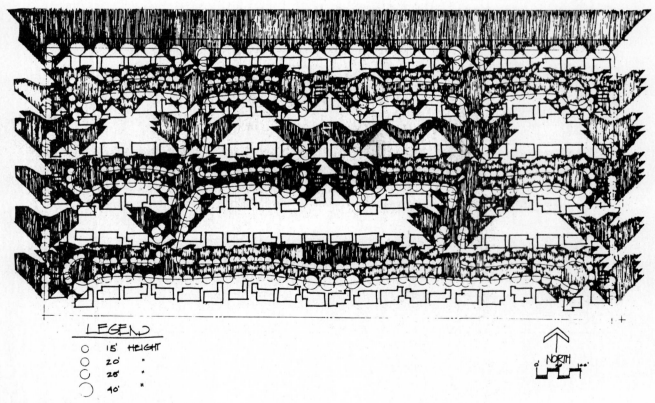

LEGEND
- O   15' HEIGHT
- O   20'   "
- O   25'   "
- ◯   40'   "

NORTH

Figure 7.63   Tree shadow plan: conventional development.

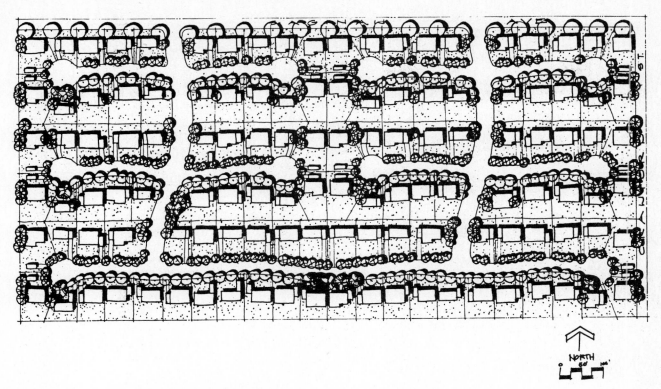

Figure 7.64 Site master plan: conventional development.

Planned Unit Development

Preliminary Site Plan The major site uses, presented in Figure 7.65, can be summarized as follows:

Housing is grouped in clusters, each cluster having a focal open space.

A centralized community open space and recreation center is linked to the clusters.

The tallest building, a mid-rise apartment, is located at the north edge of the site.

Auto circulation is kept to the periphery of the site, and a bike/pedestrian path is used for internal traffic.

The commercial center lies at the intersection of the two existing roads.

The highest-density housing is located closest to the existing roads.

The existing barn and fruit trees are used as an agricultural center.

The open spaces double as storm drainage and percolation areas.

Detailed Site Plan
  Streets, Lots, and Building Siting

All access streets run east/west to allow north/south lots.

Only on lots with roof clerestory collectors, buildings do not run north/south. Clerestories gain solar access through their roofs.

When possible, car circulation is kept to the periphery in order to concentrate and integrate community and cluster open spaces.

Layout of individual buildings and trees is based on shadow patterns.

All the general housing types used in the conventional neighborhood are used in the PUD, with the additions shown in Figure 7.66 as follows:

Apartment F: Two-story, four-unit apartment building. South wall requires solar access.

Apartment G: Low-rise attached town houses. Most are two-story, although some one-story units are included. All require solar access to the base of the south wall.

Apartment H: Attached low-rise apartments. Solar access is through the roof for passive space heating. South walls receive sun for half the day in winter. Private yard space is on the east or west.

Apartment I: Four-story, mid-rise apartment building. This is a single-loaded exterior corridor plan to allow south access and cross ventilation to all units. Patios are provided on the south side.

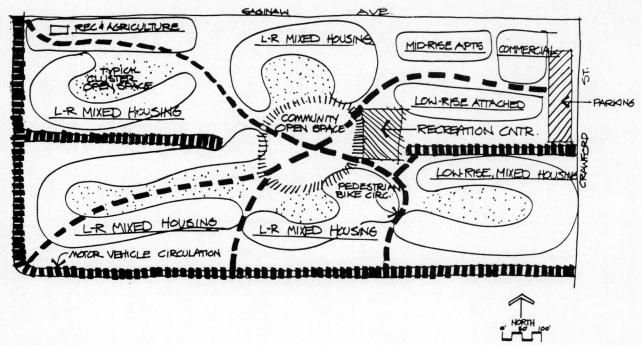

Figure 7.65 Major land uses: PUD

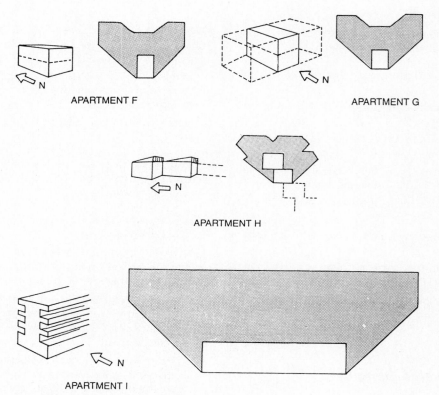

APARTMENT F

APARTMENT G

APARTMENT H

APARTMENT I

Figure 7.66 Apartment types: PUD

Figure 7.67 shows an alternative housing layout that achieves south-wall and, in many cases, south-yard solar access for all dwellings. Carports generally are located on the opposite side of the street along the southern edge of the neighborhood to maximize solar access to south yards. Figures 7.67 to 7.70 are used to determine best placements of carports:

Alternative 1 in Figure 7.67 shows carports adjacent to the houses. Solar access to the south yard is blocked by the carport in winter, but the street is half-shaded in summer.

Alternative 2 shows the carport on the south (opposite) side of the street. The shadow of the deciduous street trees allows increased winter sun to the larger south yard, but the street has minimal summer shade.

Alternative 3 shows that alternating street trees and carports results in optimum sun and shade patterns.

Shadow patterns are developed for each of the housing types; the buildings and shadow patterns are arranged on the site plan in various ways that are consistent with the land-use diagram developed earlier. Building placement is optimal when all south walls obtain the best solar access and are not shaded by adjacent structures. Figure 7.68 shows this arrangement.

Landscaping   As for trees, a process similar to that used in the conventional development is used for the PUD — the tree types are identified and shadow patterns developed. The trees and shadow patterns are then organized on the site plan to accomplish the major development objective — summer shading of yard areas and west walls of structures and winter sun access to south walls and rooftops. Figure 7.69 shows the resulting tree plan.

Finally, as in the conventional development, all the elements are combined into a final site master plan. Buildings and trees are located to achieve the best ciruulation, land use, and solar access development objectives. Figure 7.70 shows the completed PUD development, fully planned for both optimal solar access and conventional development goals.

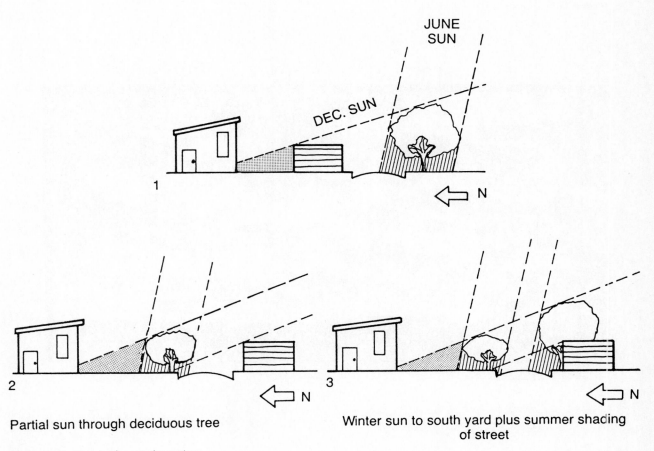

Partial sun through deciduous tree

Winter sun to south yard plus summer shading of street

Figure 7.67   Housing layout alternatives.

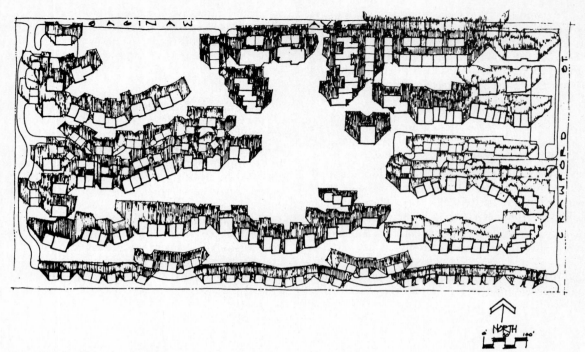

Figure 7.68 Building shadow plan: PUD

LEGEND

○ 15' HEIGHT
○ 20' "
○ 25' "
○ 40' "

Figure 7.69 Shadow plan: PUD

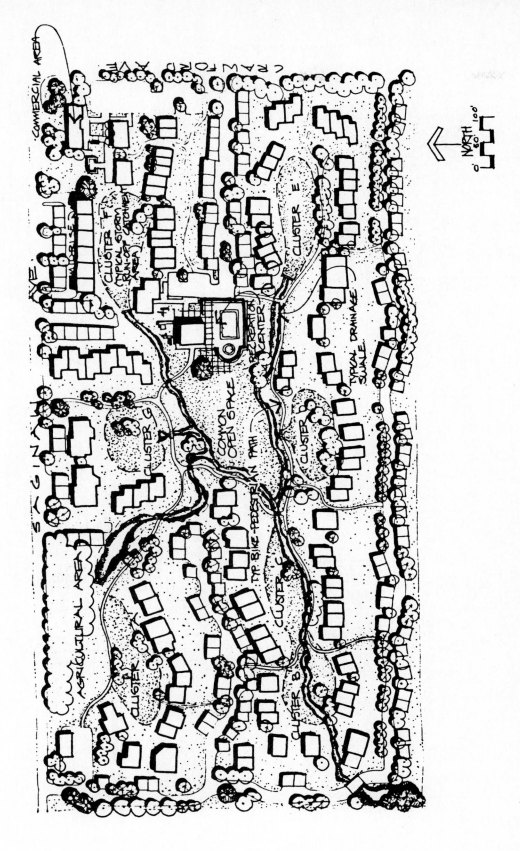

Figure 7.70 Site master plan: PUD

7.9 Site Plans—School

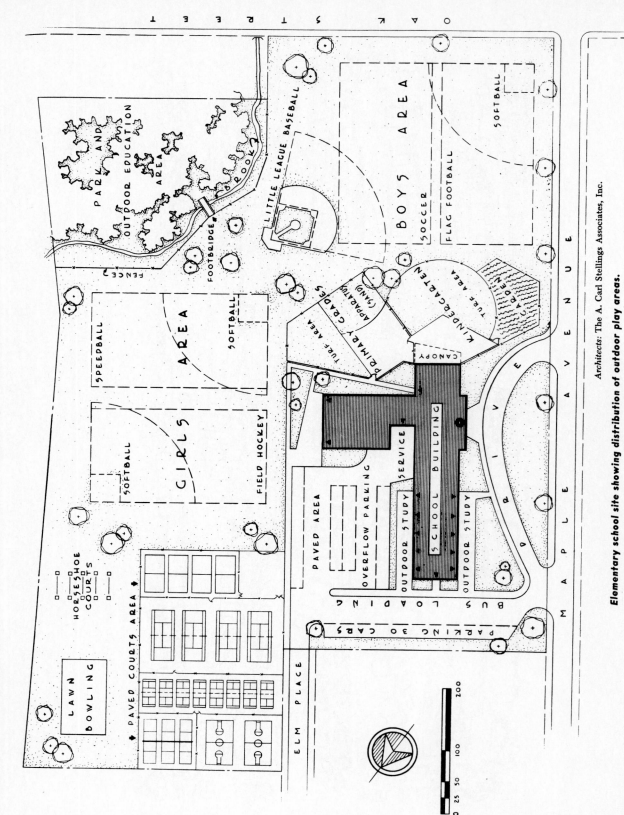

Architects: The A. Carl Stellings Associates, Inc.

Elementary school site showing distribution of outdoor play areas.

Figure 7.71

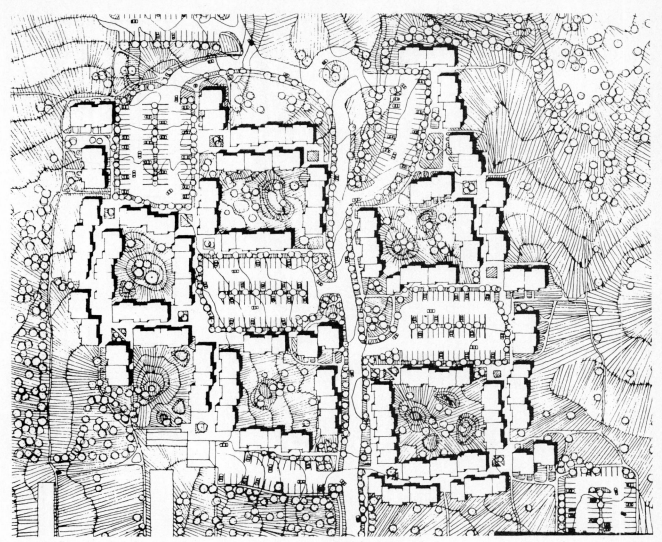

Figure 7.72   Example of a low-rise suburban development on a moderately sloping site.

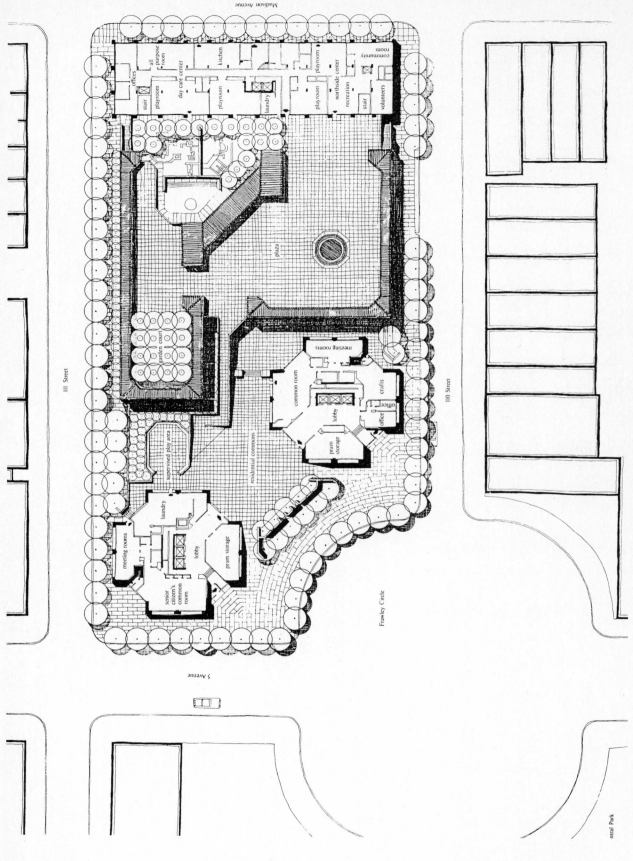

Figure 7.73 High-use urban residential development, Arthur Schomberg Plaza, New York City; architect, Gruzen & Partners.

Madison Avenue

III Street

110 Street

5 Avenue

Frawley Circle

Central Park

offices

stair

playroom

day care center

kitchen

playroom

laundry

all purpose room

playroom

playroom

recreation

northside center

community room

stair

volunteers

garden court

supervised play area

plaza

laundry

meeting rooms

lobby

pram storage

residential commons

senior citizen's common room

common room

meeting rooms

lobby

pram storage

office

crafts

office

7.10 Site Plans—Residential

The 80-unit family housing complex in Figure 7.74 is located opposite a new elementary school in a neighborhood fabric of medium-density housing, and will accommodate about 330 persons. A tri-level system of organization is utilized on the 3-acre site. This enables automobiles to be parked out of sight under the living units, and allows for the separation of general pedestrian areas from both the lower vehicular circulation and the higher town-house patios.

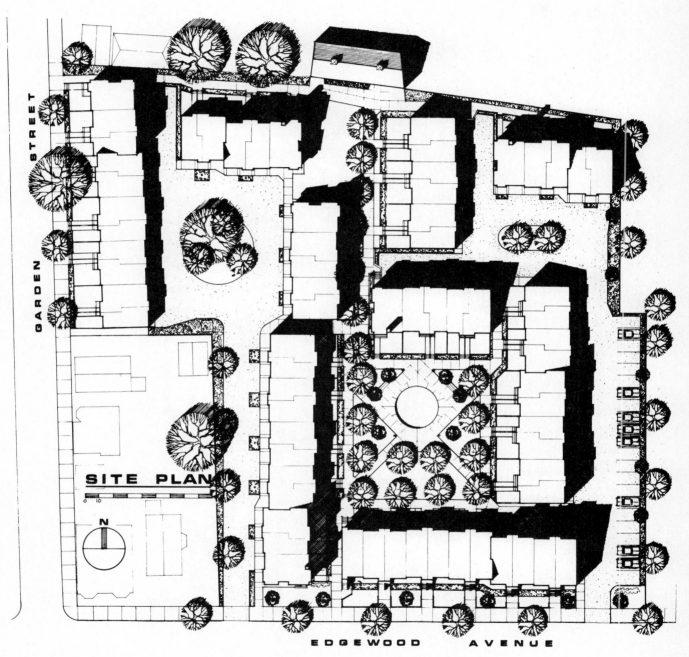

Figure 7.74 Dwight Co-Operative Town Houses, New Haven, Connecticut; architect, Gilbert Switzer, AIA, and Associates.

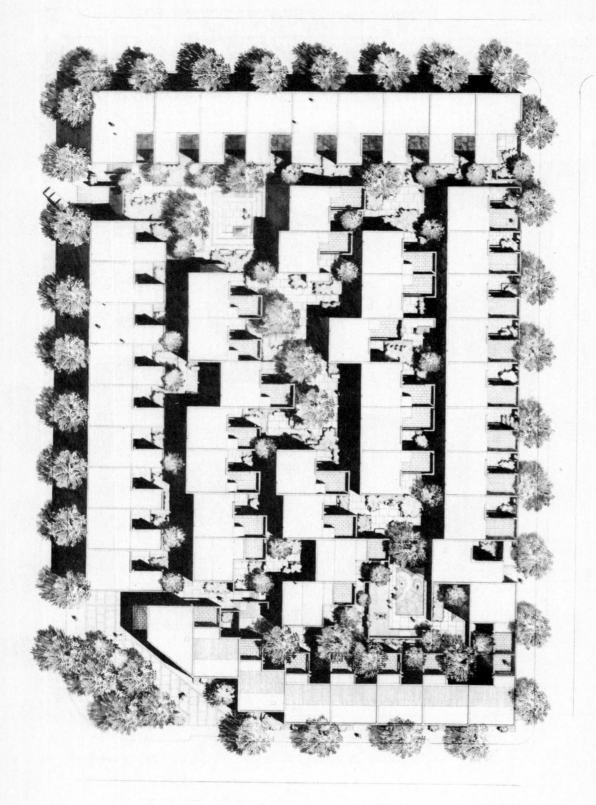

Figure 7.75   Site plan. A well-protected inner courtyard can be entered at only three secured places; architect, Louis Sauer Associates.

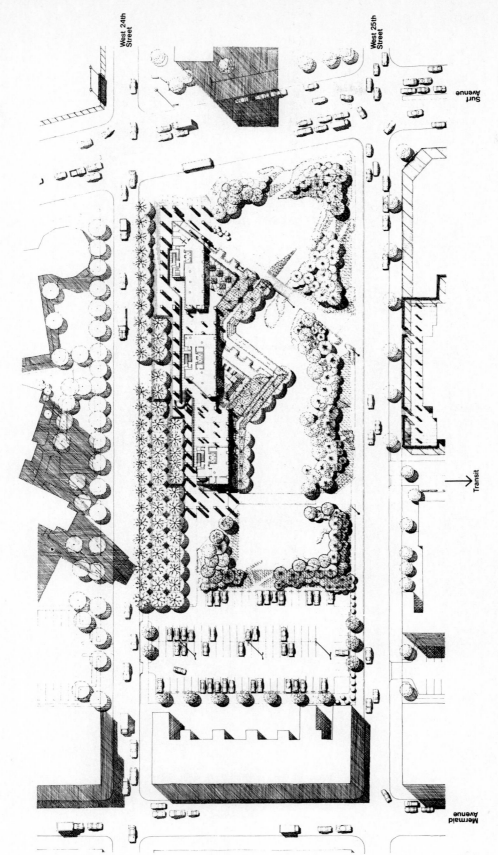

Figure 7.76 Site plan of an urban built-up development.

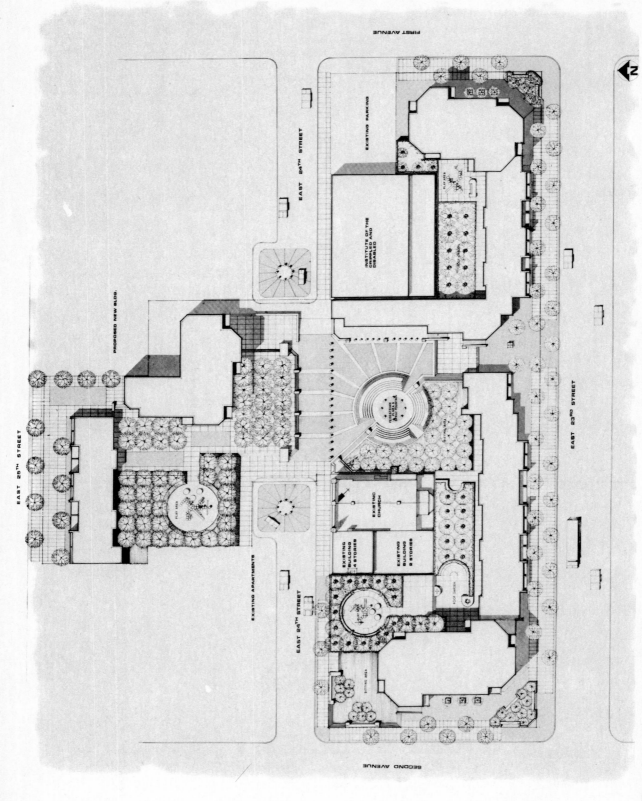

Figure 7.77 Site plan of an urban, high-use housing development, East Midtown Plaza, New York City; architect, Davis Brody & Associates.

Figure 7.78 Subdivision of site plan. Proposed residential development—Transcona, Manitoba.

Credits

Section 1

Pages 2-6: *Planning for Wildlife and Man,* U.S. Department of the Interior, Fish and Wildlife Service, Washington, D.C., 1974.

Page 6, Figs. 1.1a and 1.1b; page 7, Figs. 1.2, 1.3, and 1.4: *Penfield Road Study,* Schnadelbach Associates.

Pages 7-14: Allen Carroll, *Developer's Handbook,* Department of Environmental Protection, State of Connecticut.

Pages 15-20: John S. Hoag, *Fundamentals of Land Measurements,* Chicago Title Insurance Company, Chicago, 1971.

Pages 20-24: Guy D. Smith and Andrew R. Aandahl, "Soil Classification and Survey" in *Soil: The Yearbook of Agriculture,* U.S. Department of Agriculture, Washington, D.C., 1957.

Pages 24-30: *Engineering Field Manual for Conservation Practices,* U.S. Department of Agriculture, Soil Conservation Service, Washington, D.C., 1969.

Page 30-34: *Geological Investigations,* U.S. Army Corps of Engineers, Washington, D.C., 1960.

Pages 34-36; page 40, Fig. 1.20: *Code Manual,* New York State Building Code Commission, Albany.

Page 37, Fig. 1.17; page 90, Fig. 1.21: Donald T. Graf, *Graf's Data Sheets,* Reinhold Publishing Corporation, New York, 1949.

Pages 38, 39; page 77, Fig. 1.54: *Manual of Acceptable Practices,* Department of Housing and Urban Development, Washington, D.C., 1973.

Pages 41-45: *Soil Mechanics, Foundations, and Earth Structures,* Navfac-DM-7, U.S. Department of the Navy, Washington, D.C.

Pages 46, 47: *Minimum Property Standards,* Federal Housing Administration, Washington, D.C.

Pages 47-50: D. R. Coote and P. J. Zwerman, *Surface Drainage of Flat Lands,* New York State College of Agriculture at Cornell University, Ithaca, 1970.

Pages 50-53 (soil erosion control): *Controlling Erosion on Construction Sites,* U.S. Department of Agriculture, Soil Conservation Service, Washington, D.C., 1970.

Pages 53, 54: *Michigan Soil Erosion and Sedimentation Control Guidebook,* Department of Natural Resources, State of Michigan, 1975.

Pages 55, 56: *Planning and Design of Roads, Airbases and Heliports,* Departments of the Army and Air Force, Washington, D.C.

Page 56, Fig. 1.34: *Landscape Development,* Field Technical Office, U.S. Department of the Interior, Littleton, Colorado.

Pages 58-61: *Mapnotes,* New York State Department of Transportation, Albany, 1971.

Page 59, Fig. 1.35: De Chiara and Koppelman, *Planning Design Criteria,* Van Nostrand Reinhold Company, New York, 1969.

Pages 61-64: *The Sanborn Map,* Sanborn Map Company, Inc., Pelham, New York.

Page 65, Fig. 1.41: *Easement Planning for Utility Services,* Detroit Edison Company-Michigan Bell Telephone Co., 1962.

Page 66, Fig. 1.42; page 67, Fig. 1.43: *Substation Site Selection and Development,* Johnson, Johnson, and Roy.

Pages 67-76: *Land Planning Tools,* Environmental Systems Research Institute, Redlands, California.

Section 2

Pages 86-95: *Climates of the United States,* National Weather Service, U.S. Department of Commerce, Washington, D.C., 1974.

Page 96: Publication 204, U.S. Department of Agriculture, Washington, D.C.

Pages 97-99: *Site Planning for Solar Access,* Department of Housing and Urban Development, Washington, D.C., 1979.

Pages 100, 101 (earthquake zones): *Home Builder's Guide for Earthquake Design,* Department of Housing and Urban Development, Washington, D.C., 1980.

Page 102, Fig. 2.16 (termite zones); page 107 (frost penetration); page 108 (drainage design): *Manual of Acceptable Practices,* Department of Housing and Urban Development, Washington, D.C., 1973.

Page 103, Fig. 2.17, through page 106, Fig. 2.22: *Regulation for Flood Plains,* Report 277, American Society of Planning Officials, Planning Advisory Service, Chicago, 1972.

Page 108, Fig. 2.24; pages 108-112: *Floodplain Management Handbook,* U.S. Water Resources Council, Washington, D.C., 1981.

CREDITS

Pages 112–118 (flood plain management): *Flood Plain Regulations and Flood Plain Management,* Department of the Army, Washington, D.C., 1976.

Pages 118, 119 (floodproofing): *Floodproofing Regulations,* Office of the Chief of Engineers, U.S. Army, 1972.

Pages 119–125: *Introduction to Floodproofing,* Center for Urban Studies, University of Chicago, 1967.

Page 126: *Minimum Property Standards,* Federal Housing Administration, Washington, D.C.

Pages 126–128 (drainage systems): *Michigan State Soil Erosion and Sedimentation Control Guidebook,* State of Michigan.

Pages 128–132 (utility systems): *System Requirements for Underground Utility Installations,* Oak Ridge National Laboratory, Oak Ridge, Tennessee, 1971.

Pages 132–142 (water supply); page 143, Table: *Manual of Individual Water Supply Systems,* Environmental Protection Agency, Water Supply Division, Washington, D.C., 1973.

Page 143 (types of wells); page 143, Table: *Minimum Design Standards for Community Water Supply Systems,* Department of Housing and Urban Development, Washington, D.C., 1968.

Pages 148–160 (sanitary landfill): *Sanitary Landfill — Design and Operation,* U.S. Environmental Protection Agency, Washington, D.C., 1972.

Pages 161–164 (residental site selection): *Planning the Neighborhood,* Public Administration Service, Chicago, 1960.

Pages 164–166: Samuel Paul, *Apartments,* Van Nostrand Reinhold Company, New York, 1967.

Pages 167–169 (selection of school building sites): Nickolas L. Engelhardt, *Complete Guide for Planning New Schools,* Parker Publishing Co., Inc., Englewood Cliffs, New Jersey, 1970.

Pages 166, 168–174 (industrial site selection): Ruddell Reed, Jr., *Plant Layout,* Richard D. Irwin, Inc., Homewood, Illinois, 1961.

Pages 174–181 (site design for wildlife): *Planning for Wildlife in Cities and Suburbs,* Fish and Wildlife Service, U.S. Department of the Interior, Washington, D.C., 1978.

Pages 181–191 (waterfront considerations of site development): *Planning for Urban Fishing and Waterfront Recreation,* U.S. Department of the Interior, Washington, D.C., 1981.

Pages 191–194 (conservation measures): Soil Conservation Service, U.S. Department of Agriculture, Washington, D.C.

Pages 194, 195: Proceedings of the Environmental Assessment and Impact Statements Conference, Environmental Studies Institute, Center for Urban Research and Environmental Studies, Drexel University, Philadelphia, 1973.

Pags 195–199: A. J. Rutledge, *Anatomy of a Park,* copyright © 1971 by McGraw-Hill, Inc., used with permission of McGraw-Hill Book Company, New York.

Section 3

Pages 200–202 (sun orientation): *Intermediate Minimum Property Standards,* Department of Housing and Urban Development, Washington, D.C., 1977.

Page 203, Fig. 3.7; page 215, Fig. 3.26; page 225, Figs. 3.43 and 3.44: *Landscape Development,* Field Technical Office, U.S. Department of the Interior, Littleton, Colorado.

Page 202 (landscaping for thermal control); page 205; page 206, Figs. 3.11 and 3.12; page 238 (landscaping for noise control): John E. Flynn and Arthur W. Segil, *Architectural Interior Systems,* Van Nostrand Reinhold Company, New York, 1970.

Page 204: Federal Housing Administration, Washington, D.C.

Pages 206–208 (building solar orientation): *Solar Dwelling Design Concepts,* Department of Housing and Urban Development, Washington, D.C., 1976.

Pages 208–214: *Site Planning for Solar Access,* Department of Housing and Urban Development, Washington, D.C., 1979.

Pages 214–216 (wind orientation); pages 245–249 (landscaping elements): *Plants, People, and Environmental Quality,* National Park Service, U.S. Department of the Interior, Washington, D.C., 1972.

Pages 216–218 (topography, slopes); pages 218–221: Federal Housing Administration, Washington, D.C.

Page 221, Table: *Minimum Property Standards,* Federal Housing Administration, Washington, D.C.

Page 222: *Traffic Engineering for Better Roads,* Military Traffic Management Command, Newport News, Virginia, 1980.

Pages 226–230: Ronald L. Mace, *Accessibility Modifications,* North Carolina Department of Insurance, Raleigh, 1976.

Pages 231–238: *Noise Assessment Guidelines,* Department of Housing and Urban Development, Washington, D.C., 1971.

Pages 238–242 (spatial structure): *Planning for Wildlife and Man,* Fish and Wildlife Service, U.S. Department of the Interior, Washington, D.C., 1974.

Pages 242–244 (site accessibility): *Barrier Free Site Design,* Department of Housing and Urban Development, Washington, D.C., 1975.

Page 244, Figs. 3.69 and 3.70; page 245, Fig. 3.71: *Access America,* Architectural and Transportation Compliance Board, Washington, D.C., 1980.

Pages 250–261; pages 295–297: Kenneth L. Schellie and David Rogier, *Site Utilization and Rehabilitation Practices for Sand and Gravel Operations,* National Sand and Gravel Association, Silver Springs, Maryland, 1963.

Pages 262–265: *Landscape Architecture,* Boy Scouts of America, New Brunswick, New Jersey, 1970.

Pages 266–281: Marketing Department, Exxon Company, U.S.A.

Pages 282–294: *Living Screens for North America,* American Association of Nurserymen, Washington, D.C.

Pages 297–302 (site security): *A Design Guide for Improving Residential Security,* Center for Residential Security Design, New York; U.S. Department of Housing and Urban Development, Office of Policy Development and Research, Washington, D.C., 1973.

Pages 303–315: *Planning for Housing Security,* Department of Housing and Urban Development, Washington, D.C., 1979.

Pages 315–332: *Time-Saver Standards,* first edition, F. W. Dodge Company, New York, 1946.

Page 318, Fig. 3.122a: *Code Manual,* New York State Building Code Commission, Albany.

Page 319, Fig. 3.122b: *Minimum Property Standards,* Federal Housing Administration, Washington, D.C.

Page 330, Fig. 3.130 and 3.131: *Code Manual,* New York State Building Commission, Albany.

Section 4

Pages 334–339: *Land Use Intensity Rating,* Land Planning Bulletin 7, Federal Housing Administration, Washington, D.C., 1966.

Pages 339, 340 (neighborhood planning): Annual Report – 1957, Philadelphia City Planning Commission, Philadelphia.

Page 341, Fig. 4.5; pages 386–389: *Residential Site Development,* Canada Mortgage and Housing Corporation, Ottawa, Ontario.

Pages 340, 341 (street classification); page 340, Fig. 4.4; pages 398–400: *Control of Land Subdivision,* New York State Department of Commerce, Albany.

Pages 341–349; page 348, Figs. 4.20 through 4.25; page 349, Figs. 4.26 and 4.27: From *Earth Sheltered Community Design,* by the Underground Space Center, University of Minnesota, copyright © 1981 by the University of Minnesota. Reprinted by permission of Van Nostrand Reinhold Company, New York.

Page 344, Figs. 4.10 through 4.13; pages 352, 353 (elements of residential streets); page 355, Figs. 4.40 through 4.44; page 356, Figs. 4.45 and 4.46: *Traffic Engineering for Better Roads,* Military Traffic Management Command, Washington, D.C., 1978.

Pages 350–352, 381, 382, and 383, Fig. 4.95: *Principles of Small House Grouping,* Central Mortgage and Housing Corporation, Ottawa, Canada, 1957.

Page 353, Fig. 4.37: *Land Subdivision Manual,* Planning Commission, Monterey County, California, 1957.

Page 352 (roads); page 354, Fig. 4.38; page 355, Fig. 4.39: National Park Service, U.S. Department of the Interior, Washington, D.C.

Page 356, Fig. 4.47; page 357, Fig. 4.48; page 379: *Site Improvement Handbook for Multifamily Housing,* Housing Research and Development, University of Illinois, Urbana, 1974.

Pages 357–365: *Subdivision Regulations Manual,* Sioux City, Iowa.

Pages 365–367 (design considerations): *Subdivision Planning Standards,* Land Planning Division, Federal Housing Administration, Washington, D.C.

Page 367, Fig. 4.59: New York Regional Survey of New York and its Environs, 1929.

Page 368, Table: *Guidelines for Developing Public Recreation Facility Standards,* Ministry of Culture and Recreation, Ontario, Canada, 1976.

Page 369: *Planning and Design Workbook for Community Participation,* prepared for the New Jersey Department of Community Affairs, by Research Center for Urban and Environmental Planning, School of Architecture, Princeton University, Princeton, New Jersey.

Page 368; pages 370–375: *Plazas for People,* Department of City Planning, City of New York, 1976.

Pages 376, 377: *Recreational Community Gardening,* Bureau of Outdoor Recreation, Department of the Interior, Washington, D.C., 1976.

Page 378: *Low Rise Housing for Older People,* Department of Housing and Urban Development, Washington, D.C., 1978.

Pages 380, 381: *Concepts in Building Fire Safety,* M. David Egan, John Wiley & Sons, New York, 1978.

Page 383, Fig. 4.96; page 384: *Manual of Acceptable Practices,* Department of Housing and Urban Development, Washington, D.C.

Page 385, Fig. 4.101; page 386, Fig. 4.102: *Minimum Property Standards for 1 and 2 Living Units,* Federal Housing Administration, Washington, D.C., 1974.

Page 390, Fig. 4.111: Julian Harris Solomon, *Campsite Development,* Girl Scouts of the U.S.A., New York, 1959.

Page 391, Fig. 4.112: *Cleaning Up the Water,* Maine Department of Environmental Protection, Augusta, Maine.

Pages 392–396: *Cluster Development,* Pratt Institute, School of Architecture, Brooklyn, New York.

Page 397: *Flood Plain Regulations for Flood Plain Management,* Department of the Army, Washington, D.C., 1976.

Pages 400–420: From *Site Planning for Cluster Housing,* by R. Untermann and R. Small, copyright © 1977, Van Nostrand Reinhold Company. Reprinted by permission of the publisher.

Section 5

Pages 424, 425: *Standards and Definitions of Terms,* Ontario Department of Education, Community Programs Division.

Pages 425–456, 462–465, 474, 511: Office of the Chief of Engineers, U.S. Department of the Army, Washington, D.C.

Pages 457, 458: Temple R. Jarrell, *Bikeways,* National Recreation and Park Association, Arlington, Virginia, 1974.

Pages 459–461: *Recreational Buildings and Facilities,* U.S. Department of Agriculture, Washington, D.C., 1972.

Pages 465, 466, 468, 469; page 470, Fig. 5.76: *Children's Play Areas and Equipment,* U.S. Departments of the Army, the Navy, and the Air Force, 1969.

Page 467, Fig. 5.75: Randall Park, Village of Freeport, New York, Planning Associates, West Hempstead, New York.

Page 471: *Public Housing Design,* National Housing Agency, Federal Public Housing Authority, Washington, D.C., 1946.

Page 473, Table: H. S. Conover, *Public Grounds Maintenance Handbook,* Tennessee Valley Authority, Knoxville, Tennessee, 1953.

Pages 475–479, 481–483, 488, 489; page 490, Fig. 5.92: *Planning Areas and Facilities for Health, Physical Education, and Recreation,* Athletic Institute and American Association for Health, Physical Education, and Recreation, Washington, D.C., 1965.

Pages 479–481: *Design Guide for Home Safety,* Department of Housing and Urban Development, Washington, D.C., 1972.

Pages 484–487: Boy Scouts of America.

Page 491, Fig. 5.93; page 492, Fig. 5.94: *Recreation on Water Supply Reservoirs,* Council on Environmental Quality, Washington, D.C.

Pages 490–495: *Small Craft Harbors: Design, Construction, and Operation,* U.S. Army Corps of Engineers, Ft. Belvoir, Virginia, 1974.

Page 496, Fig. 5.97: Division of State Parks, Minnesota.

Page 498, Fig. 5.99: *Outdoor Recreation Facilities,* Department of the Army, Washington, D.C., 1975.

Page 499, Fig. 5.100: Outboard Boating Club of America and Socony Mobile Oil Company, Chicago.

Page 500, Fig. 5.101: Charles A. Chaney, *Marinas: Recommendations for Design, Construction, and Maintenance,* second edition, National Association of Engine and Boat Manufacturers, New York, 1961.

Page 500, Fig. 5.102; page 501, Table: *Marinas,* National Association of Engine and Boat Manufacturers, New York, 1947.

Pages 501–503: *Ponds,* Soil Conservation Service, U.S. Department of Agriculture, Washington, D.C., 1971.

Pages 503–506: Rees L. Jones and Guy L. Rando, *Golf Course Developments,* Technical Bulletin 70, Urban Land Institute, Washington, D.C., 1974.

Pages 504, 507, and 508: *Planning and Building the Golf Course,* National Golf Foundation, Chicago, 1967.

Pages 508–510, Fig. 5.115 through 5.119: *House and Home,* June 1968.

Pages 512–516: National Park Service, Washington, D.C.

Pages 517–522: Geoffrey Baker and Bruno Funaro, *Parking,* copyright © by Litton Educational Publishing. Reprinted by permission of Van Nostrand Reinhold Company, New York.

Page 522; page 523, Table: *Guidelines for Developing Public Recreation Facility Standards,* Ministry of Culture and Recreation, Ontario, Canada, 1976.

Page 523, Fig. 5.131; pages 524–532: *Guide to Designing Accessible Outdoor Recreation Facilities,* U.S. Department of the Interior, Ann Arbor, Michigan, 1980.

Page 524, Fig. 5.132: *Barrier-Free Site Design,* Department of Housing and Urban Development, Washington, D.C., 1975.

Pages 532–534: *Play Areas for Low-Income Housing,* Housing Research and Development, University of Illinois, Urbana-Champaign, 1972.

Section 6

Pages 536, 537, 543, 544, 548–551, 562, 563; page 564, Fig. 6.54; pages 623–625; page 626, Fig. 6.138; pages 644, 676; page 679, Fig. 6.222; pages 714–716, 752–754: *Barrier-Free Design,* Department of Housing and Urban Development, Washington, D.C., 1975.

Page 537, Fig. 6.2 (walks); page 546, Fig. 6.18: *Design for the Physically Handicapped,* Office of the Chief of Engineers, Department of the Army, Washington, D.C.

Page 538, Fig. 6.3; page 539, Fig. 6.4; page 546, Fig. 6.19; pages 647, 678: *An Illustrated Handbook of the Handicapped Section of the North Carolina State Building Code,* Ronald L. Mace, North Carolina Department of Insurance, State of North Carolina, 1977.

Pages 538, 539: *General Provisions and Geometric Design for Roads, Streets, Walks, and Open Storage Areas,* Departments of the Army, the Navy, and the Air Force, Washington, D.C., 1977.

Pages 540–542, 693, 697, 707; page 708, Fig. 6.275; pages 729, 730: *Landscape and Site Development,* Department of Public Works, Ottawa, Canada, 1972.

Page 545: *Access America,* Architectural and Transportation Barriers Compliance Board, Washington, D.C., 1980.

Pages 552, 558; page 559, Figs. 6.43 and 6.44; pages 637, 689, 690: *Time-Saver Standards,* first edition, F. W. Dodge Company, New York, 1946.

Pages 553–557; page 565, Fig. 6.57; pages 571; 638, 639, 645, and 646; page 647, Figs. 6.179 and 6.180; page 691; page 695, Fig. 6.255; pages 699, 717, 720; page 721, Fig. 6.296: *Time-Saver Standards for Building Types,* by Joseph De Chiara and John Callender, first edition, McGraw-Hill Book Company, New York, 1973.

Page 561, Figs. 6.48 and 6.49; pages 642, 684–686; page 708, Fig. 6.276; page 709: *Manual of Acceptable Practices,* Department of Housing and Urban Development Washington, D.C.

Pages 565, 566: *Design Guide for Home Safety,* Department of Housing and Urban Development, Washington, D.C., 1972.

Pages 568, 569: H. S. Conover, *Public Grounds Maintenance Handbook,* Tennessee Valley Authority, Knoxville, Tennessee, 1953.

Pages 570, 594, 622, 640, 641, 692, 698, 702–706, 728: *Landscape Development,* Field Technical Office, U.S. Department of the Interior, Littleton, Colorado.

Page 572: Department of Recreation and Conservation, Province of British Columbia.

Page 573: Fort Lauderdale, Florida.

Page 575: Central New York State Parks Commission.

Page 576: *Substation Site Selection and Development,* Consumer Power Company, Jackson, Michigan.

Pages 581–585, 602–605, 696: *Junkyards—the Highway and Visual Quality,* Federal Highway Administration, U.S. Department of Transportation, Washington, D.C., 1979.

Pages 585–587, 592; page 593, Fig. 6.99; pages 612–614, 618–621; page 711, Fig. 6.280: *Spon's Landscape Handbook,* Derek Lovejoy and Partners, E. & F. N. Spon, Ltd, London, England, 1972.

Page 578, Fig. 6.74; pages 588–590: *The Outdoor Room Plans Book,* Western Wood Products Association, Portland, Oregon.

Page 591: Highway Commission, Wisconsin.

Page 593, Fig. 6.100: Department of Parks, Seattle, Washington.

Page 595: U.S. Steel Corporation.

Pages 596–598: Bureau of Yards and Docks, Department of the Navy, Washington, D.C.

Page 599: Game, Fish, and Parks Department, Colorado.

Pages 600, 601: *Fences for the Farm and Rural Home,* Farmers Bulletin 2247, U.S. Department of Agriculture, Washington, D.C., 1971.

Page 608: Conservation Department, Indiana.

Page 609: Department of Public Works and Buildings, Illinois.

Page 610: National Park Service, U.S. Department of the Interior, Washington, D.C.

Page 626 (trails); page 627, Fig. 6.139, 6.140, and 6.141: *Bart/Trails,* U.S. Department of Transportation, Washington, D.C.

Page 627, Fig. 6.142: State Park System, Nevada.

Page 628; page 629, Fig. 6.146: Temple R. Jarrell, *Bikeways,* National Recreation and Park Association, Arlington, Virginia, 1974.

Page 629, Fig. 6.147: *Outdoor Recreation Facilities,* Department of the Army, Washington, D.C., 1975.

Pages 630–632: Byron Ashbaugh and Raymond Kordish, *Trail Planning and Layout,* National Audubon Society, New York, 1971.

Pages 632–636, 668: *Time-Saver Standards,* fourth edition, John Callender, editor. McGraw-Hill Book Company, New York, 1966.

Page 643: Department of Transportation, New York City.

Page 647, Fig. 6.181; page 648, Figs. 6.182 and 6.183; page 649, Figs. 6.184 and 6.185: *Traffic Engineering for Better Roads,* Military Traffic Management Command, Washington, D.C., 1978.

Pages 650, 651: *Urban Transportation Planning for Goods and Services,* Federal Highway Administration, U.S. Department of Transportation, Washington, D.C., 1979.

Pages 652, 653, 658–660, 669, 673: *Zoning, Parking, and Traffic,* Eno Foundation for Transportation, Saugatuck, Connecticut, 1973.

Page 655, Fig. 6.194; page 656, Figs. 6.195 and 6.196: *Planning and Design of Roads, Airbases, and Heliports,* Departments of the Army and the Air Force, Washington, D.C., 1968.

Pages 656, 657: The Eno Foundation for Highway Traffic Control, Saugatuck, Connecticut, 1977.

Page 661, Fig. 6.200; page 662, Figs. 6.201, 6.202, and 6.203: *Eastwick New House Study,* Philadelphia Redevelopment Authority, Philadelphia, 1957.

Pages 662–667, 674: *Parking Guide for Cities,* U.S. Department of Commerce, Bureau of Public Roads, Washington, D.C., 1956.

Pages 670, 671, 675: Naval Facilities Engineering Command, Department of the Navy, Washington, D.C.

Page 676, Fig. 6.218; pages 753, 755–763: *Manual of Uniform Traffic Control Devices for Streets and Highways,* Federal Highway Administration, Department of Transportation, Washington, D.C., 1971.

Page 677: *Design for the Physically Handicapped,* Military Construction Civil Works, Department of the Army, Washington, D.C., 1976.

Pages 680, 681: *Handbook for Design: Specially Adapted Housing,* Veterans Administration, Washington, D.C., 1978.

Pages 682, 683; page 684, Fig. 6.230: National Parks Service, Washington, D.C.

Page 686, Fig. 6.233; page 687, Fig. 6.234 and Table; pages 750–752: *Code Manual,* New York State Building Code Commission, Albany, 1973.

Page 687, Fig. 6.235; page 688: *Roads, Runways, and Miscellaneous Pavements,* War Department, Washington, D.C.

Page 695, Fig. 6.254: Schnadelbach Braun Partnership.

Page 711, Fig. 6.279: *Trees for New York City,* New York City Planning Commission, New York.

Page 718: Conservation Department, Indiana.

Page 721; page 722, Fig. 6.298: *Park Practice Program,* National Conference on State Parks and American Institute of Park Executives.

Page 722; page 723, Fig. 6.299: *Minimum Property Standards,* Federal Housing Administration, Washington, D.C.

Page 723, Fig. 6.300; page 726: *National Engineering Handbook,* Soil Conservation Service, U.S. Department of Agriculture, Washington, D.C., 1971.

Pages 724, 725: Division of Sewers, Queens, New York.

Pages 730–732 (cisterns): Reproduced by permission from *Private Water Systems,* MWPS-14 Midwest Plan Service, Ames, Iowa.

Pages 733–743: *Shore Protection Manual,* vol. II, U.S. Army Corps of Engineers, Coastal Engineering Research Center, Fort Belvoir, Virginia, 1973.

Pages 744–749: *Restoration of Fish Habitat in Relocated Streams,* U.S. Department of Transportation, Washington, D.C., 1979.

Pages 764–766: Bureau of Yards and Docks, Department of the Navy, Washington, D.C.

Page 767: *Memo to Architects,* New York City Housing Authority, New York.

Section 7

Pages 770–776: *Townhouse Development Process,* Michigan State Housing Development Authority, State of Michigan, 1970.

Pages 777–791: *Planned Unit Development,* New York City Planning Commission, New York.

Page 792: *Residential Site Development,* Canada Mortgage and Housing Corporation, Ottawa, Ontario, Canada.

Pages 793–796: A. J. Rutledge, *Anatomy of a Park,* copyright © 1971, by McGraw-Hill, Inc. Used with permission of McGraw-Hill Book Company, New York.

Pages 797–802: Planning Associates, Bohemia, New York.

Page 803: Land Development Division, Mobile Home Manufacturers Association.

Page 804; 835: R. T. Schnadelbach, Landscape and Ecological Consultant.

Pages 805–819: *Solar Dwelling Design Concepts,* Department of Housing and Urban Development, Washington, D. C., 1976.

Pages 819–829: *Site Planning for Solar Access,* Department of Housing and Urban Development, Washington, D.C., 1979.

Page 831: Grasslands Housing, Westchester Development Corp., Pokorny & Pertz/Architects.

Page 834: Penn Landing Square, architects, Louis Sauer & Associates.

Page 837: *Planning Residential Subdivisions,* J. Kostka, Appraisal Institute of Canada.

Index

About the Authors

Joseph De Chiara is an architect and an urban planner. During twenty-five years of practice he has dealt with many and varied types of site planning problems.

Lee E. Koppelman is a landscape architect and an urban planner. He has had a private practice in landscape design, has been a planning consultant for many projects, and is now the Planning Director of the Nassau-Suffolk Regional Planning Board.

De Chiara and Koppelman have previously collaborated on several books in this field, including *Planning Design Criteria, Urban Planning and Design Criteria,* and *Manual of Housing/Planning and Design Criteria.*